Molecular Response Mechanisms of *Apis cerana cerana* to Abiotic Stress

Baohua Xu   Xingqi Guo

The publication of this book was financially supported by Funds of Shandong Province "Double Tops" Program (2016-2020)

Molecular Response Mechanisms of *Apis cerana cerana* to Abiotic Stress

Baohua Xu

Xingqi Guo

ISBN 978-7-03-058987-3
Science Press, Beijing

Copyright © 2019 by Science Press
All rights reserved.

Responsible Editors: Haiguang Wang   Chenyang Hao

About the Authors

Baohua Xu College of Animal Science and Technology
Shandong Agricultural University
Taian, Shandong, China
+86-0538-8241462
bhxu@sdau.edu.cn

Xingqi Guo College of Life Sciences
Shandong Agricultural University
Taian, Shandong, China
+86-0538-8245679
xqguo@sdau.edu.cn

Preface

As one of the predominant pollinators, honeybees provide an important ecosystem service to crops and wild plants and generate great economic benefit for human. Unfortunately, there is clear evidence of marked regional population decreasing in honeybee populations in recent years. In particular, from the winter of 2006 to the spring of 2007, adult bees suddenly underwent mass disappearances in outwardly healthy colonies within 2-4 weeks in parts of Asia, Europe and the United States, leading to alarming levels of colony failure, termed colony collapse disorder (CCD). For hundreds of millions of years, honeybees evolved together with angiosperms, and the number and health of honeybees directly affected plant diversity, ecosystem stability and crop production. It has been demonstrated that various environmental stresses, including both abiotic and biotic stresses, functioning singly or synergistically, are the potential drivers of colony collapse. Therefore, it is necessary to study the molecular mechanisms of stress resistance that honeybees use to defend against adverse environmental stresses.

Apis cerana cerana (*A. cerana cerana*) belongs to the major ecological group of *Apis cerana*. This group has a long apicultural history and possesses strong biological characteristics in terms of hygienic behaviors, resisting *Tropilaelaps* mites, cold and cooperative group-level defenses, and can forage for a wide range of nectars and pollens. However, a variety of environmental stimuli have severely decreased the numbers of *A. cerana cerana* in some regions in recent years. To provide research support for the rational utilization of the resistance of honeybees and to cultivate stress-resistant honeybee species, we established a stress resistance biology research group for *A. cerana cerana* when the China Agriculture Research System (No. CARS-44) was established in 2008. In our study, various biochemistry, molecular biology and cell biology methods were selected. First, PCR was used to isolate target genes, and then, bioinformatics was chosen to analyze the genomic structure of target genes, the tertiary structure of proteins and the relationship between the protein encoded by the target gene and its homologous protein. RT-PCR and Western blotting were used to detect the developmental characterization and environmental stress responses of target genes at the transcriptional level and protein level. The transcriptional level of some growth- and development-related genes and antioxidant genes, the enzyme activities of several antioxidant-related enzymes, the contents of oxidation- or antioxidation-related metabolites and the phenotype of *A. cerana cerana* were also investigated after knockdown of target genes by RNAi technology. In addition, the tissue distribution and subcellular localization characteristics of the target genes were explored by immunohistochemistry and immunoblotting, respectively.

In the last 10 years, more than 50 stress response genes have been identified; most of these genes and their corresponding proteins were conserved during evolution but also presented some differences. Structural conservation may determine the similarity of gene functions, and structural differences may be caused by the diversity of species during evolution. Furthermore, it was found that some genes played important roles in the growth and development of honeybees through growth and tissue distribution characteristics. For example, RT-PCR showed that *AccLSD-1* was expressed ubiquitously from larva to adult stages and its expression level was the highest at the brown-eyed pupae stage. Analysis of the 5′-flanking region of *AccLSD-1* revealed a number of putative *cis*-acting elements, including three PPREs. *AccLSD-1* was downregulated by CLA but upregulated by Rosi. Furthermore, the combination of CLA and Rosi remarkably rescued the suppression of *AccLSD-1* expression by CLA alone. These results suggest that *AccLSD-1* is associated with the development of *A. cerana cerana*, especially during pupal metamorphosis, and can be regulated by CLA or Rosi possibly by activating PPARs. The highest mRNA level of *AccRPL17* was detected in larvae on the fifth day. Simultaneously, immunohistochemical localization showed that *AccRPL17* was primarily concentrated in muscular tissues, stigma, body wall and surrounding the eye in fifth-instar larvae. These results indicate that *AccRPL17* may play an important role in insect development.

Although multiple target genes had different expression patterns under various environmental stress conditions, all of them could respond to stress factors to varying degrees. For example, the transcription level and protein level of *AccRBM11* were all upregulated upon simulation of abiotic stress. The expression of some genes related to stress responses was remarkably suppressed when *AccRBM11* was silenced. These results suggest that *AccRBM11* is potentially involved in the regulation of abiotic stress responses. Many potential transcription factor binding sites associated with development and stress responses were identified in the 5′-flanking region of *AccsHsp22.6*. Moreover, *AccsHsp22.6* was significantly upregulated or downregulated by abiotic and biotic stresses. Recombinant AccsHsp22.6 also exhibited significant temperature tolerance, antioxidation and molecular chaperone activity. In addition, knockdown of *AccsHsp22.6* by RNAi remarkably reduced temperature tolerance in *A. cerana cerana*. Taken together, these results suggest that *AccsHsp22.6* plays an essential role in the defense against cellular stress. *AccAK* was upregulated under abiotic stress conditions, and under some conditions, this response was very pronounced. A disc diffusion assay showed that overexpression of AccAK reduced the resistance of *Escherichia coli* cells to multiple adverse stresses. These findings indicate that *AccAK* may be involved in significant responses to adverse abiotic and biotic stresses.

Environmental stress factors, such as pesticides, cold, heat, UV and heavy metal, can give rise to oxidative stress and can induce oxidative damage in organisms. To explore the antioxidant capacity of Chinese honeybees in response to oxidative damage, some common antioxidant enzymes (thioredoxin reductase, thioredoxin peroxidase and glutathione *S*-transferase) and possible antioxidant genes (cytochrome P450, cyclin-dependent kinase and decapentaplegic genes) were selected for functional analysis. It was discovered that multiple enzymes and genes could be induced when honeybees suffer oxidative damage. Silencing some antioxidant genes reduced the levels of oxidation- or antioxidation-related metabolites. For example, *AccGrx1*, *AccGrx2* and *AccTrx1* were induced by 4, 16, and 42 ℃; H_2O_2; and pesticides (acaricide, paraquat, cyhalothrin, and phoxim) treatments and repressed by UV light. *AccGrx1* and *AccGrx2* were upregulated by $HgCl_2$ treatment, whereas *AccTrx1* was downregulated. Knockdown of these three genes all enhanced the enzymatic activities of CAT and POD, the metabolite contents of hydrogen peroxide, carbonyls and ascorbate; and the ratios of GSH/GSSG and $NADP^+$/NADPH. In addition, we analyzed the transcripts of other antioxidant genes and found that some were upregulated and others were downregulated, revealing that the upregulated genes may be involved in compensating for the knockdown of *AccGrx1*, *AccGrx2* and *AccTrx1*. Taken together, these results suggest that *AccGrx1*, *AccGrx2* and *AccTrx1* may play critical roles in antioxidant defense. The expression level of *AccCYP336A1* was upregulated by cold, heat, ultraviolet radiation, H_2O_2 and pesticides (thiamethoxam, deltamethrin, methomyl and phoxim) treatments. Recombinant AccCYP336A1 protein acted as an antioxidant that resisted paraquat-induced oxidative stress. These findings suggest that *AccCYP336A1* may play a very significant role in antioxidant defense against ROS damage.

Taken together, we explored the roles of multiple genes and related antioxidant signaling pathways of *A. cerana cerana* under adverse circumstances to discover key genes with strong resistance to environmental stresses, and systematically studied and analyzed the mechanisms of *A. cerana cerana* defense against stress factors. These results will enrich the scientific knowledge of honeybee resistance biology and provide a scientific basis for the genetic improvement of honeybees through gene-editing techniques in the future.

Baohua Xu　Xingqi Guo
Sept. 8th, 2018

Contents

1 The Abiotic Stress Response of Some Key Genes in *Apis cerana cerana* ········· 1
   1.1 Two Small Heat Shock Protein Genes in *Apis cerana cerana*: Characterization, Regulation, and Developmental Expression ········· 4
   1.2 *sHsp22.6*, an Intronless Small Heat Shock Protein Gene, Is Involved in Stress Defence and Development in *Apis cerana cerana* ········· 23
   1.3 Molecular Cloning and Characterization of *Hsp27.6*: the First Reported Small Heat Shock Protein from *Apis cerana cerana* ········· 44
   1.4 Molecular Characterization and Immunohistochemical Localization of a Mitogen-activated Protein Kinase Gene, *Accp38b*, from *Apis cerana cerana* ········· 62
   1.5 *AccERK2*, A Map Kinase Gene from *Apis cerana cerana*, Plays Roles in Stress Responses, Developmental Processes, and the Nervous System ········· 73
   1.6 A Typical RNA Binding Protein Gene (*AccRBM11*) in *Apis cerana cerana*: Characterization of *AccRBM11* and Its Possible Involvement in Development and Stress Responses ········· 87
   1.7 Isolation of Arginine Kinase from *Apis cerana cerana* and Its Possible Involvement in Response to Adverse Stress ········· 110
   1.8 Molecular Cloning and Characterization of Two Nicotinic Acetylcholine Receptor β Subunit Genes from *Apis cerana cerana* ········· 131
   1.9 Molecular Cloning and Characterization under Different Stress Conditions of Insulin-like Peptide 2 (*AccILP-2*) Gene from *Apis cerana cerana* ········· 146
   1.10 Molecular Cloning, Expression and Stress Response of the Estrogen-related Receptor Gene (*AccERR*) from *Apis cerana cerana* ········· 160
   1.11 The Initial Analysis of a Serine Proteinase Gene (*AccSp10*) from *Apis cerana cerana*: Possible Involvement in Pupal Development, Innate Immunity and Abiotic Stress Responses ········· 177
   1.12 Cloning, Expression Patterns and Preliminary Characterization of *AccCPR24*, a Novel RR-1 Type Cuticle Protein Gene from *Apis cerana cerana* ········· 195
   1.13 Molecular Identification and Stress Response of the Apoptosis-inducing Factor Gene 3 (*AccAIF3*) from *Apis cerana cerana* ········· 210

2 The Enzymatic Antioxidant Molecular Mechanisms of *Apis cerana cerana* ········· 227
   2.1 Identification and Characterization of an *Apis cerana cerana* Delta Class Glutathione *S*-transferase Gene (*AccGSTD*) in Response to Thermal Stress ········· 230
   2.2 Characterization and Mutational Analysis of Omega-class GST (GSTO1) from *Apis cerana cerana*, a Gene Involved in Response to Oxidative Stress ········· 247

2.3　A Novel Omega-class Glutathione *S*-transferase Gene in *Apis cerana cerana*: Molecular Characterisation of *GSTO2* and Its Protective Effects in Oxidative Stress ·· 268

2.4　Functional and Mutational Analyses of an Omega-class Glutathione *S*-transferase (*GSTO2*) That Is Required for Reducing Oxidative Damage in *Apis cerana cerana* ····· 288

2.5　Characterization of a Sigma Class Glutathione *S*-Transferase Gene in the Larvae of the Honeybee (*Apis cerana cerana*) on Exposure to Mercury ················ 310

2.6　A Glutathione *S*-Transferase Gene Associated with Antioxidant Properties Isolated from *Apis cerana cerana* ·· 325

2.7　The Identification and Oxidative Stress Response of a Zeta Class Glutathione *S*-transferase (GSTZ1) Gene from *Apis cerana cerana* ·· 344

2.8　Molecular Cloning, Expression and Oxidative Stress Response of a Mitochondrial Thioredoxin Peroxidase Gene (*AccTpx-3*) from *Apis cerana cerana* ····················· 362

2.9　A Novel 1-Cys Thioredoxin Peroxidase Gene in *Apis cerana cerana*: Characterization of *AccTpx4* and Its Role in Oxidative Stresses ···························· 380

2.10　Identification and Characterisation of a Novel 1-Cys Thioredoxin Peroxidase Gene (*AccTpx5*) from *Apis cerana cerana* ··· 395

2.11　Glutaredoxin 1, Glutaredoxin 2, Thioredoxin 1, and Thioredoxin Peroxidase 3 Play Important Roles in Antioxidant Defense in *Apis cerana cerana* ···················· 414

2.12　Molecular Cloning, Expression and Antioxidant Characterisation of a Typical Thioredoxin Gene (*AccTrx2*) in *Apis cerana cerana* ··· 436

2.13　Characterization of a Mitochondrial Manganese Superoxide Dismutase Gene from *Apis cerana cerana* and Its Role in Oxidative Stress ································· 453

2.14　Cloning and Molecular Identification of Triosephosphate Isomerase Gene from *Apis cerana cerana* and Its Role in Response to Various Stresses ······················· 475

2.15　Identification and Characterization of a Novel Methionine Sulfoxide Reductase B Gene (*AccMsrB*) from *Apis cerana cerana* (Hymenoptera: Apidae) ················ 489

2.16　Molecular Cloning, Characterization and Expression of Methionine Sulfoxide Reductase A Gene from *Apis cerana cerana* ·· 503

3　The Nonenzymatic Antioxidant Molecular Mechanisms of *Apis cerana cerana* ······· 519

3.1　Characterization of an *Apis cerana cerana* Cytochrome P450 Gene (*AccCYP336A1*) and Its Roles in Oxidative Stresses Responses ··························· 522

3.2　Roles of a Mitochondrial *AccSCO2* Gene from *Apis cerana cerana* in Oxidative Stress Responses ··· 538

3.3　Characterization of the CDK5 Gene in *Apis cerana cerana* (*AccCDK5*) and a Preliminary Identification of Its Activator Gene, *AccCDK5r1* ····························· 558

3.4　Isolation of Carboxylesterase (esterase FE4) from *Apis cerana cerana* and Its Role in Oxidative Resistance during Adverse Environmental Stress ······················ 579

3.5 Characterization of a Decapentapletic Gene (*AccDpp*) from *Apis cerana cerana* and Its Possible Involvement in Development and Response to Oxidative Stress···· 600

3.6 Molecular Cloning, Expression and Oxidative Stress Response of the Vitellogenin Gene (*AccVg*) from *Apis cerana cerana* ··· 622

4 The Molecular Mechanisms of Growth and Development of *Apis cerana cerana* ··· 637

4.1 Characterization of the Response to Ecdysteroid of a Novel Cuticle Protein R&R Gene in the Honeybee, *Apis cerana cerana* ··· 640

4.2 Characterization of the TAK1 Gene in *Apis cerana cerana* (*AccTAK1*) and Its Involvement in the Regulation of Tissue-specific Development ···················· 656

4.3 Cloning, Structural Characterization and Expression Analysis of a Novel Lipid Storage Droplet Protein-1 (LSD-1) Gene in Chinese Honeybee (*Apis cerana cerana*) ········ 666

4.4 Ribosomal Protein L11 is Related to Brain Maturation During the Adult Phase in *Apis cerana cerana* (Hymenoptera, Apidae) ·· 682

1
The Abiotic Stress Response of Some Key Genes in *Apis cerana cerana*

As an Asian species, the *Apis cerana cerana* brings considerable economic benefits to the apiculture industry and plays a critical role in maintaining biodiversity. However, recently, the population of *Apis cerana cerana* has severely declined, which can be attributed to many abiotic stresses that exist in the environment, such as excessive pesticide usage, climate changes that include extreme heat and cold, the presence of heavy metal and exposure to ultraviolet radiation. It is essential to identify specific genes and their corresponding proteins and to reveal their expression characteristics and related biological functions in stress responses.

In the study, we cloned and characterized a series of genes that have been confirmed to be involved in stress responses in other insects, and we detected their levels of expression at various developmental stages. In the experiment, we mimicked common abiotic stress conditions encountered by *Apis cerana cerana* during its life to examine protein expression profiles. qRT-PCR, Western blotting, RNAi and disc diffusion assay were used to investigate the functions of these genes in response to abiotic stresses. Through this study, we found that although the expression patterns of multiple target genes were different under various environmental stress conditions, they could respond to different environmental stresses to a certain extent.

Small heat shock proteins (sHsps) play an important role in protecting against stress-induced cell damage and fundamental physiological processes. *AccsHsp22.6* was significantly upregulated by abiotic and biotic stresses. Recombinant AccsHsp22.6 exhibited significant temperature tolerance and antioxidant and molecular chaperone activity. Knockdown of *AccsHsp22.6* by RNA interference remarkably reduced temperature tolerance in *A. cerana cerana*. The recombinant AccHsp24.2 and AccHsp23.0 proteins were shown to have molecular chaperone activity by the thermal aggregation assay of the malate dehydrogenase; the expression of the two genes could be induced by cold shock and heat shock in an analogous manner, and *AccHsp24.2* was more susceptible than *AccHsp23.0*. The mRNA level of *AccHsp27.6* was induced by exposure to heat shock, H_2O_2, different chemicals, and the microbes *Staphylococcus aureus* and *Micrococcus luteus* and was repressed by CO_2, the pesticides pyriproxyfen and cyhalothrin, and the microbes *Bacillus subtilis* and *Pseudomonas aeruginosa*.

The MAPK family is composed of three major groups: extracellular signal-regulated kinases (ERKs), C-Jun N-terminal kinases (JNKs) and p38 MAPKs. ERK, a mitogen-activated protein kinase (MAPK), plays roles in a variety of cellular responses. The expression of *AccERK2* was induced by abiotic stresses, including heat, ion irradiation, oxidative stress, and heavy metal ions. p38 MAPK was originally identified as a 38 kDa protein that was rapidly activated by stressors; the gene expression of *Accp38b* was induced by multiple stressors. Immunohistochemistry showed significant positive staining of Accp38b in sections from the brain, eyes, fat body, and midgut of *A. cerana cerana*.

RBPs play roles not only in genome organization, growth, and development but also in stress responses through the regulation of post-transcriptional mechanisms. The expression of *AccRBM11* was upregulated under the simulation of abiotic stresses; the expression of some genes connected with development or stress responses was remarkably suppressed when *AccRBM11* was silenced. Arginine kinases (AKs) in invertebrates play the same role as creatine kinases in vertebrates. The gene expression of *AccAK* was upregulated under several abiotic and biotic stresses, and Western blotting and enzyme activity assays confirmed the results. In addition, disc diffusion assay showed that overexpression of AccAK reduced the resistance of *Escherichia coli* cells to multiple adverse stresses.

Corticotropin-releasing hormone-binding protein (CRH-BP) is an essential secreted glycoprotein for coordinating the neuroendocrine responses to stress by binding either CRH or its related peptides. The gene expression of *AccCRH-BP* in the brain was induced by exposure to environmental stresses. Analysis of the 5′-flanking region of *AccCRH-BP* revealed a number of putative development and stress transcription factor binding sites. Nicotinic acetylcholine receptors (nAChRs) mediate fast cholinergic synaptic transmission in the insect nervous system and are important targets for insecticides. It was observed that *Accβ1* and *Accβ2* were differentially affected by environmental stresses and that, in most cases, *Accβ2* was expressed at the higher level.

Insulin-like peptide 2 is an important gene in the insulin/insulin-like signaling pathway and plays a key role in metabolism, growth, reproduction, stress resistance and aging. The gene expression of *AccILP-2* was induced by environmental stressors. Estrogen-related receptor (ERR), which belongs to the nuclear receptor superfamily, has been implicated in diverse physiological processes involving the estrogen signaling pathway. The expression of *AccERR* was induced by cold (4℃), heat (42℃), ultraviolet light (UV), $HgCl_2$, and various types of pesticides.

Glutamine synthetase (GS) is an essential detoxification enzyme that plays an important role in stress responses. The expression of the *AccGS* was highly regulated by environmental stresses. Disc diffusion assay showed that the recombinant AccGS protein confers resistance to mercuric chloride ($HgCl_2$) stress in *E. coli*. Serine proteinases play important roles in innate immunity and insect development. In adult worker bees, the gene and protein expression of AccSp10 was upregulated by treatments of mimicking harmful environments. Disc diffusion assay results indicated that recombinant AccSp10 accelerated *E. coli* cell death during stimulation with harmful substances.

Cuticular proteins (CPs) are key components of the insect cuticle, a structure that plays a pivotal role in insect development and defenses. The gene expression of *AccCPR24* was influenced by environmental stressors. In addition, disc diffusion assay results showed that AccCPR24 enhanced the ability of bacterial cells to resist multiple stresses. Apoptosis-inducing factors (AIFs) play vital roles in bioenergetic and redox metabolism. The transcription level of *AccAIF3* was significantly increased by exposure to cold, $HgCl_2$, $CdCl_2$, UV, pesticides and *Ascosphaera apis*.

1.1 Two Small Heat Shock Protein Genes in *Apis cerana cerana*: Characterization, Regulation, and Developmental Expression

Zhaohua Liu, PengboYao, Xingqi Guo, Baohua Xu

Abstract: In the present study, we identified and characterized two small heat shock protein genes from *Apis cerana cerana*, named *AccHsp24.2* and *AccHsp23.0*. An alignment analysis showed that *AccHsp24.2* and *AccHsp23.0* share high similarity with other members of the α-crystallin/sHsp family, all of which contain the conserved α-crystallin domain. The recombinant AccHsp24.2 and AccHsp23.0 proteins were shown to have molecular chaperone activity by the thermal aggregation assay of the malate dehydrogenase. Three heat shock elements were detected in the 5'-flanking region of *AccHsp24.2* and eleven in *AccHsp23.0*, and two *Drosophila* Broad-Complex (BR-C) genes for ecdysone steroid response sites were found in each of the genes. The presence of these elements suggests that the expression of these genes might be regulated by heat shock and ecdysone, which was confirmed by quantitative RT-PCR (qRT-PCR). The results revealed that the expression of the two genes could be induced by cold shock (4℃) and heat shock (37℃ and 43℃) in an analogous manner, and *AccHsp24.2* was more susceptible than *AccHsp23.0*. In addition, the expression of the two genes was induced by high concentrations of ecdysone *in vitro* and *in vivo*. The accumulation of *AccHsp24.2* and *AccHsp23.0* mRNA was also detected at different developmental stages and in various tissues. In spite of the differential expression at the same stage, these genes shared similar developmental patterns, suggesting that they are regulated by similar mechanisms.

Keywords: *Apis cerana cerana*; Expression; Gene cloning; Heat shock; Ecdysone

Introduction

The Chinese honeybee (*Apis cerana cerana*), a subspecies of the Eastern honeybee or *A. cerana*, is an important indigenous species that plays a critical role in agricultural economic development as a pollinator of flowering plants, particularly winter-flowering plants. *A. cerana* is highly resistant to cold, pests, and diseases and also plays an important role in combating soil degradation by pollinating wild plants, thereby ensuring that more biomass may be returned to the soil. Therefore, *A. cerana cerana* is of great significance to the ecological environment of China. However, because of deforestation, excessive pesticide usage, environmental pollution, and other reasons, there is concern with regard to the survival of *A. cerana cerana* [1].

Heat shock proteins (Hsps) are protein chaperones that are ubiquitously expressed in most organisms; these proteins are also involved in some processes of embryogenesis, diapause, and morphogenesis [2]. Traditionally, the Hsp superfamily has been classified based on the molecular weights of the proteins, with four types of Hsps, including Hsp90, Hsp70, Hsp60, and a family of small Hsps (sHsps) [3]. There are numerous studies concerning sHsps in bacteria, algae, plants, amphibians, birds, and mammals. The four primary sHsps of *Drosophila melanogaster* (Hsp22, Hsp23, Hsp26, and Hsp27) have been studied in detail: these proteins share significant sequence similarity and are coordinately expressed following stress yet have distinct developmental expression patterns and intracellular localizations [4]. The sHsps range in molecular weight from 15-43 kDa and exhibit the greatest variation in sequence, size, and function compared to other Hsp members [5]. Despite these differences, two or three conserved domains are present in almost all sHsps, the most prominent of which is the α-crystallin domain.

In the higher-order structure, the α-crystallin domain forms a conserved β-sandwich composed of two antiparallel β-sheets [6]. This conserved structure facilitates the assembly of sHsps into oligomeric complexes of up to 800 kDa, which is crucial for their primary function as molecular chaperones, binding to partially denatured proteins to prevent irreversible protein aggregation during stresses such as extreme temperature, oxidation, ultraviolet radiation, heavy metal, and chemical intoxication [7]. In addition to stress responses, the chaperone function of sHsps is also involved in cell growth, differentiation, apoptosis [8], membrane fluidity [9], diapause [10], and lifespan [11].

According to previous reports, the function of Hsp90, Hsp70, and Hsp40 in the maturation process of molting steroid hormone receptors [12, 13], and the genes encoding small Hsps may be regulated by ecdysone. The regulation of the *D. melanogaster Hsp23* gene by ecdysterone has been analyzed by measuring the activities of hybrid *Hsp-Escherichia coli* β-galactosidase genes in transfected hormone-sensitive *D. melanogaster* cells [14]. Salivary gland culture experiments have shown that the *Hsp27* gene from *Ceratitis capitata* can also be regulated by 20-hydroxyecdysone (20E) [15]. However, there is limited information on the ecdysone regulation of the genes encoding small Hsps in other insects.

We previously cloned and characterized one small Hsp gene from *A. cerana cerana* [16]. In the present study, another two small Hsp genes were cloned, and the regulation of these genes by heat shock and ecdysone was examined. Moreover, the recombinant proteins showed molecular chaperone activity *in vitro* by inhibiting the thermal aggregation of the mitochondrial malate dehydrogenase enzyme.

Experimental Procedures

Animals and treatments

Worker bees of *A. cerana cerana* were obtained from the Technology Park of Shandong

Agricultural University. The entire bodies of the fourth (L4)- and the sixth (L6)-instar larvae, white (Pw)-, pink (Pp)-, and dark (Pd)-eyed pupae from the hive, and adult worker bees (2 and 10 d after emergence) were obtained by labeling newly emerged bees with paint. The head (HE), thorax (TH), midgut (MG), muscle (MS), epidermis (EP), and fat body (FA) of newly emerged bees were dissected on ice, frozen immediately in liquid nitrogen, and stored at −80℃. Adult bees (12 d) were also divided into groups (each group contains 30 individuals) and subjected to different temperatures.

The second (L2)-instar larvae were maintained in a 24-well plate and reared on an artificial diet at 34℃ and 95% humidity. After treated with different artificial diets for 96 h, the larvae were flash-frozen in liquid nitrogen at the indicated time points and stored at −80℃.

RNA extraction, cDNA synthesis, and DNA isolation

Total RNA was extracted by using the Trizol reagent (Invitrogen, Carlsbad, CA, USA) according to the instructions of the manufacturer. The RNA was digested with DNase I (RNase-free) to remove any genomic DNA, and cDNA was synthesized by using a reverse transcription system (TransGen Biotech, Beijing, China). Genomic DNA was isolated by using an EasyPure Genomic DNA Extraction Kit in accordance with the instructions of the manufacturer (TransGen Biotech, Beijing, China).

PCR primers and conditions

The primers used in this study are listed in Table 1. The amplification conditions are shown in Table 2.

Table 1 Primers used in this study

Abbreviation	Primer sequence (5'—3')	Description
P1	ATGTCATTGTTGCCGCTG	cDNA sequence primer of *AccHsp24.2*, forward
P2	GCTCTTCATGCTGATTCG	cDNA sequence primer of *AccHsp24.2*, reverse
P3	ATGTCTCTGATTCCATTG	cDNA sequence primer of *AccHsp23.0*, forward
P4	TCACCATCGAGCAGACGGG	cDNA sequence primer of *AccHsp23.0*, reverse
5P1	GATCGTCCACATCGTATC	5' RACE reverse primer of *AccHsp24.2*, outer
5P2	ATGTCATTGTTGCCGCTG	5' RACE reverse primer of *AccHsp24.2*, inner
3P1	GCTCTTCATGCTGATTCG	3' RACE forward primer of *AccHsp24.2*, outer
3P2	GCAAACAGCTGACGAGAAG	3' RACE forward primer of *AccHsp24.2*, inner
5P3	CCTCATCGACTACTTGATC	5' RACE reverse primer of *AccHsp23.0*, outer
5P4	ATGTCTCTGATTCCATTG	5' RACE reverse primer of *AccHsp23.0*, inner
3P3	GGAGCAACCAAAGATCCAG	3' RACE reverse primer of *AccHsp23.0*, outer

Continued

Abbreviation	Primer sequence (5'—3')	Description
3P4	TCACCATCGAGCAGACGGG	3' RACE reverse primer of *AccHsp23.0*, inner
AAP	GGCCACGCGTCGACTAGTAC(G)14	Abridged anchor primer
AUAP	GGCCACGCGTCGACTAGTAC	Abridged universal amplification primer
B26	GACTCTAGACGACATCGA(T)18	3' RACE universal adaptor primer
B25	GACTCTAGACGACATCGA	3' RACE universal primer
QP1	GTTGAACTTCAAACATTGG	Full-length cDNA primer of *AccHsp24.2*, forward
QP2	CTATGTTAATATGTATAGACTG	Full-length cDNA primer of *AccHsp24.2*, reverse
QP3	CACTCGAGAGAGAAGCTTGAC	Full-length cDNA primer of *AccHsp23.0*, forward
QP4	GTGTTCCTGCTTCTTACTGC	Full-length cDNA primer of *AccHsp23.0*, reverse
GP1	GTTGAACTTCAAACATTGG	Genomic sequence primer of *AccHsp24.2*, forward
GP2	CTATGTTAATATGTATAGACTG	Genomic sequence primer of *AccHsp24.2*, reverse
GP3	CACTCGAGAGAGAAGCTTGAC	Genomic sequence primer of *AccHsp23.0*, forward
GP4	GTGTTCCTGCTTCTTACTGC	Genomic sequence primer of *AccHsp23.0*, reverse
PS1	CAACAGAGCTATGTTGGAT	Inverse PCR forward primer, outer of *AccHsp24.2*
PX1	CACATCGCATCTGGGATC	Inverse PCR reverse primer, outer of *AccHsp24.2*
PS2	GATTGGTCGTGGTAGAGG	Inverse PCR forward primer, inner of *AccHsp24.2*
PX2	TTCTGCCTGGTGGGCAG	Inverse PCR reverse primer, inner of *AccHsp24.2*
PS3	GCAACCAAAGATCCAGAGCG	Inverse PCR forward primer, outer of *AccHsp23.0*
PX3	GGTTTGGGAATATATCCCG	Inverse PCR reverse primer, outer of *AccHsp23.0*
PS4	CGGCATTGAAGGAGAACACG	Inverse PCR forward primer, inner of *AccHsp23.0*
PX4	CTGATTGGTGGGAAGACCTG	Inverse PCR reverse primer, inner of *AccHsp23.0*
RP1	GATTGGTCGTGGTAGAGG	Real-time PCR primer of *AccHsp24.2*, forward
RP2	GCTCTTCGTGCTGATTCG	Real-time PCR primer of *AccHsp24.2*, reverse
RP3	CTGATTGGTGGGAAGACCTG	Real-time PCR primer of *AccHsp23.0*, forward
RP4	CTTGGGGTGAACTTCTGCG	Real-time PCR primer of *AccHsp23.0*, reverse
β-actin-s	GTTTTCCCATCTATCGTCGG	Standard control primer, forward
β-actin-x	TTTTCTCCATATCATCCCAG	Standard control primer, reverse
EP1	GGTACCATGTCATTATTGCCG	Protein expression primer, forward of *AccHsp24.2*
EP2	AAGCTTTTTTTCTCTGGCATC	Protein expression primer, reverse of *AccHsp24.2*
EP3	GGTACCATGTCTTTGATTCCA	Protein expression primer, forward of *AccHsp23.0*
EP4	AAGCTTATTTTCTTCTTTCTTC	Protein expression primer, reverse of *AccHsp23.0*

Table 2 PCR amplification conditions used in this study

Primer pair	Amplification conditions
P1 + P2/P3 + P4	6 min at 94℃, 40 s at 94℃, 40 s at 52℃, 1 min at 72℃ for 35 cycles, 10 min at 72℃
5P1 + B26/5P3 + B26	6 min at 94℃, 40 s at 94℃, 40 s at 53℃, 30 s at 72℃ for 28 cycles, 5 min at 72℃
5P2 + B25/5P4 + B25	6 min at 94℃, 40 s at 94℃, 40 s at 58℃, 30 s at 72℃ for 35 cycles, 5 min at 72℃
3P1 + AAP/3P3 + AAP	6 min at 94℃, 40 s at 94℃, 40 s at 49℃, 40 s at 72℃ for 28 cycles, 5 min at 72℃
3P2 + AUAP/3P4 + AUAP	6 min at 94℃, 40 s at 94℃, 40 s at 55℃, 40 s at 72℃ for 35 cycles, 5 min at 72℃
QP1 + QP2/QP3 + QP4	6 min at 94℃, 40 s at 94℃, 40 s at 52℃, 1 min at 72℃ for 35 cycles, 10 min at 72℃
GP1 + GP2/GP3 + GP4	6 min at 94℃, 40 s at 94℃, 40 s at 52℃, 1.5 min at 72℃ for 35 cycles, 10 min at 72℃
PS1 + PX1/PS3 + PX3	6 min at 94℃, 50 s at 94℃, 50 s at 50℃, 2 min at 72℃ for 30 cycles, 5 min at 72℃
PS2 + PX2/PS4 + PX4	6 min at 94℃, 50 s at 94℃, 50 s at 50℃, 2 min at 72℃ for 30 cycles, 5 min at 72℃
EP1 + EP2/EP3 + EP4	6 min at 94℃, 40 s at 94℃, 40 s at 53℃, 1.5 min at 72℃ for 28 cycles, 10 min at 72℃

Isolation of the *AccHsp24.2* and *AccHsp23.0* genes

The primer pairs P1 + P2 and P3 + P4 were designed based on a conserved DNA sequence among homologous genes from *Apis mellifera*, *Nasonia vitripennis*, and *Macrocentrus cingulum* and used to clone the internal conserved fragments of *AccHsp24.2* and *AccHsp23.0* by RT-PCR. The 5P1, 5P2, 3P1, and 3P2 primers and 5P3, 5P4, 3P3, and 3P4 primers were designed based on these internal fragment sequences for 5′ and 3′ RACE (rapid amplification of cDNA ends), respectively. Following RACE-PCR, the full-length cDNA sequences were derived through the assembly of the 5′- and 3′-end sequences and the internal fragment. Subsequently, The full-length cDNA sequence of the two genes was amplified by the primer pairs QP1 + QP2 and QP3 + QP4. The genomic DNA was also cloned (GP1 + GP2 and GP3 + GP4) by using genomic DNA as the template. All the PCRs were performed as previously described [17].

Cloning of the 5′-flanking regions of *AccHsp24.2* and *AccHsp23.0*

To clone the 5′-flanking regions of *AccHsp24.2* and *AccHsp23.0*, the restriction endonucleases *Dra* Ⅰ and *Ssp* Ⅰ, respectively, were used in inverse PCR. Genomic DNA was digested with the indicated restriction endonuclease at 37℃ overnight, ligated by using T4 DNA ligase (TaKaRa, Dalian, China), and then used as templates. The PS1 + PX1 and PS2 + PX2 primers and the PS3 + PX3 and PS4 + PX4 primers were designed based on the genomic sequence of *AccHsp24.2* and *AccHsp23.0* to obtain the 5′-flanking regions. The PCR conditions are shown in Table 2. The PCR products were cloned into the pMD18-T vector and sequenced. The transcription factor database TRANSFAC R.3.4 [18] and MatInspector were used to search for transcription factor binding sites in the 5′-flanking region.

Bioinformatic and phylogenetic analysis

Conserved domains in *AccHsp24.2* and *AccHsp23.0* were detected by using the NCBI bioinformatics tools (http://blast.ncbi.nlm.nih.gov/Blast.cgi). In addition, the DNAMAN software was used to search for open reading frames (ORFs) and to perform multiple sequence alignments. The PSIPRED Protein Sequence Analysis Workbench (http://bioinf.cs.ucl.ac.uk/psipred/) was used to predict secondary structures, and the Protein Data Bank (http://www.pdb.org/) was searched for protein 3D structure models. Phylogenetic and molecular evolutionary analyses were performed by using the neighbor-joining method with the Molecular Evolutionary Genetics Analysis (MEGA version 4.1) software.

Expression analysis by using real-time PCR

To determine the expression profile of *AccHsp24.2* and *AccHsp23.0*, real-time PCR was performed with the SYBR® PrimeScript™ RT-PCR Kit (TaKaRa, Dalian, China) by using a CFX 96™ real-time system (Bio-Rad, USA). The *AccHsp24.2* and *AccHsp23.0* transcripts were amplified by using two pairs of specific primers: RP1 + RP2 and RP3 + RP4, respectively. The primers β-actin-s and β-actin-x were used to amplify the *β-actin* transcripts from *A. cerana cerana* (GenBank accession number: XM_640276). The following real-time PCR amplification conditions were used: an initial denaturation at 95℃ for 30 s; 41 cycles of 95℃ for 5 s, 53℃ for 15 s, and 72℃ for 15 s; and a single melting cycle from 65℃ to 95℃. The relative expression was analyzed by using the comparative CT method ($2^{-\Delta\Delta C_t}$ method). Three individual samples were prepared from each sample, and all the samples were analyzed in triplicate.

Construction of expression plasmids, recombinant protein expression, and purification

To express the recombinant AccHsp24.2 and AccHsp23.0 proteins in BL21 cells, specific primers (EP1 + EP2 and EP3 + EP4, respectively) were designed to amplify the entire ORF. Each ORF sequence was cloned into pEASY-T3 and digested with the restriction endonucleases *Kpn* I and *Hin*dIII; the sequences were then ligated into the expression vector pET-30a (+), which was digested with the same restriction endonucleases. We next transformed each expression plasmid, pET-30a (+)-AccHsp24.2 and pET-30a (+)-AccHsp23.0, into the BL21 (DE3) *Escherichia coli* strain. After intermediate culture and isopropylthio-β-D-galactoside (IPTG) induction, the bacterial cells were harvested by centrifugation, resuspended, and sonicated. After centrifugation of the lysate, the pellet was resuspended in PBS and subjected to 15% SDS-PAGE for expression analysis. The recombinant proteins were purified by using MagneHis™ Protein Purification System (Promega, Madison, WI, USA) according to the instructions of the manufacturer, and the purified proteins were examined by 15% SDS-PAGE.

Ecdysone treatment *in vitro*

New pupae during the Pw stage were dissected in sterilized Ringer's saline for the extraction of the thoracic integument. The adhering muscle and fat body were removed, and pools of five or six thoracic integuments were incubated in 5 ml of commercially available Grace's culture medium specifically for insect tissues (Invitrogen, Carlsbad, CA, USA). The integument was incubated in the presence of 1, 0.1, or 0.001 μg/ml 20E of culture medium for a period of 3 h. After this process, the integuments were washed for 3 h in hormone-free medium [19]. Total RNA was extracted from the incubated integuments and analyzed by RT-qPCR.

Larval rearing in the laboratory

A. cerana cerana larvae were collectively reared in 24-well plates (Costar, Corning Incorporated, USA) and placed in a desiccator at a constant temperature (34℃) and humidity (96%); a 10% glycerol solution was used inside the desiccator. The larvae were initially fed an artificial diet [20] and, once reaching the third instar, were randomly transferred to artificial diets containing 1, 0.1, and 0.01 μg/ml 20E. 24-well plates were filled with 48 larvae (two larvae/well), and four 24-well plates were treated with the same concentration of 20E. The larvae that were still alive after 72 h of feeding on the treatment diets were separated into groups for RNA extraction and analyzed by RT-PCR.

Molecular chaperone activity of the recombinant proteins

The *in vitro* molecular chaperone activity of AccHsp24.2 and AccHsp23.0 was evaluated by measuring their capacity to suppress the thermal aggregation of mitochondrial malate dehydrogenase (MDH) from pig heart (EC 1.1.1.37; Amresco). Four samples (A, MDH; B, MDH + BSA; C, MDH + AccHsp24.2; and D, MDH + AccHsp23.0) were incubated at 43℃. BSA (bovine serum albumin) was used as a control to exclude nonspecific chaperone activity. The absorbance was monitored at 360 nm by using an Ultrospec 3000 UV/visible spectrophotometer (Pharmacia Biotech) at regular intervals (10 min) for 1 h [6].

Results

Cloning and sequence analysis of *AccHsp24.2* and *AccHsp23.0*

A sequence analysis indicated that the cDNA sequence of *AccHsp24.2* (GenBank accession number: JF411009) obtained from *A. cerana cerana* is 868 bp in length and contains a 146 bp 5′ untranslated region (UTR) and a 74 bp 3′ UTR. The 648 bp ORF encodes a protein of 216 amino acids with a predicted molecular weight of 24.2 kDa and an estimated pI of 4.93. The cDNA sequence of *AccHsp23.0* (GenBank accession number: JF411010) is 780 bp in length and contains a 124 bp 5′ UTR and a 53 bp 3′ UTR. The 603 bp ORF encodes a protein of 201 amino acids with a predicted molecular weight of 23.0 kDa and an estimated pI of 5.21. The genomic sequences of these two genes were also obtained, and no introns were present.

By Using the NCBI Conserved Domain (CD) search, the α-crystallin domains (ACDs) of metazoan α-crystallin-type sHsps were found in the deduced amino acid sequences of AccHsp24.2 and AccHsp23.0, suggesting that the two genes might belong to the α-crystallin/sHsp family. An analysis of the secondary structure indicated that the helical contents of AccHsp24.2 and AccHsp23.0 are predicted to be 3.26% and 8.5%, respectively, with strand contents of 20.93% and 24.5%, respectively. β-sheets were found to occur frequently in the α-crystallin domain, which is consistent with De Jong et al., who determined that the secondary structure of sHsps is rich in β-sheets [5]. In addition, there were no cysteine residues in the entire protein sequence of AccHsp24.2 and only two in AccHsp23.0. These findings are consistent with the research of Fu et al. [21], who showed that cysteine residues were rarely found in the sequences of molecular chaperones compared to other protein families. Moreover, we also analyzed the tertiary structures of the two gene products, and both proteins shared the best template for 3D homology modeling of the deduced protein, a solid-state NMR structure of the α-crystallin domain in α-β-crystallin oligomers (PDB code, 2KLR; E-value, 5.5127E-25 for AccHsp24.2 and 5.7653E-26 for AccHsp23.0) (Fig. 1). This model consisted of a single polypeptide chain of approximately 175 amino acids [22].

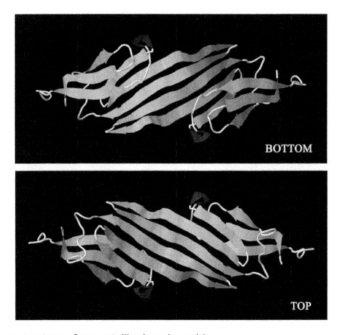

Fig. 1 The tertiary structure of α-crystallin domain residues.

Protein sequence alignment and phylogenetic analysis of AccHsp24.2 and AccHsp23.0

To further validate the conserved domain, a multiple-sequence alignment was performed by using the Clustal X program (Fig. 2). The predicted amino acid sequence of AccHsp24.2

exhibited 43.64%, 42.00%, 52.63%, 48.33%, and 55.72% similarity to Hsp27.6 from *A. cerana cerana* (GenBank accession number: GQ254650), Hsp20.7 from *Locusta migratoria* (GenBank accession number: DQ355965), sHsp from *Macrocentrus cingulum* (GenBank accession number: EU624206), Hsp21.7 from *Nasonia vitripennis* (GenBank accession number: XP001604512), and AccHsp23.0, respectively. AccHsp23.0 was 44.07%, 49.74%, 62.62%, 64.43%, and 55.72% identical to AccHsp27.6, LmHsp20.7, McsHsp, NvHsp21.7, and AccHsp24.2, respectively. The protein sequence comparison also showed that two conserved regions are present in these sHsps. The first 27 amino acids constitute the first conserved region, which is highly hydrophobic and might be involved in modulating oligomerization, subunit dynamics, and substrate binding [23]. The second region, which includes the α-crystallin domain, is prominent and determines the molecular chaperone function of the sHsps.

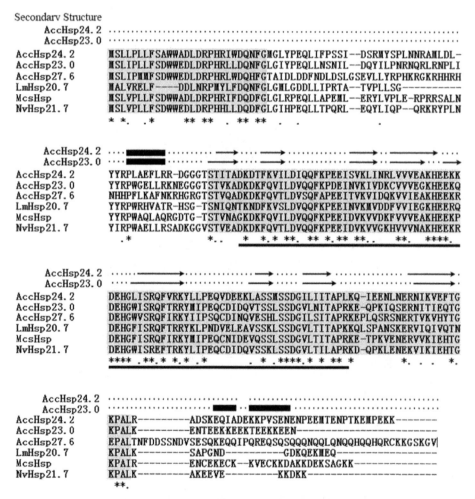

Fig. 2 Amino acid sequence comparison of sHsp homologs. The asterisks, double dots, and single dots denote fully, strongly, and weakly conserved residues, respectively. The gray sections show the two conserved regions, and the α-crystallin domain is underlined. The secondary structure assignments of AccHsp24.2 and AccHsp23.0 are shown above the sequences; the arrowheads, ribbons, and dotted lines denote helices, strands, and coils, respectively.

As shown in Fig. 3, the two sHsp genes cloned appeared in the same cluster and were orthologous to the sHsp genes from hymenopteran species (McsHsp, NvHsp21.7 and AccHsp27.6). there was an orthologous cluster containing several sHsps from different insect orders, which is showed in Fig. 3. This orthologous cluster suggests that these sHsps may have evolved prior to the divergence of the species [15].

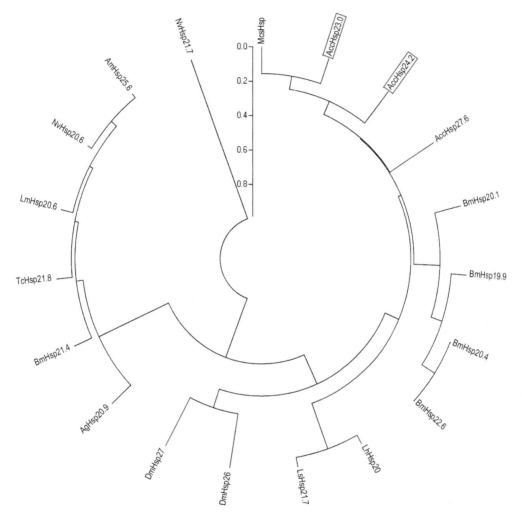

Fig. 3 Phylogenetic analysis of sHsp amino acid sequences from different insect species (Diptera, Hymenoptera, and Lepidoptera). The full names of the species and the accession numbers of the genes are designated with the following abbreviations: DmHsp26 (*Drosophila melanogaster*, NM079273), DmHsp27 (*Drosophila melanogaster*, NM079276), LhHsp20 (*Liriomyza huidobrensis*, DQ452370), LsHsp21.7 (*Liriomyza sativa*, DQ452372), McsHsp (*Macrocentrus cingulum*, EU624206), NvHsp21.7 (*Nasonia vitripennis*, XP001604512), NvHsp20.6 (*Nasonia vitripennis*, XM001607625), AccHsp27.6 (*Apis cerana cerana*, GQ254650), AccHsp24.2 (*Apis cerana cerana*, JF411009), AccHsp23.0 (*Apis cerana cerana*, JF411010), BmHsp20.1 (*Bombyx mori*, AB195971), BmHsp20.4 (*B. mori*, AF315318), BmHsp22.6 (*Bombyx mori*, FJ602788), BmHsp21.4 (*Bombyx mori*, AB195972), AgHsp20.9 (*Anopheles gambiae*, XM560153), TcHsp21.8 (*Tribolium castaneum*, XM968592), AmHsp25.6 (*Apis mellifera*, XM392405), BmHsp19.9 (*Bombyx mori*, NM001043519.1), and LmHsp20.6 (*Locusta migratoria*, DQ355964). AccHsp23.0 and AccHsp24.2 are in two boxes.

Characterization of the 5′-flanking region of *AccHsp24.2* and *AccHsp23.0*

We identified the 5′-flanking regions to investigate the regulation of the transcription of *AccHsp24.2* and *AccHsp23.0*. By using inverse PCR, we obtained a 727 bp and an 885 bp DNA fragment upstream of the translation start sites of *AccHsp24.2* and *AccHsp23.0*, respectively. As shown in Fig. 4, three heat shock elements (HSEs) were detected in *AccHsp24.2* and eleven were detected in *AccHsp23.0*. In addition, a TATA-box, representing the putative core promoter element upstream of the transcription start site, was found in both genes. We also searched for transcription factor binding sites by using MatInspector, and two Broad-Complex (BR-C) genes for ecdysone steroid response elements were found in each gene. BR-C is a key regulator of metamorphosis and primary ecdysone response gene, which play an important role in insect metamorphosis [24, 25]. The specific expression of this transcription factor in epidermal tissue is related to the development of the pupal skin [26]. The presence of DBRC sites suggested that *AccHsp24.2* and *AccHsp23.0* are related to the regulation of the ecdysone cascade.

Fig. 4 The nucleotide sequence and putative transcription factor binding sites of the 5′-flanking regions of *AccHsp24.2* (A) and *AccHsp23.0* (B). The transcription and translation start sites are indicated with arrowheads. The underlined sections indicate the DBRC binding sites. The HSEs and the TATA-box are highlighted in light gray.

Developmental and tissue-specific expression patterns of *AccHsp24.2* and *AccHsp23.0*

As shown in Fig. 5, the trends of the developmental expression patterns of the *AccHsp24.2* and *AccHsp23.0* genes were similar. The amount of *AccHsp24.2* mRNA increased rapidly from the fourth-instar larval stage to the sixth-instar larval stage, decreased from the sixth-instar larval stage to the pink-eyed pupal stage, increased slightly at the dark-eyed pupal stage, and showed an increasing trend at the adult stage. The amount of *AccHsp23.0* mRNA reached a peak at the fourth-instar larval stage and decreased at the pink-eyed pupal stage, and, similar to *AccHsp24.2*, then increased slightly at the dark-eyed pupal stage, increasing at the adult stage. The tissue-specific expression analysis indicated high transcription levels of the two genes predominantly in the head and muscle.

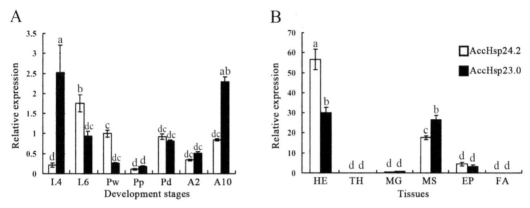

Fig. 5 Expression patterns of *AccHsp24.2* and *AccHsp23.0* at different developmental stages and in various tissues. A. Expression patterns of *AccHsp24.2* and *AccHsp23.0* at different developmental stages. Total RNA was extracted from whole honeybees at the fourth-instar larval stage(L4) and the sixth-instar larval stages (L6), the successive pupal stages, including the white-eyed pupal phase (Pw), pink-eyed pupal phase (Pp), and dark-eyed pupal phase (Pd), and 2- and 10-day-old adult worker bees (A2 and A10). B. Expression patterns of *AccHsp24.2* and *AccHsp23.0* in various tissues. Total RNA was extracted from the dissected head (HE), thorax (TH), midgut (MG), muscle (MS), epidermis (EP), and fat body (FA) of newly emerged bees. The *A. cerana cerana β-actin* gene was used as a reference house-keeping gene to normalize the expression level of the investigated gene in the RT-qPCR experiments. The histograms indicate the relative expression levels. The data represent the means ± SE of three independent experiments. The different letters above the columns indicate significant differences (A, $P < 0.05$; B, $P < 0.01$).

Heat stress and cold stress regulation of *AccHsp24.2* and *AccHsp23.0*

The expression levels of the two genes under different temperature conditions were evaluated by using RT-qPCR, and the results indicated that the two genes could be induced by both heat shock and cold shock. The expression patterns of the two genes were similar, but the level of *AccHsp24.2* was notably higher than *AccHsp23.0*. As shown in Fig. 6, after incubation at 37℃, the amount of mRNA of both genes showed no significant change but increased rapidly during the recovery period at 34℃. In response to 43℃, the expression of the two genes increased rapidly during the first 30 min and then decreased to a low level; during the recovery period at 34℃, the expression remained high and showed increasing trends. In addition to heat stress, we also examined the expression patterns of the two genes under cold shock. As shown in Fig. 6, the expression patterns of the two genes rapidly reached their peak levels in the first 30 min and then decreased to their lowest levels, even during recovery at 34℃.

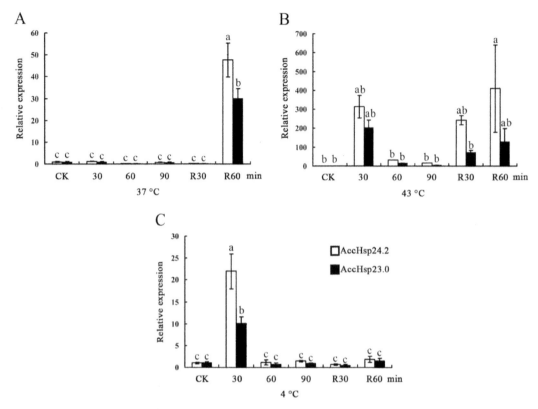

Fig. 6 Expression patterns of *AccHsp24.2* and *AccHsp23.0* in response to different temperatures. Total RNA was extracted from adult bees (12 days) treated for the indicated times at 37℃ (A), 43℃ (B) and 4℃ (C). The *A. cerana cerana β-actin* gene was used as a reference house-keeping gene to normalize the expression level of the investigated gene in the RT-qPCR experiments. The histograms indicate the relative expression levels. The data represent the means ± SE of three independent experiments. The different letters above the columns indicate significant differences ($P < 0.01$). R30 min, recovery at 34℃ for 30 min. R60 min, recovery at 34℃ for 60 min.

Ecdysone regulation of the *AccHsp24.2* and *AccHsp23.0* genes

According to Zhou and Riddiford, the expression of BR-C mRNA under the control of ecdysone is one of the first molecular events underlying the pupal commitment of the epidermis [26]. As we detected two BR-C sites in the 5′-flanking regions of both *AccHsp24.2* and *AccHsp23.0*, we speculated that the expression of the two genes might be regulated by ecdysone. To verify this hypothesis, we studied the expression pattern of the two genes in the pupal epidermis following induction by ecdysone. As shown in Fig. 7A, the results of real-time PCR indicated that the two genes exhibited similar expression patterns. The expression levels of the two genes decreased significantly after the removal of 0.1 μg/ml and 0.01 μg/ml ecdysone; however the removal of 0.001 μg/ml ecdysone did not cause a notable change in expression.

We also detected the ecdysone regulation of *AccHsp24.2* and *AccHsp23.0 in vivo*. We added three different titers of ecdysone to the diet of L2 to L6 larvae and examined the expression of the two genes by using real-time PCR. As shown in Fig. 7B, all three titers of ecdysone could upregulate the two genes, though different expression patterns were observed. The expression of *AccHsp24.2* only slightly responded to 0.1 μg/ml ecdysone in the ration, which was the lowest of the three titers, whereas the expression of *AccHsp23.0* reached extremely high levels in response to this dose.

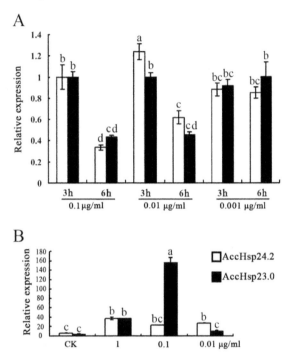

Fig. 7 Expression patterns of *AccHsp24.2* and *AccHsp23.0* in response to different titers of ecdysone *in vitro* and *in vivo*. A. Expression patterns of *AccHsp24.2* and *AccHsp23.0* in response to different titers of ecdysone *in vitro*. Total RNA was extracted from the epidermis of new pupae that were incubated *in vitro* in an incubation medium containing different titers of ecdysone (0.1, 0.01, and 0.001 μg/ml) for 3 h and were then washed for 3 h. B. Expression patterns of *AccHsp24.2* and *AccHsp23.0* in response to different titers of ecdysone *in vivo*. Total RNA was extracted from L6 bees fed an artificial diet containing different titers of ecdysone (0, 0.01, 0.1, and 1 μg/ml) for 96 h. The *A. cerana cerana β-actin* gene was used as a reference house-keeping gene to normalize the expression level of the investigated gene in the RT-qPCR experiments. The histograms indicate the relative expression levels. The data represent the means ± SE of three independent experiments. The different letters above the columns indicate significant differences ($P < 0.01$).

Molecular chaperone activity of recombinant AccHsp24.2 and AccHsp23.0 proteins

The two genes were expressed in *E. coli* BL21 cells as 6 × His-tagged fusion proteins. After induction with IPTG at 37℃, an SDS-PAGE analysis detected bands corresponding to the two 6 × His fusion proteins (Fig. 8A, lanes 1 and 4); no protein induction was observed in the non-induced controls (Fig. 8A, lanes 2 and 3). The results indicated that the two genes were expressed, and the fusion proteins were then purified (Fig. 8B, lanes 1 and 2).

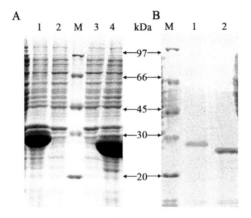

Fig. 8 Expression in *E. coli* and the purification of recombinant AccHsp24.2 and AccHsp23.0 proteins as analyzed by SDS-PAGE. Lane A1, induced AccHsp24.2 recombinant protein; lanes A2 and A3, non-induced; lane A4, induced AccHsp23.0 recombinant protein; lane B1, purified AccHsp24.2 recombinant protein; lane B2, purified AccHsp23.0 recombinant protein; lane M, molecular mass standard.

The α-crystallin domain is involved in the structure formation of sHsp and not in the determination of the chaperone function. Indeed, some sHsps that carry the α-crystallin domain have no chaperone activity [27].

Therefore, the molecular chaperone activity of the purified recombinant AccHsp24.2 and AccHsp23.0 proteins were investigated *in vitro* by using the MDH thermal aggregation assay. As shown in Fig. 9, MDH aggregated when the solution was incubated at 43 ℃, as indicated

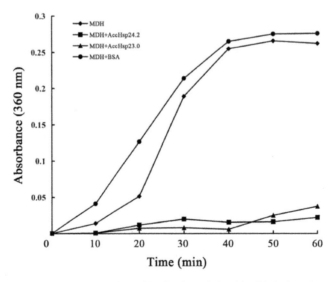

Fig. 9 Molecular chaperone activity of AccHsp24.2 and AccHsp23.0 demonstrated by using the malate dehydrogenase thermal aggregation assay. MDH incubated at 43 ℃ is represented by filled diamonds. MDH incubated with AccHsp24.2 in a molar ratio 1∶1 is represented by filled squares. MDH incubated with AccHsp23.0 in a molar ratio 1∶1 is represented by filled triangles, and MDH incubated with BSA in a molar ratio 1∶1 is represented by filled circles. The absorbance was monitored at 360 nm at regular intervals (10 min) for 1 h.

by an increase in dispersed light at 320 nm. In addition, MDH aggregated faster when mixed with BSA, which has no chaperone activity. However, the addition of an equivalent amount of AccHsp24.2 or AccHsp23.0 to MDH almost completely suppressed MDH aggregation (Fig. 9). Therefore, the two recombinant proteins can act as efficient molecular chaperones *in vitro*.

Discussion

In a previous study, we cloned and characterized the first sHsp gene (*AccHsp27.6*) in the Chinese honeybee (*A. cerana cerana*). In the present study, we isolated two other sHsp genes, and the amino acid sequence, structure, presence of the α-crystallin domain, and *in vitro* molecular chaperone activity of the recombination proteins indicated that the two genes belong to the α-crystallin/sHsp family. A homologous sequence alignment showed that these two sHsp genes share a high sequence homology to *AccHsp27.6* (Fig. 2). Accordingly, the genes were named *AccHsp24.2* and *AccHsp23.0*. The protein sequence comparison also showed that two conserved regions are present in the sHsps. The first 27 amino acids constitute the first conserved region, which is highly hydrophobic and might be involved in modulating oligomerization, subunit dynamics, and substrate binding [23]. The second region, which includes the α-crystallin domain, is prominent and determines the molecular chaperone function of the sHsps. While in the phylogenetic analysis of the two genes, the orthologous cluster suggests that these sHsps may have evolved prior to the divergence of the species [15]. These two genes also had similar developmental and heat shock expression profiles. In addition, both were induced by ecdysone, suggesting that the two genes might be regulated by ecdysone. Furthermore, we also expressed the recombinant AccHsp24.2 and AccHsp23.0 proteins and detected a remarkable *in vitro* molecular chaperone activity.

Heat shock elements (HSEs), which are recognized by heat shock factors (HSFs), are the most critical elements in Hsp gene transcription. We detected three putative HSEs in the 5'-flanking region of *AccHsp24.2* and eleven in *AccHsp23.0*. According to a report on *Drosophila Hsp22*, three functional HSEs in the 5'-flanking region are required for the proper expression of Hsp22 following stress [28]. However, the HSEs responsible for the heat induction of *AccHsp24.2* and *AccHsp23.0* have not yet been determined.

According to previous reports, sHsps can prevent the caspase-dependent apoptosis that occurred during heat and cold stresses in *Drosophila* cells [29, 30]. Some studies have analyzed the expression of the sHsps in response to heat shock in insects. For example, the expression of the *Drosophila Hsp23* gene could be induced several times by heat, over a broad temperature range (30-37°C), reaching its maximum level at 35°C [31]. However, only a few studies have investigated the expression of sHsps in response to cold stress. In non-diapausing flesh flies, the expression of *Hsp23* was induced by both heat and cold shock (43 and −10°C) [32]. In the present study, we observed the notable upregulation of *AccHsp24.2* and *AccHsp23.0* in response to 37°C during recovery for 60 min. Similarly, we also found a remarkable upregulation during incubation at 43°C and throughout the recovery process. These results suggested that sHsps could markedly improve the thermal tolerance of the insect. The upregulation of *AccHsp24.2* and *AccHsp23.0* was only observed at 30 min in response to 4℃, in agreement with a previous report that showed that there was no modulation of *Hsp23* transcription during the recovery from a short cold stress at 0℃ for 3 h in *Drosophila* [33]. However, Qin et al. [34] showed that a cold stress of only 2 h at 0℃ resulted in the upregulation of *Hsp23* and *Hsp26* in *D. melanogaster* during a 30 min recovery phase; according to Colinet et al. [35], the four sHsp genes of *D. melanogaster* were all upregulated after cold stress at 0℃ for 9 h. In addition to their molecular chaperone functions, sHsps are involved in various physiological processes, some of which are important for the cold tolerance of the insect. For example, sHsps help to protect the integrity of the actin cytoskeleton and microfilaments, which is important for the cold tolerance of the insect because increasing evidence has shown that cytoskeletal components are involved in the cold tolerance of the insect [36, 37].

In addition to heat induction, we also observed the expression of *AccHsp24.2* and *AccHsp23.0* during normal development in this study. The two genes exhibited similar expression patterns during development, suggesting that they are controlled by the same regulatory mechanisms, except at the fourth-instar larvae. *AccHsp23.0* RNA reached its peak level at the fourth-instar larvae, whereas the RNA level of *AccHsp24.2* was very low at the fourth-instar larvae and peaked at the sixth-instar larvae. After the larval stage, the two genes had different maximum RNA levels, one of which was at the dark-eyed pupal stage and the other in 10-day post-emergent adults. According to Kokolakis et al. [31], medfly *Hsp23* RNA levels peaked at the third-instar larvae and remained very high during most of larval development, decreased significantly at the end of the larval stage, increased again at puparium formation and the prepupal stages, and increased during the aging of adults. The expression patterns of *AccHsp24.2* and *AccHsp23.0* were similar to the two medfly *Hsp23* genes, suggesting analogous regulation of these genes in the two species during development.

In *Drosophila*, the expression of the *Hsp27* gene during embryonic, late larval, and prepupal development is correlated with the three major peaks in the titer of the molting hormone, ecdysone. This finding suggests that the hormone may be responsible for most of the developmental expressions of this gene [15]. In our study, we detected two BR-C sites in the 5′-flanking regions of *AccHsp24.2* and *AccHsp23.0*. Broad-Complex is an early gene induced by ecdysone and is the key transcription factor in ecdysone cascade regulation. The specific expression of Broad-Complex in the epidermal tissue is related to the formation of the pupal epidermis. It has been reported that mutated *Drosophila* larvae without all BR-C isomers could only develop to 3 days old and could not form chrysalises [13]. According to Ireland et al. [38], the activation of *Drosophila* small Hsp genes in the late larval and early pupal stages may be regulated by the steroid hormone ecdysterone. Therefore, we examined how ecdysone affected the transcription of the *AccHsp24.2* and *AccHsp23.0* genes *in vitro* and *in vivo* and found that a high titer of ecdysone could significantly upregulate the expression of the two small Hsp genes in the pupal epidermis *in vitro*. This result was consistent with Mestril et al. [14], who showed that all four *Drosophila* small Hsps (*Hsp22*, *Hsp23*, *Hsp26*, and *Hsp27*) were synthesized in elevated quantities during puparium formation, a stage when the ecdysterone titer is high. However, treatment with 0.001 μg/ml ecdysone did not improve the expression of *AccHsp24.2* and *AccHsp23.0*, suggesting that the residual ecdysone levels in the epidermis were approximately 0.001 μg/ml. In the larva-rearing experiment, both genes were upregulated by all the titers of ecdysone tested, though different patterns of expression were observed, suggesting the existence of different regulatory mechanisms in the larval stage.

In this study, we isolated and characterized two small heat shock protein genes, *AccHsp24.2* and *AccHsp23.0*, from *A. cerana cerana* and focused on their regulation by heat shock and ecdysone. However, it remains to be determined whether there is a correlation between the heat shock regulation and ecdysone regulation.

References

[1] Yang, G. Harm of introducing the western honeybee *Apis mellifera* L. to the Chinese honeybee *Apis cerana* F. and its ecological impact. Acta Entomol Sin. 2005, 3, 401-406.

[2] Hendrick, J. P., Hartl, F. U. Molecular chaperone functions of heat-shock proteins. Annu Rev Biochem. 1993, 62, 349-384.

[3] Mizrahi, T., Heller, J., Goldenberg, S., Arad, Z. Heat shock proteins and resistance to desiccation in congeneric land snails. Cell Stress and Chaperones. 2010, 15, 351-363.

[4] Michaud, S., Morrow, G., Marchand, J., Tanguay, R. M. *Drosophila* small heat shock proteins: cell and organelle-specific chaperones? Prog Mol Subcell Biol. 2002, 28, 79-101.

[5] De Jong, W. W., Caspers, G. J., Leunissen, J. A. Genealogy of the alpha-crystallin-small heat-shock protein superfamily. Int J Biol Macromol. 1998, 22, 151-162.

[6] Pérez-Morales, D., Ostoa-Saloma, P., Espinoza, B. *Trypanosoma cruzi* SHSP16: characterization of an α-crystallin small heat shock protein. Exp Parasitol. 2009, 123, 182-189.

[7] Reineke, A. Identification and expression of a small heat shock protein in two lines of the endoparasitic wasp *Venturia canescens*. Comp Biochem Physiol A: Mol Integr Physiol. 2005, 141, 60-69.

[8] Arrigo, A. P. Small stress proteins: chaperones that act as regulators of intracellular redox state and programmed cell death. Biol Chem. 1998, 379, 19-26.

[9] Tsvetkova, N. M., Horvath, I., Torok, Z., et al. Small heat-shock proteins regulate membrane lipid polymorphism. Proc Natl Acad Sci USA. 2002, 99, 13504-13509.

[10] Gkouvitsas, T., Kontogiannatos, D., Kourti, A. Differential expression of two small Hsps during diapause in the corn stalk borer *Sesamia nonagrioides* (Lef.). J Insect Physiol. 2008, 54, 1503-1510.

[11] Morrow, G., Samson, M., Michaud, S., Tanguay, R. M. Overexpression of the small mitochondrial *Hsp22* extends *Drosophila* life span and increases resistance to oxidative stress. FASEB J. 2004, 18, 598-599.

[12] Brandt, G. E., Blagg, B. S. Alternate strategies of *Hsp90* modulation for the treatment of cancer and other diseases. Curr Top Med Chem. 2009, 9, 1447-1461.

[13] Zheng, W. W., Yang, D. T., Wang, J. X., Song, Q. S., Gilbert, L. I., Zhao, X. F. Hsc70 binds to ultraspiracle resulting in the upregulation of 20-hydroxyecdsone-responsive genes in *Helicoverpa armigera*. Mol Cell Endocrinol. 2010, 315, 282-291.

[14] Mestril, R., Schiller, P., Amin, J., Klapper, H., Ananthan, J., Voellmy, R. Heat shock and ecdysterone activation of the *Drosophila melanogaster hsp23* gene; a sequence element implied in developmental regulation. EMBO J. 1986, 5, 1667-1673

[15] Kokolakis, G., Tatari, M., Zacharopoulou, A., Mintzas, A. C. The *hsp27* gene of the Mediterranean fruit fly, *Ceratitis capitata*: structural characterization, regulation and developmental expression. Insect Mol Biol. 2008, 17, 699-710.

[16] Liu, Z., Xi, D., Kang, M., Guo, X., Xu, B. Molecular cloning and characterization of *Hsp27.6*: the first reported small heat shock protein from *Apis cerana cerana*. Cell Stress and Chaperones. 2012, 17, 539-551.

[17] Wang, J., Wang, X., Liu, C., Zhang, J., Zhu, C., Guo, X. The *NgAOX1a* gene cloned from *Nicotiana glutinosa* is implicated in the response to abiotic and biotic stresses. Biosci Rep. 2008, 28, 259-266.

[18] Tsunoda, T., Takagi, T. Estimating transcription factor bindability on DNA. Bioinformatics. 1999, 15, 622-630.

[19] Sun, R., Zhang, Y., Xu, B. Characterization of the response to ecdysteroid of a novel cuticle protein R&R gene in the honey bee, *Apis cerana cerana*. Comp Biochem Physiol B: Biochem Mol Biol. 2013, 166 (1), 73.

[20] Silva, I. C., Message, D., Cruz, C. D., Campos, L. A., Sousa-Majer, M. J. Rearing Africanized honey bee (*Apis mellifera* L.) brood under laboratory conditions. Genet Mol Res. 2009, 8, 623-629.

[21] Fu, X., Li, W., Mao, Q., Chang, Z. Disulfide bonds convert small heat shock protein Hsp16.3 from a chaperone to a non-chaperone: implications for the evolution of cysteine in molecular chaperones. Biochem Biophys Res Commun. 2003, 308, 627-635.

[22] Jehle, S., Rajagopal, P., Bardiaux, B., et al. Solid-state NMR and SAXS studies provide a structural basis for the activation of αB-crystallin oligomers. Nat Struct Mol Biol. 2010, 17, 1037-1042.

[23] Sun, Y., MacRae, T. H. Small heat shock proteins: molecular structure and chaperone function. Cell Mol Life Sci. 2005, 62, 2460-2476.

[24] Von Kalm, L., Crossgrove, K., Von Seggern, D., Guild, G. M., Beckendorf, S. K. The Broad-Complex directly controls a tissue-specific response to the steroid hormone ecdysone at the onset of *Drosophila* metamorphosis. EMBO J. 1994, 13, 3505-3516.

[25] Uhlirova, M., Foy, B. D., Beaty, B. J., Olson, K. E., Riddiford, L. M., Jindra, M. Use of Sindbis virus-mediated RNA interference to demonstrate a conserved role of Broad-Complex in insect metamorphosis. Proc Natl Acad Sci USA. 2003, 100, 15607-15612.

[26] Zhou, B., Riddiford, L. M. Hormonal regulation and patterning of the Broad-Complex in the epidermis and wing discs of the tobacco hornworm, *Manduca sexta*. Dev Biol. 2001, 231, 125-137.

[27] Kokke, B. P., Leroux, M. R., Candido, E. P., Boelens, W. C., De Jong, W. W. *Caenorhabditis elegans* small heat-shock proteins Hsp12.2 and Hsp12.3 form tetramers and have no chaperone-like activity. FEBS Lett. 1998, 433, 228-232.

[28] Morrow, G., Battistini, S., Zhang, P., Tanguay, R. M. Decreased lifespan in the absence of expression of the mitochondrial small heat shock protein *Hsp22* in *Drosophila*. J Biol Chem. 2004, 279, 43382-43385.

[29] Concannon, C. G., Gorman, A. M., Samali, A. On the role of Hsp27 in regulating apoptosis. Apoptosis. 2003, 8, 61-70.

[30] Yi, S. X., Moore, C. W., Lee Jr, R. E. Rapid cold-hardening protects *Drosophila melanogaster* from cold-induced apoptosis. Apoptosis. 2007, 12, 1183-1193.

[31] Kokolakis, G., Kritsidima, M., Tkachenko, T., Mintzas, A. C. Two *hsp23* genes in the Mediterranean fruit fly, *Ceratitis capitata*: structural characterization, heat shock regulation and developmental expression. Insect Mol Biol. 2009, 18, 171-181.

[32] Yocum, G. D., Joplin, K. H., Denlinger, D. L. Upregulation of a 23 kDa small heat shock protein transcript during pupal diapause in the flesh fly, *Sarcophaga crassipalpis*. Insect Biochem Mol Biol. 1998, 28, 677-682.

[33] Sinclair, B. J., Gibbs, A. G., Roberts, S. P. Gene transcription during exposure to, and recovery from, cold and desiccation stress in *Drosophila melanogaster*. Insect Mol Biol. 2007, 16, 435-443.

[34] Qin, W., Neal, S. J., Robertson, R. M., Westwood, J. T., Walker, V. K. Cold hardening and transcriptional change in *Drosophila melanogaster*. Insect Mol Biol. 2005, 14, 607-613.

[35] Colinet, H., Lee, S. F., Hoffmann, A. Temporal expression of heat shock genes during cold stress and recovery from chill coma in adult *Drosophila melanogaster*. FEBS J. 2010, 277, 174-185.

[36] Colinet, H., Nguyen, T. T., Cloutier, C., Michaud, D., Hance, T. Proteomic profiling of a parasitic wasp exposed to constant and fluctuating cold exposure. Insect Biochem Mol Biol. 2007, 37, 1177-1188.

[37] Kim, M., Denlinger, D. L. Decrease in expression of beta-tubulin and microtubule abundance in flight muscles during diapause in adults of *Culex pipiens*. Insect Mol Biol. 2009, 18, 295-302.

[38] Ireland, R. C., Berger, E., Sirotkin, K., Yund, M. A., Osterbur, D., Fristrom, J. Ecdysterone induces the transcription of four heat-shock genes in *Drosophila* S3 cells and imaginal discs. Dev Biol. 1983, 93, 498-507.

1.2 *sHsp22.6*, an Intronless Small Heat Shock Protein Gene, Is Involved in Stress Defence and Development in *Apis cerana cerana*

Yuanying Zhang, Yaling Liu, Xulei Guo, Yalu Li, Hongru Gao, Xingqi Guo, Baohua Xu

Abstract: Small heat shock proteins (sHsps) play an important role in protecting against stress-induced cell damage and fundamental physiological processes. In this study, we identified an intronless sHsp gene from *Apis cerana cerana* (*AccsHsp22.6*). The open reading frame of *AccsHsp22.6* was 585 bp and encoded a 194 amino acid protein. Furthermore, a 2,064 bp 5′-flanking region was isolated, and potential transcription factor binding sites associated with development and stress response were identified. Quantitative PCR and Western blotting analysis demonstrated that *AccsHsp22.6* was detected at higher levels in the midgut than in other tissues tested, and it is highly expressed during the shift to different developmental stages. Moreover, *AccsHsp22.6* was significantly upregulated by abiotic and biotic stresses, such as 4, 16 and 42℃, cyhalothrin, pyridaben, H_2O_2, UV, $CdCl_2$, 20-hydroxyecdysone and *Ascosphaera apis* treatments. However, *AccsHsp22.6* was slightly repressed by other stresses, including 25℃, phoxim, paraquat and $HgCl_2$ treatments. The recombinant AccsHsp22.6 also exhibited significant temperature tolerance, antioxidation and molecular chaperone activity. In addition, we found that knockdown of *AccsHsp22.6* by RNA interference remarkably reduced temperature tolerance in *A. cerana cerana*. Taken together, these results suggest that *AccsHsp22.6* plays an essential role in the developmental stages and defence against cellular stress.

Keywords: *Apis cerana cerana*; *sHsp*; Expression analysis; Stress response; RNA interference

Introduction

Living cells are inevitably challenged by adverse conditions such as extreme temperature, oxidative stress, toxic substances or infectious agents. Protein aggregations resulting from these stresses pose a dominating threat to the cells [1]. To maintain cellular protein homeostasis, cells have evolved a "protein quality control" network, which is mainly mediated by autophagy, the proteasome, and heat shock proteins (Hsps) [2].

Hsps belong to a large protein family which is classified into several superfamilies, including Hsp100, Hsp90, Hsp70, Hsp60, Hsp40 and sHsps, based on their molecular weight and sequence homology[3]. The sHsps are ATP-independent molecular chaperones that comprise a less conserved, more diverse and small molecular weight (15-30 kDa) group of proteins[1]. sHsps share common structural characteristics: a conserved α-crystallin domain containing 80-100 amino acids, an extremely flexible, variable C-terminus and a disorganized N-terminus [4, 5]. It has been suggested that the conserved α-crystallin domain is important for maintaining molecular chaperone activity and other functions of sHsp in cells whereas the diverse C-terminal and N-terminal sequences may be involved with the diverse expressions, functions, and evolutionary patterns of sHsps [6, 7]. Thus, the sequences and number of sHsps vary between and within species.

Similar to other chaperones, sHsps can bind partially denatured proteins and facilitate correct refolding, which protects cells against protein aggregation after exposure to various stresses [3]. Thus, sHsps are ubiquitous and crucial components of protein quality control networks. In addition to their central role as chaperones, sHsps participate in diverse fundamental cellular and physiological processes, such as cell proliferation and differentiation, cytoskeletal organisation, redox homeostasis, apoptosis and the immune response [3, 4]. The varied expression pattern of sHsps is also involved in many human diseases, such as cataracts, neurodegenerative disorders, cardiovascular diseases, and cancers [8].

Although the heat shock response was first observed in *Drosophila*, the understanding of sHsps in insects is not as comprehensive and profound as in bacteria, plants, and mammals. Apart from the fundamental chaperone function, recent studies found that sHsps might be involved in insect development, particularly in larval-pupal metamorphosis. Huang et al. found that in *Liriomyza sativa* the highest expression levels of three sHsps were observed in the pupal stage, but the expression of three sHsps was upregulated gradually with development [9]. Sonoda et al. also reported similar results in the diamondback moth, *Plutella xylostella* [10]. The increased levels of sHsps were speculated to facilitate cell proliferation and the regeneration of new tissues and organs during metamorphosis [9].

The Chinese honeybee (*A. cerana cerana*) is an important pollinator that plays a critical role in maintaining the balance of regional ecologies in China. Compared with the Western honeybee (*A. mellifera*), it has some advantages, such as higher cold tolerance and disease resistance, and the capability of long-distance flight [11]. However, the population of the Chinese honeybee has declined severely in recent decades, which is attributed to an epidemic of honeybee diseases and the deterioration of its environment, apart from invasion of *A. mellifera* [11, 12]. To date, sHsps have been widely considered as the first line of defence against adverse conditions. Therefore, the study of sHsps in the Chinese honeybee is necessary. Here, we isolated and characterized an intronless *sHsp* gene from *A. cerana cerana* (*AccsHsp22.6*) and evaluated its expression profiles at different developmental stages and in various tissues. We also investigated the response of *AccsHsp22.6* to various stresses. Moreover, the recombinant AccsHsp22.6 protein displayed significant temperature tolerance, antioxidation and molecular chaperone activity. In addition, we knocked down *AccsHsp22.6* in *A. cerana cerana* and evaluated its protective role in cold or heat stress. Our results provide new insights into the molecular characterization and functions of insect sHsps.

Materials and Methods

Insects and various treatments

The Chinese honeybees (*A. cerana cerana*) were maintained in the experimental apiary of Technology Park of Shandong Agricultural University (Taian, China). The egg, the first- to the fifth-instar (L1-L5) larvae, pupae [prepupae (PP), white-eyed (Pw), pink-eyed (Pp), brown-eyed (Pb) and dark-eyed (Pd) pupae] and newly emerged adults (A1) were identified according to the criteria of Michelette and Soares and collected from the hive [13]. The 15 days post-emergence (A15) and 30 days post-emergence adults (A30) were collected at the entrance of the hive after marking newly emerged bees with paint 15 and 30 days earlier. The 15 days post-emergence adults were divided into groups (*n*=40/group), fed an adult diet of water, 70% powdered sugar and 30% honey and maintained in an incubator at a constant humidity (70%) and temperature (34℃) under a 24 h dark regimen [14]. Groups 1-4 were placed at different temperatures (4, 16, 25 and 42℃). Groups 5-12 were treated with the following conditions: ultraviolet (UV)-light (30 MJ/cm^2); H_2O_2 (0.5 μl of a 2 mmol/L dilution was injected between their first and second abdominal segments by using a sterile microscale needle); $HgCl_2$ (3 mg/ml added to food); $CdCl_2$ (0.5 mg/L added to food); or pesticides (0.5 μl of cyhalothrin, phoxim, pyridaben, and paraquat, which were applied to the thoracic notum of worker bees and the final concentrations were 20 μg/L, 1 μg/ml, 10 μmmol/L, and 10 μmmol/L, respectively). The untreated adults, as the control group, were fed normal food. Bees in group 6 injected with phosphate buffered saline (PBS) (0.5 μl/worker) were the injection controls. The larvae were collectively reared as previously described [15, 16]. For 20-hydroxyecdysone (20E) treatment, once the larvae reached their third instar, they were transferred to an artificial diet containing different doses of 20E: 1, 0.1, 0.01 and 0.001 μg/ml. The larvae were reared in 24-well plate (Costar, Corning Incorporated, USA) which were filled with 48 larvae/plate (2 larvae/well) and three plates were treated with one concentration of 20E. Ethanol was the 20E solvent and was used in the 20E controls. Its final concentration was 0.1 mg/ml. The living larvae were collected after 96 h of feeding on the treatment diets. For microbe treatment, once the larvae reached their second instar they were inoculated with *Ascosphaera apis*. One 24-well plate was filled with 48 larvae/plate (2 larvae/well) and three plates were treated. The control larvae were fed only artificial diets. The living larvae were collected after treatment at 96 h. All of the bee specimens were immediately frozen in liquid nitrogen at the indicated time points and stored at −80℃. For the tissue-specific expression analysis, the brain, epidermis, midgut and muscle of adults (15 d) were dissected on ice, flash-frozen in liquid nitrogen and stored at −80℃. These experiments were performed in triplicate.

Cloning of *AccsHsp22.6* cDNA

The extraction of total RNA and the isolation of the full-length cDNA sequence were performed as previously described [17].

Amplification of the genomic sequence and promoter of *AccsHsp22.6*

The genomic DNA was extracted according to Yan et al. [17]. The *AccsHsp22.6* promoter region was obtained by using inverse PCR (I-PCR) as previously described [18]. All of the PCR primers used are listed in Table S1. The MatInspector database (http://www.genomatix.de/matinspector.html) was used to predict transcription factor binding sites in the *AccsHsp22.6* promoter region. All of the PCR amplification conditions are listed in Table S2.

Bioinformatic analysis

The bioinformatics tools available at the NCBI server (http://blast.ncbi.nlm.nih.gov/Blast.cgi) were used to detect the *AccsHsp22.6* conserved domains. The DNAMAN software program version 5.2.2 (Lynnon Biosoft, Quebec, Canada) was used to identify the *AccsHsp22.6* ORF and align its homologues. The theoretical isoelectric point (pI) and molecular mass were predicted by the Expasy-Sib Bioinformatics Resource Portal (http://web.expasy.org/compute_pi/). The phylogenetic tree was constructed by the neighbour-joining method using the Molecular Evolution Genetics Analysis (MEGA version 5.2) software program. I-TASSER, an online protein structure prediction tool server (http://zhanglab.ccmb.med.umich.edu/I-TASSER), was used to predict the tertiary structure of AccsHsp22.6 [19].

Protein expression, purification, and antisera preparation

To construct the pET-30a (+)-AccsHsp22.6 plasmid, the coding region of *AccsHsp22.6* was amplified with primers (EP1/EP2) containing *Bam*H I and *Hin*dIII restriction enzyme sites and inserted into the expression vector pET-30a (+) (Novagen, Darmstadt, Germany) after digestion with the same restriction endonucleases. Then, this expression plasmid was transformed into BL21 (DE3) cells, and the cells were induced with 0.2 mmol/L isopropylthio-β-D-galactoside (IPTG) at 30℃ for 8 h during log phase growth. The recombinant AccsHsp22.6 protein was purified by using a method from Zhang et al. [18], and examined by 12% SDS-PAGE. To generate antisera, white mice were immunized subcutaneously by using the purified protein as previously described [18].

Real-time quantitative PCR (qPCR) and Western blotting analysis

The qPCR and Western blotting analysis were performed as described previously [18]. In brief, the SYBR® PrimeScript™ RT-PCR Kit (TaKaRa, Dalian, China) and the CFX96™ Real-time System (Bio-Rad, Hercules, CA, USA) were used for the qPCR analysis. The *A. cerana cerana β-actin* gene (GenBank accession number: HM_640276) was selected as an internal control. The gene transcriptional level was evaluated in three independent biological replicates for each group with three technical repeats for each primer pair. For Western blotting analysis, three different developmental stages (1 day larvae, pink-eyed pupae, and newly emerged adults) and four different tissues (brain, muscle, epidermis, and midgut) were selected to extract their total protein, respectively. The anti-AccsHsp22.6 serum was served as the primary antibody at a dilution of 1 : 500 (V/V). Peroxidase-conjugated goat anti-mouse IgG (Dingguo, Beijing, China) was served as the secondary antibody at a dilution of 1 : 2,000 (V/V). The SuperSignal® West Pico Trial Kit (Thermo Scientific Pierce, IL, USA) was employed to detect the proteins.

Thermotolerance and cold tolerance assay of *Escherichia coli* overexpressing AccsHsp22.6

A temperature tolerance assay was performed as previously reported from *Artemia franciscana* [20], and 4℃ was used for the cold tolerance assay. In brief, the cells with the pET-30a (+)-AccsHsp22.6 plasmid were induced with 0.2 mmol/L IPTG at 30℃ for 8 h. Immediately, the cells were incubated at 4℃ for 12, 24, 36, and 48 h and at 50℃ for 0, 15, 30, 45, and 60 min, respectively. Then, the cells were diluted to 5×10^6 cells/ml in cold LB-kanamycin broth and 50 μl cultures were plated in triplicate on LB agar plates. Colonies were counted after overnight at 37℃ to calculate the survival rate (the ratios of viable cell numbers per plate in the treated samples and untreated samples). In addition, the recombinant AccGSTO2 (*A. cerana cerana* Omega-class glutathione *S*-transferase) from our previous study, a construct overexpressing non-Hsp protein, was served as a negative control[18]. These experiments were performed in triplicate.

In vitro molecular chaperone activity

To evaluate the chaperone activity of AccsHsp22.6, its capacity to suppress aggregation of the pig heart mitochondrial malate dehydrogenase (MDH, Sigma, USA) was measured as previously described [21]. For the aggregation protection assay, 6 μmol/L MDH was incubated at 45℃ for 60 min, with AccsHsp22.6 (molar ratio 1 : 1).

Disc diffusion assay

To determine the protection capacity of AccsHsp22.6 under oxidative stress, a disc diffusion assay was implemented with a method from Burmeister et al. [22]. Different concentrations of cumene hydroperoxide (0, 25, 50, 80, or 100 mmol/L) were used in this study.

RNA interference (RNAi)

The dsRNAs were synthesised as previously described [23]. The primer sequences (HI1/HI2) are listed in Table S1. The newly emerged adult honeybees were used for the RNAi experiments and were divided into 4 groups (n=50/group). They were injected with 0.5 μl (6 μg) of ds*AccsHsp22.6*, or with an equal concentration of a ds*GFP* into their thoracic muscle. The other two groups were kept untreated, or were injected with 0.5 μl of nuclease-free water. The healthy bees were subsequently sampled each day. This experiment was repeated three times.

Immunohistochemistry (IHC)

To detect the effect of RNAi, IHC was implemented as previously described [16]. The AccsHsp22.6-antiserum (1 ∶ 50 dilution) was used as the primary antibody in this study.

Temperature tolerance assays under RNAi treatment

The newly emerged adult honeybees were divided into 4 groups (n=100/group) with an above described RNAi treatment for 1 day. Subsequently, each group was equally divided into two subgroups (n=50/subgroup), maintained, respectively, in an incubator at a constant temperature (4 or 50 ℃) and humidity (75%) for 1-5 h. Their survival rate was evaluated each hour. A bee that did not respond to stimulus was identified as dead.

Results and Discussion

Isolation and molecular characterization of *AccsHsp22.6*

The full-length 904 bp cDNA sequence of *AccsHsp22.6* (GenBank accession number: KF150016) was obtained by RT-PCR and RACE-PCR, which included a 115 bp 5' untranslated region (UTR), a 585 bp complete open reading frame (ORF) and a 204 bp 3' UTR. The predicted ORF encoded a 194 -amino acid protein with computed theoretical pI (isoelectric point) of 5.88 and M_r (molecular mass) of 22.605 kDa. In addition, AccsHsp22.6 has 35 glutamine (Gln) and glutamic acid (Glu) residues which maintain protein stability under elevated temperatures by providing additional electrostatic forces (Fig. 1) [24].

Multiple sequence alignments revealed 71%-99% sequence identity between AccsHsp22.6 and other insect sHsps including *Apis mellifera*, *A. florae*, *Bombus impatiens*, *Megachile rotundata*, *Nasonia vitripennis* and *Macrocentrus cingulum*, suggesting that the sHsps are highly conserved across Hymenoptera insects (Fig. 1A). Moreover, the deduced AccsHsp22.6 shares typical features of the sHsp family: the N-terminal domain contains the partially conserved sequence PSRLFDQXFGEXLL, which might be required for modulating sHsp oligomerisation and chaperoning [4], it has the α-crystallin domain (ACD) of metazoan ACD-type sHsps, contains 79 amino acid residues (75-153) consisting of β-sheets (β2-β5) that are essential for dimer formation [6], and has a variable C-terminal extension presenting the IXI motif, which drives higher-order assembly [5]. The analysis suggested that the AccsHsp22.6 belongs to the typical sHsp family. In addition, 8 residues that compose the putative dimer interface were predicted by the NCBI server, and the conserved protein kinase C (PKC) phosphorylation sites (Ser^{139} and Thr^{170}) were indicated by the PROSITE database.

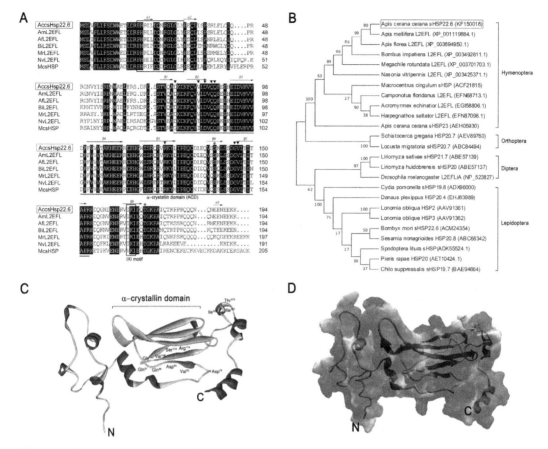

Fig. 1 Molecular properties of AccsHsp22.6. A. The amino acid sequence alignment of AccsHsp22.6 and other sHsps. The putative secondary structure of AccsHsp22.6 is shown. Identical amino acids are shaded in blue. The α-crystallin domain (ACD) is underlined. The IXI motif is boxed. The putative dimer interface and PKC phosphorylation sites are denoted by (▼) and (❖). B. Phylogenetic relationships on the basis of the amino acid sequences between sHsps from different insect species. The four insect species are shown, and AccsHsp22.6 is boxed. C. The tertiary structure of AccsHsp22.6. The putative dimer interface residues (Asp^{75}, Val^{82}, Asp^{84}, Gln^{86}, Gln^{87}, Arg^{124}, Gly^{145} and Val^{146}), PKC phosphorylation site residues (Ser^{139} and Thr^{170}), IXI motif (Ile^{165} and Ile^{167}), N-terminus, and C-terminus are shown. D. The surface electrostatic potential distributions in the structure models of AccsHsp22.6. Negative and positive charges are represented in blue and red, respectively. These images were coloured by secondary structure succession and using Swiss-PdbViewer. (For color version, please sweep QR Code in the back cover)

A phylogenetic tree of the sHsps was constructed on the basis of the amino acid sequences to investigate the evolutionary relationships among the different insect sHsps (Fig. 1B). Of the four insect species sHsps examined, AccsHsp22.6 was categorized into the Hymenoptera order, where it clustered with the predicted *A. mellifera* lethal (2) essential for life (AmL2EFL), a small heat-shock-homologous protein [25]. Thus, we named the putative protein AccsHsp22.6 by using its predicted M_r.

The three-dimensional structure of AccsHsp22.6 was also predicted by I-TASSER using the human αB-crystallin (PDB 2yjdA) as the best template as it contains all 194 amino acids and shares 39% sequence identity with AccsHsp22.6. As shown in Fig. 1C, the conserved ACD was enriched for β-strands generating a β-sheet sandwich, which is the basic unit for the dimerisation and oligomerisation of sHsps [8]. In addition, the β-sheet sandwich surface was surrounded by negative charges (Fig. 1D) and might be related to substrate binding [26].

Genomic structure and putative *cis*-acting elements of *AccsHsp22.6*

Based on chromosome location and intron number, Li et al. had roughly grouped the sHsp genes into two types in silkworms: an orthologous cluster and a species-specific cluster [7]. The former, involved in mostly metabolic processes, has at least one intron and the latter, associated with responses to environmental stresses, has no intron. To further elucidate the genomic structure of *AccsHsp22.6*, the genomic DNA sequence of honeybees was used as a template for PCR amplification. The result revealed that *AccsHsp22.6* had no intron, which is consistent with the homology DNA sequences of other insects identified (Fig. 2). Therefore, we speculated that *AccsHsp22.6* may be mainly involved in environmental stress responses.

A 2,064 bp fragment (GenBank accession number: KF150016) located in the 5'-flanking region of *AccsHsp22.6* was obtained by I-PCR. As shown in Fig. 2, except for some sequences involved in embryo or tissue development, including CF2-II, Dfd, Broad-Complex (BR-C), NIT2, and CdxA [18], several important transcription factors associated with environmental stress and immune response, such as heat shock factors (HSF), activating protein-1 (AP-1), cAMP response element binding protein (CREB), nuclear factor kappa B (NF-κB), p53 and X-box binding protein-1 (XBP-1), were also found in the region [27]. In addition, five GATA motifs which participated in the development of the endoderm and gene expression in terminally differentiated endodermal tissue, were detected in the promoter region [22]. The promoter region also revealed an alcohol dehydrogenase gene regulator 1 (ADR1), a positive regulator of transcription of genes encoding peroxisomal proteins [28], and a nuclear factor-erythroid 2-related factor 2 (Nrf2), which is a master switch controlling the expression of a series of stress response genes critical in mounting a cellular defence against electrophiles or ROS/RNS (reactive oxygen species/reactive nitrogen species) [29]. These putative transcription factor binding sites imply that *AccsHsp22.6* may be involved in not only environmental stress responses but also developmental regulation.

Temporal and spatial expression profiles of *AccsHsp22.6*

qPCR and Western blotting analysis were performed to determine the expression patterns of *AccsHsp22.6*. As shown in Fig. 3A, the mRNA levels of *AccsHsp22.6* fluctuated greatly during the developmental stages. The highest expression level was found in the newly emerged adults, and a relatively high expression of was observed during pupal development. Interestingly, we noticed that a significant increase of *AccsHsp22.6* expression always correlated with the transition of different developmental stages, including the egg-larval, the larval-pupal, and the pupal-adult transition. In holometabolous insects, the shift is always accompanied by dramatic changes in gene expression. Gala et al. reported that the expression

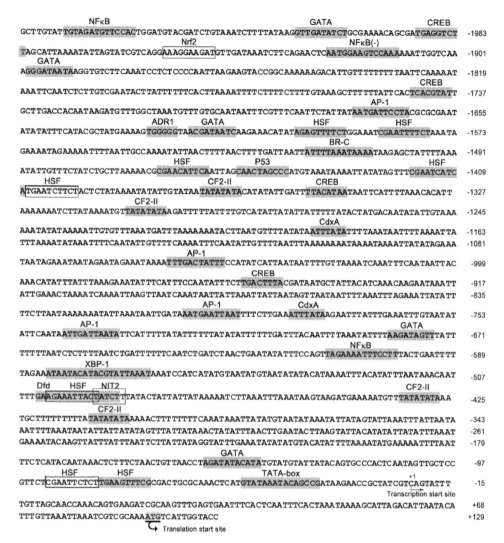

Fig. 2 The nucleotide sequence and putative cis-acting elements of the *AccsHsp22.6* regulatory region. The translation and transcription start sites are marked with arrows. The cis-acting elements are shaded in black or boxed. At least 85% matrix similarity (i.e. 15% mismatches) and 100% similarity to the core similarity were allowed in the identification of putative transcription factor binding sites.

of 65 proteins and 34 phosphoproteins changed across the shift of eggs to larvae, and the newly hatched larvae enhanced the expression of proteins related to protein folding, cytoskeleton, and metabolism to ensure the rapid growth of *A. mellifera ligustica* [30]. During metamorphosis, the natural life transition is mediated by two hormones, juvenile hormone and ecdysone (mainly 20E) which can bind a canonical ecdysone response element (EcRE) to control insect development [31, 32]. *AccsHsp22.6* transcription was shown to be associated with peaks in 20E titre during metamorphosis [33], suggesting a role in 20E response as observed for other *sHsps* [34, 35]. Moreover, BR-C, an ecdysone-responsive key regulator of metamorphosis during third-instar stage and early prepupal development [36], had been found in the regulatory region of *AccsHsp22.6*. In addition, newly emerged adults must still perform considerable metabolic, morphological, and neural maturation to prepare for energetically intensive flight [37]. Therefore, the significant increase of *AccsHsp22.6* expression at these transitions implied that *AccsHsp22.6* might be involved in honeybee development through controlling protein folding.

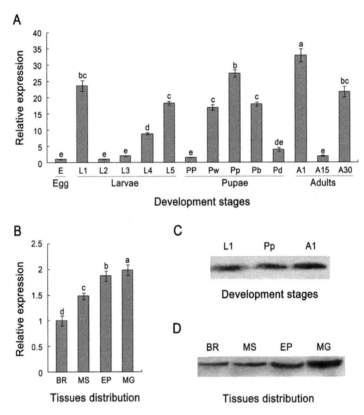

Fig. 3 Expression profile of *AccsHsp22.6* as determined by qPCR (A, B) and Western blotting analysis (C, D). The relative expression of *AccsHsp22.6* at different developmental stages (A) and in different tissues (B) is shown. The data are the means ± SE of three independent experiments. The different letters above the columns indicate significant differences ($P < 0.01$) according to Duncan's multiple range tests. C. Lanes 1-3 were loaded with an equivalent amount of protein: L1, 1 day larvae; Pp, pink-eyed pupae; A1, newly emerged adults. D. Lanes 1-4 were loaded with an equivalent amount of protein: BR, brain; MS, muscle; EP, epidermis; MG, midgut.

The expression of different *sHsps* in a specific tissue most likely reflects the particular need of that tissue. As shown in Fig. 3B, among the adult tissues tested, the highest expression of *AccsHsp22.6* appeared in the midgut, followed by the epidermis. The midgut plays important roles in the detoxification of xenobiotics and protection from oxidative stress [38], and the epidermis acts as the exoskeleton to confer physical stability and protect against various environmental stresses [39]. The tissue-specific expression of *AccsHsp22.6* indicated that the gene may play a potential role in protecting against the damage of xenobiotics. In addition, the accumulation of the AccsHsp22.6 protein was consistent with its transcription level at different developmental stages and in various tissues (Fig. 3C, D), suggesting that no noticeable variation of gene expression occurred at the protein level.

Expression profiles of *AccsHsp22.6* under abiotic stresses

Previous studies have indicated that *sHsps* could be involved in the response to heat shock and other stresses, such as UV radiation, pesticides, oxidation, chemical intoxication and microbe infection [7, 40, 41]. However, most but not all sHsps are stress-induced molecules in cells. To elucidate the potential function of *AccsHsp22.6*, its transcription under abiotic stresses, including temperature (4, 16, 25 and 42 ℃), pesticides (cyhalothrin, phoxim, pyridaben, paraquat), oxidative stresses (UV, H_2O_2), and heavy metals ($CdCl_2$ and $HgCl_2$), was examined by qPCR (Fig. 4).

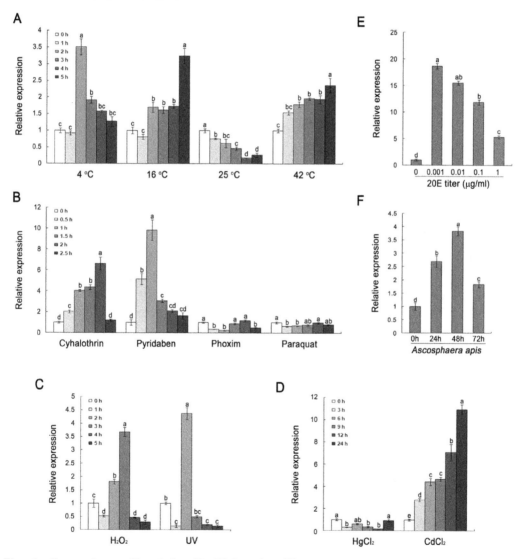

Fig. 4 Expression profile of *AccsHsp22.6* under different stress conditions. These conditions included temperature (A), pesticides (B), H_2O_2 and UV (C), heavy metal (D), 20E (E), and *Ascosphaera apis* (F). Untreated adult worker bees (Lane 0) were used as controls. The data are the means ± SE of three independent experiments. The different letters above the columns indicate significant differences ($P < 0.01$) according to Duncan's multiple range tests.

Temperature, an important environmental factor, can induce physiological changes and oxidative stress in organisms [42]. When treated with temperature stress, *AccsHsp22.6* exhibited different expression profiles and degrees of response (Fig. 4A). *AccsHsp22.6* transcription was upregulated at 4, 16 and 42°C, whereas the 25°C stress did not upregulate the transcriptional level of *AccsHsp22.6*, suggesting that this treatment might not produce enough stimuli to induce *AccsHsp22.6* expression. Alterations to *sHsps* expression induced by heat stress have been extensively reported, but only a few studies have analysed *sHsps* expression under cold stress in insects. In *D. melanogaster*, Colinet et al. reported that five *sHsp* genes in adults could be induced by cold shocks, and the upregulation of *Hsp23* and *Hsp26* transcriptions were observed during a recovery phase after a cold stress [43, 44]. Recently, Waagner et al. also found that rapid cold hardening (RCH) induced significant upregulation of *Hsp40*, *Hsp23*, and *Hsp10* genes in *Folsomia candida*, indicating that the chill sensitive collembolan relies on protein protection from a subset of Hsps during RCH [45]. These observations provide clues for the role of *AccsHsp22.6* in the cold stress response.

Pesticides are ubiquitous environmental contaminants and include insecticides, fungicides, herbicides and others. A previous study suggested that Hsps were helpful as biomarkers to explore the impact of xenobiotics on invertebrate organisms [46]. The induction of *sHsp* genes after exposure to various pesticides has been reported in *D. melanogaster* [47-49]. However, there are limited studies on the alteration of *sHsps* expression in response to pesticides. In the present study, *AccsHsp22.6* expression was slightly changed with phoxim and paraquat treatments, whereas the application of cyhalothrin and pyridaben significantly induced its expression (Fig. 4B). Phoxim, an organophosphate insecticide, acts on the nervous system of the organisms as inhibitor of acetylcholinesterase that hydrolyzes acetylcholine to terminate the transmission of nervous signals. Many organophosphates are too toxic as antiparasitic agents. However, phoxim exhibits a rather low toxicity to mammals due to its lowest rate of uptake through the cuticula [50]. Paraquat is one of the most widely used herbicides in the world. Paraquat, when ingested, is highly toxic to animals. Instead, diluted paraquat, when used for spraying, is less toxic [51]. Thus, we speculated that little effect on *AccsHsp22.6* expression from phoxim and paraquat might attribute to less absorption through the cuticula. Cyhalothrin, a lipophilic insecticide, can hinder phospholipid orientation and lead to changes in membrane fluidity [52]. Pyridaben, a new lipophilic acaricide and insecticide discovered, is an inhibitor of complex I of the mitochondrial electron transport chain [53]. In general, the lipophilic molecules pass very easily through the cell membrane into the cell. Hsp induction by certain toxicants is always connected with the cytotoxic events that affect cellular integrity [49]. Therefore, *AccsHsp22.6* induction by cyhalothrin and pyridaben suggested a possible protective role in the physiological dysfunctions evoked by the two pesticides.

sHsps are known to be important modulators in the cellular response to oxidative stress [54]. Previous studies have demonstrated that oxidative stress can be induced by many environmental conditions, including temperature, pesticides, UV radiation and heavy metal [55, 56]. In this study, some antioxidant response elements, such as HSF, AP-1, ADR1 and Nrf2, were predicted in the 5'-flanking region of *AccsHsp22.6*. To elucidate the potential role of *AccsHsp22.6* in antioxidant defence, transcriptions of *AccsHsp22.6* after H_2O_2, UV, $HgCl_2$, and $CdCl_2$ challenges were investigated by qPCR. The oxidative stress resulting from H_2O_2 and UV significantly induced *AccsHsp22.6* expression (Fig. 4C). However, compared with the time-dependent and significant induction upon $CdCl_2$ stress, the transcription level of *AccsHsp22.6* showed fluctuating and suppressed pattern during $HgCl_2$ stress (Fig. 4D). In general, the cellular injury from heavy metal challenge is mediated by direct protein denaturation and indirect ROS generation [57]. Similar to our findings, several studies have demonstrated that cadmium highly induced *sHsp* expression in other animal models [58, 59]. Nevertheless, mercury was also reported to dose-dependently induce *HdHsp20* expression in abalone [27], suggesting that low-dose mercury might have relatively low toxicity to honeybees and hence fail to induce *AccsHsp22.6* expression.

Expression profiles of *AccsHsp22.6* in response to biotic stress

Similar to abiotic stress, biotic stress also plays a significant role in the regulation of *sHsp* response. As presented in Fig. 4E, *AccsHsp22.6* transcription in honeybee larvae was significantly induced by 20E. This observation provides further evidence for the role of *AccsHsp22.6* in the control of honeybee development as described above. *AccsHsp22.6* expression was also significantly upregulated and peaked at 48 h post *Ascosphaera apis* challenge in honeybee larvae (Fig. 4F). *Ascosphaera apis*, which results in chalkbrood, is a common and widespread fungal pathogen that severely impacts the survival of honeybee larvae and overall colony health. Many studies have demonstrated that Hsps were involved in disease defence mechanisms, but the role of sHsps in microbe infection is not fully understood. Polla and Cossarizza [60] reported that Hsps can prevent protein denaturation arising from excessive ROS generation and provide a protective role in pathogen infection. In the midgut of silkworms, Wu et al. [61], found that *sHsp23.7* expression was upregulated after cytoplasmic polyhedrosis virus (CPV) infection, which might be connected to anti-BmCPV immunity. Wan et al. [27] also reported that HdHsp20 expression was significantly induced after bacteria and virus challenge. Notably, our previous study indicated that microbial infection may regulate *AccHsp27.6* expression, and the analysis of the recombinant AccHsp27.6 protein provided direct antimicrobial evidence [40]. Therefore, several immune response elements found in the 5'-flanking region of *AccsHsp22.6* and the abundant expression of *AccsHsp22.6* after *Ascosphaera apis* infection revealed that *AccsHsp22.6* might be involved in the immune response of the honeybee in chalkbrood.

Expression, purification and characterization of recombinant AccsHsp22.6 protein

AccsHsp22.6 was overexpressed in *E. coli* BL21 (DE3) cells as a His-tag fusion protein after IPTG induction (Fig. 5). An SDS-PAGE analysis revealed that the recombinant protein was soluble and had a molecular mass of approximately 30 kDa, consistent with the predicted molecular mass of 29.505 kDa (containing cleavable N- and C-terminal His-tags of approximately 7 kDa). The soluble recombinant protein was further purified by using a HisTrap™ FF column, and the concentration of the purified AccsHsp22.6 was approximately 1.67 mg/ml.

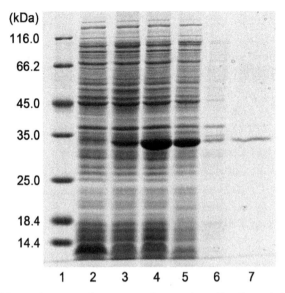

Fig. 5 An SDS-PAGE analysis on the expression and purification of AccsHsp22.6. Lane 1, low-molecular weight protein marker; Lane 2, induced overexpression of pET-30a (+) in *E. coli* BL21 cells; Lanes 3 and 4, non-induced and induced overexpression of pET-30a (+)-*AccsHsp22.6* in cells, respectively; Lanes 5 and 6, suspension and pellet of sonicated recombinant AccsHsp22.6, respectively; Lane 7, purified recombinant AccsHsp22.6.

We have observed the accumulation of *AccsHsp22.6* mRNA after exposure to heat or cold stress (Fig. 4A). To provide direct evidence that AccsHsp22.6 protected cells from extreme temperature injury *in vivo*, we further examined the effect of heat and cold stresses on the survival rate of *E. coil* BL21 (DE3) cells transformed with a vector expressing AccsHsp22.6. The recombinant AccGSTO2 from our previous work, a construct overexpressing non-hsp protein, was used as a negative control in this study [18]. As shown in Fig. 6, *E. coil* expressing AccsHsp22.6 displayed a higher resistance to heat and cold stresses than cells containing only the pET-30a (+) vector. After 48 h cold (4℃) exposure, the survival rate of cells with pET-30a (+) dropped to approximately 16.9%, and that of the cells containing pET-30a (+)-AccsHsp22.6 was 30.6% (Fig. 6A). Moreover, with the exposure time increased at 50℃, the survival rate of the cells with the vector only showed a dramatic decline and no colonies were detected, whereas the survival of the cells with pET-30a (+)-AccsHsp22.6 was 42.7% after 1 h of heat exposure (Fig. 6B). Instead, *E. coil* cells expressing AccsGSTO2 could not show a higher cellular survivability at 4℃ and 50℃ exposure than cells containing only the pET-30a (+) vector. These results indicate that AccsHsp22.6 expression in *E. coil* cells enhanced cellular thermotolerance. The honeybee is an ectotherm, and its distribution and survival are influenced by ambient temperature. Several sHsps have been reported to accumulate to over 1% of the total cellular protein after heat or cold stress, maintaining approximately one-third of cytoplasmic proteins in a soluble state [62, 63]. Therefore, the induction of *AccsHsp22.6* and the higher survival rate in *E. coil* expressing *AccsHsp22.6* during heat and cold stresses implied a similar functional characteristic to other sHsps.

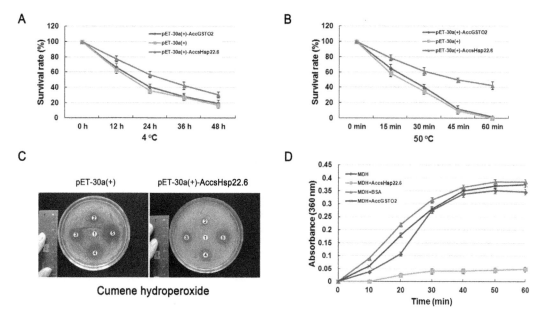

Fig. 6 The functional characterization of recombinant AccsHsp22.6 protein. A and B. The effect of cold and heat stresses, respectively, on the survival rate of E. coli BL21 (DE3) cells transformed with a vector expressing AccsHsp22.6. C. Disc diffusion assays by using E. coli overexpressing AccsHsp22.6. D. Molecular chaperone activity of AccsHsp22.6 shown by the thermal aggregation assay of the malate dehydrogenase.

In Fig. 6C, the killing zones on the plates containing E. coli overexpressing AccsHsp22.6 were smaller than in the control bacteria after overnight exposure to cumene hydroperoxide, suggesting a high tolerance to oxidative stress in cells of overexpressing AccsHsp22.6. The model external pro-oxidant, cumene hydroperoxide, was used because it is more stable than H_2O_2 under the implemented incubation conditions [22]. A disc diffusion assay also provided evidence *in vivo* for the protective roles of AccsHsp22.6 in oxidative stress.

The molecular chaperone activity is a key characteristic of Hsps. Several sHsps have been shown to possess the activity *in vitro*, preventing thermal aggregation of various model substrates [21, 40, 64]. As shown in Fig. 6D, purified AccsHsp22.6 effectively protected malate dehydrogenase (MDH) from heat-induced denaturation. During incubation at 43℃, MDH was denatured, and BSA and the purified AccGSTO2 could not prevent MDH aggregation, which was demonstrated by the notable increase of the absorbance at 360 nm [21]. However, MDH aggregation was almost entirely suppressed when mixed with an equivalent amount of AccsHsp22.6. This observation demonstrated that the AccsHsp22.6 protein could recognize and bind unfolded proteins and possessed significant chaperone activity *in vitro*.

RNAi-mediated silencing of *AccsHsp22.6* reduces the temperature tolerance of the honeybee

Consistent with gene overexpression, RNAi mediated by dsRNA has become a promising strategy for gene function study, especially in insects for which transgenesis is not available [65]. Here, the *AccsHsp22.6* gene was successfully silenced in the adults.

Compared with the no injection group, *AccsHsp22.6* mRNAs in the water or ds*GFP* injection groups were significantly increased 1 day after injection and then reverted back to a level similar to that of the no injection group after 3 days. However, significant transcriptional suppression could be observed in the ds*AccsHsp22.6*-injected group, especially at 1 day after injection (Fig. 7A). To eliminate the interference of the injection stimulus, after RNAi treatment for 3 days, these samples were used to further examine the effect of *AccsHsp22.6* silencing by histological sections. As shown in Fig. 8, the sections from ds*AccsHsp22.6*-treated honeybees were more slightly stained than those of the controls, although without visible morphologic alteration, especially in the midgut and fat body. Additionally, in the temperature tolerance assays after RNAi treatment for 3 days, we found that the *AccsHsp22.6*-RNAi honeybees had lower survival rate than that of the control honeybees during 4 or 50℃ treatment (Fig. 7B, C), meaning they were more sensitive to the temperature challenge. This finding provides further evidence that *AccsHsp22.6* has a protective role in extreme temperature stress.

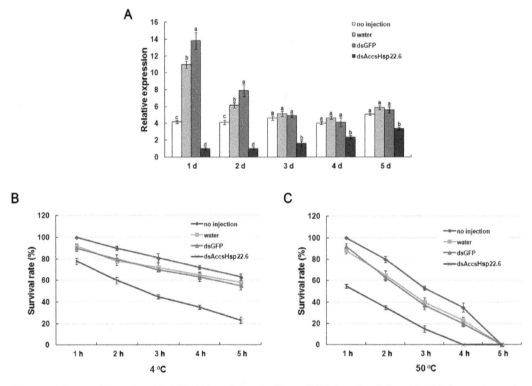

Fig. 7 Effects of *AccsHsp22.6* RNAi in adults. A. The mRNA levels of *AccsHsp22.6* are shown in RNAi tests. The survival rate of *AccsHsp22.6* RNAi adults compared with controls in 4℃ (B) and 50℃ (C). Each value is given as the mean (SD.) of three replicates. The different letters above the bars indicate significant differences ($P < 0.01$), as determined by Duncan's multiple range tests using SAS software version 9.1.

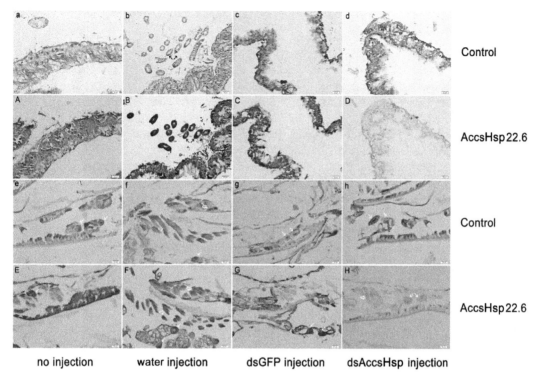

Fig. 8 Immunohistochemistry analyses on the effects of *AccsHsp22.6* knockdown in adults. a-h are negative controls for A-H, respectively.(a-d and A-D. midgut, Malpighian tube (black arrows); e-h and E-H. epidermis, muscle (white and hollow arrows), and fat body (white and solid arrows). Scale bar, 20 μm. (For color version, please sweep QR Code in the back cover)

Conclusion

In this study, we identified and characterized an intronless *sHsp* gene from *A. cerana cerana*. This protein (AccsHsp22.6) possesses the conserved structural and functional domains of the sHsp superfamily. The expression pattern of *AccsHsp22.6* at different developmental stages implied that *AccsHsp22.6* might participate in honeybee development through controlling protein folding. The induction of *AccsHsp22.6* in response to adverse environmental stress indicated that *AccsHsp22.6* may play an important role in stress resistant mechanisms. Recombinant AccsHsp22.6 was shown to enhance the tolerance of *E. coli* to extreme temperatures, antioxidation and molecular chaperone activity *in vitro*, and a disc diffusion assay provided evidence for a protective effect against oxidative stress. Finally, the knockdown of *AccsHsp22.6* also significantly reduced the survival rate of *A. cerana cerana* after exposure to heat or cold challenges. In conclusion, our results suggested that the *AccsHsp22.6* gene may be involved in stress-related defence mechanisms in *A. cerana cerana*, and its expression may enhance the tolerance of *A. cerana cerana* to harsh conditions and protect *A. cerana cerana* from stress-induced injuries.

References

[1] Basha, E., O'Neill, H., Vierling, E. Small heat shock proteins and α-crystallins: dynamic proteins with flexible functions. Trends Biochem Sci. 2012, 37, 106-117.

[2] Liberek, K., Lewandowska, A., Zietkiewicz, S. Chaperones in control of protein disaggregation. EMBO J. 2008, 27, 328-335.

[3] Garrido, C., Paul, C., Seigneuric, R., Kampinga, H. H. The small heat shock proteins family: the long forgotten chaperones. Int J Biochem Cell Biol. 2012, 44, 1588-1592.

[4] Kostenko, S., Moens, U. Heat shock protein 27 phosphorylation: kinases, phosphatases, functions and pathology. Cell Mol Life Sci. 2009, 66, 3289-3307.

[5] Bagnéris, C., Bateman, O. A., Naylor, C. E., Cronin, N., Slingsby, C. Crystal structures of α-crystallin domain dimers of αB-crystallin and Hsp20. J Mol Biol. 2009, 392, 1242-1252.

[6] Hayes, D., Napoli, V., Mazurkie, A., Stafford, W. F., Graceffa, P. Phosphorylation dependence of hsp27 multimeric size and molecular chaperone function. J Biol Chem. 2009, 284, 18801-18807.

[7] Li, Z. W., Li, X., Yu, Q. Y., Zhang, Z. The small heat shock protein (sHSP) genes in the silkworm, *Bombyx mori*, and comparative analysis with other insect sHSP genes. BMC Evol Biol. 2009, 9, 215.

[8] Sun, Y., MacRae, T. H. The small heat shock proteins and their role in human disease. FEBS J. 2005, 272, 2613-2627.

[9] Huang, L. H., Wang, C. Z., Kang, L. Cloning and expression of five heat shock protein genes in relation to cold hardening and development in the leafminer, *Liriomyza sativa*. J. Insect Physiol. 2009, 55, 279-285.

[10] Sonoda, S., Ashfaq, M., Tsumuki, H. Cloning and nucleotide sequencing of three heat shock protein genes (*hsp90*, *hsc70*, and *hsp19.5*) from the diamondback moth, *Plutella xylostella* (L.) and their expression in relation to developmental stage and temperature. Arch Insect Biochem Physiol. 2006, 62, 80-90.

[11] Li, D. C., Yang, F., Lu, B., Yang, W. J. Thermotolerance and molecular chaperone function of the small heat shock protein HSP20 from hyperthermophilic archaeon, *Sulfolobus solfataricus P2*. Cell Stress Chaperones. 2012, 17, 103-108.

[12] Gegear, R. J., Otterstatter, M. C., Thomson, J. D. Bumble-bee foragers infected by a gut parasite have an impaired ability to utilize floral information. Proc Biol Sci. 2006, 273, 1073-1078.

[13] Michelette, E. R., Soares, A. E. Characterization of preimaginal developmental stages in Africanized honey bee workers (*Apis mellifera* L.). Apidologie. 1993, 24, 431-440.

[14] Alaux, C., Ducloz, F., Crauser, D., Le Conte, Y. Diet effects on honeybee immunocompetence. Biol Lett. 2010, 6, 562-565.

[15] Sun, Y., MacRae, T. H. Small heat shock proteins: molecular structure and chaperone function. Cell Mol Life Sci. 2005, 62, 2460-2476.

[16] Yu, X., Sun, R., Yan, H., Guo, X., Xu, B. Characterization of a sigma class glutathione *S*-transferase gene in the larvae of the honeybee (*Apis cerana cerana*) on exposure to mercury. Comp Biochem Physiol B: Biochem Mol Biol. 2012, 161, 356-364.

[17] Yan, H., Jia, H., Gao, H., Guo, X., Xu, B. Identification, genomic organization, and oxidative stress response of a sigma class glutathione *S*-transferase gene (*AccGSTS1*) in the honey bee, *Apis cerana cerana*. Cell Stress Chaperones. 2013, 18, 415-426.

[18] Zhang, Y., Yan, H., Li, Y., Guo, X., Xu, B. A novel Omega-class glutathione *S*-transferase gene in *Apis cerana cerana*: molecular characterisation of GSTO2 and its protective effects in oxidative stress. Cell Stress Chaperones. 2013, 18, 503-516.

[19] Roy, A., Kucukural, A., Zhang, Y. I-TASSER: a unified platform for automated protein structure and function prediction. Nat Protoc. 2010, 5, 725-738.

[20] Crack, J. A., Mansour, M., Sun, Y., MacRae, T. H. Functional analysis of a small heat shock/alpha-crystallin protein from *Artemia franciscana*. Oligomerization and thermotolerance. Eur J

Biochem. 2002, 269, 933-942.
[21] Pérez-Morales, D., Ostoa-Saloma, P., Espinoza, B. Trypanosoma cruzi *SHSP16*: characterization of an alpha-crystallin small heat shock protein. Exp Parasitol. 2009, 123, 182-189.
[22] Burmeister, C., Lüersen, K., Heinick, A., Hussein, A., Domagalski, M., Liebau, E. *Oxidative stress in Caenorhabditis elegans*: protective effects of the Omega class glutathione transferase (*GSTO-1*). FASEB J. 2008, 22, 343-354.
[23] Elias-Neto, M., Soares, M. P., Simões, Z. L., Hartfelder, K., Bitondi, M. M. Developmental characterization, function and regulation of a Laccase 2 encoding gene in the honey bee, *Apis mellifera* (Hymenoptera, Apinae). Insect Biochem Mol Biol. 2010, 40, 241-251.
[24] Jacobsen, J. V., Shaw, D. C. Heat-stable proteins and abscisic acid action in barley aleurone cells. Plant Physiol. 1989, 91, 1520-1526.
[25] Kurzik-Dumke, U., Lohmann, E. Sequence of the new *Drosophila melanogaster* small heat-shock-related gene, lethal (2) essential for life [l(2)efl], at locus 59F4,5. Gene. 1995, 154, 171-175.
[26] Collin, V., Issakidis-Bourguet, E., Marchand, C., Hirasawa, M., Miginiac-Maslow, M. The *Arabidopsis* plastidial thioredoxins: new functions and new insights into specificity. J Biol Chem. 2003, 278, 23747-23752.
[27] Wan, Q., Whang, I., Lee, J. Molecular and functional characterization of *HdHSP20*: a biomarker of environmental stresses in disk abalone *Haliotis discus discus*. Fish Shellfish Immunol. 2012, 33, 48-59.
[28] Simon, M., Adam, G., Rapatz, W., Spevak, W., Ruis, H. The *Saccharomyces cerevisiae* ADR1 gene is a positive regulator of transcription of genes encoding peroxisomal proteins. Mol Cell Biol. 1991, 11, 699-704.
[29] Hayes, J. D., McMahon, M. NRF2 and KEAP1 mutations: permanent activation of an adaptive response in cancer. Trends Biochem Sci. 2009, 34, 176-188.
[30] Gala, A., Fang, Y., Woltedji, D., Zhang, L., Han, B., Feng, M., Li, J. Changes of proteome and phosphoproteome trigger embryo-larva transition of honeybee worker (*Apis mellifera ligustica*). J Proteomics. 2013, 78, 428-446.
[31] Rebecca, M., Hsing-Lin, W., Jun, G., Srinivas, I., Rashi, I. Impact of physicochemical properties of engineered fullerenes on key biological responses. Toxicol Appl Pharmacol. 2009, 234, 58-67.
[32] Li, Y., Zhang, Z., Robinson, G. E., Palli, S. R. Identification and characterization of a juvenile hormone response element and its binding proteins. J Biol Chem. 2007, 282, 37605-37617.
[33] Sun, R., Zhang, Y., Xu, B. Characterization of the response to ecdysteroid of a novel cuticle protein R&R gene in the honey bee, *Apis cerana cerana*. Comp Biochem Physiol B: Biochem Mol Biol. 2013, 166, 73-80.
[34] Shen, Y., Gu, J., Feng, Q. L., Kang, L. Cloning and expression analysis of six small heat shock protein genes in the common cutworm, *Spodoptera litura*. J Insect Physiol. 2011, 57, 908-914.
[35] Huet, F., Lage, J. L., Ruiz, C., Richards, G. The role of ecdysone in the induction and maintenance of *hsp27* transcripts during larval and prepupal development of *Drosophila*. Dev Genes Evol. 1996, 206, 326-332.
[36] Von Kalm, L., Crossgrove, K., Von Seggern, D., Guild, G. M., Beckendorf, S. K. The Broad-Complex directly controls a tissue-specific response to the steroid hormone ecdysone at the onset of *Drosophila* metamorphosis. EMBO J. 1994, 13, 3505-3516.
[37] Skandalis, D. A., Roy, C., Darveau, C. A. Behavioural, morphological, and metabolic maturation of newly emerged adult workers of the bumblebee, *Bombus impatiens*. J Insect Physiol. 2005, 57, 704-711.
[38] Enayati, A. A., Ranson, H., Hemingway, J. Insect glutathione transferases and insecticide resistance. Insect Mol Biol. 2011, 14, 3-8.
[39] Marionnet, C., Bernerd, F., Dumas, A., Dubertret, L. Modulation of gene expression induced in human epidermis by environmental stress *in vivo*. J Invest Dermatol. 2003, 121, 1447-1458.

[40] Liu, Z., Xi, D., Kang, M., Guo, X., Xu, B. Molecular cloning and characterization of *Hsp27.6*: the first reported small heat shock protein from *Apis cerana cerana*. Cell Stress Chaperones. 2012, 17, 539-551.

[41] Reineke, A. Identification and expression of a small heat shock protein in two lines of the endoparasitic wasp *Venturia canescens*. Comp Biochem Physiol A: Mol Integr Physiol. 2005, 141, 60-69.

[42] An, M. I., Choi, C. Y. Activity of antioxidant enzymes and physiological responses in ark shell, *Scapharca broughtonii*, exposed to thermal and osmotic stress: effects on hemolymph and biochemical parameters. Comp Biochem Physiol B: Biochem Mol Biol. 2010, 155, 34-42.

[43] Colinet, H., Lee, S. F., Hoffmann, A. Temporal expression of heat shock genes during cold stress and recovery from chill coma in adult *Drosophila melanogaster*. FEBS J. 2010, 277, 174-185.

[44] Qin, W., Neal, S. J., Robertson, R. M., Westwood, J. T., Walker, V. K. Cold hardening and transcriptional change in *Drosophila melanogaster*. Insect Mol Biol. 2005, 14, 607-613.

[45] Waagner, D., Holmstrup, M., Bayley, M., Sørensen, J. G. Induced cold-tolerance mechanisms depend on duration of acclimation in the chill-sensitive *Folsomia candida* (Collembola). J Exp Biol. 2013, 216, 1991-2000.

[46] De Pomerai, D. Heat-shock proteins as biomarkers of pollution. Hum Exp Toxicol. 1996, 15, 279-285.

[47] Mukhopadhyay, I., Nazir, A., Mahmood, K., Saxena, D. K., Chowdhuri, D. K. Toxicity of argemone oil: effect on *hsp70* expression and tissue damage in transgenic *Drosophila melanogaster* (*hsp70-lacZ*) Bg9. Cell Biol Toxicol. 2002, 18, 1-11.

[48] Gupta, S. C., Siddique, H. R., Saxena, D. K., Chowdhuri, D. K. Comparative toxic potential of market formulation of two organophosphate pesticides in transgenic *Drosophila melanogaster* (*hsp70-lacZ*). Cell Biol Toxicol. 2005, 21, 149-162.

[49] Gupta, S. C., Siddique, H. R., Saxena, D. K., Chowdhuri, D. K. Hazardous effect of organophosphate compound, dichlorvos in transgenic *Drosophila melanogaster* (*hsp70-lacZ*): induction of *hsp70*, anti-oxidant enzymes and inhibition of acetylcholinesterase. Biochim Biophys Acta. 2005, 1725, 81-92.

[50] Brimer, L., Henriksen, S. A., Gyrd-Hansen, N., Rasmussen, F. Evaluation of an *in vitro* method for acaricidal effect. Activity of parathion, phosmet and phoxim against *Sarcoptes scabiei*. Vet Parasitol. 1993, 51, 123-135.

[51] Raghu, K., Mahesh, V., Sasidhar, P., Reddy, P. R., Venkataramaniah, V., Agrawal, A. Paraquat poisoning: a case report and review of literature. J Family Community Med. 2013, 20, 198-200.

[52] Vaalavirta, L., Tähti, H. Astrocyte membrane Na^+, K^+-ATPase and Mg^{2+}-ATPase as targets of organic solvent impact. Life Sci. 1995, 57, 2223-2230.

[53] Sugimoto, N., Osakabe, M. Cross-resistance between cyenopyrafen and pyridaben in the two-spotted spider mite *Tetranychus urticae* (Acari: Tetranychidae). Pest Manag Sci. 2014, 70(7):1090-1096 [54] Christians, E. S., Ishiwata, T., Benjamin, I. J. Small heat shock proteins in redox metabolism: implications for cardiovascular diseases. Int J Biochem Cell Biol. 2012, 44, 1632-1645.

[55] Lushchak, V. I. Environmentally induced oxidative stress in aquatic animals. Aquat Toxicol. 2011, 101, 13-30.

[56] Kottuparambil, S., Shin, W., Brown, M. T., Han, T. UV-B affects photosynthesis, ROS production and motility of the freshwater flagellate, *Euglena agilis* Carter. Aquat Toxicol. 2012, 122-123, 206-213.

[57] Bertin, G., Averbeck, D. Cadmium: cellular effects, modifications of biomolecules, modulation of DNA repair and genotoxic consequences (a review). Biochimie. 2006, 88, 1549-1559.

[58] Choi, Y. K., Jo, P. G., Choi, C. Y. Cadmium affects the expression of heat shock protein 90 and metallothionein mRNA in the Pacific oyster, *Crassostrea gigas*. Comp Biochem Physiol C: Toxicol Pharmacol. 2008, 147, 286-292.

[59] Zhang, L., Wang, L., Song, L., Zhao, J., Qiu, L., Yang, G. The involvement of *HSP22* from bay scallop *Argopecten irradians* in response to heavy metal stress. Mol Biol Rep. 2010, 37, 1763-1771.

[60] Polla, B. S., Cossarizza, A. Stress proteins in inflammation. EXS. 1996, 77, 375-391.
[61] Wu, P., Wang, X., Qin, G. X., Li, M. W., Guo, X. J. Microarray analysis of the gene expression profile in the midgut of silkworm infected with cytoplasmic polyhedrosis virus. Mol Biol Rep. 2011, 38, 333-341.
[62] Derocher, A. E., Helm, K. W., Lauzon, L. M., Vierling, E. Expression of a conserved family of cytoplasmic low molecular weight heat shock proteins during heat stress and recovery. Plant Physiol. 1991, 96, 1038-1047.
[63] Pacheco, A., Pereira, C., Almeida, M. J., Sousa, M. J. Small heat-shock protein Hsp12 contributes to yeast tolerance to freezing stress. Microbiology. 2009, 155, 2021-2028.
[64] Li, J., Qin, H., Wu, J., Sadd, B. M., Chen, Y. The prevalence of parasites and pathogens in Asian honeybees *Apis cerana* in China. PLoS One. 2012, 7, e47955.
[65] Huvenne, H., Smagghe, G. Mechanisms of dsRNA uptake in insects and potential of RNAi for pest control: a review. J Insect Physiol. 2010, 56, 227-235.

1.3 Molecular Cloning and Characterization of *Hsp27.6*: the First Reported Small Heat Shock Protein from *Apis cerana cerana*

Zhaohua Liu, Dongmei Xi, Mingjiang Kang, Xingqi Guo, Baohua Xu

Abstract: Small heat shock proteins (sHsps) play an important role in the cellular defense of prokaryotic and eukaryotic organisms against a variety of internal and external stressors. In this study, a cDNA clone encoding a member of the α-crystallin/sHsp family, termed *AccHsp27.6*, was isolated from *Apis cerana cerana*. The full-length cDNA is 1,014 bp in length and contains a 708 bp open reading frame encoding a protein of 236 amino acids with a calculated molecular weight of 27.6 kDa and an isoelectric point of 7.53. Seven putative heat shock elements and three NF-κB binding sites were present in the 5′-flanking region, suggesting a possible function in immunity. A semi-quantitative RT-PCR analysis indicated that *AccHsp27.6* was expressed in all tested tissues and at different developmental stages. Furthermore, expression of the *AccHsp27.6* transcript was induced by exposure to heat shock, H_2O_2, a number of different chemicals (including SO_2, formaldehyde, alcohol, acetone, chloroform, and the pesticides phoxim and acetamiprid), and the microbes *Staphylococcus aureus* and *Micrococcus luteus*. In contrast, the mRNA expression could be repressed by CO_2, the pesticides pyriproxyfen and cyhalothrin, and the microbes *Bacillus subtilis* and *Pseudomonas aeruginosa*. Notably, the recombinant AccHsp27.6 protein exhibited significant *in vitro* molecular chaperone activity and antimicrobial activity. Taken together, these results suggest that *AccHsp27.6* might play an important role in the response to abiotic and biotic stresses and in immune reactions.

Keywords: *Apis cerana cerana*; Expression; Gene cloning; Immunological function; Small heat shock proteins

Introduction

Members of the heat shock protein (Hsp) superfamily are ubiquitously present in most organisms. The Hsps play important roles as protein chaperones [1] and are also involved in a number of biological processes, including embryogenesis, diapause, and morphogenesis. Based on their molecular weights, Hsps have been classified into four categories, including

Hsp90, Hsp70, Hsp60, and a family of small Hsps (sHsps) [2]. The sHsps, also called small stress proteins, range in molecular weight of 15-43 kDa. Compared to other types of Hsps, the sHsps exhibit the greatest variation in sequence, size, and function [3]. However, there is a conserved domain of 80-100 residues, called the α-crystallin domain that is bordered by variable amino- and carboxy-terminal extensions [4]. In the higher-order structure, the α-crystallin domain forms a conserved β-sandwich composed of two antiparallel β-sheets [5]. This conserved structure facilitates the assembly of the sHsps into oligomeric complexes of up to 800 kDa, which is crucial for their primary function as molecular chaperones that bind partially denatured proteins to prevent irreversible protein aggregation during stresses, such as extreme temperatures, oxidation, UV irradiation, heavy metal, and chemical intoxication [6]. The chaperone function of the sHsps is not only involved in stress conditions but also in normal development. Indeed, these proteins are also involved in cell growth, differentiation, apoptosis [7], membrane fluidity [8], diapause [9], and lifespan [10].

There have been numerous studies concerning sHsps in bacteria, algae, plants, amphibians, birds, and mammals. The four primary sHsps of *Drosophila melanogaster*, Hsp22, Hsp23, Hsp26, and Hsp27, have been studied in detail. They share significant sequence similarity and coordinately expressed following stress, but they have distinct developmental expression patterns and intracellular localizations [11]. Recently, a few members of the α-crystallin/sHsp family have been cloned from other insect species, including *Plutella xylostella* [12], *Mamestra brassicae* [13], *Ceratitis capitata* [14], *Sesamia nonagrioides* [9], *Liriomyza sativa* [15], and *Macrocentrus cingulum* [16].

The Chinese honeybee, *Apis cerana cerana*, is an important indigenous species that plays a critical role in agricultural economic development as the pollinator of flowering plants [17]. To date, no sHsp has been identified in the Chinese honeybee. In this study, a cDNA clone, *AccHsp27.6*, encoding a putative sHsp gene was isolated and characterized. Our results indicate that the expression of *AccHsp27.6* responds to a number of chemical and biological inducers. Moreover, a recombinant AccHsp27.6 protein exhibited significant *in vitro* molecular chaperone activity and antimicrobial activity. We speculate that *AccHsp27.6* might play an important role in regulating biotic and abiotic stress responses and immune reactions.

Experimental Procedures

Animals and treatments

The worker bees of *A. cerana cerana* obtained from the Technology Park of Shandong Agricultural University were reared on an artificial diet at 34 °C and 80% humidity. The entire bodies of the second (L2)-, the fourth (L4)-, the fifth (L5)-, and the sixth (L6)-instar larvae with white (Pw)-, pink (Pp)-, and dark (Pd)-eyed pupae were obtained from the hive, and the adult workers (1 day and 10 days after emergence) were collected at the entrance of the hive upon their return to the colony after foraging [18]. The adult bees (12 days) were divided into groups (n=40) and exposed to abiotic and biotic stresses. The abiotic stress consisted of heat shock (37 °C), UV-induced oxidative stress (30 MJ/cm^2), H$_2$O$_2$ treatment (30% H$_2$O$_2$ was delivered in 0.5 μl of distilled water to the thoracic notum of the worker bees at a final concentration of 2 mmol/L), and chemical stress through treatment with four different pesticides (imidacloprid, cyhalothrin, pyriproxyfen and phoxim; the pesticides were delivered in 0.5 μl of distilled water to the thoracic notum of the worker bees at final concentrations of 10, 12.5, 20, and 20 mg/L, respectively), CO_2, SO_2, formaldehyde, alcohol, acetone, and chloroform (a gas generator in the incubator was used to generate the CO_2 and SO_2, and the organic reagents were volatilized and added to the incubator containing the honeybees). The biotic stress was microbial infection. The 3-day-old larvae were maintained in a 24-well plate and inoculated with microbes (*Staphylococcus aureus*, *Micrococcus luteus*, *Bacillus subtilis*, and *Pseudomonas aeruginosa*). At 6 days after inoculation, the bees were flash-frozen in liquid nitrogen at the indicated time points and stored at −80℃.

RNA extraction, cDNA synthesis, and DNA isolation

Total RNA was extracted by using Trizol reagent (Invitrogen, Carlsbad, CA, USA) according to the instructions of the manufacturer. To remove genomic DNA, the RNA was treated with RNase-free DNase I. Subsequently, the cDNA was synthesized by using a reverse transcription system (TransGen Biotech, Beijing, China). The genomic DNA was isolated by using an EasyPure Genomic DNA Extraction Kit in accordance with the instructions of the manufacturer (TransGen Biotech).

PCR primers and conditions

The primers used in this study are listed in Table 1. The amplification conditions are shown in Table 2.

Table 1 The primers used in this study

Abbreviation	Primer sequence (5'—3')	Description
HP1	TTCGGAAAGTGGGGAAACATTC	cDNA sequence primer of *AccHsp27.6*, forward
HP2	AAGGGCAGTAAAGGTGTATAAGTCA	cDNA sequence primer of *AccHsp27.6*, reverse
5P1	CATAAACGCGGCAAAAGACATCA	5' RACE reverse primer of *AccHsp27.6*, outer
5P2	GGATCAGCATTTTGGCACAGC	5' RACE reverse primer of *AccHsp27.6*, inner
3P1	GATCAAATGAGAGAACAGTGAAGG	3' RACE forward primer of *AccHsp27.6*, outer
3P2	AAGGGCAGTAAAGGTGTATAAGTCA	3' RACE forward primer of *AccHsp27.6*, inner
AAP	GGCCACGCGTCGACTAGTAC(G)$_{14}$	Abridged anchor primer
AUAP	GGCCACGCGTCGACTAGTAC	Abridged universal amplification primer
B26	GACTCTAGACGACATCGA(T)$_{18}$	3' RACE universal adaptor primer
B25	GACTCTAGACGACATCGA	3' RACE universal primer
QP1	TTCGGAAAGTGGGGAAACATTC	Full-length cDNA primer, forward
QP2	GACATGCAATAAATATTGTATTAATTATTTC	Full-length cDNA primer, reverse
GP1	TTCGGAAAGTGGGGAAACATTC	Genomic sequence primer of *AccHsp27.6*, forward
GP2	AAGGGCAGTAAAGGTGTATAAGTCA	Genomic sequence primer of *AccHsp27.6*, reverse
PS1	CAGCAAAATCAGCAATTAC	Inverse PCR forward primer, outer
PX1	CGTCATAATCATCATCC	Inverse PCR reverse primer, outer

Continued

Abbreviation	Primer sequence (5'—3')	Description
PS2	CAACAACATCAGCGTTGT	Inverse PCR forward primer, inner
PX2	CGATCTAGATTCACTTGG	Inverse PCR reverse primer, inner
QS1	GCATTTGCATGTAAATAGATG	Promoter special primer, forward
QS2	GTGGGGAAACATTCCTAAC	Promoter special primer, reverse
RP1	GATGTTTTCCGACTGGTGG	RT-PCR primer of *AccHsp27.6*, forward
RP2	GAACATGGTTGGGTATCCAG	RT-PCR primer of *AccHsp27.6*, reverse
β-actin-s	GTTTTCCCATCTATCGTCGG	Standard control primer, forward
β-actin-x	TTTTCTCCATATCATCCCAG	Standard control primer, reverse
EP1	GGTACCTTCGGAAAGTGGGGAAACATTC	Protein expression primer, forward
EP2	AAGCTTTGACTTATACACCTTTACTGCCCTT	Protein expression primer, reverse

Table 2 PCR amplification conditions in the study

Primer pair	Amplification conditions
HP1/HP2	6 min at 94℃, 40 s at 94℃, 40 s at 52℃, 1 min at 72℃ for 35 cycles, 10 min at 72℃
5P1/B26	6 min at 94℃, 40 s at 94℃, 40 s at 53℃, 30 s at 72℃ for 28 cycles, 5 min at 72℃
5P2/B25	6 min at 94℃, 40 s at 94℃, 40 s at 58℃, 30 s at 72℃ for 35 cycles, 5 min at 72℃
3P1/AAP	6 min at 94℃, 40 s at 94℃, 40 s at 49℃, 40 s at 72℃ for 28 cycles, 5 min at 72℃
3P2/AUAP	6 min at 94℃, 40 s at 94℃, 40 s at 55℃, 40 s at 72℃ for 35 cycles, 5 min at 72℃
QP1/QP2	6 min at 94℃, 40 s at 94℃, 40 s at 52℃, 1 min at 72℃ for 35 cycles, 10 min at 72℃
GP1/GP2	6 min at 94℃, 40 s at 94℃, 40 s at 52℃, 1.5 min at 72℃ for 35 cycles, 10 min at 72℃
PS1/PX1	6 min at 94℃, 50 s at 94℃, 50 s at 50℃, 2 min at 72℃ for 30 cycles, 5 min at 72℃
PS2/PX2	6 min at 94℃, 50 s at 94℃, 50 s at 50℃, 2 min at 72℃ for 30 cycles, 5 min at 72℃
QS1/QS2	6 min at 94℃, 40 s at 94℃, 40 s at 52℃, 2 min at 72℃ for 35 cycles, 10 min at 72℃
RP1/RP2	6 min at 94℃, 40 s at 94℃, 40 s at 52℃, 40 s at 72℃ for 28 cycles, 5 min at 72℃
β-actin-s/β-actin-x	6 min at 94℃, 30 s at 94℃, 30 s at 53℃, 30 s at 72℃ for 28 cycles, 5 min at 72℃
EP1/EP2	6 min at 94℃, 40 s at 94℃, 40 s at 53℃, 1.5 min at 72℃ for 28 cycles, 10 min at 72℃

Isolation of the *AccHsp27.6* gene

The internal conserved fragment was cloned by RT-PCR using the primers HP1 and HP2, which were designed based on the conserved DNA sequence among the homologous genes from *Apis mellifera*, *Nasonia vitripennis*, and *Macrocentrus cingulum*. Using the sequence of the internal fragment, two sets of primers, 5P2 and 5P1 and 3P2 and 3P1, were designed and synthesized for 5' and 3' RACE (rapid amplification of cDNA ends), respectively. Following RACE-PCR, the 5'- and 3'-end sequences and the internal fragment were assembled, and the full-length cDNA sequence was derived. Subsequently, the specific primers QP1 and QP2 were used to amplify the full-length cDNA sequence by using end-to-end PCR. The genomic DNA was cloned by using the primers GP1 and GP2 and used as a template. All PCRs were performed as previously described [19].

Cloning of the 5′-flanking region of *AccHsp27.6*

The sequence of the 5′-flanking region was cloned by using inverse PCR and the restriction endonuclease *Ssp* I. Genomic DNA digested with *Ssp* I at 37°C overnight and ligated by using T4 DNA ligase (TaKaRa, Dalian, China) was used as the template. Based on the genomic sequence of *AccHsp27.6*, two pairs of primers (PS1 + PX1 and PS2 + PX2) were designed to obtain the 5′-flanking region, and two specific sequencing primers (QS1 and QS2) were designed to confirm the amplified product. The PCR conditions are shown in Table 2. The PCR products were cloned into the pMD18-T vector and sequenced. The TFBind software (http://tfbind.hgc.jp/) was used to search for transcription factor binding sites in the 5′-flanking region [20].

Bioinformatic and phylogenetic analysis

The NCBI bioinformatics tools (available at http://blast.ncbi.nlm.nih.gov/Blast.cgi) were used to detect conserved domains in *AccHsp27.6*. In addition, the DNAMAN software was used for the prediction of secondary structure and to search for open reading frames (ORFs). An online protein 3D structure prediction tool (http://swissmodel.expasy.org/) was also used in the structural analysis. Multiple sequence alignments were performed by using the Clustal X program with its default parameters [14]. Phylogenetic and molecular evolutionary analyses were performed by using the neighbor-joining method with the Molecular Evolutionary Genetics Analysis (MEGA version 4.1) software.

Expression analysis by using semi-quantitative RT-PCR

A pair of specific primers was designed to amplify a fragment of *AccHsp27.6*. The house-keeping gene *β-actin* (GenBank accession number: XM640276) was used to estimate equal amounts of RNA among the samples. RT-PCR of *β-actin* was performed by using the primers RP1 and RP2 and the same conditions used to amplify the *AccHsp27.6* fragment. The PCRs were repeated three times, and the electrophoresis results were normalized against the *β-actin* results by using the Quantity-One™ image analysis software implemented in VersaDoc 4000 (Bio-Rad, Hercules, CA, USA).

Construction of expression plasmids, recombinant protein expression and purification

To express the recombinant AccHsp27.6 protein in BL21 cells, a pair of specific primers (EP1 and EP2) was designed to amplify the 708 bp fragment encoding the entire ORF. The ORF sequence was cloned into pEASY-T3 and digested with the restriction endonucleases *Kpn* I and *Hin*dIII. After extraction and purification by using the Gel Extraction Kit (Solarbio, Beijing, China), the ORF sequence was ligated into the expression vector pET-30a (+), which was digested with the same restriction endonucleases. The expression plasmid pET-30a (+)-AccHsp27.6 was then transformed into the *E. coli* BL21 (DE3) cells. After intermediate culture and isopropylthio-β-D-galactoside (IPTG) induction at a final concentration of 0.6 mmol/L, the bacterial cells were harvested by centrifugation, resuspended, sonicated, and centrifuged again. The final pellet was resuspended in PBS and subjected to 15% SDS-PAGE. The AccHsp27.6 in the medium was purified by using the 6 × His-Tagged Protein Purification Kit (CW Biotechnology, Beijing, China) according to the instructions of the manufacturer, and the purified protein was examined by 15% SDS-PAGE. In addition, AccCPR26-6 × His fusion protein (*A. cerana cerana* cuticle protein R&R 26) was purified by the same method.

Molecular chaperone activity and antimicrobial activity of the recombinant AccHsp27.6 protein

The capacity of AccHsp27.6 to suppress the thermal aggregation of the MDH (mitochondrial malate dehydrogenase) from pig heart (EC 1.1.1.37; Amresco) was examined. Three samples (A, MDH, B, MDH + AccHsp27.6, and C, MDH + BSA) were incubated at 43 ℃. The BSA (bovine serum albumin) was used as control to eliminate the effect of nonspecific chaperone activity. Absorbance was monitored at 360 nm at regular intervals (10 min) for 1 h [5].

Sterile paper disks were saturated with a solution of recombinant AccHsp27.6 protein and dried at 30 ℃. Separate tubes of LB broth were inoculated with overnight cultures of four bacterial strains, *M. luteus*, *S. aureus*, *B. subtilis*, and *P. aeruginosa*, to give an initial bacterial density of 10^6 cells/ml and overlaid onto agarose medium plates. The AccHsp27.6-saturated paper disks were placed on the plates and incubated at 37 ℃ for 24 h.

Results

AccHsp27.6 cloning and sequence analysis

Based on the conserved DNA sequence of the sHsps from *A. mellifera*, *N. vitripennis*, and *M. cingulum*, a fragment of approximately 785 bp was amplified by using the primers HP1 and HP2 (Fig. 1). Specific primers were designed according to the obtained internal sequence and subsequently used for the amplification of 3' (primers 3P1 and 3P2) and 5'-end (primers 5P1 and 5P2) cDNA, resulting in 244 bp and 161 bp fragments corresponding to the 3'- and 5'-end sequences, respectively. The full-length *AccHsp27.6* cDNA sequence of *AccHsp27.6* was deduced and amplified by RT-PCR using the full-length cDNA primers QP1 and QP2 and confirmed by sequencing. The full-length cDNA sequence of *AccHsp27.6* (GenBank accession number: GQ254650) was 1,014 bp, with a 76 bp 5' untranslated region (UTR), a 223 bp 3' UTR and a 708 bp ORF encoding a protein of 236 amino acids with a calculated molecular weight of 27.6 kDa and an isoelectric point of 7.53.

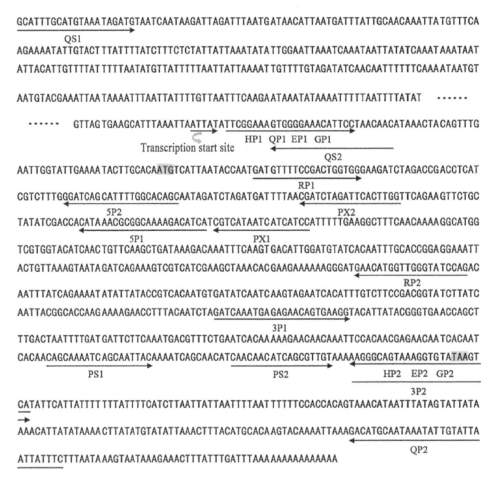

Fig. 1 Nucleotide sequence of *AccHsp27.6* and primers. The cDNA sequence is indicated on the top line, and the names of the primers and their corresponding orientations are shown on the second line.

The NCBI Conserved Domain (CD) search revealed that the α-crystallin domain (ACD) of metazoan α-crystallin-type sHsps was present in the deduced amino acid sequence of AccHsp27.6, suggesting that the *AccHsp27.6* gene might belong to the sHsp family. The analysis of the secondary structure indicated that the helix content was 10.55%, which is characteristic of the structure of sHsps [21]. β-sheets occurred frequently throughout the structure of the protein, which is consistent with previous reports that the secondary structure of the sHsps is rich in β-sheets [3]. In addition, there were only two cysteines in the entire protein sequence, which is consistent with previous data showing that cysteine residues were rarer in the sequences of molecular chaperones than in other protein families [22]. Moreover, the tertiary structure was also analyzed (Fig. 2). The human αB crystallin (PDB code is 2WJ7) was the best template for homology modeling of the deduced protein. This model consisted of a single polypeptide chain of approximately 94 amino acids from residue 67 to residue 157. In the 3D structure, the conserved 188-residue α-crystallin domain was enriched in β-strands that were organized into a β-sheet sandwich. These sheets are the basic structural unit of sHsps [4] and help facilitate dimer formation and the subsequent oligomerization of sHsps.

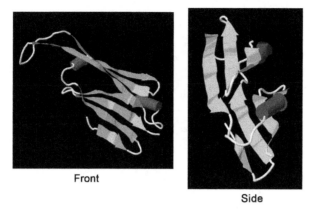

Front   Side

Fig. 2 The tertiary structure of α-crystallin domain 67-157 residues. (For color version, please sweep QR Code in the back cover)

The predicted amino acid sequence of AccHsp27.6 exhibited 43.88% and 44.07% similarity to the sHsp from *M. cingulum* (GenBank accession number: EU624206) and Hsp21.7 from *N. vitripennis* (GenBank accession number: XP001604512), respectively. The protein sequence comparison showed that AccHsp27.6 contains two prominent conserved regions (Fig. 3). In addition to the one including the α-crystallin domain, the region of the initial 27 amino acids in the amino-terminal section is of relatively high homology and this domain is highly hydrophobic and might be involved in modulating oligomerization, subunit dynamics, and substrate binding. sHsp oligomerization and chaperoning are influenced by the amino-terminus in several, but not all sHsps [4]. Usually, the amino-terminal extension is variable in length and sequence. However, according to Kokolakis et al. [14], the 14-amino acid domain of Hsp27, at the very amino-terminal region of the three proteins, is very hydrophobic and shows 57% amino acid sequence identity among *Drosophila*, medfly, and *Sarcophaga crassipalpis*. And analogous hydrophobic domain is also conserved in *Drosophila* Hsp23 and Hsp26 but not in *Drosophila* Hsp22 [23]. The relatively higher homology of AccHsp27.6 in the amino-terminus suggests the conservation of sHsp in the structure and function.

To investigate the regulation of AccHsp27.6 expression, we cloned its 5′-flanking region and obtained a 1,376 bp DNA fragment upstream of the translation start site by using inverse PCR (GenBank accession number: JF317273). The TFBIND software was used to search the transcription factor binding sites. Seven HSEs and three NF-κB binding sites were detected. In addition, a TATA-BOX, which represents the putative core promoter element upstream of the transcription start site, was found (Fig. 4).

Phylogenetic analysis

To study the evolutionary relationships among different sHsps from various insects, a phylogenetic tree was constructed by using the MEGA (version 4.1) software, and all of the amino acid sequences used were obtained from GenBank. As shown in Fig. 5, the sHsps from insects of the same order were phylogenetically closer than homologous sHsps from other insect orders. However, an orthologous cluster containing several sHsps from different insect orders was found, suggesting that these sHsps evolved before the divergence of the species [14].

52 | Molecular Response Mechanisms of Apis cerana cerana to Abiotic Stress

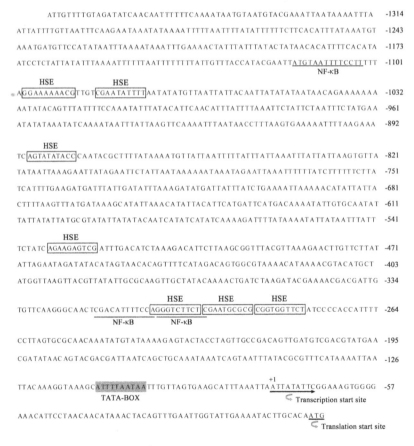

Fig. 3 Amino acid sequence comparison of sHsp homologs from *Apis cerana cerana*, *Macrocentrus cingulum*, and *Nasonia vitripennis*. The asterisks, double dots, and single dots denote fully, strongly, and weakly conserved residues, respectively. The conserved regions are shown in gray. The α-crystallin domain is underlined. The degree of identity relative to AccHsp27.6 is shown at the end of each sequence. The secondary structure assignment of the AccHsp27.6 is shown above the sequence.

Fig. 4 The nucleotide sequence and putative transcription factor binding sites of the 5'-flanking region of *AccHsp27.6*. The transcription and translation start sites are indicated with an arrowhead. The underlined sections indicate the putative NF-κB binding sites, and the boxes indicate the putative HSEs. The TATA-box is highlighted in light gray.

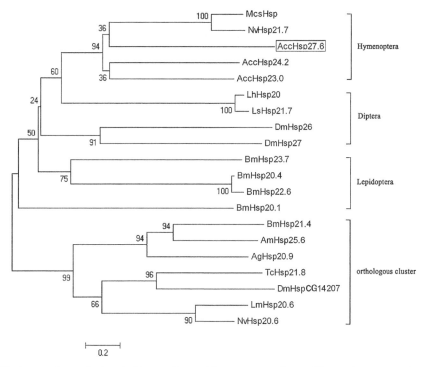

Fig. 5 Phylogenetic analysis of sHsp amino acid sequences from different insect species (Diptera, Hymenoptera, and Lepidoptera). The full names of species and the accession numbers of the genes are designated by the following abbreviations: AccHsp24.2 (*Apis cerana cerana*, JF411009), AccHsp23.0 (*Apis cerana cerana*, JF411010) DmHsp26 (*Drosophila melanogaster*, NM079273), DmHsp27 (*Drosophila melanogaster*, NM079276), DmHspCG14207 (*Drosophila melanogaster*, NM134482), LhHsp20 (*Liriomyza huidobrensis*, DQ452370), LsHsp21.7 (*Liriomyza sativa*, DQ452372), McsHsp (*Macrocentrus cingulum*, EU624206), NvHsp21.7 (*Nasonia vitripennis*, XP001604512), NvHsp20.6 (*Nasonia vitripennis*, XM001607625), AccHsp27.6 (*Apis cerana cerana*, GQ254650), BmHsp20.1 (*Bombyx mori*, AB195971), BmHsp23.7 (*Bombyx mori*, AB195973), BmHsp20.4 (*Bombyx mori*, AF315318), BmHsp22.6 (*Bombyx mori*, FJ602788), BmHsp21.4 (*Bombyx mori*, AB195972), AgHsp20.9 (*Anopheles gambiae*, XM560153), TcHsp21.8 (*Tribolium castaneum*, XM968592), AmHsp25.6 (*Apis mellifera*, XM392405), LmHsp20.6 (*Locusta miratoria*, DQ355964).

Developmental and tissue-specific expression patterns of *AccHsp27.6*

To determine the developmental expression patterns of the *AccHsp27.6* gene, semi-quantitative RT-PCR was employed to detect mRNA accumulations at different developmental stages and in various tissues. As shown in Fig. 6A, the expression level of *AccHsp27.6* showed a gradual decrease from the second to the fifth-instar larval stages followed by a remarkable increase at the sixth stage. A similar expression pattern was observed at the pupal stage, with a rapid decline from the white to the pink-eyed pupae followed by a significant increase in the dark-eyed pupae. In newly emerged adults, the expression of *AccHsp27.6* was low, and it peaked in 10-day-old adults. Analysis of the spatial expression profiles of *AccHsp27.6* indicated that its transcripts were abundant in the thorax and abdomen and scarce in the head (Fig. 6B).

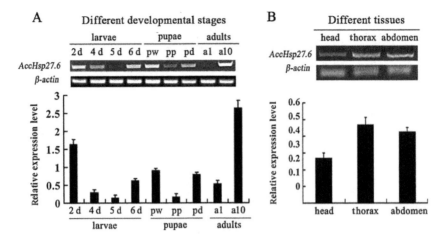

Fig. 6 Expression patterns of *AccHsp27.6* at different developmental stages and in various tissues. A. Expression patterns of *AccHsp27.6* at different developmental stages. Total RNA was isolated from larvae (2 d. second instars, 4 d. fourth instars, 5 d. fifth instars, and 6 d. sixth instars), pupae (pw. white-eyed pupae, pp. pink-eyed pupae, pd. dark-eyed pupae), and adults (a1. one day post-emergence adults, a10. ten days post-emergence adults). The transcripts of *AccHsp27.6* were detected throughout the entire development process. B. Expression patterns of *AccHsp27.6* in various tissues. Total RNA was extracted from the brain, thorax and abdomen of the adult bees (12 d) at the indicated times, and semi-quantitative RT-PCR was used to examined the transcription level of *AccHsp27.6* under different treatments. The *β-actin* gene was used as an internal control. The histograms indicate the relative expression levels. Values on the y-axis are the normalization units relative to *β-actin* levels. In addition, the values on the x-axis represent the lanes. The signal intensities (INT/mm^2) of the bands were assigned by using the Quantity-One^{TM} image analysis software implemented in VersaDoc 4000 (Bio-Rad). Each value is given as the mean (SD) of three replicates.

Expression profiles of *AccHsp27.6* in response to abiotic stress

The sHsps are products of heat shock and other stress responses. Previous studies have reported that sHsps provide a protective function under conditions of high temperature, oxidation, UV irradiation, and chemical intoxication [6]. This idea prompts us to examine the responses of *AccHsp27.6* to different environmental stressors. Twelve-day-old bees were subjected to heat shock, oxidative stresses (UV and H_2O_2), and exposure to a number of chemicals (pesticides, CO_2, SO_2, formaldehyde, alcohol, acetone, and chloroform). As shown in Fig. 7, when treated with heat stress, *AccHsp27.6* transcripts quickly accumulated to a maximum level at 2 h and then declined at 4 h. UV treatment induced only a mild increase in *AccHsp27.6* expression at 4 h. Under H_2O_2 stress conditions, an increase in the accumulation of *AccHsp27.6* transcripts was detected 1 h after treatment, and the transcription levels reached a maximum at 2 h. Interestingly, the *AccHsp27.6* gene exhibited different responses to the application of various pesticides. Treatments with pyriproxyfen and cyhalothrin downregulated *AccHsp27.6* expression, whereas the application of phoxim and acetamiprid greatly upregulated gene expression, with maximum expression levels at 3 h and 1 h, respectively. The application of other chemicals, including SO_2, formaldehyde, alcohol, acetone, and chloroform, induced the expression of *AccHsp27.6* to various degrees. In contrast, the CO_2 challenge suppressed its transcription to nearly zero.

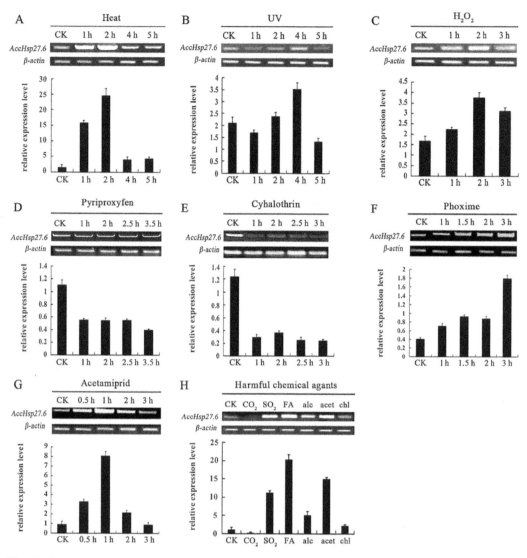

Fig. 7 Expression profiles of *AccHsp27.6* under abiotic stress. Total RNA was extracted from adult bees (12 d) treated at the indicated times with heat (A), UV irradiation (B), H_2O_2 (C), pesticides (D, pyriproxyfen; E, cyhalothrin; F, phoxim; G, acetamiprid), and hazardous chemicals (H). Semi-quantitative RT-PCR was used to examine the transcription level of *AccHsp27.6* under different treatments. FA, formaldehyde; alc, alcohol; acet, acetone; chl, chloroform. The *β-actin* gene was used as an internal control. The line CK indicates the control. The histograms show the relative expression levels. The values on the *y*-axis are the normalization units relative to *β-actin* levels. In addition, the values on the *x*-axis represent the lanes. The signal intensities (INT/mm^2) of the bands were assigned by using the Quantity-One^{TM} image analysis software implemented in VersaDoc 4000 (Bio-Rad). Each value is given as the mean (SD) of three replicates.

AccHsp27.6 mRNA accumulation under biotic stress

Bacterial diseases of honeybee larvae seriously affect the growth of the honeybee, the quality of the bee products, and the beekeeping industry. *S. aureus*, *P. aeruginosa*, and *M.*

luteus are opportunistic pathogens that are widely distributed in the natural environment, and *B. subtilis* could stimulate the development of animal immune organs. To investigate whether *Hsp27.6* is involved in the immune system, these four bacterial strains were injected into the body surface of 3-day-old larvae. Three days later, RT-PCR analysis revealed that *AccHsp27.6* expression was induced by *S. aureus* and *M. luteus* and suppressed by *B. subtilis* and *P. aeruginosa* (Fig. 8). These results suggest an important role for *AccHsp27.6* in honeybee microbe resistance.

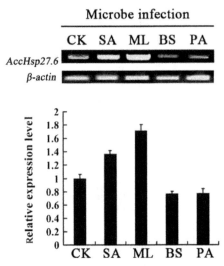

Fig. 8 Expression profiles of *AccHsp27.6* in response to microbe infection. SA, *Staphylococcus aureus*; ML, *Micrococcus luteus*; BS, *Bacillus subtilis*; PA, *Pseudomonas aeruginosa*. The 3-day-old larvae fed in the 24-well plates were inoculated with microbes and maintained until they were 6 days old. Total RNA was extracted at the indicated times, and semi-quantitative RT-PCR was used to examine the transcription level of *AccHsp27.6*. The *β-actin* gene was used as an internal control. The line CK indicates the control. The histograms show the relative expression levels. The values on the y-axis are the normalization units relative to *β-actin* levels. In addition, the values on the x-axis represent the lanes. The signal intensities (INT/mm^2) of the bands were assigned by using the Quantity-One™ image analysis software implemented in VersaDoc 4000 (Bio-Rad). Each value is given as the mean (SD) of three replicates.

Expression and antimicrobial activity of a recombinant AccHsp27.6 protein

The *AccHsp27.6* gene was expressed in *E. coli* strain BL21 cells as a 6 × His-tagged fusion protein. After induction with IPTG at 37℃, SDS-PAGE analysis detected a band corresponding to the HSP27.6-6 × His fusion protein (Fig. 9, lane 3). No protein induction was observed in the non-induced control (Fig. 9, lane 2). The data indicated that the *AccHsp27.6* gene was expressed, and a purified HSP27.6-6 × His fusion protein was obtained (Fig. 9, lane 1).

The molecular chaperone activity *in vitro* of the recombinant AccHsp27.6 was investigated by the thermal aggregation assay of the malate dehydrogenase. As shown in Fig. 10, MDH aggregated when incubation at 43℃, which was revealed by the significant increase of the absorbance at 360 nm [5]. In addition, MDH aggregated faster when mixed with BSA that has no chaperone activity. However, the addition of an equivalent of AccHsp27.6 to MDH almost totally suppressed MDH aggregation (Fig. 10). Therefore, the recombinant protein could have significant molecular chaperone activity *in vitro*.

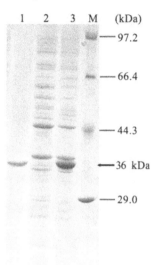

Fig. 9 Expression in *E. coli* and purification of the recombinant AccHsp27.6 protein were analyzed by SDS-PAGE. Lane 1, purified recombinant AccHsp27.6 protein; Lane 2, non-induced; Lane 3, induced; Lane M, molecular mass standard.

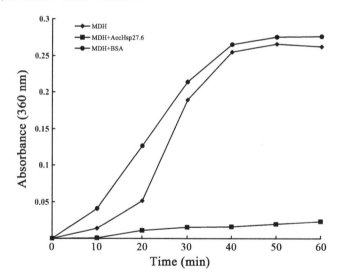

Fig. 10 Molecular chaperone activity of AccHsp27.6 shown by the thermal aggregation assay of the malate dehydrogenase. MDH was heated at 43℃ (♦); mixed with AccHsp27.6 in a molar ratio 1∶1 (■); or mixed with BSA (1∶1) (●). Absorbance was monitored at 360 nm at regular intervals (10 min) for 1 h.

We investigated the antimicrobial activity of the bioactive recombinant AccHsp27.6 protein against several bacteria by using the disk diffusion method. As shown in Fig. 11, the protein exhibited significant antimicrobial activity against *S. aureus*, *M. luteus*, *B. subtilis*, and *P. aeruginosa*. In addition, AccCPR26 purified by the same method with AccHsp27.6 was used as the control to exclude the effect of the 6 × His-tag and the toxicity that accumulate during the purification steps. And as shown in Fig. 11 in enclosure, AccCPR26 had no clearly antimicrobial activity.

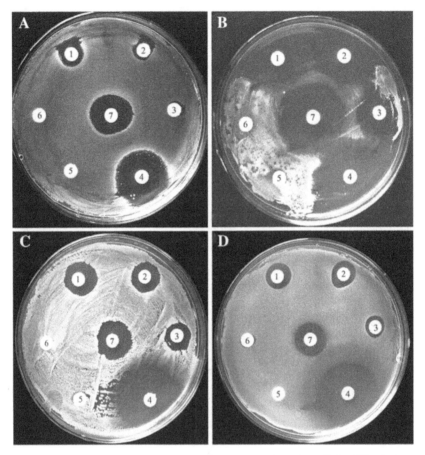

Fig. 11 Antimicrobial activity of AccHsp27.6 shown by the disk diffusion method. A, *Staphylococcus aureus*; B, *Micrococcus luteus*; C, *Bacillus subtilis*; D, *Pseudomonas aeruginosa*. 1, two-fold dilution of the AccHsp27.6 protein solution; 2, five-fold dilution of the AccHsp27.6 protein solution; 3, ten-fold dilution of the AccHsp27.6 protein solution; 4, positive control (kanamycin, 250 mg/ml); 5, negative control (the total protein from *E. coli* BL21 (DE3) cells, 0.5 mg/ml); 6, PBS; 7, the original concentrated solution of purified AccHsp27.6 protein (0.5 mg/ml).

Discussion

Although the molecular and biochemical characterization of *sHsp* genes in insects has been extensively studied, most studies are performed in model insect species, such as *D. melanogaster*. However, the expression of *sHsps* in the Chinese honeybee (*A. cerana cerana*) has not been previously reported. This is the first study to investigate the existence of sHsps in the Chinese honeybee and their possible involvement in response to abiotic and biotic stresses. The results of experiments involving the cloning, characterization, and expression of a Chinese honeybee sHsp gene suggests the possibility of an *sHsp*-mediated resistance against adverse environments in the honeybee.

The HSEs are DNA sequences recognized by heat shock factors (HSFs), transcription factors that are responsible for the expression of Hsps following stress. HSEs are the most critical elements in Hsp gene transcription. We detected seven putative HSEs in the 5'-flanking region of *AccHsp27.6*. Previous studies on the *Drosophila Hsp22* promoter revealed three functional HSEs that are required for the proper expression of *Hsp22* following stress [24]. The HSE responsible for the heat-induction of the *AccHsp27.6* gene has not yet been determined.

The sHsps function as molecular chaperones not only protect proteins from denaturation under high temperature stress but also develop protective functions under other stress conditions, such as cold, drought, oxidation, hypertonic stress, heavy metal, and even high population densities [25]. In the present study, the expression of *AccHsp27.6* in response to abiotic stress was analyzed by RT-PCR. The results showed that *AccHsp27.6* could be markedly induced by heat shock and H_2O_2 treatments as well as the application of pesticides (phoxim and acetamiprid) and several other chemicals (including SO_2, formaldehyde, alcohol, acetone, and chloroform). However, gene expression was suppressed by CO_2 and the application of pyriproxyfen and cyhalothrin.

Interestingly, the expression of the *AccHsp27.6* gene exhibited different responses to the application of various microbes. *S. aureus* and *M. luteus* upregulated *AccHsp27.6* expression, whereas *B. subtilis* and *P. aeruginosa* downregulated the gene expression. These data suggest that AccHsp27.6 is related to the immune system.

Insects lack an acquired immune system, and their defense mechanisms mainly rely on innate immune responses. Extensive work over the past 10 years suggests that the Hsp family might be potential activators of the innate immune system [26]; however, information concerning immunity and the potential role of the sHsps is relatively limited and has emerged only recently. An sHsp gene from shrimp, *Hsp21*, was downregulated during a white spot syndrome viral infection, which might be related to host cell apoptosis [27]. The genome-wide analysis of host gene expression in silkworm cells infected with *Bombyx mori* nucleopolyhedrovirus revealed that the expression of *Hsp20.1* was decreased to approximately 50% at 2 h post-infection, which suggested that silkworm *Hsp20.1* is involved in anti-BmNPV immunity at the early phase of infection[28]. More recently, an sHsp gene from the bloody clam (*Tegillarca granosa*) was reported to be involved in the immune response against *Vibrio parahaemolyticus* and lipopolysaccharides [29]. A microarray analysis of the gene expression profile in the midgut of silkworms infected with cytoplasmic polyhedrosis virus revealed that a number of Hsps (including *sHsp23.7*) were upregulated, suggesting that these Hsps might be involved in anti-BmCPV immunity [26]. However, previous studies have focused primarily on expression analysis at the transcriptional level. In this study, several lines of evidence indicate that *AccHsp27.6* might be involved in immune reactions. First, three nuclear factor-κB (NF-κB) binding sites were observed in the 5'-flanking region of *AccHsp27.6*. NF-κB is a ubiquitous transcription factor that plays critical roles in host defense and chronic inflammatory diseases by regulating the expression of multiple inflammatory and immune genes [30]. Second, the transcription of *AccHsp27.6* was greatly upregulated or downregulated upon microbial infection. Direct evidence came from the analysis of the recombinant AccHsp27.6 protein, which showed that the protein exhibited antimicrobial activity *in vitro*. Therefore, it is reasonable to speculate that AccHsp27.6 could be related to the immune response of the honeybee.

In this study, we only examined microbial infections in the larvae because bacterial diseases of the honeybee larvae seriously affect the growth of the honeybee, the quality of bee products and the beekeeping industry. Moreover, larvae may have the least developed immune systems, thus requiring HSP protection and eliciting a strong response to microbial challenge. Therefore, it would have been useful to examine the effects of a honeybee-specific microbe, such as *Paenibacillus larvae*, which causes American foulbrood. Further investigation is of great importance to uncover a more comprehensive picture of the functional roles and mechanisms of AccHsp27.6 in the immune response.

References

[1] Hendrick, J. P., Hartl, F. U. Molecular chaperone functions of heat-shock proteins. Annu Rev Biochem. 1993, 62, 349-384.

[2] Mizrahi, T., Heller, J., Goldenberg, S., Arad, Z. Heat shock proteins and resistance to desiccation in congeneric land snails. Cell Stress Chaperon. 2010, 15, 351-363.

[3] De Jong, W. W., Caspers, G. J., Leunissen, J. A. Genealogy of the alpha-crystallin-small heat-shock protein superfamily. Int J Biol Macromol. 1998, 22, 151-162.

[4] Sun, Y., Macrae, T. H. Small heat shock proteins: molecular structure and chaperone function. Cell Mol Life Sci. 2005, 62, 2460-2476.

[5] Pérezmorales, D., Ostoasaloma, P., Espinoza, B. *Trypanosoma cruzi* SHSP16: characterization of an α-crystallin small heat shock protein. Exp Parasitol. 2009, 123, 182.

[6] Reineke, A. Identification and expression of a small heat shock protein in two lines of the endoparasitic wasp *Venturia canescens*. Comp Biochem Physiol A: Mol Integr Physiol. 2005, 141, 60-69.

[7] Arrigo, A. P. Small stress proteins: chaperones that act as regulators of intracellular redox state and programmed cell death. Biol Chem. 1998, 379, 19-26.

[8] Tsvetkova, N. M., Horváth, I., Török, Z., et al. Small heat-shock proteins regulate membrane lipid polymorphism. Proc Natl Acad Sci USA. 2002, 99, 13504-13509.

[9] Gkouvitsas, T., Kontogiannatos, D., Kourti, A. Differential expression of two small Hsps during diapause in the corn stalk borer *Sesamia nonagrioides* (Lef.). J Insect Physiol. 2008, 54, 1503-1510.

[10] Morrow, G., Battistini, S., Zhang, P., Tanguay, R. M. Decreased lifespan in the absence of expression of the mitochondrial small heat shock protein *Hsp22* in *Drosophila*. J Biol Chem. 2004, 279, 43382-43385.

[11] Michaud, S., Morrow, G., Marchand, J., Tanguay, R. M. *Drosophila* small heat shock proteins: cell and organelle-specific chaperones? Prog Mol Subcell Biol. 2002, 28, 79-101.

[12] Sonoda, S., Ashfaq, M., Tsumuki, H. Cloning and nucleotide sequencing of three heat shock protein genes (*hsp90*, *hsc70*, and *hsp19.5*) from the diamondback moth, *Plutella xylostella* (L.) and their expression in relation to developmental stage and temperature. Arch Insect Biochem. 2006, 62, 80-90.

[13] Sonoda, S., Ashfaq, M., Tsumuki, H. A comparison of heat shock protein genes from cultured cells of the cabbage armyworm, *Mamestra brassicae*, in response to heavy metals. Arch Insect Biochem. 2007, 65, 210-222.

[14] Kokolakis, G., Tatari, M., Zacharopoulou, A., Mintzas, A. C. The *hsp27* gene of the Mediterranean fruit fly, *Ceratitis capitata*: structural characterization, regulation and developmental expression. Insect Mol Biol. 2008, 17, 699-710.

[15] Huang, L. H., Wang, C. Z., Kang, L. Cloning and expression of five heat shock protein genes in relation to cold hardening and development in the leafminer, *Liriomyza sativa*. J Insect Physiol. 2009,

55, 279-285.
[16] Xu, P., Xiao, J., Liu, L., Li, T., Huang, D. Molecular cloning and characterization of four heat shock protein genes from *Macrocentrus cingulum* (Hymenoptera: Braconidae). Mol Biol Rep. 2010, 37, 2265-2272.
[17] Yang, G. H. Harm of introducing the western honeybee *Apis mellifera* L. to the Chinese honeybee *Apis cerana* F. and its ecological impact. Acta Entomol Sin. 2005, 48, 401-406.
[18] Bitondi, M. M., Nascimento, A. M., Cunha, A. D., Guidugli, K. R., Nunes, F. M., Simoes, Z. L. Characterization and expression of the *Hex 110* gene encoding a glutamine-rich hexamerin in the honey bee, *Apis mellifera*. Arch Insect Biochem. 2006, 63, 57-72.
[19] Wang, J., Wang, X., Liu, C., Zhang, J., Zhu, C., Guo, X. The *NgAOX1a* gene cloned from *Nicotiana glutinosa* is implicated in the response to abiotic and biotic stresses. Bioscience Rep. 2008, 28, 259-266.
[20] Tsunoda, T., Takagi, T. Estimating transcription factor bindability on DNA. Bioinformatics (Oxford, England). 1999, 15, 622-630.
[21] Augusteyn, R. C. α-crystallin: a review of its structure and function. Clin Exp Optom. 2004, 87, 356-366.
[22] Fu, X., Li, W., Mao, Q., Chang, Z. Disulfide bonds convert small heat shock protein Hsp16.3 from a chaperone to a non-chaperone: implications for the evolution of cysteine in molecular chaperones. Biochem Bioph Res Co. 2003, 308, 627-635.
[23] Southgate, R., Ayme, A., Voellmy, R. Nucleotide sequence analysis of the *Drosophila* small heat shock gene cluster at locus 67B. J Mol Biol. 1983, 165, 35-57.
[24] Morrow, G., Samson, M., Michaud, S., Tanguay, R. M. Overexpression of the small mitochondrial *Hsp22* extends *Drosophila* life span and increases resistance to oxidative stress. FASEB J: Official Publication of the Federation of American Societies for Experimental Biology. 2004, 18, 598-599.
[25] Li, Z. W., Li, X., Yu, Q. Y., Xiang, Z. H., Kishino, H., Zhang, Z. The small heat shock protein (sHSP) genes in the silkworm, *Bombyx mori*, and comparative analysis with other insect sHSP genes. BMC Evol Biol. 2009, 9, 215.
[26] Wu, P., Wang, X., Qin, G. X., Liu, T., Jiang, Y. F., Li, M. W., Guo, X. J. Microarray analysis of the gene expression profile in the midgut of silkworm infected with cytoplasmic polyhedrosis virus. Mol Biol Rep. 2011, 38, 333-341.
[27] Huang, P. Y., Kang, S. T., Chen, W. Y., Hsu, T. C., Lo, C. F., Liu, K. F., Chen, L. L. Identification of the small heat shock protein, HSP21, of shrimp *Penaeus monodon* and the gene expression of HSP21 is inactivated after white spot syndrome virus (WSSV) infection. Fish Shellfish Immun. 2008, 25, 250-257.
[28] Sagisaka, A., Fujita, K., Nakamura, Y., Ishibashi, J., Noda, H., Imanishi, S., Mita, K., Yamakawa, M., Tanaka, H. Genome-wide analysis of host gene expression in the silkworm cells infected with *Bombyx mori* nucleopolyhedrovirus. Virus Res. 2010, 147, 166-175.
[29] Bao, Y., Wang, Q., Liu, H., Lin, Z. A small HSP gene of bloody clam (*Tegillarca granosa*) involved in the immune response against *Vibrio parahaemolyticus* and lipopolysaccharide. Fish Shellfish Immun. 2011, 30, 729-733.
[30] Barnes, P. J. Molecules in focus nuclear factor-κB. Int J Biochem Cell Biol. 1997, 29, 867-870.

1.4 Molecular Characterization and Immunohistochemical Localization of a Mitogen-activated Protein Kinase Gene, *Accp38b*, from *Apis cerana cerana*

Liang Zhang, Fei Meng, Yuzhen Li, Mingjiang Kang, Xingqi Guo, Baohua Xu

Abstract: The p38 mitogen-activated protein kinase (MAPK) is involved in various processes, including stress responses, development, and differentiation. However, little information on p38 MAPK in insects is available. In this study, a p38 MAPK gene, *Accp38b*, was isolated from *Apis cerana cerana* and characterized. The quantitative real-time PCR (qPCR) analysis revealed that *Accp38b* was induced by multiple stressors. Notably, the expression of *Accp38b* was relatively higher in the pupae phase than in other developmental phases. During the pupae phase, *Accp38b* expression was higher in the thorax than in the head and abdomen and higher in the fat body than in the muscle and midgut. Immunohistochemisty showed significant positive staining of Accp38b in sections from the brain, eyes, fat body, and midgut of *A. cerana cerana*. These results suggest that *Accp38b* may play a crucial role in stress responses and have multiple aspects function during development.

Keyword: *Apis cerana cerana*; Expression analysis; Immunohistochemistry; Stress response; p38 MAPK

Introduction

Mitogen-activated protein kinases (MAPKs) are a family of serine/threonine protein kinases that regulate a number of cellular responses, including proliferation, differentiation, nervous system development and immunity [1, 2]. Generally, the MAPK family is composed of three major groups: the extracellular regulated kinases (ERKs), the C-Jun N-terminal Kinases (JNKs) and the p38 MAPKs [3, 4].

As a member of a highly conserved subfamily of MAPKs, the p38 MAPK was originally identified as a 38 kDa protein that was rapidly activated by stressors [5]. In mammals, the p38 MAPK subfamily consists of four isoforms (p38α, p38β, p38γ, and p38δ) based on the conserved Thr-Gly-Tyr (TGY) motif and high homology at the amino acid level [4, 6]. Increasing evidence has indicated that p38 MAPK could be activated by various environmental stressors, such as ultraviolet (UV) irradiation, oxidative stress, heat or osmotic shock [7, 8]. The activation of p38 MAPK is performed via dual phosphorylation of both the serine/threonine and tyrosine residues in the conserved TGY motif by its upstream MAPK kinase [8]. The activated p38 MAPK plays important roles in many biological processes, including proliferation, differentiation, wound healing, inflammatory responses, and energy metabolism [8, 9].

In *Drosophila melanogaster*, the genome encodes three p38 orthologs, which are D-p38a (CG5475), D-p38b (CG7393), and D-p38c (CG3338) [10]. Similar to mammalian p38 MAPKs, D-p38 MAPKs are involved in the response to environmental stressors and immunity [11]. D-p38b has been reported to be involved in transforming growth factor beta (TGF-β) superfamily signal transduction in *Drosophila* wing morphogenesis [12]. Recently, D-p38b was reported to regulate intestinal stem cell (ISC) proliferation and differentiation in the midgut [2]. Although they have high homology with mammalian p38 MAPKs, the roles of p38 MAPKs in insects have not been fully investigated and should be further determined.

The Chinese honeybee, *Apis cerana cerana*, is an important indigenous species that plays a critical role in agricultural economic development as a pollinator of flowering plants. As the environment deteriorates, an increasing number of *A. cerana cerana* are disappearing. The p38 MAPKs play crucial roles in immunity and stress responses [11]. However, to date, no information is available on the p38 MAPK from *A. cerana cerana*. In this study, we isolated and characterized a p38 MAPK isoform gene named *Accp38b* from *A. cerana cerana*. Our results suggest that *Accp38b* may play an important role in stress responses and the development of the organism.

Materials and Methods

Insect materials and treatments

A. cerana cerana were bred at the experimental apiary of Shandong Agricultural University, Taian, China. Larvae, pupae and adult worker bees were identified by age. The worker bees were collected upon emergence from combs of outdoor hives. For the environmental stress analysis, the worker bees were divided into groups of 40 individuals, which were kept at a constant temperature, and treated with UV (30 MJ/cm^2), heat (43℃), cold (4℃) and strong light [250 μmol/(m^2·s)] for the indicated times. For the H$_2$O$_2$ treatment, 30% H$_2$O$_2$ was delivered in 0.5 μl of distilled water to the thoracic notum of the worker bees at a final concentration of 2 mmol/L. For the pesticide treatment, imidacloprid was delivered in 0.5 μl of distilled water to the thoracic notum of the worker bees at a final concentration of 20 mg/L. To analyze the influence of p38 MAPK effectors on the expression of *Accp38b*, A pollen and sucrose solution containing LPS (1 μg/ml) or SB203580 (10 μmol/L) was fed to the adult workers. Untreated larvae, pupae and adult worker bees were collected for gene expression analysis at different developmental periods. Five individual bees were used for each period. The treated materials were immediately frozen in liquid nitrogen and stored at −80℃ for RNA extraction.

RNA extraction, cDNA synthesis, and DNA isolation

The total RNA was extracted with the Trizol reagent (Invitrogen, Carlsbad, CA) according to the protocol of the manufacturer. The total RNA was digested with RNase-free DNase I (Promega, Madison, USA) and used for cDNA synthesis with the reverse transcriptase system (TransGen Biotech, Beijing, China). The genomic DNA was extracted from the fifth-instar larvae by using the EasyPure Genomic DNA Extraction Kit (TransGen Biotech, Beijing, China) according to the instructions of the manufacturer.

Isolation of the cDNA, genomic sequence and the 5'-flanking region of *Accp38b*

To determine the cDNA and the genomic DNA sequence of *Accp38b*, cloning procedures were performed as described previously [25]. The 5'-flanking region of *Accp38b* was amplified according to the genomic sequence of *A. mellifera*. The primer sequences used were provided in Supplementary Table 3. The MatInspector database (http://www.cbrc.jp/research/db/TFSEA RCH.html) was used to predict putative *cis*-acting element and transcription factor binding sites in the 5'-flanking region of *Accp38b*.

General bioinformation and phylogenetic analysis

The conserved domains in *Accp38b* were identified by using a tool on the NCBI server (http://blast.ncbi.nlm.nih.gov/Blast.cgi). The theoretical isoelectric point and molecular weight were predicted with PeptideMass (http://us.expasy.org/tools/peptide-mass.html). Multiple alignment analysis of p38 MAPKs was conducted by using DNAMAN software, version 5.2.2 (Lynnon Biosoft Company, Quebec, Canada). The phylogenetic tree was constructed by the NJ (Neighbor-Joining) method using MEGA software, version 4.1.

Protein expression, purification, and Western blotting analysis

The Accp38b coding region was ligated into the pET-30a (+) expression vector after digestion with the restriction endonucleases *Kpn* I and *Eco*R I. The recombinant plasmid pET-30a (+)-*Accp38b* was transformed into *Escherichia coli* strain BL21 (DE3) cells. The expression and purification of the fusion protein were performed as described previously [25]. Western blotting was performed according to the procedure of Tufail et al. [26] with minor modifications. Briefly, the purified protein was separated by sodium dodecylsulfate-polyacrylamide gel electrophoresis (SDS-PAGE) (12%). Anti-Accp38b serum was used as the primary antibody at a dilution of 1∶500 (*V/V*). Goat anti-rat immunoglobulin G (IgG) (Dingguo, Beijing, China) conjugated to horseradish peroxidase (HRP) was used as the secondary anti-body (1∶150). The HRP was visualized by using a chemiluminescent HRP substrate. The anti-Accp38b replaced by phosphate buffered saline (PBS, pH 7.2) in the reaction was used as the negative control.

Preparation of antisera and immunohistochemistry

The purified Accp38b protein (100 μg/ml) mixed with an equal volume of Freund's complete adjuvant was used to subcutaneously immunize white mice. The mice were boosted three times with 25 µg/ml of a mixture of purified protein and Freund's incomplete adjuvant (1 ∶ 1) at one week intervals. Three days after the last injection, blood was collected, and the serum was separated by centrifugation at 3,000 r/min for 10 min and stored at −80℃. For immunohistochemistry, the head, abdomen and midgut tissues from the bees were separated and fixed in 4% paraformaldehyde. The fixed tissues were dehydrated with an ethanol gradient, cleared in xylene, and embedded in paraffin. The paraffin sections were dewaxed, blocked with 10% normal goat serum for 30 min and incubated with anti-Accp38b anti-body (1 ∶ 50) at 4℃ overnight. Subsequently, the sections were washed in PBS, incubated with HRP-conjugated goat anti-rat IgG (1 ∶ 150) at 37℃ for 1 h, washed with PBS, and incubated with HRP-streptavidin (Solarbio, Beijing, China) at a dilution of 1 ∶ 200 for 40 min. The color development was performed with 3, 3′-diaminobenzidine solution (DAB, 0.25 mg/ml). The material was counterstained with the Harris hematoxylin, dehydrated, mounted with neutral balsam and analyzed by using a light microscope (BX51, Olympus, Japan). The normal serum was used instead of the anti-Accp38b serum as the negative control.

Quantitative real-time PCR

The cDNA synthesis was performed as described above. The *Accp38b* primer pairs Q1/Q2 and *A. mellifera β-actin* (XM623-378) primer pairs β1/β2 were used for qPCR with the SYBR®PrimeScript™ RT-PCR Kit (TaKaRa, Dalian, China) in a CFX96™ Real-time System (Bio-Rad) with the following conditions: 95℃ for 30 s, 40 amplification cycles (95℃ for 10 s, 53℃ for 20 s, and 72℃ for 20 s) and a melting cycle from 65℃ to 95℃. The data were analyzed by using the CFX Manager software, version 1.1, and significant differences were determined by using the Statistical Analysis System (SAS) software, version 9.1. All reactions were performed with three technical replicates.

Results

Isolation and sequence analysis of *Accp38b* cDNA

The full-length cDNA sequence of *Accp38b* (GU321334) consisted of 1,465 nucleotides, containing a 1,083 bp open reading frame (ORF), a 148 bp 5′-untranslated region (UTR) and a 234 bp 3′ UTR. *Accp38b* was predicted to encode a protein of 360 amino acid residues with a putative molecular weight of 41.34 kDa and an isoelectric point of 6.15. Multiple sequence alignments showed that *Accp38b* contains 11 conserved subdomains, a catalytic loop (activation-loop), and a phosphorylation motif (TGY motif) (Fig. 1A). The TGY motif between subdomains Ⅶ and Ⅷ is the typical feature of p38 MAPKs, which is different from the TPY motif in JNK family and TEY motif in ERK family (Fig. 1B). Additionally, kinase interaction motif (KIM) docking sites and putative substrate binding sites were predicted by using a bioinformatics tool available on the NCBI server. Accp38b shares 99.72% sequence similarity with Amp38b from *Apis mellifera*, 91% with Nvp38 from *Nasonia vitripennis*, 83% with Bmp38 from *Bombyx mori*, and 76.63% and 76.36% with D-p38a and D-p38b from *D. melanogaster*, respectively. A phylogenetic analysis revealed that Accp38b is closely related to the p38 MAPKs from insects, such as *N. vitripennis*, *B. mori* and *D. melanogaster* (Fig. 1C).

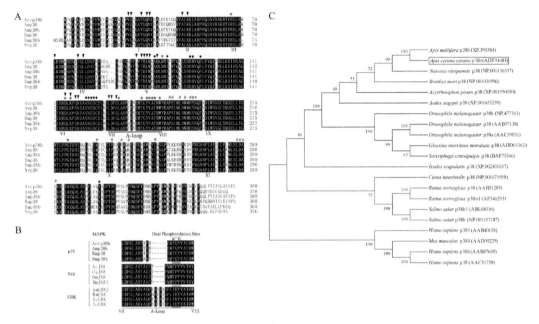

Fig. 1 Multiple amino acid alignments and phylogenetic analysis. A. The alignment of the amino acid sequence of Accp38b with other p38 MAPKs is shown. Identical amino acids are shaded in black. The protein kinase subdomains are shown with numerals (I-XI) at the bottom of the sequences, and the activation loop (A-loop) is underlined. The putative ATP binding sites and lipid binding sites of p38 MAPKs are denoted by triangle (▼) and hash symbol (#), respectively. The kinase interaction motif (KIM) docking sites are indicated by dots (•). The asterisk (*) over the sequences indicate the putative substrate binding sites. B. A comparison of the phosphorylation motif between subdomains VII and VIII of p38, JNK and ERK is shown. The motifs are boxed, and the dual phosphorylation sites are marked with asterisks. C. A phylogenetic tree was constructed from different species by the neighbor-joining method using MEGA 4.1 software. The numbers above or below the branches indicate the bootstrap values (>50%) from 500 replicates. All of the gene messages from various species are listed in Supplementary Table 1.

Identification of the genomic structure of *Accp38b*

The length of *Accp38b* in the genome (GU321334) is 4,038 bp, including 10 exons separated by 9 introns (Supplementary Fig. 1). The number of bases and the GC content (%) of each exon of *Accp38b* were similar to those of *Amp38b*, with the GC content (%) ranging from 26.32% to 44.23%. However, the number of bases and the GC content (%) of each intron of *Accp38b* were different from those of *Amp38b* (Supplementary Table 1).

To analyze the organization of the regulatory region of *Accp38b*, a 649 bp fragment (GU295943) upstream of the transcription start site was isolated. Many *cis*-acting elements were predicted, such as the DNA replication-related element (DRE) and cAMP-responsive element (CRE) (Supplementary Fig. 2). The DRE is the binding site for the DRE binding factor (DREF), which plays an important role in the regulation of DNA replication, cell proliferation-related genes and the homeostasis of the adult midgut in *D. melanogaster* [13]. The CRE is the binding site for the CRE binding protein (CREB), which regulates gene expression during cell differentiation, proliferation, and apoptosis [14, 15]. In addition, binding sites for the CCAAT/enhancer binding protein (CEBP), glucocorticoid responsive and related element (GREF) and cell cycle homology regulatory factor (CHRF) were also found (Supplementary Fig. 2). These data suggest that *Accp38b* may have multiple functions related to cell growth and organism development.

Differential expression of *Accp38b* at different developmental phases and in various tissues

To investigate the expression patterns of *Accp38b* at different developmental stages, qPCR was performed. As shown in Fig. 2A, the mRNA level of *Accp38b* in the pupae and adult was 13.09- and 3.94-fold higher than that in the larvae, respectively. The mRNA level in the thorax and abdomen was 2.61- and 1.73-fold higher than that in the head, respectively (Fig. 2B). Tissue-specific expression analysis indicated that the expression of *Accp38b* normalized to that in the midgut was the highest in the fat body and muscle (Fig. 2C). These results suggest that *Accp38b* may have various functions at different developmental phases and in various tissues.

Expression patterns of *Accp38b* under environmental stressors and effectors treatments

To detect the expression patterns of *Accp38b* under different environmental stress conditions, the adults were exposed to a series of stressors. The expression of *Accp38b* was induced after 1 h of UV light treatment and reached the highest level after 4 h of UV light treatment (Fig. 2D). Under the heat condition, the expression of *Accp38b* was significantly increased after 2 h of UV light treatment and then gradually decreased from 3 h to 4 h (Fig. 2E). A similar expression pattern for *Accp38b* was found under the cold, light and H_2O_2 conditions (Fig. 2F-H). Honeybees are often poisoned by pesticides used in agricultural production. When the adults were treated with imidacloprid, the expression of *Accp38b* was increased at 1 h and reached the highest level at 2 h (Fig. 2I). However, the expression was significantly decreased at 4 h when the expression level was even lower than that without any stress, which suggested that the pesticide might have some depression effect on the expression of *Accp38b*. Consistent with previous studies, the transcription of *Accp38b* could be induced by LPS and inhibited by SB203580 (Fig. 2J, K). These results indicate that *Accp38b* may be involved in responses to stressors.

Immunohistochemical localization of Accp38b

To localize Accp38b, paraffin sections of head, eye, fat body and midgut from the pupae were stained with an anti-Accp38b antibody. The results showed that Accp38b was expressed in the brain, especially in the esophagus and the periphery of the median calyx of the mushroom

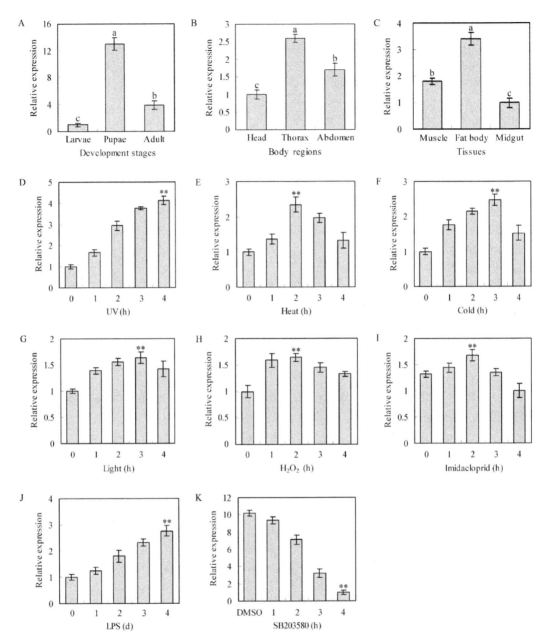

Fig. 2 Expression profile of *Accp38b* by quantitative real-time PCR. The relative expression of *Accp38b* at different developmental phases (A) and in different body regions (B) and tissues (C) is shown. Data are the means ± SE of three independent experiments. Different letters above the columns indicate significant differences ($P < 0.01$) according to Duncan's multiple range tests. The relative expression of *Accp38b* under different stress conditions, such as UV (30 MJ/cm^2) (D), heat (43°C) (E), cold (4°C) (F), light [250 μmol/(m^2·s)] (G), H_2O_2 (2 mmol/L) (H), imidacloprid (20 mg/L) (I), LPS (1 μg/ml) (J), and SB203580 (10 μmol/L) (K), is shown. The expression levels were normalized to that of untreated bees (0 h) or the dimethyl sulfoxide (DMSO) control (H). Data are the means ± SE of three independent experiments. Significant differences determined by Duncan's multiple range tests are indicated with two asterisks (**$P < 0.01$, comparing the control and the highest level).

body (Fig. 3A). Notably, Accp38b was widely distributed in the photoreceptor cells in the retina and compound eye (Fig. 3B). Furthermore, significant positive staining was found in fat body sections, especially the surrounding fat particles, and in the midgut wall (Fig. 3C, D). These findings imply that Accp38b may play important roles in the nervous, visual, and digestive systems and in metabolic regulation.

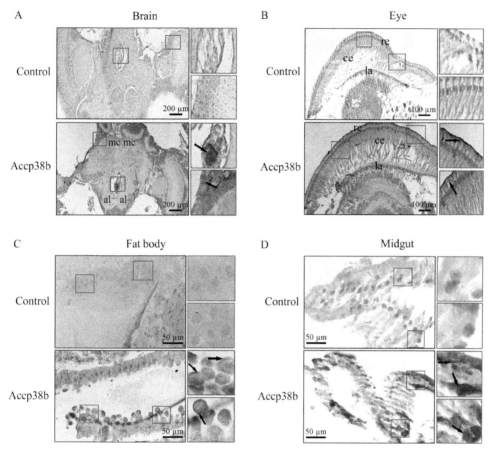

Fig. 3 Immunocytochemical analysis for the tissue distribution of Accp38b. The expression of Accp38b is represented by positive staining in the brain (A), eye (B), fat body (C), and midgut (D). The arrows indicate the positive staining. Normal serum was used instead of the anti-Accp38b serum as the negative control. The abbreviations indicate the following: al, antennal lobe; mc, median calyx of the mushroom body; la, lamina; ce, compound eye; and re, retina. (For color version, please sweep QR Code in the back cover)

Discussion

p38 MAPK plays important roles in many biological processes [8]. However, research concerning p38 MAPKs has mainly focused on mammals, and there is little information on the role of p38 MAPKs in insects. In this study, we described the cloning, characterization and immunohistochemical localization of a novel member of the p38 MAPKs from *A. cerana cerana*. qPCR analysis revealed that *Accp38b* was induced by many cellular stressors, such as

UV, heat, cold, and imidacloprid treatment (Fig. 2). Consistent with these results, the p38 MAPK in *Drosophila* is activated by a variety of environmental stimuli, including osmotic shock, heat, and UV irradiation [16, 17]. These results suggest that the function of p38 MAPKs in insects is conserved. Recently, the p38 MAPK from *Drosophila* was reported to participate in the defense response against Cry toxins from *Bacillus thuringiensis*, pathogenic bacteria and fungi [10, 18]. This observation provides clues for the role of *Accp38b* in the defense response; however, it should be further explored.

The expression analysis revealed that *Accp38b* had the highest expression in pupae (Fig. 2A). At the pupal stage, cells become highly differentiated, and many organs tend to mature. Therefore, it is reasonable to deduce that *Accp38b* may be involved in differentiation or development. Furthermore, *Accp38b* mRNA could be detected in the head, thorax and abdomen, and the highest relative expression of *Accp38b* was in the thorax (Fig. 2B). Because the primary locomotoriums are located in the thorax and *Accp38b* expression was also observed in muscle, we hypothesize that *Accp38b* is involved in the movement of *A. cerana cerana*. In insects, the energy for movement or other activity is primarily stored in the adipose cells. Interestingly, there was relatively higher expression of *Accp38b* in the fat body (Fig. 2C). Immunohistochemistry also showed significant staining of Accp38b in the fat body (Fig. 3C). Recently, it was reported that the p38 MAPK from *Drosophila* could activate activating transcription factor (ATF)-2, which is a member of the ATF/CREB family of transcription factors that regulates fat metabolism [19]. The expression of *Accp38b* could be detected in the abdomen, which contains abundant adipose tissue, and digestive organs, such as the midgut. As expected, the significant staining of Accp38b was observed in the section of midgut (Fig. 3D). In *Drosophila*, D-p38b has been reported to play a crucial role in the balance between intestinal stem cell proliferation and differentiation in the adult midgut [2]. In the brain section, the significant expression of Accp38b was observed in the esophagus and the periphery of the median calyx of the mushroom body by immunohistochemistry (Fig. 3A). Similarly, p38α and p38β, human orthologs of Accp38b, were abundantly expressed in brain tissues, and p38α plays an important role in the central nervous system [20, 21]. Additionally, the expression analysis revealed that *Accp38b* was responsive to light (Fig. 2G), and there was marked positive staining of Accp38b in the photoreceptor cells of the retina and compound eye (Fig. 3B). Therefore, it is plausible that *Accp38b* is involved in signal transduction in the visual system. A similar observation has been reported for the rat p38 MAPK, which functions in the developing visual system [22]. To our knowledge, this data offer the first insight into the function of p38 MAPK in light-related signal transduction in the visual system. These results suggest that Accp38b may be involved in the nervous, visual, and digestive systems and in metabolic and developmental regulation.

In the 5′-flanking region of *Accp38b*, several DREF binding sites were predicted (Supplementary Fig. 2). DREF play an important role in regulation of DNA replication and cell proliferation-related genes by binding to the DRE of target genes [23]. The DRE/DREF system is required for the homeostasis in the adult midgut [24]. In *Drosophila*, *D-p38b* gene expression is regulated by DREF in the adult midgut [13]. This result provides further support for a function of Accp38b in the midgut. In addition, other elements involved in the regulation of gene expression were also found, such as CEBP,

CREF and PARF binding sites. These results further suggest that the signal transduction function of *Accp38b* in multiple aspects.

In conclusion, the characterization of *Accp38b* strongly suggests that Accp38b may be involved in the nervous, visual, and digestive systems as well as in metabolic and developmental regulation. These results broaden our knowledge regarding the role of p38 MAPKs in insects.

References

[1] Davis, M. M., Primrose, D. A., Hodgetts, R. B. A member of the p38 mitogen-activated protein kinase family is responsible for transcriptional induction of dopa decarboxylase in the epidermis of *Drosophila melanogaster* during the innate immune response. Mol Cell Biol. 2008, 28, 4883-4895.
[2] Park, J. S., Kim, Y. S., Yoo, M. A. The role of p38b MAPK in age-related modulation of intestinal stem cell proliferation and differentiation in *Drosophila*. Aging. 2009, 1, 637-651.
[3] Raman, M., Chen, W., Cobb, M. H. Differential regulation and properties of MAPKs. Oncogene. 2007, 26, 3100-3112.
[4] Ono, K., Han, J. The p38 signal transduction pathway: activation and function. Cell Signal. 2000, 12, 1-13.
[5] Han, J., Bibbs, L., Ulevitch, R. J. A MAP kinase targeted by endotoxin and hyperosmolarity in mammalian cells. Science. 1994, 265, 808-811.
[6] Zarubin, T., Han, J. Activation and signaling of the p38 MAP kinase pathway. Cell Res. 2005, 15, 11-18.
[7] An, H., Lu, X., Liu, D., Yarbrough, W. G. LZAP inhibits p38 MAPK (p38) phosphorylation and activity by facilitating p38 association with the wild-type p53 induced phosphatase 1 (WIP1). PLoS One. 2011, 6, e16427.
[8] Gong, X., Ming, X., Deng, P., Jiang, Y. Mechanisms regulating the nuclear translocation of p38 MAP kinase. J Cell Biochem. 2010, 110, 1420-1429.
[9] Roux, P. P., Blenis, J. ERK and p38 MAPK-activated protein kinases: a family of protein kinases with diverse biological functions. Microbiol Mol Biol Rev. 2004, 68, 320-344.
[10] Chen, J., Xie, C., Tian, L., Hong, L., Wu, X., Han, J. Participation of the p38 pathway in *Drosophila* host defense against pathogenic bacteria and fungi. Proc Natl Acad Sci USA. 2010, 107, 20774-20779.
[11] Craig, C. R., Fink, J. L., Yagi, Y., Ip, Y. T., Cagan, R. L. A *Drosophila* p38 orthologue is required for environmental stress responses. EMBO Rep. 2004, 5, 1058-1063.
[12] Adachi-Yamada, T., Nakamura, M., Irie, K., Tomoyasu, Y., Sano, Y., Mori, E., Goto, S., Ueno, N., Nishida, Y., Matsumoto, K. p38 mitogen-activated protein kinase can be involved in transforming growth factor beta superfamily signal transduction in *Drosophila* wing morphogenesis. Mol Cell Biol. 1999, 19, 2322-2329.
[13] Park, J. S., Kim, Y. S., Kim, J. G., Lee, S. H., Park, S. Y., Yamaguchi, M., Yoo, M. A. Regulation of the *Drosophila p38b* gene by transcription factor DREF in the adult midgut. Biochim Biophys Acta. 2010, 1799, 510-519.
[14] Ishida, M., Mitsui, T., Yamakawa, K., Sugiyama, N., Takahashi, W., Shimura, H., Endo, T., Kobayashi, T., Arita, J. Involvement of cAMP response element-binding protein in the regulation of cell proliferation and the prolactin promoter of lactotrophs in primary culture. Am J Physiol Endocrinol Metab. 2007, 293, E1529-E1537.
[15] Carlezon Jr, W. A., Duman, R. S., Nestler, E. J. The many faces of CREB. Trends Neurosci. 2005, 28, 436-445.
[16] Han, Z. S., Enslen, H., Hu, X., Meng, X., Wu, I. H., Barrett, T., Davis, R. J., Ip, Y. T. A conserved p38 mitogen-activated protein kinase pathway regulates *Drosophila* immunity gene expression. Mol

Cell Biol. 1998, 18, 3527-3539.
- [17] Han, S. J., Choi, K. Y., Brey, P. T., Lee, W. J. Molecular cloning and characterization of a *Drosophila* p38 mitogen-activated protein kinase. J Biol Chem. 1998, 273, 369-374.
- [18] Cancino-Rodezno, A., Alexander, C., Villasenor, R., Pacheco, S., Porta, H., Pauchet, Y., Soberón, M., Gill, S.S., Bravo, A. The mitogen-activated protein kinase p38 is involved in insect defense against Cry toxins from *Bacillus thuringiensis*. Insect Biochem Mol Biol. 2010, 40, 58-63.
- [19] Okamura, T., Shimizu, H., Nagao, T., Ueda, R., Ishii, S. ATF-2 regulates fat metabolism in *Drosophila*. Mol Biol Cell. 2007, 18, 1519-1529.
- [20] Borders, A. S., De Almeida, L., Van Eldik, L. J., Watterson, D. M. The p38α mitogen-activated protein kinase as a central nervous system drug discovery target. BMC Neurosci. 2008, 9 Suppl 2, S12.
- [21] Takeda, K., Ichijo, H. Neuronal p38 MAPK signalling: an emerging regulator of cell fate and function in the nervous system. Genes Cells. 2002, 7, 1099-1111.
- [22] Oliveira, C. S., Rigon, A. P., Leal, R. B., Rossi, F. M. The activation of ERK1/2 and p38 mitogen-activated protein kinases is dynamically regulated in the developing rat visual system. Int J Dev Neurosci. 2008, 26, 355-362.
- [23] Matsukage, A., Hirose, F., Yoo, M. A., Yamaguchi, M. The DRE/DREF transcriptional regulatory system: a master key for cell proliferation. Biochim Biophys Acta. 2008, 1779, 81-89.
- [24] Park, S. Y., Kim, Y. S., Yang, D. J., Yoo, M. A. Transcriptional regulation of the *Drosophila* catalase gene by the DRE/DREF system. Nucleic Acids Res. 2004, 32, 1318-1324.
- [25] Meng, F., Kang, M., Liu, L., Luo, L., Xu, B., Guo, X. Characterization of the TAK1 gene in *Apis cerana cerana* (*AccTAK1*) and its involvement in the regulation of tissue-specific development. BMB Rep. 2011, 44, 187-192.
- [26] Tufail, M., Naeemullah, M., Elmogy, M., Sharma, P. N., Takeda, M., Nakamura, C. Molecular cloning, transcriptional regulation, and differential expression profiling of vitellogenin in two wing-morphs of the brown planthopper, *Nilaparvata lugens* Stal (Hemiptera: Delphacidae). Insect Mol Biol. 2010, 19, 787.

1.5 *AccERK2*, A Map Kinase Gene from *Apis cerana cerana*, Plays Roles in Stress Responses, Developmental Processes, and the Nervous System

Yuzhen Li, Liang Zhang, Mingjiang Kang, Xingqi Guo, Baohua Xu

Abstract: Extracellular signal-regulated kinase (ERK), a mitogen-activated protein kinase (MAPK), plays roles in a variety of cellular responses. However, limited information is available on the relationship between ERKs and environmental stress. In this report, an ERK gene, *AccERK2*, was cloned and characterized from *Apis cerana cerana*. Polypeptide sequence alignment revealed that the single-copied AccERK2 shares high identity with other known ERKs and contains the typical conserved Thr-Glu-Tyr (TEY) motif in its activation loop. Genomic sequence analysis revealed that the seven exons of *AccERK2* are interrupted by six introns, and the seventh intron is located in the 3′ untranslated region. Semi-quantitative reverse transcription (RT-PCR) indicated that *AccERK2* was expressed at higher levels in the larval and pupal stages than in the adult stage. *AccERK2* was also most highly expressed in the brain. The expression of *AccERK2* was induced by abiotic stresses, including heat, ion irradiation, oxidative stress, and heavy metal ions. Based on these results, it appears that *AccERK2* in *A. cerana cerana* participates in developmental processes, the nervous system, and responses to environmental stressors.

Keywords: *Apis cerana cerana*; ERK; Abiotic stress; Developmental processes; Nervous system

Introduction

Mitogen-activated protein kinase (MAPK) cascades are evolutionarily conserved signaling pathways that transduce a wide range of extracellular stimuli to regulate cellular activities, such as proliferation, differentiation, metabolism, and programmed cell death [1]. This signal transduction module includes three-tiered sequential elements: MAPK kinase kinases (MAPKKKs), MAPK kinases (MAPKKs), and MAPKs [2-4]. As the primary endpoints of the cascades, MAPKs play roles in regulating specific cellular responses by modulating the functions of target transcription factors or cytoplasmic proteins [5]. So far, the identified MAPKs have been classified into four subfamilies: extracellular signal-regulated kinases (ERKs), stress-activated protein kinases/c-Jun N-terminal kinases (SAPKs/JNKs), p38

MAPKs, and ERK5 [4, 5]. Each subfamily has its own specific biological functions [6].

The ERKs, the oldest known members of MAPKs, contain at least six isoform family members. ERK1 and ERK2 (also named p44 MAPK and p42 MAPK, respectively) are the typical members of ERKs, and their activation requires dual phosphorylation of the threonine and tyrosine residues in the Thr-Glu-Tyr (TEY) motif by their upstream activators [1]. Mitogenic stimuli, such as growth factors and phorbol esters, are well-identified inducers of ERKs [7, 8], and few environmental stresses, such as high temperature and oxidative stress, can also activate ERKs. For example, Fujiwara and Denlinger reported that temperature elevation activates ERK and that activated ERK acts in breaking pupal diapause in flesh flies [9]. Substantial activation of ERK in response to oxidative stress enhances the survival of mammalian 3T3 cells [10]. In *Drosophila*, ERKs are strongly activated by hydrogen peroxide but weakly activated by osmotic stress [11]. To date, however, there is no good explanation for this result. Thus, the precise role of ERKs under various environmental stresses remains to be determined.

Studies indicated that ERK1/2 play roles in cellular proliferation, differentiation, and the prevention of apoptosis [7]. Emerging evidence suggests that ERK1/2 are related to the central nervous system. Guo et al. reported that ERK1/2 are abundantly and ubiquitously expressed throughout the rat central nervous system and further supported the functional importance of ERKs-mediated signaling pathways in the processing of the negative consequences of pain associated with sensory, emotional, and cognitive dimensions [12]. The roles of ERK1/2 in the purine-mediated neuroprotection of clonal rat hypoxic pheochromocytoma (PC12) cells and primary rat cerebellar granule neurons were confirmed [13]. ERK1/2 also have a link with the higher functions of learning and memory [14]. Those reports about the functions of ERK1/2 in the nervous system mainly focus on the mammalian; however, the functions of ERK1/2 in other species remain unknown.

The Chinese honeybee, *Apis cerana cerana*, is an economically indigenous species with many unique characteristics, such as the pollination of flowering crops at low temperatures and the resistance to ectoparasitic mites via a specific grooming behavior [15]. Therefore, we chose to investigate the putative function of ERK by using *A. cerana cerana* as an experimental insect. Here, we report on the outcomes of our investigation.

Materials and Methods

Biological specimens and treatments

Foraging *A. cerana cerana* (Chinese honeybee) workers were acquired from the experimental apiary of Shandong Agricultural University (Taian, China). Emerging worker bees were collected from the combs of outdoor hives. To detect ERK2 developmental and tissue expression patterns, bees of different ages (larval, pupal, and adult bees) were collected from a large nest. Selected tissues, including the brain, muscle, fat body, midgut, and epidermis, were isolated from anesthetized bees with CO_2. For testing the effects of temperature, ionic rays, and oxidative stress, worker bees were divided into groups of 40 individuals and separately treated with heat (43℃), cold (4℃), ultraviolet (UV)-light (30 MJ/cm^2), and H_2O_2 (30%; Kaitong, Tianjin, China). To analyze the influence of ERK effectors, bacterial lipopolysaccharide (LPS; 1 μl/ml; Sigma, St. Louis, MO) and U0126 (10 μmol/L; Promega, Madison, WI) were separately mixed with the pollen to feed the adult workers. In addition, NaCl (0.3 mol/L; Kaitong, Tianjin, China) and $MnCl_2$ (400 pmol/L; Bodi, Tianjin, China) were fed to workers to determine the expression profiles of *AccERK2* under osmotic and heavy metal ion conditions. All of the bees and tissues were flash frozen in liquid nitrogen and immediately stored at −80℃.

RNA extraction, cDNA synthesis, and genomic DNA isolation

Total RNA was extracted from the worker bees by using the Trizol reagent (Invitrogen,Carlsbad, CA) following the protocol of the manufacturer and digested with RNase-free Dnase Ⅰ (Promega) to remove the potential genomic DNA contamination. cDNA synthesis was carried out using the reverse transcription kit (TransGen Biotech, Beijing, China). Total genomic DNA was isolated from worker bees by using the EasyPure Genomic DNA Extraction Kit according to the instructions of the manufacturer (TransGen Biotech).

Isolation of the full-Length cDNA sequence of *AccERK2*

The full-length cDNA sequence of *AccERK2* was amplified by reverse transcription-PCR (RT-PCR) and rapid amplification of cDNA end-PCR (RACE-PCR). Based on the amino acid and nucleotide sequences of ERKs from other species, primers (IP1 and IP2) were designed to obtain the internal conserved fragment (all primers are listed in Table 1).The 5′ RACE and 3′ RACE were carried out with the specific primers (5P1/5P2 and3P1/3P2) and universal primers (AAP/AUAP and B25/B26). By aligning the sequences from 5′ and 3′ RACE reactions and the amplified fragment, we deduced the full-length cDNA sequence of *AccERK2*. Using this constructed sequence, the specific primers, QP1and QP2, were designed to amplify the complete coding sequence of *AccERK2*. All of the PCR products were separated by 1% agarose gel electrophoresis, purified, and cloned into pEASY-T3 vector (TransGen Biotech). The recombinant DNA molecules were transformed into *Escherichia coli* competent cells and the selected positive clones were sequenced.

Amplification of the genomic sequence of *AccERK2*

To amplify the *AccERK2* genomic sequence, primers, GP1 and GP2, designed based on the *AccERK2* cDNA sequence were employed. All of the PCR products were separated, purified, cloned into pEASY-T3 vector (TransGen Biotech), and sequenced.

Bioinformatic and phylogenetic analysis

Sequences homologous to *AccERK2* were retrieved by using the bioinformatics tools on the NCBI server (http://blast.ncbi.nlm.nih.gov/Blast.cgi) and aligned by using DNAMAN version 5.2.2 (Lynnon Biosoft, Quebec, Canada). The theoretical isoelectric point and molecular weight of AccERK2 was predicted by PeptideMass (http://web.expasy.org/peptide_mass/).

The phylogenetic analysis was conducted by using Molecular Evolutionary Genetics Analysis (MEGA version 4.1) with the neighbor-joining method.

Table 1 The primers used in this study

Primer name	Description	Primer sequence (5'—3')
IP1	Internal fragment primer, forward	CGAAGTTGGTCCACGATACA
IP2	Internal fragment primer, reverse	CCGGTTCATCAGCAGGATCAT
5P1	5' RACE reverse primer, primary	CACCCTCACCCATGTATGAC
5P2	5' RACE reverse primer, nested	CATAAGCAGATACGACCATTCCG
AAP	Abridged anchor primer	GGCCACGCGTCGACTAGTAC(G)$_{14}$
AUAP	Abridged universal amplification primer	GGCCACGCGTCGACTAGTAC
3P1	3' RACE reverse primer, primary	GGGAGTCCTTGGATCACCATC
3P2	3' RACE reverse primer, nested	CGGATTGTTGTAGAGGATGC
B25	Universal primer, nested	GACTCTAGACGACATCGA
B26	Universal primer, primary	GACTCTAGACGACATCGA(T)$_{18}$
FP1	Full-length sequence primer, forward	ATGGCGGGTGAAGGTTCTG
FP2	Full-length sequence primer, reverse	ATAGCGTTTTCTTGATGATTC
RT1	RT-PCR primer, forward	CGGAATGGTCGTATCTGCTTATG
RT2	RT-PCR primer, reverse	GCATCCTCTACAACAATCCG
β-1	Standard control primer, forward	GTTTTCCCATCTATCGTCGG
β-2	Standard control primer, reverse	TTTTCTCCATATCATCCCAG
GP1 (FP1)	Genomic sequence primer, forward	ATGGCGGGTGAAGGTTCTG
GP2	Genomic sequence primer, reverse	TCTGACATTATATTTTCGCTTCC

Southern blotting hybridization

Total genomic DNA extracted from worker bees was digested with the restriction endonucleases, *Eco*R I, *Eco*R V, and *Hin*d III (TaKaRa, Dalian, China), at 37℃ overnight, separated by 1% agarose gel electrophoresis and transferred onto hybond-N$^+$ nylon membranes (Amersham, Pharmacia, England) by capillary blotting. The probe was amplified by using the FP1 and FP2 primers and labeled with [α^{32}P]-dCTP by using the Primer-a-Gene Labeling System (Promega) according to the instructions of the manufacturer. The Southern blotting procedure was performed as described previously [16].

Semi-Quantitative RT-PCR

To determine the expression patterns of the *AccERK2* gene, semi-quantitative RT-PCR was carried out. The primers RT1 and RT2 were designed to detect the expression profiles of *AccERK2* under different treatments, and the β-1 and β-2 primers were used to amplify a house-keeping gene, *β-actin*, which was used as the standard control. The samples were run in triplicate.

Western blotting analysis

The complete open reading frame (ORF) of *AccERK2* was inserted into the pET-30a (+) vector (Novagen, Madison, WI), and the recombinant plasmid pET-30a (+)-AccERK2 was transformed into *E. coli* BL21 (DE3) (Novagen). The cultures were induced with 0.2 mmol/L isopropylthio-β-D-galactoside (IPTG; Promega) for 4 h to get the purified AccERK2 protein with the MagneHis™ Protein Purification System (Promega). Then, the antisera of AccERK2 were obtained by injecting the inclusion AccERK2 protein (100 μg/ml) subcutaneously into white mice. The total protein extracted from *A. cerana cerana* according to Li et al. was separated by 12% sodium dodecyl sulphate polyacrylamide gel electrophoresis (SDS-PAGE) and subsequently electrotransferred onto a polyvinylidene fluoride membrane (Millipore, Bedford, MA) by using a Semi-dry Transfer Conit [17]. Western blotting was performed according to the procedure of Meng et al. [18]. The anti-AccERK2 serum was used as the primary antibody at a dilution of 1∶500 (*V/V*). Peroxidase-conjugated goat anti-mouse immunoglobulin G (Dingguo, Beijing, China) was used as the secondary antibody at a dilution of 1∶2,500 (*V/V*). The results of the binding reactions were visualized by using horseradish peroxidase (HRP) substrate. Phosphate-buffered saline (PBS, pH 7.2) instead of the anti-AccERK2 was used as the negative control in the reaction.

Results

Isolation and characterization of the full-length cDNA of *AccERK2*

The full-length cDNA of *AccERK2* (also named p42 MAPK; GenBank accession number: GU321335) was obtained by RT-PCR and RACE-PCR. It contains 1,271 bp nucleotides comprising a 1,098 bp ORF, a 64 bp 5′-untranslated region (UTR), and a 109 bp 3′ UTR. The ORF encodes a polypeptide of 365 amino acids in length with the predicted molecular mass of 41.99 kDa and the isoelectric point of 5.81.

Protein sequence alignment and phylogenetic analysis

Compared with other known ERKs, AccERK2 is highly identical to *Apis mellifera*, *Bombyx mori*, *Sarcophaga crassipalpis*, and *Tribolium castaneum*. Several conserved sequences, 11 conserved subdomains and a phosphorylation motif (TEY motif) located in the catalytic loop (activation-loop) were identified (Fig. 1A). These conserved sequences indicated that AccERK2 is a typical ERK subfamily member. The TEY motif between subdomains Ⅶ and Ⅷ, which is necessary for MAPK activation, is a characteristic feature of ERKs, whereas TPY and TGY motifs are typical of JNKs and p38s, respectively (Fig. 1B).

To investigate the evolutionary relationship between AccERK2 and other ERKs, a neighbor-joining phylogenetic tree was generated by MEGA 4.1 (Fig. 2). The phylogenetic tree showed that AccERK2 is clustered with Arthropoda.

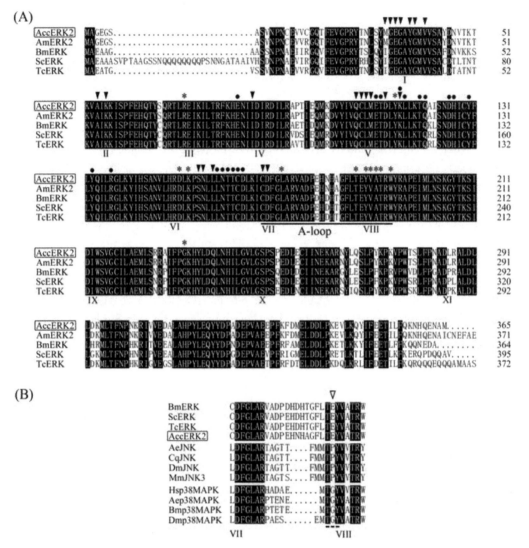

Fig. 1 Sequence analysis of AccERK2. (A) Alignment of the amino acid sequences of AccERK2 (ADT91684) and other ERKs. The protein features are indicated as follows: numerals (Ⅰ-Ⅺ), the MAP kinase subdomains; arrowheads (▼), the putative ATP-binding sites; asterisks (*), the substrate-binding sites; dots (●), the kinase interaction motif (KIM)-docking sites; black line (—), the activation loop (A-loop). The selected amino acid sequences are *Apis mellifera* ERK2 (XP_393029), *Bombyx mori* ERK (NP_001036921), *Sarcophaga crassipalpis* ERK (BAF75365), and *Tribolium castaneum* ERK (XP_966833). Identical amino acid residues in this alignment are indicated with the black background. (B) Comparison of the specific phosphorylation motifs between subdomains Ⅶ and Ⅷ of ERKs, JNKs, and p38 MAPKs. The conserved phosphorylation motifs are marked by the dashed line (- - -) and the notable distinction is indicated by triangle (▽). The partial amino acid sequences of MAPKs are as follows: BmERK, ScERK, TcERK, AccERK, *Ae* JNK (XP_001653365), *Culex quinquefasciatus* JNK (XP_001842827), *Drosophila melanogaster* JNK (AAB51187), *Mus musculus* JNK3 (BAA85877), *Harpegnathos saltator* p38 MAPK (EFN78769), *A. aegypti* p38 MAPK (XP_001653239), *B. mori* p38 MAPK (NP_001036996), and *D. melanogaster* p38 MAPK (AAB97138).

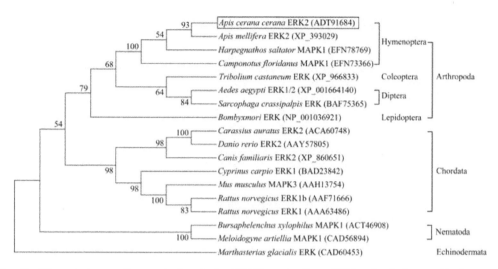

Fig. 2 The phylogenetic analysis of AccERK2 and other ERKs. A phylogenetic tree was constructed by using MEGA 4.1 software with the neighbor-joining method. The names of the protein and GenBank IDs are provided. The numbers above or below the branches indicate the bootstrap values (> 50%) of 500 replicates.

Structural and copy number analyses of the genomic sequence of AccERK2

A 2,513 bp genomic fragment of *AccERK2* (GenBank accession number: GU321337) was identified from total genomic DNA by using the specific primers, GP1 and GP2. Comparison of the genomic sequence and cDNA sequence of *AccERK2* showed that there are six introns separating the seven exons. An additional intron is located in the 3′ UTR, which is different from other ERK genes. All of the introns in the genomic sequence of *AccERK2* have the typical characteristics of introns, being both abundant in A+T (Table 2) and flanked by the conserved 5′-GT and AG-3′ splice site consensus sequences. To further analyze the structure of the *AccERK2* genomic sequence, the comparison analysis was carried out on selected *ERKs* (Fig. 3). Similar to *A. cerana cerana ERK2*, *Bombus terrestris ERK* also has an additional intron in the 3′ UTR. However, all of the introns of *A. mellifera ERK2* and *T. castaneum MAPK1* are located in the ORF sequence, and *AmERK2* has one more intron than *TcMAPK1*. The sizes of the introns of *TcMAPK1* are greater than the ERK genes of the other species and the largest is 34,055 bp. Thus, the numbers, sizes, and relative positions of all the introns of *ERKs* in different species are different.

Table 2 Exon and intron sizes and A+T content in the *AccERK2* gene

Exon no.	Size (bp)	AT content (%)	Intron no.	Size (bp)	AT content (%)
1	125	52.80	1	743	77.52
2	127	73.23	2	67	82.09
3	111	72.97	3	76	85.53
4	252	67.06	4	120	89.17
5	168	63.69	5	84	90.48
6	189	66.67	6	76	82.89
7	126	67.46	7	79	91.14

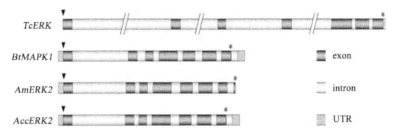

Fig. 3 Schematic representation of the DNA structures of *ERKs*. The initiation codons (ATG) and the terminal codons are marked by arrowheads (▼) and asterisks (*), respectively. The legend displays the pattern of exons, introns, and untranslated regions. Notably, some introns of *Tribolium castaneum* ERK have been partially abridged in the schematic representation due to their sizes. The sequences are from *AccERK2* (GU321337), *AmERK2* (AADG05008776), *Bombus terrestris* MAPK1 (NC_015764), and *TcERK* (NC_007424).

To estimate the copy numbers of *AccERK2* in the *A. cerana cerana* genome, Southern blotting analysis was performed and only a single band was identified by *AccERK2* probe in each of the three different restriction enzyme-digested DNA lanes (Fig. 4).

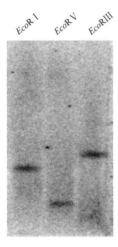

Fig. 4 Southern blotting analysis of *AccERK2*. The restriction enzymes used in digestion of the genomic DNA are shown above the panel. [α^{32}P] labeled *AccERK2* cDNA was used as the probe.

Developmental and tissue expression profiles of *AccERK2*

To analyze the developmental expression patterns of *AccERK2* in *A. cerana cerana*, semi-quantitative RT-PCR was performed. The expression level of *AccERK2* was relatively higher in the larvae and pupae than in the adults (Fig. 5A). The pupae were dissected into head, thorax, and abdomen sections to determine the expression of *AccERK2* in the different body regions. No significant differences of the transcription levels of *AccERK2* were found between the head and abdomen, whereas a slightly lower expression level of *AccERK2* was detected in the thorax (Fig. 5B). The highest transcription level was detected predominantly in the brain and the lowest in muscle, but no significant differences were found among the fat body, midgut, and epidermis (Fig. 5C).

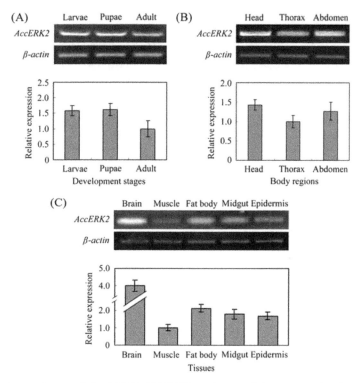

Fig. 5 Expression patterns of *AccERK2*. RT-PCR analyses of the expression levels of *AccERK2* for different developmental phases (A), body regions (B), and tissues (C). Vertical bars represent the mean ± SE (*n* = 3).

Expression patterns of *AccERK2* under different environmental stresses

Following the heat treatment, the transcription level of *AccERK2* was slightly elevated (Fig. 6A), yet no significant alteration of the *AccERK2* expression was observed after the cold treatment (Fig. 6B). Following the UV and H_2O_2 treatments, the expression of *AccERK2* was rapidly induced, reaching the highest level at 2 h, and then gradually decreasing between 3 and 4 h (Fig. 6C and D). No marked change in the transcription level of *AccERK2* was observed under the NaCl treatment (Fig. 6E).

Worker bees were separately fed with $MnCl_2$ (400 pmol/L), LPS (1 μg/ml), and U0126 (10 μmol/L), respectively. Following exposure to Mn^{2+}, the expression of *AccERK2* gradually decreased over time (Fig. 6F). Similarly, *AccERK2* expression was significantly inhibited by U0126, an inhibitor of the ERK pathway, the expression of *AccERK2* was nearly undetectable at 16 h after the treatment (Fig. 6G). In contrast, the membrane component of gram-negative bacteria, LPS, thought to be an activator of the MAPK signal transduction pathway, caused an accumulation of the AccERK2 transcript (Fig. 6H).

To conform that AccERK2 actually exists in the bees, Western blotting was performed. The expression of AccERK2 in *A. cerana cerana* could be detected by using anti-AccERK2. Following the UV treatment, the changes of the total AccERK2 protein were consistent with its transcriptional level (Fig. 7).

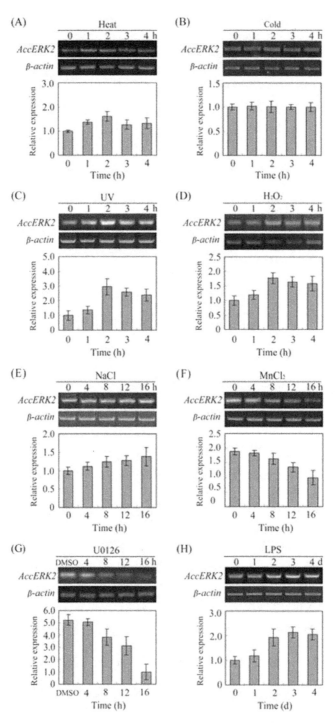

Fig. 6 Expression patterns of *AccERK2* under different stresses. The expression levels of *AccERK2* were detected by RT-PCR following heat (43℃) (A), cold (4℃) (B), UV (30 MJ/cm^2) (C), H$_2$O$_2$ (30%) (D), NaCl (0.3 mol/L) (E), MnCl$_2$ (400 pmol/L) (F), U0126 (10 μmol/L) (G), and LPS (1 μg/ml) (H) treatments. The *β-actin* gene was used as an internal control, and the conditional control was normalized to bees without any stress treatment (0 h) for (A-F, H) or the dimethyl sulfoxide (DMSO, 0.1%) treatment for (G). The vertical bars represent the mean ± SE (*n* = 3).

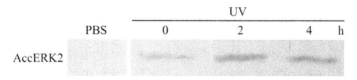

Fig. 7 Western blotting analysis of AccERK2 changes following UV treatment. The total protein was immunoblotted with anti-AccERK2. The signal of the binding reactions was visualized with HRP substrates. PBS instead of anti-AccERK2 was used as the negative control.

Discussion

The classical MAPKs, ERK1/2, are widely expressed and have been implicated in a wide range of cellular functions, including the control of cell proliferation, differentiation, and survival [19]. Studies demonstrated that ERK1/2 play roles in the development of synaptic plasticity, learning, and memory [20-22]. The role and significance of ERK1/2 in response to abiotic stress have recently begun to emerge through indirect evidence. In this study, we identified and characterized the *ERK2* gene from the Chinese honeybee, *A. cerana cerana*. According to the typical features of the ERK superfamily, it was named *AccERK2* and was used for further investigation.

Genomic analysis displayed that the numbers, sizes, and relative positions of all the introns of ERKs in different species are variable, suggesting that the introns of ERKs are not conserved. Southern blotting hybridization indicated that there is only one copy of *AccERK2* in the *A. cerana cerana* genome. We infer that ERKs may play a role in the evolution of different insects.

We examined the expression patterns of *AccERK2*. The results indicated that the transcription level of *AccERK2* was higher in the larval and pupal stages than in the adult stage. ERK2 has a key role in mesoderm formation during embryonic development [23]. We infer that *AccERK2* may play a role during development. In *Drosophila*, ERK was thought to be involved in regulating vein and/or intervein gene expression during wing tissue development in the larval and pupal stages [8]. This is a possible explanation for the relatively higher expression of *AccERK2* in the larval and pupal stages. We suggest the influence of *AccERK2* on development mainly occurs in these stages.

The MAPK signaling pathway can respond to extracellular stimuli and coordinate appropriate responses [2]. In this study, the transcription level of *AccERK2* was slightly elevated under high temperature. Some reports indicate that temperature changes influence honeybee behavior. Honeybees fan their wings to reduce the temperature and protect the brood when the ambient temperature exceeds 36°C [18]. *Apis cerana cerana* has the unique characteristic of low temperature tolerance [15], thus, it is reasonable to conclude that *AccERK2* has a protective role against low temperature.

AccERK2 transcripts rapidly accumulated after UV irradiation and H_2O_2 treatment. These oxidative stresses enhance the generation of the endogenous reactive oxygen species (ROS) [16, 24]. ROS have been considered to attack neurons in the brain and cause neurodegenerative diseases. Hyperbaric oxygen significantly induces deficits of cognitive performance in rat [25]. Indeed, several reports indicate that ERK2 plays roles in regulating neuronal plasticity [22, 26, 27]. In mice, ERK2 is necessary for the transition from oligodendrocyte progenitor cells (OPCs) to mature oligodendrocytes, which play a critical role in the brain and spinal cord by generating myelin sheaths [28]. Consistent with this suggestion, we found that *AccERK2* was predominantly expressed in the brain. Taken together, we infer that *AccERK2* acts in oxidative stress- or UV-related neuronal transmissions.

In present study, the influences of osmotic stress and heavy metal ions on *AccERK2* were also detected and demonstrated inverse effects. The transcription levels of *AccERK2* were slightly increased under the osmotic stress. Similarly, in *Drosophila*, ERK was weakly activated by osmotic stress induced by NaCl or sorbitol [11]. The expression of *AccERK2* gradually decreased over time when exposed to manganese, a neurotoxin, demonstrating a connection between ERK2 and the nervous system. Walowitz and Roth demonstrated that ERK1/2 are required for manganese-induced PC12 differentiation and neurite outgrowth [29]. Although manganese is an essential trace mineral required for normal growth and development [30, 31], it also has a direct negative effect on neurons and glia within the central nervous system. Its toxicity is connected with dopaminergic neurons, globus pallidus, and striatum in brain areas [32]. Thus, manganese would presumably damage the nervous system. ERK2 translocates from the cytoplasm into the nucleus driven by ROS, where it directly interacts with p53. The activated p53 induces the upregulation of manganese superoxide dismutase, which represents a possible protective mechanism of NB4 cells to counteract oxidative stress [33]. Taken together, *AccERK2* might be involved in responding to multiple abiotic stresses and might further indirectly affect the nervous system.

References

[1] Nishimoto, S., Nishida, E. MAPK signalling: ERK5 versus ERK1/2. EMBO Rep. 2006, 7, 782-786.
[2] Strniskova, M., Barancik, M., Ravingerova, T. Mitogen-activated protein kinases and their role in regulation of cellular processes. Gen Physiol Biophys. 2002, 21, 231-255.
[3] Johnson, G. L. Defining MAPK interactomes. ACS Chem Biol. 2011, 6, 18-20.
[4] Manna, P. R., Stocco, D. M. The role of specific mitogen-activated protein kinase signaling cascades in the regulation of steroidogenesis. J Signal Transduct. 2011, 2011, 821615.
[5] Imajo, M., Tsuchiya, Y., Nishida, E. Regulatory mechanisms and functions of MAP kinase signaling pathways. IUBMB Life. 2006, 58, 312-317.
[6] Biondi, R. M, Nebreda, A. R. Signalling specificity of Ser/Thr protein kinases through docking-site-mediated interactions. Biochem J. 2003, 372, 1-13.
[7] Junttila, M. R., Li, S. P., Westermarck, J. Phosphatase-mediated crosstalk between MAPK signaling pathways in the regulation of cell survival. FASEB J. 2008, 22, 954-965.
[8] Mouchel-Vielh, E., Rougeot, J., Decoville, M., Peronnet, F. The MAP kinase ERK and its scaffold protein MP1 interact with the chromatin regulator Corto during *Drosophila* wing tissue development. BMC Dev Biol. 2011, 11, 17.
[9] Fujiwara, Y., Denlinger, D. L. High temperature and hexane break pupal diapause in the flesh fly, *Sarcophaga crassipalpis*, by activating ERK/MAPK. J Insect Physiol. 2007, 53, 1276-1282.
[10] Martindale, J. L., Holbrook, N. J. Cellular response to oxidative stress: signaling for suicide and survival. J Cell Physiol. 2002, 192, 1-15.

[11] Liu, W., Silverstein, A. M., Shu, H., Martinez, B., Mumby, M. C. A functional genomics analysis of the B56 isoforms of *Drosophila* protein phosphatase 2A. Mol Cell Proteomics. 2007, 6, 319-332.

[12] Guo, S., Liu, M. G., Long, Y. L., et al. Region- or state-related differences in expression and activation of extracellular signal-regulated kinases (ERKs) in naive and pain-experiencing rats. BMC Neurosci. 2007, 8, 53.

[13] Tomaselli, B., Nedden, S. Z., Podhraski, V., Baier-Bitterlich, G. p42/44 MAPK is an essential effector for purine nucleoside-mediated neuroprotection of hypoxic PC12 cells and primary cerebellar granule neurons. Mol Cell Neurosci. 2008, 38, 559-568.

[14] Peng, S., Zhang, Y., Zhang, J., Wang, H., Ren, B. ERK in learning and memory: a review of recent research. Int J Mol Sci. 2010, 11, 222-232.

[15] Li, H. L., Zhang, Y. L., Gao, Q. K., Cheng, J. A., Lou, B. G. Molecular identification of cDNA, immunolocalization, and expression of a putative odorant-binding protein from an Asian honey bee, *Apis cerana cerana*. J Chem Ecol. 2008, 34, 1593.

[16] Wang, M., Kang, M. J., Guo, X. Q., Xu, B. H. Identification and characterization of two phospholipid hydroperoxide glutathione peroxidase genes from *Apis cerana cerana*. Comp Biochem Physiol C: Toxicol Pharmacol. 2010, 152, 75.

[17] Li, J., Zhang, L., Feng, M., Zhang, Z., Pan, Y. Identification of the proteome composition occurring during the course of embryonic development of bees (*Apis mellifera*). Insect Mol Biol. 2009, 18, 1-9.

[18] Meng, F., Zhang, L., Kang, M., Guo, X., Xu, B. Molecular characterization, immunohistochemical localization and expression of a ribosomal protein L17 gene from *Apis cerana cerana*. Arch Insect Biochem Physiol. 2010, 75, 121-138.

[19] Kurada, P., White, K. Ras promotes cell survival in *Drosophila* by downregulating hid expression. Cell. 1998, 95, 319-329.

[20] Morozov, A., Muzzio, I. A., Bourtchouladze, R., et al. Rap1 couples cAMP signaling to a distinct pool of p42/44MAPK regulating excitability, synaptic plasticity, learning, and memory. Neuron. 2003, 39, 309-325.

[21] Sweatt, J. D. Mitogen-activated protein kinases in synaptic plasticity and memory. Curr Opin Neurobiol. 2004, 14, 311-317.

[22] Thomas, G. M., Huganir, R. L. MAPK cascade signalling and synaptic plasticity. Nat Rev Neurosci. 2004, 5, 173-183.

[23] Yao, Y., Li, W., Wu, J., Germann, U. A., Su, M. S., Kuida, K., Boucher, D. M. Extracellular signal-regulated kinase 2 is necessary for mesoderm differentiation. Proc Natl Acad Sci USA. 2003, 100, 12759-12764.

[24] Yasui, H., Sakurai, H. Chemiluminescent detection and imaging of reactive oxygen species in live mouse skin exposed to UVA. Biochem Biophys Res Commun. 2000, 269, 131-136.

[25] Ohwada, K., Takeda, H., Yamazaki, M., Isogai, H., Nakano, M., Shimomura, M., Fukui, K., Urano, S. Pyrroloquinoline quinone (PQQ) prevents cognitive deficit caused by oxidative stress in rats. J Clin Biochem Nutr. 2008, 42, 29-34.

[26] Wei, F., Zhuo, M. Activation of Erk in the anterior cingulate cortex during the induction and expression of chronic pain. Mol Pain. 2008, 4, 28.

[27] Liu, M. G., Wang, R. R., Chen, X. F., Zhang, F. K., Cui, X. Y., Chen, J. Differential roles of ERK, JNK and p38 MAPK in pain-related spatial and temporal enhancement of synaptic responses in the hippocampal formation of rats: multi-electrode array recordings. Brain Res. 2011, 1382, 57-69.

[28] Fyffe-Maricich, S. L., Karlo, J. C., Landreth, G. E., Miller, R. H. The ERK2 mitogen-activated protein kinase regulates the timing of oligodendrocyte differentiation. J Neurosci. 2011, 31, 843-850.

[29] Walowitz, J. L., Roth, J. A. Activation of ERK1 and ERK2 is required for manganese-induced neurite outgrowth in rat pheochromocytoma (PC12) cells. J Neurosci Res. 1999, 57, 847-854.

[30] Roth, J. A., Horbinski, C., Higgins, D., Lein, P., Garrick, M. D. Mechanisms of manganese-induced rat pheochromocytoma (PC12) cell death and cell differentiation. Neurotoxicology. 2002, 23, 147-157.

[31] Bowman, A. B., Kwakye, G. F., Herrero, H. E., Aschner, M. Role of manganese in neurodegenerative diseases. J Trace Elem Med Biol. 2011, 25, 191-203.

[32] Stanwood, G. D., Leitch, D. B., Savchenko, V., Wu, J., Fitsanakis, V. A., Anderson, D. J., Stankowski, J. N., Aschner, M., McLaughlin, B. Manganese exposure is cytotoxic and alters dopaminergic and GABAergic neurons within the basal ganglia. J Neurochem. 2009, 110, 378-389.

[33] Li, Z., Shi, K., Guan, L., Cao, T., Jiang, Q., Yang, Y., Xu, C. ROS leads to MnSOD upregulation through ERK2 translocation and p53 activation in selenite-induced apoptosis of NB4 cells. FEBS Lett. 2010, 584, 2291-2297.

1.6 A Typical RNA Binding Protein Gene (*AccRBM11*) in *Apis cerana cerana*: Characterization of *AccRBM11* and Its Possible Involvement in Development and Stress Responses

Guilin Li, Haihong Jia, Hongfang Wang, Yan Yan, Xingqi Guo, Qinghua Sun, Baohua Xu

Abstract: RNA binding motif proteins (RBMs) belong to RNA binding proteins that display extraordinary post-transcriptional gene regulation roles in various cellular processes, including development, growth and stress responses. Nevertheless, only a few examples of the roles of RBMs are known in insects, particularly in *Apis cerana cerana*. In the present study, we characterized the novel RNA binding motif protein 11 from *Apis cerana cerana*, which was named AccRBM11 and whose promoter sequence included abundant potential transcription factor binding sites that are connected to responses to adverse stress and early development. Quantitative PCR results suggested that *AccRBM11* was expressed at the highest levels in 1-day post-emergence worker bees. *AccRBM11* mRNA and protein levels were higher in the poison gland and the epidermis than in other tissues. Moreover, the transcription levels of *AccRBM11* were upregulated upon all the simulation of abiotic stresses. Furthermore, Western blotting analysis indicated that the expression levels of AccRBM11 protein could be induced under some abiotic stressors, a result that did not completely in agree with the qRT-PCR results. It is also noteworthy that the expression of some genes that connected with development or stress responses were remarkably suppressed when *AccRBM11* was silenced, which suggested that *AccRBM11* might play a similar role in development or stress reactions with the above genes. Taken together, the data presented here provide evidence that AccRBM11 is potentially involved in the regulation of development and some abiotic stress responses. We expect that this study will promote future research on the function of RNA binding proteins.

Keywords: *Apis cerana cerana*; RNA binding protein; Abiotic stress; Expression patterns

Introduction

As an Asian species, the Chinese honeybee brings huge economic benefits to the apiculture industry and plays a critical role in maintaining biodiversity. Compared to western honeybees, *Apis cerana cerana* has an acute sense of smell, a strong resistance to mites, and can forage a wide range of nectars and pollens; these advantages are irreplaceable [1-3]. However, recently, the population of *Apis cerana cerana* has severely declined, which can be attributed to the many abiotic stresses that exist in the environment, such as excessive pesticide usage, climate changes of extreme heat and cold and the presence of heavy metal and ultraviolet radiation. Publication of the *Apis mellifera* genomic sequence in 2006 powerfully facilitated honeybee research [4], and the report of the genomic sequence of *Apis cerana cerana* in 2015 provided a wealth of information for better understanding the natural biology and complex behaviours of the Asian honeybee [5]. However, it remains essential to identify specific genes and their corresponding proteins and to reveal their expression characteristics and related biological functions in stress responses.

Cellular response to environmental stress is complex. Cells contain multiple regulatory mechanisms that are generally considered to have protective functions. The regulation can cause specific gene regulation or activation as well as post-transcriptional and translational events. With regards to post-transcriptional regulation, diverse RNA binding proteins (RBPs) are the central post-transcriptional regulators of RNA metabolism. Typical RBPs are characterized by the presence of one or more RNA-recognizing domains (RRMs, also known as CS-RBD, RNP, or RBD domains), which are the largest parts of the protein and are composed of 75-85 amino acids [6]. Large-molecular-weight RBPs contain a nuclear localization signal and can combine with nascent mRNAs to be responsible for their export from the nucleus. The structure of RBPs may be related to their function. In recent years more and more studies have begun to explore the functions of RBPs.

RBPs not only play a role in genome organization, growth and development but also in stress responses through the regulation of post-transcriptional mechanisms. RBPs were implicated in low oxygen level [7] and could respond to H_2O_2 stress in HeLa cells [8]. A glycine-rich RBP could be induced by cold stress and mediate cold-inducible suppression of cell growth [9]. In the Pashmina goat, RNA binding motif protein 3 (RBM3) was downregulated under deep hypothermic conditions [10]. RBMs belong to RBPs. Guo et al. [11] proposed that *RBM4*, *RBM7*, and *RBM11* had strong homology in the RBD family according to GenBank. The role of *RBM4* could be modulated by stressful cellular conditions [12]. The phosphorylation of human RBM7 by the p38(MAPK)/MK2 is involved in stress-dependent modulations of the noncoding RNA metabolism [13]. *RBM11* displayed dynamic movement between the speckle and the nucleoplasm when cells were exposed to genotoxic and oxidative stresses [14].

Although the functions of RBPs have been explored in other species, there is limited knowledge on the role of RBPs in honeybees, particularly in *Apis cerana cerana*. In this study, to gain insight into the role of the Chinese honeybee RNA binding motif protein 11 gene (*AccRBM11*), we characterized the *RBM11* gene from *Apis cerana cerana* and investigated its mRNA levels in different tissues and at different developmental stages. We also simulated common abiotic stress conditions encountered by *Apis cerana cerana* during its life to examine *AccRBM11* mRNA and protein expression profiles. To our knowledge, this is the first report to examine the role of RBPs in stress responses in honeybees.

Experimental Procedures

Biological specimens and various treatments

The Chinese honeybees (*Apis cerana cerana*) used in this study were maintained in artificial beehives at Shandong Agricultural University (Taian, China). Honeybees in various developmental stages were classified according to the criteria of Michelette and Soares [15]. The egg, larvae, pre-pupal phase pupae, pupae, 1-day post-emergence worker bees, non-egg-laying queen bee, and egg-producing queen bee were collected from the hive, whereas 19-day post-emergence worker bees and 30-day post-emergence worker bees were collected from the hive entrance by marking 1-day post-emergence worker bees with paint 19 and 30 days earlier. Drones were also collected from the hive. The 19-day worker bees were collected randomly from the hive and were kept in a thermostat-regulated environment (34℃) with 70% relative humidity under a 24 h dark regimen. Then, they were randomly divided into ten groups (n=60/group), and each group was treated with different stress conditions (Supplementary Table 1), which were simulated and similar to those that *Apis cerana cerana* are subjected to during its life. All of the control groups (untreated 19-day worker bees) were fed a normal adult diet and were incubated at 34℃ with 70% relative humidity. The controls for the group that was injected with H_2O_2 were injected with phosphate buffered saline (0.5 μl/honeybee). The honeybees were collected at the appropriate time after treatment (Supplementary Table 1). The wing, honey sac, muscle, epidermis, poison gland, midgut, haemolymph, tentacle, and rectum of 19-day worker bees were dissected on the ice to detect tissue-specific expression. All of the samples were frozen in liquid nitrogen and stored at −70℃ until used.

Total RNA extraction, cDNA synthesis and genomic DNA preparation

Total RNA extraction, cDNA synthesis and preparation of genomic DNA were performed by using Trizol reagent (Invitrogen, Carlsbad, CA, USA) and EasyScript First-Strand cDNA Synthesis (TransGen Biotech, Beijing, China) and EasyPure Genomic DNA Extraction Kits (TransGen Biotech, Beijing, China), respectively, according to the instructions of each manufacturer.

Acquisition of *AccRBM11* cDNA, 5′-flanking region, and genomic DNA sequence

The genome information for the Chinese honeybee was uncovered in 2015 [5], however, it was not released. Therefore, the open reading frame (ORF), 5′-flanking region (partial sequence of the promoter) and the genomic DNA sequence of *AccRBM11* were obtained by using specific primers, which are designed based on the *RBM11*-like sequence from *Apis mellifera* (a honeybee shares high homology with *Apis cerana cerana*). The 5'-untranslated region (5' UTR) and 3'-untranslated region (3' UTR) of *AccRBM11* were cloned by 5'- and 3'-rapid amplification of the cDNA end (RACE).

Bioinformatics analysis

The AccRBM11 homologous sequences used in this paper were downloaded from the NCBI servers (http://blast.ncbi.nlm.nih.gov/Blast.cgi). The physical and chemical properties of AccRBM11 were predicted with the ProtParam tool (http://www.expasy.ch/tools/protparam.html) and DNAMAN version 5.22 (Lynnon Biosoft, Quebec, Canada). Multiple alignments were performed in DNAMAN version 5.22. The phylogenetic tree was generated with the neighbour-joining method by using Molecular Evolutionary Genetics Analysis (MEGA version 4.1). The putative *cis*-acting elements of the *AccRBM11* 5'-flanking sequence were predicted by the MatInspector database (http://www.cbrc.jp/research/db/TFSEARCH.html). A PSORT II server was used to predict the nuclear localization of AccRBM11. Putative AccRBM11 antimicrobial peptides were predicted by using the following website: http://aps.unmc.edu/AP/design/design_improve.php.

Fluorescent real-time quantitative PCR

Fluorescent real-time quantitative PCR (qRT-PCR) was carried out by using the SYBR® PrimeScript™ RT-PCR Kit (TaKaRa, Dalian, China) and a CFX96™ Real-Time PCR Detection System (Bio-Rad, Hercules, CA, USA) to determine the *AccRBM11* transcription profile. mRNA expression levels were normalized to the house-keeping gene *β-actin* (GenBank accession number: XM-640276), which was stably expressed and could be considered as a reference gene [16-19]. A 25 µl amplification reaction volume was used to perform the qRT-PCR reaction, which contained 12.5 µl Takara SYBR® premix Ex *Taq*™, 9.5 µl double distilled water, 2 µl cDNA, and 0.5 µl of each primer. The following qRT-PCR protocol was performed: (1) 30 s at 95℃ for pre-denaturation, (2) 40 cycles of amplification (5 s at 95℃ for denaturation, 15 s at 54℃ for annealing, and 15 s at 72℃ for extension), and (3) a single melting cycle from 65℃ to 95℃. Relative *AccRBM11* gene expression was analysed with the CFX Manager software (version 1.1), using the 2 (−delta delta C(T)) method [20]. Statistical Analysis System (SAS) software (version 9.1) was used to determine the statistical significance. All of the experimental conditions were implemented in three independent biological replicates, and the PCR reactions were performed in triplicate.

Protein expression, purification and preparation of anti-AccRBM11

The ORF of *AccRBM11* was cloned into the *Bam*H Ⅰ and *Sac* Ⅰ sites of the expression vector pET-30a (+) (Novagen, Madison, WI). Next, the expression plasmid pET-30a (+)-AccRBM11 was transformed into Transetta (DE3) chemically competent cells (a type of *Escherichia coli*; TransGen Biotech, Beijing, China). Next, the bacteria were cultured in 500 ml Luria-Bertani (LB) with 250 µl kanamycin (100 mg/ml) until the cell density reached

0.4-0.6 at 600 nm. Next, the cultures were induced with 0.1 mmol/L isopropylthio-β-D-galactoside (IPTG, Promega) at 37℃ for 12 h. Finally, a HisTrap™ FF column (GE Healthcare, Uppsala, Sweden) was used to purify the recombinant AccRBM11. The target SDS-PAGE albumen glue was added moderate sodium chloride injection (0.9%) (Cisen Pharmaceutical, Jining, China) and benzylpenicillin sodium for injection (Lukang Pharmaceutical, Jining, China) and ground in a mortar with a pestle. The appropriate amount of ground sample was injected into the body of a white mouse (Taibang, Taian, China). The remaining experimental procedures were carried out as described in Meng et al. [21]. The obtained AccRBM11 antibody was hybridized to a blot containing the overexpressed protein of AccRBM11 in order to detect the specificity of the anti-AccRBM11.

Western blotting analysis

Total protein was extracted from *Apis cerana cerana* following the instructions of the manufacturer that provided by a tissue protein extraction kit (ComWin Biotech, Beijing, China), and quantified with a total protein assay kit (using a standard BCA method; ComWin Biotech, Beijing, China). A 12% SDS-PAGE was used to separate equal concentration of total protein from each sample of the same treatment. Then, the albumin glue that contained target proteins was subsequently electrotransferred onto a PVDF membrane (ComWin Biotech, Beijing, China) by using a wet transfer method. The anti-AccRBM11 serum was used as the primary antibody at a dilution of 1 : 500 (V/V). Peroxidase-conjugated goat anti-mouse immunoglobulin G (Jingguo Changsheng Biotechnology, Beijing, China) was used as the secondary antibody at a dilution of 1 : 2,000 (V/V). α-tubulin was used as a control at a dilution of 1 : 50,000 (V/V); its protein levels is usually not change in the body of an organism [22-24]. The tubulin antibody (1 : 50,000) purchased from Beyotime (China) recognizes the C-terminal region of α-tubulin. The results of the binding reaction were visualized by using a SuperSignal™ West Dura Extended Duration Substrate (Thermo Fisher Scientific, Shanghai, China).

RNA interference mediated silencing of the *AccRBM11* gene and transcriptional levels of some other genes after *AccRBM11* knockdown

RNA interference experiment was used to knock down *AccRBM11* transcripts. The non-conserved sequence region in the *AccRBM11* ORF was chosen to design the specific primers. The primers with a T7 polymerase promoter at their 5′-end were used to synthesize a linear DNA template. *AccRBM11* double stranded RNA (dsRNA) was produced by using RiboMAX T7 large-scale RNA production systems as per the instructions of the manufacturer [25]. The green fluorescent protein gene (*gfp*, synthetic construct, GenBank accession number: U87974) was used as a control [25]. *Apis cerana cerana* genes do not share homology with dsRNA-GFP, so an RNAi response will be not triggered by dsRNA-GFP in the body of *Apis cerana cerana*. The 19-day post-emergence worker bees were selected for the RNAi treatments and were divided into three groups (n = 40/group). Using a microsyringe, 6 μg of dsRNA-AccRBM11 or dsRNA-GFP was injected between the first and the second abdominal segments into each adult. The last group was untreated. The above three groups were maintained in an incubator (at 34℃ with 60% relative humidity and under a 24 h dark regimen). The healthy adult bees were sampled each day and stored at −70℃. qRT-PCR was performed to detect *AccsHsp22.6*, *AccSOD2*, *AccTpx4*, *AccAK*, *AccRPL11*, *AccTpx1*, *Accp38b*, *AccGSTO1*, and *AccTrx1* expression profiles when *AccRBM11* was knocked down. These experiments were carried out in three independent biological replicates.

Disc diffusion assay

Cells overexpressing AccRBM11 were plated onto LB-kanamycin agar plates and incubated at 37℃ for 50 min. Then, filter discs (6 mm diameter) were placed on the surface of the agar plates. The filter discs were soaked with 2.5 μl of different concentrations of cumene hydroperoxide (0, 50, 100, 200, and 400 mmol/L), paraquat (0, 100, 200, and 300 mmol/L), $HgCl_2$ (0, 10, 20, 50, and 70 mg/ml) and $CdCl_2$ (0, 8, 16, 23, and 42 mg/ml). The bacteria were incubated at 37℃ for 10 h, and the killing zones around the discs were measured. Transetta (DE3) chemically competent cells containing the pET30-a (+) vector were used as control.

Primers and amplification conditions

All of the primer pairs and polymerase chain reaction (PCR) amplification procedures used in this paper are presented in Supplementary Table 2 and Supplementary Table 3, separately. An ideal primer pair used for qRT-PCR should have an efficiency (E) value of 90%-110%, a coefficient (R^2) of more than 0.980, and single peak melting curves [26, 27]. The primers used for qRT-PCR in this paper were designed on the basis of the principle of quantitative primer design; the E value, R^2, and numbers of melting curve peaks for each qRT-PCR primer pair are listed in Supplementary Table 4, all of which complied with the qRT-PCR requirements. All of the primers used in this study were synthesized by the Sangon Biotechnological Company (Shanghai, China).

GenBank accession characteristic of the genes used in this study

A lot of genes were used to execute bioinformatics analysis, and their GenBank accession number and species name were listed in Supplementary Table 5.

Results

Characterization of *AccRBM11*

The full-length cDNA sequence of *AccRBM11* (GenBank accession number: KR827409) is 1,205 bp, which contains a 116 bp 5′ UTR, a 561 bp complete ORF, and a 528 bp 3′ UTR (Fig. 1). A typical polyadenylation signal sequence (AATAA) was found in the 3′ UTR (Fig. 1). Fig. 2a revealed that AccRBM11 was most homologous to its ortholog protein in *Apis mellifera*

(99.46%), and the N-terminus was more highly conserved than the C-terminus. The ORF of *AccRBM11* encodes a 186 amino acids protein with a predicted molecular mass of 22.115 kDa and a predicted theoretical pI of 9.10. AccRBM11 included an RRM-RBM7-like (RNA recognition motif in the RBP7 and similar proteins) specific hit that contained 74 residues (6-79) (Fig. 2a), which was predicted by the NCBI Conserved Domain Database Search. The secondary structure of AccRBM11 is represented as $\beta_1\alpha_1\beta_2\beta_3\alpha_2\beta_4\alpha_3$ and the secondary structure of the RRM domain of AccRBM11 as $\beta_1\alpha_1\beta_2\beta_3\alpha_2\beta_4$ (Fig. 2a).

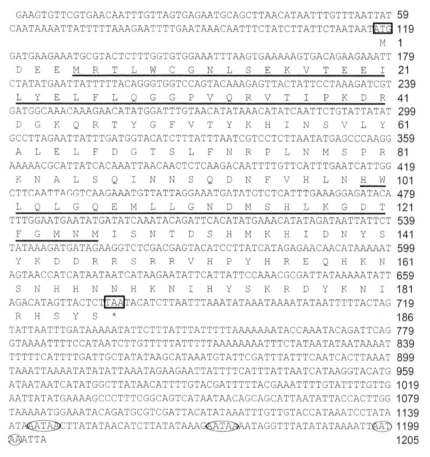

Fig. 1 *AccRBM11* nucleotide sequence and deduced amino acid sequence. The cDNA sequence is shown on the top line, and the deduced amino acid sequence is shown on the second line. The boxes mark the ATG (start codon) and the TAA (stop codon), and the asterisk denotes the stop codon. The predicted antimicrobial peptides are underlined. Ellipses show the possible polyadenylation signal. The sequence of *AccRBM11* was deposited into GenBank with the accession number KR827409.

Phylogenetic tree analysis revealed that AccRBM11 was more closely related to AmRBM11-like protein than to other species (Fig. 2b). Fig. 2c showed the possible three-dimensional structure of AccRBM11 that might be important for understanding AccRBM11 protein structure. In addition, the NLS prediction showed that AccRBM11 was most likely localized in the nucleus, suggesting that it might play a role in the nucleus.

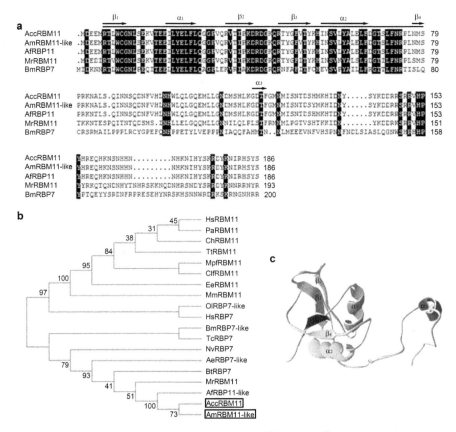

Fig. 2 Characterization of RNA-binding proteins (RBPS) from different species and the tertiary structure of AccRBM11. Protein sequences were acquired from NCBI (Supplementary Table 5). a. Multiple alignment of AccRBM11 amino acid sequences with other RNA-binding proteins. Identical regions are shaded in black. The putative RRM domain of AccRBM11 is marked by a straight line, and the predicted secondary structure of AccRBM11 is labeled with arrows. b. Phylogenetic analysis of RBPs from various species; AfRBM11-like and AccRBM11 are boxed. C. The tertiary structure of AccRBM11. The helices, sheets, and coils are marked with different colors.

Analysis of the *AccRBM11* genetic sequence structure

To further study the *AccRBM11* genomic locus, a 1,552 bp full-length segment of DNA was obtained, which contained one intron and two exons (GenBank accession number: KR827410). As shown in Fig. 3 and Supplementary Table 6, the lengths and number of exons had been conserved; however, the sizes of the introns, the 3′ UTR, and the 5′ UTR were not conserved during the evolutionary period. Supplementary Table 6 revealed that the A+T content was richer than the content of G+C and that the A+T contents of the exons were much lower than in the introns, which was consistent with the results from a previous study [5].

Putative transcription factor binding sites (TFBs) on the *AccRBM11* promoter

A 1,464 bp promoter sequence (GenBank accession number: KR827410) was isolated to investigate the transcription regulatory regions of *AccRBM11*. Many putative TFBs were predicted by the software MatInspector in both the sense strand (Fig. 4) and the antisense strand (data not shown). These included sites for heat shock factor (HSF), CdxA, Nkx-II, NIT2, BR-C, DRI, CF2-II, and ovo homolog-like transcription factor. Moreover, many possible TFBs were also detected in the 5′ UTR of *AccRBM11*. TATA box was also identified in the promoter region of *AccRBM11*, which can combine with TFII to initiate gene transcription.

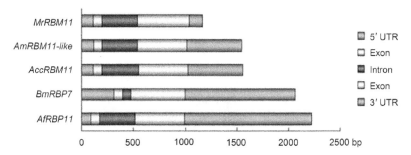

Fig. 3 Schematic representation of DNA structures. The pattern of untranslated regions, introns, and exons in AccRBM11, AmRBM11-like, AfRBP11, MrRBM11, and BmRBP7 are presented according to the scale shown below. The genomic DNA from the above genes (except AccRBM11) was loaded from the NCBI database, and their GenBank accession numbers are presented in Supplementary Table 5.

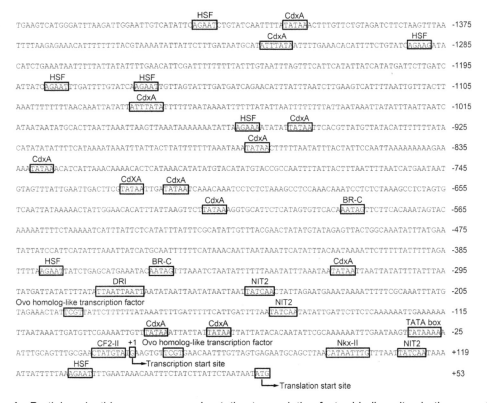

Fig. 4 Partial nucleotide sequences and putative transcription factor binding sites in the promoter of *AccRBM11*. The transcription start and translation sites are labeled with an arrow, and the putative TFBs are marked with boxes.

Expression and characterization of recombinant AccRBM11

AccRBM11 was overexpressed in Transetta cells as a histidine fusion protein. SDS-PAGE result showed that the recombinant had a molecular mass of less than 29 kDa (Fig. 5), which was a little smaller than the predicted molecular mass of 29.946 kDa (including cleavable C- and N-terminal His-tags of approximately 7.831 kDa). The recombinant protein was presented in the form of an inclusion body in the Transetta cells, and the nearly insoluble recombinant protein could not be purified with a HisTrap™ FF column, even at low temperature (data not shown).

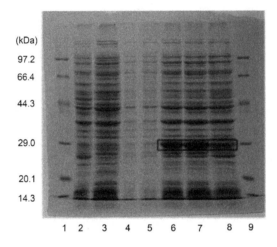

Fig. 5 The expression of the recombinant AccRBM11 protein. Recombinant AccRBM11 was separated by SDS-PAGE. Lanes 1 and 9, protein molecular weight marker; lane 2, induced overexpression of pET-30a (+) in Transetta cells; lanes 3 and 8, uninduced and induced overexpression of pET-30a (+)-AccRBM11, respectively, in Transetta cells; lanes 4 and 5, suspension of sonicated recombinant AccRBM11; and lanes 6 and 7, pellets from sonicated recombinant AccRBM11. The site of recombinant AccRBM11 is marked with a box.

Specific detection of anti-AccRBM11

Total protein from overexpressed AccRBM11 in Transetta cells was used to detect the specificity of the anti-AccRBM11, whereas total protein from uninduced overexpression of pET-30a (+)-AccRBM11 in Transetta cells was used as a control. As shown in Supplementary Fig. 1, only the lane from the cells overexpressing AccRBM11 had a band, and it was a single band, which suggested that the anti-AccRBM11 was specific for overexpressed AccRBM11.

Temporal and spatial *AccRBM11* profiles

qRT-PCR was used to examine the expression patterns of *AccRBM11* during various developmental stages and in different tissues. As shown in Fig. 6a, *AccRBM11* was expressed at all stages and displayed the highest transcriptional levels in the 1-day post-emergence worker bees (A1). Among the tissues examined, the most well-defined expression was observed in the poison gland (Pg), followed by the epidermis (Ep) and midgut (Mi) (Fig. 6b). Western blotting was also performed to examine AccRBM11 temporal and spatial protein expression. As shown in Fig. 6c, AccRBM11 expression was higher in the Ep than in the Pg, whereas AccRBM11 proteins could not be detected in the Mi and Re. The AccRBM11 protein contents were higher in the L5 stage, Pp and Pb, followed by the stage of L3 and A19 (Fig. 6d), which was not completely consistent with the result of qRT-PCR.

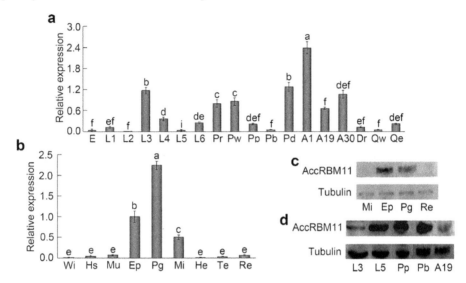

Fig. 6 *AccRBM11* expression profile at different developmental stages and in various tissues. a. *AccRBM11* expression levels at diverse developmental stages including egg (E), 1- to 6-day larvae (L1-L6), pre-pupal phase (Pr), white-eyed (Pw), pink-eyed (Pp), brown-eyed (Pb) and dark-eyed (Pd) pupae, 1-day post-emergence worker bees (A1), 19-day post-emergence worker bees (A19), 30-day post-emergence worker bees (A30), drone (Dr), the non-egg-laying queen bee (Qw), and the egg-producing queen bee (Qe). b. The transcription levels of *AccRBM11* in the wing (Wi), honey sac (Hs), muscle (Mu), epidermis (Ep), poison gland (Pg), midgut (Mi), hemolymph (He), tentacle (Te), and rectum (Re). The *β-actin* gene is shown for comparison. Vertical bars and letters above vertical bars represent the mean ± SE of three different samples and significant differences ($P < 0.000,1$), respectively. c. The expression level of AccRBM11 protein in different tissues. Tubulin was used as an internal control. d. The amount of AccRBM11 protein at different developmental stages. Tubulin was chosen as an internal control.

AccRBM11 expression profiles under different types of abiotic stress

Though the mRNA and protein levels of *AccRBM11* in the 19-day post-emergence worker bees were not high. Older than 18-day post-emergence worker bees are usually responsible for foraging nectar, pollen, and water, so they were most likely to be subject to environmental stress. Thus, the 19-day post-emergence worker bees were chosen to be treated with different stresses to explore the possible response of *AccRBM11* at the transcript and protein levels. qRT-PCR was used to characterize the mRNA levels of *AccRBM11* after exposure to UV, H_2O_2, VC, heavy metals ($HgCl_2$ and $CdCl_2$), pesticides (acaricide, cyhalothrin, and paraquat), and extreme temperatures (4 and 44 ℃). Expression levels were normalized to those in the control adult bees.

As shown in Fig. 7a, the transcription level of *AccRBM11* was upregulated under UV treatment, and peaked at 0.5 h. The mRNA level of *AccRBM11* was increased 2.26-fold at 4 h and peaked at 6 h following $HgCl_2$ treatment (Fig. 7b). The mRNA level of *AccRBM11* climbed 22.77-fold compared with the control group after treatment with $CdCl_2$ (Fig. 7c). Cyhalothrin, paraquat and acaricide also upregulated *AccRBM11* expression (Fig. 7d-f); however, there were differences in the degrees of the increase and in the transcriptional patterns. Experiments involving exposure to different temperatures were presented in Fig. 7g, h, which indicated that the expression levels of *AccRBM11* were markedly upregulated following exposure to 4 and 44 ℃ temperatures. According to the qRT-PCR analysis, The mRNA level of *AccRBM11* were upregulated after treatment with H_2O_2 (Fig. 7i) and VC (Fig. 7j) and reached maximums at 3 and 24 h, respectively.

Western blotting analysis

Western blotting analysis was performed to further explore AccRBM11 expression patterns under adverse abiotic stress conditions. AccRBM11 was detected by using an anti-AccRBM11, which was specific to AccRBM11 (Supplemental Fig. S1). After exposure to 44 ℃ (Fig. 8a) and 4 ℃ (Fig. 8b) for 1, 3, and 5 h, AccRBM11 expression was increased several fold. VC treatment caused clear increases in AccRBM11 expression levels (Fig. 8c). As shown in Fig. 8d and Fig. 8e, the amount of AccRBM11 was the highest at 9 and 4 h of $CdCl_2$ and $HgCl_2$ treatment, respectively. Following exposure to H_2O_2, the protein levels of AccRBM11 rose gradually from 0 to 1.5 h (Fig. 8f). In contrast, AccRBM11 expression patterns did not display any significant changes when treated with acaricide (Supplemental Fig. S2a), paraquat (Supplemental Fig. S2b), cyhalothrin (Supplemental Fig. S2c), or UV (Supplemental Fig. S2d).

The knockdown of *AccRBM11* transcript and the expression profiles of other genes after *AccRBM11* silencing

RNAi experiment was carried out to explore the role of *AccRBM11* in development and stress responses. As shown in Fig. 9a, the *AccRBM11* gene was successfully silenced compared with the control groups, and the lowest transcription levels of *AccRBM11* were discovered at one and two days post-injection of dsRNA-AccRBM11. The qRT-PCR results showed that *AccSOD2*, *AccTpx4*, *AccAK*, *AccTpx1*, *AccGSTO1*, *AccTrx1*, *AccsHsp22.6*, *AccRPL11*, and *Accp38b* were all suppressed when *AccRBM11* was silenced (Fig. 9b-j).

Disc fusion assay of recombinant AccRBM11 in response to diverse stressors

The recombinant AccRBM11 contained a His-tag and AccRBM11. Interestingly, it required much more time for the *Escherichia coli* (*E. coli*) cells with AccRBM11 to reach the same optical density at 600 nm compared with that of the *E. coil* cells without AccRBM11. After exposure to different concentrations of reagents ($HgCl_2$, $CdCl_2$, paraquat, and cumene hydroperoxide), the halo diameters around the filters on agar plates that cultured the bacteria containing overexpression AccRBM11 were larger than around the controls, even though the death zones were distinctly different under the different treatment condition (Fig. 10). The online database APD2 was used to explore the mechanism indicated by the experimental results. At least two antimicrobial peptides were identified in the AccRBM11 amino acid sequence (Fig. 1). The peptides have neutral charges and may display antimicrobial activities.

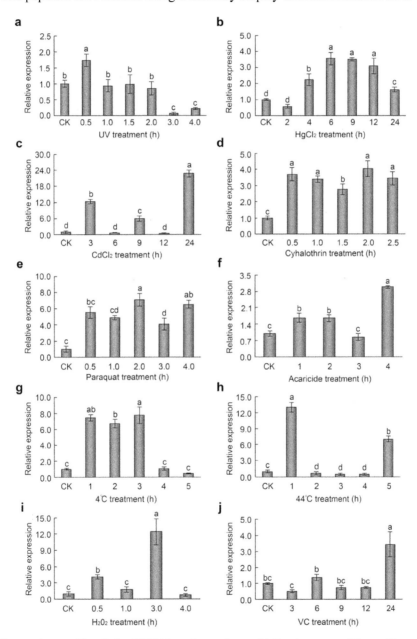

Fig. 7 Expression profile of *AccRBM11* under various abiotic stress conditions. These stresses included UV (a), $HgCl_2$ (b), $CdCl_2$ (c), cyhalothrin (d), paraquat (e), acaricide (f), 4℃ (g), 44℃ (h), H_2O_2 (i), and VC (j). The *β-actin* gene was used as an internal control. Control check is abbreviated CK. The data represent the mean ± SE of independent experiments (n = 3). The letters above the vertical bars indicate significant differences (P < 0.000,1) by using Duncan's multiple range tests.

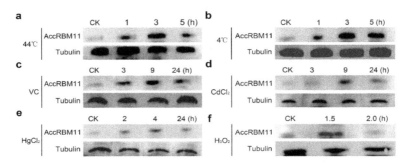

Fig. 8 Western blotting analysis of AccRBM11. Nineteen-day-old worker bees were treated with 44℃ (a), 4℃ (b), VC (c), CdCl$_2$ (d), HgCl$_2$ (e), and H$_2$O$_2$ (f). An equivalent concentration of extracted protein from the same treatment at different time was loaded in different lane and anti-AccRBM11 was used to immunoblot the targeted protein. Tubulin was used as an internal control.

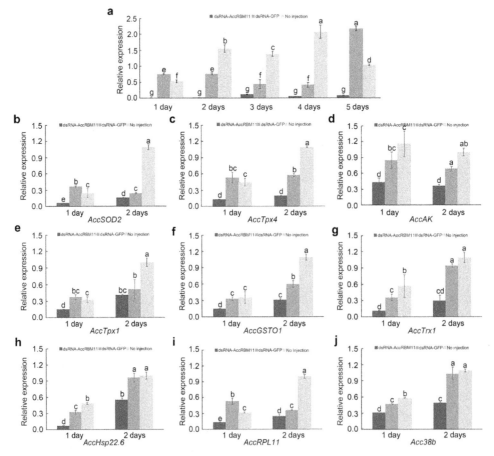

Fig. 9 RNA interference on the transcription levels of *AccRBM11* in adults and the effect of *AccRBM11* silencing on other genes in *Apis cerana cerana*. a. Each 19-day post-emergence worker bees was injected with 6 μg of dsRNA-AccRBM11. An equivalent volume of dsRNA-GFP was also injected for the control group. Untreated adults were also chosen as controls. b-j. Expression profiles of other development or stress response genes when *AccRBM11* was knocked down. The *β-actin* gene was used as an internal control. All of the data are presented as the mean ± SE of three independent experiments. Various letters above the bars indicate significant differences ($P < 0.000,1$) by using Duncan's multiple range tests.

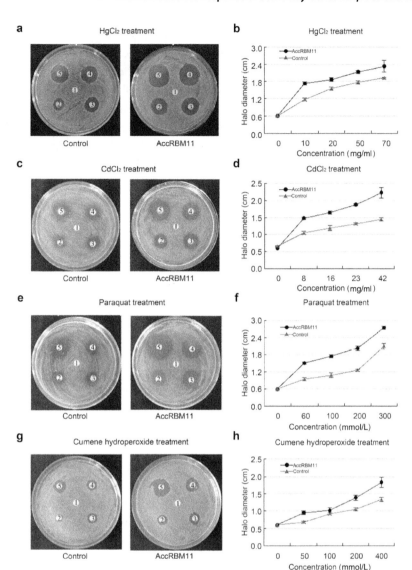

Fig. 10 Resistance of bacteria cells overexpressing AccRBM11 to various stressors. A total of 5 × 10⁸ cells were inoculated onto LB agar plates. AccRBM11 was overexpressed in bacterial cells, and the *E. coli* cells that were transfected with pET30a (+) were used as a control. Killing zone diameters were compared in the line chart. The data are the mean ± SE of three replicates. a and b. The $HgCl_2$ concentrations on filter disks 1-5 were 0, 10, 20, 50, and 70 mg/ml, respectively. c and d. The $CdCl_2$ concentrations on filter disks 1-5 were 0, 8, 16, 23, and 42 mg/mL, respectively. e and f. The paraquat concentrations on filter disks 1-5 were 0, 60, 100, 200, and 300 mmol/L, respectively. g and h. The cumene hydroperoxide concentrations on filter disks were 0, 50, 100, 200, and 400 mmol/L, respectively.

Discussion

RNA binding proteins are well-known for post-transcriptionally regulating of RNA metabolism. Recently, an increasing number of studies have begun to investigate the functions of RBPs in response to different stresses in plants [28]. Studies have also indicated a pivotal role for RBPs in animal under adverse types of stresses, such as in *Homo sapiens* and *Mus musculus* [8-10, 29, 30]. However, the roles of RBPs in stress response are poorly understood in insects. The present study was undertaken to characterize a novel RBP gene (*AccRBM11*) identified in *Apis cerana cerana* and to investigate its response to different abiotic stresses. Our findings suggested that *AccRBM11* could be induced by some stress agents.

AccRBM11 has a typical RRM domain, which specifically belongs to the RRM-RBM7-like subfamily. This subfamily is characterized by the RRM in *RBM7*, *RBM11* and their eukaryotic homologues. An AccRBM11 phylogenetic tree was generated to analyze the evolutionary relationships among different species, and the result showed that AccRBM11 shared a high degree of homology with RBP11 and RBP7 from other species; however, its greatest homology was to RBM11-like from *Apis mellifera* (Fig. 2b), which was consistent with the results predicted by multiple protein alignment. We therefore named the newly acquired gene from the Chinese honeybee *AccRBM11*. Kenan et al. [31] deemed that RRM domain could form a $\beta_1\alpha_1\beta_2\beta_3\alpha_2\beta_4$ secondary structure and that the direction of the two α helixes was perpendicular to the direction of the β-sheet. It is possible that the *AccRBM11* RRM domain has this $\beta_1\alpha_1\beta_2\beta_3\alpha_2\beta_4$ structure (Fig. 2a). Thus, the secondary structure characteristics of AcRBM11 might be involved in AccRBM11 functions.

To better understand the potential roles of *AccRBM11*, a 1,464 bp 5'-flanking region that contained many TFBs connected with environmental stress and development was cloned (Fig. 4). Interestingly, the *AccRBM11* 5' UTR also included some TFBs (Fig. 4), which indicated that the 5' UTR might also participate in the transcriptional regulation of *AccRBM11*. Among these, CF2-II is known to be related to development and growth [32, 33]. BR-C can regulate the development of the third instars and early pre-pupal larvae [34]. NIT2 has been described as the major nitrogen regulatory gene [35]. A previous study suggested that RBPs could be developmental regulators [36]. It showed that *AccRBM11* could be expressed in all *Apis cerana cerana* stages and its expression displayed a distinct increase during the transition from egg to larva, larva to pupa, and pupa to adult (Fig. 6a). The Western blotting results were not completely consistent with the results of qRT-PCR (Fig. 6d). Recent studies have shown that the transitions between developmental stages are often followed by immense changes in gene expression and that the expression of definite qualitative and quantitative proteins that implement specific functions are required in each stage [37]. This finding suggested that *AccRBM11* might be necessary for the transitions in *Apis cerana cerana* between developmental stages.

CdxA, Nkx-2, DRI, and Ovo homolog-like transcription factor binding sites were also found in the 5'-flanking region of *AccRBM11*. CdxA participates in tissue-specific expression [38]. Nkx-2 is related to organogenesis [39]. DRI determines the development of neural and gut cells [40], and Ovo homolog-like transcription factor is necessary for epidermal and germline differentiation [41]. Tissue-specific analysis showed that the highest expression of *AccRBM11* appeared in the poison gland, followed by the epidermis and midgut (Fig. 6b), which did not agree with the Western blotting findings (Fig. 6c). This particular expression of *AccRBM11* in different tissues most likely reflects the specific demand of that tissue. Moreover, the poison gland plays a key role in self-defence. The epidermis plays an important role in imparting physical stability and in taking part in responses to environmental stress. The midgut is associated with the detoxification of exogenous substances and protection from oxidative stress [42, 43]. These findings indicated that *AccRBM11* might be involved in protecting *Apis cerana cerana* against harm from environmental stress.

In addition to the TFBs mentioned above, a higher amount of HSF which is involved in heat-induced transcriptional activation [44], was found in the 5′-flanking sequence of *AccRBM11*. High temperature may be an important environmental factor that influences development because it can induce physiological changes [45]. The behaviour of the honeybee can be influenced by temperature changes, and *Apis cerana cerana* are more sensitive to heat stress [46]. qRT-PCR was performed to detect whether *AccRBM11* responded to heat. As shown in Fig. 7h, the mRNA level of *AccRBM11* reached their highest levels at 1 h, which suggested that *AccRBM11* might play a role in heat stress. In organisms, cold pressure suppresses protein synthesis rate, changes cellular membrane lipid composition and ultimately inhibits cell growth [47, 48]. RBPs have been known to play important roles in the cellular response under cold stress, and can be induced by low temperature [9, 10, 49]. The qRT-PCR results indicated that *AccRBM11* was increased several fold (Fig. 7g) under 4℃ treatment. The data suggested that *AccRBM11* might play a role in avoiding cold lesions.

In other species, many previous studies have demonstrated that RBPs could respond to environmental stress [7, 28, 50-53]. Therefore, we also detected the expression levels of *AccRBM11* when *Apis cerana cerana* was subjected to other abiotic stresses, such as pesticides, heavy metal, UV, H_2O_2, and VC. Pesticides are important factors that lead to environmental pollution. These include herbicides, insecticides, fungicides and others, which damage the physiological and biochemical functions of lymphocytes and erythrocytes and result in the lesions to lipid biomembranes [54]. The transcription levels of *AccRBM11* were all induced by cyhalothrin (Fig. 7d), paraquat (Fig. 7e) and acaricide (Fig. 7f) treatment, suggesting a role for *AccRBM11* in protecting honeybees against pesticides. Heavy metal can influence the normal development of insects. Mercury (Hg) and cadmium (Cd) are well-known to be the most poisonous heavy metals in the natural world [55] and can inactivate the function of proteins by directly binding to enzyme metal ion sites. Worker honeybees may contact heavy metal when they forage for pollen and nectar outside. The expression levels of *AccRBM11* were enhanced by $HgCl_2$ (Fig. 7b) and $CdCl_2$ (Fig. 7c) treatment by 3.6-fold and 22.7-fold, respectively, indicating that *AccRBM11* might function to avoid injury under $HgCl_2$ and $CdCl_2$ stresses. As an environmental stress, UV light influences insect habits and living [56]. qRT-PCR findings revealed that UV induced the mRNA level of *AccRBM11* (Fig. 7a). The UV-inducible RBP A18 has been shown to protect human cells against genotoxic stress by translocating to the cytosol and stabilizing specific transcripts related to cell survival [29]. *AccRBM11* might function similarly to avoid UV injury. Hydrogen peroxide (H_2O_2) is a main type oxidant that can cause oxidative damage. We detected high expression levels of *AccRBM11* after 3 h of H_2O_2 treatment (Fig. 7i). Recent studies had indicated that RBPs could be involved in oxidative responses [8, 51, 57]. *AccRBM11* might play a role in oxidative stress. Fig. 7j showed that *AccRBM11* was upregulated by VC. VC is a typical antioxidant. However, VC can also induce the decomposition of lipid hydroperoxide to endogenous genotoxins and can lead to DNA oxidative damage [58]. Thus, in this study, the dose of VC might have been sufficient to induce *AccRBM11* to take part in the response to oxidative injury.

Moreover, Western blotting was carried out to investigate the protein levels of *AccRBM11* when Chinese honeybees were subjected to various abiotic stresses. When treated with 44℃ (Fig. 8a), 4℃ (Fig. 8b), VC (Fig. 8c), $CdCl_2$ (Fig. 8d), $HgCl_2$ (Fig. 8e), and H_2O_2 (Fig. 8f), the expression level of AccRBM11 protein increased in some capacity, even though they were upregulated to various degrees and at different times. However, no clear response of the protein levels of *AccRBM11* was observed after treatment with acaricide (Supplemental Fig. S2a), paraquat (Supplemental Fig. S2b), cyhalothrin (Supplemental Fig. S2c), and UV (Supplemental Fig. S2d). It is worth noting that the upregulated degree and time points for *AccRBM11* showed distinct differences in the transcript and protein levels. The mRNA and its corresponding protein for a gene are both present in a living body; however, usually only the protein is functional. Changes in AccRBM11 protein levels when subjected to abiotic stress in this study indicated that AccRBM11 might be implicated in responses to some abiotic pressure.

Regarding the transcriptional patterns of *AccRBM11* that were not completely consistent with its protein levels, the following explanations should be taken into account. First, stress treatments may be enough to induce gene transcription, but not enough to necessarily affect its translation. Second, the accumulation and degradation of protein are existed in organisms. For instance, when the transcription of a gene is inhibited, increased protein expression levels could be due to protein accumulation. Third, the transcription and translation of a gene could be regulated by various signal transduction pathways under environmental pressure, such as post-transcriptional regulation. Recently, Mitobe et al. [52] reported that *invE* mRNA levels were easily detectable, though translation of the protein was tightly inhibited, which is similar for post-transcriptional regulation through RBP Hfq. Last but not least, several RNAs can participate in the transcription and translation of mRNAs, such as circRNAs and miRNAs. Previous studies had shown that circRNAs are implicated in regulating transcription, post-transcription, splicing, and protein activation [59, 60]. A recent study revealed that miRNAs were involved in regulating many pivotal processes during amelogenesis by influencing translation and mRNA stability in rat incisors [61]. The differences in the expression patterns of *AccRBM11* and its protein may be a result of its regulation by circRNAs and miRNAs.

RNAi experiment was performed to further investigate the function of *AccRBM11*. RNA silencing works through post-transcriptional gene regulation mechanisms [62, 63]. RNAi-mediated knockdown of endogenous target gene expression has become a popular strategy determining gene function, as transgenesis is hard to achieve in some species, especially in animals [63, 64]. This experimental method also has been used in honeybees, and studies have proven that gene knockdown is important for studying of gene function [65, 66]. Here, the mRNA levels of *AccRBM11* were successfully knocked down by RNAi, especially at one and two days after injection of dsRNA-AccRBM11 (Fig. 9a). Compared with the controls, *AccRBM11* silencing markedly downregulated *AccSOD2*, *AccTpx4*, *AccAK*, *AccTpx1*, *AccGSTO1*, *AccTrx1*, *AccsHsp22.6*, *AccRPL11*, and *Accp38b* transcripts to different degrees (Fig. 9b-j). Among these genes, *AccSOD2*, *AccTpx4*, *AccAK*, *AccTpx1*, *AccGSTO1*, and

AccTrx1 had been demonstrated to be involved in different environment stress responses[66-71]. *AccsHsp22.6*, *AccRPL11*, and *Accp38b* could not only function in stress defence but also in development [72-74]. Thus, *AccRBM11* might have the similar functions in development and stress responses in *Apis cerana cerana* with the above genes.

We also used a disc diffusion assay to examine the ability of AccRBM11 to resist abiotic stress. Recent evidence had shown that the human YB-1 protein can bind to RNA and act as an RNA chaperone to play a role in cell survival under environmental stress. When introduced into *E. coli*, the *YB-1* gene conferred high resistance to bacterial cells to environmental stress [75]. However, after exposure to four different treatment conditions, AccRBM11 overexpression in *E. coli* cells displayed low resistance to these disparate stresses (Fig. 10). This might be because of potential antimicrobial peptides that were identified in the AccRBM11 amino acid sequence [69], which might inhibit the growth of bacteria [76]. The inhibitory effect is relatively weak (not enough to completely restrain the growth of *E. coli* cells), and the cells still can express recombinant AccRBM11 protein [77, 78].

In conclusion, our study indicated that the expression of *AccRBM11* could be enhanced by some abiotic stresses both at the transcription and protein levels. The increased expression pattern possibly relates to the changes in ROS levels. AccRBM11 might play an important role in the development of *Apis cerana cerana* and in stress response challenges and might cause the elevated resistance of honeybees to environmental stress. These findings may contribute to further inquiry into the function of RNA binding proteins in honeybees and other insects.

References

[1] Peng, Y. S., Fang, Y., Xu, S., Ge, L. The resistance mechanism of the Asian honey bee, *Apis cerana* Fabr., to an ectoparasitic mite, *Varroa jacobsoni* Oudemans. J Invertebr Pathol. 1987, 49 (1), 54-60.

[2] Cheng, S. L. The Apicultural Science in China. China Agriculture Press, Beijing. 2001.

[3] Oldroyd, B., Wongsiri, S. Asian Honey Bees: Biology, Conservation, and Human Interactions. Harvard University Press, Cambridge. 2006.

[4] The Honeybee Genome Sequencing Consortium. Insights into social insects from the genome of the honeybee *Apis mellifera*. Nature. 2006, 443 (7114), 931-949.

[5] Park, D., Jung, J. W., Choi, B. S., et al. Uncovering the novel characteristics of Asian honey bee, *Apis cerana*, by whole genome sequencing. BMC Genomics. 2015, 16 (1).

[6] Norbury, C. J. Cytoplasmic RNA: a case of the tail wagging the dog. Nat Rev Mol Cell Biol. 2013, 4 (10), 643-653.

[7] Kang, W. H., Park, Y. D., Hwang, J. S., Park, H. M. RNA-binding protein Csx1 is phosphorylated by LAMMER kinase, Lkh1, in response to oxidative stress in *Schizosaccharomyces pombe*. FEBS Letters. 2007, 581 (18), 3473-3478.

[8] Mironova, A. A., Barykina, N. V., Zatsepina, O. V. Cytological analysis of the response of nucleolar RNA and RNA-binding proteins to oxidative stress in HeLa cells. Tsitologiia. 2014, 8 (6), 461-472.

[9] Nishiyama, H., Itoh, K., Kaneko, Y., Kishishita, M., Yoshida, O., Fujita, J. A glycine-rich RNA-binding protein mediating cold-inducible suppression of mammalian cell growth. J Cell Biol. 1997, 137 (4), 899-908.

[10] Zargar, R., Urwat, U., Malik, F., et al. Molecular characterization of RNA binding motif protein 3 (*RBM3*) gene from Pashmina goat. Res Vet Sci. 2015, 98, 51-58.

[11] Guo, T. B., Boros, L. G., Chan, K. C., Hikim, A. P. S., Hudson, A. P., Swerdloff, R. S., Mitchell, A.

P., Salameh, W. A. Spermatogenetic expression of RNA-binding motif protein 7, a protein that interacts with splicing factors. J Androl. 2003, 24 (2), 204-214.

[12] Lin, J. C., Hsu, M., Tarn, W. Y. Cell stress modulates the function of splicing regulatory protein RBM4 in translation control. Proc Natl Acad Sci USA. 2007, 104 (7), 2235-2240.

[13] Tiedje, C., Lubas, M., Tehrani, M., Menon, M. B., Ronkina, N., Rousseau, S. Cohen, P., Kotlyarov, A., Gaestel, M. p38MAPK/MK2-mediated phosphorylation of RBM7 regulates the human nuclear exosome targeting complex. RNA. 2014, 21 (2), 262-278.

[14] Pedrotti, S., Busa, R., Compagnucci, C., Sette, C. The RNA recognition motif protein RBM11 is a novel tissue-specific splicing regulator. Nucleic Acids Res. 2012, 40 (3), 1021-1032.

[15] Michelette, E. D. F., Soares, A. E. E. Characterization of preimaginal developmental stages in Africanized honey bee workers (*Apis mellifera* L). Apidologie. 1993, 24 (4), 431-440.

[16] Lourenço, A. P., Mackert, A., dos Santos Cristino, A., Simões, Z. L. P. Validation of reference genes for gene expression studies in the honey bee, *Apis mellifera*, by quantitative real-time RT-PCR. Apidologie. 2008, 39 (3), 372-385.

[17] Li, Y., Zhang, L., Kang, M., Guo, X., Xu, B. *AccERK2*, a map kinase gene from *Apis cerana* cerana, plays roles in stress responses, developmental processes, and the nervous system. Arch Insect Biochem Physiol. 2012, 79 (3), 121-134.

[18] Scharlaken, B., De Graaf, D. C., Goossens, K., Peelman, L. J., Jacobs, F. J. Differential gene expression in the honeybee head after a bacterial challenge. Dev Comp Immunol. 2008, 32 (8), 883-889.

[19] Umasuthan, N., Revathy, K. S., Lee, Y., Whang, I., Choi, C. Y., Lee, J. A novel molluscan sigma-like glutathione S-transferase from *Manila clam, Ruditapes philippinarum*: Cloning, characterization and transcriptional profiling. Comp Biochem Physiol C: Toxicol Pharmacol. 2012, 155 (4), 539-550.

[20] Livak, K. J, Schmittgen, T. D. Analysis of relative gene expression data using real-time quantitative PCR and the 2(-Delta Delta C(T)) method. Methods. 2001, 25 (4), 402-408.

[21] Meng, F., Zhang, L., Kang, M., Guo, X., Xu, B. Molecular characterization, immunohistochemical localization and expression of a ribosomal protein L17 gene from *Apis cerana cerana*. Arch Insect Biochem Physiol. 2010, 75 (2), 121-138.

[22] Liu, Y., Zhao, Y., Dai, Z., Chen, H., Yang, W. Formation of diapause cyst shell in brine shrimp, *Artemia parthenogenetica*, and its resistance role in environmental stresses. J Biol Chem. 2009, 284 (25), 16931-16938.

[23] Zhang, Z., Wang, Q., Li, P., Zhou, Y., Li, S., Yi, W., Chen, A., Kong, P., Hu, C. Overexpression of the interferon regulatory factor 4-binding protein in human colorectal cancer and its clinical significance. Cancer Epidemiol. 2009, 33 (2), 130-136.

[24] Tang, W. J., Wang, L. F., Xu, X. Y., Zhou, Y., Jin, W. F., Wang, H. F., Gao, J. Autocrine/paracrine action of vitamin D on *fgf23* expression in cultured rat osteoblasts. Calcif Tissue Int. 2010, 86 (5), 404-410.

[25] Elias-Neto, M., Soares, M. P. M., Simões, Z. L. P., Hartfelder, K., Bitondi, M. M. G. Developmental characterization, function and regulation of a laccase 2 encoding gene in the honey bee, *Apis mellifera*, (Hymenoptera, Apinae). Insect Biochem Mol Biol. 2010, 40 (3), 241-251.

[26] Bustin, S. A., Benes, V., Garson, J. A. et al. The miqe guidelines: minimum information for publication of quantitative real-time pcr experiments. Clin Chem. 2009, 55 (4), 611-622.

[27] Yang, J., Kemps-Mols, B., Spruyt-Gerritse, M., Anholts, J., Claas, F., Eikmans, M. The source of SYBI green master mix determines outcome of nucleic acid amplification reactions. BMC Res Notes. 2016, 9, 292.

[28] Lorković, Z. J. Role of plant RNA-binding proteins in development, stress response and genome organization. Trends Plant Sci. 2009, 14 (4), 229-236.

[29] Yang, C., Carrier, F. The UV-inducible RNA-binding protein A18 (A18 hnRNP) plays a protective role in the genotoxic stress response. J Biol Chem. 2001, 276 (50), 47277-47284.

[30] Aoki, K., Matsumoto, K., Tsujimoto, M. Xenopus cold-inducible RNA-binding protein 2 interacts

with ElrA, the *Xenopus* homolog of HuR and inhibits deadenylation of specific mRNAs. J Biol Chem. 2003, 278 (48), 48491-48497.

[31] Kenan, D. J., Query, C. C., Keene, J. D. RNA recognition: towards identifying determinants of specificity. Trends Biochem Sci. 1991, 16, 214-220.

[32] Štanojević, D., Hoey, T., Levine, M. Sequence-specific DNA-binding activities of the gap proteins encoded by hunchback and Krüppel in *Drosophila*. Nature. 1989, 341 (6240), 331-335.

[33] Gogos, J. A., Hsu, T., Bolton, J., Kafatos, F. C. Sequence discrimination by alternatively spliced isoforms of a DNA binding zinc finger domain. Science. 1992, 257 (5078), 1951-1955.

[34] Spokony, R. F., Restifo, L. L. Anciently duplicated broad complex exons have distinct temporal functions during tissue morphogenesis. Dev Genes Evol. 2007, 217 (7), 499-513.

[35] Fu, Y. H., Marzluf, G. A. Characterization of nit-2, the major nitrogen regulatory gene of *Neurospora crassa*. Mol Cell Biol. 1987, 7 (5), 1691-1696.

[36] Bandziulis, R. J., Swanson, M. S., Dreyfuss, G. RNA-binding proteins as developmental regulators. Genes Dev. 1989, 3 (4), 431-437.

[37] Gala, A., Fang, Y., Woltedji, D., Zhang, L., Han, B., Feng, M., Li, J. Changes of proteome and phosphoproteome trigger embryo-larva transition of honeybee worker (*Apis mellifera ligustica*). J Proteomics. 2013, 78, 428-446.

[38] Ericsson, A., Kotarsky, K., Svensson, M., Sigvardsson, M., Agace, W. Functional characterization of the CCL25 promoter in small intestinal epithelial cells suggests a regulatory role for caudal-related homeobox (Cdx) transcription factors. J Immunol. 2006, 176 (6), 3642-3651.

[39] Briscoe, J., Sussel, L., Serup, P., Hartigan-O'Connor, D., Jessell, T. M., Rubenstein, J. L. R., Ericson J. Homeobox gene *Nkx2.2* and specification of neuronal identity by graded sonic hedgehog signalling. Nature. 1999, 398 (6728), 622-627.

[40] Gregory, S. L., Kortschak, R. D., Kalionis, B., Saint, R. Characterization of the dead ringer gene identifies a novel, highly conserved family of sequence-specific DNA-binding proteins. Mol Cell Biol. 1996, 16 (3), 792-799.

[41] Delon, I., Chanut-Delalande, H., Payre, F. The Ovo/Shavenbaby transcription factor specifies actin remodelling during epidermal differentiation in *Drosophila*. Mech Dev. 2003, 120 (7), 747-758.

[42] Marionnet, C., Bernerd, F., Dumas, A., et al. Modulation of gene expression induced in human epidermis by environmental stress *in vivo*. J Invest Dermatol. 2003, 121 (6), 1447-1458.

[43] Enayati, A. A., Ranson, H., Hemingway, J. Insect glutathione transferases and insecticide resistance. Insect Mol Biol. 2005, 14(1), 3-8.

[44] Santoro, N., Johansson, N., Thiele, D. J. Heat shock element architecture is an important determinant in the temperature and transactivation domain requirements for heat shock transcription factor. Mol Cell Biol. 1998, 18 (11), 6340-6352.

[45] An, M. I., Choi, C. Y. Activity of antioxidant enzymes and physiological responses in ark shell, *Scapharca broughtonii*, exposed to thermal and osmotic stress: effects on hemolymph and biochemical parameters. Comp Biochem Physiol B: Biochem Mol Biol. 2010, 155 (1), 34-42.

[46] Tautz, J., Maier, S., Groh, C., Rössler, W., Brockmann, A. Behavioral performance in adult honey bees is influenced by the temperature experienced during their pupal development. Proc Natl Acad Sci USA. 2003, 100 (12), 7343-7347.

[47] Rao, P. N., Engelberg, J. HeLa cells: effects of temperature on the life cycle. Science. 1965, 148 (3673), 1092-1094.

[48] Burdon, R. H. Temperature and animal cell protein synthesis. Symp Soc Exp Biol. 1986, 41, 113-133

[49] Bae, W., Jones, P. G., Inouye, M. CspA, the major cold shock protein of *Escherichia coli*, negatively regulates its own gene expression. J Bacteriol. 1997, 179 (22), 7081-7088.

[50] Mazan-Mamczarz, K., Galbán, S., de Silanes, I. L., Martindale, J. L., Atasoy, U., Keene, J. D., Gorospe, M. RNA-binding protein HuR enhances p53 translation in response to ultraviolet light irradiation. Proc Natl Acad Sci USA. 2003, 100 (14), 8354-8359.

[51] Abdelmohsen, K., Kuwano, Y., Kim, H. H., Gorospe, M. Posttranscriptional gene regulation by

[51] RNA-binding proteins during oxidative stress: implications for cellular senescence. Biol Chem. 2008, 389 (3), 243-255.

[52] Mitobe, J., Morita-Ishihara, T., Ishihama, A., Watanabe, H. Involvement of RNA-binding protein Hfq in the osmotic-response regulation of *invE* gene expression in *Shigella sonnei*. BMC Microbiology. 2009, 9 (1), 110.

[53] Tanaka, M., Sasaki, K., Kamata, R., Hoshino, Y., Yanagihara, K., Sakai, R. A novel RNA-binding protein, Ossa/C9orf10, regulates activity of Src kinases to protect cells from oxidative stress-induced apoptosis. Mol Cell Biol. 2009, 29 (2), 402-413.

[54] Narendra, M., Bhatracharyulu, N. C., Padmavathi, P., Varadacharyulu, N. C. Prallethrin induced biochemical changes in erythrocyte membrane and red cell osmotic haemolysis in human volunteers. Chemosphere. 2007, 67 (6), 1065-1071.

[55] Rashed, M. N. Monitoring of environmental heavy metals in fish from Nasser Lake. Environ Int. 2001, 27 (1), 27-33.

[56] Mazza, C. A., Izaguirre, M. M., Zavala, J., Scopel, A. L., Ballaré, C. L. Insect perception of ambient ultraviolet-B radiation. Ecol Lett. 2002, 5 (6), 722-726.

[57] Brégeon, D., Sarasin, A. Hypothetical role of RNA damage avoidance in preventing human disease. Mutat Res. 2005, 577 (1), 293-302.

[58] Lee, S. H., Oe, T., Blair, I. A. Vitamin C-induced decomposition of lipid hydroperoxides to endogenous genotoxins. Science. 2001, 292 (5524), 2083-2086.

[59] Memczak, S., Jens, M., Elefsinioti, A., et al. Circular RNAs are a large class of animal RNAs with regulatory potency. Nature. 2013, 495 (7441), 333-338.

[60] Ashwal-Fluss, R., Meyer, M., Pamudurti, N. R., et al. circRNA biogenesis competes with pre-mRNA splicing. Mol Cell. 2014, 56 (1), 55-66.

[61] Yin, K., Hacia, J. G., Zhong, Z., Paine, M. L. Genome-wide analysis of miRNA and mRNA transcriptomes during amelogenesis. BMC Genomics. 2014, 15 (1), 998.

[62] Ding, S. W. RNA-based antiviral immunity. Nat Rev Immunol. 2010, 10 (9), 632-644.

[63] Goic, B., Vodovar, N., Mondotte, J. A., et al. RNA-mediated interference and reverse transcription control the persistence of RNA viruses in the insect model *Drosophila*. Nat Immunol. 2013, 14 (4), 396-403.

[64] Huvenne, H., Smagghe, G. Mechanisms of dsRNA uptake in insects and potential of rnai for pest control: a review. J Insect Physiol. 2010, 56 (3), 227-235.

[65] Wilson, M. J., Kenny, N. J., Dearden, P. K. Components of the dorsal-ventral pathway also contribute to anterior-posterior patterning in honeybee embryos (*Apis mellifera*). Evodevo. 2014, 5 (1), 1419-1419.

[66] Yao, P., Chen, X., Yan, Y., Feng, L., Zhang, Y., Guo, X., Xu, B. Glutaredoxin 1, glutaredoxin 2, thioredoxin 1, and thioredoxin peroxidase 3 play important roles in antioxidant defense in *Apis cerana cerana*. Free Radic Biol Med. 2014, 68 (3), 335-346.

[67] Yu, F., Kang, M., Meng, F., Guo, X., Xu, B. Molecular cloning and characterization of a thioredoxin peroxidase gene from *Apis cerana cerana*. Insect Mol Biol. 2011, 20 (3), 367-378.

[68] Jia, H., Sun, R., Shi, W., Yan, Y., Han, L., Guo, X., Xu, B. Characterization of a mitochondrial manganese superoxide dismutase gene from *Apis cerana cerana*, and its role in oxidative stress. J Insect Physiol. 2014, 60 (1), 68-79.

[69] Meng, F., Zhang, Y., Liu, F., Guo, X., Xu, B. Characterization and mutational analysis of Omega-class GST (GSTO1) from *Apis cerana cerana*, a gene involved in response to oxidative stress. PLoS One. 2014, 108 (4), 927-939.

[70] Chen, X., Yao, P., Chu, X., Hao, L., Guo, X., Xu, B. Isolation of arginine kinase from *Apis cerana cerana* and its possible involvement in response to adverse stress. Cell Stress Chaperones. 2015, 20 (1), 169-183.

[71] Huaxia, Y., Fang, W., Yan, Y., Feng, L., Wang, H., Guo, X., Xu, B. A novel 1-Cys thioredoxin peroxidase gene in *Apis cerana cerana*: characterization of *AccTpx4* and its role in oxidative stresses.

Cell Stress Chaperones. 2015, 20 (4), 1-10.
[72] Meng, F., Lu, W., Yu, F., Kang, M., Guo, X., Xu, B. Ribosomal protein l11 is related to brain maturation during the adult phase in *Apis cerana cerana*, (Hymenoptera, Apidae). Naturwissenschaften. 2012, 99 (5), 343-352.
[73] Zhang, L., Meng, F., Li, Y., Kang, M., Guo, X., Xu, B. Molecular characterization and immunohistochemical localization of a mitogen-activated protein kinase, Acc38b, from *Apis cerana cerana*. BMB Rep. 2012, 45 (5), 293-298.
[74] Zhang, Y., Liu, Y., Guo, X., Li, Y., Gao, H., Guo, X., Xu, B. *sHsp22.6*, an intronless small heat shock protein gene, is involved in stress defence and development in *Apis cerana cerana*. Insect Biochem Mol Biol. 2014, 53 (2), 160-165.
[75] Li, Z., Wu, J., DeLeo, C. J. RNA damage and surveillance under oxidative stress. IUBMB Life. 2006, 58 (10), 581-588.
[76] Xu, P., Shi, M., Chen, X. X. Antimicrobial peptide evolution in the Asiatic honey bee *Apis cerana*. PLoS One. 2009, 4 (1), e4239.
[77] Yu, Y., Yu., Y. Huang, H., et al. A short-form C-type lectin from amphioxus acts as a direct microbial killing protein via interaction with peptidoglycan and glucan. J Immunol. 2007, 179 (12), 8425-8434.
[78] Guo, X., Jin, X., Li, S., et al. A novel C-type lectin from eriocheir sinensis functions as a pattern recognition receptor with antibacterial activity. Fish Shellfish Immunol. 2013, 35 (5), 1554-1565.

1.7 Isolation of Arginine Kinase from *Apis cerana cerana* and Its Possible Involvement in Response to Adverse Stress

Xiaobo Chen, Pengbo Yao, Xiaoqian Chu, Lili Hao, Xingqi Guo, Baohua Xu

Abstract: Arginine kinases (AK) in invertebrates play the same role as creatine kinases in vertebrates. Both proteins are important for energy metabolism, and previous studies on AK focused on this attribute. In this study, the arginine kinase gene was isolated from *Apis cerana cerana* and was named *AccAK*. A 5′-flanking region was also cloned and shown to contain abundant putative binding sites for transcription factors related to development and response to adverse stress. We imitated several abiotic and biotic stresses suffered by *A. cerana cerana* during their life, including heavy metal, pesticides, herbicides, heat, cold, oxidants, antioxidants, ecdysone, and *Ascosphaera apis* and then studied the expression patterns of *AccAK* after these treatments. *AccAK* was upregulated under all conditions, and, in some conditions, this response was very pronounced. Western blotting and AccAK enzyme activity assays confirmed the results. In addition, a disc diffusion assay showed that overexpression of AccAK reduced the resistance of *Escherichia coli* cells to multiple adverse stresses. Taken together, our results indicated that AccAK may be involved of great significance in response to adverse abiotic and biotic stresses.

Keywords: *Apis cerana cerana*; Arginine kinase; Abiotic and biotic stresses; Expression patterns; Enzyme activity

Introduction

Arginine kinases (AK) in invertebrates and creatine kinases in vertebrates play major roles in energy metabolism by functioning as phosphagen kinases [1]. AK catalyses a reversible phosphorylation reaction as follows:

Phosphoarginine+MgADP+αH$^+$↔Arginine+MgATPPhosphoarginine+MgADP+αH$^+$↔Arginine+MgATP

Arginine can be phosphorylated by MgATP to form phosphoarginine and MgADP, and during bursts of cellular activity, the reaction can be reversed to regenerate ATP [2]. This system for ATP turnover could provide a large pool of "high-energy phosphates" to refresh

ATP levels during periods of high energy demand of the host organisms through temporal and spatial energy buffering [3, 4].

AKs are ubiquitous in invertebrates, and some cDNA sequences have been obtained from a variety of invertebrates. Notably, more than 40 *AK* cDNAs have been cloned from insects, either in their full-length forms or as fragments. Among these sequences, the *AK*s from the American cockroach [5, 6], *Apis mellifera* [7], *Bombyx mori* [8], and locusts [9, 10] have been well studied. Carlson et al. [11] isolated the first insect AK from *A. mellifera*, drawing attention to AK purification, enzyme activity analysis, and tissue localization.

Earlier studies on the function of AKs from insects focused on their roles in energy metabolism and development. Schneider et al. [12] study on *Locusta migratoria* speculated that the phospho-L-arginine/L-arginine kinase system acted as a shuttle mechanism for high-energy phosphate between the mitochondria and myofibrils, as well as a high-energy phosphate buffer system. The highest *AK* mRNA expression in various *A. mellifera* tissues was found in the compound eye, while the levels of mRNA encoding *α-GPDH*, another metabolically important enzyme, were quite low. Together, these results suggested that AK contributed to the energy releasing mechanism in the visual system, which has high and fluctuating energy demands [7]. The expression of the AK gene in the red imported fire ant *Solenopsis invicta* and its AK protein activity were developmentally and tissue specifically regulated. This regulation was dependent on the different demands for energy-consumption and production in the different castes and was linked to their different labor and physiological activities in the colonies [13].

However, there is emerging evidence that AK might participate in responses to adverse environmental conditions and during innate immune responses. In the gills of *Callinectes sapidus*, acclimation to salinity had no impact on AK activity, but exposure to low salinity resulted in a 1.7-fold increase in expression [14, 15]. In *Penaeus monodon*, increased AK activity was detected in different tissues after exposure to low salinity [16]. *Trypanosoma cruzi* treated with hydrogen peroxide produced an up to 10-fold increase in AK expression, while trypanosomes treated with nifurtimox, an oxidative stress generating compound, exhibited more than 2-fold increase in expression [17]. After exposure to copper sulfate ($CuSO_4$), AK was shown to be downregulated in a proteomic study of *Artemia sinica* larvae [18]. After chronic cadmium exposure, AK was identified as a downregulated protein by using a proteomic approach in the anterior gills of the Chinese mitten crab, *Eriocheir sinensis* [19]. These reports provided clues that AKs may be involved in the response to abiotic stress, even though its expression patterns may differ depending on the particular stressor. In the context of immune responses, by using two-dimensional electrophoresis and electrospray ionization mass spectrometry, AK was found to be significantly decreased in the plasma of *Fenneropenaeus chinensis* 45 min after laminarin injection, but its expression level recovered after 3 h [20]. It was reported that laminarin could act as an immunostimulant that can enhance resistance of many animals against bacterial and viral infections and could be also a source of energy to crustacea [21]. In many studies, laminarin was used as an immunostimulant [22, 23]. Furthermore, it was reported that shrimp AK could act as an allergen that caused allergic reactions in hypersensitive individuals [24, 25]. These findings suggested that AK might play a role in the process of immunization.

Apis cerana cerana is the main honeybee species in China, which has enjoyed exceptional advantages over other species, such as the longer period of collecting honey, the more resistance to disease, and the lesser cost of food. As a flowering plant pollinator, *A. cerana cerana* greatly contributes to the balance of regional ecology and agricultural development [26]. The environment in China is getting worse, which is a threat for raising the honeybees. Thus, a study examining the resistance of *A. cerana cerana* to adverse stress is of great significance, as raising this species is becoming more and more difficult. Though some work has been carried out on AKs in other insects, little is known about the role of AK in response to environment stresses and pathogen stimuli, especially in *A. cerana cerana*. To increase this knowledge, the *AK* of *A. cerana cerana* was isolated in our study, and the common stress suffered by the honeybees were imitated in order to thoroughly assess the role AK may play in response to adverse stress.

Materials and Methods

Specimens and treatments

Chinese honeybees (*A. cerana cerana*) were obtained from the artificial beehives in Shandong Agricultural University (Taian, China). The honeybees were raised for collecting honey. For analysis of *AccAK* expression during different developmental stages, the eggs; the first (L1)-, the second (L2)-, the third (L3)-, the fourth (L4)-, and the fifth (L5)-instar larvae; prepupal phase pupae; white-eyed (Pw), pink-eyed (Pp), brown-eyed (Pb), and dark-eyed (Pd) pupae; and 1-day-old worker adults were collected from the hive, while 15-day-old adults and 30-day-old adults were collected from the hive entrance. The larvae and pupae were classified according to the criteria of Michelette and Soares [27]. The newly emerged bees were marked and identified as 1-day-old adults. The 15- and 30-day-old adults were collected after 15 and 30 days, respectively. The 15-day-old adults were divided into several groups for expression analysis under various environmental stresses. The adults were maintained in an incubator at constant temperature (34°C) and humidity (70%) under a 24 h dark regimen and fed with a pollen-and-sucrose solution [28]. Group 1 was fed with $CdCl_2$ (0.5, 5, or 10 mg/L) that was added into their food for 1, 4, 5, and 6 d. Group 2 was fed with $CdCl_2$ (1 mg/ml) that was added into their food. Groups 3-6 were treated with pesticides (acaricide and pyriproxyfen at a final concentration of 20 mg/ml) and herbicides (phoxim and paraquat at a final concentration of 20 mg/ml). Groups 7-8 were placed at high (42°C) and low (4°C) temperatures. Groups 9-10 were subjected to hydrogen peroxide (H_2O_2) and vitamin C. H_2O_2 was diluted to a concentration of 2 mmol/L, and 0.5 µl of this solution was applied to the thoracic notum of the adult bees. Vitamin C was diluted to a final concentration of 20,000 mg/kg in the food. The fourth-instar larvae were divided into two groups for biotic treatments. Group 1 was treated with ecdysone diluted in the larval food at a final concentration of 0.01, 0.1, or 1 µg/ml. Group 2 was treated with *Ascosphaera apis*, a main fungal parasite of *A. cerana cerana*, for 1, 2, or 3 d, which was cultured in the larval food at a concentration of 1×10^6 CFU/ml. Three samples were used for each treatment.

Primers and PCR procedure

The primers and polymerase chain reaction (PCR) procedure used in these experiments are listed in Table 1 and Table 2, respectively.

Table 1 Primer sequences used in this research

Abbreviation	Primer sequence (5'—3')	Description
AKF	GACGTGCCTCGAGGAGTAACT	cDNA sequence primer, forward
AKR	GGATCCTCGATGTATCAAGAACGT	cDNA sequence primer, reverse
AK3RO	GGCAATCTTGATCCAGCTAAT	3' RACE forward primer, outer
AK3RI	TCGTGTAAGATGCGGTCGCT	3' RACE forward primer, inner
B26	GACTCTAGACGACATCGA(T)18	3' RACE universal primer, outer
B25	GACTCTAGACGACATCGA	3' RACE universal primer, inner
AK5RO	GTAAATACCAACGCCAGAAT	5' RACE reverse primer, outer
AK5RI	GAAGAGTAGAATCGAAGGAAGTT	5' RACE reverse primer, inner
AAP	GGCCACGCGTCGACTAGTAC(G)$_{14}$	Abridged anchor primer
AUAP	GGCCACGCGTCGACTAGTAC	Abridged universal amplification primer
AKQDZF	GTTGCATTTTGTTTTATTAAACT	Promoter-specific primer, forward
AKQDZR	GAGTTACTCCTCGAGGCACGT	Promoter-specific primer, reverse
AKRTF	GAAGACTGACGAGCACCCG	Real-time PCR primer, forward
AKRTR	GAAGGTCTTATCATCGTTGTGGT	Real-time PCR primer, reverse
β-s	AGAATTGATCCACCAATCCA	Standard control primer, forward
β-x	GGTACCATGCAGCACATATTATTG	Standard control primer, reverse
AKPETF	GGATCCATGGTTGACCAAGCTGTTT	Primers of constructing vector, forward
AKPETR	AAGCTTAAGTTCCTTTTCGAGTTTAAT	Primers of constructing vector, reverse

Table 2 Procedures used in this research

Primer pair	Amplification conditions
AKF/AKR	5 min at 94℃, 40 s at 94℃, 40 s at 50℃, 1 min at 72℃ for 35 cycles, 5 min at 72℃
AK3R1/B26	5 min at 94℃, 40 s at 94℃, 40 s at 45℃, 30 s at 72℃ for 35 cycles, 5 min at 72℃
AK3R2/B25	5 min at 94℃, 40 s at 94℃, 40 s at 52℃, 30 s at 72℃ for 35 cycles, 5 min at 72℃
AK5R1/AAP	5 min at 94℃, 40 s at 94℃, 40 s at 47℃, 40 s at 72℃ for 35 cycles, 5 min at 72℃
AK5R2/AUAP	5 min at 94℃, 40 s at 94℃, 40 s at 52℃, 40 s at 72℃ for 35 cycles, 5 min at 72℃
QDZF/QDZR	5 min at 94℃, 40 s at 94℃, 40 s at 53℃, 40 s at 72℃ for 35 cycles, 5 min at 72℃
PETF/PETR	5 min at 94℃, 40 s at 94℃, 40 s at 53℃, 1 min at 72℃ for 35 cycles, 5 min at 72℃

RNA extraction, cDNA synthesis, and DNA preparation

Trizol® reagent (Invitrogen, Carlsbad, CA, USA) was used to extract total RNA according to the instructions of the manufacturer, and RNA was treated with RNase-free Dnase Ⅰ to remove any potential genomic DNA. The first-strand cDNA was obtained by using the EasyScript cDNA Synthetic SuperMix (TransGen Biotech, Beijing, China) according to the protocol of the manufacturer. The EasyPure Genomic DNA Extraction Kit (TransGen Biotech, Beijing, China) was used to isolate genomic DNA according to the instructions of the manufacturer.

cDNA isolation of *AK*

The specific primers AKF and AKR were designed and synthesized (Sangon Biotechnological Company, Shanghai, China) based on the *AK* sequence of *A. mellifera*, an organism that shares high similarity with *A. cerana cerana*, to obtain the internal region of the *AccAK* gene. Then, the specific primers AK5RO and AK5RI were designed for the 5′ rapid amplification of cDNA end (RACE) based on the obtained sequence, while the AK3RO and AK3RI primers were used for 3′ RACE. The Abridged anchor primer (AAP) and the abridged universal amplification primer (AUAP) were used in the first and the second rounds of 5′ RACE, respectively, whereas the B26 and B25 primers were used in the first and the second rounds of 3′ RACE, respectively. The resulting PCR products were ligated into the pEASY-T3 vectors and transformed into *Escherichia coli* strain DH5α for sequencing.

Amplification of the 5′-flanking region of *AccAK*

The 5′-flanking region of *AccAK* was amplified by using specific primers AKQDZF and AKQDZR, which were designed against the *A. mellifera* promoter region sequence. The resulting PCR products were cloned into the pEASY-T3 vectors and transformed into *E. coli* strain DH5α for sequencing. The TFSEARCH database (http://www.cbrc.jp/research/db/TFSEARCH.html) was used to predict the putative transcription factor binding sites in the 5′-flanking region of *AccAK*.

Bioinformatic analysis

Bioinformatics tools available on the NCBI server (http://blast.ncbi.nlm.nih.gov/Blast.cgi) were used to retrieve the homologous AccAK protein sequence from *A. mellifera*. The open reading frame (ORF) was identified and aligned with multiple homologues by using DNAMAN software 5.2.2; additionally, the molecular mass and isoelectric point were calculated. Phylogenetic tree analysis of the AK amino acid sequences from different species was carried out by using the neighbor-joining method in Molecular Evolutionary Genetic Analysis (MEGA version 4.1).

Fluorescence real-time quantitative PCR

Quantitative PCR (qPCR) was performed with the SYBR Premix Ex *Taq* (TaKaRa) in the CFX96™ real-time PCR Detection System (Bio-Rad, Hercules, CA, USA) to measure *AccAK* transcription at different developmental stages and under adverse abiotic and biotic stresses. The specific primers AKRTF and AKRTR were designed based on the cDNA sequence of *AccAK*. The β-s and β-x primers were designed to amplify the house-keeping gene *β-actin* (GenBank accession number: HM_640276), which was used to normalize RNA levels. We have first validated the primers of *β-actin* and *AccAK*. The efficiency of *β-actin* and *AccAK* both approached 100%. The melting curves had single peaks and the correlation coefficients (R^2) were 0.997 and 0.998, respectively. The amplification parameters were performed as described in Yao et al. [29]. All the experiments were carried out in triplicate. The data were analyzed with the CFX Manager software program (version 1.1) by using the $2^{-\Delta\Delta C_t}$ comparative CT method [30]. Differences among the samples were determined by one-way ANOVA using the Statistical Analysis System (SAS) software, version 9.1. The significant differences were labelled with different letters. The same letter indicated that there was no significant difference between the two groups. The different letters indicated that there was significant difference. The overlapped letters indicated that there was difference between the two groups but the difference was not significant.

Expression and purification of recombinant AccAK

To obtain the AccAK protein, the ORF of *AccAK* was amplified by using a pair of primers containing *Bam*H I and *Hin*dIII restriction sites and was inserted into the expression vector pET-30a (+). Then the expression plasmid pET30a (+)-AccAK was transformed into *E. coli* strain Rosseta (DE3) which was reformed from BL21 for protein expression. The cells were cultured in Luria-Bertani (LB) broth with kanamycin at 37℃ until the cells density reached 0.4-0.6 OD_{600}. Then, the AccAK expression was induced with 1.0 mmol/L isopropylthio-β-D-galactoside (IPTG) at 28℃ for 8 h. The protein was purified on a HisTrap^{TM} FF column (GE Healthcare, Uppsala, Sweden) according to the instructions of the manufacturer. The expression of the target protein was analyzed by 12% sodium dodecyl sulfate polyacrylamide gel electrophoresis (SDS-PAGE).

Anti-AccAK preparation and Western blotting analysis

The anti-AccAK was obtained by injecting the purified AccAK protein subcutaneously into white mice as described by Yan et al. [31]. Total protein were extracted from *A. cerana cerana* according to Li et al. [32] and quantified with the BCA Protein Assay Kit (Thermo Scientific Pierce, IL, USA). Equal amounts of protein from each sample were separated by 12% SDS-PAGE and subsequently electrotransferred onto a polyvinylidene fluoride membrane (Millipore, Bedford, MA). Western blotting was performed according to Meng et al. [33]. The anti-AccAK serum was used as the primary antibody at a dilution of 1∶400 (*V/V*). Peroxidase-conjugated goat anti-mouse immunoglobulin G (Dingguo, Beijing, China) was used as the secondary antibody at a dilution of 1∶2,000 (*V/V*). The results were visualized by using the SuperSignal® West Pico Trial Kit (Thermo Scientific Pierce, IL, USA).

AccAK activity assay

The AK activity assay was conducted according to the procedures described in "Enzymatic Assay of AK" available on the following website: http://www.sigmaaldrich.com/img/assets/18220/Arginine_Kinase.pdf. Briefly, in the reaction mix, the final concentrations are 178 mmol/L glycine, 0.33 mmol/L 2-mercaptoethanol, 13 mmol/L magnesium sulfate, 133 mmol/L potassium chloride, 20 mmol/L phosphor (enol) pyruvate, 6.7 mmol/L adenosine 5′-triphosphate, 0.13 mmol/L β-nicotinamide adenine dinucleotide, 2 U pyruvate kinase, 3 U lactic dehydrogenase, 17 mmol/L L-arginine. The reaction solution was preincubated for 2 min at 30℃ before reaction. The AK activity was measured at $A_{340\ nm}$ immediately after adding 0.04 ml of AK enzyme solution into the 1 ml reaction mix. The progress was recorded as the decrease in the $A_{340\ nm}$ for 5 min, and the activity was recorded as the $\Delta A_{340\ nm}/min$ by using the maximum linear rate. One unit can convert 1.0 μmol/L of L-arginine and ATP to N'-phospho-L-arginine and ADP per minute at pH 8.6 at 30℃.

Disc diffusion assay

Disc diffusion assay was performed according to Zhang et al. [34]. Approximately 5×10^8 bacterial cells overexpressing AccAK were plated on LB-kanamycin agar plates and incubated at 37℃ for 1 h. Cells with pET-30a (+) vector were used as the control. Filter discs (6 mm diameter) soaked in different concentrations of $HgCl_2$ (0, 10, 20, 40, or 80 mg/ml), paraquat (0, 10, 50, 100, or 200 mmol/L), t-butyl hydroperoxide (0, 100, 250, 500, or 1,000 mmol/L), or cumene hydroperoxide (0, 10, 50, 100, or 200 mmol/L) were placed on the surface of the top agar. The cells were incubated at 37℃ for 24 h before the killing zones around the discs were measured.

Results

cDNA Isolation of *AccAK* and bioinformatics analysis

The full-length cDNA of *AccAK* was obtained from *A. cerana cerana* RNA by using reverse-transcription PCR and rapid amplification of cDNA end PCR. The full-length cDNA is 1,674 bp long and contains a 70 bp 5′ untranslated region (UTR), a 539 bp 3′ UTR, and a 1,065 bp ORF, which encodes a 355-amino acid protein with a predicted molecular mass of 39 kDa and a theoretical isoelectric point of 5.34. Multiple sequence alignments of several AK protein sequences showed that the putative AccAK shared 96.90%, 88.73%, 87.92%, and 87.32% similarity with *A. mellifera*, *Solenopsis invicta*, *Periplaneta americana*, and *Helicoverpa armigera*, respectively (Fig. 1a). As shown in Fig. 1a, AccAK contained the typical conserved residues like other AKs, including seven conserved arginine (Arg) binding residues (Ser^{62}, Gly^{63}, Val^{64}, Tyr^{67}, Glu^{224}, Cys^{270}, and Glu^{313}), five Arg residues (123, 125, 228, 279, and 308) that interact with ADP, and the active center sequence CPTNLGT (270-276), of which the first cysteine residue is the active site necessary for kinase activity. Additionally, the Asp and Arg residues at positions 61 and 192, respectively, could form an ion pair structure [35-37].

A neighbor-joining phylogenetic tree was built by using 21 AK sequences from various insects to demonstrate the putative evolutionary relationships among the AKs (Fig. 1b). The result showed that AKs were highly conserved throughout evolution. The AK in *A. cerana cerana* shared the highest similarity with the AK found in *A. mellifera*. Additionally, this sequence was more closely related to the AKs in honeybees than to the homologues present in other species.

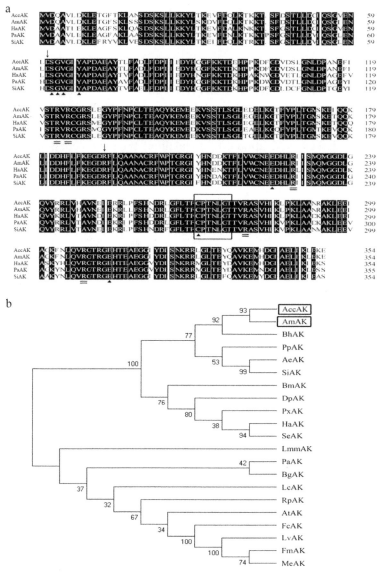

Fig. 1 Characterization and evolutionary relationship of arginine kinases (AKs) from various species. a. Multiple amino acid sequence alignment of AccAK (GenBank accession number: KF772855) from *A. cerana cerana*, AmAK (NM_001011603.1) from *A. mellifera*, HaAK from *Helicoverpa armigera* (EF600057.1), PaAK from *P. americana* (GU301882.1), and SiAK from *S. invicta* (EU817514.1). Conserved regions are shown in black. Conserved arginine binding residues are marked with black triangles. Arginine (Arg) residues that interact with ADP are marked with double lines. The active center sequence CPTNCGT is boxed. The Asp and Arg residues at positions 61 and 192, respectively, which can form an ion pair structure are marked with arrows. b. Phylogenetic analysis of AK amino acid sequences from various species. The sequences were obtained from the NCBI database. The homologous AKs are from the following species: *A. mellifera* (NM_001011603.1), *Bombus hypocrite* (JF751027.1), *Pteromalus puparum* (FJ882065.1), *Acromyrmex echinatior* (GL888624.1), *Solenopsis invicta* (EU817514.1), *B. mori* (EU327675.1), *Danaus plexippus* (AGBW01011584.1), *Papilio xuthus* (AK401168.1), *H. armigera* (EF600057.1), *Spodoptera exigua* (GQ379235.1), *L. migratoria manilensis* (DQ513322.1), *P. americana* (GU301882.1), *Blattella germanica* (FJ855501.1), *Lucilia cuprina* (JQ088101.1), *Riptortus pedestris* (AK417004.1), *Anasa tristis* (JQ266746.1), *F. chinensis* (AY661542.1), *Litopenaeus vannamei* (EU346737.1), *Fenneropenaeus merguiensis* (FJ895112.1), and *Metapenaeus ensis* (EU497674.1).

Analyses of partial potential *cis*-acting elements in the 5′-flanking region of *AccAK*

To predict the putative mechanism of AccAK regulation, a 1,207 bp 5′-flanking region of *AccAK* was isolated, and the putative transcription factor binding sites were predicted. Representative portions of these sequences are shown in Fig. 2. In this region, nearly 100 heat shock factor (HSF) binding sites were found. This transcription factor plays a crucial role in heat-induced transcriptional activation [38]. In addition, the region contains a large number of binding sites for other transcription factors, including those for caudal-related homeobox (Cdx) protein, which participates in tissue-selective development [39, 40], fork head domain protein crocodile (Croc), which is a brain development-specific transcriptional activator [41], and Broad-Complex, which is a key ecdysone-responsive regulator of metamorphosis [42]. Additionally, some binding sites for transcription factors related to early stage tissue development and growth or involved in the response to adverse stress, such as Hunchback (Hb), Dfd, NIT2, cAMP-responsive element binding protein (CREB), and CCAAT/enhancer binding protein (C/EBP), were also found.

Fig. 2 The nucleotide sequence and the partial putative transcription factor binding sites in the 5′-flanking region. The transcription start site is marked with an arrow, and the putative transcription factor binding sites are boxed.

Developmental expression levels of *AccAK*

qPCR was performed to investigate the expression levels of *AccAK* at different developmental stages (Fig. 3). The results showed that the lowest expression level was during the egg stage. In the pupae, the expression levels in dark-pigmented phase pupae were somewhat higher than in the pupae of other phases. The highest level of expression appeared at the fourth-instar larvae, while the levels in the other larval instar stages and the prepupal phase pupae were not as high, although were still higher than the expression observed in pupae. The second highest expression level was observed in the 15- and the 30-day-old adult worker bees.

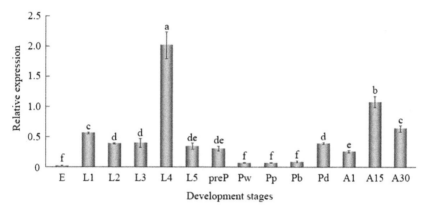

Fig. 3 The relative expression of *AccAK* during different development stages. The data are the mean ± SE of three replicates. The letters above the bar indicate significant differences ($P < 0.000,1$) as determined by Duncan's multiple range tests using SAS software version 9.1.

Expression profiles of *AccAK* under adverse abiotic stress

The mRNA transcription levels under various environmental stresses were quantified by using qPCR and were normalized to an untreated control group. Surprisingly, *AccAK* was induced by all of the tested treatments. *AccAK* robustly responded to $CdCl_2$ treatment, which caused an 86-fold increase in expression after treatment with 5 mg/L for 4 days compared with the control group (Fig. 4b). Similarly, $CdCl_2$ treatment with different concentrations and for different times induced different patterns of *AccAK* expression (Fig. 4a, c, and d). Although *AccAK* expression changed in response to all tested pesticides and herbicides, it seemed that the induction varied by treatment type, as paraquat and pyriproxyfen caused a much more intense increase in expression than acaricide and phoxim (Fig. 5a-d). Extreme temperatures that honeybees may experience (42 ℃ and 4 ℃) were also examined. The results showed that *AccAK* expression was enhanced more drastically after exposure to the high temperature (Fig. 6a, b). We also analyzed the expression level after treatments with H_2O_2 and vitamin C, as H_2O_2 is a type of oxidant and vitamin C is an antioxidant. The two treatments both caused an increase in *AccAK* expression, but H_2O_2 caused a drastic increase, while vitamin C treatment resulted in only a slight increase (Fig. 6c, d).

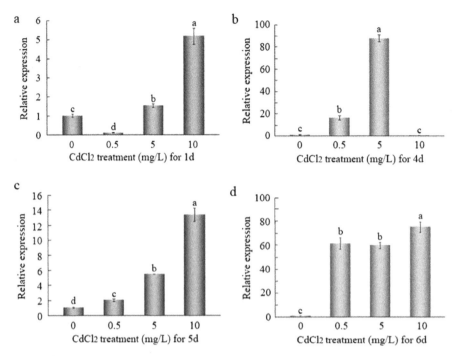

Fig. 4 Expression profiles of *AccAK* under heavy metal stress. The treatments are as follows: CdCl$_2$ (24 h) (a), CdCl$_2$ (4 days) (b), CdCl$_2$ (5 days) (c), and CdCl$_2$ (6 days) (d). The data are the mean ± SE of three replicates. The letters above the bar indicate significant differences ($P < 0.000,1$) as determined by Duncan's multiple range tests using SAS software version 9.1.

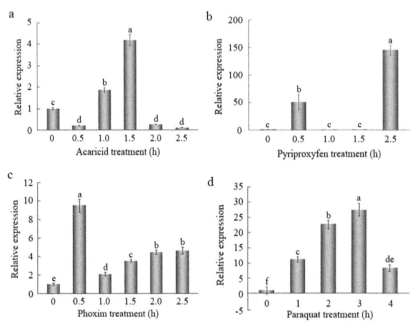

Fig. 5 Expression profiles of *AccAK* under pesticide and herbicide stresses. These stresses are as follows: acaricide (a), pyriproxyfen (b), phoxim (c), and paraquat (d). The data are the mean ± SE of three replicates. The letters above the bar indicate significant differences ($P < 0.000,1$) as determined by Duncan's multiple range tests using SAS software version 9.1.

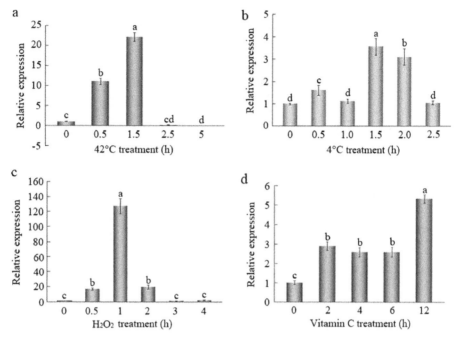

Fig. 6 Expression profiles of *AccAK* under extreme temperature, oxidant, and antioxidant stresses. These stresses are 42℃ (a), 4℃ (b), H_2O_2 (c), and vitamin C (d). The data are the mean±SE of three replicates. The letters above the bar indicate significant differences ($P<0.000,1$) as determined by Duncan's multiple range tests using SAS software version 9.1.

Expression profiles of *AccAK* under biotic stress

To further understand the role *AccAK* plays in honeybees, we treated the fourth-instar larvae with different concentrations of ecdysone. As described above, qPCR was performed to examine the accumulation of *AccAK* transcripts. As shown in Fig. 7a, transcription was similarly induced under all ecdysone concentrations. Additionally, we diluted *A. apis* in the larval food. The expression of *AccAK* initially decreased after treatment with *A. apis* for 1 day, but increased immediately on the second day. Then, on the third day, its expression decreased again but remained higher than that of the control group (Fig. 7b).

Western blotting analysis

To further detect AccAK in honeybees under adverse stress, Western blotting was performed. Anti-AccAK was used to detect AccAK. Following 42℃ treatments for 0.5 and 1.5 h, the expressions of AccAK were induced; especially after 1.5 h, the protein level was several times induced (Fig. 8a). After exposure to phoxim for 0.5 and 1.0 h, the expressions of AccAK were also induced (Fig. 8b), although not that high as they were after 42℃ treatment. The results at the protein level were consistent with the expressions of AccAK at the mRNA level.

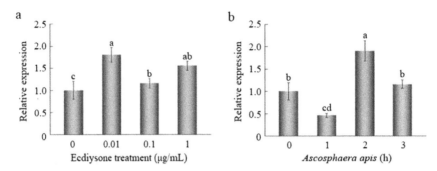

Fig. 7 Expression profiles of *AccAK* under various biotic stresses. The stresses included ecdysone (a) and *A. apis* treatments (b). The data are the mean ± SE of three replicates. The letters above the bar indicate significant differences ($P < 0.000,1$) as determined by Duncan's multiple range tests using SAS software version 9.1.

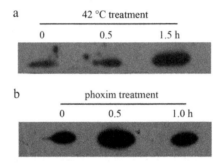

Fig. 8 Western blotting analysis of AccAK changes after 42℃ (a) and phoxim (b) treatments. The total protein was immunoblotted with anti-AccAK. The signal of the binding reactions was visualized with HRP substrates.

AccAK enzyme activities analyses under adverse stress

AccAK enzyme activities were examined under various stresses. After 42℃ treatments for 0.5, 1.0, and 2.5 h, the AccAK enzyme activities increased several times compared with the untreated group and reached the peak at 1.0 h (Fig. 9a). After phoxim treatment for 0.5, 1.0, and 1.5 h, the enzyme activity increased constantly though only slightly (Fig. 9b). Following H_2O_2 treatment for 10, 20, and 30 min and $CdCl_2$ (1 mg/ml) treatment for 2, 4, and 6 h respectively, the enzyme activity drastically increased (Fig. 9c, d). The enzyme activities all increased after the four adverse stresses, though not strictly consistent with the results at the mRNA level which might result from the accumulation and degradation of protein.

Disc fusion assay under various stresses

The target protein contained the AccAK protein and the His-tag. SDS-PAGE showed that the target protein could be induced by IPTG, and only the amount of the target protein increased constantly as time increased (Fig. 10). The protein was detected soluble (data not shown). It took longer time for the bacteria containing AccAK to produce the same number of cells as the bacteria without AccAK. After overnight exposure to various stressors, including

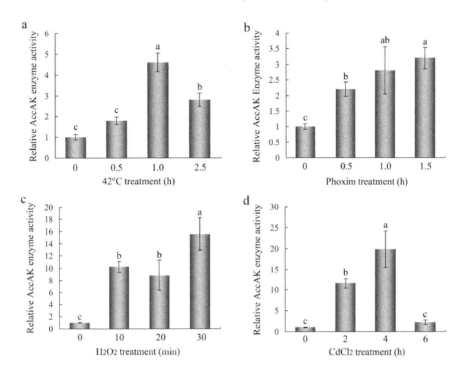

Fig. 9 Relative enzyme activities of AccAK under various stresses. These stresses are 42℃ (a), phoxim (b), H_2O_2 (c), and $CdCl_2$ (d). The data are the mean ± SE of three replicates. The letters above the bar indicate significant differences ($P<0.001$) as determined by Duncan's multiple range tests using SAS software version 9.1.

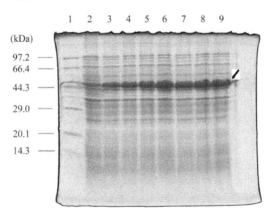

Fig. 10 The expression of the recombinant AccAK protein. Lane 1, protein molecular weight marker; lane 2, expression of AccAK without IPTG induction; lanes 3-9, expression of AccAK after IPTG induction for 2, 3, 4, 5, 6, 7, and 8 h.

$HgCl_2$, paraquat, t-butyl hydroperoxide, and cumene hydroperoxide, the bacteria overexprssing AccAK were more susceptible to the adverse stress, as the killing zones of the bacteria were bigger on the plates compared with the control bacteria. The sensitivity of the bacteria overexpressing AccAK varied from the different stressors, as the killing zones around $HgCl_2$, paraquat, and t-butyl hydroperoxide-soaked filters were much bigger than the control group compared with those around cumene hydroperoxide-soaked filters (Fig. 11). To explore the mechanism of the inhibiting effect of the AccAK to the growth of bacteria, the putative antimicrobial motifs were predicted on the website: http://aps.unmc.edu/AP/, and the result showed that there is at least one antimicrobial motif existing in the AccAK (Fig. 12). The motif has neutral charge, and may have antimicrobial activity.

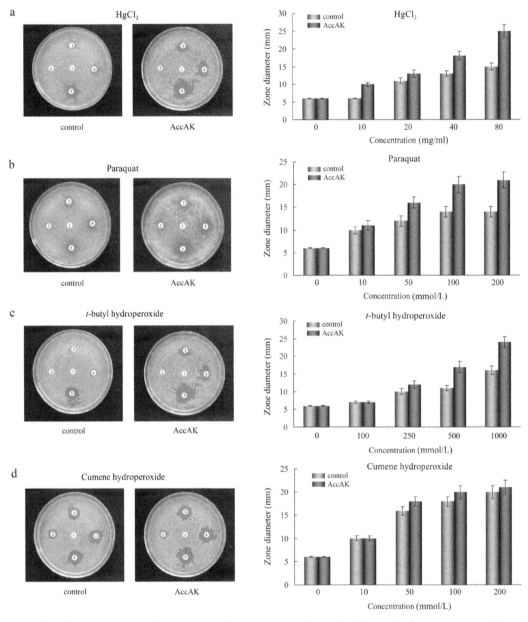

Fig. 11 The resistance of bacteria cells overexpressing AccAK to $HgCl_2$, paraquat, t-butyl hydroperoxide, and cumene hydroperoxide. The halo diameters of the killing zones were compared in the histograms. a.3 The $HgCl_2$ concentrations of discs 1-5 are 0, 10, 20, 40, and 80 mg/ml respectively. b. The paraquat concentrations of discs 1-5 are 0, 10, 50, 100, and 200 mmol/L, respectively. c. The t-butyl hydroperoxide concentrations are 0, 100, 250, 500, and 1,000 mmol/L, respectively. d. The cumene hydroperoxide concentrations are 0, 10, 50, 100, and 200 mmol/L, respectively. The data are the mean ± SE of three replicates.

MVDQAVLDKLETGFTKLANSDSKSLLKKYLTKEVFDQLKTKKTSFDSTL
LDCIQSGVENLDSGVGIYAPDAEAYTLFADLFDPIIEDYHGGFKKTDEHP
PKDFGDVDSLGNLDPANE\[FIVSTRVRCGRSLEGYPFNPCLTEAQYKEME
EK\]VSSTLSGLEDELKGTFYPLTGMSKEIQQKLIDDHFLFKEGDRFLQAAN
ACRFWPTGRGIYHNDDKTFLVWCNEEDHLRIISMQMGGDLGQVYRRLV
HAVNEIERRLPFSHNDRLGFLTFCPTNLGTTVRASVHIKLPKLAANRAKL
EEIAGKFNLQVRGTRGEHTEAEGGIYDISNKRRLGLTEYQAVKEMHDGI
AELIKLEKEL

Fig. 12 The putative antimicrobial peptide motif. The amino acids sequence is boxed.

Discussion

Previous studies demonstrated that AK plays a major role in energy metabolism and contributes to growth and development in invertebrates. An increasing number of studies have been conducted on the roles of AK in response to adverse stress. However, no studies had systematically imitated the various types of stresses that animals may experience during their life.

In this study, we imitated several stresses that *A. cerana cerana*, the main species used in Chinese farming, may encounter during their lifespan, and we explored the resulting changes in AK expression at the mRNA and protein level as well as the AK enzyme activity. Also, we explored the resistance of the bacteria overexpressing AccAK to determine the role that AK may play. The honeybees used in this study were obtained from artificial beehives, and they were raised for collecting honey. So, they were always subjected to wild environment.

To achieve this goal, we first isolated a predicted AK gene from *A. cerana cerana* and named it *AccAK*. Similar to other typical AKs found in the NCBI database, AccAK contained conserved AK residues. Through sequence alignment, it was shown that AccAK showed a high degree of similarity with several typical AK amino acid sequences. Furthermore, its phylogenetic tree indicated that AK sequences have been conserved throughout evolution, and AccAK is in the typical AK cluster, sharing the highest degree of homology with the AK of *A. mellifera*. Taken together, these results indicate that the gene we isolated is a typical *AK*.

A promoter determines the timing and level of gene expression, both of which depends on the role the gene plays. To predict the putative roles of AccAK, a 1,207 bp 5′-flanking region was cloned. The region contained a large number of putative binding sites for transcription factors involved in development and response to environmental stress. This result drove us to study the putative roles that AccAK potentially played. To achieve this goal, we analyzed *AccAK* expression at different growth stages and under various abiotic and biotic stresses. Stage-specific expression analysis indicated that the highest accumulation of *AccAK* was detected in the fourth-instar larvae. The fourth-instar larvae are at the stage when protein synthesis is the highest [43], and these larvae are the most vulnerable to *A. apis* infection and the easiest to die of chalkbrood disease [44]. Although lower than in the fourth-instar larvae, *AccAK* expression is still very high in adults, as the adult workers' main duty is to work outside where they consume more energy and are exposed to worse environmental stress. The prepupal phase also showed a high level of *AccAK* expression. In *Drosophila*, the prepupal phase is the stage at which AK mostly accumulated and when the most ecdysone was secreted [45]. Ecdysone plays a crucial role in coordinating moulting, metamorphosis, and reproduction in insects [46]. It is reasonable to infer that AKs play an ecdysone-related role in the growth and developmental progress of insects [47]. Similarly, at the egg and pupal stages, when the bees stay in the hives and have little exposure to adverse environments, *AK* expression was the lowest. The stage-specific expressions of AccAK provided clues that AccAK might play roles when the honeybees were exposed to adverse stress.

In addition to stage-specific expression analysis, the expression patterns of *AccAK* were also characterized after various treatments imitating adverse environments. When the honeybees were treated with a low concentration of $CdCl_2$ for a short time, *AccAK* expression was either lower or only slightly higher than that of the control group. Increasing the concentration or extending the treatment time resulted in increased expression. A high concentration of $CdCl_2$ for a short time could induce an obvious increase of AccAK enzyme activity. A lot of heavy metals including $CdCl_2$ [48] can induce the formation of endogenous reactive oxygen species (ROS), which include superoxide anions, hydrogen peroxide, and hydroxyl radicals. Endogenously generated ROS should be kept in balance, as high ROS concentration can damage DNA, protein, and lipids [49]. H_2O_2 and vitamin C, a typical oxidant and antioxidant, respectively, both cause oxidative damage. H_2O_2-induced ROS elevation and resulted in cell death caused by oxidative stress [50, 51]. It was reported that vitamin C, a well-known antioxidant, could induce decomposition of lipid hydroperoxides to endogenous genotoxins and resulted in DNA oxidative damage [52]. *AccAK* expression was upregulated after these two treatments, and the enzyme activity also obviously increased after H_2O_2 treatment. It is noticeable that the *AccAK* expressions and enzyme activities all increased drastically after $CdCl_2$ and H_2O_2 treatments compared with other treatments. These findings indicated that AccAK might play a prior role in the response to oxidative damage. Bad weather is a common adverse situation experienced by *A. cerana cerana*. There is a large number of putative HSF binding sites in the *AccAK* 5'-flanking region that participate in heat-induced transcriptional activation. Previous studies showed that heat and cold could both upregulate *AK* expression in *Helicoverpa assulta* [37]. Similarly, our results demonstrated that *AccAK* expression increased under both high and low temperatures. Western blotting and enzyme activity assay of AccAK after 42°C treatment both confirmed the result, suggesting that AccAK might play a role in response to extreme temperature. Pesticides and herbicides are the main threat to a honeybee's life. The two treatments are thought to influence the development process of the insect. Although different patterns of AK expression were observed, both pesticide and herbicide treatments resulted in an increase in the expression of *AccAK*. In addition, phoxim treatment resulted in increase in both protein expression and enzyme activity of AccAK, which may play roles in the developmental processes of *A. cerana cerana*. All the results indicated that the AccAK would be induced and activated after abiotic stress which indicated that AccAK quite possibly plays a significant role in the response of *A. cerana cerana* to adverse environmental situations. Quite a lot of genes that involved in response to adverse stress have been well studied in *A. cerana cerana*. The suite of stressors would result in changes of these genes expression in different patterns. For example, high temperature (42°C) treatment inhibited the expression of *AccTrx2* [29]. $HgCl_2$ treatment inhibited the expression of *Acctpx-3* [53]. Pyriproxyfen treatment inhibited the expression of

AccSOD2 [54]. Though some of these stressors also cause increases of gene expressions in *A. cerana cerana*, the levels and patterns are quite different from that of *AccAK*. These findings showed that the suit of stressors would not result in induction of all other genes in *A. cerana cerana*, and the *AccAK* mRNA is specifically upregulated during stress.

In addition to the abiotic treatments, we also imitated biotic situations. In *B. mori*, the expression of *AK* is consistent with the accumulation of ecdysone [55]. As mentioned above, *AccAK* was highly expressed during the prepupal phase when ecdysone levels were also the highest. BRCZ is an ecdysone-regulated transcription factor and can regulate gene expression at a low ecdysone concentration [56]. In the *AccAK* 5'-flanking region, there are plenty of potential binding sites for the BRCZ, which suggests that AccAK might be regulated by ecdysone. The qPCR results confirmed this prediction. It has been shown that AKs participate in immune responses. After injection of *Vibrio alginolyticus*, *AK* expression in *Portunus trituberculatus* peaked after 3 h with a 5.01-fold increase [57]. By using the mRNA differential display technique, the *AK* level in *Penaeus stylirostris* was found to be upregulated following white spot virus injection [58]. Similarly, treatment with *A. apis*, which is a main fungal parasite of *A. cerana cerana*, resulted in an increase in *AccAK* expression on the second and the third days. This result confirmed the role AK played in immunization.

To perform the disc diffusion assay, we cultured the bacteria with AccAK and the control bacteria to achieve the same cell density. It is worth noting that it took longer time for the bacteria with AccAK to reach the same OD_{600} which provides a clue that AccAK may be an inhibitor for the growth of bacteria. After exposure to four adverse treatments, overexpression of AccAK reduced the resistance of the bacteria cells to the adverse stress. Then the amino acid sequence was input on website for prediction of antimicrobial activity, and there is a putative anti-microbial peptide in the AccAK protein. The result suggested that AccAK possibly acts as an inhibitor of the bacterial growth and protects honeybees from bacterial damage.

In conclusion, we identified the *AK* gene in *A. cerana cerana*. Then, to predict its role, we cloned the 5'-flanking region and predicted putative binding sites for transcription factors. Next, the AK expression patterns during different developmental stages and under different adverse stresses were examined at the transcriptional level. Then the expression patterns at the protein level and the enzyme activities were examined to confirm the results. At last, a disc fusion assay indicated the inhibiting effect of AccAK to the growth of bacteria. Taken together, our research provides evidence that AccAK is potentially involved in response to adverse stress. However, these findings just raised the possibility that AccAK plays a role in response to adverse stress. To further understand the characteristics of AccAK, more studies that directly address the functional significance of the gene and its products should be carried out.

References

[1] Uda, K., Fujimoto, N., Akiyama, Y., Mizuta, K., Tanaka, K., Ellington, W. R., Suzuki, T. Evolution of the arginine kinase gene family. Comp Biochem Physiol D: Genomics Proteomics. 2006, 1 (2), 209-218.

[2] Ellington, W. R. Phosphocreatine represents a thermodynamic and functional improvement over other muscle phosphagens. J Exp Biol. 1989, 143 (1), 177-194.

[3] Kammermeier, H., Seymour, A. M. L. Meaning of energetic parameters. Basic Res Cardiol. 1993, 88

(5), 380-384.

[4] Sauer, U., Schlattner, U. Inverse metabolic engineering with phosphagen kinase systems improves the cellular energy state. Metab Eng. 2004, 6 (3), 220-228.

[5] Brown, A. E., France, R. M., Grossman, S. H. Purification and characterization of arginine kinase from the American cockroach (*Periplaneta americana*). Arch Insect Biochem Physiol. 2004, 56 (2), 51-60.

[6] Sookrung, N., Chaicumpa, W., Tungtrongchitr, A., Vichyanond, P., Bunnag, C., Ramasoota, P., Tongtawe, P., Sakolvaree, Y., Tapchaisri, P. *Periplaneta americana* arginine kinase as a major cockroach allergen among Thai patients with major cockroach allergies. Environ Health Perspect. 2006, 114 (6), 87.

[7] Kucharski, R., Maleszka, R. Arginine kinase is highly expressed in the compound eye of the honey-bee, *Apis mellifera*. Gene. 1998, 211 (2), 343-349.

[8] Liu, Z., Xia, L., Wu, Y., Xia, Q., Chen, J., Roux, K. H. Identification and characterization of an arginine kinase as a major allergen from silkworm (*Bombyx mori*) larvae. Int Arch Allergy Immunol. 2009, 150 (1), 8-14.

[9] Li, M., Wang, X. Y., Bai, J. G. Purification and characterization of arginine kinase from locust. Protein Pept Lett. 2006, 13 (4), 405-410.

[10] Wu, Q. Y., Li, F., Zhu, W. J., Wang, X. Y. Cloning, expression, purification, and characterization of arginine kinase from *Locusta migratoria manilensis*. Comp Biochem Physiol B: Biochem Mol Biol. 2007, 148 (4), 355-362.

[11] Carlson, C. W., Fink, S. C., Brosemer, R. W. Crystallization of glycerol 3-phosphate dehydrogenase, triosephosphate dehydrogenase, arginine kinase, and cytochrome c from a single extract of honeybees. Arch Biochem Biophys. 1971, 144 (1), 107-114.

[12] Schneider, A., Wiesner, R. J., Grieshaber, M. K. On the role of arginine kinase in insect flight muscle. Insect Biochem. 1989, 19 (5), 471-480.

[13] Wang, H., Zhang, L., Zhang, L., Lin, Q., Liu, N. Arginine kinase: differentiation of gene expression and protein activity in the red imported fire ant, *Solenopsis invicta*. Gene. 2009, 430 (1), 38-43.

[14] Kinsey, S. T., Lee, B. C. The effects of rapid salinity change on *in vivo* arginine kinase flux in the juvenile blue crab, *Callinectes sapidus*. Comp Biochem Physiol B: Biochem Mol Biol. 2003, 135 (3), 521-531.

[15] Holt, S. M., Kinsey, S. T. Osmotic effects on arginine kinase function in living muscle of the blue crab *Callinectes sapidus*. J Exp Biol. 2002, 205 (12), 1775-1785.

[16] Shekhar, M. S., Kiruthika, J., Ponniah, A. G. Identification and expression analysis of differentially expressed genes from shrimp (*Penaeus monodon*) in response to low salinity stress. Fish Shellfish Immunol. 2013, 35 (6), 1957-1968.

[17] Miranda, M. R., Canepa, G. E., Bouvier, L. A., Pereira, C. A. *Trypanosoma cruzi*: oxidative stress induces arginine kinase expression. Exp Parasitol. 2006, 114 (4), 341-344.

[18] Zhou, Q., Wu, C., Dong, B., Li, F., Liu, F., Xiang, J. Proteomic analysis of acute responses to copper sulfate stress in larvae of the brine shrimp, *Artemia sinica*. Chin J Oceanol Limnol. 2010, 28, 224-232.

[19] Silvestre, F., Dierick, J. F., Dumont, V., Dieu, M., Raes, M., Devos, P. Differential protein expression profiles in anterior gills of *Eriocheir sinensis* during acclimation to cadmium. Aquat Toxicol. 2006, 76 (1), 46-58.

[20] Yao, C. L., Wu, C. G., Xiang, J. H., Dong, B. Molecular cloning and response to laminarin stimulation of arginine kinase in haemolymph in Chinese shrimp, *Fenneropenaeus chinensis*. Fish Shellfish Immunol. 2005, 19 (4), 317-329.

[21] Johansson, M. W., Keyser, P., Sritunyalucksana, K., Söderhäll, K. Crustacean haemocytes and haematopoiesis. Aquaculture. 2000, 191 (1), 45-52.

[22] Awad, E. M., Osman, O. A. Laminarin enhanced immunological disorders of septicimeric albino rats infected with *Aeromonas hydrophila*. Egypt J Immunol Egypt Assoc Immunol. 2002, 10 (2), 49-56.

[23] Vargas-Albores, F., Yepiz-Plascencia, G. Beta glucan binding protein and its role in shrimp immune response. Aquaculture. 2000, 191 (1), 13-21.

[24] Leung, P. S., Chu, K. H. Current molecular immunological perspectives on seafood allergies. Recent Res Dev Allergy Clin Immunol. 2001, 2, 183-195.

[25] Yu, C. J., Lin, Y. F., Chiang, B. L., Chow, L. P. Proteomics and immunological analysis of a novel shrimp allergen, Pen m 2. J Immunol. 2003, 170 (1), 445-453.

[26] Yang, G. Harm of introducing the western honeybee *Apis mellifera* L. to the Chinese honeybee *Apis cerana* F. and its ecological impact. Acta Entomol Sin. 2005, 48 (3), 401.

[27] Michelette, E. D. F., Soares, A. E. E. Characterization of preimaginal developmental stages in Africanized honey bee workers, *Apis mellifera*. Apidologie. 1993, 24 (4), 431-440.

[28] Yao, P. B., Chen, X. B., Yan, Y., Liu, F., Zhang, Y. Y., Guo, X. Q., Xu, B. H. Glutaredoxin 1, glutaredoxin 2, thioredoxin 1 and thioredoxin peroxidase 3 play an important role in antioxidant defense in *Apis cerana cerana*. Free Radic Biol Med. 2014, 68, 335-346.

[29] Yao, P. B., Hao, L. L., Wang, F., Chen, X. B., Yan, Y., Guo, X. Q., Xu, B. H. Molecular cloning, expression and antioxidant characterisation of a typical thioredoxin gene (*AccTrx2*) in *Apis cerana cerana*. Gene. 2013, 527 (1), 33-41.

[30] Livak, K. J., Schmittgen, T. D. Analysis of relative gene expression data using real-time quantitative PCR and the $2^{-\Delta\Delta C_t}$ method. Methods. 2001, 25 (4), 402-408.

[31] Yan, H., Jia, H., Wang, X., Gao, H., Guo, X., Xu, B. Identification and characterization of an *Apis cerana cerana* delta class glutathione S-transferase gene (AccGSTD) in response to thermal stress. Naturwissenschaften. 2013, 100 (2), 153-163.

[32] Li, X., Zhang, X., Zhang, J., Zhang, X., Starkey, S. R., Zhu, K. Y. Identification and characterization of eleven glutathione S-transferase genes from the aquatic midge *Chironomus tentans* (Diptera: Chironomidae). Insect Biochem Mol Biol. 2009, 39 (10), 745-754.

[33] Meng, F., Zhang, L., Kang, M., Guo, X., Xu, B. Molecular characterization, immunohistochemical localization and expression of a ribosomal protein L17 gene from *Apis cerana cerana*. Arch Insect Biochem Physiol. 2010, 75 (2), 121-138.

[34] Zhang, Y., Yan, H., Lu, W., Li, Y., Guo, X., Xu, B. A novel Omega-class glutathione S-transferase gene in *Apis cerana cerana*: molecular characterisation of GSTO2 and its protective effects in oxidative stress. Cell Stress Chaperones. 2013, 18 (4), 503-516.

[35] Fujimoto, N., Tanaka, K., Suzuki, T. Amino acid residues 62 and 193 play the key role in regulating the synergism of substrate binding in oyster arginine kinase. FEBS Lett. 2005, 579 (7), 1688-1692.

[36] Takeuchi, M., Mizuta, C., Uda, K., Fujimoto, N., Okamoto, M., Suzuki, T. Unique evolution of Bivalvia arginine kinases. Cell Mol Life Sci Cmls 2004, 61 (1), 110-117.

[37] Zhang, Y. C., An, S. H., Li, W. Z., Guo, X. R., Luo, M. H., Yuan, G. H. Cloning and mRNA expression analysis of arginine kinase gene from *Helicoverpa assulta* (Guenée) (Lepidoptera: Noctuidae). Acta Entomol Sin. 2011, 54 (7), 754-761.

[38] Fernandes, M., Xiao, H., Lis, J. T. Fine structure analyses of the *Drosophila* and *Saccharomyces* heat shock factor-heat shock element interactions. Nucleic Acids Res. 1994, 22 (2), 167-173.

[39] Ericsson, A., Kotarsky, K., Svensson, M., Sigvardsson, M., Agace, W. Functional characterization of the CCL25 promoter in small intestinal epithelial cells suggests a regulatory role for caudal-related homeobox (Cdx) transcription factors. J Immunol. 2006, 176 (6), 3642-3651.

[40] Hsu, T., Gogos, J. A., Kirsh, S. A., Kafatos, F. C. Multiple zinc finger forms resulting from developmentally regulated alternative splicing of a transcription factor gene. Science. 1992, 257 (5078), 1946-1950.

[41] Jeffrey, P. L., Capes-Davis, A., Dunn, J. M., Tolhurst, O., Seeto, G., Hannan, A. J., Lin, S. L. CROC-4: a novel brain specific transcriptional activator of c-fos expressed from proliferation through to maturation of multiple neuronal cell types. Mol Cell Neurosci. 2000, 16 (3), 185-196.

[42] Von Kalm, L., Crossgrove, K., Von Seggern, D., Guild, G. M., Beckendorf, S. K. The Broad-Complex directly controls a tissue-specific response to the steroid hormone ecdysone at the onset of *Drosophila* metamorphosis. EMBO J. 1994, 13 (15), 3505.

[43] Li, J., Li, H., Zhang, Z., Pan, Y. Identification of the proteome complement of high royal jelly

producing bees (*Apis mellifera*) during worker larval development. Apidologie. 2007, 38 (6), 545-557.

[44] Bailey, L. The effect of temperature on the pathogenicity of the fungus *Ascosphaera apis* for larvae of the honeybee, *Apis mellifera*. Insect Pathology and Microbial Control. North Holland Publishing Co., Amsterdam, 2002, 162-167.

[45] Jiang, C., Lamblin, A. F. J., Steller, H., Thummel, C. S. A steroid-triggered transcriptional hierarchy controls salivary gland cell death during *Drosophila* metamorphosis. Mol Cell. 2000, 5 (3), 445-455.

[46] Liu, Y., Xu, P., Li, Y., Shu, H., Huang, D. Progress in ecdysone receptor (EcR) and insecticidal mechanisms of ecdysteroids. Acta Entomol Sin. 2007, 50 (1), 67.

[47] James, J. M., Collier, G. E. Early gene interaction during prepupal expression of *Drosophila* arginine kinase. Dev Genet. 1992, 13 (4), 302-305.

[48] Liu, J., Qu, W., Kadiiska, M. B. Role of oxidative stress in cadmium toxicity and carcinogenesis. Toxicol Appl Pharmacol. 2009, 238 (3), 209-214.

[49] Narendra, M., Bhatracharyulu, N. C., Padmavathi, P., Varadacharyulu, N. C. Prallethrin induced biochemical changes in erythrocyte membrane and red cell osmotic haemolysis in human volunteers. Chemosphere. 2007, 67 (6), 1065-1071.

[50] Goldshmit, Y., Erlich, S., Pinkas-Kramarski, R. Neuregulin rescues PC12-ErbB4 cells from cell death induced by H_2O_2 regulation of reactive oxygen species levels by phosphatidylinositol 3-kinase. J Biol Chem. 2001, 276 (49), 46379-46385.

[51] Casini, A. F., Ferrali, M., Pompella, A., Maellaro, E., Comporti, M. Lipid peroxidation and cellular damage in extrahepatic tissues of bromobenzene-intoxicated mice. Am J Pathol. 1986, 123 (3), 520.

[52] Lee, S. H., Oe, T., Blair, I. A. Vitamin C-induced decomposition of lipid hydroperoxides to endogenous genotoxins. Science. 2001, 292 (5524), 2083-2086.

[53] Yao, P., Lu, W., Meng, F., Wang, X., Xu, B., Guo, X. Molecular cloning, expression and oxidative stress response of a mitochondrial thioredoxin peroxidase gene (*AccTpx-3*) from *Apis cerana cerana*. J Insect Physiol. 2013, 59 (3), 273-282.

[54] Jia, H., Sun, R., Shi, W., Yan, Y., Li, H., Guo, X., Xu, B. Characterization of a mitochondrial manganese superoxide dismutase gene from *Apis cerana cerana* and its role in oxidative stress. J Insect Physiol. 2014, 60, 68-79.

[55] Wang, H. B., Xu, Y. S. cDNA cloning, genomic structure and expression of arginine kinase gene from *Bombyx mori* (L.). Sci Agric Sin. 2006, 11, 27.

[56] Karim, F. D., Guild, G. M., Thummel, C. S. The *Drosophila* Broad-Complex plays a key role in controlling ecdysone-regulated gene expression at the onset of metamorphosis. Development. 1993, 118 (3), 977-988.

[57] Song, C., Cui, Z., Liu, Y., Li, Q., Wang, S. Cloning and expression of arginine kinase from a swimming crab, *Portunus trituberculatus*. Mol Biol Rep. 2012, 39 (4), 4879-4888.

[58] Astrofsky, K. M., Roux, M. M., Klimpel, K. R., Fox, J. G., Dhar, A. K. Isolation of differentially expressed genes from white spot virus (WSV) infected Pacific blue shrimp (*Penaeus stylirostris*). Arch Virol. 2002, 147 (9), 1799-1812.

1.8 Molecular Cloning and Characterization of Two Nicotinic Acetylcholine Receptor β Subunit Genes from *Apis cerana cerana*

Xiaoli Yu, Mian Wang, Mingjiang Kang, Li Liu, Xingqi Guo, Baohua Xu

Abstract: Nicotinic acetylcholine receptors (nAChRs) mediate fast cholinergic synaptic transmission in the insect nervous system and are important targets for insecticides. In this study, we identified and characterized two novel β subunit genes (*Accβ1* and *Accβ2*) from *Apis cerana cerana*. Homology analysis indicated that *Accβ1* and *Accβ2* possess characteristics that are typical of nAChR subunits although *Accβ2* was distinct from *Accβ1* and the other nAChR subunits, due to its unusual transmembrane structure and uncommon exon-intron boundary within the genomic region encoding the TM1 transmembrane domain. Analysis of the 5′-flanking regions indicated that *Accβ1* and *Accβ2* possess different regulatory elements, suggesting that the genes might exhibit various expression and regulatory patterns. RT-PCR analysis demonstrated that *Accβ2* was expressed at a much higher level than *Accβ1* in the tissues of adult bees. During development, *Accβ1* was highly expressed at the pupal stages, whereas *Accβ2* was abundantly expressed at the larval stages. Furthermore, *Accβ1* and *Accβ2* were both induced by exposure to various insecticides and environmental stresses although *Accβ2* was more responsive than *Accβ1*. These results indicate that *Accβ1* and *Accβ2* may have distinct roles in insect growth and development and that they may belong to separate regulatory pathways involved in the response to insecticides and environmental stresses. This report is the first description of the differences between the nAChR β subunit genes in the Chinese honeybee and establishes an initial foundation for further study.

Keywords: Nicotinic acetylcholine receptor β subunit; *Apis cerana cerana*; Promoter; Semi-quantitative RT-PCR; Insecticide

Introduction

Nicotinic acetylcholine receptors (nAChRs) are important excitatory neurotransmitter receptors in the central nervous system of insects. They belong to the Cys-loop ligand-gated ion channel (LGIC) superfamily, which includes the receptors for γ-aminobutyric acid, glycine, glutamate and 5-hydroxytryptamine [1]. The nAChRs are composed of five homologous subunits, typically two α and three non-α subunits, arranged around a central, cation-permeable pore [2, 3]. In insects, these subunits are separated into α or non-α (β) subunits. The α subunits contain two adjacent cysteine residues that are important for acetylcholine (ACh) binding, whereas the non-α subunits lack these two residues [4]. Each subunit includes an extracellular domain containing the ligand-binding site, a transmembrane domain comprising four helices (TM1-4) and an intracellular domain for ion channel modulation and cellular localization [5]. In addition, alternative splicing and RNA A-to-I editing [6, 7] generate a diverse population of nAChR subtypes that influence the various pharmacological functions of nAChRs in insects [8]. The fast actions of ACh at the synapses are mediated by nAChRs, and insecticides such as neonicotinoids act as agonists to their molecular targets, the nAChRs [9].

Currently, only limited information exists on the β subunits of the insect nAChR. The β subunits have been reported to contribute to the formation of nAChRs *in vivo* [10]. [3H] epibatidine studies in mice have indicated that the β2 subunit is required for the upregulation of nAChRs *in vivo* [11]. Functional ion channel activity or ligand-binding activity is clearly observed when any of the four α subunits (Dα1-Dα4) is co-expressed to form a hybrid receptor with chick β2, rat β2 or rat β4 subunits [12]. In addition, the co-expression of certain α subunits from *Drosophila melanogaster* with β subunits from *Xenopus laevis* oocytes results in functional nAChRs [13]. Immunoprecipitation studies have demonstrated that Dα1 and Dα2 form hetero-oligomers with Dβ2 *in vivo*, while Dα3 is in the same receptor complex as Dβ1 [14, 15]. The β2 subunit often combines with the α4 subunit to form α4β2 pentameric channels, which play an important role in the central nervous system [16]. Previous studies have reported that non-α subunits are essential for the pharmacological properties of the insect nAChRs [17]. Moreover, each insect exhibits at least one highly divergent subunit that may perform species-specific functions [18]. Notably, one of the divergent subunits in the honeybee is the β2 subunit [9].

To date, several complete insect nAChR subunit genes have been identified. The model genetic organism *D. melanogaster* has 10 nAChR subunits [10], as does the malaria mosquito, *Anopheles gambiae* [13]. The honeybee, *Apis mellifera*, has 11 genes encoding nAChR subunits [9], and *Bombyx mori*, a member of the order Lepidoptera, has 12 subunits [19]. The insect nAChRs have long been recognized as potential targets for insecticides, but the widespread use of insecticides can also affect useful, non-target insects, such as the honeybee. The Chinese honeybee, *Apis cerana cerana*, is an important beneficial insect in agriculture that is constantly exposed to environmental pollution, including insecticides [20]. This species exists widely in China, but limited information on nAChRs in the Chinese honeybee is available. Thus, *A. cerana cerana* was selected as the subject of this study to obtain data on the changes in the regulatory patterns of the β subunit genes induced by insecticides and other environmental stresses.

In the present study, we isolated the *Accβ1* and *Accβ2* genes from *A. cerana cerana* and provide, for the first time, a comparative analysis of the expression patterns of *Accβ1* and *Accβ2* under various conditions. The study of these two β genes in Chinese honeybee will provide a useful reference for further studies of the function of the β subunits with respect to the effects of insecticides and environmental stresses on nAChRs and may help in the development of improved insecticides to protect the Chinese honeybee from injury.

Materials and Methods

Insects and treatments

The Chinese honeybees, *A. cerana cerana*, were maintained at Shandong Agricultural University, China. The colonies were fed in incubators at a constant temperature (33℃) and humidity (80%). The entire body of the second (L2) larval instars, the fourth (L4) larval instars, prepupal (PP1 and PP2) phase pupae, earlier pupal (Pw) phase pupae, brown (Pb) phase pupae and dark-pigmented (Pbd) phase pupae were obtained from the hive, and the adult workers (2 d and 12 d after emergence) were collected at the entrance of the hive [21]. The adults (12 d) were divided into 11 groups (n=10-15/group). Groups 1-6 were treated with imidacloprid, acetamiprid, lambda-cyhalothrin, phoxim, pyriproxyfen and cyhalothrin, respectively. All insecticides were purchased from Huayang Technology Co., Ltd. (Shandong, China) and were dissolved in a pollen-and-sucrose solution that was employed as food for the bees. The final effective concentrations were determined according to the instructions of the manufacturer (see Supplementary Table 1). For the stress treatments, groups 7 and 8, which consisted of adult bees, were subjected to ultraviolet (UV) light (30 MJ/cm^2) and heat (42℃) [22], respectively, and groups 9 and 10 were fed with pollen-and-sucrose solutions containing $HgCl_2$ (1.0 μg/ml) [23] and nystatin (25 μg/ml), respectively. Control bees from group 11 were fed only the pollen-and-sucrose solution. Three bees were harvested at the appropriate times (see Supplementary Table 2) under each condition and stored at −80℃ until used; each experiment was performed in triplicate.

RNA extraction, cDNA synthesis and DNA isolation

Total RNA was extracted by using the Trizol reagent (Invitrogen, Carlsbad, CA) according to the instructions of the manufacturer and kept at −80℃ until used. To remove potentially contaminating genomic DNA, total RNA was digested with RNase-free Dnase I (Promega, Madison, WI). Single-stranded cDNA was then synthesised with a reverse transcriptase system (TransGen Biotech, Beijing, China) with an adaptor primer oligo d(T)$_{18}$ at 42℃ for 50 min. Genomic DNA was isolated by using an EasyPure Genomic DNA Extraction Kit according to the instructions of the manufacturer (TransGen Biotech).

Isolation of cDNA and genomic sequences of *Accβ1* and *Accβ2*

Single-stranded cDNA was synthesized by using a reverse transcriptase system (TransGen Biotech). PCR was performed to amplify the interval fragments containing *Accβ1* and *Accβ2*. The forward primers F1 and P1 were designed based on the conserved amino acid sequences of the TM1 region in other insects, including *B. mori*, *A. mellifera* and *A. gambiae*. Similarly, the reverse primers F2 and P2 were designed based on the conserved amino acid sequences in the TM4 region. To obtain the full-length cDNAs of *Accβ1* and *Accβ2*, 5′ and 3′ rapid amplification of cDNA end (5' and 3' RACE) were performed. All PCR products were cloned into the pEASY-T3 vector (TransGen Biotech) transformed into *Escherichia coli* strain DH5α and then sequenced by Jinsite Science and Technology (Nanjing) Co., Ltd., China. The primers and PCR amplification conditions are outlined in Supplementary Tables 3 and 4.

To obtain the genomic sequences of *Accβ1* and *Accβ2*, two pairs of primers (M1/M2 and N1/N2, respectively) were designed and synthesized based on the cDNA sequences of *Accβ1* and *Accβ2*. By using *A. cerana cerana* genomic DNA as a template, PCR was employed to amplify the genomic sequences of the *Accβ1* and *Accβ2* genes. The PCR products were then cloned into the pMD18-T vector (TaKaRa, Kyoto, Japan), transformed into *E. coli* strain DH5α and sequenced by Jinsite Science and Technology (Nanjing) Co., Ltd. The PCR amplification conditions are summarized in Supplementary Tables 3 and 4.

Sequence and general bioinformatic analysis

The nucleotide and deduced amino-acid sequences of Accβ1 and Accβ2 were analyzed and compared by using the BLAST search program (http://blast.ncbi.nlm.nih.gov/Blast.cgi). Open reading frames (ORFs) were predicted by using DNAMAN version 5.2.2 software (Lynnon Biosoft Company, Quebec, Canada). The transmembrane segments and topology of Accβ1 and Accβ2 were predicted by TMHMM 2.0 (http://www.cbs.dtu.dk/services/TMHMM-2.0/). Multiple protein sequence alignments were performed by using the Clustal X program. The phylogenetic tree and molecular evolutionary analyses were performed by using MEGA version 4.1 software. Transcription factor-binding sites in the 5'-flanking regions were predicted by using the MatInspector database (http://www.cbrc.jp/research/db/TFSEARCH.html).

Amplification of the 5'-flanking regions of *Accβ1* and *Accβ2*

To obtain the 5'-flanking region of *Accβ1*, an inverse polymerase chain reaction (I-PCR) was performed by using the restriction endonucleases *Eco*R I and *Bcl* I and two primer pairs (PYS1/PYX1 and PYS2/PYX2) that were designed based on the genomic sequence of *Accβ1* (Supplementary Table 3). The nested PCR conditions are listed in Supplementary Table 4. The PCR products were cloned into the pEASY-T3 vector and sequenced by Jinsite Science and Technology (Nanjing) Co., Ltd. A similar method was used to clone the 5'-flanking region of *Accβ2* by using the restriction endonuclease *Cla* I and the primer sets PZS1/PZX1 and PZS2/PZX2.

Expression analysis of *Accβ1* and *Accβ2* by semi-quantitative RT-PCR

To further characterize *Accβ1* and *Accβ2*, two specific primer pairs were employed to amplify an mRNA fragment from each. *A. cerana cerana β-actin* gene (GenBank accession number: XM640276) which is a house-keeping gene was selected for normalization of RNA levels. The primers and reaction conditions are provided in Supplementary Tables 3 and 4, and each reaction was performed in triplicate. Quantitative data obtained from the stained PCR bands were normalized to *β-actin* by using the Quantity-One™ image analysis software (Bio-Rad, Hercules, CA). The ratios of *Accβ1* and *Accβ2* to *β-actin* were then calculated.

Results

Isolation and sequence analysis of *Accβ1* and *Accβ2*

The cDNA sequence of *Accβ1* (GenBank accession number: HM049634) obtained from *A. cerana cerana* was 2,039 bp in length, containing a 154 bp 5' untranslated region (UTR) and a 325 bp 3' UTR. The 1,560 bp ORF encoded a protein of 519 amino acids with a predicted molecular weight of 59.64 kDa and an estimated pI of 8.24. The cDNA sequence of *Accβ2* (GenBank accession number: GU722330) was 1,810 bp in length and contained a 116 bp 5' UTR and a 419 bp 3' UTR. The 1,275 bp ORF encoded a protein of 424 amino acids with a predicted molecular weight of 49.17 kDa and an estimated pI of 5.67.

Protein sequence alignment and the evolutionary relationship of nAChR subunits

To better understand the sequence similarities among Accβ1, Accβ2 and other nAChR subunits, a multiple-sequence alignment was performed by using the Clustal X program (Fig. 1). Accβ1 and Accβ2 both exhibited characteristics that are common to members of the Cys-loop LGIC superfamily, including an N-terminal signal peptide sequence, four transmembrane hydrophobic, conserved segments (TM1-TM4), a highly variable intracellular region between TM3 and TM4 and an extracellular N-terminal region with six distinct regions (loops A-F) that were involved in ligand binding. Notably, the prediction of membrane topology by using TMHMM 2.0 identified a special transmembrane structure spanning the amino acids 1-50 in Accβ2 that was not present in Accβ1 (Fig. 2). Similar to other nAChR subunits, Accβ1 and Accβ2 possessed numerous phosphorylation sites for the cAMP-dependent protein and protein kinase C and/or putative N-glycosylation sites within the extracellular N-terminal region.

To determine the phylogenetic position of Accβ1 and Accβ2, an evolutionary tree of various insect nAChR subunits was generated (Fig. 3). Accβ1 appeared to be orthologous to Amelβ1 and Accβ2 to Amelβ2. Surprisingly, Accβ2 shared little homology with Accβ1 or any of the other insect β2 subunits and was, instead, more closely related to Amelα9.

Genomic structure analysis of *Accβ1* and *Accβ2*

To further elucidate the differences between the Accβ1 and Accβ2 subunits, the genomic sequences of these two genes were determined. The genomic sequence of *Accβ1* (GenBank accession number: HM049635) spans 2,725 bp and includes five exons and four introns; *Accβ2* (GenBank accession number: GU722331) spans 2,633 bp and includes seven exons and six introns. In addition, all of the introns of *Accβ1* and *Accβ2* exhibited typical intron features such as being AT-rich and flanked by the 5' splice donor GT and the 3' splice acceptor AG sequences. The genomic structures of *Accβ1* and *Accβ2* were then compared by using

BLAST (Fig. 4). Notably, *Accβ2* exhibited an uncommon exon-intron boundary within TM1, but not TM2-TM4, whereas *Accβ1* did not exhibit this boundary.

```
                                                                              LpD
Accβ1   .......MHFCSRLARILLIS...AVFCVGLCSEDEE.RLVRDLFRGYNKLIRPVQNNTEKVHVNFGIAFVQLINVNEKNQIMKSNVWLR    79
Accβ2   MLNMKNIFPVLFLIINVSIHGQVICFICKEVTSTSALHRLKQYLFCDYDRDIIPEEK..NNDKIDFGLSIQHYN.VDEYSHTVDFHVLLK    87
Amelβ1  .......MHNICSRLGRILLIS...AVFCVGLCSEDEE.RLVRDLFRGYNKLIRPVQNNTEKVHVNFGIAFVQLINVNEKNQIMKSNVWLR    80
Amelβ2  MLNMKNIFPVLFVIINVLIHGQVICFVCKDITSTSALYRLKLYLFCDYDRDIIPEQK..NNIKIDFGLSIQHYN.VDEYSHTVDFHVMLK    87
Bmβ1    MYPHPAVPAAMSCVSRAFIATTLLAILYSGWCSEDEE.RLVRDLFRGYNKLIRPVQNNTQKVDVRFGLAFVQLINVNEKNQIMKSNVWLR    89
Bmβ2    ...MSLLFGLLLLLVRGSAQD...CVIDHRIPEDAWEQKLHTDLIN..TGSLEPLCNNTEPLDVYIRFTLRYFEYLSEES.TFNIYTRVH    81
Amelα1  .........MATAISCLVAP.....FPGASANSEAK.RLYDDLLSNYNRLIRPVGNNSDRLTVKMGIRLSQLIDVNLKNQIMTTNVWVE    74
                            LpA                 LpE
                                                       *                   *
Accβ1   FIWTDYQLQWDEADYGGIGVLRLPP.DKVWKPDIVLFNNADGNYEVRY.KSNVLIYPNGDVLWVPPAIYQSSCTIDVTYFPFDQQTCIMK   167
Accβ2   LMWKQNHLTWKSSEFDSINSIRVKS.YEIWPDIVMHSVTSVGIDFEMPPVECIVFNSGTVLCVPFTTYTPVCEFIHTWWPYGIINCHIH   176
Amelβ1  FIWTDYQLQWDEADYGGIGVLRLPP.DKVWKPDIVLFNNADGNYEVRY.KSNVLIYPNGDVLWVPPAIYQSSCTIDVTYFPFDQQTCIMK   168
Amelβ2  LMWEQSHLTWKSSEFDSINSIRVKS.YEIWPDIVMHSVTSVGIDLEMPSVECIVFNSGTILCVPFTTYTPVCEFIHTWWPYDIINCHIH   258
Bmβ1    LVWMDYQIMWDEADYGGIGVLRLPP.DKVWKPDIVLFNNADGNYEVRY.KSNVLIYPNGEVLWVPPAIYQSSCTIDVTYFPFDQQTCIMK   177
Bmβ2    ISWTHDRLTWDPKDYGGIEETVLVSGTEMWYQTFRVLNSPETDDIVHHYNVPCQVAHTGRVVCVPRFNDPAICVPDLTDWPYDRQTCSLL   171
Amelα1  QEWNDYKLKWNPDDYGGVDTLHVPS.EHIWLPDIVLYNNADGNYEVTI.MTKAILHHTGKVVWKPPAIYKSFCEIDVEYFPFDEQTCFMK   162
        LpB                 LpF                   LpC                               TM1
Accβ1   FGSWTFNGDQVSLALYNNKN.......FVDLSDYWKSGTWDIINVPAYIN...TYKGDFPTETDITFYIIRRKTLFYTVNLILPTVLIS   247
Accβ2   IASWSHGSNEIKLNSLDTEQ........ILDDMYNNNNTEWEIVHMSHSERTIDSKFGSGFTSDLLSYNILLRRHYSMNSNTYVTLTIVLM   258
Amelβ1  FGSWTFNGDQVSLALYNNKN.......FVDLSDYWKSGTWDIINVPAYIN...TYKGDFPTETDITFYIIRRKTLFYTVNLILPTVLIS   248
Amelβ2  IASWSHGSNEIKLNSLDTEQ........ILDDMYNNNNTEWEIVHMSHSESTIDSKFGLGFTTDLLSYNILLRRHYSMNSNTYVTLTIVLM   258
Bmβ1    FGSWTFNGDQVSLALYNNKN.......FVDLSDYWKSGTWDIIEVPAYIN...IYEGNHPTETDITFYIIRRKTLFYTVNLILPTVLIS   257
Bmβ2    IAPPDHNTGGANKIKLSFGG.....RAITMFGAEYGAEWMIIDYLQTNG..........UTTLNMTEVMERHGEGLAAVVVFPGVMLS   244
Amelα1  FGSWTYDGYTVDLRHLAQTEDSNQIEVGIDLTDYYISVEWDIIKVPAVRNEA.FYICTEEPYPDIVFNTTLRRKTLFYTVNLIIPCVGIS   251
                              TM2                             TM3
Accβ1   FLCVLVFYLPAEAGEKVTLGISILLSLVVFLLLVSKILPPTSLVLPLIAKYLLFTFIMNTVSILVTVIIINWNFRGPRTHRMPQLIRKIF   337
Accβ2   TMTIMTLWLEPSSTERMIIANLNFIIHLFCLLDVQWRIPFNGIQIPNLMVFYEKSLALAAFSLMLTSILR...YLQELHVDAPTWISSVT   345
Amelβ1  FLCVLVFYLPAEAGEKVTLGISILLSLVVFLLLVSKILPPTSLVLPLIAKYLLFTFIMNTVSILVTVIIINWNFRGPRTHRMPQLIRKIF   338
Amelβ2  TMTIMTLWLEPSSTERMIIANLNFIIHLFCLLDVQWRIPFNGIQMPNLMVFYEKSLALAAFSLMLTSILR...YLQELHVDAPTWISSVT   345
Bmβ1    FLCVLVFYLPAEAGEKVTLGISILLSLVVFLLLVSKILPPTSLVLPLIAKYLLFTFIMNTVSILVTVIIINWNFRGPRTHRMPLWIRSVF   347
Bmβ2    ALTLTALALDPRQSTRASILGFSVAAHVYFINQLRAMAPLHQLASVPSLVYYRGSLLATLVALVSSLCLG..ALCRKSSPAPRWLHELQ   332
Amelα1  FLSVLVFYLPSDSGEKVSLSISILLSLTVFFLLIAEIIPPTSLTVPLLGKYLLFTMVLVTLSVVVTIAVLNVNFRSPVTHRMARWVRVVF   341

Accβ1   LKYLPTIIMMRRPKK.........................................TRLRWMMEIPNVTL   366
Accβ2   ESVL.................................................................   349
Amelβ1  LKYLPTIIMMRRPKK.........................................TRLRWMMEIPNVTL   367
Amelβ2  ESVL.................................................................   349
Bmβ1    LHYLPAMLLMRRPRK.........................................TRLRWMMEMPGMGA   376
Bmβ2    AAVG.................................................................   336
Amelα1  IQVLPRFLLIERPKKDEDEEEEEVVVVGRNGAGVGAMNANGEAVGEVDEDDIDGDDDDDDVEAANGKPAEGMLTDVFHVQETDKYDA   431

Accβ1   PTSTYSGSPTELPKHLPSS.LTKSKMEVMELSDIHHPNCKINRKVHHTTSSSS....AAGGEGIAD.RRGSESSDSVLLSPEASKATEAV   450
Accβ2   ................................KSKVGQVFLITILDS................................KVSARI   370
Amelβ1  PTSTYSGSPTELPKHLPTS.LTKSKMEVMELSDIHHPNCKINRKVHHTTSSSS....AAGGEGIAD.RRGSESSDSVLLSPEASKATEAV   451
Amelβ2  ................................KSKIGQVFLITILDS................................KVSARI   370
Bmβ1    PP..HAAAPHDLPKHISAIGAKQSKMEAMELSDIHHPNCKINRAAGGGEVGALGDLGALGGLGLGGERRSESSDSLLLSPEAAKATEAV   464
Bmβ2    ................SHWTRLLIP.............................................................   345
Amelα1  YYGNRFSGEYEIPAHGLPPSATRYDLGAVATVGTTVAPCFEEPLPSLPLPGADDD..LFGPASPAYVHEDVSPTFEKPLVREIEKTIDDA   519
                                                             TM4
Accβ1   EFIAEHLRNEDLYIQTRELWKYVAMVIDRIQLYIFFLVTTAGTIGILMDAPHIFEYVD.........QDHIIEIYRGK..   519
Accβ2   EMNEDDNTSLVSLDRKQYTWRYTSVLIGWSAFLCISFVYII..MLIIFIPRNIYES........................   424
Amelβ1  EFIAEHLRNEDLYIQTRELWKYVAMVIDRIQLYIFFLVTTAGTIGILMDAPHIFEYVD.........QDHIIEIYRGK..   520
Amelβ2  EMNEDDNTSLVSLDKKQYTWRHTSVLIGWSAFLCISFVYII..MLIIFIPRNIYENFIS.....................   427
Bmβ1    EFIAEHLRNEDLYIQTRELWKYVAMVIDRIQLYIFFIVTTAGTVGILMDAPHIFEYVD.........QDRIIEIYRGK..   533
Bmβ2    RWECGETNN...TERDTEEWTNVANIANAACLLIFPVVYVS...MYFSLVPQFYPITN......................   397
Amelα1  RFIAQHARNKDKFESVEEDWKYVAMVLDRIFLWIFTVACVLGTVLIILQAPSLYDTTKPIDIKYSKIAKKKMMLMNMGPEE   600
```

Fig. 1 Multiple alignments of insect β1 and β2 subunits by using the Clustal X program. The amino acid sequences are from *Apis mellifera* (Amelβ1, NP001073028; Amelβ2, NP001091699; Amelα1, NP001091690) and *Bombyx mori* (Bmβ1, ABV72692; Bmβ2, ABV72693). The two cysteines forming the Cys-loop are denoted by asterisks. The vicinal cysteines characteristic of α-subunits are boxed, and the six loops (loop A-F) involved in ACh binding and the transmembrane domains (TM1-4) are indicated by lines. Potential N-glycosylation sites are indicated by black shading, and phosphorylation sites indicated by gray shading.

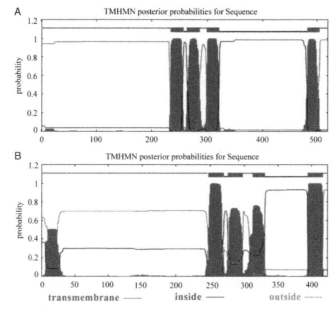

Fig. 2 Predicted membrane topology of Accβ1 and Accβ2. A. The predicted membrane topology of Accβ1. B. The predicted membrane topology of Accβ2. The transmembrane segments and topology were predicted by using TMHMM 2.0 software.

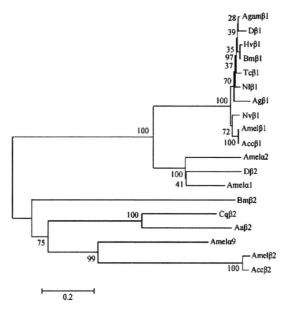

Fig. 3 Phylogenetic relationship of Accβ1 and Accβ2 to other nAChR subunits from different insects as determined by neighbor-joining distance analysis. The numbers at each branch node mark the confidence level of posterior probability. The amino acid sequences are from the following organisms: *Anopheles gambiae* (Agamβ1, AAU12514), *Drosophila melanogaster* (Dβ2, CAA39211; Dβ1, CAA27641), *Heliothis virescens* (Hvβ1, AAD09810), *Bombyx mori* (Bmβ1, ABV72692; Bmβ2, ABV72693), *Apis mellifera* (Amelβ1, NP001073028; Amelβ2, NP001091699; Amelα1, NP001091690; Amelα9, NP001091694; Amelα2, NP001011625), *Tribolium castaneum* (Tcβ1, NP001103418), *Nilaparvata lugens* (Nlβ1, ACJ07013), *Aphis gossypii* (Agβ1, AAM94384), *Culex quinquefasciatus* (Cqβ2, XP001862560), and *Aedes aegypti* (Aaβ2, EAT43659).

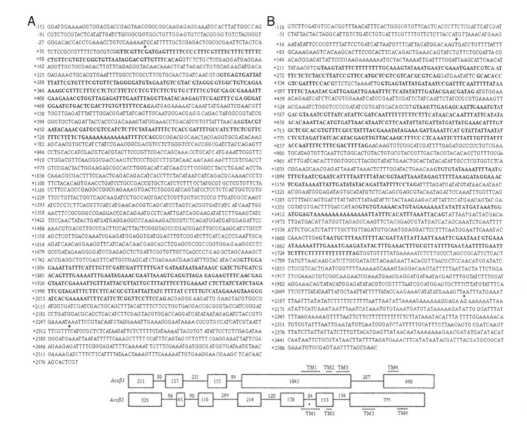

Fig. 4 Schematic representation of genomic structures of *Accβ1* and *Accβ2* genes. A. Genomic DNA sequence of *Accβ1*. B. Genomic DNA sequence of *Accβ2*. The introns are bold. The ATG and TAA are both indicated with *. Boxes and lines represent exons and introns, respectively. The sizes of exons are indicated in the boxes and the sizes of introns are indicated above the lines. The uncommon exon-intron boundary within TM1 of *Accβ2* is designated by asterisks.

Identification of putative transcription factor-binding sites in the 5'-flanking regions of *Accβ1* and *Accβ2*

To understand the mechanism that determines the expression and regulation of these two genes, IPCR was employed to amplify a 1,057 bp DNA fragment upstream of the *Accβ1* translation start site. A 483 bp DNA fragment in the 5'-flanking region of *Accβ2* was obtained by using a similar method. The web-based MatInspector program was employed to predict the putative transcription factor-binding sites in the genome upstream of both genes (Fig. 5). In the 5'-flanking regions of *Accβ1* and *Accβ2*, predicted binding sites for heat shock factors (HSFs) were identified, which played important roles in heat-induced transcriptional activation [24], suggesting that *Accβ1* and *Accβ2* were responsive to heat. Predicted binding sites for the caudal-related homeobox (Cdx) protein, which contributes to tissue-selective expression [25], were found in the 5'-flanking region of *Accβ2*. In addition, predicted binding sites for Nkx-2, which is involved in organogenesis [26], and CF2-II, which is required for embryogenesis [27], were observed in the 5'-flanking region of *Accβ1*.

1 The Abiotic Stress Response of Some Key Genes in *Apis cerana cerana* | 139

Fig. 5 Nucleotide sequence and putative transcription factor-binding sites of the 5'-flanking regions of *Accβ1* and *Accβ2*. A. The 5'-flanking region of *Accβ1*. B. The 5'-flanking region of *Accβ2*. The translation (ATG) and transcription start sites are marked with arrows and boxed and underlined, respectively. The transcription factor-binding sites are boxed.

Expression patterns of *Accβ1* and *Accβ2* in various tissues and developmental stages

To investigate the expression patterns of *Accβ1* and *Accβ2* in various tissues and developmental stages, semi-quantitative RT-PCR was performed. The expression profiles of *Accβ1* and *Accβ2* were different in the tissues examined. Although both *Accβ1* and *Accβ2* were expressed in the same tissues, the expression levels of *Accβ2* were approximately six-fold higher than those of *Accβ1* (Fig. 6A), with the highest levels of *Accβ2* mRNA being observed in the abdomen. During larval development, the levels of *Accβ1* decreased gradually and were initially higher than those of *Accβ2* (L2, L4), whereas the levels of *Accβ2* remained relatively steady until PP2, when they increased dramatically. During the successive pupal stages (Pw, Pb, Pbd), the transcription level of *Accβ1* increased and, by Pb, reached levels higher than those of *Accβ2*, which decreased over time. In adults (2- and 12-day-olds), the transcription level of *Accβ1* decreased, whereas those of *Accβ2* increased somewhat (Fig. 6B). These results suggest that *Accβ1* might play a key role in the pupation of the bee, but *Accβ2* mainly affects larval development.

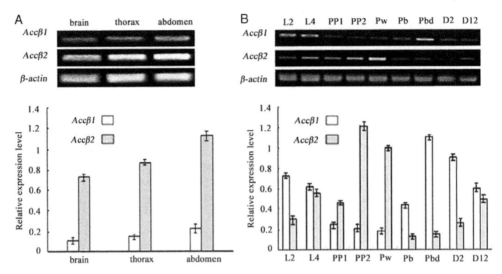

Fig. 6 Expression patterns of *Accβ1* and *Accβ2*. A. Distribution of *Accβ1* and *Accβ2* in brain, thorax and abdomen. B. Expression of *Accβ1* and *Accβ2* at the 2nd (L2) and the 4th (L4) larval instars, prepupal (PP1 and PP2), earlier pupal (Pw), brown (Pb) and dark-pigmented (Pbd) phase pupae, and adults (2 d and 12 d after emergence). The *β-actin* gene was used for normalization. The histograms show the relative expression levels of *Accβ1* and *Accβ2*. The y-axis values are the normalization units relative to *β-actin* levels, and the x-axis values are the different lanes. The signal intensities (INT/mm^2) of the bands were assigned by using the Quantity-One^{TM} image analysis software. Each value is the mean (SD) of three replicates.

Expression profiles of *Accβ1* and *Accβ2* in the presence of insecticides and environmental stresses

Semi-quantitative RT-PCR was used to determine the effects of insecticides on the expression levels of *Accβ1* and *Accβ2* after exposure to imidacloprid, acetamiprid, lambda-cyhalothrin, phoxim, pyriproxyfen and cyhalothrin. Treatment with imidacloprid resulted in a gradual decrease in *Accβ1* and *Accβ2*, with the lowest levels observed at 2 h, followed by an increase to pretreatment expression levels (Fig. 7A). Treatment with acetamiprid significantly upregulated the expression of *Accβ1* and *Accβ2* compared with that of the untreated controls (Fig. 7B). Treatment with lambda-cyhalothrin resulted in the increased transcription of both *Accβ1* and *Accβ2*, which peaked at 2 h post-treatment and declined thereafter (Fig. 7C). Treatment with phoxim caused the levels of both *Accβ1* and *Accβ2* mRNA to increase significantly and then to decrease after reaching peak levels at 2 and 1.5 h post-treatment, respectively (Fig. 7D). Treatment with pyriproxyfen, which did not appear to have an effect on the expression of *Accβ2*, induced the expression of *Accβ1*, which reached a peak at 1 h post-treatment and decreased thereafter (Fig. 7E). Treatment with cyhalothrin resulted in the increased expression of *Accβ1* and *Accβ2*, with peaks for both at 4 h post-treatment (Fig. 7F). However, after various insecticide treatments, the bees all manifested a nervous convulsion and subsequently died. These results suggest that *Accβ1* and *Accβ2* might be differently associated with the regulation of these insecticides.

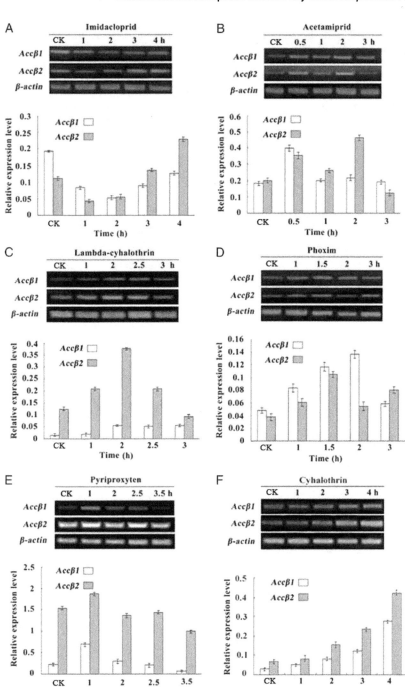

Fig. 7 Expression patterns of Accβ1 and Accβ2. A-F. The expression patterns of Accβ1 and Accβ2 in response to exposure to different insecticides in 12-day-old adult bees. The expression levels were determined at different times after treatment with imidacloprid (A), acetamiprid (B), lambda-cyhalothrin (C), phoxim (D), pyriproxyfen (E) and cyhalothrin (F). CK denotes the control. The β-actin gene was used for normalization. The histograms show the relative expression levels of Accβ1 and Accβ2. The y-axis values are the normalization units relative to β-actin levels, and the x-axis values are the different gel lanes. The signal intensities (INT/mm^2) of the bands were assigned by using the Quantity-One^{TM} image analysis software. Each value is the mean (SD) of three replicates.

To investigate whether *Accβ1* and *Accβ2* are involved in the response to environmental stresses, including UV light, heat, heavy metal ($HgCl_2$) and nystatin, semi-quantitative RT-PCR was performed. Exposure to $HgCl_2$ resulted in increased levels of *Accβ1*, but decreased levels of *Accβ2* 16 h post-treatment (Fig. 8A). Exposure to either UV light or heat (42℃) resulted in the upregulation of *Accβ2* expression and the downregulation of *Accβ1* expression in a time-dependent manner (Fig. 8B, C). Exposure to nystatin induced a slight increase in the expression of *Accβ1* and *Accβ2* (Fig. 8D). However, the toxic effect of environmental stresses resulted in the deaths of bees. Taken together, these results suggest that these two genes might play distinct roles in the responses to various stresses.

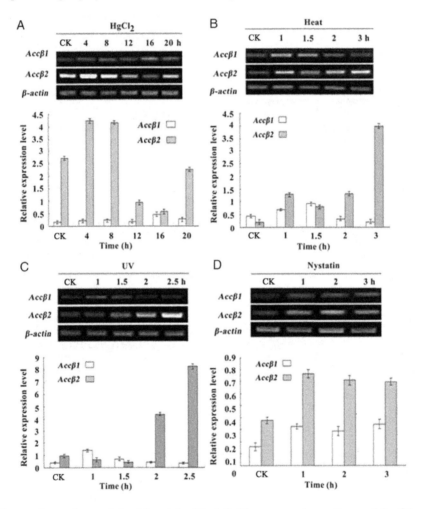

Fig. 8 Expression patterns of *Accβ1* and *Accβ2*. A-D. The expression patterns of *Accβ1* and *Accβ2* in response to environmental stress in 12-day-old adult bees. The expression levels were determined at different times after treatment with $HgCl_2$ (A), heat (B), UV (C) and nystatin (D). CK denotes the control. The *β-actin* gene was used for normalization. The histograms show the relative expression levels of *Accβ1* and *Accβ2*. The y-axis values are the normalization units relative to *β-actin* levels, and the x-axis values are the different gel lanes. The signal intensities (INT/mm^2) of the bands were assigned by using the Quantity-One™ image analysis software. Each value is the mean (SD) of three replicates.

Discussion

In this study, we identified and characterized the *Accβ1* and *Accβ2* genes from the Chinese honeybee, *A. cerana cerana*. A protein sequence alignment indicated that Accβ1 and Accβ2 possess the typical features of the Cys-loop LGIC superfamily, sharing high homology with the nAChR β-subunits of *A. mellifera*. A phylogenetic analysis suggested that Accβ2 and Amelα9 are closely related. In *A. mellifera*, the *Amelβ2* and *Amelα9* subunit genes encode divergent subunits but are located within 10 kb of each other in the honeybee genome and may represent a recent evolutionary duplication event of a single gene [9]. Notably, Accβ2 was predicted to have a transmembrane structure distinct not only from that of Accβ1, but also from other cloned nAChR subunits. In addition, *Accβ2* was demonstrated to have uncommon exon-intron boundaries within TM1 that are not present in *Accβ1*. These results indicate that *Accβ2* and *Accβ1* encode different nAChR subunits that might also exhibit distinct functions and that further study of *Accβ2* may offer promising targets in the future for the rational design of insecticides that target specific pest insects while sparing beneficial species.

The nAChR proteins are widely dispersed in the brain along with acetylcholinesterases [28], and to play roles in learning, memory, olfactory signaling and visual processing. Eleven subunits in *A. mellifera* are transcribed in each developmental stage and tissue tested [9]. In this study, *Accβ1* and *Accβ2* mRNA transcripts were expressed, not only in the brain, but also in the thorax and abdomen, with *Accβ2* generally expressed at higher levels. The binding of transcription factors, such as the Cdx transcription factors, to sites in the 5′-flanking region of the *Accβ2* gene might contribute to the varied expression rates. We hypothesize that the *Accβ1* and *Accβ2* genes might play important roles in the regulation of the growth and development of the tissues. The expression patterns of both genes throughout the stages of development suggest that *Accβ1* might serve to modulate the regulation of pupal development, whereas *Accβ2* might be involved in the regulation of larval development. The analysis of Nkx-2, which is involved in organogenesis [26], and CF2-II, which is required for embryogenesis [27], also suggests that *Accβ1* is involved in development. However, further studies to better understand the various roles of *Accβ1* and *Accβ2* during growth and development are required.

Most commercial insecticides are neurotoxins that act on nAChRs in the insect nervous system [29]. Previous studies have indicated that the neonicotinoid compound interacts with a cationic subsite within insect nAChRs [30], with the interactions of imidacloprid with the nitroimine group and of the bridgehead nitrogen with particular amino-acid residues providing relevant examples [31]. In this study, semi-quantitative RT-PCR analysis demonstrated that *Accβ1* and *Accβ2* could be regulated by imidacloprid and acetamiprid. To obtain more direct evidence, the expression profiles of both genes following treatments with lambda-cyhalothrin, phoxim, pyriproxyfen and cyhalothrin were determined. These four insecticides induce the transcription of *Accβ1* and *Accβ2*, although at different levels, with *Accβ2* being more sensitive in general. These results indicate that both β subunit genes in the honeybee are differently involved in the response to a variety of insecticides. Further study of the different roles played by the *Accβ1* and *Accβ2* genes in the response to insecticides may reveal promising targets in the future for the rational design of insecticides that will not affect the honeybee. In addition, we investigated, for the first time, whether *Accβ1* and *Accβ2* participate in the response to environmental stresses such as UV light, heat, heavy metal and nystatin [32], any of which could significantly influence the bees during their foraging activity. Environmental stresses have been demonstrated to mediate a wide variety of toxic effects, such as DNA damage and genotoxicity, and to contribute to pathological conditions [33]. In this study, it was observed that *Accβ1* and *Accβ2* were differently affected by several environmental stresses and that, in most cases, *Accβ2* was expressed at higher levels. These results suggest that these two genes might play important, but distinct roles in stress-response regulation. In addition, binding sites for putative transcription factors, such as the HSFs, were identified in the 5'-flanking regions of both *Accβ1* and *Accβ2*. However, many other insecticide-responsive elements, particularly transcription-factor binding sites, may exist in the honeybee genome. Our future research will involve transgenic studies of the *Accβ1* and *Accβ2* promoters to better understand the functions of both genes in response to insecticides and environmental stresses.

References

[1] Millar, N. S., Harkness, P. C. Assembly and trafficking of nicotinic acetylcholine receptors (Review). Mol Membr Biol. 2008, 25, 279-292.

[2] Unwin, N. Refined structure of the nicotinic acetylcholine receptor at 4A resolution. J Mol Biol. 2005, 346, 967-989.

[3] Sine, S. M., Engel, A. G. Recent advances in Cys-loop receptor structure and function. Nature. 2006, 440, 448-455.

[4] Rinkevich, F. D., Scott, J. G. Transcriptional diversity and allelic variation in nicotinic acetylcholine receptor subunits of the red flour beetle, *Tribolium castaneum*. Insect Mol Biol. 2009, 18, 233-242.

[5] Tsetlin, V., Hucho, F. Nicotinic acetylcholine receptors at atomic resolution. Curr Opin Pharmacol. 2009, 9, 306-310.

[6] Stetefeld, J., Ruegg, M. A. Structure and functional diversity generated by alternative mRNA splicing. Trends Biochem Sci. 2005, 30, 515-521.

[7] Jin, Y., Tian, N., Cao, J., Liang, J., Yang, Z., Lv, J. RNA editing and alternative splicing of the insect nAChR subunit alpha6 transcript: evolutionary conservation, divergence and regulation. BMC Evol Biol. 2007, 7, 98.

[8] Millar, N. S. Assembly and subunit diversity of nicotinic acetylcholine receptors. Biochem Soc Trans. 2003, 31, 869-874.

[9] Jones, A. K., Raymond-Delpech, V., Thany, S. H., Gauthier, M, Sattelle, D. B. The nicotinic acetylcholine receptor gene family of the honey bee, *Apis mellifera*. Genome Res. 2006, 16, 1422-1430.

[10] Sattelle, D. B., Jones, A. K., Sattelle, B. M., Matsuda, K., Reenan, R. Edit, cut and paste in the nicotinic acetylcholine receptor gene family of *Drosophila melanogaster*. BioEssays. 2005, 27, 366-376.

[11] McCallum, S. E., Collins, A. C., Paylor, R., Marks, M. J. Deletion of the beta 2 nicotinic acetylcholine receptor subunit alters development of tolerance to nicotine and eliminates receptor upregulation. Psychopharmacology. 2006, 184, 314-327.

[12] Tomizawa, M., Casida, J. E. Structure and diversity of insect nicotinic acetylcholine receptors. Pest Manag Sci. 2001, 57, 914-922.

[13] Jones, A. K., Grauso, G., Sattelle, D. B. The nicotinic acetylcholine receptor gene family of the malaria mosquito, *Anopheles gambiae*. Genomics. 2005, 85, 176-187.

[14] Chamaon, K., Schulz, R., Smalla, K. H., Seidel, B., Gundelfinger, E. D. Neuronal nicotinic acetylcholine receptors of *Drosophila melanogaster*: the alpha-subunit dalpha3 and the beta-type subunit ARD co-assemble within the same receptor complex. FEBS Lett. 2000, 482, 189-192.

[15] Chamaon, K., Smalla, K. H., Thomas, U., Gundelfinger, E. D. Nicotinic acetylcholine receptors of

Drosophila: three subunits encoded by genomically linked genes can co-assemble into the same receptor complex. J Neurochem. 2002, 80, 149-157.

[16] Bondarenko, V., Tillman, T., Xu, Y., Tang, P. NMR structure of the transmembrane domain of the n-acetylcholine receptor β2 subunit. Biochimica Biophysica Acta. 2010, 1798, 1608-1614.

[17] Tomizawa, M., Millar, N. S., Casida, J. E. Pharmacological profiles of recombinant and native insect nicotinic acetylcholine receptors. Insect Biochem Mol Biol. 2005, 35, 1347-1355.

[18] Jones, A. K., Brown, L. A., Sattelle, D. B. Insect nicotinic acetylcholine receptor gene families: from genetic model organism to vector, pest and beneficial species. Invert Neurosci. 2007, 7, 67-73.

[19] Shao, Y. M., Dong, K., Zhang, C. X. The nicotinic acetylcholine receptor gene family of the silkworm, Bombyx mori. BMC Genomics. 2007, 8, 324.

[20] Rabea, E. I., Nasr, H. M., Badawy, M. E. I. Toxic effect and biochemical study of chlorfluazuron, oxymatrine, and spinosad on honey bees (Apis mellifera). Arch Environ Contam Toxicol. 2010, 58, 722-732.

[21] Bitondi, M. M. G., Nascimento, A. M., Cunha, A. D., Guidugli, K. R., Nunes, F. M. F., Simões, Z. L. P. Characterization and expression of the hex 110 gene encoding a glutamine-rich hexamerin in the honey bee, Apis mellifera. Arch Insect Biochem. 2006, 63, 57-72.

[22] Meng, F., Zhang, L., Kang, M. J., Guo, X. Q., Xu, B. H. Molecular characterization, immunohistochemical localization and expression of a ribosomal protein L17 gene from Apis cerana cerana. Arch Insect Biochem Physiol. 2010, 75, 121-138.

[23] Wang, M., Kang, M. J., Guo, X. Q., Xu, B. H. Identification and characterization of two phospholipid hydroperoxide glutathione peroxidase genes from Apis cerana cerana. Comp Biochem Physiol C: Toxicol Pharmacol. 2010, 152, 75-83.

[24] Fernandes, M., Xiao, H., Lis, J. T. Fine structure analyses of the Drosophila and Saccharomyces heat shock factor-heat shock element interactions. Nucleic Acids Res. 1994, 22, 167-173.

[25] Ericsson, A., Kotarsky, K., Svensson, M., Sigvardsson, M., Agace, W. Functional characterization of the CCL25 promoter in small intestinal epithelial cells suggests a regulatory role for Caudal-Related Homeobox (Cdx) transcription factors. J Immunol. 2006, 176, 3642-3651.

[26] Durocher, D., Charron, F., Warren, R., Schwartz, R. J., Nemer, M. The cardiac transcription factors Nkx2-5 and GATA-4 are mutual cofactors. EMBO J. 1997, 16, 5687-5696.

[27] Hsu, T., Gogos, J. A., Kirsh, S. A., Kafatos, F. C. Multiple zinc finger forms resulting from developmentally regulated alternative splicing of a transcription factor gene. Science. 1992, 257, 1946-1950.

[28] Thany, S. H., Crozatier, M., Raymond-Delpech, V., Gauthier, M., Lenaers, G. Apisa2, Apisa7-1 and Apisa7-2: three new neuronal nicotinic acetylcholine receptor a-subunits in the honeybee brain. Gene. 2005, 344, 125-132.

[29] Millar, N. S., Denholm, I. Nicotinic acetylcholine receptors: targets for commercially important insecticides. Invert Neurosci. 2007, 7, 53-66.

[30] Tomizawa, M., Zhang, N., Durkin, K. A., Olmstead, M. M., Casida, J. E. The neonicotinoid electronegative pharmacophore plays the crucial role in the high affinity and selectivity for the Drosophila nicotinic receptor: an anomaly for the nicotinoid cation-π interaction model. Biochem. 2003, 42, 7819-7827.

[31] Matsuda, K., Buckingham, S. D., Kleier, D., Rauh, J. J., Grauso, M., Sattelle, D. B. Neonicotinoids: insecticides acting on insect nicotinic acetylcholine receptor. Trends Pharmacol Sci. 2001, 22, 573-580.

[32] Hac-Wydro, K., Kapusta, J., Jagoda, A., Wydro, P., Dynarowicz-Latka, P. The influence of phospholipid structure on the interactions with nystatin, a polyene antifungal antibiotic: A Langmuir monolayer study. Chem Phys Lipids. 2007, 150, 125-135.

[33] Franco, R., Sánchez-Olea, R., Reyes-Reyes, E. M., Panayiotidis, M. I. Environmental toxicity, oxidative stress and apoptosis: Ménage à Trois. Mutat Res. 2009, 674, 3-22.

1.9 Molecular Cloning and Characterization under Different Stress Conditions of Insulin-like Peptide 2 (*AccILP-2*) Gene from *Apis cerana cerana*

Xuepeng Chi, Wei Wei, Weixing Zhang, Hongfang Wang, Zhenguo Liu, Baohua Xu

Abstract: Insulin-like peptide 2 is an important gene in the insulin/insulin-like signaling pathway. The insulin-like peptide 2 gene (*ILP-2*) appears to play a key role in metabolism, growth, reproduction, stress resistance and aging. In this study, we isolated and characterized an *ILP-2* gene from *Apis cerana cerana* known as *AccILP-2*. The full-length cDNA of *AccILP-2* is 494 bp and contains a 222 bp open reading frame (ORF) encoding a protein of 73 amino acids with a calculated molecular weight of 8.706 kDa and an isoelectric point of 9.02. Multiple sequence alignment revealed that *AccILP-2* shares high identity with *ILP* genes from *Apis mellifera*, *Apis florea*, *Bombus terrestris*, *Eufriesea mexicana* and *Dufourea novaeangliae* and contains a typical IlGF-insulin-bombyxin-like domain. Quantitative PCR analysis indicated that the greatest expression of *AccILP-2* was observed in the muscles of adult workers. Expression analysis results also indicated that the expression of *AccILP-2* was the greatest during the dark-eye developmental period. Furthermore, expression of *AccILP-2* transcript was induced by environmental stressors, including exposure to 4℃, ultraviolet light, H_2O_2, heavy metals ($HgCl_2$ and $CdCl_2$) and pesticides (dichlorvos, paraquat and cyhalothrin). However, the expression of *AccILP-2* was downregulated at 44℃. In conclusion, these results suggested that *AccILP-2* might play an important role in the response to abiotic stress. We expect that our study will promote future research on the function of the insulin/insulin-like signaling pathway.

Keywords: *Apis cerana cerana*; Insulin-like peptides; Gene cloning; Gene expression

Introduction

Insulin, a protein hormone produced by pancreatic islet β cells, is induced by endogenous or exogenous substances [1]. In previous studies, some scholars postulated that invertebrates may possess this hormone. Until 1984, bombyxin was isolated from the head of *Bombyx mori* by scientists. Structure analysis showed that bombyxin and insulin had the same chemical precursors. The insulin-like peptides (ILPs) of invertebrates are central life-history regulators and are functionally homologous to insulin and insulin-like growth factor (IGF) ligands in mammals [2]. The *ILP* genes are also important components of the insulin/insulin-like signaling (IIS) pathway. The IIS pathway, along with the parallel acting target-of-rapamycin (TOR) pathway, is a fine example of such evolutionary conservation [3-5]. Previous studies have shown that the IIS pathway is an evolutionarily conserved module that controls body size and is correlated with organ growth in metazoans [6]. Other studies have indicated that the IIS pathway has an evolutionarily conserved role in nutrient regulation and it plays an important role in determining growth rate and size, metabolic traits, and various forms of stress resistance [7].

The application of molecular techniques and genomics in recent decades allowed Nagasawa et al. to separate and purify the neuropeptide hormone bombyxin, which shares a common chemical precursor with insulin, from peptides derived from *Bombyx* heads [8]. After 1984, homologous insulin genes were successfully extracted from *Samia cynthia*, A*grius convoluvuli*, *Lymnaea stagnalis*, *Locusta migratoria* and *Caenorhabditis elegans* [1]. However, although insulin-like peptides are known to play a role in stress resistance, this activity has only been demonstrated in *Drosophila* [9]. The roles of insulin-like peptides in the stress response are poorly understood in other insects.

In 2006, whole genome sequencing of *A. mellifera* was successfully completed, and the results revealed two insulin-like peptides genes, *AmILP1* and *AmILP2*, and two insulin receptor genes, *AmInR1* and *AmInR2* [10]. Other studies showed that during critical stages of caste development in queen and worker insect larvae, *AmILP-2* is the predominantly transcribed ILP in both castes and is more highly expressed in the workers than in the queens. Expression analysis of *AmInR-1* and *AmInR-2* in the ovaries of queen and worker larvae indicates relative independence in tissue-specific versus overall IIS pathway activity [6]. Corona et al. demonstrated that old queens had lower head expression of insulin-like peptide and its putative receptors than old workers and that queen longevity was related to decreased IIS expression [11]. Ihle et al. found that *AmILP1* plays a role in nutritional homeostasis [12].

Honeybees inevitably experience environmental stresses, such as that resulting from cold, heat, ultraviolet (UV) light, heavy metal and pesticide exposure. The honeybee is an excellent model organism for studying the effects of the stress response on a particular gene [13]. While studies of *ILP-2* in some species have mainly focused on the modulation of different pathways, little is known about the role of *ILP-2* in *A. cerana cerana*. Whether the functions of *AccILP-2* gene during stress resistance in honeybees is unknown. Identifying the functions of the *ILP-2* gene would enable us to better understand the stress resistance mechanisms of *ILP-2* in *A. cerana cerana*. To this end, we isolated *AccILP-2* from *A. cerana cerana*. *AccILP-2* mRNA was detected at different developmental stages and in various tissues, suggesting that *AccILP-2* plays an important role during all developmental stages of honeybee. Furthermore, accumulation of *AccILP-2* mRNA was induced by exposure to different temperatures, $CdCl_2$, $HgCl_2$, UV light, H_2O_2, paraquat, cyhalothrin, and dichlorvos. *AccILP-2* regulation was immediately apparent after mild stress. Based on these results, we speculated that *AccILP-2* might play an important role in alleviating stress.

Materials and Methods

Biological specimens and various treatments

The Chinese honeybees (*A. cerana cerana*) used in this study were maintained in an experimental apiary at Shandong Agricultural University (Taian, China). We chose 6 healthy colonies for our experiment. The bees were fed in incubators maintained at 33℃ and 60% relative humidity with a photoperiod of 24 h dark [13]. Honeybees at various life stages were classified as larvae, pupae, or adults according to age [14] and pupal eye color [15]. The honeybees were further divided into the following eleven stages: the first (L1), the second (L2), the third (L3)-, the fourth (L4)-, and the fifth (L5)-instar larvae; prepupae (Pp) and white-eyed (Pw), pink-eyed (Pp), and dark-eyed (Pd) pupae; and 1-day adult workers (A1) and 15-day adult workers (A15). We recorded egg positions on the comb and the time of oviposition. Thereafter, samples from each stage were collected from the hive. The 15-day worker (A1 workers) bees were collected from the hive entrance by marking the 1-day worker (A1 workers) bees with paint 15 days earlier. All the 15-day workers from different colonies were randomly divided into several groups of 40 each (n=40), fed a diet of water and 50% sucrose solutions and maintained in an incubator at a constant humidity (70%) [16]. Bees incubated at 33℃ served as a control. Groups 1 and 2 were kept at 4℃ and 44℃, respectively, for 4 h. For stress treatments, groups 3, 4, 5 and 6 were treated with ultraviolet (UV) oxidative stress (280-315 nm, 30 MJ/cm^2), H_2O_2 (1 µl 30% H_2O_2 was injected between their first and second abdominal segments by using a sterile microscale needle), $CdCl_2$ (1 mg/ml added to the 50% sucrose feed solutions) and $HgCl_2$ (0.5 mg/ml added to the 50% sucrose feed solutions), respectively. For pesticide exposure, groups 7, 8 and 9 were treated with paraquat (0.1 µl added to 10 ml of the 50% sucrose feed solutions), cyhalothrin (1 µl added to 10 ml of the 50% sucrose feed solutions) and dichlorvos (1 µl added to 1 L of the 50% sucrose feed solutions), respectively. None of the treatments resulted in bee death during the sampling period. However, when greater concentrations of pesticides or chemical additives were investigated, the worker bees died during the sampling period. The control group remained untreated and was only fed the 50% sucrose solutions. The samples were collected at the appropriate time after treatment. For the UV and pesticide exposure groups, we collected samples every hour for 6 h. For H_2O_2 treatment, we collected samples at 0.25, 0.5, 1, 1.5, 2 and 3 h after treatment. For the two heavy metal treatments, we collected samples at 1.5, 3, 4.5, 6, 12 and 24 h after treatment. For the different temperature treatments, we collected the bees at 0.25, 0.5, 1, 2, 3 and 4 h during processing. At each sampling time point, 5 workers were collected, and one worker was used for RNA extraction at each sampling time point. The thorax muscle, thorax and abdomen epidermis, venom gland, midgut, head and rectum from 15-day ten workers were dissected on ice to detect tissue-specific gene expression. All of the samples were immediately frozen in liquid nitrogen and stored at −80℃ until used. At least three biological replicates were performed for each treatment.

Total RNA extraction, cDNA synthesis, and genomic DNA preparation

The RNA extraction was performed by using the Total RNA Kit II (OMEGA, Norcross,

GA, USA), and RNA concentration was determined after extraction. After dilution, the first-strand cDNA was synthesized with 1 μg of total RNA by using the Transcriptor First Strand cDNA Synthesis Kit (Roche, Mannheim, Germany) at 25 ℃ for 10 min, 55 ℃ for 30 min and 85 ℃ for 5 min. The genomic DNA preparation was extracted by using a QIAamp DNA Investigator Kit in accordance with the instructions of the manufacturer (QIAGEN GmbH, Hilden, Nordrhein-Westfalen, Germany).

PCR primers and conditions

All primers used in this study are listed in Table 1, and the amplification conditions are listed in Table 2.

Table 1 The primers used in this study

Abbreviation	Primer sequence (5'—3')	Description
IP1	ATGCACCAATATTATGGTCGAACTTTG	cDNA sequence primer, forward
IP2	TCCCTCATTAACGGGCACCG	cDNA sequence primer, reverse
5P1	GGATATAGATCGTAACCATAAAATGCC	5'RACE reverse primer, outer
5P2	TCATCCATCTCCATTTCTTGG	5' RACE reverse primer, inner
3P1	AACTCTAAAGGAATTCATGAAG	3' RACE forward primer, outer
3P2	TTGCACAACAGAGGAGTT	3' RACE forward primer, inner
AAP	GGCCACGCGTCGACTAGTACGGGGGGGGGGGGGG	Abridged anchor primer
AUAP	GGCCACGCGTCGACTAGTAC	Abridged universal amplification primer
B26	GACTCTAGACGACATCGATTTTTTTTTTTTTTTTT	3' RACE universal adaptor primer
B25	GACTCTAGACGACATCGA	3' RACE universal primer
INP1	GGATATAGATCGTAACCATAAAATGCC	Full-length cDNA primer, forward
INP2	TTGCACAACAGAGGAGTT	Full-length cDNA primer, reverse
β-actin-s	TTATATGCCAACACCTGTCCTTT	Standard control primer, forward
β-actin-x	AGAATTGATCCCACCAATCCA	Standard control primer, reverse
RP1	TAGGAGCGCAACTCCTCTGT	Real-time PCR primer, forward
RP2	CCAAGAAATGGAGATGGATGA	Real-time PCR primer, reverse

Table 2 PCR amplification conditions in this study

Primer pair	Amplification conditions
IP1, IP2	10 min at 94 ℃, 40 s at 94 ℃, 40 s at 51.3 ℃, 30 s at 72 ℃ for 35 cycles, and 10 min at 72 ℃
3P1/B26	10 min at 94 ℃, 40 s at 94 ℃, 40 s at 43 ℃, 30 s at 72 ℃ for 35 cycles, and 10 min at 72 ℃
3P2/B25	10 min at 94 ℃, 40 s at 94 ℃, 40 s at 47.3 ℃, 30 s at 72 ℃ for 35 cycles, and 10 min at 72 ℃
5P1AAP	10 min at 94 ℃, 40 s at 94 ℃, 40 s at 49.3 ℃, 30 s at 72 ℃ for 35 cycles, and 10 min at 72 ℃
5P2AUAP	10 min at 94 ℃, 40 s at 94 ℃, 40 s at 49.6 ℃, 30 s at 72 ℃ for 35 cycles, and 10 min at 72 ℃
INP1/ INP2	10 min at 94 ℃, 40 s at 94 ℃, 40 s at 43 ℃, 30 s at 72 ℃ for 35 cycles, and 10 min at 72 ℃
RP1/ RP2	10 min at 94 ℃, 40 s at 94 ℃, 40 s at 49.6 ℃, 30 s at 72 ℃ for 35 cycles, and 10 min at 72 ℃
β-actin-s/β-actin-x	10 min at 94 ℃, 40 s at 94 ℃, 40 s at 49.6 ℃, 30 s at 72 ℃ for 35 cycles, and 10 min at 72 ℃

Isolation of the *AccILP-2* gene

Whole genome sequencing of *Apis cerana cerana* has been completed [17]. The internal conserved fragment was cloned by RT-PCR by using the primers IP1 and IP2, which were designed based on the conserved DNA sequence from *A. mellifera*. By using the internal conserved fragment, two set of primers, 5P1 and 5P2 and 3P1 and 3P2, were designed for 5′ and 3′ RACE (rapid amplification of cDNA end), respectively. Following RACE-PCR, the 5′ and 3′ end sequences and the internal fragment were assembled, and the full-length cDNA sequence was derived. Based on the cDNA sequence of *A. mellifera* and the internal conserved fragment, the specific primers INP1 and INP2 were used to amplify the full-length cDNA sequence.

Bioinformatics analysis

The nucleotide sequence of *AccILP-2* and amino acid sequences of AccILP-2 were analyzed by using NCBI bioinformatics tools (http://blast.ncbi.nlm.nih.gov/Blast.cgi). The physical and chemical properties of *AccILP-2* were predicted by using the ProtParam tool (http://www.expasy.ch/tools/protparam.html) and DNAMAN version 5.22 (Lynnon Biosoft, Quebec, Canada). Multiple alignment analysis, the theoretical isoelectric point, and molecular weight prediction were also conducted by using DNAMAN software. To explore the phylogenetic relationships among species, a phylogenetic tree was constructed by using the neighbor-joining method and the Molecular Evolutionary Genetics Analysis program (MEGA version 4.1). A PSORT II server was used to predict the nuclear localization of ILP-2. Lastly, the tertiary structure was inferred by using the online protein structure prediction tool SWISS-MODEL (http://swissmodel.expasy.org/).

Expression analysis by using fluorescent real-time quantitative PCR

A pair of specific primers was designed and synthesized by Sangon Biotech (Shanghai, China) to amplify a fragment from *AccILP-2*. Fluorescent real-time quantitative PCR (qRT-PCR) was accomplished by using the SYBR Green Master (Rox) (Roche, Indianapolis, USA) and a 7500 Real Time PCR System (Applied Biosystems, ABI, USA) to determine the transcription profile of *AccILP-2*. Expression of the house-keeping gene *β-actin*, which was stably expressed and could be considered as a reference gene, was used to normalize mRNA expression levels [18-22]. qRT-PCR was performed in a 20 µl amplification reaction volume containing 10 µl of SYBR, 7 µl of double distilled water, 2 µl of cDNA, and 0.5 µl of each primer. The qRT-PCR was performed according to the following program: 2 min at 50℃, 10 min at 95℃, 15 s at 95℃ and 1 min at 60℃ for 40 cycles. The relative expression ratio was calculated from the real-time PCR efficiencies and the crossing point deviation of an unknown sample versus a control. Relative *AccILP-2* gene expression was analyzed with a new mathematical model [23]. SPSS software (version 16.0) was used to determine the statistical significance of the gene expression profiles. All experiments included at least three independent biological replicates, and the PCRs were performed in triplicate.

Statistical analysis

Error bars in the figures denote the SE of the mean (SEM) from three independent experiments. Significant differences were determined by Tukey's HSD test using SPSS software (version 16.0, SPSS Inc., Chicago, Illinois). A P value of 0.05 was used to determine statistical significance.

Results

Sequence analysis of the full-length cDNA of *ILP-2*

The full-length cDNA of *ILP-2* was 494 bp and contained a 112-bp 5′ untranslated region (UTR), a 222 bp complete open reading frame (ORF), and a 160 bp 3′ untranslated region (UTR) (Fig. 1). The ORF of *ILP-2* encodes a 73-amino acid polypeptide that contains a 15-amino acid signal peptide and has a predicted molecular mass of 8.706 kDa and a theoretical isoelectric point (pI) of 9.02. In addition, a typical polyadenylation signal sequence (AATAA) was observed in the 3′ UTR.

```
GGCCACGCGTCGACTAGTACGATCGGATGTGTTTCAGTGTGGCCAAAAGGGACAGACA   58
GTGACAGAAATGCACCAATATTGTG GTCGAACTTTGTCAAGTACATTACAAATA ATG TGC  118
                                                         M   C   2
GGTAGTGTTTATAATTCTCGATTTAAAAAAAGTAACCAAGAAATG GAGATGGATGATTAT  178
 G   S   V   Y   N   S   R   F   K   K   S   N   Q   E   M   E   M   D   D   Y   22
ATGGCATTTTATG GTTACGATCTATATCCTTATAAATCTATTAAAAATG CGAAAAAAATG  238
 M   A   F   Y   G   Y   D   L   Y   P   Y   K   S   I   K   N   A   K   K   M   42
ATACGATTTCGAAGAAACTCTAAAGGAATTCATGAAGAATGTTGTCTAAAATCTTG CACA  298
 I   R   F   R   R   C   S   K   G   I   H   E   E   C   C   L   K   S   C   T   62
ACAGAGGAGTTGC GCTCCTATTGC GGTGCCCGT TAA TGAGAGAATTTTTTTAAAATTTTA  358
 T   E   E   L   R   S   Y   C   G   A   R   *                                73
AACATAAATTTAAGAAAATATTATTTTTTC CTATTTAC TG ATTATATAATTTTTAACAAT  418
TTATGATATATTT TC AATAA TTTTATTTTTCATTCA ATGTAACAATTCAGTAAAATCAAA  478
ATTACGAATATTTTGT                                             494
```

Fig. 1 *AccILP-2* nucleotide sequence and deduced amino acid sequence. The nucleotide sequence is shown on the top line and the amino acid sequence is shown on the second line. The ATG (start codon) and the TAA (stop codon) are marked by using boxes, and the asterisk denotes the stop codon. The ellipses show the possible polyadenylation signal.

Fig. 2A reveals that AccILP-2 was most homologous to its ortholog protein in *A. mellifera*, and the C-terminus was more highly conserved than the N-terminus. To analyze the possible evolutionary relationships among *ILP-2* genes, phylogenetic tree analysis was conducted by using MEGA4, which revealed that *AccILP-2* was closely related to AfLIRP-like, AmILP-2 and AcLIRP-like (Fig. 2B). To further characterize AccILP-2, its potential tertiary structure was constructed by using the SWISS-MODEL server (Fig. 2C).

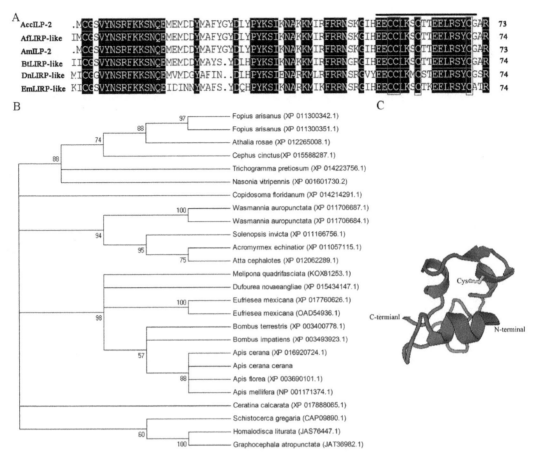

Fig. 2 Characterization of insulin peptides like gene from various species and the tertiary structure of AccILP-2. Protein sequences were acquired from NCBI. A. Multiple alignment of AccILP-2 amino acid sequences with other similar species sequence. Identical regions are shaded in black. The sequences used in analysis were from *A. cerana cerana, A. mellifera, A. florea, Bombus terrestris, Eufriesea mexicana* and *Dufourea novaeangliae*. The putative domain of AccILP-2 is marked by a straight line. B. Phylogenetic analysis of ILP-2 from various species. The numbers above or below the branches indicate the bootstrap values (> 50%) of 500 replicates. C. The tertiary structure of AccILP-2.

Analysis of genomic structure

To further study the characteristics of *AccILP-2*, a 695 bp full-length DNA segment was amplified by using genomic DNA. The results demonstrated that this fragment contained two exons and one intron, and the intron was within the ORF. Furthermore, the nucleotide sequence was consistent with the canonical GT - AG rule at the splice junctions. Additionally, a comparison of the genomic sequence of *AccILP-2* with that from other insects of the IlGF-superfamily such as *A. florea*, *A. mellifera* and *Bombus terrestris* revealed variations in the intron numbers and gene lengths (Fig. 3). Notably, in comparison with *A. florea* and *B. terrestris*, the DNA structure of *AccILP-2* was more similar to that of *AmILP-2* in terms of intron and exon numbers as well as DNA length.

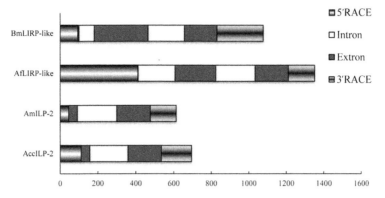

Fig. 3 Schematic representation of the DNA structures. The pattern of untranslated regions, introns, and exons in *AccILP-2*, *AmILP-2*, *A. florea* LIRP-like (*AfLIRP-like*) and *Bombus* LIRP-like *BmLIRP-like* are presented according to the scale shown below. The genomic DNA from the above genes (except *AccILP-2*) was loaded from the NCBI database.

Temporal and spatial profiles of *AccILP-2*

To determine the developmental expression patterns of *AccILP-2*, in this study, we researched the relative expression levels of *AccILP-2* among different life stages and different tissues (Fig. 4A, B). During different developmental stages, the results showed that the relative expression level of *AccILP-2* was nearly equal during the larva period, except at L2 ($P > 0.05$), and decreased rapidly in the prepupal stage ($P < 0.05$). *AccILP-2* expression then gradually increased during the pupal period and reached a peak in the dark-eyed pupal stage ($P < 0.05$). In newly emerged adults, *AccILP-2* expression was less than that observed during the dark-eyed pupal stage ($P < 0.05$). *AccILP-2* expression increased in the 15-day adult workers than in the 1-day adult workers. Analysis of *AccILP-2* spatial expression profiles indicated that transcripts were most abundant in the muscle ($P < 0.05$) followed by the epidermis. The expression levels in the head, midgut, rectum and venom gland were not significant ($P > 0.05$). *AccILP-2* was minimally expressed in the head.

Expression profiles of *AccILP-2* under different abiotic stresses

Many previous studies have investigated the insulin/insulin-like signaling (IIS) pathway; however, little is known about the functions of *AccILP-2*. To verify whether the expression level of *AccILP-2* was influenced by abiotic stress, total RNA was extracted from 15-day-old adults that had been exposed to different environmental stresses. qRT-PCR was used to characterize the mRNA levels of *AccILP-2* after exposure to UV, H_2O_2, heavy metals ($HgCl_2$ and $CdCl_2$), pesticides (dichlorvos, paraquat and cyhalothrin) and different temperature treatments (4 ℃ and 44 ℃) [13, 24]. Gene expression levels were normalized to those in the control adult bees. As shown in Fig. 5A, *AccILP-2* transcription was slightly downregulated following UV exposure but reached a maximum at 3 h of UV exposure. The expression level declined slightly and then remained stable ($P > 0.05$). When treated with H_2O_2, *AccILP-2* transcripts quickly accumulated to a maximum at 0.25 h after which the transcription level was decreased (Fig. 5B). Under $HgCl_2$ stress conditions, *AccILP-2* transcription increased and

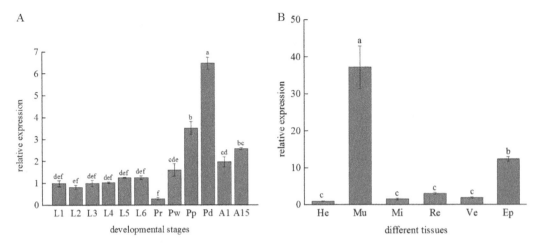

Fig. 4 The expression profile of *AccILP-2* in different developmental stages and different tissues in *Apis cerana cerana*. A. The expression level of *AccILP-2* at diverse developmental stages including 1- to 6-day larvae (L1-L6), pre-pupal (Pr), white-eyed (Pw), pink-eyed (Pp), dark-eyed (Pd), 1-day worker bees (A1) and 15-day worker bees (A15). B. *AccILP-2* transcript in head (He), muscle (Mu), midgut (Mi), rectum (Re), venom gland (Ve) and epidermis (Ep). The β-actin gene was used as an internal control. Control check are 1-day larvae (L1) and head (He), respectively. Each value is given as the mean (SD) of three replicates. Significant differences across mRNA expression are indicated with letters a, b, c, d, etc., as determined by the Tukey's HSD test ($P < 0.05$).

peaked at 1.5 h before declining (Fig. 5C). Treatment with $CdCl_2$ increased the accumulation of *AccILP-2* transcripts as detected at 1.5 h after treatment, and the transcription levels reached a maximum at 3 h. After 3 h, the transcription level was downregulated (Fig. 5D). Interestingly, the *AccILP-2* gene exhibited different responses to the application of various pesticides. Treatments with dichlorvos and paraquat upregulated *AccILP-2* expression, with maximum expression levels observed at 6 h, respectively. In these experiments, the expression levels increased gradually before reaching maximum levels (Fig. 5E, F). However, the application of cyhalothrin induced a mild decrease in *AccILP-2* expression followed by upregulation to maximum expression at 5 h ($P < 0.05$) (Fig. 5G). Experiments involving exposure to different temperatures are presented in Fig. 5H, which indicated that *AccILP-2* expression level was upregulated following exposure to 4 ℃ and reached a maximum at 2 h. However, the transcription levels were downregulated following treatment at 44 ℃ for 0.25 h ($P < 0.05$); this downregulation was followed by a gradual increase in expression. The transcription levels reached a maximum at 3 h and were nearly equal to the expression levels observed in the control adult bees (Fig. 5I).

Discussion

Previous studies of the insulin-like peptides in honeybees have mostly investigated the insulin signaling pathway, and few have demonstrated the functions of the insulin-like peptide genes [12]. Here, we report the cloning and characterization of the *AccILP-2* gene from *A. cerana cerana* for the first time and investigate its response to various abiotic stresses. Our results suggest that *AccILP-2* expression could be induced by some environment stressors.

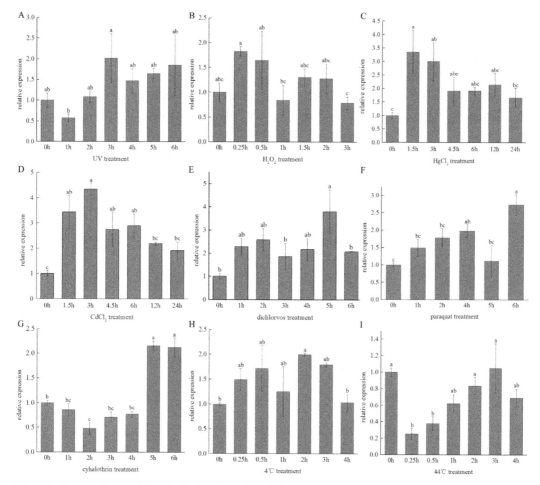

Fig. 5 Expression profile of *AccILP-2* under different abiotic stress conditions. These stresses included UV (A), H_2O_2 (B), $HgCl_2$ (C), $CdCl_2$ (D), dichlorvos (E), paraquat (F), cyhalothrin (G), 4℃ (H), and 44℃ (I). The *β-actin* gene was used as an internal control. Control check is abbreviated 0 h. Each value is given as the mean (SD) of three replicates. Significant differences across mRNA expression are indicated with letters a, b, c, d, etc., as determined by the Tukey's HSD test ($P < 0.05$).

After we obtained the amino acid sequence of *AccILP-2*, a multiple sequence alignment was built with genes from other species, including those from *A. florea*, *A. mellifera*, *B. terrestris*, *Dufourea novaeangliae* and *Eufriesea mexicana*. The alignment results indicated that the amino acid sequence of *AccILP-2* shares high sequence similarity with that from *A. mellifera*. The conserved domain was predicted by NCBI. We found that *AccILP-2* has a typical IlGF-insulin-bombyxin-like domain, which is characteristic of the IlGF-like superfamily. Typically, the active forms of these peptide hormones from the superfamily are composed of two chains (A and B), which are linked by two disulfide bonds. The arrangement of four cysteines is conserved in the "A" chain; Cys1 is linked by a disulfide bond to Cys3, and Cys2 and Cys4 are linked by interchain disulfide bonds to cysteines in the "B" chain [25-28]. From the amino acid sequence, we detected the four Cys displayed in the red box of Fig. 2. An AccILP-2 phylogenetic tree was then generated to analyze the evolutionary relationships among the genes from different species, and the results indicated that *AccILP-2* had a high degree of homology with the *ILP-2* gene from *A. mellifera*, *A. cerana* and *A. florea*. To better understand the genetic relationships among *A. cerana cerana*, *A. mellifera* and *A. florea*, we cloned the intron of *AccILP-2*. The result indicated that the DNA sequence of *AccILP-2* has only one intron and two exons, which was similar to the genetic structure of *A. mellifera*. However, the numbers, sizes, and relative positions of all the introns of ILP-2 in different species are different, suggesting that the introns of ILP-2 are not conserved. The results of the multiple sequence alignment and phylogenetic tree confirmed that *A. cerana cerana* had a high degree of similarity with *A. mellifera*.

Although the structure and homology of the *AccILP-2* gene were previously known, the function of this gene in honeybee remains unknown. Our analysis of *AccILP-2* gene expression during various developmental stages showed that *AccILP-2* may regulate the growth of *A. cerana cerana*. *AccILP-2* was most highly expressed at the dark-eye pupal stage rather than in adults. This finding may be because the bees' cells become highly differentiated and many organs mature when they begin to pupate [13]. The high expression at these stages indicated that *AccILP-2* might play an important role in the transition from pupae to adult. These results were consistent with those reported by de Azevedo et al. [6]. The expression level in A1 was higher than in A15, and we deduced that this may be due to environmental changes, such as exposure to UV radiation, heat, and other sources of handicap [29].

Our analysis of the relative expression levels of *AccILP-2* in different tissues indicated that the highest mRNA accumulation occurred in the muscles followed by the epidermis. This particular expression pattern of *AccILP-2* among different tissues most likely reflects the specific demands of each tissue. Muscle tissues undergo significant oxidative stress due to high oxygen consumption and therefore serve an important role in resistance to oxidative stressors [13]. In addition, the epidermis, it is a barrier to external stressors and participates in resistance to oxidative stress.

Temperature was thought to be a dominant abiotic cause of physiological changes in organisms. High temperatures may be an important environmental factor that influences development because exposure to extreme temperatures can induce physiological changes [30]. The behavior of the honeybees can be influenced by temperature changes, and *A. cerana cerana* are more sensitive to heat stress [31]. In our study, when *A. cerana cerana* were kept at 44℃, *AccILP-2* transcription was downregulated, which suggested that *AccILP-2* might play a role in heat stress regulation. The increased thermotolerance and accumulation of heat-shock proteins observed in IIS-mutant worms is consistent with this hypothesis [9], and thus, we predicted that the transcription level of *AccILP-2* would be upregulated. The result was not consistent with this prediction. This response to high temperature is worth further study. In organisms, cold pressure suppresses the rate of protein synthesis, changes the cellular membrane lipid composition, and ultimately inhibits cell growth [32, 33]. the transcription levels of *AccILP-2* were upregulated by exposure to 4℃. Stress induced by temperature changes has been associated with enhanced reactive oxygen species (ROS) generation and oxidative stresses [34]. We deduced that when the temperature exceeded a certain level, the transcription levels would be downregulated as a response to the environmental and oxidative stresses.

As an environmental stress, UV light influences insect habits and living conditions [35]. Previous studies have shown that insulin-like growth factor Ⅰ can protect fibroblasts from apoptosis induced by UV-B light [36]. H_2O_2 concentration inside the bodies of bees increased dramatically when *A. cerana cerana* were exposed to UV radiation [34]. Since the initial observation that insulin induces the production of H_2O_2 in adipose cells, the intracellular generation of this molecule has been detected in various nonphagocytic cells [37]. UV and H_2O_2 exposure can increase ROS production [29, 38], and our study indicated that the expression of *AccILP-2* was upregulated following exposure to UV and H_2O_2 to scavenge ROS, which suggested that *AccILP-2* plays a crucial role in oxidation resistance. The same reaction also proved the relationship between UV and H_2O_2.

It is well-known that mercury (Hg) and cadmium (Cd) are the most poisonous heavy metalsin the natural world [39] and that they can inactivate protein function by directly binding to enzyme metal ion sites [24]. Metal ions are also well-known to induce oxidative stress by stimulating ROS production [40]. When the worker bees forage for the pollen and nectar outside, they may encounter heavy metal. Our study suggested that *AccILP-2* expression was enhanced by $CdCl_2$ and $HgCl_2$ treatments, indicated that *AccILP-2* may prevent injury under conditions of $HgCl_2$ and $CdCl_2$ stresses. And the result were consistent with other stresses that can increase ROS production.

The overuse of pesticides is an important factor in environmental pollution. Previous studies have shown that pesticides can damage the physiological and biochemical functions of lymphocytes and erythrocytes, which can cause lipid biomembrane lesions [41]. Several studies have also described an increase in oxidative stress due to insecticide exposure [42-44]. In our study, all pesticides including dichlorvos, paraquat and cyhalothrin induced the expression of *AccILP-2*, indicated that *AccILP-2* plays an important role in protecting honeybees against the damaging effects of pesticides.

In conclusion, *AccILP-2* transcription in response to a variety of environmental stresses indicated the importance of *AccILP-2* in the development of *A. cerana cerana* and in stress responses. The *AccILP-2* gene may enhance honeybees resistance to environmental stressors. However, as is well-known, *ILP-2* is an important gene in the insulin/insulin-like signaling (IIS) pathway and how *AccILP-2* cooperates with other genes in this pathway is unknown. In addition, the protein expression levels of ILP-2 and overexpression of *AccILP-2* or silencing of *AccILP-2* will be the focus of future studies. Further examination of the mechanisms involved in honeybees gene regulation is needed.

References

[1] Zheng, X., Lu, Y., Zhang, P., Zhang, L., Li. C., Xue, Y., Gong. L. Effect of inhibiting the expression of insulin-like peptide gene BBX-B8 on development and reproduction of silkworm, *Bombyx mori*. Afr J Biotechnol. 2012, 11 (10), 2548-2554.

[2] Nilsen, K. A., Ihle, K. E., Frederick, K., Fomdrk, M. K., Smedal, B., Hartfelder, K., Amdam, G. V. Insulin-like peptide genes in honey bee fat body respond differently to manipulation of social behavioral physiology. J Exp Biol. 2011, 214, 1488-1497.

[3] Brogiolo, W., Stocke, H., Ikeya, T., Rintelen, F., Fernandez, R., Hafen, E. An evolutionarily conserved function of the *Drosophila* insulin receptor and insulin-like peptides in growth control. Curr Biol. 2001, 11 (4), 213-221.

[4] Oldham, S., Hafen, E. Insulin/IGF and target of rapamycin signaling, a TOR de force in growth control. Trends Cell Biol. 2003, 13(2), 79-85.

[5] Edgar, B. A. How flies get their size: genetics meets physiology. Nat Rev Genet. 2006, 7 (12), 907-916.

[6] de Azevedo, S. V., Hartfelder, K. The insulin signaling pathway in honey bee (*Apis mellifera*) caste development – differential expression of insulin-like peptides and insulin receptors in queen and worker larvae. J Insect Physiol. 2008, 54 (6), 1064-1071.

[7] Wu, Q., Brown, M. R. Signaling and function of insulin-like peptides in insects. Annu Rev Entomol. 2006, 51, 1-24.

[8] Nagasawa, H., Kataoka, H., Hori, Y., Isogai, A., Tamura, S., Suzuki, A. Isolation and some characterization of the prothoracicotropic hormone from *Bombyx mori*. Gen Comp Endocrinol. 1984, 53 (1), 143-152.

[9] Broughton, S. J., Piper, M. D., Ikeya, T., Bass, T. M., Jacobson, J., Driege, Y., Maartinez, P. Longer lifespan, altered metabolism, and stress resistance in *Drosophila* from ablation of cells making insulin-like ligands. Proc Natl Acad Sci USA. 2005, 102 (8), 3105-3110.

[10] Weinstock, G. M., Robinson, G. E., Gibbs, R. A. Insights into social insects from the genome of the honeybee *Apis mellifera*. Nature. 2006, 443 (7114), 931-949.

[11] Corona, M., Velarde, R. A., Remolina, S., Moran-Lauter, A. Vitellogenin, juvenile hormone, insulin signaling, and queen honey bee longevity. Proc Natl Acad Sci USA. 2007, 104 (17), 7128-7133.

[12] Ihle, K. E., Baker, N. A., Amdam, G. V. Insulin-like peptide response to nutritional input in honey bee workers. J Insect Physiol. 2014, 69, 49-55.

[13] Liu, F., Gong, Z., Zhang, W., Wang, Y., Ma, L. Identification and characterization of a novel methionine sulfoxide reductase B gene (*AccMsrB*) from *Apis cerana cerana* (Hymenoptera, Apidae). Ann Entomol Soc Am. 2015, 4 (108), 575-584.

[14] Robinson, G. E. Regulation of division of labor in insect societies. Annu Rev Entomol. 1992, 37, 637-665.

[15] Kerr, W. E., Laidlaw, H. H. General genetics of bees. Adv Genet. 1956, 8, 109-153.

[16] Alaux, C., Ducloz, F., Crauser, D., Conte, Y. L. Diet effects on honeybee immunocompetence. Biol Lett. 2010, 6, 562-565.

[17] Park, D., Jung, J. W., Choi, B. S., Jayakodi, M., Lee, J., Lim, J. Uncovering the novel characteristics of Asian honey bee, *Apis cerana*, by whole genome sequencing. BMC Genomics. 2015, 16 (1), 1.

[18] Lourenço, A. P., Mackert, A., Cristino, A. D. S., Simões, Z. L. P. Validation of reference genes for gene expression studies in the honey bee, *Apis mellifera*, by quantitative real-time RT-PCR. Apidologie. 2008, 39 (3), 372-385.

[19] Scharlaken, B., Graaf, D. C., Goossens, D. K., Peelman, L. J., Jacobs, F. J. Differential gene expression in the honeybee head after a bacterial challenge. Dev Comp Immunol. 2008, 32 (8), 883-889.

[20] Li, Y., Zhang, L., Kang, M., Guo, X., Xu, B. *AccERK2*, a map kinase gene from *Apis cerana cerana*, plays roles in stress responses, developmental processes, and the nervous system. Arch Insect Biochem Physiol. 2012, 79 (3), 121-134.

[21] Umasuthan, N., Revathy, K. S., Lee, Y., Whang, I., Choi, C. Y., Lee, J. A novel molluscan sigma-like glutathione S-transferase from Manila clam, *Ruditapes philippinarum*, cloning, characterization and transcriptional profiling. Comp Biochem Physiol C: Toxicol Pharmacol. 2012, 155 (4), 539-550.

[22] Li, D. C., Yang, F., Lu, B., Chen, D. F., Yang, W. J. Thermotolerance and molecular chaperone function of the small heat shock protein *HSP20* from hyperthermophilic archaeon, *Sulfolobus solfataricus* P2. Cell Stress and Chaperones. 2012, 17 (1), 103-108.

[23] Pfaffl, M. W. A new mathematical model for relative quantification in real-time RT-PCR. Nucleic Acids Res. 2001, 29 (9), e45.

[24] Li, G., Jia, H., Wang, H., Yan, Y., Guo, X., Sun, Q., Xu, B. A typical RNA-binding protein gene (*AccRBM11*) in *Apis cerana cerana*: characterization of *AccRBM11* and its possible involvement in development and stress responses. Cell Stress and Chaperones. 2016, 21 (6), 1005-1019.

[25] Marchlerbauer, A., Bryant, S. H. CD-Search: protein domain annotations on the fly. Nucleic Acids Res. 2004, 32, W327-W331.

[26] Marchlerbauer, A., Anderson, J. B., Chitsaz, F., Derbyshire, M. K., Deweesescott, C. CDD: specific functional annotation with the Conserved Domain Database. Nucleic Acids Res. 2009, 37, D205-D210.

[27] Marchlerbauer, A., Lu, S., Anderson, J. B., Chitsaz, F., Derbyshire, M. K., Deweesescott, C. CDD: a Conserved Domain Database for the functional annotation of proteins. Nucleic Acids Res. 2011, 39, D225-D229.

[28] Marchlerbauer, A., Derbyshire, M. K., Gonzales, N. R., Lu, S., Chitsaz, F. CDD: NCBI's conserved domain database. Nucleic Acids Res. 2015, D222-D226.

[29] Kottuparambil, S., Shin, W., Brown, M. T., Han, T. UV-B affects photosynthesis, ROS production and motility of the freshwater flagellate, *Euglena agilis* Carter. Aquat Toxicol. 2012, 122, 206-213.

[30] An, M., Choi, C. Y. Activity of antioxidant enzymes and physiological responses in ark shell, *Scapharca broughtonii*, exposed to thermal and osmotic stress: effects on hemolymph and biochemical parameters. Comp Biochem Physiol B: Biochem Mol Biol. 2010, 155 (1), 34-42.

[31] Tautz, J., Maier, S., Groh, C., Rössler, W., Brockmann, A. Behavioral performance in adult honey bees is influenced by the temperature experienced during their pupal development. Proc Natl Acad Sci USA. 2003, 100 (12), 7343-7347.

[32] Rao, P. N., Engelberg, J. HeLa cells: effects of temperature on the life cycle. Science. 1965, 148(3673), 1092-1094.

[33] Burdon, R. H. Temperature and animal cell protein synthesis. Symp Soc Exp Biol. 1987, 41, 113-133.

[34] Yan, Y., Zhang, Y., Huaxia, Y., Wang, X., Yao, P., Guo, X., Xu, B. Identification and characterisation of a novel 1-Cys thioredoxin peroxidase gene (*AccTpx5*) from *Apis cerana cerana*. Comp Biochem Physiol B: Biochem Mol Biol. 2014, 172-173, 39-48.

[35] Mazza, C. A., Izaguirre, M. M., Zavala, J., Scopel, A. L., Ballaré, C. L. Insect perception of ambient ultraviolet-B radiation. Ecol Lett. 2002, 5 (6), 722-726.

[36] Kulik, G., Klippel, A., Weber, M. J. Antiapoptotic signalling by the insulin-like growth factor I receptor. Mol Cell Biol. 1997, 17 (3), 1595-1606.

[37] Rhee, S. G., Bae, Y. S., Lee, S. R., Kwon, J. Hydrogen peroxide: a key messenger that modulates protein phosphorylation through cysteine oxidation. Sci Stke. 2000, (53), pe1.

[38] Lushchak, V. I. Environmentally induced oxidative stress in aquatic animals. Aquat Toxicol. 2011, 101 (1), 13-30.

[39] Rashed, M. N. Monitoring of environmental heavy metals in fish from Nasser Lake. Environ Int. 2001, 27 (1), 27-33.

[40] Cove, D. J. Cholorate toxicity in *Aspergillus nidulans*: the selection and characterisation of chlorate resistant mutants. Heredity (Edinb). 1976, 36 (2), 191-203.

[41] Narendra, M., Bhatracharyulu, N. C., Padmavathi, P., Varadacharyulu, N. C. Prallethrin induced biochemical changes in erythrocyte membrane and red cell osmotic haemolysis in human volunteers. Chemosphere. 2007, 67 (6), 1065-1071.

[42] Piner, P., Sevgiler, Y., Under, N. *In vivo* effects of fenthion on oxidative processes by the modulation of glutathione metabolism in the brain of *Oreochromis niloticus*. Environ Toxicol. 2007, 22 (6), 605-612.

[43] Monteiro, D. A., Rantin, F. T., Kalinin, A. L. The effects of selenium on oxidative stress biomarkers in the freshwater characid fish matrinxã, *Brycon cephalus* (Günther,1869) exposed to organophosphate insecticide Folisuper 600 BR® (methyl parathion). Comp Biochem Physiol C: Toxicol Pharmacol. 2009, 149, 40-49.

[44] Thomaz, J. M., Martins, N. D., Monteiro, D. A., Rantin, F. T., Kalinin, A. L. Cardio-respiratory function and oxidative stress biomarkers in *Nile tilapia* exposed to the organophosphate insecticide trichlorfon (NEGUVON®). Ecotoxicol Environ Safety. 2009, 72, 1413-1424.

1.10 Molecular Cloning, Expression and Stress Response of the Estrogen-related Receptor Gene (*AccERR*) from *Apis cerana cerana*

Weixing Zhang, Ming Zhu, Ge Zhang, Feng Liu, Hongfang Wang, Xingqi Guo, Baohua Xu

Abstract: Estrogen-related receptor (ERR), which belongs to the nuclear receptor superfamily, has been implicated in diverse physiological processes involving the estrogen signaling pathway. However, little information is available on ERR in *Apis cerana cerana*. In this report, we isolated the *ERR* gene and investigated its involvement in antioxidant defense. Quantitative real-time polymerase chain reaction (qPCR) revealed that the highest mRNA expression occurred in eggs during different developmental stages. The expression levels of *AccERR* were the highest in muscle, followed by the rectum. The predicted transcription factor binding sites in the promoter of *AccERR* suggested that *AccERR* potentially functions in early development and in environmental stress responses. The expression of *AccERR* was induced by cold (4 ℃), heat (42 ℃), ultraviolet light (UV), $HgCl_2$, and various types of pesticides (phoxim, deltamethrin, triadimefon, and cyhalothrin). Western blotting was used to measure the expression levels of AccERR protein. These data suggested that *AccERR* might play a vital role in abiotic stress responses.

Keywords: *Apis cerana cerana*; Estrogen-related receptor; Biochemical properties; qPCR; Oxidative stress

Introduction

Organisms are exposed to various unfavorable abiotic stresses, including temperature changes and exposure to heavy metal, ultraviolet radiation, and different pesticides, all of which induce the formation of reactive oxygen species (ROS) [1]. High ROS concentrations induce cellular senescence and apoptosis by oxidizing nucleic acids, proteins, and lipid membranes [2].

Estrogen-related receptors (ERRs) compose a group of nuclear receptors that were originally identified based on amino acid sequence similarity to the estrogen receptors (ER), particularly in the DNA-binding domain (DBD) [3, 4]. The ERRs include ERRα (NR3B1), ERRβ (NR3B2) and ERRγ (NR3B3) [5]. ERRs comprise a subgroup of steroid hormone receptors. ERRs have been demonstrated to share both structural and functional attributes with ERs, including binding to synthetic estrogenic ligands [6]. ERRs are localized to the nucleus and are expressed in multiple cell types in animals. ERRs can function as gene repressors or gene activators in association with coregulators, depending on the cellular and genetic context [7, 8]. One of the crucial roles of ERRs in animals is in various facets of cardiac physiology. A healthy and functional heart is necessary for survival and requires efficient ATP generation and the maintenance of intracellular calcium homeostasis [9]. ERRs regulate many target genes to maintain heart functions, including genes involved in energy metabolism, apoptosis, cell cycle regulation and cardiac development. In osteoblasts and osteoclasts, the high expression of ERRα controls growth and differentiation *in vitro* and *in vivo*. During osteoblastogenesis, which is characterized by upregulated bone sialoprotein and downregulated osteopontin messenger RNA (mRNA) levels, silencing ERRα in human mesenchymal stem cells (hMSCs) increases osteoblastic differentiation. These results suggested essential and unexpected roles for ERRα in regulating bone development and differentiation [10-12]. Collectively, previous studies have highlighted an important role for ERRs in several aspects of energy metabolism in both normal and cancer cells [13, 14]. It has been reported that ERR plays a central role in carbohydrate metabolism and directs a critical metabolic transition during *Drosophila* development [15]. ERRs do not bind to endogenous estrogens; however, in the absence of estrogens or ligands, ERRs can constitutively transactivate target gene promoters or response element-containing reporter genes [4, 16, 17]. *In vitro*, ERRα can form a heterodimer with ERα via direct protein-protein interactions (PPIs). ERRs can regulate their own target genes or modulate the transcription of genes that are regulated by ERs or other orphan nuclear receptors in different target cells [18, 19]. Based on these reports, we hypothesized that ERRs might be involved in ER-mediated signaling pathways.

As a pollinator of flowering plants, *Apis cerana cerana* plays an important role in maintaining the balance of regional ecologies and agriculture economic development [20]. Although ERRs have been studied in various species, little is known about their functions and expression in insects. Therefore, we chose *A. cerana cerana* to ascertain the tissue distribution and transcription of AccERR in response to environmental stress. In this study, we isolated the full-length cDNA corresponding to AccERR from *A. cerana cerana* and evaluated its expression during different developmental stages and various tissues. In addition, we investigated the *in vivo* activity of AccERR under various oxidative stress conditions, such as temperature change and exposure to heavy metal ($HgCl_2$), ultraviolet radiation, or insecticides (triadimefon, phoxim, deltamethrin, cyhalothrin and imidacloprid). This study provided new evidence in the response to oxidative stress.

Materials and Methods

Biological specimens and tissues preparation

Worker bees, *A. cerana cerana* were collected from hives of the Experimental Apiary of the Department of Animal Nutrition and Feed Science, University of Shandong Agricultural University (Taian, China).

To obtain larvae of uniform age, queens were periodically confined for 12 h on combs without brood, where they laid eggs. At appropriate age, worker larvae were removed from the brood frames and maintained in an incubator (34℃ and 80% relative humidity), where development progressed normally. The larvae and pupae were identified according to the criteria of Michelette and Soares [21]. The different developmental stages including egg, the first to the fifth instars, prepupae, white-eyed pupae (unpigmented cuticle), pink-eyed pupae (pigmented cuticle), brown-eyed pupae (unpigmented cuticle), brown-eyed pupae (dark pigmented cuticle), and newly emerged. Newly emerged adult workers were collected from outdoor hives. They were caged in groups of 50 individuals and kept in incubators (32℃ and 60% relative humidity) until 15 days before they were treated. Those 15-day post-emergence adult worker bees were treated with different conditions. Groups 1 and 2 were placed at 4℃ and 42℃, and collected at the proper time. Groups 3 and 4 were treated with ultraviolet light (UV) (254 nm, 30 MJ/cm^2) and HgCl$_2$ (3 mg/ml mixed in food). Groups 5-9 were treated with pesticides (thriadimefon, phoxim, deltamethrin, cyhalothrin and imidacloprid, which were diluted to 20 mg/L and sprayed gauze). The control group kept in incubators (32℃ and 60% relative humidity) without treatment. The worker bees were collected at the proper time after treatment, and then the samples were flash-frozen in liquid nitrogen.

The brain, epidermis, muscle, midgut, haemolymphe and rectum were isolated separately from 15-day post-emergence adult workers. These tissues were frozen in liquid nitrogen immediately and stored at −80℃ until used.

RNA extraction, cDNA synthesis, and DNA isolation

RNAiso Plus (TaKaRa, Dalian, China) was used to extract total RNA according to the instructions of the manufacturer. Total RNAs were treated with DNase I. The RNA integrity was checked by electrophoresis on a 1.0% agarose gel in TAE buffer and visualized under UV light. Spectrophotometer was used to quantify the absorbance ratio at 260/280 nm of RNA samples.

The first-strand cDNA was generated by using Transcript® All-in-One First-Strand cDNA Synthesis SuperMix for qPCR (TransGen Biotech, Beijing, China) according to the protocol of the manufacturer. Total RNAs were digested with genomic DNA (gDNA) remover to eliminate potential DNA contamination.

gDNA was isolated by using the EasyPure Genomic DNA Kit (TransGen Biotech, Beijing, China) according to the protocol of the manufacturer.

qPCR transcriptional analysis

cDNA was synthesized as described above. *AccERR* expression at different environmental stages was analyzed by qPCR. The *β-actin* gene was selected as the control gene to normalize the qPCR data because it has been validated as among the most stably expressed genes [22, 23]. Negative controls without the addition of RT enzyme were run to check for contamination by gDNA. The sequences for the specific primer pair RTP1/RTP2 and the *actin* gene primers actin-f/ actin-r are listed in Table 1. The reverse transcription PCR (RT-PCR) mix included 10.0 μl of TransStart® Tip Green qPCR SuperMix (TransGen Biotech, Beijing, China), 1.6 μl of cDNA, 0.4 μl of each primer (10 mmol/L each), and 7.6 μl of ddH$_2$O in a final volume of 20 μl. RT-PCR was performed by using a CFX96^{TM} Real-time System (Bio-Rad) under the following cycling conditions: 94℃ for 30 s followed by 40 cycles of 94℃ for 5 s, 55℃ for 15 s, and 72℃ for 10 s. A melting cycle was performed from 65℃ to 95℃. All the samples were prepared from at least three individuals and were assayed in triplicate. Individual ERR mRNA level was log2-transformed and relative quantities were calculated according to the $2^{-\triangle\triangle C_t}$ method (Applied Biosystems, user bulletin no. 2 [24]). We log2-transformed the data in order to approximate normality as is often done with gene expression data sets, as these data are non-linear and the variance is often very unequal across the samples [25, 26].

Table 1 PCR primers in this study

Abbreviation	Primer sequence (5'→3')	Description
ER1	TGCGTGCTTTATTCGTAGTCTTC	cDNA sequence primer, forward
ER2	TACAGTCTCCGTTTCCTTTCTC	cDNA sequence primer, reverse
5P1	CGGGCATCATACAAACCACAT	5' RACE reverse primer, outer
5P2	TGGTGGTCGGTGAACAAACT	5' RACE reverse primer, inner
AAP	GGCCACGCGTCGACTAGTAC(G)14	Abridged anchor primer
AUAP	GGCCACGCGTCGACTAGTAC	Abridged universal amplification primer
3P1	ATCAAATGCGGCTTCTGC	3' RACE forward primer, outer
3P2	GGCTTCAAAGATTGGGATT	3' RACE forward primer, inner
B26	GACTCTAGACGACATCGA(T)18	3' RACE universal primer, outer
B25	GACTCTAGACGACATCGA	3' RACE universal primer, inner
ERR1	GTGTATACATACTGGTATGTGTGC	Full-length cDNA primer, forward
ERR2	CTATCACTCACTGTCTTCTGC	Full-length cDNA primer, reverse
IP1	ATACGCCAGTGAAACCAG	Inverse PCR forward primer, outer
IP2	GAAATAAATAAACGCAGGAG	Inverse PCR reverse primer, outer
IP3	CCGACATGCTGCAAGTAT	Inverse PCR forward primer, inner
IP4	CAGATTATGGTGGAGGCG	Inverse PCR reverse primer, inner
QS1	CCATCGCTACCATCGTCA	Promoter special primer, forward
QS2	CCATGATCGTCTCTTTTGTCC	Promoter special primer, reverse
RTP1	CCGACCACTACACTTCAGCA	Real-time PCR primer, forward
RTP2	GCCTTCCTCAGACGTTCCTA	Real-time PCR primer, reverse
actin-f	CCGTGATTTGACTGACTACCT	Standard control primer, forward
actin-r	AGTTGCCATTTCCTGTTC	Standard control primer, reverse
PR1	GAATTCTTAGGAACGTCTGAGGAAG	Protein expression primer, forward
PR2	CTCGAGTCGATAATAAGCAGCTTCT	Protein expression primer, reverse

Cloning the full-length cDNA of *AccERR*

To determine the cDNA sequence of *AccERR*, primers ER1 and ER2 were designed based on the conserved regions of *ERR* from other insects and were then synthesized (Sangon Biotech Company, Shanghai, China). The 5'-flanking region of *AccERR* was amplified as previously described [27]. To determine the 5′ and 3′ untranslated region (UTR) sequences, 5′ and 3′ rapid amplification of cDNA end (RACE) were performed by using specific primers (5P1/5P2 and 3P1/3P2) designed based on the sequence of the internal fragment. All the PCR products were cloned into the pEASY-T3 vector (TransGen Biotech, Beijing, China), and the resulting plasmids were transformed into competent *Escherichia coli* DH5α cells for sequencing.

Primers and amplification condition

The primers used in the present study were listed in Table 1. The PCR amplification conditions were shown in Supplemental data 1.

Isolation of genomic DNA and the 5′-flanking region of *AccERR*

gDNA was extracted from whole adult bees by using the EasyPure® Genomic DNA Kit (TransGen Biotech, Beijing, China) according to the instructions of the manufacturer. To better understand the organization of the regulatory elements, primers (IP1/IP2 and IP3/IP4) were used to obtain the 5′-flanking region of *AccERR*, by using inverse PCR (IPCR). Briefly, the total genomic DNA was digested with the restriction endonuclease *Sca* I at 37℃ overnight, self-ligated to form circles by using T4 DNA ligase (TaKaRa, Dalian, China), and used as a template for I-PCR. Finally, specific primers (QS1 and QS2) were designed to amplify the 5′-flanking region of *AccERR*. The TFSEARCH database (www.cbrc.jp/research/db/TFSEARCH.html) was used to analyze the promoter region of *AccERR*.

Bioinformatic analysis

The AccERR nucleotide sequence was analyzed and compared by using DNAMAN version 5.2.2 (Lynnon Biosoft, Quebec, Canada). The amino acid sequence was aligned with homologous ERR protein sequences from various species by using DNAMAN software 5.2.2 (Lynnon Biosoft, Quebec, Canada). The online protein structure prediction tool SWISS-MODEL (http://swissmodel.expasy.org/) was used to predict the tertiary structure of AccERR. We also used ExPASy online software to predict the molecular mass and theoretical isoelectric point (pI). The phylogenetic tree was constructed by using Molecular Evolutionary Genetics Analysis (MEGA version 4.0) software by using the neighbor-joining method.

AccERR protein expression, purification and antibodies preparation

To construct the plasmid for expression in *E. coli* Transetta (DE3) Chemically Competent Cells (TransGen Biotech, Beijing, China), a 1,032 bp cDNA fragment of *AccERR* encoding the partial open reading frame (ORF) was amplified with a specific primer pair (PR1/PR2) that incorporated an *Eco*R I site in the forward primer and an *Xho* I site in the reverse primer. After cutting the PCR product with *Eco*R I and *Xho* I, the AccERR partial fragment was amplified and inserted into the pET-21a (+) expression plasmid. Next, the recombinant plasmid pET-21a (+)-AccERR was transformed into *E. coli* Transetta (DE3) Chemically Competent Cells (TransGen Biotech, Beijing, China) for protein expression. The bacteria were grown in 10 ml of Luria-Bertani (LB) medium containing 100 mg/ml kanamycin at 37℃. Recombinant protein expression was induced by the addition of isopropylthio-β-D-galactoside (IPTG, 5 mmol/L final concentration) until the bacterial density reached 0.4-0.6 OD_{600}, and the cells were grown at 37℃ for 7 h. Next, the bacterial cells were harvested and centrifuged at 5,000 g at 4℃ for 10 min. The pellet was resuspended in lysis buffer, sonicated and centrifuged again. Finally, the supernatant and the pellet solubilized in PBS were analyzed by 10% SDS-polyacrylamide gel electrophoresis (SDS-PAGE) to determine the expression level of the target recombinant protein. The recombinant AccERR protein was purified by using a HisTrap^{TM} FF column (GE Healthcare, Uppsala, Sweden). The purified protein is more than 95% pure by SDS-PAGE gel analysis. The purified protein was injected subcutaneously into rabbit for generation of antibodies [28].

Western blotting

Total protein extracts were prepared by splitting the whole workers in 1 ml of RIPA lysis buffer for 3 min. The lysates were clarified by centrifugation at 10,000 g for 10 min at 4℃, and the supernatants were used to assay protein levels. The total proteins were quantified with the BCA Protein Assay Kit (Beyotime Institute of Biotechnology, JingSu, China). Total protein extracts (80 μg) were separated by electrophoresis on 7.5% SDS-PAGE gels by using a slab gel apparatus and then transferred to polyvinylidene fluoride (PVDF) nitrocellulose membranes. After transfer, the membranes were blocked in blocking buffer for 1.5 h at room temperature. The membranes were then incubated with primary antibodies against ERR, tubulin at 4℃ overnight, washed three times with TBST buffer, and incubated with the appropriate HRP conjugated secondary antibodies (Cwbiotech, Beijing, China) for 4 h at 4℃. Proteins were visualized by using BeyoECL Plus (Beyotime, Jiangsu, China), and then the blots were quantified by using a FusionCapt Advance system (Oriental Science, Beijing, China) and the analysis software Fusion (Oriental Science, Beijing, China).

Data analysis

The gene expression results are presented as the mean±SD (standard deviation) from three independent experiments (n = 3). The significance was determined by using Duncan's multiple range test and analysis of variance (ANOVA). The calculations were performed by using Statistical Analysis System (SAS) version 9.1. $P < 0.01$ were considered statistically significant. The different letters (a-f) above bars denote significant differences between the treatment and control groups.

Results

Sequence analysis of full-length cDNA of *AccERR* and purification of recombinant AccERR protein

Based on RT-PCR and rapid amplification of cDNA end PCR (RACE-PCR), we cloned the full-length cDNA corresponding to *AccERR*. The full-length cDNA of *AccERR* (GenBank accession number:1789032) was 1,671 bp long, including 1,305 bp for the open reading frame (ORF), 154 bp in the 5' untranslated region (UTR) and 212 bp in the 3' untranslated region (UTR). The ORF encoded a polypeptide of 434 amino acids with a predicted molecular mass of 49.131 kDa and a pI of 7.43. The full-length nucleotide and amino acid sequences of AccERR are shown in Fig. 1.

```
1                               TAGTGTATACATACTGGTATGTGTGCTTTATTCGTAGTCTTCATGAGTGTTTCAATTCTTGACA
65    TTGTGGTCTTTTTCGCTTTTCATCTAGGAAAAGTATAATAGGAAAATAAGAATACAGTGTTCTTATATCATTTGGACAAAAGAGACGATC
1     ATGGATTCATGGATGTACGATGTGGTTTGTATGATGCCCGGTGGCGGCACCGAAAATATGATTGGAAATAATAGGACTATGGCAACATT
1      M  D  S  W  M  Y  D  V  V  C  M  M  P  G  G  G  T  E  N  M  I  G  N  N  R  T  M  A  N  I
91    AAACAAGAAATTGAAAATCCTACAACGCCTACGCAAAATTATCAAGTTTGTTCACCGACCACTACACTTCAGCATCAAGAGGTGATTTGT
31     K  Q  E  I  E  N  P  T  T  P  T  Q  N  Y  Q  V  C  S  P  T  T  T  L  Q  H  Q  E  V  I  C
181   AGTAAAATAGAAGTTCCGCCAGATTATGGTGGAGGCGAAGGTAGTCCTGGAAGTCCAGAAATGCACCATTGTTCGTCAACTACGCAACCT
61     S  K  I  E  V  P  P  D  Y  G  G  G  E  G  S  P  G  S  P  E  M  H  H  C  S  S  T  T  Q  P
271   TTAGGAACGTCTGAGGAAGGCGTTAAAGAGGAAGATATGATACCTAGAAGGCTTTGCCTAGTCTGTGGTGACGTTGCAAGTGGATTTCAT
91     L  G  T  S  E  E  G  V  K  E  E  D  M  I  P  R  R  L  C  L  V  C  G  D  V  A  S  G  F  H
361   TATGGAGTTGCATCTTGCGAAGCGTGTAAAGCTTTTTTCAAAAGAACCATACAAGGTAATATCGAGTATACATGTCCCGCAAATGGGGAA
121    Y  G  V  A  S  C  E  A  C  K  A  F  F  K  R  T  I  Q  G  N  I  E  Y  T  C  P  A  N  G  E
451   TGTGAAATAAATAAACGCAGGAGAAAAGCGTGTCAGGCATGTAGGTTTCAAAAGTGTCTTAGACAAGGCATGTTAAAAGAGGGAGTCCGA
151    C  E  I  N  K  R  R  R  K  A  C  Q  A  C  R  F  Q  K  C  L  R  Q  G  M  L  K  E  G  V  R
541   TTGGATCGTGTTCGAGGAGGGAGACAAAAATATAGAAGATCCACCGATCCCTATACGCCAGTGAAACCAGCTCCCTTGGAAGATAATAAA
181    L  D  R  V  R  G  G  R  Q  K  Y  R  R  S  T  D  P  Y  T  P  V  K  P  A  P  L  E  D  N  K
631   ATGCTGGAAGCTTTGGCTGCCTGTGAACCCGACATGCTGCAAGTATCAAATATCTCTCATACTCTTGACACAGATCAAAGAGTACTTGGA
211    M  L  E  A  L  A  A  C  E  P  D  M  L  Q  V  S  N  I  S  H  T  L  D  T  D  Q  R  V  L  G
721   CAATTATCTGATTTATATGATCGGGAATTAGTCGGAATAATTGGCTGGGCAAAACAGATTCCAGGATTCAGCAGTTTAGCTTTAAACGAT
241    Q  L  S  D  L  Y  D  R  E  L  V  G  I  I  G  W  A  K  Q  I  P  G  F  S  S  L  A  L  N  D
811   CAAATGCGGCTTCTGCAAAGTACGTGGGCTGAGATATTGACGTTTAGTCTCGCGTGGAGAAGCATGCCAAATAATGGTAGATTAAGGTTT
271    Q  M  R  L  L  Q  S  T  W  A  E  I  L  T  F  S  L  A  W  R  S  M  P  N  N  G  R  L  R  F
901   GCACAAGATTTTACTTTAGACGAACGACTTGCCCGCGAATGTCATTGTACAGAGTTATACACGCATTGTATCCAAATAGTAGAAAGGCTT
301    A  Q  D  F  T  L  D  E  R  L  A  R  E  C  H  C  T  E  L  Y  T  H  C  I  Q  I  V  E  R  L
991   CAAAGATTGGGATTAACCAGAGAAGAATATTATGTGTTGAAGGCATTGATATTAGCAAACAGTGACGCAAGATCAGACGAACCACAGGCA
331    Q  R  L  G  L  T  R  E  E  Y  Y  V  L  K  A  L  I  L  A  N  S  D  A  R  S  D  E  P  Q  A
1081  TTATACCGCTTTCGAGATTCTATCTTAAATTCATTATCAGACTGTATGGCCGCAGTAAGACCAGGACAAGCGCTTCGTGCCACTCAAAAT
361    L  Y  R  F  R  D  S  I  L  N  S  L  S  D  C  M  A  A  V  R  P  G  Q  A  L  R  A  T  Q  N
1171  ATGTTCCTAGTGTTACCCAGTCTCAGGCAGGTCGATGGAATTGTTAGACGATTTTGGTCTAGTGTGTATCGAACGGGAAAAGTACCAATG
391    M  F  L  V  L  P  S  L  R  Q  V  D  G  I  V  R  R  F  W  S  S  V  Y  R  T  G  K  V  P  M
1261  AATAAGCTCTTTGTAGAAATGTTAGAAGCTGCTTATTATCGATGAAATATCAACTCGAACTCGATTTATTAATATCTATATTGAGGACTT
421    N  K  L  F  V  E  M  L  E  A  A  Y  Y  R  *
46    CATAGAATCGCTGTTATTAAGTAACACTCGCACATTAAGAATGACTTAAAGGCACGCATTTTGTTGTGTTCCTAAGATATAAGCAGAAGA
136   CAGTGAGTGATAGTGAAATGAGAAAAAGAAACGGAGACTGTACAGTTTTATTCATTTTCTGTAAGAGAAAGACTTAT
```

Fig. 1 The nucleotide sequence of AccERR and the deduced amino acid sequence. The nucleotide sequence is shown on the top line, and the deduced amino acid sequence is shown on the second line. The start coden (ATG) and the stop coden (TGA) are indicated with boxes. The stop coden is indicated with an asterisk.

The multiple sequence alignments revealed that the deduced amino acid sequence of AccERR was closely related to that of *Apis mellifera* ERR (NP_001155988.1, 100% protein sequence identity), *Drosophila melanogaster* ERR (NP_729340.1, 45.9% protein sequence identity), *Tribolium castaneum* ERR (NP_001135406.1, 53.45% protein sequence identity), and *Aedes aegypti* ERR (XP_00163736.1, 50.21% protein sequence identity). The highly conserved region (DBD) was shown in Fig. 2A.

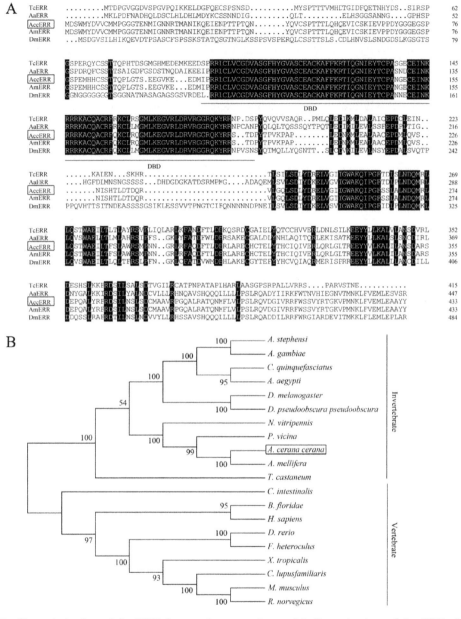

Fig. 2 Characterization of AccERR from various species and tertiary structure of AccERR. A. The alignment of the deduced AccERR amino acid sequence with other known ERR. B. Phylogenetic analysis of ERR protein sequences from several species. The protein sequences of ERR were obtained from NCBI as follows: *B. floridae* (*Branchiostoma floridae*, AAU88062.1), *H. sapiens* (*Homo sapiens*, NP_004443.2), *D. melanogaster* (*Drosophila melanogaster*, NP_729340.1), *A. mellifera* (*Apis mellifera*, NP_001155988.1), *A. stephensi* (*Anopheles stephensi*, BAE96770.1), *D. pseudoobscura pseudoobscura* (*Drosophila pseudoobscura pseudoobscura*, XP_001354210.2), *N. vitripennis* (*Nasonia vitripennis*, XP_001604033.2), *C. quinquefasciatus* (*Culex quinquefasciatus*, XP_001862599.1), *A. aegypti* (*Aedes aegypti*, EAT34188.1), *P. vicina* (*Polyrhachis vicina*, ABR88112.1), *C. intestinalis* (*Ciona intestinalis*, NP_001071700.1), *M. musculus* (*Mus musculus*, NP_031979.2), *D. rerio* (*Danio rerio*, NP_998120.1), *F. heteroclitus* (*Fundulus heteroclitus*, ABB80450.1), *X. tropicalis* [*Xenopus* (*Silurana*) *tropicalis*, NP_001072756.1], *R. norvegicus* (*Rattus norvegicus*, NP_001008511.2), *A. gambiae* (*Anopheles gambiae*, XP_321343.4), *C. lupus familiaris* (*Canis lupus familiaris*, NP_001002936.1) and *T. castaneum* (*Tribolium castaneum*, NP_001135406.1).

The predicted protein sequence of AccERR and representative sequences from ERRs in other species were used to build a phylogenetic tree by using the neighbor-joining method in Molecular Evolutionary Genetics Analysis (MEGA) version 4.0 software (Fig. 2B).

The soluble recombinant ERR protein was purified by using HisTrap™ FF columns, and the eluted protein was analyzed by SDS-PAGE. The SDS-PAGE analysis revealed that the recombinant AccERR protein was mainly expressed in soluble fraction, with very little expression in the inclusion bodies. The SDS-PAGE analysis also showed that the recombinant protein had a molecular mass of approximately 42 kDa (Supplementary data 2).

Identification of partial putative transcription factor binding sites within the 5'-flanking region of AccERR

To further understand the molecular mechanism of transcriptional regulation of AccERR, a 1,304 bp sequence upstream of the transcription start site was cloned by PCR using two specific primers, QS1 and QS2. To analyze the *AccERR* promoter sequence, we used TFSEARCH, which predicted putative *cis*-acting elements. As shown in Fig. 3, this analysis identified Broad-Complex (BR-C), heat shock factor (HSF), cell factor 2-II (CF2-II), ovo homolog-like transcription factor (TF), Caudal-related homeobox (CdxA), NIT2, Sox-5, and ADRI. We identified binding sites for HSFs, which play important roles in heat-induced transcriptional activation in *Drosophila* [29]. A TATA box and a CAAT box were found in the promoter; these elements are critical for transcription. In addition, TF binding sites involved in germline and epidermis differentiation were also predicted. Previous studies have shown that the transcription factor BR-C plays important roles in the tissue-specific response to ecdysone, in muscle fiber development and in the regulation of development [30]. Various important transcription factors involved in tissue development and early growth were identified, including binding sites for CF2-II, CdxA, Sox-5 and the protein encoded by the nitrogen regulatory (NIT2) gene [27, 31-33].

Expression of AccERR at different developmental stages and in various tissues

To determine the expression of the *AccERR* gene during different developmental stages, qPCR was performed by using total RNA extracted from egg, larvae, pupae and adults. As shown in Fig. 4A, *AccERR* was most highly expressed in eggs; *AccERR* expression decreased rapidly from the egg to the first-instar larval stage and slightly increased in prepupae. *AccERR* expression was higher in pupae than in larvae, but there were no significant differences in expression between the pupae and adult stages.

Previous research has demonstrated that *AccERR* is expressed in different adult tissues [33]. Therefore, to further determine the expression levels of *AccERR* in the *A. cerana cerana* adult, total RNA from the head, muscle, midgut, epidermis, hemolymph, wings and rectum was extracted for qPCR analysis. As shown in Fig. 4B, the highest expression was detected in muscle, followed by rectum, and the lowest expression was in the epidermis.

```
                    IF                              CAAT box
CCATCGCTACCATCGTCACCACTGTCTTAGCCGCCGCCAGTGCTTTCAATATAGCCACAGCCTCCGCCC    -1236
CCGCCTCCTATCTTCCACCTCCCACCTTCCATCTTCCACTTCTCTCCTCCCTCTTATTCTCACCCTTCC   -1167
        IF
ATAGCAATATATATGTTGCACGCATTACGCCAGTAACGTATCAACTTAGGTCCTATTTAGGTCCTGTTT   -1098
         ADRI                                                    CAAT box
CAACTATGGGGGTAAAGGGAGAGTTTTGGAGCTCGATCGATCTACCTGCTAGCCTGAAAATTCAATTGG    -1029
         CAAT box
CTTGTATATGGACTCAATGCCATTTTATACGTTTCTTTGTTCACGTATTTTTATCATATTTATTCTGTG   -960
                                          IF              CAAT box
TAAAAAACCATTCATATAACATTTATTTTCTTCTACTTCGTAGAATCATTTATTCAATTAACATCTTTT   -891
       IF                                          HSF
CCTGTTTCTTCATCGTGTTTAGAATACTGAAGTAAAAAGAGGTTCACTTTATAGAAAGACCTGCTTCAG   -822
BR-C
AATAGCTTCACGGCCAAGAGTTAATGACCGATTGGTCATGACCATGATAATGTTCACGGTAATTTTATC   -753
    IF     CdxA
AATTTCGTTTATATTTATACAACATGGTAACTATTATGCTTTAAATTGAAGTTGATGATTTTATCATCA   -684
TCGATAAATACCGATCTAATTCTTTGCAAATGCAAATCTGTTGCAAATGCACTCTGTTGTAAGATATCT   -614
TGAAGATATTAATGTTTCTGAAAAATTTACTTATATTCTATACATACATATAATATATGTTTATAATAT   -546
                   TATA box
TAATATTAATATTATATAAAATATAATGTTTTTAATCTCAAGAACATTATTTGTTTATATTTAATTGAA   -477
  HSF                    Sox-5
ATAGAAATTATTTTTGGAATGTTTTTATAATAACAATAAATCAGATTTATTAATAATAAAAAAAATTT    -409
     CdxA         CAAT box          TATA box
ATTAATAACATTTATATTTAATACAACAATAAATATATAATATTATAAATATTATATAATAATAAATTT   -340
      CF2-II            TATA box
ATTAACAATATATATATAATAATAATAAATCATTATATAAATGTTGAAATTTTTATCTATTCATGAGAA   -271
                          TATA box
ATTATTTTATAACAGATCATGTGTAGATATATAAATCATAATCATAAAAAATTGAATATCTGTTGATCA   -202
ATTAGATAAAATATTTGACATATTTTGTCTGAAGATTCGGTTAAATTCCTTAAACGATAAGGTAATTAG   -133
   HSF
CATAAGAAATATATATTTTGCGAATTTGTGGCGCCTGGTATTAGGATCGGCCATCTCCGTGTATTACCG   -64
                                IF                       ADRI
TATCCTCAGAATTGCCGTCAGTCGCATACGGCTCGTCGCGCCGTCAACGTCGTGGTAGGGGGATAGTGT   +6
                                                              →+1
                                                         transcription start site
ATACATACTGGTATGTGTGCTTTATTCGTAGTCTTCATGAGTGTTTCAATTCTTGACATTGTGGTCTTT   +75
TTCGCTTTTCATCTAGGAAAAGTATAATAGGAAAATAAGAATACAGTGTTCTTATATCATTTGGACAAA   +144
AGAGACGATCATGGATTCAT                                                    +164
         → translation start site
```

Fig. 3 Nucleotide sequence of the 5'-flanking region of *AccERR*. The translation and Transcription start sites were marked with arrowhead. The transcription factor binding sites are boxed.

Expression profiles of *AccERR* in response to various environmental stresses

Previous studies have reported that the expression of *AccERR* is induced by abiotic stress, such as low temperature [34]. To investigate whether the expression of *AccERR* was influenced by other environmental stresses, we exposed worker bees (15 days post-emergence) to temperature changes, heavy metal, pesticides and ultraviolet light (UV). qPCR was carried out to analyze the expression of *AccERR* after various environmental stresses. The expression level was normalized to control untreated bees. Worker bees were exposed to low (4℃) and high (42℃) temperatures. As shown in Fig. 5A, the expression of *AccERR* increased rapidly, dropped between 0.25 h and 0.5 h, and then increased slightly. In response to exposure to high temperature, the expression of *AccERR* significantly decreased compared with the control (Fig. 5B). To characterize the effect of $HgCl_2$, worker bees were fed $HgCl_2$ mixed with sugar water. The expression of *AccERR* increased rapidly at 1 h compared with the control group and then decreased slightly (Fig. 5D). The expression of *AccERR* peaked 0.5 h after UV treatment. Interestingly, there were differences in the expression of *AccERR* after treatment with the five pesticides (Fig. 5E-I). Together, these results strongly suggested that *AccERR* may be involved in responses to various oxidative stresses.

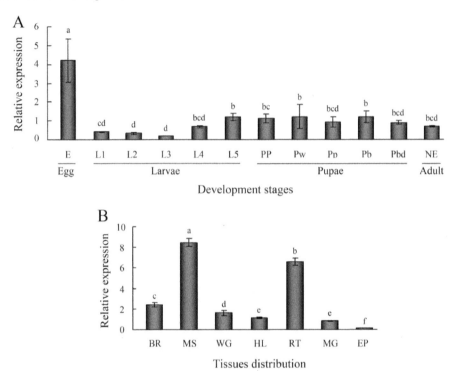

Fig. 4 The expression profile of *AccERR* as determined by qPCR. The relative expression of *AccERR* at different developmental stages and tissues were shown in A and B, respectively. The different developmental stages including egg, larvae (L1-L5, from the first to the fifth instars), pupae (PP, prepupae, Pw, white-eyed pupae, unpigmented cuticle, Pp, pink-eyed pupae, unpigmented cuticle, Pb, brown-eyed pupae, unpigmented cuticle, Pbd, brown-eyed pupae, dark pigmented cuticle), and adult (NE, newly emerged). The various tissues were brain (BR), muscle (MS), wings (WG), haemolymphe (HL), rectum (RT), midgut (MG), epidermis (EP). The different letters above the columns represent the significant differences ($P < 0.01$), as determined by Duncan's multiple range tests.

Western blotting analysis

To further verify the expression patterns of AccERR in response to different abiotic stresses, samples with UV and deltamethrin treatments were subjected to Western blotting. As shown in Fig. 6A, expression of AccERR protein was enhanced by UV treatment and reached the highest level at 0.5 h. Deltamethrin treatments caused the expression of AccERR protein to increase significantly (Fig. 6B). The results revealed that the expression of AccERR was consistent with that at the transcriptional level.

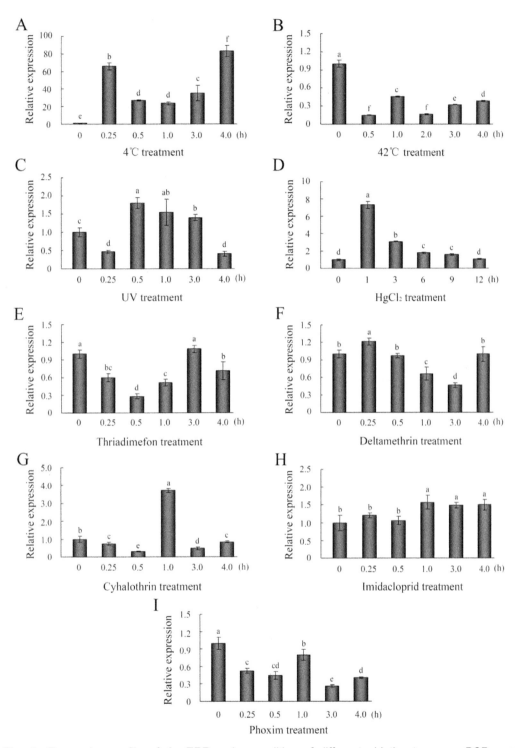

Fig. 5 Expression profile of *AccERR* under condition of different abiotic stresses. qPCR was performed on RNA extracted from adult bees at the marked times after treatment with 4 ℃ (A), 42 ℃ (B), UV (C), $HgCl_2$ (D), thriadimefon (E), deltamethrin (F), cyhalothrin (G), imidacloprid (H) and phoxim (I). The different letters above the columns represent the significant differences ($P<0.01$), as determined by Duncan's multiple range tests.

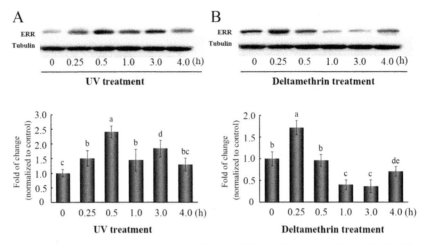

Fig. 6 Western blotting analysis of AccERR. SDS-PAGE (7.5%) was performed with 80 μg of total protein. Tubulin expression is used as an internal control.

Discussion

ERR research has mainly been concentrated in model species, such as *Mus musculus* and *D. melanogaster*; the expression and function of AccERR in *A. cerana cerana* in response to exposure to oxidative stress has not been previously reported. In this study, we identified ERR from *A. cerana cerana* based on its high sequence homology to *A. mellifera* ERR and evaluated the response of AccERR to different abiotic stresses. Our present study is the first to show that AccERR plays an important role in protecting *A. cerana cerana* from oxidative stress.

A sequence analysis revealed that the cloned *AccERR* transcript shared 94% homology at the cDNA level and 100% identity at the amino acid level with *A. mellifera* ERR, suggesting that ERRs have been highly conserved during evolution. In particular, the DNA-binding domain (DBD) has been conserved during evolution.

The qPCR analysis revealed that AccERR expression was the highest in the egg during all the developmental stages, and AccERR expression was higher in pupae than in larvae. Therefore, we speculated that AccERR might play an important role in early development. It is revealed that an ERR family member coordinates metabolism with growth, indicate that a proliferative metabolic program is used in the context of normal development [14]. Studying the tissue distribution of AccERR mRNA can provide a better understanding of physiology. The tissue-specific expression analysis revealed that AccERR mRNA was most highly expressed in the muscle, followed by the rectum (Fig. 4). The thorax is the major movement center and contains substantial muscles that are required for collecting pollen and clearing comb. The presence of AccERR expression in muscular tissues suggests that the function of AccERR is related to movement. This observation is consistent with the high expression of ERRα in mammalian skeletal muscle [35-37]. Furthermore, the 5'-flanking region of *AccERR* contains many predicted transcription factor binding sites, such as a BR-C binding site and an ovo homolog-like TF binding site, that are involved in the early stages of tissue development and growth.

Temperature is an abiotic factor that causes physiological changes in organisms [38]. When the temperature exceeds 36°C, bees fan their wings to reduce the temperature and protect the brood [39, 40]. *A. cerana cerana* are more sensitive to heat shock stress [41]. Moreover, previous results have demonstrated that oxidative stress can be induced by temperature [1, 42]. In this study, the expression of AccERR increased compared to the control group when adults were exposed to a temperature of 4°C. This result is consistent with the high expression of ERRα in the muscle and brown adipose tissue of mammalian organisms [35]. When adults were exposed to temperatures of 42°C, AccERR expression was low compared to the control group. A possible explanation for this result is that AccERR might play a different role in the resistance to heat.

UV is one of the most ubiquitous abiotic stresses for most animals; UV can cause oxidative stress through the generation of ROS, which in turn damage membrane lipids and proteins [43, 44]. Previous studies have found that UV-B (280-320 nm) can affect insect habits and life [45-47]. Perhaps UV-B inhibits a specific DNA repair pathway and interferes with functional protein activity. In addition, UV-A (320-400 nm) radiation damages the photoreceptive cells of the compound eyes of the butterfly species *Papilio xuthus* and *Pieris napi* [48]. In the present study, the qPCR analysis revealed that AccERR was induced by UV radiation. This result corroborated the hypothesis that AccERR plays a crucial role in protecting organisms against abiotic stress.

Previous studies have shown that many heavy metals, such as Cd and Hg, cause the intracellular formation of ROS, which promote cellular oxidative stress [49, 50]. In this study, AccERR expression increased rapidly, peaking at 1 h, and then decreased slowly in response to heavy metal exposure. The increased AccERR expression following $HgCl_2$ exposure could enhance the tolerance of *Apis cerana cerana* to oxidative stress. Considering our result, we hypothesized that heavy metal (such as $HgCl_2$) could cause stress and induce the expression of AccERR through distinct pathways.

Previous studies have demonstrated that insecticides are a significant threat to insects; insecticides impair biochemical and physiological functions by inducing the oxidation of lipid biomembranes [51]. At a later stage of differentiation, ERRα has been shown to be important for the oxidative capacity of myotubes and for skeletal muscle regeneration [52-54]. In this study, we found that when the adults were exposed to triadimefon, deltamethrin, cyhalothrin, and imidacloprid (Fig. 5E-H), the transcripts of AccERR increased before decreasing. The increased AccERR expression following insecticide stressors could increase the tolerance of bees to abiotic stress. A possible explanation for the decreasing may be that with increasing stress, other stress-related genes might play crucial roles in cellular stress responses, and the role of AccERR may be minimized. However, we found that AccERR was repressed during the exposure to phoxim (Fig. 5I), indicating that AccERR might play different roles in resistance to phoxim treatments or might be involved in different signal transduction processes. The expression changes of AccERR mRNA in response to treatment with different insecticides suggested that AccERR might play important roles in the response to abiotic stress.

In conclusion, we identified AccERR from *A. cerana cerana*, analyzed the molecular characteristics of AccERR and determined the AccERR expression levels at various developmental stages. In addition, the expression patterns of AccERR in response to different abiotic stresses suggest that AccERR may play an important role in oxidative stress-related defense mechanisms. All of these results provide useful information for further understanding the expression, regulation, and function of AccERR in *A. cerana cerana*, which may be important for the survival of these insects. The roles of AccERR in other domains are worthy of further inquiry.

References

[1] Kottuparambil, S. UV-B affects photosynthesis, ROS production and motility of the freshwater flagellate, *Euglena agilis* Carter. Aquatic Toxicology. 2012, 122-123 (3), 206-213.

[2] Lee, K. S. Characterization of a silkworm thioredoxin peroxidase that is induced by external temperature stimulus and viral infection. Insect Biochemistry and Molecular Biology. 2005, 35 (1), 73-84.

[3] Committee, N. R. N. A unified nomenclature system for the nuclear receptor superfamily. Cell. 1999, 97 (2), 161-163.

[4] Hong, H. Hormone-independent transcriptional activation and coactivator binding by novel orphan nuclear receptor ERR3. Journal of Biological Chemistry. 1999, 274 (32), 22618-22626.

[5] Tremblay, A. M., Giguère, V. The NR3B subgroup: an ovERRview. Nuclear Receptor Signaling. 2007, 5, e009.

[6] Giguère, V. To ERR in the estrogen pathway. Trends in Endocrinology and Metabolism Tem. 2002, 13 (5), 220-225.

[7] Giguere, V. Orphan nuclear receptors: from gene to function. Endocrine Reviews. 1999, 20 (5), 689-725.

[8] Coward, P., Lee, D., Hull, M. V., Lehmann, J. M. 4-hydroxytamoxifen binds to and deactivates the estrogen-related receptor gamma. Proceedings of the National Academy of Sciences. 2001, 98 (15), 8880-8884.

[9] Taegtmeyer, H., Golfman, L., Sharma, S., Razeghi, P., van Arsdall, M. Linking gene expression to function: metabolic flexibility in the normal and diseased heart. Annals of the New York Academy of Sciences. 2004, 1015, 202-213.

[10] Bonnelye, E., Merdad, L., Kung, V, Aubin, J. E. The orphan nuclear estrogen receptor-related receptor alpha (ERRalpha) is expressed throughout osteoblast differentiation and regulates bone formation *in vitro*. Journal of Cell Biology. 2001, 153 (5), 971-983.

[11] Delhon, I., Gutzwiller, S., Morvan, F., Rangwala, S., Wyder, L., Evans, G., Studer, A., Kneissel, M., Fournier, B. Absence of estrogen receptor-related-α increases osteoblastic differentiation and cancellous bone mineral density. Endocrinology. 2009, 14 (10), 4463-4472.

[12] Teyssier, C., Gallet, M., Monfoulet. L., et al. Absence of ERRalpha in female mice confers resistance to bone loss induced by age or estrogen-deficiency. PLoS One. 2009, 4 (11), 130-142.

[13] Giguère, V. Transcriptional control of energy homeostasis by the estrogen-related receptors. Endocrine Reviews. 2008, 29 (6), 677-696.

[14] Deblois, G., Giguère, V. Oestrogen-related receptors in breast cancer: control of cellular metabolism and beyond. Nature Reviews Cancer. 2013, 13 (1), 27-36.

[15] Tennessen, J. M., Baker, K. D., Lam, G., Evans, J., Thummel, C. S. The *Drosophila* estrogen-related receptor directs a metabolic switch that supports developmental growth. Cell Metabolism. 2011, 13, 139-148.

[16] Xie, W. Constitutive activation of transcription and binding of coactivator by estrogen-related receptors 1 and 2. Molecular Endocrinology. 2000, 13 (12), 2151-2162.

[17] Zhang, Z., Teng, C. Estrogen receptor-related receptor alpha 1 interacts with coactivator and

constitutively activates the estrogen response elements of the human lactoferrin gene. Journal of Biological Chemistry. 2000, 275 (27), 20837-20846.

[18] Vanacker, J. M. Activation of the osteopontin promoter by the orphan nuclear receptor estrogen receptor related alpha. Cell Growth and Differentiation: the Molecular Biology Journal of the American Association for Cancer Research. 1998, 9 (12), 1007-1014.

[19] Lu, D., Kiriyama, Y., Lee, K. Y., Giguère, V. Transcriptional regulation of the estrogen-inducible pS2 breast cancer marker gene by the ERR family of orphan nuclear receptors. Cancer Research. 2001, 61 (18), 6755-6761.

[20] Consortium, H. G. S. Insights into social insects from the genome of the honeybee *Apis mellifera*. Nature. 2006, 443, 931-949.

[21] Michelette, E. D. F., Soares, A. Characterization of preimaginal developmental stages in Africanized honey bee workers (*Apis mellifera* L). Apidologie. 1993, 24 (4), 431-440.

[22] Lourenco, A. P., Mackert, A., Cristino, A. D. S., Simões, Z. L. P. Validation of reference genes for gene expression studies in the honey bee, *Apis mellifera*, by quantitative real-time RT-PCR. Apidologie. 2008, 39 (3), 372-385.

[23] Wang, Y., Baker, N., Amdam, G. V. RNAi-mediated double gene knockdown and gustatory perception measurement in honey bees (*Apis mellifera*). Jove. 2013, 77 (77), e50446.

[24] Livak, K. J. Analysis of relative gene expression data using real-time quantitative PCR and the 2(-Delta Delta C (T)) Method. Methods. 2001, 25 (4), 402-408.

[25] Ballman, K. V. Genetics and genomics: gene expression microarrays. Circulation. 2008, 118, 1593-1597

[26] Rieu, I., Powers, S. J. Real-time quantitative RT-PCR: design, calculations, and statistics. Plant Cell. 2009, 21, 1031-1033.

[27] Yu, F. Molecular cloning and characterization of a thioredoxin peroxidase gene from *Apis cerana cerana*. Insect Molecular Biology. 2011, 20 (3), 367-378.

[28] Bitondi, M. M. G., Simões, Z. L. P. The relationship between level of pollen in the diet, vitellogenin and juvenile hormone titres in Africanized *Apis mellifera* workers. Journal of Apicultural Research. 1996, 35, 27-36.

[29] Ding, Y. Characterization of the promoters of Epsilon glutathione transferases in the mosquito *Anopheles gambiae* and their response to oxidative stress. Biochemical Journal. 2005, 387 (3), 879-888.

[30] Von, K. L., Crossgrove, K., Von, S. D., Guild, G. M., Beckendorf, S. K. The Broad-Complex directly controls a tissue-specific response to the steroid hormone ecdysone at the onset of *Drosophila metamorphosis*. Embo Journal. 1994, 13 (15), 3505-3516.

[31] Štanojević, D. Sequence-specific DNA-binding activities of the gap proteins encoded by hunchback and Krüppel in *Drosophila*. Nature. 1989, 341 (6240), 331-335.

[32] Rørth, P. Specification of C/EBP function during *Drosophila* development by the bZIP basic region. Science. 1994, 266 (5192), 1878-1881.

[33] Ericsson, A., Kotarsky, K., Svensson, M., Sigvardsson, M., Agace, W. Functional characterization of the CCL25 promoter in small intestinal epithelial cells suggests a regulatory role for Caudal-related homeobox (Cdx) transcription factors. Journal of Immunology. 2006, 176 (6), 3642-3651.

[34] Schreiber, S. N., Knutti, D., Brogli, K., Uhlmann, T., Kralli, A. The transcriptional coactivator Pgc-1 regulates the expression and activity of the orphan nuclear receptor estrogen-related receptor A (Err). Journal of Biological Chemistry. 2003, 278 (11), 9013-9018.

[35] Bonnelye, E. The ERR-1 orphan receptor is a transcriptional activator expressed during bone development. Molecular Endocrinology. 1997, 11 (7), 905-916.

[36] Bonnelye, E., Vanacker, J. M., Spruyt, N., Alric, S., Fournier, B., Desbiens, X., Laudet, V. Expression of the estrogen-related receptor 1 (ERR-1) orphan receptor during mouse development. Mechanisms of Development. 1997, 65, 1-85.

[37] Shi, H., Shigeta, H., Yang, N., Fu, K., O'Brian, G., Teng, CT. Human estrogen receptor-like 1

(ESRL1) gene: genomic organization, chromosomal localization, and promoter characterization. Genomics. 1997, 44 (1), 52-60.

[38] Mi, A., Choi, C. Y. Activity of antioxidant enzymes and physiological responses in ark shell, *Scapharca broughtonii*, exposed to thermal and osmotic stress: effects on hemolymph and biochemical parameters. Comparative Biochemistry and Physiology Part B: Biochemistry and Molecular Biology. 2010, 155 (1), 34-42.

[39] Dyer, F. C., Seeley, T. D. Interspecific comparisons of endothermy in honeybees (*Apis cerana cerana*): deviations from the expected size-related patterns. J Exp Biol. 1987, 127, 1-26.

[40] Tautz, J., Maie, S., Groh, C., Rossler, W., Brockmann, A. Behavioral performance in adult honey bees is influenced by the temperature experienced during their pupal development. Proceedings of the National Academy Sciences of the United States of America. 2003, 100 (12), 7343-7347.

[41] Yang, M. X., Wang, Z. W., Li, H., Zhang, Z. Y., Tan, K., Radloff, S. E., Hepburn, H. R. Thermoregulation in mixed-species colonies of honeybees (*Apis cerana cerana* and *Apis mellifera*). Journal of Insect Physiology. 2010, 56 (7), 706-709.

[42] Lushchak, V. I. Environmentally induced oxidative stress in aquatic animals. Aquatic Toxicology. 2010, 101 (1), 13-30.

[43] Ravanat, J. L., Douki, T., Cadet, J. Direct and indirect effects of UV radiation on DNA and its components. Journal of Photochemistry and Photobiology B: Biology. 2001, 63 (1-3), 88-102.

[44] Cadet, J., Sage E., Douki, T. Ultraviolet radiation-mediated damage to cellular DNA. Mutation Research/Fundamental and Molecular Mechanisms of Mutagenesis. 2005, 571 (1-2), 3-17.

[45] Mazza, C. A., Izaguirre, M. M., Jorge, Z., Scopel, A. L., Ballaré, C. L. Insect perception of ambient ultraviolet-B radiation. Ecology Letters volume. 2002, 5 (6), 722-726.

[46] Nguyen, T. T. A proteomic analysis of the aphid *Macrosiphum euphorbiae* under heat and radiation stress. Insect Biochemistry and Molecular Biology. 2009, 39 (1), 20-30.

[47] Jung, I, Kim., T. Y., Kim-Ha, J. Identification of Drosophila SOD3 and its protective role against phototoxic damage to cells. Febs Letters. 2011, 585 (12), 1973-1978.

[48] Meyer-Rochow, V. B. Selective photoreceptor damage in four species of insects induced by experimental exposures to UV-irradiation. Micron. 2002, 33 (1), 23-31.

[49] Park, E. J., Park, K. Induction of reactive oxygen species and apoptosis in BEAS-2B cells by mercuric chloride. Toxicology *in vitro* an International Journal Published in Association with BIBRA. 2007, 21 (5), 789-794.

[50] Tang, T. Molecular cloning and expression patterns of copper/zinc superoxide dismutase and manganese superoxide dismutase in *Musca domestica*. Gene. 2012, 505 (2), 211-220.

[51] Narendra, M., Bhatracharyulu, N. C., Padmavathi, P., Varadacharyulu, N. C. Prallethrin induced biochemical changes in erythrocyte membrane and red cell osmotic haemolysis in human volunteers. Chemosphere. 2007, 67 (6), 1065-1071.

[52] Shao, D, Liu., Y, Liu, X. J., et al. PGC-1β-regulated mitochondrial biogenesis and function in myotubes is mediated by NRF-1 and ERRα. Mitochondrion. 2010, 10, 516-527.

[53] Wang, S. C. M., Myers, S., Dooms, C., Capon, R., Muscat, G. E. O. An ERRβ/γ agonist modulates GRα expression, and glucocorticoid responsive gene expression in skeletal muscle cells. Molecular and Cellular Endocrinology. 2010, 315, 146-152.

[54] Murray, J. Estrogen-related receptor α regulates skeletal myocyte differentiation via modulation of the ERK MAP kinase pathway. American Journal of Physiology - Cell Physiology. 2011, 301 (3), C630.

1.11 The Initial Analysis of a Serine Proteinase Gene (*AccSp10*) from *Apis cerana cerana*: Possible Involvement in Pupal Development, Innate Immunity and Abiotic Stress Responses

Lijun Gao, Hongfang Wang, Zhenguo Liu, Shuchang Liu, Guangdong Zhao, Baohua Xu, Xingqi Guo

Abstract: Serine proteinases play important roles in innate immunity and insect development. We isolated a serine proteinase gene, designated *AccSp10*, from the Chinese honeybees (*Apis cerana cerana*). RT-qPCR and Western blotting analysis at different pupal development stages indicated that *AccSp10* might be involved in melanin formation in pupae and promote pupal development. In adult workers, the expression of *AccSp10* was upregulated by treatments of mimicking harmful environments such as the presence of *Bacillus bombysepticus*, different temperatures (4, 24 and 42 ℃), $HgCl_2$, H_2O_2, and paraquat; the exception was treatment with VC (vitamin C), which did not upregulate *AccSp10* expression. Western blotting confirmed the results. A disc diffusion assay indicated that recombinant AccSp10 accelerated *E. coli* cell death during stimulation with harmful substances ($HgCl_2$, paraquat and cumene hydroperoxide). These findings suggest that *AccSp10* may be involved in the pupal development of Chinese honeybees and protection against microorganisms and abiotic harms.

Keywords: *Apis cerana cerana*; Serine proteinase; Microorganism and abiotic stresses; Expression analysis; Antibacterial activity

Introduction

Insects are the most widely distributed species in the world, often living in environments with many pathogens (for example, fungi and Gram-negative bacteria) and various unfavourable environmental stressors (extreme temperature, heavy metal, pesticides), which can cause serious oxidative damage [1, 2]. All of these factors pose a threat to the lives of insects, including the Chinese honeybees (*Apis cerana cerana*), which is an important species in maintaining the balance of regional ecologies in China.

When pathogens enter insects, insects can initiate a type of humoral immunity, which is an important innate immune response involving serine proteinases that includes coagulation, melanization and Toll pathway signalling [3-5]. These serine proteinase cascade pathways can rapidly amplify responses to infection and stimulate pathogen killing [6]. The coagulation mechanism is well known based on studies in horseshoe crab; plasmozyme is activated by serine proteinase cascade reactions, leading to the coagulation of haemolymph for pathogen clearance [3, 4]. In the melanization pathway, after pathogen-associated molecular patterns such as fungal β-1, 3-glucan or bacterial peptidoglycan are recognized by pattern recognition proteins in insects [7, 8], a series of serine proteinases in haemolymph are activated, which leads to proPO-activating proteinase (PAP) activation and the transformation of proPO (prophenoloxidase) into PO (phenoloxidase) [9, 10]. PO is the key proteinase that catalyses the formation of melanin. This pathway plays an important role in microbe killing and wound healing [5, 11]. The Toll molecule was originally shown to be a type 1 transmembrane receptor that controls dorsal-ventral patterning of the *Drosophila* embryo [12-14] and later was identified to be involved in host resistance against pathogens [15-17]. Many serine proteinases with clip-domains have been found in *Drosophila*, *Manduca sexta* and *Bombyx mori*, which participate in Toll pathway activation and stimulate the synthesis of antimicrobial peptides (metchnikowin thanatin, magainin) [6, 16, 18-22].

Additional studies have found that extracellular serine proteases are required for dorsal-ventral growth of the *Drosophila* embryo [23, 24]. Ahola [25] reported that the serine protease family was associated with larval growth rate, development time and pupal weight in the Glanville fritillary butterfly. In mammalian preimplantation embryos, a serine protease (Hepsin) was also found to be expressed [26]. These imply that serine proteases might play a role in the development of various organisms.

These serine proteinases that participate in immunity and insect development are conserved proteinases in terms of their structure, which contains a conservative catalytic triad with the His, Asp and Ser residues [27-29]. Additionally, many serine proteinases have clip-domains at their amino terminus, but little is known about the functions of clip-domains at present [30-32]. Serine proteinases exist in haemolymph in the form of zymogens, which are cut and activated at specific sites and then activate downstream proteins, when pathogens invade insects [5, 29, 33]. At present, functional serine proteinases have been found in *Bombyx mori*, *Drosophila melanogaster*, *Anopheles gambiae*, *Tenebrio molitor*, *Holotrichia diomphalia* and *Manduca sexta*, particularly in *Drosophila melanogaster* and *Manduca sexta* [6, 9, 34]. However, the functions of most clip-domain proteases are unknown, even in well-studied insect species.

All of the previous data indicate that serine proteinases play a crucial role in innate immunity and insect development. However, the melanization reaction pathway that these enzymes involved has not understood well [6, 35], particularly in the Chinese honeybees (*A. cerana cerana*). Furthermore, little is known about the functions of serine proteinases when insects are subjected to abiotic stress. Thus, the study of serine proteinases in the Chinese honeybees (*A. cerana cerana*) is necessary. Here, we identified a gene (*AccSp10*) from the Chinese honeybees (*A. cerana cerana*) for the first time and predicted that it is a serine proteinase gene that may be involved in Chinese honeybee development and protection against microorganisms and environmental harms.

Materials and Methods

Specimens

Insects used in the following experiments were Chinese honeybees (*A. cerana cerana*) from Shandong Agricultural University (Taian, China). According to the criteria of Michelette and Soares [36], the honeybees in the experiments were used at the following stages: eggs; larvae (L1-L6), pupae including prepupae (P0), white eyes (Pw), pink eyes (Pp), brown eyes (Pb) and dark eyes (Pd); and adults (A1, 1-day post-emergence; A15, 15-day post-emergence; and A30, 30-day post-emergence). The whole bodies of eggs, larvae, pupae and 1-day-old adult workers were collected from the hive. The A15 or A30-day-old adult workers were obtained by marking newly emerged bees with paint. These honeybees were breed at a constant humidity (70%) and temperature (34℃) and under a 24 h dark regimen [37]. The whole bodies of bees were flash-frozen in liquid nitrogen and stored at −80℃.

Bacillus bombysepticus was preserved in our laboratory.

Treatments

The 1-day-old post-emergence adult workers were divided into groups ($n = 40$). Groups 1-3 were exposed to 4, 24 and 42℃, respectively. Groups 4-6 were treated with *Bacillus bombysepticus* ($OD_{600} \approx 100$), $HgCl_2$ (3 mg/ml) and VC (0.02 mg/ml), which was added to food. Group 7 was injected with H_2O_2 (0.5 μl of a 2 mmol/L dilution) between the first and the second abdominal segments. Paraquat (10 μmol/L) was daubed on the thoracic notum of worker bees in Group 8. At specific times, specimens were collected and immediately frozen in liquid nitrogen and stored at −80℃.

Preparation of RNA and cDNA

The extraction of total RNA and complementary DNA (cDNA) synthesis were performed as previously described [2].

PCR amplification conditions

Primers are shown in Table 1. The PCR amplification conditions were 10 min at 94℃, 40 s at 94℃, 40 s at 47℃, 1 min 20 s at 72℃ for 35 cycles, 10 min at 72℃ (primers SC1 and SC2 (listed in Table 1)) and 10 min at 94℃, 40 s at 94℃, 40 s at 50℃, 1 min 30 s at 72℃ for 35 cycles, 10 min at 72℃ [primers RP1 and RP2 (listed in Table 1)] respectively.

Table 1 The primers used in this study

Primers	Primer sequence (5′–3′)	Description
SC1	TGAAGGCATGTCTTGATTATC	Synthesizing cDNA primer, forward
SC2	CCTTTGTATTATTAATCTGGCC	Synthesizing cDNA primer, reverse
SQ1	CAATGAATGGATAAGACCCAGTTG	Q-PCR primer, forward

Continued

Primers	Primer sequence (5'–3')	Description
SQ2	GTCCACCACTATCTCCCTG	Q-PCR primer, reverse
β-s	GTTTTCCCATCTATCGTCGG	Standard control primer, forward
β-x	TTTTCTCCATATCATCCCAG	Standard control primer, reverse
RP1	GGATCCTTAGATAAAGGCGAAGCATG BamH I	Primer of protein expression, forward
RP2	TCCGAGATCTGGCCAAACTATATTCTC Xho I	Primer of protein expression, reverse

Cloning of open reading frame of AccSp10

The internal fragment of *AccSp10* cDNA was obtained by using primers SC1 and SC2, which were designed based on the conserved sequence of *Sp10* from *Apis mellifera*, *Apis dorsata*, and *Apis florea*. Subsequently. The PCR product [open reading frame (ORF) without the 5' and 3' cDNA end] was ligated into the pEASY-T3 vector and transformed into *Escherichia coli* strain DH5α for sequencing.

Bioinformatics analysis

The bioinformatics tools (http://blast.ncbi.nlm.nih.gov/Blast.cgi) at the National Center for Biotechnology Information (NCBI) were used to analyze conserved sequences of *AccSp10*. The homologous sequences of AccSp10 in other species from NCBI were aligned by using the DNAMAN software. The theoretical isoelectric point and molecular mass of the predicted protein were predicted with the online tool PeptideMass (http://us.expasy.org/tools/peptide). SWISS-MODEL was used to predict the advanced structure of the AccSp10 protein. Antibacterial activity loci were predicted by using the website http://aps. unmc.edu/AP/.

RT-qPCR analysis

Reverse transcription quantitative polymerase chain reaction (RT-qPCR) was used to identify the expression patterns of the *AccSp10* gene. Total RNA of different samples was extracted separately and the cDNA was synthesized as described previously [2]. A 278 bp fragment of *AccSp10* was obtained by using primers SQ1 and SQ2 (listed in Table 1). The house-keeping gene *β-actin* from *A. cerana cerana* (XM640276) was used as a control gene, because it has been verified as the most stably expressed gene in honeybees [38, 39]. SYBR Premix Ex *Taq* (TaKaRa, Dalian, China) was used in a 25 μl volume on a CFX96™ Real-Time System (Bio-Rad, Hercules, CA, USA). We have first validated the primers of *AccSp10*. The melting curves had single peak. The amplification efficiency (E) and the correlation coefficients (R^2) were 95.1% and 0.998, respectively. The PCR cycling conditions were as follows: 95℃ for 30 s and then 40 cycles (95℃ for 5 s, 55℃ for 15 s and 72℃ for 15 s). Lastly, there was a melting cycle from 65℃ to 95℃. Each sample was repeated in triplicate, and at least three samples were used for each primer pair. The linear relation, amplification efficiency and data analysis were performed by using the CFX Manager Software version 1.1. The analysis of the significant differences was conducted by using Duncan's multiple range tests with the Statistical Analysis System (SAS) software version 9.1.

Expression and purification of recombinant AccSp10 protein

The ORF fragment of *AccSp10* cDNA (without signal sequence), amplified by using a pair of primers containing *Bam*H I and *Xho* I (primers RP1 and RP2), was inserted into the expression vector pET-21a (+) and was then transformed into *E. coli* BL21 cells. The cells were cultured in Luria-Bertani (LB) broth with kanamycin at 37℃ until the cell density reached 0.4-0.6 OD_{600}. Then, the *AccsSp10* expression was induced with 0.5 mmol/L isopropylthio-β-D-galactoside (IPTG) at 28℃ for 6 h. The bacterial cells were harvested by centrifugation at 5,000 g for 10 min at 4℃, resuspended in a loading buffer.

As described previously, the induced protein was collected and then purified by using a HisTrap^{TM} FF column (GE Healthcare, Uppsala, Sweden) according to the instructions of the manufacturer. 12% sodium dodecyl sulfate polyacrylamide gel electrophoresis (SDS-PAGE) was used to detect the target protein.

Recombinant AccSp10 protein identification

SDS-PAGE was performed to characterize the recombinant AccSp10 protein. A target band was picked up and further identified by matrix-assisted laser desorption/ionization-time-of-flight tandem mass spectrometry (MALDI-TOF/TOF, ultrafleXtreme^{TM}, Brucher, Germany). In brief, the protein sample was digested with trypsin (Promega, Madison, WI) after by DTT reduction and iodine generation acetamide alkylation process. The extracted peptide mixtures were analyzed by MALDI-TOF/TOF MS. All peptide spectra were matched with the online Mascot program (http://www.matrixscience.com) against the NCBIprot databases. Carbamidomethyl of Cys was chosen as fixed modification; oxidation of Met was chosen as dynamic modification; and two missing tryptic cleavage were allowed.

Antibody preparation and Western blotting analysis

The purified recombinant AccSp10 (200 μg) was mixed with an equal volume of Freund's complete adjuvant (Sigma, USA) and was injected into the New Zealand rabbit. The second injection was the same with the first after 2 weeks. Subsequently, three successive injections of 100 μg of protein mixed with an equal volume of Freund's incomplete adjuvant were administered at 2-week intervals. Fourteen days after the final injection, blood was collected and centrifuged at 3,000 r/min for 5 min. The supernatant antibodies were stored at −70℃, for further application.

To analyze the expression of the *AccSp10* gene at the protein level, total protein was extracted from *A. cerana cerana* at four different pupal developmental stages following treatment with microorganisms and at 24℃. Anti-AccSp10 serum was used as the primary antibody at a dilution of 1∶400 (*V/V*), and peroxidase-conjugated goat anti-rabbit IgG was used as the secondary antibody at a dilution of 1∶2,000 (*V/V*). Anti-β-actin (Sigma, USA) was used as a control. The proteins were detected by using a SuperSignal® West Pico Trial Kit (Thermo Science Pierce, IL, USA).

Antibacterial assay of recombinant AccSp10 protein *in vitro*

A modified disc diffusion assay from Burmeister et al. [40] was performed. After overexpressing AccSp10 in *E. coli* BL21, the protein was incubated on LB-kanamycin agar plates at 37°C for 1 h, and filter discs (6 mm diameter) with different concentrations of $HgCl_2$, paraquat and cumene hydroperoxide were placed on the agar surface for 12 h at 37°C. Then, the inhibition zones were measured. *E. coli* BL21 cells with empty pET-21a (+) were used as the negative control.

ProPO activation assay

Recombinant proteins were mixed with tissue protein extract from prepupae, which had a low basal PO activity. The mixtures were incubated at room temperature for 10 min. Protein concentration was measured by using the BCA Protein Assay Kit (Jiancheng, Nanjing, China). 40 μl tissue protein extract with recombinant proteins (30 μg) in 500 μl of phosphate-buffered saline containing protease inhibitors was mixed with 1,500 μl of phosphate-buffered saline saturated with L-3,4-dihydroxiphenylalanine (Sigma, USA). After incubation at room temperature for 30 min, the absorbance at 470 nm of the samples was measured. One unit of PO activity was defined as the unit amount of enzyme producing an increase in absorbance (ΔA_{470}) of 0.001 μg/min. Each experiment was repeated at least three times.

Statistical analysis

Statistics were performed by using the mean ± SD from triplicate experiments ($n = 3$). Statistical significance was determined by Duncan's multiple range test using the SAS version 9.1 software program (SAS Institute, Cary, NC, USA). Significance was set at $P < 0.05$.

Results

Characterization of *AccSp10*

The ORF of *AccSp10* (GenBank accession number: AKT73552) is 1,164 bp and encodes a 387-amino-acid polypeptide with a predicted theoretical isoelectric point (pI) of 7.60 and a molecular mass of 43.39 kDa, as shown in Fig. 1A. Alignment of the AccSp10 protein was performed with homologous proteins from other species, specifically Apis mellifera Sp10 (XP_001120043.2), *Manduca sexta* HP6 (AAV91004.1), *Drosophila melanogaster* persephone (NP_573297.1), *Drosophila melanogaster* snake, isoform A (NP_524338.2) and *Drosophila melanogaster* spatzle-processing enzyme (NP_651168.1). To understand the relationship between structure and function for the AccSp10 protein, the tertiary structures of AccSp10 and cleaved AccSp10 were predicted; the catalytic triad and clip-domain of AccSp10 are identified in blue, as shown in Fig. 1B, C.

Transcription of *AccSp10* and Western blotting analysis at pupal development

The RT-qPCR results showed that the expression of *AccSp10* was low in larvae and the highest in the 30-day-old adult workers. In the pupae, the transcription level of *AccSp10* was the highest in the dark eye stage compared to the other pupal stages (Fig. 2A). From white eyes (Pw) to dark eyes (Pd), pupal colour slowly become black, particularly the eyes, indicating that melanization might be very rich; thus, we selected pupal stages to evaluate the expression of *AccSp10* at the protein level. In the Western blotting assay, anti-AccSp10 was used to detect AccSp10. The results indicated that the protein level of AccSp10 increases from white eyes (Pw) to dark eyes (Pd) in the pupae (Fig. 2B).

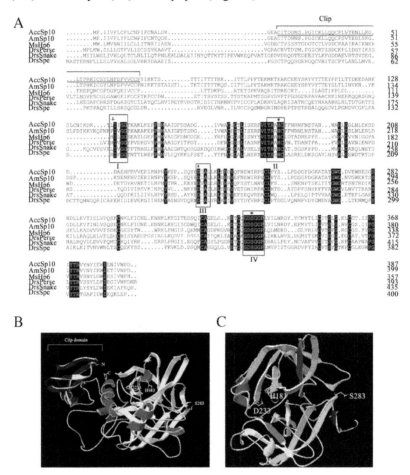

Fig. 1 A. Multiple amino acid sequence alignment of AccSp10. The three conserved regions (II, III and IV) are marked with black boxes. The three amino acid residues (His[183], Asp[233] and Ser[283]) of the catalytic triad of AccSp10 are indicated with dots. The protein active site (Leu[138] and Ile[139]) is indicated with arrows. The clip-domain is underlined. B. The tertiary structure of AccSp10. The tertiary structure of AccSp10 was built by using homology modelling in SWISS-MODEL. The catalytic triad of AccSp10 (His[183], Asp[233] and Ser[283]), N-terminus and C-terminus are shown. The structure of the clip-domain is indicated in blue. C. The tertiary structure of cleaved AccSp10. The catalytic triad (His[183], Asp[233] and Ser[283]) is shown. (For color version, please sweep QR Code in the back cover)

Transcriptional expression of *AccSp10* and Western blotting analysis under microorganism and abiotic conditions

The RT-qPCR results indicated that the expression of *AccSp10* was upregulated under *Bacillus bombysepticus* and abiotic conditions (4℃, 24℃, 42℃, $HgCl_2$, H_2O_2 and paraquat) but not under VC treatment (Fig. 3A, B and Fig. 4A-C). During the Western blotting assay, anti-AccSp10 was used to detect AccSp10. Under microorganism treatment and at 24℃, the protein levels of AccSp10 first declined and then rose at different treatment times (Fig. 3C, D).

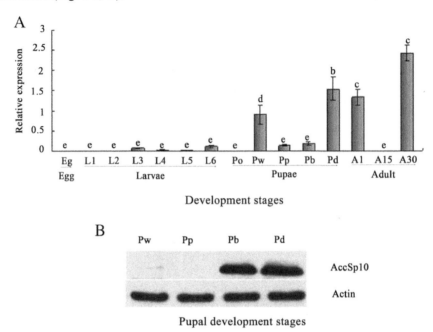

Fig. 2 Expression profiles of *AccSp10* and Western blotting analysis of AccSp10 at different developmental stages. A. Transcriptional expression levels of *AccSp10* at different developmental stages: egg, larvae from the first to the sixth instars (L1-L6), pupae (Po, prepupae; Pw, white eyes; Pp, pink eyes; Pb, brown eyes; Pd, dark eyes), and adults (A1, 1-day post-emergence; A15, 15-days post-emergence; A30, 30-days post-emergence). The data represent the means ± SE (n = 3). Letters above the columns indicate significant differences (P < 0.001). B. Western blotting analysis of AccSp10 at different pupal developmental stages. The β-actin protein was used as an internal control. All lanes were loaded with equivalent amounts of protein.

Overexpression and purification of AccSp10

The ORF fragment of *AccSp10* cDNA (without the signal sequence) was expressed in *E. coli* BL21, and the recombinant AccSp10 protein was purified with HisTrap^{TM} FF columns and is shown in lanes 6 and 7 (Fig. 5). The concentration of the soluble recombinant protein was approximately 1.8 μg/μl.

Recombinant AccSp10 protein identification

The target strip, as shown in Fig. S1A (provided as Supplemental data), was picked up and further identified by MALDI-TOF/TOF MS. The results showed that the protein score (130) was greater than 93 (P < 0.05) and significant; the matched protein was serine proteinase 10 (GenBank accession number: AKT73552.1, *Apis cerana cerana*), as shown in Fig. S1B (provided as Supplemental data). The matched peptides were shown in Fig. S1C (provided as Supplemental data). These results indicated that the recombinant protein was the same protein coding by the sequence.

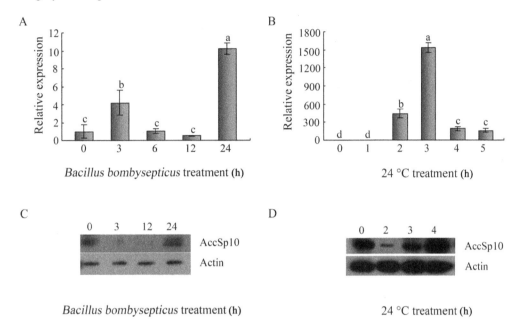

Fig. 3 The expression profiles of *AccSp10* and the Western blotting analysis of AccSp10 with *Bacillus bombysepticus* ($OD_{600} \approx 100$) and 24℃ treatments. A, B. Transcriptional expression levels of *AccSp10* with *Bacillus bombysepticus* ($OD_{600} \approx 100$) and 24℃ treatments. The data are the means ± SE (n = 3). Letters above the columns indicate significant differences ($P < 0.001$). C, D. The Western blotting analysis of AccSp10 with *Bacillus bombysepticus* ($OD_{600} \approx 100$) and 24℃ treatments. The β-actin protein was used as an internal control. All lanes (each condition) were loaded with equivalent amounts of protein.

Antibacterial assay with recombinant AccSp10 protein

To evaluate the capacity of recombinant AccSp10 as a serine protease *in vitro*, a disc diffusion assay was performed. *E. coli* cells with overexpressing AccSp10 were exposed to $HgCl_2$, paraquat and cumene hydroperoxide. The results showed that the death zones of *E. coli* with overexpressing AccSp10 were larger in diameter on experimental plates than control plates (Fig. 6).

After culturing the bacteria with AccSp10 6, 7 and 8 h, the absorbance of bacteria was measured at 600 nm, respectively; results showed that the absorbance of bacteria with AccSp10 was less than the control (without AccSp10) (Fig. 7).

ProPO activation assay

The PO activity of tissue protein extract with recombinant AccSp10 was higher than control (Fig. 8).

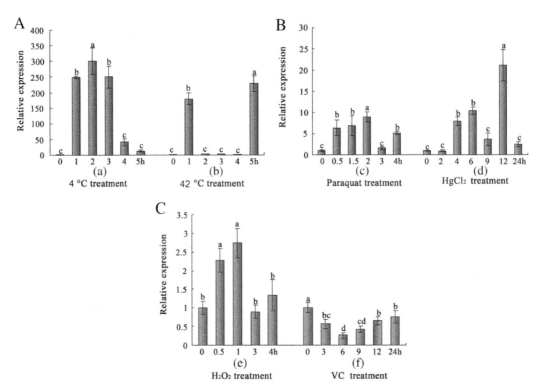

Fig. 4 The expression profiles of *AccSp10* under different abiotic environmental conditions. These conditions were temperature (a), paraquat (b-c), $HgCl_2$ (b-d), H_2O_2 (c-e) and VC (c-f). The *β-actin* gene was used as an internal control. The data represent the means ± SE (n = 3). Letters above the columns indicate significant differences (P < 0.001).

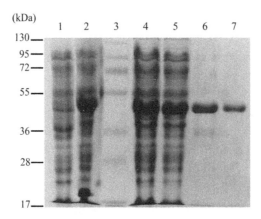

Fig. 5 An SDS-PAGE analysis of the expression and purification of recombinant AccSp10 protein. Lanes 1 and 2, non-induced and induced overexpression of recombinant AccSp10 protein (10 μl, respectively); lane 3, protein molecular weight marker; lanes 4 and 5, suspension of sonicated recombinant AccSp10 protein (10 μl, respectively); lanes 6 and 7, purified recombinant AccSp10 protein (18 μg and 10.8 μg, respectively).

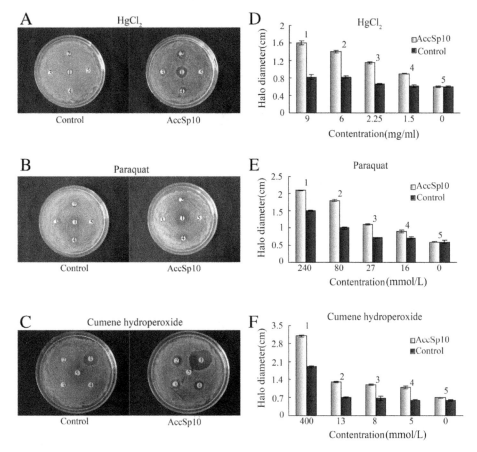

Fig. 6 Antibacterial activity assay with recombinant AccSp10 protein employing the disc diffusion method. LB agar plates were inoculated with 5×10^8 cells. *E. coli* BL21 cells transfected with empty pET-21a (+) vector were used as controls. The filter discs labelled 1 represent the negative controls. Filter discs labelled 2, 3, 4 and 5 represent different concentrations of $HgCl_2$ (A, D), paraquat (B, E) and cumene hydroperoxide (C, F). The diameters of the death zones were measured. The data shown are the means ± SE ($n = 3$).

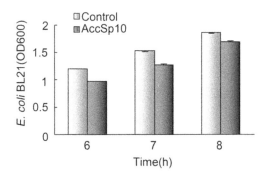

Fig. 7 The AccSp10 inhibition for the growth of *Escherichia. coli* BL21 cells. The inhibition for the growth of *E. coli* BL21 cells. The absorbance was measured after 6, 7 and 8 h, respectively. *E. coli* BL21 cells transfected with empty pET-21a (+) vector were used as controls.

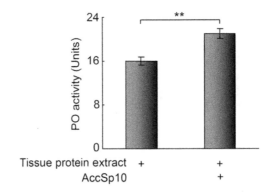

Fig. 8 The proPO activation assay. PO activity was measured by using dopamine as a substrate. The tissue protein extract without recombinant AccSp10 was used as the control. The data represent the means ± SE (n = 3). Asterisks indicate a significant difference from the control at P < 0.001.

Discussion

In insects, serine proteinases with serine in their active centres are likely to be involved in development and innate immunity physiology [30, 41, 42]. According to reports, many serine proteinases with clip-domains were shown to participate in melanism reactions to help clear pathogens and initiate the Toll pathway, in which antimicrobial peptides are produced [5, 6, 43, 44]. These studies were also performed in *Manduca sexta* and *Drosophila melanogaster*. In this study, a serine proteinase gene from *A. cerana cerana*, designated *AccSp10*, was cloned and preliminarily analyzed under simulated environments for the first time. The results suggest that *AccSp10* takes part in stress responses and likely plays a part in honeybee development and protection against harmful damage.

A sequence analysis indicates that *AccSp10* is a typical serine proteinase with three highly conserved sequences (TAAHC, DIAL and GDSGGP) in which the residues His^{183}, Asp^{233} and Ser^{283} form a catalytic triad that is essential for a serine protease to maintain catalytic functions [29]. The protein active site (Leu^{138} and Ile^{139}) was predicted from which AccSp10 was cleaved and activated with a surplus molecular mass of 27.78 kDa. After AccSp10 is activated, it may then activate downstream interacting proteins. These features of serine proteinases have been identified in many studies [5, 7]. In addition, a clip-domain (residues from Cys^{25} to Pro^{70}) was predicted in the *AccSp10* sequence. Studies have indicated that serine proteinases with clip-domains play a significant role in embryonic development and immune responses; [4, 6, 16, 23, 45]. Based on previous studies, it appears that AccSp10 is a typical serine protease with a clip-domain and might possess the same functions as other serine proteases such as *MsHp6* or *DrsPersephone* [6, 46].

Melanin synthesis and deposition in the cuticle occur at specific times during insect development. Phenoloxidase is an important enzyme in the biosynthesis of melanin and exists as an inactive precursor, prophenoloxidase. When phenoloxidase is activated by a cascade of serine proteases, it can catalyse reactions that lead to melanin and cuticle sclerotization [47-49]. Bitondi [50] reported that cuticular pigmentation of *Apis mellifera* correlated with phenoloxidase activation. The process of cuticular pigmentation begins at the pupal phase in

Apis mellifera [51]. This study indicated that serine proteases activate phenoloxidase and give rise to melanin during pupal development. To further probe the function of *AccSp10* in development, the expression pattern of the *AccSp10* gene was evaluated at different developmental stages. The transcription levels of the *AccSp10* gene were measured by using qPCR and exhibited large differences at all experimental stages, as shown in Fig. 2A. The transcription expression of *AccSp10* was the highest in the dark eyes (Pd) stage than the other pupal stages. The Western blotting assay showed that the AccSp10 protein was upregulated gradually with pupal growth, and there was greater accumulation at the brown eyes (Pb) and dark eyes (Pd) stages. Pw (white-eyed) is the junction from larvae to pupae, which is a period during which cells become highly differentiated, and many organs tend to formation, for example, formation of the head, thorax, abdomen and legs; the result of *AccSp10* for RT-qPCR is not in line with Western blotting at Pw stage, as shown in Fig. 2, suggesting that the AccSp10 may quickly participate in other development processes besides melanization reaction or the translator process of AccSp10 may be regulated after transcription. These results indicated that *AccSp10* was likely involved in the production of melanin and promoted the development of pupae, especially changes in body surface colour, which might facilitate improved insect immunity.

Studies have shown that serine proteinases can initiate cascade reactions to resist pathogen invasion and promote wound healing [3, 5, 52]. To further understand the functions of *AccSp10*, external environment conditions were imitated. In this study, we found that the transcriptional expression of *AccSp10* was upregulated when adult workers were treated with *Bacillus bombysepticus*, as shown in Fig. 3A. Western blotting showed that the protein levels of AccSp10 first declined and then rose, as shown in Fig. 3C. This suggested that *AccSp10* may take part in the elimination of *Bacillus bombysepticus* via lash-up and stress reactions. When accepting the external stimulus, organisms could occur lash-up and stress reactions (lash-up reactions showed quick reflexes, but stress reactions slow), *AccSp10* expressed less in the early during lash-up reactions because of instinct, for example, during bacterial immune challenge at 3 h, at the same time AccSp10 protein was largely used to resist abiotic stresses; then, with the emergence of stress reactions, the expression of *AccSp10* was greatly upregulated by hormone; so the transcription of *AccSp10* was lower at 3 h and higher at 24 h, the result of Western blotting was gradually higher from 3 to 24 h and, the protein was less at 3 h than the control because of being consumed. This phenomenon could be observed under other stresses. The expression of *AccSp10* was also upregulated at extreme temperatures (4, 24 and 42°C) and under harmful abiotic conditions at different time points (treated with $HgCl_2$, H_2O_2 and paraquat) (Fig. 4A-C, E). An and Choi (2010) found that temperature is an abiotic environmental factor that induces physiological changes in organisms. When the temperature is higher than a certain range, it can induce damage in insects, including water imbalance, changes in the ion concentration in cells, membrane destruction, and structural changes to DNA and proteins [53-58]. Cold injury globally influenced biological structures and processes and was a primary cause of insect death [59-61]. The best survival temperature of Chinese honeybees was approximately 34°C, so 4, 24 and 42°C were extreme temperatures. When adult workers were exposed to extreme temperatures (4, 24 and 42°C), *AccSp10* mRNA accumulation was greater than controls and higher especially at 24 than 4 and 42°C (Fig. 3B and Fig. 4A). The results indicated that Chinese honeybees had stronger ability to regulate expression of *AccSp10* to pull through abiotic harms at 24 than 4 and 42℃. Upregulation of the AccSp10 protein was similarly shown in the Western blotting assay at 24℃ (Fig. 3D). These results might imply some other potential role of AccSp10 in response to extreme temperatures, suggesting that this gene may play a role in overwintering and oversummering similar with the Hsps [62]. Li [63] reported that heavy metal could destroy the activity of the insect antioxidant enzyme system and cause cell damage. H_2O_2, a typical oxidant, can cause oxidative damage, resulting in cell death [64]. Paraquat, one of the most widely employed herbicides in the world, can produce super oxygen anion free radicals and damage the chloroplast membranes of plant. The upregulated *AccSp10* at different times after exposure to extreme temperatures, $HgCl_2$, H_2O_2 and paraquat suggested that this gene play a potential role in environmental adaptation. However, we also found that *AccSp10* was downregulated when honeybees were treated with VC (Fig. 4C-F). Etlik [65] reported that vitamins protected red blood cells from lipoperoxidation induced by sulphur dioxide. This suggested that the workers likely enhance immunity against oxidative damage under these conditions and maintain life through other processes rather than the *AccSp10* pathway. All of the previous data indicate that *AccSp10* might be involved in pathogen clearance, wound healing and protection against microorganisms and abiotic harm.

To further demonstrate the role of *AccSp10* in pathogen elimination, an *in vitro* disc diffusion assay was performed. The killing zones for *E. coli* overexpressing AccSp10 were larger in diameter on the experimental plates than on control plates (Fig. 6), indicating that AccSp10 plays a role in killing *E. coli* cells after activation by different concentrations of $HgCl_2$, paraquat and cumene hydroperoxide. In addition, while culturing the bacteria with *AccSp10*, we also found that it took longer for the bacteria with *AccSp10* to reach the same OD_{600} (Fig. 7); above indicated that AccSp10 might inhibit bacterial growth. The MIC (minimum inhibitory concentration) assay could further measure the antibacterial specificity of recombinant AccSp10, but because the recombinant protein we got was easy to form precipitation, we felt regretted extremely that we could not obtain a MIC value of recombinant AccSp10. To further explore the anti-bacterial mechanism exerted by AccSp10, the antibacterial activity loci of AccSp10 were predicted by using the website http://aps.unmc.edu/AP/. We found that there were two antimicrobial loci (showed in blue and red colours) present in activated AccSp10, as shown in Fig. S2a and S2b (provided as Supplemental data). This indicated that AccSp10 might play an antimicrobial role like other antimicrobial peptides [21]. All of the previous data suggest that AccSp10 is likely involved in the clearance of *E. coli* cells to protect honeybees against bacterial damage. To further verify this view, we performed the proPO activation assay; result showed that tissue protein extract had low basal PO activity that could be activated dramatically by addition of recombinant AccSp10 (Fig. 8). This powerfully indicated that AccSp10 could kill pathogen by melanization pathway.

In conclusion, we identified a gene from *A. cerana cerana*, *AccSp10*; predicted its physiological course, including its participation during different developmental stages and then analyzed its functions in the elimination of harmful microorganisms and resistance to abiotic stress. Furthermore, new studies will test whether *AccSp10* is really involved in the melanism response or Toll pathway signalling to protect honeybees against pathogen damage. Furthermore, this information will provide a foundation for the prevention and cure of honeybee diseases.

References

[1] Halliwell B., Gutterridge J. M. C. Production Against Radical Damage: Systems with Problems, Second Ed. Free Radical Biology and Medicine Clarendon Press, Oxford. 1989.

[2] Yao, P., Lu, W., Meng, F., Wang, X., Xu, B., Guo, X. Molecular cloning, expression and oxidative stress response of a mitochondrial thioredoxin peroxidase gene (*Acctpx-3*) from *Apis cerana cerana*. Journal of Insect Physiology. 2013, 59, 273-282.

[3] Muta, T., Iwanaga, S. The role of hemolymph coagulation in innate immunity. Curr Opin Immunol. 1996, 8, 41-47.

[4] Iwanaga, S., Kawabata, S., Muta, T. New types of clotting factors and defense molecules found in horseshoe crab hemolymph: their structures and functions. Journal of Biological Chemistry. 1998, 123, 1-15.

[5] Jiang, H., Vilcinskas, A., Kanost, M. R. Immunity in lepidopteran insects. Advances in Experimental Medicine and Biology. 2010, 34, 89-100.

[6] An, C., Ishibashi, J., Ragan, E. J., Jiang, H., Kanost, M. R. Functions of *Manduca sexta* hemolymph proteinases HP6 and HP8 in two innate immune pathway. Journal of Biological Chemistry. 2009, 284, 19716-19726.

[7] Gupta, S., Wang, Y., Jiang, H. *Manduca sexta* prophenoloxidase serine proteinase homologs (SPHs) simultaneously. Insect Biochemistry and Molecular Biology. 2005, 35, 241-248.

[8] Hughes, A. L. Evolution of the betaGRP/GNBP/beta-1,3-glucanase family of insects. Immunogenetics. 2012, 64, 549-558.

[9] Gorman, M. J., Wang, Y., Jiang, H., Kanost, M. R. *Manduca sexta* hemolymph proteinase 21 activates prophenoloxidase-activating proteinase 3 in an insect innate immune response proteinase cascade. Journal of Biological Chemistry. 2007, 282, 11742-11749.

[10] Wang, Y., Jiang, H. Reconstitution of a branch of the *Manduca sexta* prophenoloxidase activation cascade *in vitro*: snake-like hemolymph proteinase 21 (HP21) cleaved by HP14 activates prophenoloxidase-activating proteinase-2 precursor. Insect Biochemistry and Molecular Biology. 2007, 37, 1015-1025.

[11] Mavrouli, M. D., Tsakas, S., Theodorou, G. L., Lampropoulou, M., Marmaras, V. J. Map kinases mediate phagocytosis and melanization via prophenoloxidase activation in medfly hemocytes. Biochimica Et Biophysica Acta-biomembranes. 2005, 1774, 145-156.

[12] Anderson, K. V., Bokla, L., Nüssleinvolhard, C. Establishment of dorsal-ventral polarity in the *Drosophila* embryo: the induction of polarity by the Toll gene product. Cell. 1985, 42, 791-798.

[13] Anderson, K. V., Jürgens, G., Nüsslein-Volhard, C. Establishment of dorsal-ventral polarity in the *Drosophila* embryo: genetic studies on the role of the Toll gene product. Cell. 1985, 42, 779-789.

[14] Hashimoto, C., Hudson, K. L., Kv, A. The Toll gene of *Drosophila*, required for dorsal-ventral embryonic polarity, appears to encode a transmembrane protein. Cell. 1988, 52, 269-279.

[15] Lemaitre, B., Nicolas, E., Michaut, L., Reichhart, J. M., Hoffmann. J. A. The dorsoventral regulatory gene cassette spatzle/Toll/cactus controls the potent antifungal response in *Drosophila* adults. Cell. 1996, 86, 973-983.

[16] Ligoxygakis, P., Pelte, N., Hoffmann, J. A., Reichhart, J. M. Activation of *Drosophila* Toll during fungal infection by a blood serine protease. Science. 2002, 297, 114-116.

[17] Michel, T., Reichhart, J. M., Hoffmann, J. A., Royet, J. *Drosophila* Toll is activated by Gram-positive bacteria through a circulating peptidoglycan recognition protein. Nature. 2001, 414, 756-759.

[18] Kim, C. H., Kim, S. J., Roh, K. B., Lee, B. L. A three-step proteolytic cascade mediates the activation

of the peptidoglycan-induced toll pathway in an insect. Journal of Biological Chemistry. 2008, 283, 1799-7607.

[19] Roh, K. B., Kim, C. H., Lee, H., Park, J. W., Lee, B. L. Proteolytic cascade for the activation of the insect toll pathway induced by the fungal cell wall component. Journal of Biological Chemistry. 2009, 284, 19474-19481.

[20] Kambris, Z., Brun, S., Jang, I. H., Nam, H. J., Lemaitre, B. *Drosophila*, immunity: a large-scale *in vivo* RNAi screen identifies five serine proteases required for Toll activation. Current Biology. 2006, 16, 808-813.

[21] Bulet, P., Hetru, C., Dimarcq, J. L., Hoffmann, D. Antimicrobial peptides in insects; structure and function. Developmental and Comparative Immunology. 1999, 23, 329-344.

[22] Lehrer, R. I., Ganz, T. Antimicrobial peptides in mammalian and insect host defence. Current Opinion in Immunology. 1999, 11, 23-27.

[23] Belvin, M. P., Anderson, K. V. A conserved signalling pathway: the *Drosophila* Toll-dorsal pathway. Annal Review of Cell and Developmental Biology. 1996, 12, 393-416.

[24] Trapasso, L. M., Simpson, R. M. Positive and negative regulation of Easter, a member of the serine protease family hat controls dorsal-ventral patterning in the *Drosophila* embryo. Development. 1998, 125, 1261-1267.

[25] Ahola, V., Koskinen, P., Paulin, L., Lehtonen, R., Hanski, I. Temperature- and sex-related effects of serine protease alleles on larval development in the Glanville fritillary butterfly. Journal of Evolutionary Biology. 2015, 28(12), 2224-2235.

[26] Vu, T. K., Liu, R. W., Haaksma, C. J., Tomasek, J. J., Howard, E. W. Identification and cloning of the membrane-associated serine protease, hepsin, from mouse preimplantation embryos. Journal of Biological Chemistry. 1997, 272, 31315-31320.

[27] Carter, P., Wells, J. A. Dissecting the catalytic triad of a serine protease. Nature. 1988, 332, 564-568.

[28] Polgár, L. The catalytic triad of serine peptidases. Cellular and Molecular Life Sciences Cmls. 2005, 62, 2161-2172.

[29] Jiang, H., Kanost, M. R. The clip-domain family of serine proteinases in arthropods. Insect Biochemistry and Molecular Biology. 2000, 30, 95-105.

[30] Zou, Z., Lopez, D. L., Kanost, M. R., Evans, J. D., Jiang, H. Comparative analysis of serine protease-related genes in the honeybees genome: possible involvement in embryonic development and innate immunity. Insect Molecular Biology. 2006, 15, 603-614.

[31] Ross, J., Jiang, H., Wang, Y. Serine proteases and their homologs in the *Drosophila melanogaster* genome: an initial analysis of sequence conservation and phylogenetic relationships. Gene. 2003, 304, 117-131.

[32] Veillard, F., Troxler, L., Reichhart, J. M. *Drosophila melanogaster* clip-domain serine proteases: structure, function and regulation. Biochimie. 2015, 122, 255-269.

[33] James, M. N., Sielecki, A. R. Molecular structure of an aspartic proteinase zymogen, porcine pepsinogen, at 1.8 A resolution. Nature. 1986, 319, 33-38.

[34] An, C., Budd, A., Michel, K. Characterization of a regulatory unit that controls melanization and affects longevity of mosquitoes. Cellular and Molecular Life Sciences Cmls. 2011, 68, 1929-1939.

[35] Tang, H., Kambris, Z., Lemaitre, B., Hashimoto, C. Two proteases defining a melanization cascade in the immune system of *Drosophila*. Journal of Biological Chemistry. 2006, 281, 28097-28104.

[36] Michelette, E., Soares, A. Characterization of preimaginal developmental stages in Africanized honey bee workers (*Apis mellifera* L.). Apidologie. 1993, 24, 431-440.

[37] Alaux, C., Ducloz, F., Crauser, D., Le, C. Y. Diet effects on honeybees immunocompetence. Biology Letters. 2010, 6, 562-565.

[38] Zhou, Y., Wang, F., Liu, F., Wang, C., Yan, Y., Guo, X., Xu, B. Cloning and molecular identification of triosephosphate isomerase gene from *Apis cerana cerana* and its role in response to various stresses. Apidologie. 2016, 1, 1-13.

[39] Scharlaken, B., De Graaf, D. C., Goossens, K., Brunain, M., Jacobs, F. J. Reference gene selection for

insect expression studies using quantitative real-time PCR, the honeybees, *Apis mellifera*, head after a bacterial challenge. Journal of Insect Science. 2008, 8, 1-10.

[40] Burmeister, C., Luërsen, K., Heinick, A., Hussein, A., Liebau, E. Oxidative stress in *Caenorhabditis elegans*: protective effects of the Omega class glutathione transferase (GSTO-1). Faseb Journal. 2008, 22, 343-354.

[41] Krem, M. M., Di, C. E. Evolution of enzyme cascades from embryonic development to blood coagulation. Trends in Biochemical Sciences. 2002, 27, 67-74.

[42] Rawlings, N. D., Barrett, A. J. Evolutionary families of peptidases. Journal of Cellular Biochemistry. 1993, 290, 205-218.

[43] Aggarwal, K., Silverman, N. Positive and negative regulation of the *Drosophila* immune response. Bmb Reports. 2008, 41, 267-277.

[44] Gabriella, F., Ioannis, E., István, V. A serine proteinase homologue, SPH-3, plays a central role in insect immunity. Journal of Immunology. 2011, 186, 4828-4834.

[45] Kanost, M. R., Jiang, H., Yu, X. Q. Innate immune responses of a lepidopteran insect, *Manduca sexta*. Immunological Reviews. 2004, 198, 97-105.

[46] Ming, M., Obata, F., Kuranaga, E., Miura, M. Persephone/Spatzle pathogen sensors mediate the activation of Toll receptor signalling in response to endogenous danger signals in apoptosis-deficient *Drosophila*. Journal of Biological Chemistry. 2014, 289, 7558-7568.

[47] Andersen, S. O. 2-sclerotization and tanning of the cuticle. *In*: Kerkut, G. A., Gilbert, L. I. Comprehensive Insect Physiology Biochemistry and Pharmacology, Vol. 3. Pergamon Press, Oxford, 1985, 59-74.

[48] Hiruma, K., Riddiford, L. M. Granular phenoloxidase involved in cuticular melanization in the tobacco hornworm: regulation of its synthesis in the epidermis by juvenile hormone. Developmental Biology. 1988, 130, 87-97.

[49] Yoshida, H., Ashida, M. Microbial activation of two serine enzymes and prophenoloxidase in the plasma fraction of hemolymph of the silkworm, *Bombyx mori*. Insect Biochemistry. 1986, 16, 539-545.

[50] Bitondi, M., Mora, I. M., Simões, Z., Figueiredo, V. L. C. The *Apis mellifera*, pupal melanization program is affected by treatment with a juvenile hormone analogue. Journal of Insect Physiology. 1998, 44, 499-507.

[51] Rembold, H., Caste specific modulation of juvenile hormone titers in *Apis mellifera*. Insect Biochemistry. 1987, 17, 1003-1006.

[52] Lavine, M. D., Strand, M. R. Surface characteristics of foreign targets that elicit an encapsulation response by the moth *Pseudoplusia includens*. Journal of Insect Physiology. 2001, 47, 965-974.

[53] An, M. I., Choi, C. Y., Activity of antioxidant enzymes and physiological responses in ark shell, *Scapharca broughtonii*, exposed to thermal and osmotic stress: effects on hemolymph and biochemical parameters. Comparative Biochemistry & Physiology Part B: Biochemistry & Molecular Biology. 2010, 155, 34-42.

[54] Yoder, J. A., Denlinger, D. L. Water balance in flesh fly pupae and water vapor absorption associated with diapause. Journal of Experimental Biology. 1991, 157, 273-286.

[55] Hallman, G. J., Denlinger, D. L. Temperature Sensitivity in Insects and Application in Integrated Pest Management. Westview Press, Oxford. 1998, 29.

[56] Walter, M. F. Biessmann, H., Petersen, N. S. Heat shock causes the collapse of the intermediate filament cytoskeleton in *Drosophila* embryos. Developmental Genetics. 1990, 11, 270-279.

[57] Jaenicke, R. Protein stability and molecular adaptation to extreme conditions. European Journal of Biochemistry. 1991, 202, 715-728.

[58] Greenspan, R. J., Finn, J. A., Hall, J. C. Acetylcholinesterase mutants in drosophila and their effects on the structure and function of the cental nervous system. Journal of Comparative Neurology. 1980, 189, 741-774.

[59] Hodkova, M., Hodek, I. Photoperiod, diapause and cold-hardiness. European Journal of Entomology.

2004, 101, 445-458.

[60] Koštál, V., Vambera, J., Bastl, J. On the nature of pre-freeze mortality in insects: water balance, ion homeostasis and energy charge in the adults of *Pyrrhocoris apterus*. Journal of Experimental Biology. 2004, 207, 1509-1521.

[61] Koštál, V., Renault, D., Mehrabianová, A., Bastl, J. Insect cold tolerance and repair of chill-injury at fluctuating thermal regimes: role of ion homeostasis. Comparative Biochemistry and Physiology Part A: Molecular and Integrative Physiology. 2007, 147, 231-238.

[62] Yang, X. Q., Zhang, Y. L., Gao, P., Jia, L. Y. Characterization of multiple heat-shock protein transcripts from *Cydia pomonella*: their response to extreme temperature and insecticide exposure. Journal of Agricultural and Food Chemistry. 2016, 64, 4288-4298.

[63] Li, L., Liu, X., Guo, Y., Ma, E. Activity of the enzymes of the antioxidative system in cadmium-treated *Oxya chinensis* (Orthoptera Acridoidae). Environmental Toxicology and Pharmacology. 2005, 20, 412-416.

[64] Goldshmit, Y., Erlich, S., Pinkas-Kramarski, R. Neuregulin rescues PC12-ErbB4 cells from cell death induced by H_2O_2. Regulation of reactive oxygen species levels by phosphatidylinositol 3-kinase. Journal of Biological Chemistry. 2001, 276, 46379-46385.

[65] Etlik, O., Tomur, A., Tuncer, M., Ridvanağaoğlu, A. Y., Andaç, O. Protective effect of antioxidant vitamins on red blood cell lipoperoxidation induced by SO_2 inhalation. Journal of Basic and Clinical Physiology and Pharmacology. 1997, 8, 31-43.

1.12 Cloning, Expression Patterns and Preliminary Characterization of *AccCPR24*, a Novel RR-1 Type Cuticle Protein Gene from *Apis cerana cerana*

Xiaoqian Chu, Wenjing Lu, Rujiang Sun, Yuanying Zhang, Xingqi Guo, Baohua Xu

Abstract: Cuticular proteins (CPs) are key components of insect cuticle, a structure that plays a pivotal role in insect development and defense. In this study, we cloned the full-length cDNA of a CP gene from *Apis cerana cerana* (*AccCPR24*). An amino acid sequence alignment indicated that AccCPR24 contains the conserved Rebers and Riddiford consensus sequence and shares high similarity with the genes from other hymenopteran insects. We then isolated the genomic DNA and found that the first intron, which is present in other CP genes, is absent in *AccCPR24*. Real-time quantitative polymerase chain reaction (qPCR) analysis revealed that *AccCPR24* is highly expressed in the late pupal stage and midgut. Expression was inhibited by an exogenous ecdysteroid *in vitro* but was enhanced by this hormone *in vivo*; environmental stressors, such as heavy metal and pesticides, also influenced gene expression. In addition, a disc diffusion assay showed that AccCPR24 enhanced the ability of bacterial cells to resist multiple stresses. We infer from our results that AccCPR24 acts in honeybee development and in protecting these insects from abiotic stress.

Keywords: *AccCPR24*; Cuticle protein; Development; Defense

Introduction

The arthropod epidermis is composed of an epithelium that anchors the apical cuticle and a layered extracellular matrix that protects the insect against pathogens and dehydration and functions as an exoskeleton [1]. Although rigid, the epidermis possesses flexibility to facilitate movement [2]. During insect development, prominent changes occur in the cuticle, which is composed of cuticular proteins (CPs) and chitin, particularly during molting and metamorphosis. Additionally, the physical properties of the cuticle are mainly determined by its protein components and the degree of networking between its various constituents, that is, chitin and CPs [3].

In contrast to the uniformity of chitin [4], CPs are divided into many types. Willis [5] described 12 CP families, including CPR, CPF, CPFL, TWDL, CPLCA, CPLCG, CPLCW,

CPLCP, CPG, apidermin, CPA1 and CPA3. CPs have been widely studied since Rebers and Riddiford [6] first defined the R&R consensus sequence. Although only 38 complete CP sequences were included in the first review of the topic [3], the number has increased by approximately three-fold (to 139) in 10 years [7], and hundreds of CPs now have been annotated [5]. CPs are popular research topics, mainly because of their expression characteristics during molting and because they provide a good model to study the molting control mechanism in insects.

The current studies of CPs are mainly focused on analyses of bioinformatics, developmental expression patterns, and functional roles as a barrier to the environment [5, 8]. The determination of insect genome sequences has provided a vast amount of bioinformatic information. Of the 12 families mentioned above, CPR, named for its conserved R&R consensus sequence, is the largest [5]. The original R&R sequence was G-x(8)-G-x(6)-Y-x- A-x-E-x-GY-x(7)-P-x-P, where x represents any amino acid and the numerals present the number of intervening amino acids. However, the sequence was modified in subsequent studies and extended N-terminal to form the "extended R&R Consensus" [9]. This 68-residue motif was proposed to adopt a β-sheet conformation that was considered to act in the binding to chitin [9-11]. This was confirmed by related experiments performed in *Bombyx mori* [12]. Based on sequences, the R&R consensus motif can be divided into three categories: RR-1 and RR-2 (the most common) and RR-3 (the least common) [13, 14]. RR-1 is mainly present in soft (flexible) cuticles, whereas RR-2 is associated with rigid (hard) cuticles [7].

Togawa et al. [15] found that most CP genes are highly expressed in the period around molting. Thus, it is necessary to consider the relationship between CP genes and hormone regulation during molting. Ecdysteroids (mainly 20E) regulate molting and are released in periodic pulses [8]. These hormones bind to their receptor to regulate the expression of transcription factors (e.g., bFTZ-F1 and DHR38) through concentration fluctuations, which then coordinate the expression of genes downstream [8]. The CPs of *Apis mellifera* and *B. mori* are subjected to direct and/or indirect regulation by 20E [16, 17].

With regard to its role as the outermost barrier, the epidermis has various physiological functions that contribute to insect resistance. Vontas et al. [18] found that a possible CP was significantly upregulated in a resistant strain of *Anopheles stephensi* that survived recurring insecticide selection pressure, demonstrating the potential role of this protein in reducing penetration. Possible CPs from *Leptinotarsa decemlineata* are involved in the response to azinphosmethyl and drought [19], and CPs are induced in *Daphnia pulex* and *Orchesella cincta* exposed to the heavy metal cadmium [20, 21]. We infer the epidermal tissue and exoskeleton act in a range of non-obvious roles in insect biology.

Apis cerana cerana is a honeybee of economic value native to China. This species is widely bred and exhibits a stronger resistance to harsh conditions in comparison to *A. mellifera* [22]. In this study, we report on the function of *AccCPR24* from *A. cerana cerana*. We infer from the results that *AccCPR24* is an RR-1-type CPR CP that participates in development and stress responses.

Materials and Methods

Insect material and treatments

Chinese honeybees were raised in an artificial hive in the experimental apiary of Shandong Agricultural University, Taian, China. Larvae, pupae and adult bees were identified by age, eye color and cuticle pigmentation. Pupal phases include the early pupal phase (Pw) and the pink (Pp) and brown (Pb) eye phases. The insects were divided into several groups of 30 and maintained in 24-well plates (Costar, Corning Incorporated, Corning, NY) in an incubator at a constant temperature (34 ℃) and humidity (70%) under a 24 h dark regimen. Selected tissues, including the brain, muscle, midgut and epidermis, were isolated from the brown (Pb)-eyed pupae. For the *in vivo* hormone treatments, the third-instar larvae that were fed a normal artificial diet [23] and served as the control; groups 1-4 were fed a diet containing 20E at the indicated concentrations. White-eyed pupae were used to test the influence of 20E on epidermization *in vitro*. The cuticle was isolated in sterilized Ringer solution, and the fat body was cleaned; the cuticle was then cultured in 1 × Grace's Insect Basal Medium (Mediatech, Incorporated, Manassas, VA). The cuticle cultured in normal medium served as the control, whereas groups 5-7 were incubated in media containing 20E at final concentrations of 1, 0.1 or 0.001 mg/ml medium. Brown (Pb)- and dark (Pd)-eyed pupae were used to detect the influence of environment stress. For the heavy metal treatment, the pupae were dipped into a heavy metal solution for 5 s, and then placed on a filter paper to dry their surface before they were returned to normal conditions; pupae dipped in water for 5 s served as the control. For the pesticide treatment, the pupae were placed in 24-well plates and covered with a piece of gauze; a drop of pesticide (20 mg/L final concentration) was then dropped onto the gauze covering each well; pupae cultured normally served as the control. All bees were frozen in liquid nitrogen at appropriate time points and stored at −80 ℃ until analysis.

Primers

The primer sequences used in this experiment are listed in Table 1.

Cloning cDNA of *AccCPR24*

Total RNA was extracted with the Trizol reagent (Invitrogen, Carlsbad, CA) according to the instructions of the manufacturer and then digested with RNase-free Dnase I to remove any contaminating DNA. The quality of total RNA was determined by a UV-visible protein nucleic acid analyzer (Thermo Scientific, Wilmington, DE). cDNA was synthesized by using a reverse transcription kit (TransGen Biotech, Beijing, China). The primers AP1 and AP2 were designed to amplify the internal conserved fragment by RT-PCR; two pairs of specific primers (5W/N and 3W/N) were generated to perform 5′ and 3′ rapid amplification of cDNA end (RACE). For 5′ RACE, the first-stand cDNA was purified and poly(C) tailed and used as the template for the first round of amplification with primer 5W and the abridged anchor primer. The resulting product was then used as the nested PCR template with primer 5N and the abridged universal amplification primer. For 3′ RACE, 3W/B26 and 3N/B25 were used in the first and the second rounds, respectively. Based on the 5′, 3′ and intermediate fragments, we isolated the full-length cDNA of *AccCPR24* by using the specific primers QP1 and QP2. The PCR products were purified and cloned into the pEASY-T3 vector (TransGen Biotech) for sequencing (Biosune Biotechnological Company, Shanghai, China). The reaction conditions are presented in Table 2.

Table 1 Primer sequences used in this research

Abbreviation	Primer sequence (5′—3′)	Description
AP1	GGACAGTATGGTGCATTTCG	cDNA sequence primer, forward
AP2	GAATCATCAGGTCCGGGT	cDNA sequence primer, reverse
3W	GCACATTTACCAACTCCTCCTC	3′ RACE forward primer, outer
3N	ATTCCAGAGGCAATTAGACGAGC	3′ RACE forward primer, inner
B26	GACTCTAGACGACATCGA(T)18	3′ RACE universal primer, outer
B25	GACTCTAGACGACATCGA	3′ RACE universal primer, inner
5W	CGAAATGCACCATACTGTCC	5′ RACE reverse primer, outer
5N	CAGCCAAATAAACTGAACACC	5′ RACE reverse primer, inner
AAP	GGCCACGCGTCGACTAGTAC(G)14	Abridged anchor primer
AUAP	GGCCACGCGTCGACTAGTAC	Abridged universal amplification primer
Q1	CAACCATCATCATGCAGCG	Full length cDNA primer, forward
Q2	ATTTGGTCGGATTATACGGC	Full length cDNA primer, reverse
N1	TCAACAACCAACAATACAACCG	Genomic sequence primer, forward
N2	CAGTTTGTGTTGGTGGAGC	Genomic sequence primer, reverse
N3	GTTGCTGAAAGTGGAAGAC	Genomic sequence primer, forward
N4	GAGGAGGAGTTGGTAAATGTG	Genomic sequence primer, reverse
P1	TTCATAATCTTATAATCTCAGTCT	Promoter specific primer, forward
P2	GTACAAATACTGCCAGCAT	Promoter specific primer, reverse
RT1	GCAATATTCAGCAACCATCATCATG	Real-time PCR primer, forward
RT2	GTCTTCCACTTTCAGCAAC	Real-time PCR primer, reverse
β-s	TTATATGCCAACACTGTCCTTT	Standard control primer, forward
β-x	AGAATTGATCCACCAATCCA	Standard control primer, reverse
E1	GGTACCATGCAGCACATATTATTG	Protein expression primer, forward
E2	GAGCTCATATCTTCGGTAAAGATTTG	Protein expression primer, reverse

Table 2 PCR reaction conditions

Primer pair	Amplification conditions
AP1/AP2	5 min at 94℃, 40 s at 94℃, 40 s at 48℃, 1 min at 72℃ for 35 cycles, 5 min at 72℃
3W/B26	5 min at 94℃, 40 s at 94℃, 40 s at 52℃, 30 s at 72℃ for 31 cycles, 5 min at 72℃
3N/B25	5 min at 94℃, 40 s at 94℃, 40 s at 52℃, 30 s at 72℃ for 35 cycles, 5 min at 72℃
5W/AAP	5 min at 94℃, 40 s at 94℃, 40 s at 48℃, 40 s at 72℃ for 31 cycles, 5 min at 72℃
5N/AUAP	5 min at 94℃, 40 s at 94℃, 40 s at 47℃, 40 s at 72℃ for 35 cycles, 5 min at 72℃
Q1/Q2	5 min at 94℃, 40 s at 94℃, 40 s at 48℃, 1 min at 72℃ for 35 cycles, 5 min at 72℃

Continued

Primer pair	Amplification conditions
N1/N2	5 min at 94℃, 40 s at 94℃, 40 s at 49℃, 1 min at 72℃ for 35 cycles, 5 min at 72℃
N3/N4	5 min at 94℃, 40 s at 94℃, 40 s at 47℃, 1 min at 72℃ for 35 cycles, 5 min at 72℃
P1/P2	5 min at 94℃, 40 s at 94℃, 40 s at 43℃, 40 s at 72℃ for 35 cycles, 5 min at 72℃
E1/E2	5 min at 94℃, 40 s at 94℃, 40 s at 50℃, 1 min at 72℃ for 30 cycles, 5 min at 72℃
RT1/RT2	5 min at 94℃, 40 s at 94℃, 30 s at 53℃, 30 s at 72℃ for 28 cycles, 5 min at 72℃
β-s/β-x	5 min at 94℃, 30 s at 94℃, 30 s at 53℃, 30 s at 72℃ for 28 cycles, 5 min at 72℃

Cloning *AccCPR24* genomic DNA and 5′-flanking region

Genomic DNA was exacted from adult bees by using the EasyPure Genomic DNA Extraction Kit (TransGen Biotech). Several pairs of primers were used to isolate *AccCPR24* genomic DNA. We generated the 5′-flanking region by using the specific primers P1/P2, which were designed based on the *A. mellifera* sequence due to the high homology between the two species.

Bioinformatic analysis

Homologous CP sequences from various insects were acquired from GenBank at the National Center for Biotechnology Information (NCBI). The theoretical isoelectric point (pI) and average mass were predicted by using PeptideMass (http://us.expasy.org/tools/peptide-mass.html). A phylogenetic analysis was performed by using Molecular Evolutionary Genetics Analysis (MEGA version 4.1) with the neighbor-joining method. The TFSEARCH database (www.cbrc.jp/research/db/TFSEARCH.html) was used to identify the transcription factor binding sites (TEBSs) in the 5′-flanking region.

qPCR

Total RNA was extracted and reverse transcribed as just described. *β-actin* was used as the reference gene. The specific primers R1/R2 and β-s/β-x were designed and synthesized based on the *AccCPR24* and *β-actin* sequences, respectively. qPCR was performed by using SYBR Premix Ex *Taq* (TaKaRa, Dalian, China) and a CFX 96™ Real-time System (Bio-Rad, Hercules, CA) in a total volume of 25 μl. The reaction conditions were as follow: 95℃ for 30 s, followed by 40 amplification cycles of 95℃ for 5 s, 55℃ for 15 s and 72℃ for 15 s and a melting cycle from 65℃ to 95℃. The data were analyzed by using CFX Manager Software (version 1.1) and the differences in *AccCPR24* expression were calculated by using the $2^{-\Delta\Delta C_t}$ comparative CT method [24]. Each experiment was repeated three times independently.

Disc diffusion assay

An expression vector was constructed and protein solubility was analyzed according to Yu et al. [25]. Briefly, The complete open reading frame of *AccCPR24* was cloned into the pET-30a (+) vector by using its *Kpn* I and *Sac* I sites and expressed in *Escherichia coli* BL21 (DE3). Luria-Bertani (LB) cultured *E. coli* was induced with a final concentration of 1 mg/ml isopropylthio-β-D-galactoside (IPTG) for 6 h. Then the cells were centrifuged at 2431 g for 5 min, and the pellet was resuspended in 20 ml phosphate-buffered saline (pH 7.2). Then the suspension was sonicated for 30 min on ice and centrifuged; the suspension and pellet were analyzed with sodium dodecyl sulphatepoly-acrylamide gel electrophoresis (SDS-PAGE). *Escherichia coli* transformed with pET-30a (+) was used as the control. For disc diffusion assay, both control bacterial cells and those expressing AccCPR24 were cultured in LB liquid medium and induced with IPTG. The cells were spread onto LB-kanamycin agar plates at 5×10^8 cells per plate. The plates were incubated at 37 ℃ for 1 h, and filter discs (8 mm diameter) soaked with the solution to be analyzed were placed on the agar surface for overnight treatment. The next day, the plates were photographed and the killing zones area were measured.

Statistical analysis

The data obtained in the experiment are presented as the means ± standard deviation, with $n = 3$. The overall differences were determined with a one-way ANOVA by using Statistical Analysis System (SAS) software, version 9.1.

Results

Isolation and characterization of AccCPR24

The cDNA of *AccCPR24* (GenBank ID: KC478604) is 638 bp, with an open reading frame of 519 bp, a 5' untranslated region (UTR) of 39 bp and a 3' UTR of 80 bp (Fig. 1A). The molecular weight of the protein is predicted to be 19.2 kDa, with pI of 6.27.

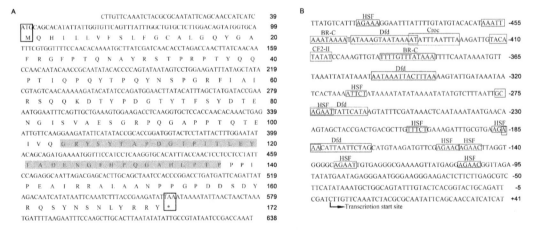

Fig. 1 The cDNA and protein sequence alignments and characterization of the 5'-flanking fragment of AccCPR24. A. The initiation codon, termination codon, and their corresponding amino acids are boxed; the R&R consensus is indicated in gray. B. The transcription start site is marked with an arrowhead. The predicted binding sites for transcription factors are boxed.

1 The Abiotic Stress Response of Some Key Genes in *Apis cerana cerana* | 201

We analyzed the protein sequence by using an online tool based on hidden Markov models (http://bioinformatics2.biol.uoa.gr/cuticleDB/index.jsp): the protein is identified in silico as a RR-1 type, with a score of 67.4 and an *E*-value of 1e-20. The score is a result of multiple alignments between AccCPR24 and proteins in the database. The protein sequence analysis indicated that AccCPR24 has a high similarity with proteins of other hymenopteran insects, including *A. mellifera* CPR24 (95.35%), *Apis florea* SgAbd-1-like (96.53%) and *Bombus impatiens* SgAbd-1-like (88.95%); the R&R consensus sequence was found in all these proteins (Fig. 2A). A neighbor-joining phylogenetic tree was constructed by using MEGA 4 to analyze the evolutionary relationship between the CPs (Fig. 2B). Based on the protein sequence alignment, AccCPR24 shares a high similarity with the CPs of other Hymenoptera.

The full-length sequence of *AccCPR24* is 1,833 bp and contains three exons separated by two introns. The two introns are A/T rich (85.5% and 87.7%, respectively), which is a typical characteristic of insect introns, and are flanked by a 5' splice donor GT and 3' splice acceptor AG. The first intron found in other bee genomes is absent in *AccCPR24* (Fig. 2C).

Additionally, a 500 bp fragment (GenBank ID: KC478605) upstream of the transcription initiation site was obtained, and we examined the regulatory region of *AccCPR24* by using this fragment (Table 3 and Fig. 1B). Several TFBSs were predicted by the web-based software program TFSEARCH. We identified several interesting TFBSs that have been studied in *Drosophila*, including binding sites for nine heat shock factors HSFs, one CF2-II, two Broad-Complex (BR-C) protein, four Dfds and one Croc. In addition, a basic TATA box is also present in this region.

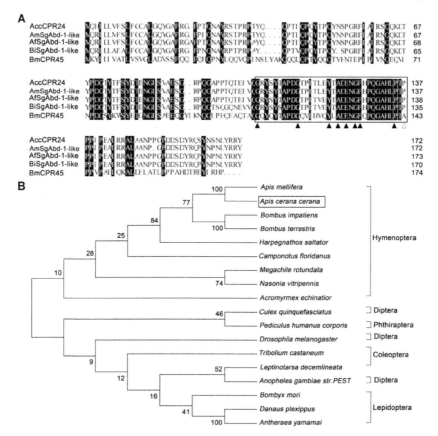

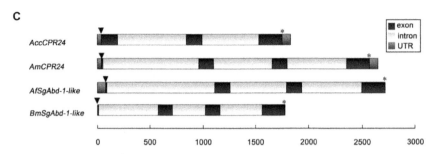

Fig. 2 Homology and evolution analysis of CPs from various species and the gene structure alignment of CPs. A. Homology comparison of CPs from *A. cerana cerana* (KC478604), *A. mellifera* (XP_001123101.2), *Bombyx mori* (NP_001036869.1), *A. florae* (XP_003695125.1), and *Bombus impatiens* (XP_003487686.1). The R&R consensus is marked with a black frame, the conserved amino acids are marked with a solid triangle, and the last proline (alanine in *B. mori*) is marked with an open triangle. B. The phylogenetic tree constructed with CPs from *A. cerana cerana* (KC478604), *A. mellifera* (XP_001123101.2), *B. impatiens* (XP_003487686.1), *Bombus terrestris* (XP_003402560.1) *Megachile rotundata* (XP_003701044.1), *Harpegnathos saltator* (EFN78519.1), *Acromyrmex echinatior* (EGI60084.1), *Camponotus floridanus* (EFN600843.1), *Nasonia vitripennis* (NP_001166266.1), *B. mori* (NP_001166719.1), *T. castaneum* (XP_974125.1), *Danaus plexippus* (EHJ74702.1), *Antheraea yamamai* (AER27818.1), *L. decemlineata* (ABZ04122.1), *Culex quinquefasciatus* (XP_001851251.1), *Pediculus humanus corporis* (XP_002432953.1), *Anopheles gambiae str. PEST* (XP_315458.4) and *Drosophila melanogaster* (NP_648420.1) All proteins are RR-1-type CPs. Most proteins cluster in the same order such as Hymenoptera and Lepidoptera. C. The exons and introns lengths are indicated by the scale underneath. The exons, introns, and UTR are indicated by black, light gray, and deep gray. The initiation codons are denoted with solid triangles, and the terminator codons are denoted with asterisks.

Table 3 Positions and sequences of potential transcription factor binding sites

Transcription factor	Sequence (5'—3')	Position
HSF	AGAAN	−489; −311; −274; −207; −187; −155; −150; −134; −106
BR-C	WWWRKAAASAWAW	−459; −393
CF2-II	RTATATRTA	−413
Dfd	NNNNNATTAMYNNNN	−445; −350; −276; −185
Croc	WANARTAAATATNNNN	−432

Note: Merger base, W represents A or T, S represents C or G; R represents A or G; K represents G or T; Y represents C or T; N represents A, C, G or T

Expression pattern of *AccCPR24* at selected development stages and in isolated tissues

Compared to the pupa stage, the gene expression level was extremely low and showed no significant difference in the larval and adult stages; in particular, it was nearly undetectable in adult stage (Fig. 3C). A stage-specific expression analysis of the different pupal stages showed that the expression levels were very low in the early and intermediate pupal stages, with an abrupt increase at the late pupal stage (Fig. 3A). Various tissues were extracted from pupae, and as shown in Fig. 3B, *AccCPR24* gene expression was detected in all of these tissues, with maximum transcript accumulation in the midgut.

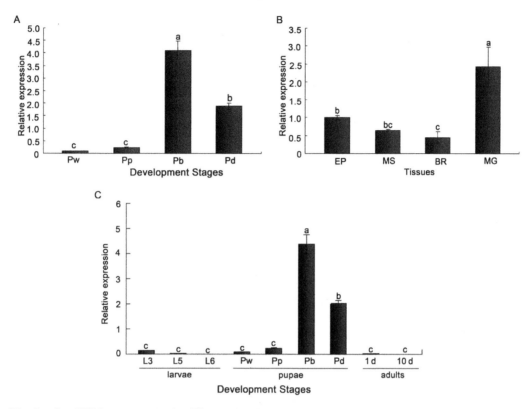

Fig. 3 *AccCPR24* expression in different development stages and different tissues. A. *AccCPR24* expression in different pupal stages, as measured by qRT-PCR. Pw, white eyes, Pp, pink eyes, Pb, brown eyes, Pd, dark eyes. B. *AccCPR24* expression in different tissues measured by qRT-PCR. EP, epidermis, MS, muscle, BR, brain, MG, midgut. The vertical bars indicated the means ± SE (n = 3). C. Different stages including the third-instar larval (L3), the fifth-instar larval (L5), the sixth-instar larval (L6), white (Pw), pink (Pp), brown (Pb) eyes and dark (Pd) eyes pupal phase, 1- and 10-day-old adult workers. The letters above the bars indicated significant differences (P < 0.05).

20E influences *AccCPR24* expression

As shown in Fig. 4A, compared to the controls, 1 and 0.1 μg/ml 20E significantly inhibited *AccCPR24* expression, and the inhibition was enhanced over time. When the concentration was reduced to 0.001 μg/ml, the inhibition was low in the first 6 h but increased over time. We also performed a similar experiment *in vivo* (Fig. 4B) in which the third-instar larvae (L3) were fed various concentrations of 20E for 4 days. In contrast to the *in vitro* results, *AccCPR24* expression was induced by the hormone, and the induction was enhanced within a certain range over time. The level of *AccCPR24* transcripts decreased when the concentration was increased to 1 μg/ml, which is similar to the level of endogenous ecdysteroids, but was still higher than the control.

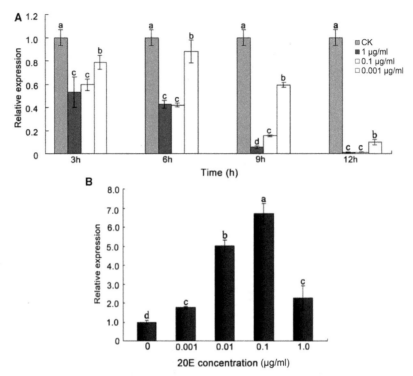

Fig. 4 Effects of 20E on *AccCPR24* expression. A. The influence on pupae of various 20E concentrations for various time periods. B. The influence of various 20E concentration on the third-instar larvae for 4 days; 1 μg/ml is regarded as the endogenous ecdysteroids level [26]. The vertical bars indicate the means ± SE ($n = 3$). The letters above the bars indicated significant differences ($P < 0.05$).

Expression of *AccCPR24* under various environmental pressures

As shown in Fig. 5, *AccCPR24* expression was inhibited by high concentrations of the heavy metals $HgCl_2$ and $CdCl_2$ but enhanced by a low metal concentration and the pesticides paraquat and deltamethrin. The heavy metal-induced downregulation of *AccCPR24* did not change significantly over time. *AccCPR24* expression increased 5.94-fold in the paraquat group compared to the control group and then decreased.

Disc diffusion assay under various pressures

SDS-PAGE analysis showed that the recombinant protein had a molecular mass of approximately 32 kDa that is equal to the predicted molecular weight of the AccCPR24 fusion protein and was produced predominantly as a soluble protein (data not shown). After overnight treatment, the killing zones around the filters soaked in the indicated reagents were much smaller on the plates with the AccCPR24-overexpressing *E. coli* compared to control bacteria (Fig. 6).

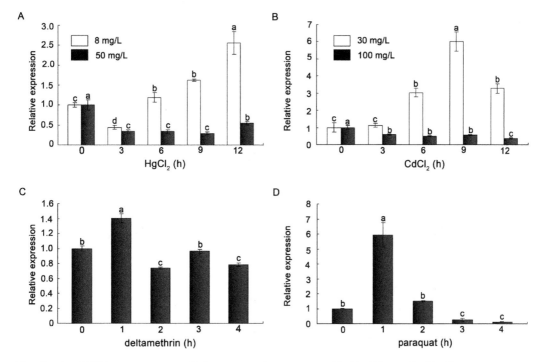

Fig. 5 *AccCPR24* expression with heavy metals exposure, HgCl$_2$ (A) and CdCl$_2$ (B) or pesticides, deltamethrin (C), and paraquat (D). The vertical bars indicated the means ± SE (n = 3). The letters above the bars indicated significant differences (P < 0.05).

Discussion

We named the novel gene described here *AccCPR24*, with the numeral "24" indicating its high similarity to CPR24 from *A. mellifera* rather than the chromosome band on which the gene is located [4]. The protein exhibits several CP characteristics. (i) The mature protein is predicted to consist of 172 residues, in accordance with small size of most CPs. (ii) The protein contains the R&R consensus, though the 24th residue in the consensus is a phenylalanine in *AccCPR24* in contrast to the initial sequence according to a previously reported [5]; this residue is widely present in other insects, such as *A. mellifera*, *B. mori* and *A. florae*, which were included in the homology comparison. (iii) AccCPR24 has aromatic residues that extend beyond the N-terminal of the R&R consensus but not enough to form the extended R&R consensus (pfam00379). (iv) The R&R consensus does not contain cysteine. We infer from these results that *AccCPR24* encodes a typical CP.

Cuticle renewal in insects tracks their developmental stage, with the cuticle in different phases exhibiting different features because of changes in the protein components. *AccCPR24* is highly expressed in the late pupa stage, consistent with the onset of the deposition of the adult cuticle; however, high expression was not maintained in the adult phase, and the early phase transcriptional repression might be due to the pulse of 20E in the pupal stage [27, 28]. The pre-ecdysial expression of *AccCPR24* indicated to us that the protein might be located in the exocuticle. AccCPR24 also exhibits the reported characteristics of an exocuticle protein, such as a low molecular weight and gene repression by 20E [29, 30]. These results indicate that the temporally specific expression of *AccCPR24* is at least partly regulated by 20E during development and that AccCPR24 plays an important role in the differentiation of the adult cuticle.

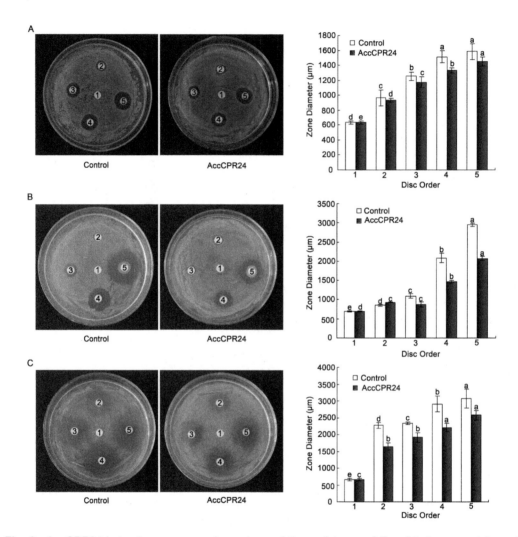

Fig. 6 AccCPR24 heterologous expression enhanced the resistance of *E. coli* to heavy metals and pesticides. The diameters of the killing zone area are compared in the histograms. A. The $HgCl_2$ concentrations of discs 1-5 are 0, 100, 200, 400 and 800 mg/ml, respectively. B. The $CdCl_2$ concentrations of discs 1-5 are 0, 0.001, 0.01, 0.1 and 1 mol/L, respectively. C. The paraquat concentrations of discs 1-5 are 0, 10, 50, 100 and 200 mol/L, respectively. The vertical bars indicated the means ± SE (n = 3). The letters above the bars indicate significant differences (P < 0.05).

BR-C transcription factors compose a protein family encoded by the BR-C genes. BR-C transcription is induced by 20E [31], and these factors mediate the process by which 20E regulates the transcription of CPs. This hypothesis has been confirmed in other insects, such as *Anopheles gambiae* [15], *Bombyx mori* [32] and *Tribolium castaneum* [33]. As expected, we also found two theoretical binding sites for BR-C in the 5'-flanking region of *AccCPR24*. The site distal from the transcriptional start site might bind BR-C Z4, whereas we predict the proximal site binds the BR-C Z1, BR-C Z3 and BR-C Z4 isoforms. This result further suggests to us that the pupa-specific expression of *AccCPR24* is regulated by 20E.

To test our hypothesis, we performed both *in vivo* and *in vitro* experiments. *AccCPR24* expression was inhibited in the presence of 1 and 0.1 μg/ml 20E, with the inhibition increasing over time. Inhibition was initially low at 0.001 μg/ml but was enhanced at 12 h. We infer *AccCPR24* is downregulated by 20E in a concentration-dependent manner. However after feeding the third-instar larvae on diets laced with 20E, the transcription levels increased with increasing concentrations. This may be due to some abnormal effects caused by continuously feeding larvae 20E for 4 days. Indeed 20E analogs have potential for insect control by causing premature molts in susceptible insects [34]. Premature moles require synthesis of more CPs, such as AccCPR24; we speculate that this need drives the upregulation of *AccCPR24*. A 20E pulse is necessary for the function of several CPs [28]. Overall, *AccCPR24* expression is regulated, at least part, by 20E.

Molting and ecdysis are influenced by heavy metal [35, 36], and the expression of genes-encoding proteins involved in cuticle formation varies in insects that are tolerant to heavy metal compared to reference animals [20, 21]. *AccCPR24* expression was repressed by mercury and cadmium. Bacteria transformed with *AccCPR24* exhibited a stronger resistance to such stress compared to controls. Similarly, gene expression was induced by insecticide treatment and a low heavy metal concentration, whereas a high heavy metal concentration suppressed gene expression. Although the gene expression results were differed among the heavy metal and insecticide treatments, the results of our disc diffusion assay were identical. Heterologous expression results showed that the protein can protect *E. coli* against both heavy metal and insecticide exposure. Here we infer a role for AccCPR24 in insect defense.

References

[1] Gangishetti, U., Veerkamp, J., Bezdan, D., Schwarz, H., Lohmann, I., Moussian, B. The transcription factor grainy head and the steroid hormone ecdysone cooperate during differentiation of the skin of *Drosophila melanogaster*. Insect Mol Biol. 2012, 21, 283-295.

[2] Gallot, A., Rispe, C., Leterme, N., Gauthier, J. P., Jaubert-Possamai, S., Tagu, D. Cuticular proteins and seasonal photoperiodism in aphids. Insect Biochem Mol Biol. 2010, 40, 235-240.

[3] Andersen, S. O., Hojrup, P., Roepstorff, P. Insect cuticular proteins. Insect Biochem Mol Biol. 1995, 25, 153-176.

[4] Karouzou, M. V., Spyropoulos, Y., Iconomidou, V. A., Cornman, R. S., Hamodrakas, S. J., Willis, J. H. *Drosophila* cuticular proteins with the R&R consensus: annotation and classification with a new tool for discriminating RR-1 and RR-2 sequences. Insect Biochem Mol Biol. 2007, 37, 754-760.

[5] Willis, J. H. Structural cuticular proteins from arthropods: annotation, nomenclature, and sequence characteristics in the genomics era. Insect Biochem Mol Biol. 2010, 40, 189-204.

[6] Rebers, J. E., Riddiford, L. M. Structure and expression of a *Manduca sexta* larval cuticle gene homologous to *Drosophila* cuticle genes. J Mol Biol. 1988, 203, 411-423.

[7] Willis, J. H., Iconomidou, V. A., Smith, R. F., Hamodrakas S. J. Cuticular proteins. *In*: Gilbert, L. I., Iatrou, K., Gill, S. S. Comprehensive Molecular Insect Science, vol. 4. Elsevier Press, Oxford. 2005, 79-109.

[8] Charles, J. P. The regulation of expression of insect cuticle protein genes. Insect Biochem Mol Biol. 2010, 40, 205-213.

[9] Iconomidou, V. A., Willis J. H., Hamodrakas S. J. Is β-pleated sheet the molecular conformation which dictates the formation of the helicoidal cuticle? Insect Biochem Mol Biol. 1999, 29, 285-292.

[10] Iconomidou, V. A., Chryssikos, G. D., Gionis, V., Willis, J. H., Hamodrakas, S. J. "Soft"-cuticle protein secondary structure as revealed by FT-Raman, ATR FT-IR and CD spectroscopy. Insect Biochem Mol Biol. 2001, 31, 877-885.

[11] Hamodrakas, S. J., Willis, J. H., Iconomidou, V. A. A structural odel of the chitin-binding domain of cuticle proteins. Insect Biochem Mol Biol. 2002, 32, 1577-1583.

[12] Togawa, T., Nakato, H., Izumi, S. Analysis of the chitin recognition mechanism of cuticle proteins from the soft cuticle of the silkworm, *Bombyx mori*. Insect Biochem Mol Biol. 2004, 34, 1059-1067.

[13] Andersen, S. O. Amino acid sequence studies on endocuticular proteins from the desert locust, *Schistocerca gregaria*. Insect Biochem Mol Biol. 1998, 28, 421-434.

[14] Andersen, S. O. Studies on proteins in post-ecdysial nymphal cuticle of locust, *Locusta migratoria*, and cockroach, *Blaberus craniifer*. Insect Biochem Mol Biol. 2000, 30, 569-577.

[15] Togawa, T., Dunn, W. A., Emmons, A. C., Nagao, J., Willis, J. H. Developmental expression patterns of cuticular protein genes with the R&R consensus from *Anopheles gambiae*. Insect Biochem Mol Biol. 2008, 38, 508-519.

[16] Wang, H. B., Moriyama, M., Iwanaga, M., Kawasaki, H. Ecdysone directly and indirectly regulates a cuticle protein gene, BMWCP10, in the wing disc of *Bombyx mori*. Insect Biochem Mol Biol. 2010, 40, 453-459.

[17] Ali, M. S., Wang, H. B., Iwanaga, M., Kawasaki, H. Expression of cuticular protein genes, *BmorCPG11* and *BMWCP5* is differently regulated at the pre-pupal stage in wing discs of *Bombyx mori*. Comp Biochem Physiol B: Biochem Mol Biol. 2012, 162, 44-50.

[18] Vontas, J., David, J. P., Nikou, D., Hemingway, J., Christophides, G. K., Louis, C., Ranson, H. Transcriptional analysis of insecticide resistance in *Anopheles stephensi* using cross-species microarray hybridization. Insect Mol Biol. 2007, 16, 315-324.

[19] Zhang, J., Goyer, C., Pelletier, Y. Environmental stresses induce the expression of putative glycine-rich insect cuticular protein genes in adult *Leptinotarsa decemlineata* (Say). Insect Mol Biol. 2008, 17, 209-216.

[20] Shaw, J. R., Colbourne, J. K., Davey, J. C., Glaholt, S. P., Hampton, T. H., Chen, C. Y., Folt, C. L., Hamilton, J. W. Gene response profiles for *Daphnia pulex* exposed to the environmental stressor cadmium reveals novel crustacean metallothioneins. BMC Genom. 2007, 8, 477.

[21] Roelofs, D., Janssens, T. K., Timmermans, M. J., et al. Adaptive differences in gene expression associated with heavy metal tolerance in the soil arthropod *Orchesella cincta*. Mol Ecol. 2009, 18, 3227-3239.

[22] Wang, Z. L., Liu, T. T., Huang, Z. Y., Wu, X. B., Yan, W. Y., Zeng, Z. J. Transcriptome analysis of the Asian honey bee *Apis cerana cerana*. PLoS One. 2012, 7, 1-11.

[23] Yu, X., Sun, R., Yan, H., Guo, X., Xu, B. Characterization of a sigma class glutathione S-transferase gene in the larvae of the honeybee (*Apis cerana cerana*) on exposure to mercury. Comp Biochem Physiol B: Biochem Mol Biol. 2012, 161, 356-364.

[24] Livak, K. J., Schmittgen, T. D. Analysis of relative gene expression data using real-time quantitative PCR and the 2(−Delta C (C)) method. Methods. 2001, 25, 402-408.

[25] Yu, F., Kang, M., Meng, F., Guo, X., Xu, B. Molecular cloning and characterization of a thioredoxin peroxidase gene from *Apis cerana cerana*. Insect Mol Biol. 2011, 20, 367-378.

[26] Pinto, L. Z., Hartfelder, K., Bitondi, M. M. G., Simões, Z. L. P. Ecdysteroid titer in pupae of highly social bees related to distinct modes of caste development. J Insect Physiol. 2002, 48, 783-790.

[27] Karouzou, M. V., Spyropoulos, Y., Iconomidou, V. A., Cornman, R. S., Hamodrakas, S. J., Willis, J. H. *Drosophila* cuticular proteins with the R&R consensus: annotation and classification with a new tool for discriminating RR-1 and RR-2 sequences. Insect Biochem Mol Biol. 2007, 37, 754-760.

[28] Noji, T., Ote, M., Takeda, M., Mita, K., Shimada, T., Kawasaki, H. Isolation and comparison of different ecdysone-responsive cuticle protein genes in wing discs of *Bombyx mori*. Insect Biochem

Mol Biol. 2003, 33, 671-679.

[29] Soares, M. P., Elias-Neto, M., SimÕes, Z. L., Bitondi, M. M. A cuticle protein gene in the honeybee: expression during development and in relation to the ecdysteroid titer. Insect Biochem Mol Biol. 2007, 37, 1272-1282.

[30] Apple, R. T., Fristrom, J. W. 20-hydroxyecdysone is required for, and negatively regulates transcription of *Drosophila* pupal cuticle protein genes. Dev Biol. 1991, 146, 569-582.

[31] Ijiro, T., Urakawa, H., Yasukochi, Y., Takeda M., Fujiwara Y. cDNA cloning, gene structure, and expression of Broad-Complex (BR-C) genes in the silkworm, *Bombyx mori*. Insect Biochem Mol Biol. 2004, 34, 963-969.

[32] Wang, H. B., Iwanaga, M., Kawasaki, H. Activation of BMWCP10 promoter and regulation by BR-C Z2 in wing disc of *Bombyx mori*. Insect Biochem Mol Biol. 2009, 39, 615-623.

[33] Parthasarathy, R., Tan, A., Bai, H., Palli, S. R. Transcription factor broad suppresses precocious development of adult structures during larval–pupal metamorphosis in the red flour beetle, *Tribolium castaneum*. Mech Dev. 2008, 125, 299-313.

[34] Bengochea, P., Christiaens, O., Amor, F., Viñuela, E., Rougé, P., Medina, P., Smagghe, G. Ecdysteroid receptor docking suggests that dibenzoylhydrazine-based insecticides are devoid of any deleterious effect on the parasitic wasp *Psyttalia concolor* (Hym. Braconidae). Pest Manag Sci. 2012, 68, 976-985.

[35] Bondgaard, M., Bjerregaard, P. Association between cadmium and calcium uptake and distribution during the moult cycle of female shore crabs, *Carcinus maenas*: an *in vivo* study. Aquat Toxicol. 2005, 72, 17-28.

[36] Nørum, U., Bondgaard, M., Pedersen, T. V., Bjerregaard, P. *In vivo* and *in vitro* cadmium accumulation during the moult cycle of the male shore crab *Carcinus maenas*-interaction with calcium metabolism. Aquat Toxicol. 2005, 2, 29-44.

1.13 Molecular Identification and Stress Response of the Apoptosis- inducing Factor Gene 3 (*AccAIF3*) from *Apis cerana cerana*

Fang Wang, Yuanying Zhang, Pengbo Yao, Xingqi Guo, Han Li, Baohua Xu

Abstract: Apoptosis-inducing factors (AIFs) play vital roles in bioenergetic and redox metabolism. Compared with studies focused on the apoptogenic function of AIF, few studies have reported on its oxidoreductase function, especially in insects. In this study, we identified a novel AIF gene *AccAIF3* that was isolated from *Apis cerana cerana*, and investigated its expression and structural features. Quantitative real-time PCR (qRT-PCR) revealed the highest mRNA level of *AccAIF3* in adult bees at 15 days. qRT-PCR results revealed that *AccAIF3* transcripts were induced by cold, $CdCl_2$, $HgCl_2$, UV, pyriproxyfen and cyhalothrin, but were downregulated by ecdysone. We also compared the expression of *AIF3* from *Apis cerana cerana* and *Apis mellifera ligustica* Spinola to *Ascosphaera apis*, and found that *AIF3s* may respond to *Ascosphaera apis* in a species-specific manner. These data suggested that *AccAIF3* may play vital roles in the response to abiotic and biotic stresses and contribute to the adaptability of honeybees to adversities.

Keywords: *Apis cerana cerana*; Biotic and abiotic stresses; Quantitative real-time PCR; *AccAIF3*

Introduction

Apoptosis-inducing factor (AIF) is a phylogenetically conserved redox-active flavoprotein that is confined to mitochondria, where it exerts a vital function in bioenergetic and redox metabolism [1, 2]. The first AIF was identified and named by Guido Kroemer and colleagues [3, 4]. The crystal structures of mouse and human AIF revealed two important regions: a domain with homology to a bacterial NADH-dependent ferredoxin reductase and a domain required for DNA fragmentation (a putative DNA-binding site) [5, 6]. In most mammals, AIF also contains a mitochondrial localization sequence (MLS) in its N-terminal and harbours two nuclear localization signals (NLS) [7]. Two additional members are found in humans, apoptosis-inducing factor-like mitochondrion-associated inducer of death (AMID), which has also been found in the cytosol and has no MLS [8, 9], and apoptosis-inducing factor-like (AIFL), which lacks a clear MLS [10]. In addition to these homologues, some alternative splice forms of AIF are known: AIFexon2a, AIFexon2b, Ash, Ash2 and AIFsh3. The alternative usage of exon 2 (exon 2a or 2b) allows for the production of AIFexon2a (AIF1) and AIFexon2b [11, 12]. Ash is a short variant that contains NLS and the pro-apoptotic segment, but lacks the MLS and redox-active domain [13]. AIFsh2 localizes to mitochondria and has redox function, but lacks the pro-apoptotic domain. AIFsh3 resembles AIFsh2, although it lacks MLS [14]. Among all the forms, AIF is most abundant, so we chose AIF as our experimental subject. As its name indicates, AIF may function in the life of the cell.

Confronted with environmental stress, in principle, cells can react in two distinct ways. To repair damage or resume normal cellular functions, some of the stressed cells can activate defence mechanisms to adapt to stressful conditions. On the other hand, programmed cell death can be activated, leading to apoptosis. The deciding factors in the choice between these two pathways are the stress intensity and cell-intrinsic parameters [15-17]. Previous studies have identified that mitochondria have a dual role in cells. Mitochondria generate ATP by oxidative phosphorylation and play a central role in metabolic pathways. In addition, mitochondria play an important role in the regulation of cell death [18, 19]. As a mitochondrial protein, AIF also functions to activate these two opposing ways.

Initial studies concerning AIFs were focused on its caspase-independent apoptotic function. AIF appears to be a caspase-independent death factor [3]. $AIF^{-/Y}$ ES cells are sensitive to various apoptotic stimuli, such as ultraviolet irradiation (UV) and anisomycin [20]. AIF is essential for the first wave of cell death during mouse morphogenesis. As discussed previously, the addition of recombinant AIF to purified nuclei causes chromatin condensation [3].

However, recent *in vivo* data indicate that in addition to its lethal activity, AIF plays a vital role in the mitochondria of healthy cells by regulating the activity of the mitochondrial respiratory chain complex Ⅰ; this form of AIF is dispensable for apoptosis and has an NADH oxidase domain [21, 22]. Other studies have shown that AIF has life-supporting mitochondrial activity [20, 23]. AIF has been proposed to act as a putative reactive oxygen species (ROS) scavenger, but this hypothesis remains to be confirmed [1]. Another subsequent work showed that the loss of AIF leads to an increase in ROS, and exogenous antioxidants can alleviate the complex Ⅰ deficiency caused by AIF depletion [24]. Compared with studies in mammals, few studies have focused on insects. *D. melanogaster* AIF (DmAIF) is highly conserved in the regions of mouse AIF (mAIF) which was critical for its oxidoreductase activity [20].

Chinese honeybees (Apis cerana cerana) play a vital role in the balance of native ecologies and economic crops. Compared with *Apis mellifera ligustica* Spinola (*A. mellifera* L.), *A. cerana cerana* has unique characteristics, such as heat hardiness and disease resistance, as well as the pollination of flowering crops at low temperature [25]. Environmental pollution threatens the existence of *A. cerana cerana*. In order to protect *A. cerana cerana*, we chose it as experimental insect to investigate its defences against oxidant stress. In this study, we cloned and characterized AIF3 from *A. cerana cerana* and detected its levels of expression at various developmental stages. We also cloned the 5'-flanking region of AccAIF3 and predicted some putative *cis*-acting elements. qRT-PCR results revealed that the AccAIF3 transcript was upregulated in response to various environmental stresses. Our study may contribute to the knowledge of AIF in insects and protect Chinese honeybees from injury.

Materials and Methods

Biological specimens and treatment

A. cerana cerana and *A. mellifera* were maintained at the experimental apiary of Shandong Agricultural University, Taian, China. Eggs were collected from the honeycomb. By using age, eye colour and shape, we divided worker honeybees into ten groups: larvae from the first to the fifth instars (L1-L5), prepupae (PP) and pupae with white eyes (Pw), pink eyes (Pp), brown eyes (Pb) and dark eyes (Pd) [26]. We labeled bees to distinguish their ages. The one-day-old adult worker honeybees (A1) were collected soon after they emerged from the honeycomb, while 15-day post-emergence (A15) and 30-day post-emergence (A30) adults were collected at the entrance of the hive [27]. The number of collected A30 groups was 10, and the number of collected A15 groups was 300 to do experiments. Worker bees collected were maintained in wooden cages (20 × 15 × 10 cm) covered with transparent plastic, and the number of bees per cage was 50. The adults were fed basic adult food containing 30% honey from the source colonies, 70% powdered sugar and water [28] in an incubator with 60% relative humidity at 34 ℃ under a 24 h dark regimen. A15 adults were divided into nine groups ($n = 25$). Groups 1-3 were treated at 4, 25 and 42 ℃. The controls for groups 1-3 were incubated at 34 ℃. Groups 4-6 were treated with pesticides (pyriproxyfen, cyhalothrin and phoxim) as food. The pesticides were dissolved in food at a final concentration of 20 mg/L and 0.5 μl were delivered to each bee in the thoracic notum. All insecticides were purchased from Huayang Technology Co., Ltd. (Shandong, China). The controls for groups 4-6 were fed normal food. Group 7 was subjected to ultraviolet (UV) light (30 MJ/cm^2). Groups 8 and 9 were fed a pollen and sucrose solution containing $HgCl_2$ (3 mg/ml) and $CdCl_2$ (3 mg/ml), and each bee was fed a mixture containing pollen and $HgCl_2$ or $CdCl_2$ (3 μg) with a micropipette. The controls for groups 8 and 9 were fed the basic adult diet. For ecdysone treatment, L3 were placed in 24-well plates and fed food dissolved in a pollen and sucrose solution at a final concentration of 1.0 μg/ml and 0.001 μg/ml. *A. cerana cerana* L3 and *A. mellifera* L3 were placed in 24-well plates and fed brood food containing *Ascosphaera apis* (spore numbers to 10^6/ml). The L3 control group was fed normal food. Three honeybees were harvested at the appropriate time (Table 1) for each condition and were stored at −80 ℃ until they were analyzed. Each treatment had three replicates.

Primers

Primers used for PCR amplification are listed in Table 2.

Extraction of total RNA, isolation of genomic DNA and cDNA synthesis

Total RNA was extracted from worker honeybees by using Trizol reagent (Invitrogen, Carlsbad, CA) and digested with RNase-free Dnase I (Promega, Madison, WI). The first stand cDNA was synthesised by using a reverse transcriptase system (TransGen Biotech, Beijing, China). Genomic DNA was isolated from worker bees by using the EasyPure Genomic DNA Extraction Kit (TransGen Biotech, Beijing, China).

Table 1 The collection time (post treatment) for each experimental group

Experiment condition	Collection time after treatment
4 ℃	0.5, 1, 1.5, 2, and 2.5 h
25 ℃	0.5, 1.5, 2.5, 3.5, and 4.5 h
42 ℃	0.5, 1, 1.5, 2, and 2.5 h
pyriproxyfen	1, 2, 3, 4, 5 and 6 h
cyhalothrin	1, 2, 3, and 4 h
phoxim	0.5, 1, 1.5, 2 and 2.5 h
UV	1, 2, 3, 4 and 5 h
$HgCl_2$	3, 6, 9 and 12 h
$CdCl_2$	3, 6, 9 and 12 h
ecdysone	3, 6, 9 and 12 h
Ascosphaera apis	1, 2 and 3 d

Table 2 The primers used in this study

Abbreviation	Primer sequence (5'—3')	Description
PU	CCACATGCAATTCGAGGTCAATC	Internal fragment primer, forward
PD	GGATCTGCCGCAACACTACTCAT	Internal fragment primer, reverse
5P1	GCTTTCACGCGTACAAAACCAGATG	5' RACE reverse primer, primary
5P2	CCTGTGTGAAGTAAAGCCCCAT	5' RACE reverse primer, nested
AAP	GGCCACGCGTCGACTAGTAC (G)$_{14}$	Abridged anchor primer
AUAP	GGCCACGCGTCGACTAGTAC	Abridged universal amplification primer
3P1	GCAGTCTCCTATCGAAATGCTACC	3' RACE reverse primer, primary
3P2	TTTCAGCTGCAATAGGTCATTATCC	3' RACE reverse primer, nested
B25	GACTCGAGTCGACATCGAT	Universal primer, nested
B26	GACTCGAGTCGACATCGA (T)$_{18}$	Universal primer, primary
QCU	CCACATGCAATTCGAGGTCAATC	Full length primer, forward
QCD	GCAATTTCATTCAATTTTCCACTTG	Full length primer, reverse
N1	GGGCGGAAAAAATTGTAAAGGACT	Genomic sequence primer, forward
N2	GCTTTCACGCGTACAAAACCAGAT	Genomic sequence primer, reverse
N3	CATCTGGTTTTGTACGCGTGAAAG	Genomic sequence primer, forward
N4	GGATCTGCTGCAACACTACTCATAGC	Genomic sequence primer, reverse
RTU	TGGGGCTTTACTTCACACAGG	qRT-PCR primer, forward
RTD	GCTTTCACGCGTACAAAACCAGATG	qRT-PCR primer, reverse
QDU	GCTGGACCTTCTTGTGTGAT	5'-flanking region primer, forward
QDD	GCCTGGTAGAGTATTAGAATTTGGT	5'-flanking region primer, nested
RTU	TGGGGCTTTACTTCACACAGG	qRT-PCR primer, forward
RTD	GCTTTCACGCGTACAAAACCAGATG	qRT-PCR primer, reverse
ARP1-s	GTTTTCCCATCTATCGTCGG	Standard control primer, forward
ARP1-x	TTTTCTCCATATCATCCCAG	Standard control primer, reverse
ARP1-s-y	TTGTATGCCAACACTGTCCTTT	*A. mellifera* standard control primer, forward
ARP1-x-y	AGAATTGACCCACCAATCCA	*A. mellifera* standard control primer, reverse
RTU-y	TGGGGCTTTACTTCATACAGG	*A. mellifera* AIF3, qRT-PCR primer, forward
RTD-y	CGCTTTCACGTGTACAAAACCAGATG	*A. mellifera* AIF3, qRT-PCR primer, reverse

Isolation of the full-length cDNA of *AccAIF3*

The full-length cDNA of *AccAIF3* was isolated by reverse transcription-PCR (RT-PCR) and rapid amplification of cDNA end-PCR (RACE-PCR). Primers PU and PD were designed to amplify the internal fragment, and primers 5P1/5P2, 3P1/3P2, AAP/AUAP and B26/B25 were designed for 5′ RACE and 3′ RACE (all primers are listed in Table 2). Finally, the full-length cDNA of *AccAIF3* was obtained by using primers QCU and QCD (primers are listed in Table 2). All products were purified, cloned into the pEASY-T3 vector (TransGen Biotech, China), transformed into the *Escherichia coli* strain DH5α and then sequenced. All sequencing was done by BioSune by using ABI 3730 XL sequenator.

Amplification of the 5′-flanking region and amplification of the genomic sequence of *AccAIF3*

The promoter region was amplified by the inverse polymerase chain reaction (I-PCR) method by using specific primers (QD51/QD31 and QD52/QD32) (primers are listed in Table 2). The template was genomic DNA digested by restriction endonuclease *Eco*R I and then self-ligated by T4 DNA ligase (TakaRa, Dalian, China). Primers QDU and QDD (primers are listed in Table 2) were used to obtain the full-length sequence of 5′-flanking region. Transcription factor binding sites (TFBS) were searched in the 5′-flanking region by using the MatInspector database (http://www.cbrc.jp/research/db/TFSEARCH.html) and TRANSFAC database (http://www.gene-regulation.com/pub/databases.html). To amplify the genomic sequence of *AccAIF3*, two pairs of specific primers (N1/N2 and N3/N4; shown in Table 2) were designed and used. All PCR products were purified, cloned into the pEASY-T3 vector (TransGen Biotech, China) and transformed into the *Escherichia coli* strain DH5α for sequencing. All sequencing was done by BioSune by using ABI 3730 XL sequenator.

Bioinformatic analysis

Conserved domains of *AccAIF3* were analyzed and compared by using the BLAST algorithm available at NCBI server (https://www.ncbi.nlm.nih.gov/Structure/cdd/wrpsb.cgi). By using DNAMAN software 5.2.2 developed by the American Lynnon Biosoft Company (USA), we identified the open reading frame (ORF) and conducted multiple alignments amongst homologues. A phylogenetic tree was generated by using Molecular Evolutionary Genetic Analysis (MEGA version 4.0). The web software program TFSEARCH (http://www.cbrc.jp/research/ db/TFSEARCH.html) was used to predict the *cis*-elements in the promoter region.

The transcription expression level of *AccAIF3* by using quantitative PCR

Total RNA was extracted from samples from different treatments, and the first-strand cDNA was synthesised as described above. qRT-PCR was performed with a modified qRT-PCR volume containing 1.6 μl of diluted cDNA from different samples, 10.0 μl of Takara SYBR® premix Ex Taq^{TM}, 0.4 μl of each primer (10 pmol/ml), and 7.6 μl of ddH$_2$O from Yao et al. [29] by using primers RTU/RTD designed based on the sequences of *AccAIF3*, primer ARP1-s/ARP1-x designed based on the *A. cerana cerana ARP1* gene (GenBank accession number: HM640276) and primers ARP1-s-y/ARP1-x-y designed based on the *A. mellifera* (GenBank accession number: GB44311). *ARP1* was selected for the normalization of RNA levels [30], and we also tested the expression stability of this endogenous control gene (date not shown). Primers RTU-y/RTd-y were designed based on the *A. mellifera AIF3* gene (GenBank accession NM_001185146.1). All primers above are listed in Table 2. The relative *AccAIF3* gene expression was analyzed by the comparative CT method ($2^{-\Delta\Delta C_t}$ method). At least three individual samples were prepared for each sample and all samples were analyzed in three replicates.

Statistical analysis

Error bars denote the standard error of the mean (SEM) from three independent experiments. Significant differences were determined by Duncan's multiple range tests by using the Statistical Analysis System (SAS) version 9.1 software programs (SAS Institute, Cary, NC, USA).

Results

Isolation and sequence analysis of the full-length cDNA of *AccAIF3*

We cloned the cDNA sequence of *AIF3* from *A. cerana cerana* (GenBank accession number: KF745895). Sequence analysis indicates that the cDNA is 2,586 bp long, containing a 1,734 bp ORF, a 622 bp 5' untranslated region (UTR) and a 230 bp 3' UTR. The ORF encodes a protein of 577 amino acids. Fig. 1a shows that AccAIF3 is most similar to *A. mellifera* AIF3 (AmAIF3, 96.37% identity, GenBank ID: XP_625035.2).

The conserved domains of AccAIF3 were identified by NCBI Conserved Domain Database Search, and included a Rieske domain and a pyridine nucleotide-disulfide oxidoreductase domain (Pyr_redox) (Fig. 1b). The Rieske domain, a [2Fe-2S] cluster binding domain, is involved in electron transfer. The Pyr_redox domain is a small NADH binding domain within a larger FAD binding domain. A phylogenetic tree of AIFs from different species was constructed by using a neighbour-joining method implemented in MEGA, and showed the possible evolutionary relationship among AIFs (Fig. 2). Phylogenetic protein sequence analysis revealed that AccAIF3 was more closely related to AmAIF3, respectively, than other homologues. In addition, this result was in agreement with the relationship predicted from the multiple sequence alignment.

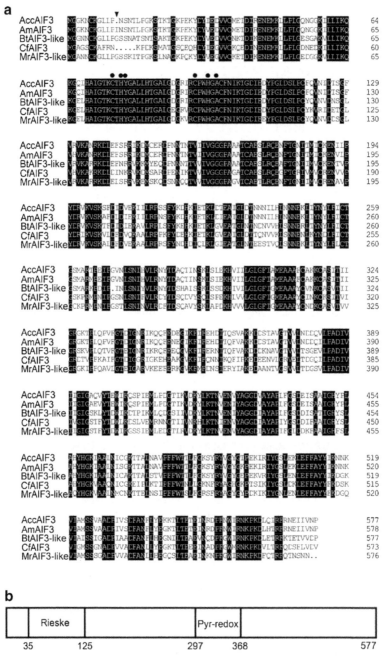

Fig. 1 a. Characterization of apoptosis-inducing factors (AIFs) from various species. An alignment of the identified AccAIF3 protein sequence (GenBank ID: KF745895) with other known insect AIF3s. The amino acid sequences used in the analysis were from *A. mellifera* (AmAIF3, GenBank ID: XP_625035.2), *Bombus terrestris* (BtAIF3-like, XP_003695310.1), *Camponotus floridanus* (Cfaif3, EFN71795.1) and *Megachile rotundata* (Mraif3-like, XP_003703654.1). Identical amino acid residues are shaded in black. A triangle was used to show the specific amino acid that exists in *A. mellifera* but does not exist in *A. cerana cerana*. Circles were used to show the [2Fe-2S] cluster binding site. b. The conserved domains of AccAIF3.

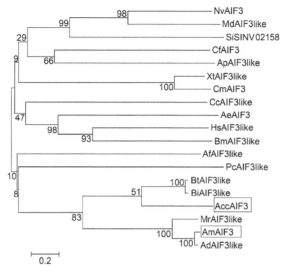

Fig. 2 A phylogenetic tree of AIF3s was constructed from different species by using the neighbour-joining method and MEGA 4.0 software. The amino acid sequences of the AIF3s obtained from NCBI are as follows: NvAIF3 (*Nasonia vitripennis*, XP_001607016.1), MdAIF3-like (*Musca domestica*, XP_005183902.1), Si SINV_02158 (*Solenopsis invicta*, EFZ18689.1), Cfaif3 (*Camponotus floridanus*, EFN71795.1), ApAIF3-like (*Acyrthosiphon pisum*, XP_003241906.1), XtAIF3-like [*Xenopus* (*Silurana*) *tropicalis*, XP_002936104.1], CmSIF3 (*Chelonia mydas*, EMP37830.1), CcAIF3-like (*Ceratitis capitata*, XP_004518444.1), AeAIF3 (*Acromyrmex echinatior*, EGI65010.1), HsAIF3 (*Harpegnathos saltator*, EFN79971.1), BmAIF3-like (*Bombyx mori*, XP_004926894.1), AfAIF3-like (*Apis florea*, XP_003695310.1), PcAIF3-like (*Pomacea canaliculata*, AFQ23946.1), BtAIF3-like (*Bombus terrestris*, XP_003695310.1), BiAIF3-like (*Bombus impatiens*, XP_003489952.1), Mraif3-like (*Megachile rotundata*, XP_003703654.1), AmAIF3 (*A. mellifera*, GenBank ID: XP_625035.2) and AdAIF3-like (*Apis dorsata*, XP_006620108.1).

The genetic structure of *AccAIF3*

The full-length DNA sequence of AccAIF3 is 4,051 bp long, containing six introns inserted within seven exons (GenBank accession number: KF745896). All introns have classical characteristics: they are flanked by the 5'-GT splice donor and 3'-AG splice donor and exhibit high AT content. To further investigate the features of AccAIF3, a comparison was performed. According to Table 3 and Fig. 3, the results indicated that the AT content of the introns was really much higher than that in exons. In addition, the length of exons or introns in AIF3s from different species was also highly conserved; especially in *A. cerana. cerana* and *A. mellifera*, which agrees with the relationship predicted from the multiple sequence alignment.

Characterization of the 5'-flanking region of *AccAIF3*

To investigate the regulatory region of AccAIF3, a 663 bp fragment upstream of the transcription start site was cloned by I-PCR. Some putative TFBS were predicted and are shown in Fig. 4. In addition, the basic TATA-box and CAAT-box, which are critical for transcription, were also found in the 5'-flanking region (data not shown). All the TFBS indicated that AccAIF3 might be involved in the regulation of stress and development at the level of transcription.

Table 3 The lengths of the exons and introns and the AT content in AccAIF3, AmAIF3 and MrAIF3-like

Exon	Gene	Length (bp)	AC content (%)	Intron	Gene	Length (bp)	AT content (%)
1	AccAIF3	70	63	1	AccAIF3	1062	84
	AmAIF3	73	64		AmAIF3	1153	84
	MrAIF3-like	73	54		MrAIF3-like	1012	72
2	AccAIF3	292	66	2	AccAIF3	61	87
	AmAIF3	293	66		AmAIF3	62	89
	MrAIF3-like	293	64		MrAIF3-like	70	79
3	AccAIF3	337	69	3	AccAIF3	66	90
	AmAIF3	337	68		AmAIF3	65	93
	MrAIF3-like	337	61		MrAIF3-like	59	75
4	AccAIF3	351	73	4	AccAIF3	87	90
	AmAIF3	351	74		AmAIF3	106	90
	MrAIF3-like	351	64		MrAIF3-like	69	76
5	AccAIF3	206	68	5	AccAIF3	99	87
	AmAIF3	206	68		AmAIF3	96	85
	MrAIF3-like	206	61		MrAIF3-like	61	76
6	AccAIF3	217	69	6	AccAIF3	90	88
	AmAIF3	217	69		AmAIF3	100	88
	MrAIF3-like	217	64		MrAIF3-like	69	76
7	AccAIF3	260	72				
	AmAIF3	260	73				
	MrAIF3-like	254	63				

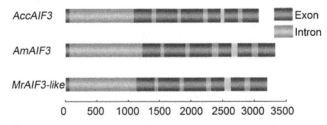

Fig. 3 Comparison of genomic structures among the AIF3s gene family. Lengths of genomic DNA sequences of AccAIF3 (KF745896) and its homologues, including AmAIF3 (NC_007076.3) and MrAIF3-like (NW_003797177.1), are shown. The exons are indicated with black boxes, and the introns are indicated with grey boxes.

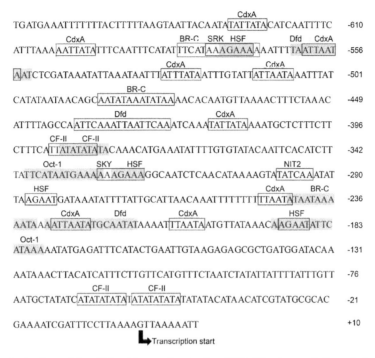

Fig. 4 Characterization of the 5'-flanking region in *AccAIF3*. The transcription start site is marked with an arrowhead. The predicted transcription factor binding sites mentioned in the text are marked as follows: HSF, CF2-II, Dfd, SKY, SRK, Oct-1, NIT2 and BR-C.

Developmental expression pattern of *AccAIF3*

Total RNA from the larvae, pupae and adults was extracted to investigate the expression pattern of AccAIF3 at different developmental stages by using qRT-PCR. In Fig. 5, the results indicate that AccAIF3 is rarely expressed in eggs. High expression levels of AccAIF3 were detected in the Pd, A1 and A30 stages, and the expression of AccAIF3 peaked at the A15 stage.

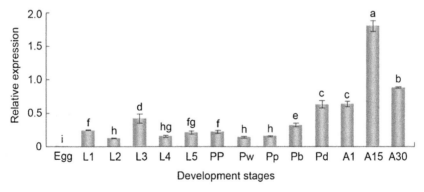

Fig. 5 The expression patterns of *AccAIF3*. mRNA expression of *AccAIF3* at different developmental stages: eggs, larvae (L1-L5), pupae (PP, Pw, Pp, Pb and Pd) and adults (A1, A10 and A30). Each value is presented as the mean (SD) of three replicates. Vertical bars represent the mean ± SE ($n = 3$). Letters above the bars indicate significant differences ($P < 0.001$) identified by SAS software version 9.1.

The expression profile of *AccAIF3* under environmental stress, and exposure to hormones and fungus

To characterize the influence of temperature on the AccAIF3 transcriptional level, qRT-PCR was performed. The expression levels were normalized to those of control worker honeybees. Under cold (4℃) treatments, the expression of AccAIF3 reached a peak after 1.0 h, followed by a sharp decrease (Fig. 6a), while heat treatment (42℃) had slight significant influence on the AccAIF3 transcriptional level (Fig. 6c). Under 25℃ treatment, the level of AccAIF3 transcription was induced within 0.5 h (Fig. 6b).

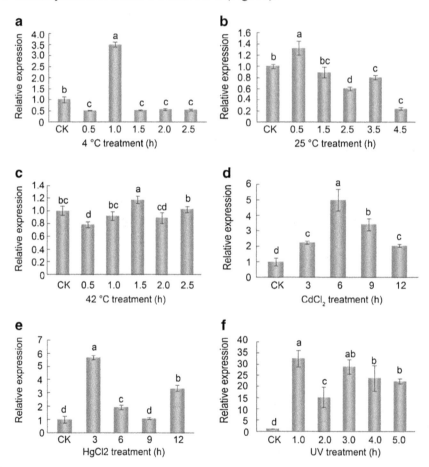

Fig. 6 The expression profiles of *AccAIF3*. a-c. The expression of *AccAIF3* under various temperature stress treatments, with the *β-actin* gene shown for comparison at the appropriate time. d and e. The transcription level of *AccAIF3* under heavy metal treatments with $CdCl_2$ and $HgCl_2$. f. *AccAIF3* mRNA expression level after exposure to UV treatment at the appropriate time. CK is the abbreviation for control check. Vertical bars represent the mean ± SE ($n = 3$). Letters above the bars indicate significant differences ($P < 0.001$) identified by SAS software version 9.1.

Under $CdCl_2$ treatment, the transcriptional rates of AccAIF3 were induced with a peak at 6 h (Fig. 6d). Different from $CdCl_2$ treatment, $HgCl_2$ treatment caused a nearly 6.0-fold relative expression increase in AccAIF3 (Fig. 6e). In addition, UV treatment caused a drastic upregulation of AccAIF3 expression (Fig. 6f), which indicated that AccAIF3 might be associated with UV-related activation.

As worker honeybees work within the environment, they contact unavoidable pesticides, which may lead to death. To study the potential role of AccAIF3 against pesticides, qRT-PCR was performed. Treatment with pyriproxyfen persistently induced a sharply increase in the level of AccAIF3 transcription, except at 4 h (Fig. 7a). The transcription level of AccAIF3 was highest at 2 h following cyhalothrin treatment (Fig. 7b). In contrast to pyriproxyfen and cyhalothrin, no significant alteration was observed after phoxim treatment (Fig. 7c). To study the response to ecdysone, which is a hormone that influences the development of larvae to pupae, 1.0 μg/ml and 0.001 μg/ml of ecdysone were used to treat larvae separately. Fig. 7d and e showed that the transcriptions of AccAIF3 were noticeably inhabited.

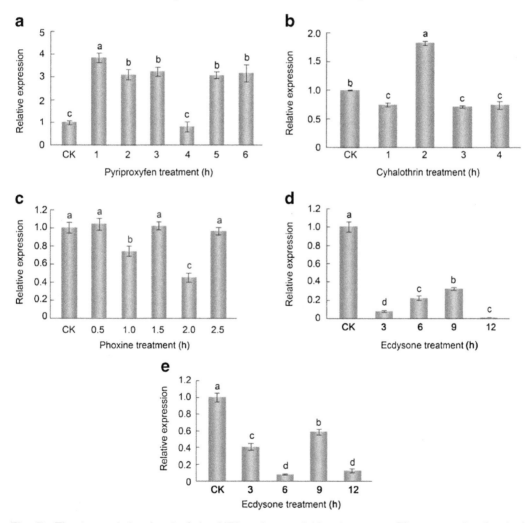

Fig. 7 The transcription level of *AccAIF3* under pesticide stress. a-c. The expression level of *AccAIF3* under pyriproxyfen, cyhalothrin and phoxim treatment. The ecdyson treatments were 1.0 μg/ml (d) and 0.001 μg/ml (e). CK is the abbreviation for control check. Each value is presented as the mean (SD) of three replicates. Letters above the bars indicate significant differences ($P < 0.001$) identified by SAS software version 9.1.

Ascosphaera apis, a primary fungus, threatens the life of the honeybee. To detect the differences in AccAIF3 and AmAIF3 in response to *Ascosphaera apis*, qRT-PCR was performed. Fig. 8 showed that the transcript rate of AccAIF3 reached a peak at approximately 5.0-fold relative to the normal, while that of AIF3 in *A. mellifera* was not significantly altered, which showed that AccAIF3 might play an important role in resistance to *Ascosphaera apis*.

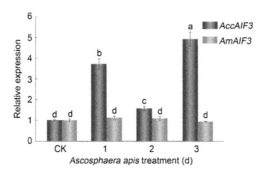

Fig. 8 Functional comparison of *AccAIF3* and *AmAIF3* after exposure *Ascosphaera apis* for 1, 2 and 3 d. CK is the abbreviation for control check. The expression levels of *AccAIF3* are indicated with grey boxes, and the transcription of *AmAIF3* is indicated with black boxes. Vertical bars represent the mean ± SE ($n = 3$). Letters above the bars indicate significant differences ($P < 0.001$) identified by SAS software version 9.1.

Discussion

Apoptosis is pivotal for the development and maintenance of healthy tissues by removing damaged and mutated cells and preventing the induction of cancer. The deregulation of apoptosis leads to the progression of tumours and malignancy due to an accumulation of gene mutations and genetic instability [31, 32]. AIFs are factors involved in apoptosis. Although many studies have focused on the mammals AIFs, few reports have investigated AIFs in insects. In this study, we reported the first cloning and characterization of the *AIF3* gene from *A. cerana cerana*, and we have studied its expression level in response to various stresses.

AmAIF3 has another name in NCBI: *A. mellifera* thioredoxin reductase 3 (AmTrxR3) (GenBank accession number: XM_625032.4). TrxRs is a flavoprotein that contains an FAD prosthetic group, an NADPH binding site and an active site [33]. Studies have demonstrated that TrxRs in their oxidized form can protect organisms against oxidant injury [34]. In addition, TrxRs also play roles in cell growth and transformation, and the recycling of ascorbate [33]. Among all known enzymatic functions of eukaryotic proteins, AIF has the strongest homology with plant semidehydroascorbate and ascorbate reductases [3]. In this respect, AccAIF3 may have a potential function such as AmTrxR3 in maintaining the redox homeostasis of a cell.

To better understand the potential roles of AccAIF3, the 5'-flanking region was isolated. Cell factor 2-II (CF2-II), related to development and growth [35], was identified. Therefore, we studied the expression pattern of AccAIF3 throughout development. Adult worker honeybees have to forage for pollen and nectar at 2-3 weeks of age [36]. We hypothesized that when worker honeybees fly to forage, the expression of AccAIF3 is needed to protect organisms against various environmental stresses. Interestingly, the qRT-PCR results demonstrated that the transcription level of AccAIF3 was the highest at A15 (Fig. 5). This stage is approximately when worker honeybees begin to forage outside.

Other TFBS heat shock factors (HSFs) related to heat-induced transcriptional activation [37] were also found in the region. Temperature can cause physiological changes in organisms and is one of the most important abiotic environmental factors [38]. To avoid harm caused by temperature, organisms have evolved many strategies, such as changing the cell membrane fluidity, expressing some proteins and seeking shelter [39]. To detect whether AccAIF3 was associated with temperature stress, the expression levels of AccAIF3 were measured. The data showed that the AccAIF3 transcription levels were significantly increased in a short time when bees exposed to 4 and 25°C, while no significant alteration was observed at 42°C (Fig. 6a-c), indicating that *A. cerana cerana* may adapt to heat conditions. These data indicated that AccAIF3 might be associated with temperature-induced transcriptional activation. In addition, binding sites for Broad-Complex (BR-C), which contributes to the tissue-specific response to ecdysone [40], were identified in the promoter. To study the responses of AccAIF3 to ecdysone, qRT-PCR was performed. Results indicated that the transcription of AccAIF3 was inhibited in response to different concentrations of ecdysone (Fig. 7d, e). The reasons that the results were opposite to the function of the putative BR-C may be due to the material, the treatment method and the collection time.

In addition to the TFBS mentioned above, CdxAs were also identified. Frumkin et al. [41] demonstrated that CdxAs play roles in gut closure, indicating that AccAIF3 may be involved in gut protection. Some diseases initially occur in the gut. For example, *Ascosphaera apis* is a filamentous fungus that causes chalkbrood [42, 43]. After ingestion, the fungal spores germinate in the larval gut, and then mycelia cross the gut lining and proliferate through the body cavity. Chalkbrood can lead to heavy losses in affected honeybee colonies [44]. At present, very few studies have focused on the molecular mechanism of honeybee immune responses to chalkbrood [45]. In this study, a comparison was performed to investigate the changes in AccAIF3 and AmAIF3 after exposure to *Ascosphaera apis*. Our data indicated that despite the high homology between AccAIF3 and AmAIF3, the responses were different (Fig. 8). AccAIF3 may be involved in the protection against *Ascosphaera apis*, while AmAIF3 may not be relevant. This phenomenon perhaps resulted from the *cis*-acting factors Dfd, which determine the specificity of homeotic gene action [46].

Furthermore, we also detected changes in the expression levels of AccAIF3 with exposure to various environmental stresses, such as heavy metal, UV, and pesticides. Heavy metal contamination remains a critical environmental problem [47]. Worker honeybees foraging for nectar and pollen outside may contact heavy metal. qRT-PCR results revealed that both $CdCl_2$ and $HgCl_2$ induce the transcription level of AccAIF3 in a time-dependent manner (Fig. 6d, e), and AccAIF3 may function in avoiding injury. UV is harmful to organisms and can destroy the structure of DNA, inhibit DNA repair pathways and interfere with the functions of proteins [48]. The transcription levels of AccAIF3 were dramatically enhanced by UV treatment (Fig. 6f), suggesting a role for AccAIF3 in the enzymatic protection against oxidative stress. The excessive use of pesticides has a negative impact on the normal development of the digestive system and glands in insects [49]. Fig. 7a-c showed that AccAIF3 might protect bees against pesticides. Murakami and Johnson [50] revealed that increased resistance to environmental stress is necessary for protection, and enhanced resistance to ROS may be efficient for the extension of life in the fruit flies.

In conclusion, we cloned the *AIF3* gene from *A. cerana cerana*, and the expression of AccAIF3 peaked at the A15 stage during development. The transcription levels of AccAIF3 were significantly increased by exposure to cold, $HgCl_2$, $CdCl_2$, UV, pesticides and *Ascosphaera apis*. The increased expression level perhaps partially corresponds to the increased levels of ROS. Taken together, AccAIF3 may have an important influence on the response to environmental stress and result in the elevated resistance of honeybees to adversities.

References

[1] Klein, J. A., Longo-Guess, C. M., Rossmann, M. P., Seburn, K. L., Hurd, R. E., Frankel, W. N., Bronson, R. T., Ackerman, S. L. The harlequin mouse mutation downregulates apoptosis-inducing factor. Nature. 2002, 419 (6905), 367-374.

[2] Urbano, A., Lakshmanan, U., Choo, P. H., Kwan, J. C., Ng, P. Y., Guo, K., Dhakshinamoorthy, S., Porter, A. AIF suppresses chemical stress-induced apoptosis and maintains the transformed state of tumor cells. EMBO J. 2005, 24 (15), 2815-2826.

[3] Susin, S. A., Lorenzo, H. K., Zamzami, N., et al. Molecular characterization of mitochondrial apoptosis-inducing factor. Nature. 1999, 397 (6718), 441-446.

[4] Kroemer, G., Reed, J. C. Mitochondrial control of cell death. Nat Med. 2000, 6 (5), 513-519.

[5] Maté, M. J., Ortiz-Lombardía, M., Boitel, B., Haouz, A., Tello, D., Susin, S. A., Penninger, J., Kroemer, G., Alzari, P. M. The crystal structure of the mouse apoptosis-inducing factor AIF. Nat Struct Mol Biol. 2002, 9 (6), 442-446.

[6] Ye, H., Cande, C., Stephanou, N. C., et al. DNA binding is required for the apoptogenic action of apoptosis inducing factor. Nat Struct Mol Biol. 2002, 9 (9), 680-684.

[7] Lorenzo, H. K., Susin, S. A., Penninger, J., Kroemer, G. Apoptosis inducing factor (AIF): a phylogenetically old, caspase-independent effector of cell death. Cell Death Differ. 1999, 6 (6), 516.

[8] Wu, M., Xu, L. G., Li, X., Zhai, Z., Shu, H. B. AMID, an apoptosis-inducing factor-homologous mitochondrion-associated protein, induces caspase-independent apoptosis. J Biol Chem. 2002, 277 (28), 25617-25623.

[9] Ohiro, Y., Garkavtsev, I., Kobayashi, S., et al. A novel p53-inducible apoptogenic gene, PRG3, encodes a homologue of the apoptosis-inducing factor (AIF). FEBS Lett. 2002, 524 (1), 163-171.

[10] Xie, Q., Lin, T., Zhang, Y., Zheng, J., Bonanno, J. A. Molecular cloning and characterization of a human AIF-like gene with ability to induce apoptosis. J Biol Chem. 2005, 280 (20), 19673-19681.

[11] Loeffler, M., Daugas, E., Susin, S. A., Zamzami, N., Metivier, D., Nieminen, A. L., Brothers, G., Penninger, J. M., Kroemer, G. Dominant cell death induction by extramitochondrially targeted apoptosis-inducing factor. FASEB J. 2001, 15, 758-767.

[12] Otera, H., Ohsakaya, S., Nagaura, Z. I., Ishihara, N., Mihara, K. Export of mitochondrial AIF in response to proapoptotic stimuli depends on processing at the intermembrane space. EMBO J. 2005, 24, 1375-1386.

[13] Delettre, C., Yuste, V. J., Moubarak, R. S., Bras, M., Lesbordes-Brion, J. C., Petres, S., Bellalou, J., Susin, S. A. AIFsh, a novel apoptosis-inducing factor (AIF) pro-apoptotic isoform with potential pathological relevance in human cancer. J Biol Chem. 2006, 281, 6413-6427.

[14] Delettre, C., Yuste, V. J., Moubarak, R. S., Bras, M., Robert, N., Susin, S. A. Identification and characterization of *AIFsh2*, a mitochondrial apoptosis-inducing factor (AIF) isoform with NADH

oxidase activity. J Biol Chem. 2006, 281, 18507-18518.

[15] Garrido, C., Schmitt, E., Cande, C., Vahsen, N., Parcellier, A., Kroemer, G. HSP27 and HSP70: potentially oncogenic apoptosis inhibitors. Cell Cycle. 2003, 2, 579-584.

[16] Mosser, D. D., Morimoto, R. I. Molecular chaperones and the stress of oncogenesis. Oncogene. 2004, 23, 2907-2918.

[17] Thompson, C. B. Apoptosis in the pathogenesis and treatment of disease. Science. 1995, 267 (5203), 1456-1462.

[18] Kroemer, G. The proto-oncogene Bcl-2 and its role in regulating apoptosis. Nat Med. 1997, 3 (6), 614-620.

[19] Green, D. R., Reed, J. C. Mitochondria and apoptosis. Science. 1998, 281, 1309-1312.

[20] Joza, N., Galindo, K., Pospisilik, J. A., et al. The molecular archaeology of a mitochondrial death effector: AIF in *Drosophila*. Cell Death Differ. 2008, 15 (6), 1009-1018.

[21] Hangen, E., Kroemer, G., Modjtahedi, N. Vital functions of apoptosis inducing factor (AIF). Gastroenterol Hepatol Bed Bench. 2009, 2 (1).

[22] Hangen, E., Blomgren, K., Bénit, P., Kroemer, G., Modjtahedi, N. Life with or without AIF. Trends Biochem Sci. 2010, 35 (5), 278-287.

[23] Miramar, M. D., Costantini, P., Ravagnan, L., et al. NADH oxidase activity of mitochondrial apoptosis-inducing factor. J Biol Chem. 2001, 276 (19), 16391-16398.

[24] Apostolova, N., Cervera, A. M., Victor, V. M., Cadenas, S., Sanjuan-Pla, A., Alvarez-Barrientos, A., Esplugues, J. V., McCreath, K. J. Loss of apoptosis-inducing factor leads to an increase in reactive oxygen species, and an impairment of respiration that can be reversed by antioxidants. Cell Death Differ. 2006, 13, 354-357.

[25] Li, H. L., Zhang, Y. L., Gao, Q. K., Cheng, J. A., Lou, B. G. Molecular identification of cDNA, immunolocalization, and expression of a putative odorant-binding protein from an Asian honey bee, *Apis cerana cerana*. J Chem Ecol. 2008, 34 (12), 1593-1601.

[26] Michelette, E. R., Soares, A. E. E. Characterization of preimaginal developmental stages in Africanized honey bee workers (*Apis mellifera* L). Apidologie. 1993, 24 (4), 431-440.

[27] Yan, H., Meng, F., Jia, H., Guo, X., Xu, B. The identification and oxidative stress response of a zeta class glutathione S-transferase (GSTZ1) gene from *Apis cerana cerana*. J Insect Physiol. 2012, 58 (6), 782-791.

[28] Alaux C., Ducloz F., Crauser D., Le Conte Y. Diet effects on honeybee immunocompetence. Biol Lett. 2010, 6, 562-565.

[29] Yao, P., Hao, L., Wang, F., Chen, X., Yan, Y. Guo, X., Xu, B. Molecular cloning, expression and antioxidant characterisation of a typical thioredoxin gene (*AccTrx2*) in *Apis cerana cerana*. Gene. 2013, 527 (1), 33-41.

[30] Yao, P., Chen, X., Yan, Y., Liu, F., Zhang, Y., Guo, X., Xu, B. Glutaredoxin 1, glutaredoxin 2, thioredoxin 1, and thioredoxin peroxidase 3 play important roles in antioxidant defense in *Apis cerana cerana*. Free Radic Biol Med. 2014, 68, 335-346.

[31] Johnstone, R. W., Ruefli, A. A., Lowe, S. W. Apoptosis: a link between cancer genetics and chemotherapy. Cell. 2002, 108 (2), 153-164.

[32] Kitada, S., Pedersen, I. M., Schimmer, A. D., Reed, J. C. Dysregulation of apoptosis genes in hematopoietic malignancies. Oncogene. 2002, 21 (21), 3459-3474.

[33] Mustacich, D., Powis, G. Thioredoxin reductase. Biochem J. 2000, 346, 1-8.

[34] Yang, H., Kang, M., Guo, X., Xu, B. Cloning, structural features, and expression analysis of the gene encoding thioredoxin reductase 1 from *Apis cerana cerana*. Comp Biochem Physiol B: Biochem Mol Biol. 2010, 156 (3), 229-236.

[35] Stanojević, D., Hoey, T., Levine, M. Sequence-specific DNA-binding activities of the gap proteins encoded by hunchback and Krüppel in *Drosophila*. Nature. 1989, 341, 331-335.

[36] Ament, S. A., Corona, M., Pollock, H. S., Robinson, G. E. Insulin signaling is involved in the regulation of worker division of labor in honey bee colonies. Proc Natl Acad Sci USA. 2008, 105,

4226-4231.

[37] Fernandes, M., Xiao, H., Lis, J. T. Fine structure analyses of the *Drosophila* and *Saccharomyces* heat shock factor-heat shock element interactions. Nucleic Acids Res. 1994, 22 (2), 167-173.

[38] An, M. I., Choi, C. Y. Activity of antioxidant enzymes and physiological responses in ark shell, *Scapharca broughtonii*, exposed to thermal and osmotic stress: effects on hemolymph and biochemical parameters. Comp Biochem Physiol B: Biochem Mol Biol. 2010, 155 (1), 34-42.

[39] Yang, L. H., Huang, H., Wang, J. J. Antioxidant responses of citrus red mite, *Panonychus citri* (McGregor) (Acari: Tetranychidae), exposed to thermal stress. J Insect Physiol. 2010, 56, 1871-1876.

[40] Von Kalm, L., Crossgrove, K., Von Seggern, D., Guild, G. M., Beckendorf, S. K. The Broad-Complex directly controls a tissue-specific response to the steroid hormone ecdysone at the onset of *Drosophila metamorphosis*. EMBO J. 1994, 13 (15), 3505.

[41] Frumkin, A., Pillemer, G., Haffner, R., Tarcic, N., Gruenbaum, Y., Fainsod, A. A role for CdxA in gut closure and intestinal epithelia differentiation. Development. 1994, 120 (2), 253-263.

[42] Williams, D. L. A veterinary approach to the European honey bee (*Apis mellifera*). Vet J. 2000, 160, 61-67.

[43] Hornitzky, M. Literature review of chalkbrood, a fungal disease of honeybees. Rural Industries Research and Development Corporation. 2001, 1, 150.

[44] Cornman, R. S., Bennett, A. K., Murray, K. D., Evans, J. D., Elsik, C. G., Aronstein, K. Transcriptome analysis of the honey bee fungal pathogen, *Ascosphaera apis*: implications for host pathogenesis. BMC Genomics. 2012, 13 (1), 285.

[45] Aronstein, K., Saldivar, E. Characterization of a honey bee Toll related receptor gene Am18w and its potential involvement in antimicrobial immune defense. Apidologie. 2005, 36 (1), 3-14.

[46] Ekker, S. C., Von Kessler, D. P., Beachy, P. A. Differential DNA sequence recognition is a determinant of specificity in homeotic gene action. EMBO J. 1992, 11 (11), 4059.

[47] Nagajyoti, P. C., Lee, K. D., Sreekanth, T. V. M. Heavy metals, occurrence and toxicity for plants: a review. Environ Chem Lett. 2010, 8 (3), 199-216.

[48] Park, S. H., Chung, Y. M., Lee, Y. S., Kim, H. J., Kim, J. S., Chae, H. Z., Yoo, Y. D. Antisense of human peroxiredoxin II enhances radiation-induced cell death. Clin Cancer Res. 2000, 6 (12), 4915-4920.

[49] Franco, R., Sánchez-Olea, R., Reyes-Reyes, E. M., Panayiotidis, M. I. Environmental toxicity, oxidative stress and apoptosis: menage a trois. Mutat Res. 2009, 674 (1), 3-22.

[50] Murakami, S., Johnson, T. E. A genetic pathway conferring life extension and resistance to UV stress in *Caenorhabditis elegans*. Genetics. 1996, 143, 1207-1218.

2

The Enzymatic Antioxidant Molecular Mechanisms of *Apis cerana cerana*

As an important indigenous species of the Chinese honeybee, *Apis cerana cerana* (*A. cerana cerana*) not only provides rich nutrition via bee products but also plays an important role in crop pollination. It has been reported that bees were one of the animals that assisted in the rapid development of the agricultural economy due to their pollination function. Insects may encounter various stresses in the environment, such as heavy metal pollution, insecticides, herbicides, temperature challenges, and bacterial infections. These environmental stresses have promoted the development of diverse protective systems.

Under normal conditions, a dynamic balance between ROS generation and scavenging exists, resulting in a relatively low level of ROS maintained within cells. However, endogenous or exogenous insults can break this balance and lead to the excessive production or accumulation of ROS (i.e., oxidative stress). To protect themselves from the damage caused by ROS, insects have evolved a complex network of enzymatic antioxidant systems including enzymatic and nonenzymatic components. The dominant components of the antioxidant system in insects include thioredoxin peroxidases (Tpxs), catalases (CATs), glutathione-*S*-transferases (GSTs) and superoxide dismutases (SODs).

In this study, we evaluated gene expression patterns in response to oxidative stress. We explored the role of the selected genes by performing gene expression analyses, disc diffusion assay, RNA interference (RNAi), and analyses of superoxide dismutase (SOD), peroxidase (POD), catalase (CAT) and COX activity *in vivo* after gene knockdown. Through analysis of physiological parameters, the results revealed that the analyzed genes played important roles in cellular stress responses and antioxidative processes in *A. cerana cerana*.

Glutathione-*S*-transferases (GSTs) are members of a multifunctional enzyme superfamily that plays a pivotal role in both insecticide resistance and protection against oxidative stress. In this study, we isolated and characterized seven GST genes (*GSTD*, *GSTO1*, *GSTO2*, *GSTS1*, *GSTs4*, *GSTT1* and *GSTZ1*) in *A. cerana cerana*. The expression profile of these genes was significantly altered in response to various oxidative stresses. *Escherichia coli* cells overexpressing AccGSTS1, AccGSTO1 and AccGSTO2 showed long-term resistance under conditions of oxidative stress. The recombinant AccGSTD, AccGSTs4 and AccGSTZ1 proteins displayed high antioxidant activity under oxidative stress. Mutations in Cys-28, Cys-70, and Cys-124 affected the catalytic activity and antioxidant activity of AccGSTO1. The knockdown of *AccGSTO2* by RNA interference triggered increased mortality. Mutants of Cys28 and Cys124 significantly affected the enzyme and antioxidant activity of AccGSTO2, and recombinant AccGSTS1 protein showed characteristic glutathione- conjugating catalytic activity toward 1-chloro-2,4-dinitrobenzene. Functional assays revealed that AccGSTS1 could remove H_2O_2, thereby protecting DNA from oxidative damage.

Thioredoxin peroxidases (Tpxs) play an important role in maintaining redox homeostasis and in protecting organisms from the accumulation of toxic reactive oxygen species (ROS). The results showed that *AccTpx3* transcription varied in response to the oxidative stress; transcription was induced by some stresses and repressed by others. *AccTpx4* was induced by various oxidative stresses, such as cold, heat, insecticides, H_2O_2, and $HgCl_2$. The expression level of *AccTpx5* was upregulated under various abiotic stresses, except UV and pyriproxyfen treatments. Recombinant AccTpx3 protein acted as a potent antioxidant that resisted paraquat-induced oxidative stress, recombinant AccTpx4 protein could efficiently degrade

H_2O_2 in the presence of DL-dithiothreitol (DTT) *in vivo* and protect cells against oxidative stress, and the purified recombinant AccTpx5 protein protected the supercoiled form of plasmid DNA from damage in the thiol-dependent mixed-function oxidation (MFO) system.

Glutaredoxins (Grxs) and thioredoxins (Trxs) play important roles in maintaining intracellular thiol-redox homeostasis by scavenging reactive oxygen species. *AccGrx1*, *AccGrx2*, and *AccTrx1* were induced by cold, heat, H_2O_2, and pesticide treatments and repressed by UV light. The knockdown of *AccGrx1*, *AccGrx2* and *AccTrx1* enhanced the enzymatic activity of CAT and POD; the metabolite contents of hydrogen peroxide, carbonyls, and ascorbate; and the ratios of GSH/GSSG and $NADP^+$/NADPH. These results suggest that *AccGrx1*, *AccGrx2*, and *AccTrx1* may play critical roles in antioxidant defense.

Thioredoxins (Trxs) are a family of small, highly conserved and ubiquitous proteins that are involved in protecting organisms against toxic reactive oxygen species (ROS). Expression analyses indicated that *AccTrx-like1* expression was induced by both H_2O_2 and low temperature. Moreover, the activity of catalase and malondialdehyde were inversely related to the expression levels of *AccTrx-like1* in honeybees injected with H_2O_2 at 30 min. These results suggest that *AccTrx-like1* is an important antioxidant gene. The expression profile of *AccTrx2* was upregulated by such abiotic stresses, namely, 4, 16 and 25℃; H_2O_2; and cyhalothrin, acaricide, paraquat, phoxim and mercury treatments. Characterization of the recombinant protein showed that the purified AccTrx2 had insulin disulfide reductase activity. These results indicate that *AccTrx2* plays an important role in the response to oxidative stress.

The expression of *AccSOD2* was induced by cold, heat, H_2O_2, ultraviolet light, $HgCl_2$, and pesticide treatments; recombinant AccSOD2 protein could play a functional role in protecting cells from oxidative stress. The expression of *AccTPI* was upregulated by some abiotic stresses, while it was downregulated by $HgCl_2$, ultraviolet light, and vitamin C treatments. Recombinant AccTPI proteins can protect cells from oxidative stress. *AccPPO* was induced by the infection of *Ascosphaera apis* and by various oxidative stress and pesticide treatments. These findings provide solid evidence for the important role of *AccPPO* in resisting *Ascosphaera apis* invasion and the adverse environment to protect *A. cerana cerana*.

The expression of *AccMsrB* was upregulated by various oxidative stresses, including 4, 16, 25, and 42℃; ultraviolet light; H_2O_2, $CdCl_2$, and $HgCl_2$; and paraquat, imidacloprid, and cyhalothrin; its expression was downregulated when exposed to phoxim. The expression of *AccMsrA* was upregulated by multiple oxidative stresses, including ultraviolet light, heat and H_2O_2. The gene expression of *AccS6K-p70* was significantly downregulated by hydrogen peroxide and pyriproxyfen. Both the juvenile hormone analog pyriproxyfen and oxidative stresses may impair development of bee larvae through downregulation of *AccS6K-p70* expression.

2.1 Identification and Characterization of an *Apis cerana cerana* Delta Class Glutathione *S*-transferase Gene (*AccGSTD*) in Response to Thermal Stress

Huiru Yan, Haihong Jia, Xiuling Wang, Hongru Gao, Xingqi Guo, Baohua Xu

Abstract: Glutathione *S*-transferases (GSTs) are members of a multifunctional enzyme superfamily that plays a pivotal role in both insecticide resistance and protection against oxidative stress. In this study, we identified a single-copy gene, *AccGSTD*, as being a Delta class GST in the Chinese honeybee (*Apis cerana cerana*). A predicted antioxidant response element, CREB, was found in the 1,492 bp 5'-flanking region, suggesting that *AccGSTD* may be involved in oxidative stress response pathways. Real-time PCR and immunolocalization studies demonstrated that *AccGSTD* exhibited both developmental- and tissue-specific expression patterns. During development, *AccGSTD* transcript was increased in adults. The *AccGSTD* expression level was the highest in the brain of the honeybee. Thermal stress experiments demonstrated that AccGSTD could be significantly upregulated by temperature changes in a time-dependent manner. It is hypothesized that high expression levels might be due to the increased levels of oxidative stress caused by the temperature challenges. Additionally, functional assays of the recombinant AccGSTD protein revealed that AccGSTD has the capability to protect DNA from oxidative damage. Taken together, these data suggest that AccGSTD may be responsible for antioxidant defense in adult honeybees.

Keywords: Glutathione *S*-transferase; *Apis cerana cerana*; Oxidative stress; Antioxidant defense

Introduction

Glutathione *S*-transferases (GSTs) are a diverse family of multifunctional enzymes that are found in almost all living organisms. These enzymes act by conjugating reduced glutathione (GSH) to the thiol group of cysteine on the electrophilic centers of endogenous and xenobiotic compounds. The addition of GSH renders these compounds less toxic and more water-soluble, allowing them to be more efficiently excreted from the body of the organism [1, 2]. Some GSTs can also catalyze the dehydrochlorination of dithiothreitol (DTT) to the noninsecticidal metabolite DDE by using reduced glutathione (GSH) as a cofactor rather than as a conjugate [3-5]. In addition to conjugation and dehydrochlorination, GSTs are also involved in intracellular transport, the biosynthesis of hormones, the degradation of tyrosine, and protection against oxidative stress [1, 6-8].

Based on their different subunit structures, isoelectric points, kinetics, and immunological properties, mammalian GSTs can be divided into three major groups: the cytosolic GSTs (including seven classes: Alpha, Mu, Pi, Theta, Sigma, Omega, and Zeta), the mitochondrial GSTs (Kappa class), and microsomal GSTs [6, 9-11]. The mitochondrial GSTs are located in mammalian mitochondria and peroxisomes and have not been found in insects to date. Studies on insect GSTs have mainly focused on cytosolic GSTs, which are the largest family of GSTs and have activities that are unique to this group of enzymes [2]. In addition to the ubiquitous Omega, Sigma, Theta, and Zeta classes, two insect-specific classes of cytosolic GSTs, Delta and Epsilon, have been identified [5, 12]. Both of these classes exist as large gene clusters in the *Anopheles gambiae* and *Drosophila melanogaster* genomes and account for over 65% of the total complement of cytosolic GSTs in the genomes of these insects [12]. The other four classes have a broader taxonomic distribution and likely play essential house-keeping roles [2, 5, 12-14]. The independent expansion and radiation of the Delta and Epsilon gene families in *A. gambiae* and *D. melanogaster* and the functional verification of their role in xenobiotic detoxification suggest that they are important for the adaptation of insects to environmental selection pressures [12]. By contrast, the number of these two GST classes is greatly reduced in honeybees. Only one Delta GST and no Epsilon GSTs [15] have been found in *Apis mellifera*, which may contribute to the sensitivity of this species to certain pesticides.

Work on insect GSTs was initially motivated by the possible involvement of these enzymes in insecticide resistance [16], and correlations between an increase in the expression and activity of GSTs with insecticide resistances have been demonstrated in insecticide-resistant strains of many insects [17, 18]. But recent studies have instead concentrated on the role of GSTs in mediating the oxidative stress response. After exposure to dietary H_2O_2, the expression levels of both Delta and Sigma GSTs in the *D. melanogaster* midgut increased, suggesting that GSTs play a role in oxidative stress [19]. In a previous report, GSTs from the *Chironomus riparius* fourth-instar larvae were upregulated upon exposure to the pro-oxidative stress inducer paraquat, suggesting their involvement in oxidative stress defense mechanisms. In addition, GSTs were induced following exposure to cadmium and silver nanoparticles, again suggesting a protective role against oxidative stress [20]. Yang et al. provided evidence that a *Solen grandis* Sigma GST might serve as an antioxidant enzyme functioning in the detoxification of both microorganism glycans and organic contaminants [21].

Although the molecular basis and mechanism of GST-related insecticide resistance have been well studied in Dipteran and some Lepidoptera, very little progress has been made in *Apis cerana cerana* (Chinese honeybee). *A. cerana cerana*, an important beneficial insect in agriculture, lacks many of the well-studied detoxification enzymes and is highly susceptible to various environmental stresses [22]. Thus, in this study, we intended to clone and characterize the single insect-specific GST (*AccGSTD*) present in *A. cerana cerana*. Our study provides new insight into the molecular characteristics of GSTs and helps to develop strategies for mitigating the insecticide resistance problem.

Materials and Methods

Insects

The colonies of *A. cerana cerana* used in the present study were maintained in the experimental apiary of Shandong Agricultural University (Taian, China). The larvae, pupae, and 1-day-old adult workers were taken from the hive, while the 2-week-old adult workers were collected at the entrance of the hive when they returned to the colony after foraging. The adults were maintained in an incubator at a constant temperature (34℃) and humidity (70%) under a 24 h dark regimen and fed a pollen-and-sucrose solution. For thermal stress experiments, adult bees were placed at 4, 15, 25, 34, or 43℃. The bees kept at 34℃ served as a control. Various tissues from the fourth-instar larvae, including the brain, fat body, epidermis, muscle, and midgut of adults (15 days) were dissected out on ice and collected while fresh. All of the honeybees in this study were harvested at the indicated time points and stored at −70℃.

Isolation of the *AccGSTD*

Total RNA was extracted from the larvae by using Trizol reagent (Invitrogen, USA) and incubated with RNase-free Dnase I (Promega, Madison, WI, USA) to eliminate potential genomic DNA contamination. The first-strand cDNA was then synthesized by using reverse transcriptase (TransGen Biotech, Beijing, China) according to the protocol of the manufacturer. Genomic DNA was isolated from the whole bodies of adult bees by using the EasyPure Genomic DNA Extraction Kit (TransGen Biotech, Beijing, China) according to the instructions of the manufacturer.

Primers D1 and D2 (Supplementary Table S1) were designed and synthesized (Shanghai Sangon Biotechnological Company, China) by using the conserved regions of the GSTs from *A. mellifera* (NP_001171499), *Nasonia vitripennis* (NP_001165913), and *Locusta migratoria* (ADR30117) to obtain the internal region of AccGSTD. Based on the sequence of the amplified fragment, specific primers 5R1/2 and 3R1/2 (Supplementary Table S1) were subsequently designed and used for 5′ and 3′ rapid amplification of cDNA end (RACE), respectively. Two other primers, QC1 and QC2 (Supplementary Table S1), were synthesized to verify the sequence of the deduced full-length cDNA of AccGSTD. All of the PCR products were purified by using a gel extraction kit (TaKaRa, Dalian, China), ligated into pEASY-T3 vectors (TransGen Biotech, China), and transformed into *Escherichia coli* strain DH5α for sequencing.

To obtain the genomic DNA sequence of *AccGSTD*, four pairs of primers (NS1/NX1, NS2/NX2, NS3/NX3, and NS4/NX4) were designed based on the full-length cDNA sequence of *AccGSTD* and synthesized. Four fragments were obtained by using the *A. cerana cerana* genomic DNA as the template. The PCR products were purified, cloned into pMD18-T, and then transformed into competent *E. coli* DH5α cells for sequence analysis.

Bioinformatic analysis

Homologous AccGSTD protein sequences from various species were retrieved by using the basic local alignment search tool (BLAST) program from the National Center for Biotechnology Information (Bethesda, MD, USA, http://www.ncbi.nlm.nih.gov/). The open reading frame (ORF) of *AccGSTD* was identified, and multiple alignments were conducted by using homologous proteins with DNAMAN version 5.2.2 (Lynnon Biosoft Company, USA). The molecular mass and isoelectric point of the relevant protein were calculated by using PeptideMass (http://web.expasy.org/cgi-bin/compute_pi/pi_tool). Phylogenetic analysis was performed by using the neighbor-joining method in Molecular Evolutionary Genetic Analysis (MEGA version 4.1). The TFSEARCH database (http://www.cbrc.jp/research/db/TFSEARCH) was used to search for transcription factor binding sites in the 5′-flanking region.

Amplification of the 5′-flanking region of *AccGSTD*

I-PCR was used to acquire the 5′-flanking region of *AccGSTD*. Two pairs of specific primers (QR1/QR2 and QF1/QF2) were designed based on the genomic DNA sequence of *AccGSTD*. Genomic DNA was completely digested with *Xba* I and self-ligated by using T4 DNA ligase (TaKaRa) to produce a template. The primers QY1 and QY2 were then designed to examine the sequence of the 5′-flanking sequence of *AccGSTD*. All of the primer sequences and PCR reaction conditions above are provided in Supplementary Table S1 and Supplementary Table S2, respectively.

Quantitative analysis

To determine the *AccGSTD* copy number, quantitative real-time PCR (qPCR) was used as described by Mason et al. [23]. *AccGSTZ1*, determined as a single copy gene by Southern blotting [24], was used as a control. Plasmids containing these two genes were extracted and the concentrations were estimated. Accurate 10-fold serial dilutions of the plasmids were prepared and utilized to obtain the standard curves necessary for relative quantification of *AccGSTD* and *AccGSTZ1*. Then, the amount of both the *AccGSTD* and *AccGSTZ1* in genomic DNA was determined. Finally, the copy number was calculated as previously described.

qPCR was also performed to assess the transcription levels of AccGSTD by using SYBR Premix Ex *Taq* (TaKaRa) in the CFX96™ Real-Time PCR Detection System (Bio-Rad, Hercules, CA, USA). The *A. cerana cerana β-actin* gene (XM640276) was used as an internal control. Details of the primers used for qPCR are given in Supplementary Table S1. The PCR cycling parameters are as follows: 95℃ for 30 s; 40 cycles at 95℃ for 5 s, 56℃ for 15 s and 72℃ for 15 s, followed by a single melting cycle from 65℃ to 95℃. Standard curves and fold change in AccGSTD expression relative to the control were developed and analyzed by CFX Manager software version 1.1 and the $2^{-\Delta\Delta C_t}$ method [25].

Expression and purification of recombinant AccGSTD

The full-length coding sequence of *AccGSTD* amplified by primers YH1, which included a

*Bam*H I site, and YH2, which included a *Sac* I site, was inserted by using those same enzymes into the pET-30a (+) binary vector. The construct was verified by DNA sequencing. The recombinant clone was transformed into *E. coli* BL21 strain (DE3) cells to express the protein inducing the cells with isopropylthio-β-D-galactoside (IPTG) at 37℃ for 4 h. The protein was then purified by using the MagneHis™ Protein Purification System (Promega). The protein purity (20 μl protein was used) was verified by 12% polyacrylamide gel electrophoresis and Coomassie staining. The protein concentration was determined by the Bradford method using the Bio-Rad protein assay dye reagent and bovine serum albumin as a standard [26].

Production of anti-AccGSTD and Western blotting analysis

The purified recombinant AccGSTD protein (50 μg) was mixed with an equal volume of Freund's complete adjuvant (a total volume of 300 μl) (Sigma, USA) and was injected into mice. Two successive injections of 30 μg of protein mixed with an equal volume of Freund's incomplete adjuvant (a total volume of 300 μl) were administered at 1-week intervals starting 1 week after the initial injection. Three days after the final injection, blood was collected and centrifuged at 3,000 r/min for 5 min. The supernatant antibodies were stored at −70℃ for further application.

The specificity of the AccGSTD antibody was confirmed by Western blotting. The purified recombinant protein mixed with sample buffer was resolved by SDS-polyacrylamide gel electrophoresis (SDS-PAGE) and subsequently electrotransferred onto a polyvinylidine difluoride membrane. After blotting for 1 h, the membrane was blocked in 5% (*m/V*) skim milk diluted with PBS buffer for 30 min and then incubated with anti-AccGSTD polyclonal antibody (1 : 400 dilution) at 4℃ overnight on a shaker. After washing with PBS for 5 min (3×), the membrane was incubated with 1 : 4,000 (*V/V*) HRP-conjugated goat anti-mouse IgG (Beijing Dingguo Changsheng Biotech Co., Ltd., China) for 2 h at 37℃. After repeated washing, the membrane was subjected to 4-chloro-1-naphthol (Solarbio, China) for detection.

Immunolocalization

The fresh whole bodies of the fourth-instar larvae were fixed overnight at 4℃ in PBS containing 4% paraformaldehyde. The larvae were dehydrated with a series of ascending concentrations of ethanol, cleared in xylene, and embedded in paraffin blocks. Tissue sections (10 μm) were cut by using a microtome, mounted on poly-L-lysine-coated glass slides, and incubated at 60℃ for 30-60 min. To enhance the immunostaining, antigen retrieval methods were used as follows: the tissue sections were dewaxed in xylene, rehydrated through graded ethanol, rinsed in PBS, and treated with a mixture of 30% H_2O_2 and distilled water (1 : 10) for 5-10 min at room temperature in the dark to block endogenous peroxidases. The slides were then washed again and placed in 500 ml of citrate buffer (2.1 g of citric acid in 1 L of distilled water, pH 6.0), microwaved for 20 min, and washed with PBS (2×). After antigen retrieval, the sections were treated with 5% BSA for 20 min at room temperature, followed by incubation with the anti-AccGSTD antibody (1 : 400 dilution) at 4℃ overnight. As a negative control, the primary antibody was omitted in some samples. The following day, the sections were rinsed in PBS (3×) and covered with HRP-conjugated goat anti-rat IgG diluted in PBS (1∶200) at 37℃ for 1 h. The slides were washed in PBS (3×) and incubated with HRP-streptavidin (Solarbio, China) at 37℃ for 20 min. The DAB Horseradish Peroxidase Color Development Kit (Wuhan Boster Biological Technology Co., Ltd., China) was then applied to produce a brown reaction product. Finally, the sections were dehydrated through graded ethanol, cleared in xylene, and mounted with distrene plasticiser xylene. Stained sections were visualized with a LSM 510 META microscope (Carl Zeiss AG, Germany).

Characterization of recombinant AccGSTD

GST activity was spectrophotometrically determined in a standard assay mixture (100 μl) containing 0.1 mol/L sodium phosphate buffer (pH 6.5) in the presence of 5 mmol/L 1-chloro-2,4-dinitrobenzene (CDNB), 5 mmol/L GSH, and enzyme. The rate of conjugation change was monitored by continuously measuring the increase in absorbance at 340 nm for 1 min. The specific activity of AccGSTD was expressed as mole of CDNB conjugated with GSH per minute per milligram of protein. The molar extinction coefficient of 9.6 $(mmol/L)^{-1} cm^{-1}$ was used to convert absorbance into moles.

The DNA cleavage assay was performed by using supercoiled pUC19 plasmid DNA (TaKaRa) as a substrate to detect the DNA-protecting function of AccGSTD. DNA damage was caused by the mixed-function oxidation (MFO) system. The reaction mixture (50 μl) containing 3 μmol/L $FeCl_3$, 10 mmol/L DTT in 100 mmol/L HEPES buffer (pH 7.0), and increasing concentrations of AccGSTD ranging from 0 to 300 mg/ml was incubated at 37℃ for 1 h. This was followed by the addition of 200 ng of supercoiled pUC19 plasmid DNA for 2 h at 37℃, and the product was then subjected to 1.2% agarose gel electrophoresis for determination of DNA cleavage.

Statistical analysis

Data are expressed as the mean ± standard deviation (SD) with $n = 3$ and analyzed by ANOVA. Significant differences between the means were determined by Duncan's multiple range test using SAS software version 9.1.

Results

Isolation and classification of *AccGSTD*

An internal fragment isolated from Chinese honeybees showed significant similarity to GST genes from other insect species. Then, 5' and 3' RACE-PCR were performed to amplify the full-length cDNA sequence. The full-length cDNA was identified as the *AccGSTD* cDNA sequence (GenBank ID: JF798573): it was 1,009 bp in length and contained an ORF of 657 bp that encodes a protein of 218 amino acid residues. The calculated molecular mass of the deduced protein is 25.359 kDa with an estimated pI of 7.24. BLAST analysis yielded the highest identity to Delta GSTs, with 54%-83% amino acid identity. Multiple alignments of derived amino acid sequences with other insect Delta GSTs revealed that the AccGSTD protein contains the key conserved residues across different insect species (Fig. 1a), including amino acid residues S10 and N48 (numbered based on DmGSTD1), which represent the catalytic pocket, and P54-L142-G150-D157, which are involved in the folding of GSTs. The phylogenetic analysis of AccGSTD along with GSTs from other species clearly supports the assignment of this GST sequence to the Delta class (Fig. 1b).

Genomic organization and copy number of AccGSTD

To investigate the structure of the *AccGSTD* genomic locus, we isolated and characterized the genomic DNA of *AccGSTD*. The complete genomic sequence of *AccGSTD* (GenBank ID: JF798572) was 5,218 bp long and contained six introns.

A comparison of the genomic structure of *AccGSTD* with other *GSTD1* sequences from *A. mellifera*, *N. vitripennis*, and *D. melanogaster* revealed variations in the intron numbers and gene lengths (Fig. 2). Interestingly, all four species have intron(s) in the 5' UTR, although the number and length are varied. Importantly, both the Chinese and western honeybees have an intron that can be as long as 3,000 bp in the 5' UTR. In addition, the three bees have a total of four introns in the coding region while the fly has none, demonstrating the diversity that arose between Hymenoptera and Diptera during evolution.

To determine the copy number of *AccGSTD* in the honeybee genome, qPCR analysis was performed by using the single copy gene *AccGSTZ1* [24] as an internal standard. The standard curves of *AccGSTZ1* and *AccGSTD* are shown in Supplementary Fig. 1, and the correlation coefficients were rather good, being 0.999 and 0.998, respectively. As shown in Supplementary Table S3, these results suggest that a single copy of *AccGSTD* is present in the *A. cerana cerana* genome.

Identification of the 5'-flanking region of AccGSTD

The 5'-flanking region (approximately 1,492 bp) of *AccGSTD* (GenBank ID: JF798572) was obtained to determine the location of the regulatory regions of the gene. Putative transcription factor binding sites were predicted by using the web-based software program TFSEARCH. Portions of the promoter sequence and putative transcription factor binding sites are shown in Fig. 3. Except for some common *cis*-acting elements, including HSF, NIT-2, SRY, SOX-5, v-Myb, and CF2-II, several binding sites for embryo or tissue developmentally regulated transcription factors, including hunchback, Oct-1, GATA-1, and Dfd were also found in the promoter region [27]. Also of interest was the identification of putative binding sites for StuAp, a transcription factor required for the development and sexual reproductive cycle [28]. In addition, the binding motif for Croc, a brain development-specific transcriptional activator, was detected [29]. Importantly, antioxidant response element CREB [30] was also found, suggesting that AccGSTD may be involved in the oxidative stress response.

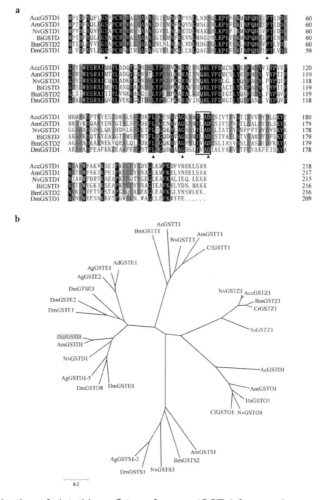

Fig. 1 Characterization of glutathione S-transferases (GSTs) from various species. a. Alignment of the deduced AccGSTD protein sequence (GenBank ID: JN637474) from *A. cerana cerana* with Delta class GST homologues. GST sequences from different species were obtained from the National Center for Biotechnology Information database: AmGSTD1 (*A. mellifera*, GenBank ID: NP_001171499), NvGSTD1 (*N. vitripennis*, NP_001165913), BiGSTD (*Bombus ignitus*, NP_001036974), BmGSTD2 (*Bombyx mori*, NP_001036974.1), and DmGSTD1 (*D. melanogaster*, ABW76195.1). The shaded regions indicate identical residues. The conserved amino acids in the catalytic pocket and protein folding region are marked by filled circles and filled triangles, respectively. The signature sequence "TIAD" of Delta class GSTs is boxed. b. Phylogenetic relationships of cytosolic GSTs from different organisms. The scale bar indicates the evolutionary distance between the groups. AccGSTD is highlighted in gray. The GenBank accession numbers of the protein sequences are AmGSTS1 (NP_001153742.1), DmGSTS1 (NP_725653.1), BmGSTS2 (NP_001036994), AgGSTS1-2 (AAM53611.1), NvGSTS3 (NP_001165920.1), AmGSTD1 (NP_001171499.1), NvGSTD1 (NP_001165913.1), AgGSTD1-5 (CAB03592.1), DmGSTD1 (ABW76195.1), DmGSTD8 (NP_524916.1), BiGSTD (ACL51929.1), BmGSTD1 (NP_001037183), BmGSTD2 (NP_001036974.1), NvGSTZ1 (NP_001165931.1), BmGSTZ1 (NP_001037418.1), CrGSTZ1 (ADK66969.1), SsGSTZ1 (NP_001230567.1), AccGSTZ1 (JN637475), AmGSTO1 (XP_624501.1), NvGSTO1 (NP_001165912.1), NvGSTO2 (NP_001165916), CfGSTO1 (EFN62827.1), HsGSTO1 (EFN81737.1), AcGSTO1 (ACY95464.1), AmGSTT1 (XP_624692.1), NvGSTT3 (NP_001165927.1), CfGSTT1 (EFN68893.1), AcGSTT1 (ACY95465.1), BmGSTT1 (NP_001108463.1), BmGSTT5 (NP_001108463), DmGSTE1 (NP_611323.1), AgGSTE1 (AAL59658.1), AdGSTE1 (ABD43203.1), DmGSTE2 (NP_611324.1), AgGSTE2 (AAG45164.1), and DmGSTE3 (NP_611325).

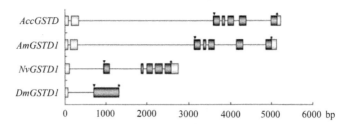

Fig. 2 Exon-intron organization of *AccGSTD* from *A. cerana cerana* and its homologs, including *AmGSTD1* from *A. mellifera* (NP_001171499), *NvGSTD1* from *N. vitripennis* (NP_001165913), and *DmGSTD1* from *D. melanogaster* (ABW76195.1). Untranslated regions and exons are indicated as light gray and gray, respectively, while the introns are shown as lines. The initiation (ATG) and termination (TAG) codons are indicated by inverted triangles and asterisks, separately.

```
                     StuAp
CAGATGCGTTACAACCTTCGATCGCGTCCCACGATATCGTCGCCGTCACACTTTCACGCTTATGTTTGCCGTGCGCTTTACGACATCGCTATGG     -1398
                                                         Dfd
CAACTCGATCTACCATTCGACCGTGTTATTTGCCTTCCCTTCCCTACCCGTATTACTCGCGAAACGAGCTTTAATCGACTCGGGCCGGTGCTTCG   -1304
                              Dfd
TTATCTTCGATGGCCGGCCAATCGGCTAATTAATTAGAGAATCGTTGACGCGTGAACGCGTGTGGAATTCGCGGTGATCTCGGATTAAATCGG     -1209
         v-Myb
GGTGAGTCGAATAACGGAACTTTTCGTTTGGAAAAATCGAGGGAAGGAGAGAGAGAGAGAAAGGTGTCTGTCGTCTTGCCACGAAAATGATA      -1116
                             GATA-1            ADR1                       NIT2
ATTAATTCCGTTTTATTAAGTTGGCGATACGTTTAATATTTGGGGGAACGTTGGAAACGGAAGTTAGATAATTCCCGCGGTATTCCATTTCGC      -1024
TGATTTTCTCCCCGCATTTTCTTCTATTTCGAAATGCTTGCTTCCCCGCATTTTGCGATGAATACCAACGATACGAAATTTACTAAGTTTTCCAAT  -930
                    Oct-1
TCGAATTCGGTGTAAGAATACGTAATGATAAACTTCAACTTCGTTGCATCTCGTCCAATGAAAATTGAATCAACGTGTATCGAGATAATTGAT    -834
                                        NIT2
AGAAAGAAAGGAAAGGAACGATCGATTTCGATAAGAAAAATTGGAAATAACGTCTTTCGATATATATATATATAAACGCGTGTATCGTAAAAT     -741
GGCATTTATTCGAAATCGCTCGTACTTTGCACCCGTAACTCTTTATACTCTTTTCGAGAGGATCTCCATAAAATTCCATATGCAATTTGGTTGATA  -649
                                                                                    HSF
ATCCTTCTCATCCATAGAATTCGAGACACAAGAGAAGAGAGACGAGTAAATTTGTATGCGGGATACGCATGCAAATACGAAGCGTAGAAATT     -554
    HSF                                               CREB                       HSF
CCAGAAATTTCCTCTATTCTTGATCTACCTCTACAAAAATCGTGCGCAGTGACGTTGAAGAGACACAAAGTAAAAAAAGAAAAGGGAGGCA      -462
   SRY    Hb    Croc                                                   ADR1
AAAGAAATTCGGGAAAAAAATAAAGAGAATACGAAAAGATGCCGATTGATTTTTATTATTGGAGTTCAGCCCGCCGTGTCGGTCGGTTTTA       -370
TTGCTGGCAAAAGCGATCGGCGTTCATTTGAAACTGAAAACGATCTCTCCTATGAAAGGGGAGCACATGAAATCCGAATTCGTGAAGGTAAT     -278
              CdxA                                        Sox-5
TCGCTCGTTTTATTCGAAGCATATATATTTATAGTAAACGTATTTCTTTTTTTAATTCATAAAAACAATTTTGATCGTGTCTCAGTTGAATCCG    -185
                                                        CF2-II
CAACACGTAATACCGACGATAGACGATAATGGCTTCGTTTTGTGCGAAAGGTACGCACACACACATATATATATGTATGTATCTTTGGATTCGC    -90
                                                                                            +1
TCTTCGATAATAACTCATTCGCGTGTTCCCAGCCGGCCGATTATGGGATACTTGGTGAGCAAATACGCCAAAAACGATTCTCTGTATCCGAAA     +3
                                                                                Transcription start
GACCCAAAGAGGAGAGGTATGGTGGATCAAATGCTGTATTTCGATATCGGTACCCTTAACGAGAACGTGGTTAAATGTTAC......TACTCGACG +161
                                                                the first intron (71 bp)
ATATTTTTGGGGGCGCACTCCTTGGACGAGGAAAAACGTACGAGCCGTGGAGAGATCGTGCGAGGTGTTGAACGCGTATCTCGGCGAGCGCGA    +253
GTACGTTGCCGGCGATACGCTCACCATTGCCGATTTTGCCATCCACACGACCATCTGCGTCCTGTTG......AAATTAAAATGCC           +3610
                                              the second intron (3277 bp)  Translation start
```

Fig. 3 Partial nucleotide sequence and characterization of the 5'-flanking region in *AccGSTD*. Putative *cis*-acting elements are boxed. Both the transcription start site and the translation start site are marked with arrows.

Developmental regulation and tissue distribution of *AccGSTD*

Quantitative analysis was performed to determine the stage-specific expression patterns of *AccGSTD* in all of the stages of development including four different larval instars (the third, the fourth, the fifth, and the sixth), pupae, and adults. *AccGSTD* showed low expression levels in earlier developmental stages especially during the pupal stage, and the fold changes in mRNA abundance in other developmental stages were normalized to the basal level in the pink eyes pupae. There were relative 20.1- and 8.3-fold differences of *AccGSTD* transcript in the 1- and 14-day-old adults than in the pink eyes pupae, respectively (Fig. 4a). The results suggest that *AccGSTD* may be involved in the regulation of development.

The distribution of *AccGSTD* mRNAs in adult tissues including the brain, fat body, epidermis, muscle, and midgut were further determined. As shown in Fig. 4b, there were noticeable variations of *AccGSTD* expression among different tissues. The highest levels of *AccGSTD* mRNA were observed in the brain, and a significant 57.6-fold change was determined compared with the muscle, indicating that both the synthesis and function of *AccGSTD* are most important in the brain.

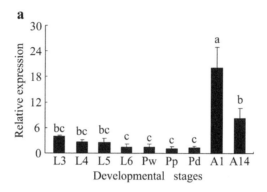

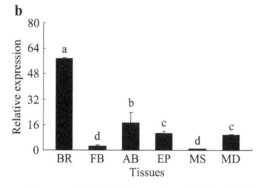

Fig. 4 The expression patterns of *AccGSTD*. a. *AccGSTD* mRNA expression at different developmental stages: larvae from the third to the sixth instars (L3-L6), pupae (Pw, white eyes, Pp, pink eyes, and Pd, dark eyes), and adults (A1, 1 day post-emergence and A14, 14-day post-emergence adults). b. The distribution of *AccGSTD* in the brain (BR), fat body (FB), abdomen (AB), epidermis (EP), muscle (MS), and midgut (MD). The *β-actin* gene was used as an internal control. Vertical bars represent the mean ± SD (n = 3). Different letters topped on the bar designated significant difference at $P < 0.01$ in SAS.

In vitro expression, purification, Western blotting, and immunohistochemistry of AccGSTD

To further evaluate the expression of AccGSTD protein in native *A. cerana cerana*, the

AccGSTD protein was overexpressed *in vitro*. SDS-PAGE showed that the recombinant protein was expressed as a 28 kDa polypeptide (Fig. 5a). This target protein was purified, and the final concentration was approximately 0.4 mg/ml. The purified protein was subcutaneously injected into mice to produce polyclonal antibodies. To examine the specificity of the anti-AccGSTD, ELISA was performed (data not shown). Simultaneously, Western blotting was also performed (Supplementary Fig. 2), and only a single band was observed, suggesting that the anti-AccGSTD antibody is relatively specific.

The whole bodies of the fourth-instar larvae were then embedded in paraffin wax and subjected to immunolocalization analysis in the presence of the anti-AccGSTD polyclonal antibody. All sections were used for DAB staining. The results revealed that AccGSTD is widely distributed in the larvae (data not shown). Notably, strong positive reactions were detected in the body wall (epidermis) and intima of the midgut (Fig. 5b), which is in agreement with the results from the mRNA localization studies, suggesting that AccGSTD might play important roles in these tissues.

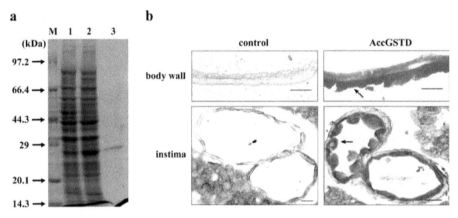

Fig. 5 Expression, purification, Western blotting, and immunostaining of AccGSTD. a. Expression and purification of AccGSTD protein by using SDS-PAGE. Lane M, protein molecular weight marker; lane 1, uninduced expression of recombinant AccGSTD protein; lane 2, overexpression of recombinant AccGSTD protein after IPTG induction; lane 3, purified recombinant AccGSTD protein. b. Immunostaining of AccGSTD in the fourth-instar larvae. The negative control group is displayed on the left, and the experimental group is shown on the right. An anti-AccGSTD polyclonal antibody was used as the primary antibody, and HRP-conjugated goat anti-mouse IgG was used as the secondary antibody. Obvious immunohistochemical staining is marked with arrows. Scale bar = 500 μm.

Expression profiles of *AccGSTD* under thermal stress

To determine the effects of different temperature challenges on *AccGSTD* expression, qPCR was carried out as described above, by using a constant hive temperature of 34 ℃ as a control. As presented in Fig. 6, after treatment for 1 h, *AccGSTD* transcript abundance was significantly induced by 15, 25, and 43 ℃ and reached a peak at 15 ℃. Notably, the strong induction by 15 ℃ lasted for 5 h, while 43 ℃ cannot enhance the expression level of AccGSTD from 3 h. On the contrary, *AccGSTD* expression can be significantly induced by low temperature (4 ℃) at 3 h, and then decreased. All of the temperatures could not upregulate the mRNA accumulation of

AccGSTD at 7 h. These results suggested that *AccGSTD* transcription was enhanced by both cold and heat treatments in a time-dependent manner ($F = 12.5$, $P < 0.0001$).

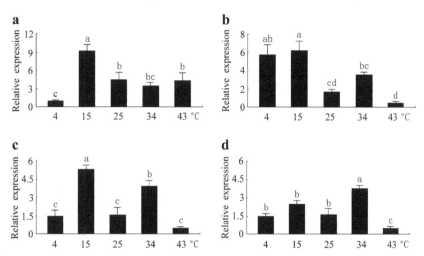

Fig. 6 The expression profiles of *AccGSTD* after temperature challenges. a-d. Expression levels after treatment (4, 15, 25, 34, 43℃) for 1, 3, 5, and 7 h, respectively. The normal hive temperature (34℃) of *A. cerana cerana* was considered as a control at every experimental group. The *β-actin* gene was used as an internal control. Vertical bars represent the mean ± SD ($n = 3$). Different letters topped on the bar designated significant difference at $P < 0.01$ in SAS.

Characterization of recombinant AccGSTD

To further understand the function of AccGSTD, purified recombinant AccGSTD protein was subjected to an enzyme-catalyzed reaction to measure the glutathione *S*-transferase activity. The results showed that the catalytic activity of AccGSTD with CDNB was 55.6 nmol/(min·mg).

Given that *AccGSTD* expression levels can be enhanced by thermal stress, a characteristic that is associated with the generation of increased reactive oxygen species (ROS) and the production of oxidative damage, we hypothesized that AccGSTD might be involved in DNA repair. To evaluate the ability of AccGSTD to protect DNA from oxidative damage, an MFO system that can produce hydroxyl radicals was chosen. As shown in Fig. 7, in the MFO system, the supercoiled form of pUC19 plasmid was converted into a nicked form by the hydroxyl radical (lane 3). However, supercoiled pUC19 was protected by purified AccGSTD in a dose-dependent manner, suggesting that AccGSTD may play a protective role in oxidative stress.

Fig. 7 Protection of DNA from oxidative damage by AccGSTD in the mixed-function oxidation system. Lane 1, pUC19 plasmid DNA only; lane 2, pUC19 plasmid DNA + FeCl$_3$; lane 3, pUC19 plasmid DNA + FeCl$_3$ + dithiothreitol (DTT); lane 4, pUC19 plasmid DNA + FeCl$_3$+ BSA; lanes 5-12, pUC19 plasmid DNA + FeCl$_3$ + DTT + purified AccGSTD (5, 10, 50, 100, 150, 200, 250, and 300 mg/ml, respectively). SF, supercoiled form; NF, nicked form.

Discussion

Insects may encounter various stresses in the environment, such as heavy metal pollution, insecticides and herbicides, temperature challenges, and bacterial infections. These environmental stresses have promoted the development of diverse protective systems. GSTs are one of the most important antioxidant enzyme families that protect organisms against oxidative stress [8, 20, 31]. It is believed that the GST enzymes play essential roles in the survival of insects. Although multiple isoforms of GSTs in insects have been identified and characterized, the number of GSTs found in Chinese honeybees is rather limited and more *AccGST* genes remain to be classified. The *A. cerana cerana* GST described in this study has been identified as belonging to the Delta class of GSTs, which is currently considered to be insect-specific.

In non-insect species, many GST enzymes are differentially regulated in response to various inducers or environmental signals in a tissue- or development-specific manner to protect tissues against oxidative damage and oxidative stress. A similarly complex regulation pattern also applies to insect GSTs [19-21, 32]. Relatively low or high temperatures are responsible for increased levels of oxidative damage and can lead to oxidative stress [33]. Our present study identified the oxidative stress response of this single insect-specific GST to thermal stress for different treatment durations. As shown in Fig. 6, the expression of *AccGSTD* was significantly induced by heat and cold exposure, which might be due to the increased levels of oxidative stress caused by thermal stress, suggesting that *AccGSTD* may act as an effective antioxidant enzyme that removes excessive ROS under oxidative stress and thus ensures the survival of the honeybee. Consistent with these findings, binding sites for CREB, which is involved in oxidative stress, were found during a preliminary analysis of the *AccGSTD* promoter. It has been reported that the increase in the DNA repair capacity correlates with increased stress resistance and might thus lead to a longer lifespan [34]. Parkes et al. demonstrated that GSTs and GSH play a vital role in the prevention and/or repair of oxidative damage in *D. melanogaster* [35]. In the present study, DNA cleavage assay demonstrated that AccGSTD can protect DNA from oxidative damage and DNA repair, leading us to propose that AccGSTD may be a potential antioxidant enzyme for defending against DNA damage to maintain the lifespan of the honeybee.

The expression of *AccGSTD* was analyzed in different life isoforms of the *A. cerana cerana* by using qPCR. The highest expression level of *AccGSTD* was found in the adult (Fig. 4a). A honeybee colony is composed of one queen and tens of thousands of workers, and the young workers progress from cell cleaning to brood feeding and other in-hive tasks as they age [36]. 2-3 weeks after emergence, the adult workers shift their activities to outside of the hive, where they forage for nectar and pollen for the remainder of their lives [37]. This is the only stage that the bees are exposed to the external environment, and they are more susceptible to excessive ROS and oxidative damage during this period. The expression of *AccGSTD* at this stage may help protect the bees from oxidative injury. Old *A. mellifera* workers, which were the individuals that likely had the highest levels of flight activity, tended to have the highest expression levels of antioxidant genes including GSTs [38]. Putative promoter analysis identified putative binding sites for StuAp, a transcription factor required for the development and sexual reproductive cycle, suggesting that *AccGSTD* may be involved in developmental stage regulation.

Different tissues and organs have different rates of metabolic activity and oxygen consumption and, thus, result in different antioxidant levels. Most of the non-insect-specific GSTs show high expression levels in various tissues, and this expression pattern is primarily related to their roles as house-keeping proteins. Conversely, a majority of GSTs from the Delta and Epsilon classes show more tissue-specific expression patterns [39]. A single specific GST in the gut of Lepidopteran larvae has been found to isomerize the jasmonic acid precursor, 12-oxophytodienoic acid (*cis*-OPDA) [40]. In the present study, qPCR showed that *AccGSTD* was mainly expressed in the brain (Fig. 4b). Interestingly, most of the Delta and Epsilon GSTs in the silkworm also show little or no constitutive expression in tissues associated with detoxification, such as fat body [39]. In *A. mellifera*, this enzyme has been detected in the brain, reproductive tissue, thorax, and abdomen [38, 41] and was postulated to play a role in protection against oxidative stress [41]. The honeybee brain, which is involved in the regulation of the social behavior associated with the transition from hive work to foraging [37], is critically sensitive to oxidative damage [42]. The highly concentrated localization of AccGSTD in the brain, along with the brain development-related transcriptional activator Croc, suggests that AccGSTD may account for the antioxidant defense of this especially sensitive organ. In *D. melanogaster*, the specific overexpression of an antioxidant enzyme (superoxide dismutase) in motor neurons led to an extension of the fly lifespans [43, 44]. Thus, our results suggest that AccGSTD might also contribute to ensure the lifespan of the honeybee. Immunohistochemistry analysis of the fourth-instar larvae indicated strong positive staining in the intima of the midgut as well as in the epidermis (Fig. 5d), which acts as the exoskeleton of the honeybee and confers physical stability and plays a role in protection. Marionnet et al. [45] provided evidence that the epidermis provides protection against a variety of environmental stresses, including oxidative stress. As noted by Sawicki et al. [31], GSTs may be relevant in the metabolism of lipid peroxidation products in specific tissues in which they are expressed, and we hypothesize that AccGSTD may be essential for the protective role of the epidermis and protect the honeybees from environmental stress and mechanical injuries.

One of the notable and unique features reported about Delta GSTs is the mosquito-specific alternative splicing, which has been confirmed in four mosquito species: *A. gambiae* [46], *Anopheles dirus* [47], *Aedes aegypti* [7], and *Culex quinquefasciatus* [48]. However, this unique phenomenon has not been reported in other insect species such as *A. mellifera* or *D. melanogaster*. In *D. melanogaster*, 20 (10 Epsilon and 10 Delta) of the 37 GST genes are intronless [31], whereas only 42 introns have been found in the 32 *A. gambiae* GSTs (1.3 introns per gene) [49]. Interestingly, the present study identified that the protein coding region of AccGSTD contains four introns. The longest intron was observed in the 5′ uncoding region and was 3,277 bp long, while the other introns ranged from 71 to 593 bp. Similar results were found by Yu et al. [39], and insect-specific classes of GST in silkworms also show more and longer introns than are found in the Dipteran genomes. In addition, they identified several repetitive sequences in the longer introns, the majority of which were transposable elements or the remainders of transposable elements. These transposable elements could be implicated in the regulation of the oxidative stress response.

In conclusion, we cloned a Delta class GST from *A. cerana cerana* and identified its tissue- and developmental-dependent expression pattern. Recombinant AccGSTD protein revealed the ability to protect DNA from oxidative damage. Strong expression induced by temperature challenges indicated that AccGSTD may participate in the resistance to oxidative stress caused by thermal stress, which is important to maintain the lifespan of the honeybee. Our results provide preliminary insights into the genomic organization and expression profiles of GSTs in the Chinese honeybee. Further functional research is needed to identify the mechanisms of antioxidant defense of the protein.

References

[1] Enayati, A. A., Ranson, H., Hemingway, J. Insect glutathione transferases and insecticide resistance. Insect Mol Biol. 2005, 14, 3-8.

[2] Grant, D. F., Matsumura, F. Glutathione *S*-transferase 1 and 2 in susceptible and resistant insecticide resistant *Aedes aegypti*. Pestic Biochem Physiol. 1989, 33, 132-143.

[3] Clark, A. G., Shamaan, N. A. Evidence that DDT-dehydrochlorinase from the housefly is a glutathione *S*-transferase. Pestic Biochem Physiol. 1984, 22, 249-261.

[4] Lumjuan, N., Mccarroll, L., Prapanthadara, L. A., Hemingway, J., Ranson, H. Elevated activity of an Epsilon class glutathione transferase confers DDT resistance in the dengue vector, *Aedes aegypti*. Insect Biochem Mol Biol. 2005, 35, 861-871.

[5] Ranson, H., Rossiter, L., Ortelli, F., Jensen, B., Wang, X. Identification of a novel class of insect glutathione *S*-transferases involved in resistance to DDT in the malaria vector *Anopheles gambiae*. Biochem J. 2001, 359, 295-304.

[6] Hayes, J. D., Flanagan, J. U., Jowsey, I. R. Glutathione transferases. Annu Rev Pharmacol Toxicol. 2005, 45, 51-88.

[7] Lumjuan, N., Stevenson, B. J., Prapanthadara, L., Somboon, P., Brophy, P. M., Ranson, H. The *Aedes aegypti* glutathione transferase family. Insect Biochem Mol Biol. 2007, 37, 1026-1035.

[8] Vontas, J. G., Small, G. J., Nikou, D. C., Ranson, H., Hemingway, J. Purification, molecular cloning and heterologous expression of a glutathione *S*-transferase involved in insecticide resistance from the rice brown planthopper, *Nilaparvata lugens*. Biochem J. 2002, 362, 329-337.

[9] Blanchette, B., Feng, X., Singh, B. R. Marine glutathione *S*-transferases. Mar Biotechnol. 2007, 23, 513-542.

[10] Sheehan, D., Meade, G., Foley, V. M., Dowd, C. A. Structure, function and evolution of glutathione transferases, implications for classification of non-mammalian members of an ancient enzyme superfamily. Biochem J. 2001, 360, 1-16.

[11] Robinson, A., Huttley, G. A., Booth, H. S., Board, P. G. Modelling and bioinformatics studies of the human Kappa-class glutathione transferase predict a novel third transferase family with homology to prokaryotic 2-hydroxychromene-2-carboxylate isomerases. Biochem J. 2004, 379, 541-552.

[12] Ranson, H., Claudianos, C., Ortelli, F., Abgrall, C., Feyereisen, R. Evolution of supergene families associated with insecticide resistance. Science. 2002, 298, 179-181.

[13] Board, P. G., Coggan, M., Chelvanayagam, G., Easteal, S., Jermiin, L. S., Schulte, G. K. Identification, characterization, structure of the Omega class glutathione transferases. J Biol Chem. 2000, 275, 24798-24806.

[14] Low, W. Y., Ng, H. L., Morton, C. J., Parker, M. W., Batterham, P., Robin, C. Molecular evolution of glutathione *S*-transferases in the genus *Drosophila*. Genetics. 2007, 177, 1363-1375.

[15] Claudianos, C., Ranson, H., Johnson, R. M., Biswas, S., Feyereisen, R., Oakeshott, J. G. A deficit of detoxification enzymes, pesticide sensitivity and environmental response in the honeybee. Insect Mol

Biol. 2006, 15, 615-636.
[16] Hemingway, J., Ranson, H. Insecticide resistance in insect vectors of human disease. Ann Rev Entomol. 2000, 45, 371-391.
[17] Li, X., Schuler, M. A., Berenbaum, M. R. Molecular mechanisms of metabolic resistance to synthetic and natural xenobiotics. Ann Rev Entomol. 2007, 52, 231-253.
[18] Lumjuan, N., Rajatileka, S., Wicheer, J., Somboon, P., Lycett, G., Ranson, H. The role of the *Aedes aegypti* Epsilon glutathione transferases in conferring resistance to DDT and pyrethroid insecticides. Insect Biochem Mol Biol. 2011, 41, 203-209.
[19] Li, H. M., Buczkowski, G., Mittapalli O., Xie, J., Pittendrigh, B. R. Transcriptomic profiles of *Drosophila melanogaster* third instar larval midgut and responses to oxidative stress. Insect Mol Biol. 2008, 17, 325-339.
[20] Nair, P. M. G., Choi, J. Identification, characterization and expression profiles of *Chironomus riparius* glutathione *S*-transferase (GST) genes in response to cadmium and silver nanoparticles exposure. Aquat Toxicol. 2011, 101, 550-560.
[21] Yang, J. L., Wei, X. M., Xu, J., Yang, D. L., Liu, X. Q., Hua, X. K. A Sigma-class glutathione *S*-transferase from *Solen grandis* that responded to microorganism glycan and organic contaminants. Fish Shellfish Immunol. 2012, 32, 1198-1204.
[22] Honeybee Genome Sequence Consortium. Insights into social insects from the genome of the honeybee *Apis mellifera*. Nature. 2006, 443, 931-949.
[23] Mason, G., Provero, P., Vaira, A. M., Accotto, G. P. Estimating the number of integrations in transformed plants by quantitative real-time PCR. BMC Biotechnol. 2002, 2, 20.
[24] Yan, H. R., Meng, F., Jia, H. H., Guo, X. Q., Xu, B. H. The identification and oxidative stress response of a Zeta class glutathione *S*-transferase (GSTZ1) gene from *Apis cerana cerana*. J Insect Physiol. 2012, 58, 782-791.
[25] Livak, K. J., Schmittgen, T. D. Analysis of relative gene expression data using real-time quantitative PCR and the $2^{-\Delta\Delta C_t}$ method. Methods. 2001, 25, 402-408.
[26] Bradford, M. M. A rapid and sensitive method for the quantitation of microgram quantities of protein utilizing the principle of protein-dye binding. Anal Biochem. 1976, 72, 248-254.
[27] Udomsinprasert, R., Pongjaroenkit, S., Wongsantichon, J., Oakley, A., Ketterman, A. J. Identification, characterization and structure of a new Delta class glutathione transferase isoenzyme. Biochem J. 2005, 388, 763-771.
[28] Dutton, J. R., Johns, S., Miller, B. L. StuAp is a sequence-specific transcription factor that regulates developmental complexity in *Aspergillus nidulans*. EMBO J. 1997, 16, 5710-5721.
[29] Jeffrey, P. L., Capes-Davis, A., Tolhurst, O., Seeto, G., Hannan, A. J., Lin, S. L. CROC-4, a novel brain specific transcriptional activator of c-fos expressed from proliferation through to maturation of multiple neuronal cell types. Mol Cell Neurosci. 2000, 16, 185-196.
[30] Shi, Y. L., Venkataraman, S. L., Dodson, G. E., Mabb, A. M., LeBlanc, S., Tibbetts, R. S. Direct regulation of CREB transcriptional activity by ATM in response to genotoxic stress. Proc Natl Acad Sci USA. 2004, 101, 5898-5903.
[31] Sawicki, R., Singh, S. P., Mondal, A. K., Benes, H., Zimniak, P. Cloning, expression and biochemical characterization of one Epsilon class (GST-3) and ten Delta-class (GST-1) glutathione *S*-transferases from *Drosophila melanogaster*, and identification of additional nine members of the Epsilon class. Biochem J. 2003, 370, 661-669.
[32] Ma, B., Chang, F. N. Purification and cloning of a Delta class glutathione *S*-transferase displaying high peroxidase activity isolated from the German cockroach *Blattella germanica*. FEBS J. 2007, 274, 1793-1803.
[33] Lopez-Martinez, G., Elnitsky, M. A., Benoit, J. B., Lee Jr, R. E., Denlinger, D. L. High resistance to oxidative damage in the Antarctic midge *Belgica antarctica*, and developmentally linked expression of genes encoding superoxide dismutase, catalase and heat shock proteins. Insect Biochem Mol Biol. 2008, 38, 796-804.

[34] Hyun, M., Lee, J., Lee, K., May, A., Bohr, V. A., Ahn, B. Longevity and resistance to stress correlate with DNA repair capacity in *Caenorhabditis elegans*. Nucleic Acids Res. 2008, 36, 1380-1389.

[35] Parkes, T. L., Hilliker, A. J., Phillips, J. P. Genetic and biochemical analysis of glutathione S-transferases in the oxygen defence system of *Drosophila melanogaster*. Genome. 1993, 36, 1007-1014.

[36] Winston, M. L. The Biology of the Honey Bee. Harvard University Press, Cambridge. 1987.

[37] Ament, S. A., Corona, M., Pollock, H. S., Robinson, G. E. Insulin signaling is involved in the regulation of worker division of labor in honey bee colonies. Proc Natl Acad Sci USA. 2008, 105, 4226-4231.

[38] Corona, M., Hughes, K. A., Weaver, D. B., Robinson, G. E. Gene expression patterns associated with queen honey bee longevity. Mech Ageing Dev. 2005, 126, 1230-1238.

[39] Yu, Q. Y., Lu, C., Li, B., Fang, S. M., Zuo, W. D., Dai, F. Y., Zhang, Z., Xiang Z. H. Identification, genomic organization and expression pattern of glutathione S-transferase in the silkworm, *Bombyx mori*. Insect Biochem Mol Biol. 2008, 38, 1158-1164.

[40] Dabrowska, P., Freitak, D., Vogel, H., Heckel, D. G., Boland, W. The phytohormone precursor OPDA is isomerized in the insect gut by a single, specific glutathione transferase. Proc Natl Acad Sci USA. 2009, 106, 16304-16309.

[41] Collins, A. M., Williams, V., Evans, J. D. Sperm storage and antioxidative enzyme expression in the honey bee, *Apis mellifera*. Insect Mol Biol. 2004, 13, 141-146.

[42] Rival, T., Soustelle, L., Strambi, C., Besson, M. T., Iche, M., Birman, S. Decreasing glutamate buffering capacity triggers oxidative stress and neuropil degeneration in *Drosophila* brain. Curr Biol. 2004, 17, 599-605.

[43] Parkes, T. L., Hilliker, A. J., Phillips, J. P. Motorneurons, reactive oxygen, and life span in *Drosophila*. Neurobiol Aging. 1999, 20, 531-535.

[44] Sohal, R. S., Agarwal, A., Agarwal, S., Orr, W. C. Simultaneous overexpression of copper- and zinc-containing superoxide dismutase and catalase retards age-related oxidative damage and increases metabolic potential in *Drosophila melanogaster*. J Biol Chem. 1995, 270, 15671-15674.

[45] Marionnet, C., Bernerd, F., Dumas, A., Verrecchia, F., Dubertret, L. Modulation of gene expression induced in human epidermis by environmental stress *in vivo*. J Invest Dermatol. 2003, 121, 1447-1458.

[46] Ranson, H., Collins, F., Hemingway, J. The role of alternative mRNA splicing in generating heterogeneity within the *Anopheles gambiae* class I glutathione S-transferase family. Proc Natl Acad Sci USA. 1998, 95, 14284-14289.

[47] Pongjaroenkit, S., Leetachewa, S., Prapanthadara, L., Ketterman, A. J. Genomic organization and putative promoters of highly conserved glutathione S-transferases originating by alternative splicing in *Anopheles dirus*. Insect Biochem Mol Biol. 2001, 31, 75-85.

[48] Kasai, S., Komagata, O., Okamura, Y., Tomita, T. Alternative splicing and developmental regulation of glutathione transferases in *Culex quinquefasciatus* Say. Pestic Biochem Physiol. 2009, 94, 21-29.

[49] Ding, Y., Ortelli, F., Rossiter, L. C., Hemingway, J., Ranson, H. The *Anopheles gambiae* glutathione transferase supergene family: annotation, phylogeny and expression profiles. BMC Genom. 2003, 4, 35.

2.2 Characterization and Mutational Analysis of Omega-class GST (GSTO1) from *Apis cerana cerana*, a Gene Involved in Response to Oxidative Stress

Fei Meng, Yuanying Zhang, Feng Liu, Xingqi Guo, Baohua Xu

Abstract: The Omega-class of GSTs (GSTOs) is a class of cytosolic GSTs that have specific structural and functional characteristics that differ from those of other GST groups. In this study, we demonstrated the involvement of the *GSTO1* gene from *A. cerana cerana* in the oxidative stress response and further investigated the effects of three cysteine residues of GSTO1 protein on this response. Real-time quantitative PCR (qPCR) showed that *AccGSTO1* was highly expressed in larvae and foragers, primarily in the midgut, epidermis, and flight muscles. The *AccGSTO1* mRNA was significantly induced by cold and heat at 1 and 3 h. The TBA (2-thiobarbituric acid) method indicated that cold or heat resulted in MDA accumulation, but silencing of *AccGSTO1* by RNAi in honeybees increased the concentration of MDA. RNAi also increased the temperature sensitivity of honeybees and markedly reduced their survival. Disc diffusion assay indicated that overexpression of AccGSTO1 in *E. coli* caused the resistance to long-term oxidative stress. Furthermore, AccGSTO1 was active in an *in vitro* DNA protection assay. Mutations in Cys-28, Cys-70, and Cys-124 affected the catalytic activity and antioxidant activity of AccGSTO1. The predicted three-dimensional structure of AccGSTO1 was also influenced by the replacement of these cysteine residues. These findings suggest that *AccGSTO1* plays a protective role in the response to oxidative stress.

Keywords: *Apis cerana cerana*; *AccGSTO1*; Oxidative stress; Mutational analysis

Introduction

Reactive oxygen species (ROS), such as oxygen radical superoxide (O_2^-) and hydroxyl (OH^-), are constantly generated during aerobic metabolism. Although the rate of oxidant scavenging maintains a dynamic balance under normal conditions, an imbalance is created when overproduction of endogenous or exogenous oxidants exceeds the intracellular antioxidant capacity, leading to excessive ROS and oxidative stress [1, 2]. As there are varied levels of ROS in living organisms, ROS can have both beneficial and harmful effects on a

---

PLoS One, DOI: 10.1371/journal.pone.0093100

diverse array of biological processes. In insects, there is compelling evidence that ROS can regulate ageing. In other words, oxidative stress can decrease their lifespan [3]. ROS can also act as modulators of signal transduction pathways [4, 5] or can accelerate the aggregation of abnormal proteins, inducing oxidative stress that is associated with many diseases such as atherosclerosis, Alzheimer's disease, and Parkinson's disease, all of which are connected to ageing and lifespan [6, 7]. Thus, to defend against oxidative damage, aerobic organisms have evolved autologous enzymatic antioxidant systems, including primary and secondary antioxidant enzymes [8].

The glutathione S-transferases (GSTs; EC 2.5.1.18) are a family of phase II enzymes that are widely found in both eukaryotic and prokaryotic cells. A prominent characteristic of these proteins is the ability to utilize glutathione in reactions, contributing to the biotransformation and degradation of various environmental xenobiotics, such as drugs, insecticides, and intracellular ROS [9, 10]. Unlike mammalian GSTs, insect GSTs can be grouped into six cytosolic classes: Delta (GSTD), Epsilon (GSTE), Omega (GSTO), Sigma (GSTS), Theta (GSTT), Zeta (GSTZ), one structurally unrelated microsomal class (GSTmic), and one unclassified class (u) [8, 11, 12]. Although some GSTs have a canonical GST structure and exhibit overlapping substrate specificities, other members are highly specific.

The Omega-class GSTs have unique structural and functional characteristics that differ from the other GST groups. The first GSTO was found through a bioinformatic analysis of human expressed sequence tags [13]. Subsequently, GSTOs have been identified in plants [14], yeast [15], insects [16], and bacteria [17], and the number of GSTO genes varies depending on the species. For instance, three Omega GST genes have been found in the human genome [18], two GSTO genes have been found in mice and rats [11], and two GSTO genes have been identified in honeybees [8]. These GSTOs also display distinct genetic organization, crystal structure, substrate specificitie, and catalytic activitie[19].

In HsGSTO1, a novel cysteine residue (Cys-32) is located in the active site, an arrangement that is distinct from the prototypical serine and tyrosine residues in the other classes of the GST superfamily [13, 20]. The crystal structure of HsGSTO1 revealed that Cys-32 can form a mixed disulfide bond with the thiol of GSH precisely over the helix axis at the N-terminal region [13]. GSTO1 exhibits a canonical cytosolic GST fold, however, it has unique N-terminal and C-terminal extensions that are not observed in the other GST classes. These two domains of GSTO1 interact to form a distinct structural unit that is crucial for substrate specificity [20]. It is therefore not surprising that GSTO1 has unique enzymatic property, most significantly thiol transferase (TTase) and dehydroascorbate reductase (DHAR) activity[21, 22]. Activities against 1-chloro-2,4-dinitrobenzene (CDNB) or ethacrynic acid (ECA), typical substrates for other classes of GSTs, are low or not detected in GSTO1 [23, 24].

In early studies, elevated GSTO1 activity was associated with resistance to radiation and cytotoxic drugs [25], and the participation of these enzymes in the biomethylation pathway of inorganic arsenic metabolism was later reported [26]. Recent evidence suggests that GSTO1 is involved in protection against oxidative stress, including platinum resistance, the heat shock response, and the response to UV-B light [23, 24, 27, 28]. The *GSTO1* gene also modifies the age-at-onset of Alzheimer's disease and Parkinson's disease [7]. Other proposed roles of GSTO1 include the modulation of Ca^{2+} ion channels [29], the participation in IL-1β activation [30], and a direct interaction in pyrimidodiazepine synthesis [31].

The activity and function of GSTO1 can be influenced by alterations in the active site. Mutation of the active site Cys-32 residue in human GSTO1 causes a strong increase in activity toward CDNB [22], whereas the Asp-140 variant has a lower thiol transferase activity [32]. The replacement of Cys-33 of GSTO1 in *Caenorhabditis elegans* elevates the worm's sensitivity to cumene hydroperoxide [23], and mutations of Cys-38 and Pro-39 in silkmoths also affect BmGSTO1 activity [24].

The Asian honeybee (*Apis cerana cerana*) is a valuable indigenous species that plays an important role in the pollination of flowering plants, especially in low temperature environments. In recent decades, global warming and excessive pesticide use have threatened the existence of this species. Although some GST genes have been identified in *A. cerana cerana*, the specific GSTO1 remains unknown. Furthermore, the influence of different GSTO1 cysteine residues on the oxidative stress response has not yet been studied. Thus, to further elucidate the protective role of GSTO1 in a multicellular organism, we isolated and characterized a *GSTO1* gene from *A. cerana cerana*, and investigated the biochemical properties and antioxidant capacity of AccGSTO1; three mutated proteins were also investigated. Our findings provide valuable insight into the function of GSTO1 in the oxidative stress response.

Experimental Procedures

Insects, tissue dissection and stress treatment

Colonies of *A. cerana cerana* were reared in beehives at the experimental apiary of Shandong Agricultural University, Taian, China. Each larval and pupal stage was identified by calculating the number of days after the queen laid eggs. Adult workers were collected at the correct time after marking the newly emerged bees with paint. The brain, epidermis, flight muscle, midgut, and mandibular glands were dissected from the adult workers on ice and stored at −80℃. For the stress treatments, the adult workers were fed water and pollen and were exposed to 75% relative humidity at different temperatures: 4, 15, 25, 34, 43, and 50℃. Thereafter, the treated bees were frozen in liquid nitrogen and stored at −80℃ until used.

RNA extraction, cDNA synthesis, and DNA isolation

Total RNA was rapidly extracted from the dissected tissues by using the Trizol reagent (Invitrogen, Carlsbad, CA, USA) according to the instructions of the manufacturer. After digestion with RNase-free DNase Ⅰ, the first-strand cDNA was synthesized by using the EasyScript First-Strand cDNA Synthesis SuperMix (TransGen Biotech, Beijing, China). Genomic DNA was isolated by using the EasyPure Genomic DNA Extraction Kit (TransGen Biotech, Beijing, China) in accordance with the protocol of the manufacturer.

Cloning of the *AccGSTO1* gene

The procedures were performed as previously described [33]. Briefly, conserved primers

GP1 and GP2 were designed to obtain an internal fragment of the *AccGSTO1* gene by PCR. RACE primers and full-length primers were subsequently generated to verify the *AccGSTO1* cDNA sequence. Several specific primers were used to isolate the *AccGSTO1* genomic DNA. All primers and their sequences are listed in Table S1.

Real-time quantitative PCR

Real-time quantitative PCR (qPCR) was performed by using the SYBR PrimeScript^{TM} RT-PCR Kit (TaKaRa, Dalian, China) in a 25 μl reaction volume by using a CFX96^{TM} Real-Time PCR Detection System. The *β-actin* gene (GenBank ID: XM640276) was selected as a reference gene due to its stable expression in honeybees [34]. The following amplification reaction protocol was used: pre-denaturation at 95℃ for 30 s; 40 cycles of denaturation at 95℃ for 5 s, annealing at 56℃ for 15 s, and extension at 72℃ for 15 s; and a single melting cycle from 65℃ to 95℃. All the conditions were analyzed in three independent biological replicates, and the PCR reactions were performed in triplicate. The data were analyzed by using the CFX Manager software, version 1.1, with the $2^{-\Delta\Delta C_t}$ method. The statistical significance was determined by a Duncan's multiple range test using Statistical Analysis System (SAS) version 9.1 software.

Determination of the malonyl dialdehyde (MDA) concentration

Lipid oxidation results in the generation of many harmful secondary products and MDA can act as a reliable indicator of lipid peroxidation to indirectly reflect cellular damage. In this assay, the total proteins from the foragers held at different temperatures were rapidly extracted by using a Tissue Protein Extraction Kit (CWBIO, Beijing, China). Each sample was digested with 990 μl of tissue protein extraction reagent with 10 μl of PMSF and incubated at 4℃ for 20 min, followed by centrifugation at 10,000 g for 10 min. The supernatant was removed and diluted at a 1 : 10 ratio (V/V). The MDA concentration was measured by using an MDA assay kits (Institute of Biological Technology of Jiancheng, Nanjing, PR China). Because TBA was used as the substrate, this method was also called the TBA method. Simply, TBA combined with MDA to form red products, and the absorbance was measured at 532 nm. The MDA concentration was expressed as nmol MDA per mg protein. Three replicates were performed in each experiment.

RNA-mediated interference (RNAi) of *AccGSTO1*

Double-stranded RNAs (dsRNAs) were synthesized as described in an established method [33]. Briefly, specific primers with a T7 polymerase promoter were produced to synthesize the ds*AccGSTO1* by using the Ribo Max^{TM} T7 Large Scale RNA Production Systems (Promega, Madison, WI, USA). After digestion with Dnase Ⅰ and isopropanol extraction, the dsRNA was annealed by gradual cooling from 95℃ to room temperature. dsGFP was selected as a control [35]. All dsRNAs were transported by oral feeding into the foragers.

Stress resistance assays

The foragers were collected from source colonies and divided into nine groups ($n=80$/group). Groups 1-3 of the foragers were fed a normal diet. For Groups 4-6 and Groups 7-9, double-stranded *GFP* or *AccGSTO1* (20 μg) were added to the normal diet, respectively. All groups were maintained in incubators at a constant temperature (50℃) and humidity (75%) for 0.5-5 h after 1 day. The survival rates of the foragers were evaluated hourly. A forager was scored as dead when it did not respond to any stimulus.

Overexpression and purification of recombinant proteins

The full-length cDNA of *AccGSTO1* was integrated into the prokaryotic expression vector pET-30a (+) and subsequently transformed into *Escherichia coli* BL21 (DE3) cells. After the cells reached an optical density $(OD)_{600}$ of 0.2-0.5, isopropylthio-β-D-galactoside (IPTG) was added to the culture at a final concentration of 0.6 mmol/L to induce protein expression. SDS-PAGE was used to analyze the homogeneity of the proteins. The protein purification was performed as previously described [36]. Briefly, cell pellets were resuspended in lysis buffer (phosphate-buffered saline containing 15% glycerol, 0.3 mol/L NaCl, and 20 mmol/L imidazole, pH 7.4). After sonication and centrifugation, the supernatant was applied to a 1 ml HisTrap™ FF column (GE Healthcare, Uppsala, Sweden) for further purification. The proteins were extensively eluted with elution buffer (phosphate-buffered saline containing 0.25 mol/L imidazole, pH 7.4 and 0.5 mol/L arginine). The protein concentration was determined by using Coomassie Brilliant Blue G-250, and bovine serum albumin was used as a standard.

Antibody preparation and immunoblot analysis

Purified AccGSTO1 protein (100 μg/ml) mixed with Freund's complete adjuvant was subcutaneously injected into six white male mice. A week later, the mice were again injected with a mixture of 25 μg/ml purified protein and Freund's incomplete adjuvant for increased immunity. In all, the immunization was performed three times at one-week intervals. Three days after the last injection, the antisera were collected by centrifugation at 3,000 r/min for 15 min and stored at −80℃ [37]. The total proteins from the foragers were rapidly extracted by using a Tissue Protein Extraction Kit (CWBIO, Beijing, China). Purified AccGSTO1 and equal amounts of total proteins were subjected to 12% SDS-PAGE and then electrotransferred to a polyvinylidene difluoride membrane (PVDF, Millipore Corporation, UK). The immunoblot was performed by using the anti-AccGSTO1 polyclonal antibody (1∶100 dilution) and peroxidase-conjugated goat anti-mouse secondary antibody (Dingguo, Beijing, China) at a dilution of 1∶2000 (*V/V*). Subsequently, the immunoblot was visualized by using the SuperSignal West Pico Trial Kit (Thermo Scientific Pierce, IL, USA).

Site-directed mutagenesis

Three mutants (C28A, C70A, and C124A) were generated to replace each of the three cysteine residues in AccGSTO1. PCR-based mutagenesis was performed by using the

Quick-Change Site-Directed Mutagenesis Kit (Agilent Technologies, Wilmington, DE, USA) according to the recommendations of the manufacturer. The recombinant plasmid containing *AccGSTO1* was used as the template, and nucleotide sequencing was performed to verify the mutations. The mutants were then transformed into DE3 cells for protein expression.

Disc diffusion assay

The *E. coli* cultures containing the target protein were plated on LB-kanamycin agar plates and incubated at 37°C for 1 h. Sterile filter discs (6 mm diameter) were saturated with 20 μl aliquots of different concentrations of compounds, including paraquat (0, 50, 100, 200, and 300 mmol/L), cumene hydroperoxide (0, 25, 50, 75, and 100 mmol/L) and *t*-butyl hydroperoxide (0, 15, 20, 25, and 30 mmol/L). All the discs were placed on the surface of the agar and incubated at 37°C for 24 h. The statistical significance of the inhibition zone was calculated by using SAS version 9.1 software.

DNA cleavage assay with the MFO system

Reaction mixtures of 25 μl, including 100 mmol/L HEPES buffer, 3 mmol/L $FeCl_3$, 10 mmol/L DTT, and increasing concentrations of protein in equal volumes, were incubated at 37°C for 30 min. Subsequently, 300 ng of supercoiled pUC19 plasmid DNA was added to the reaction mixture, which was incubated for 2.5 h at 37°C. Agarose gel electrophoresis was performed by using a 1% gel to determine DNA cleavage. The signal intensities of the bands were determined by using the Quantity-One™ Image Analysis software (Bio-Rad, Hercules, CA, USA).

Measurements of enzyme activity

The enzymatic activity toward 1-chloro-2,4-dinitrobenzene (CDNB) was measured by using spectrophotometry, as described by Habig [38]. The glutathione-dependent dehydroascorbate reductase (DHAR) activity was determined spectrophotometrically with DHA at 265 nm, as described by Wells [39]. The glutathione peroxidase activity was measured by recording at 340 nm the oxidation of NADPH in a reaction mixture containing NADPH, GSH, glutathione reductase and H_2O_2 [40]. Alternatively, cumene hydroperoxide or *t*-butyl hydroperoxide was also used instead of H_2O_2 in the reaction to detect peroxidase activity.

Optimum temperature and pH and kinetic analysis

The effects of pH and temperature on enzymatic activity were examined by using a standard DHAR assay, as described previously [36]. To determine the optimal temperature, the reaction was initiated by adding DHA to the protein mixture over a temperature range of 5 to 55°C and recording for 2 min. Citrate buffer (50 mmol/L, pH 4.0-5.0), sodium phosphate buffer (50 mmol/L, pH 6.0-7.0), and Tris-HCl buffer (50 mmol/L, pH 8.0-9.0) were used to determine the optimal pH. The apparent kinetic parameters of the recombinant proteins were determined by using a GSH range of 0.2-4.0 mmol/L and a fixed DHA concentration of 1 mmol/L. Similarly, the apparent K_m and V_{max} values for DHA were determined by using a DHA range of 0.1-2.0 mmol/L and a fixed GSH concentration of 2.25 mmol/L. All of the reactions were measured at pH 8.0 and 25℃. The Michaelis constant was calculated by using the Lineweaver-Burk method in the Hyper program [41].

Bioinformatic analysis

The amino acid sequence was aligned with homologous GSTO1 sequences by using DNAMAN version 5.2.2 (Lynnon Biosoft, Quebec, Canada). The conserved domains in GSTO1 were predicted by using the NCBI server (http://www.ncbi.nlm.nih.gov/Structure/cdd/cdd.shtml). A phylogenetic analysis was performed by using the neighbor-joining method with the Molecular Evolutionary Genetic Analysis (MEGA) software version 4.0. All the tertiary structures were drawn by using Swiss-PdbViewer.

Results

Identification and sequence analysis of *AccGSTO1*

The cDNA sequence of *AccGSTO1* (GenBank ID: KF496073) is 1,040 bp long and is composed of a 246 bp 5' untranslated region (UTR), a 68 bp 3' UTR, and a 726 bp open reading frame (ORF) that encodes a peptide with a predicted molecular mass of 27.9 kDa. A multiple sequence alignment revealed that the AccGSTO1 N-terminal region is highly conserved with other species (Fig. 1A). Residues that contribute to the H-site region, as described for HsGSTO1 [13], were either conserved or conservatively replaced. However, unlike HsGSTO1, AccGSTO1 contains two additional Cys residues, and the active site Cys-28 is not predicted to be located in the α1 helix. Furthermore, a neighbor-joining phylogenetic tree including proteins from the six cytosolic GST superfamilies of different species was generated (Fig. 1B). The tree clearly grouped the Omega family in a cluster distinct from the other five classes of GSTs. Compared to *Bombus terrestris* and *Nasonia vitripennis*, GSTO1 from *A. cerana cerana* was placed closest to *Apis mellifera* and *Apis florea*. The genomic structure of *AccGSTO1* was also analyzed and found to be 1,579 bp (GenBank ID: KF496074), containing five introns. In contrast, *AccGSTO2* (GenBank ID: JX456219) is much larger than *AccGSTO1*, encompassing 2,147 bp [36]. Although there are five introns in both *AccGSTO* genes, the first intron of *AccGSTO1* is located in the 5' UTR (Fig. 1C).

Expression profiles of *AccGSTO1*

Stage-specific expression profiling showed that the highest expression levels were detected in larvae and foragers. The amount of *AccGSTO1* mRNA decreased rapidly in older bees, and almost no significant expression was detected in dead bees that died from natural causes (Fig. 2A). The tissue-specific expression analysis showed that *AccGSTO1* mRNA was expressed in all the tissues tested, with higher expression in the midgut, epidermis, and flight muscles and a relatively low level in the brain (Fig. 2B). Conversely, the highest expression of *AccGSTO2* has been detected in the brain [36]. As shown in Fig. 2C, after different temperature treatments, the expression of *AccGSTO1* was significantly induced at 5, 15, and 43 ℃ compared with the expression level at 34 ℃ ($P < 0.01$). The highest expression of *AccGSTO1* was detected after 1 h of temperature treatment. The overall expression profiles of *AccGSTO1* were different from those previously obtained for *AccGSTO2*, but cold and heat treatment have also been shown to induce *AccGSTO2* in a similar way [36]. These results reveal that the expression of *AccGSTO1* may be linked to the lifespan of the honeybee and the response to abiotic stress.

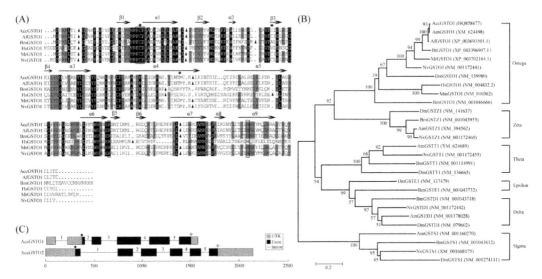

Fig. 1 Sequence analysis of the AccGSTO1. (A) Alignment of AccGSTO1 with the Omega-class GSTs from *Apis florea* (AfGSTO1), *Bombyx mori* (BmGSTO1), *Homo sapiens* (HsGSTO1), *Megachile rotundata* (MrGSTO1) and *Nasonia vitripennis* (NvGSTO1). Identical amino acids are shaded in black and similar amino acids are shown in gray. The putative secondary structures are displayed, and the H-site regions are boxed. The Cys residues in AccGSTO1 are marked by asterisks. The intron/exon junctions are indicated by triangles. (B) Phylogenetic relationships among GSTO1 and other GST families. Six main groups are shown. Acc, *Apis cerana cerana*; Af, *Apis florea*; Am, *Apis mellifera*; Bm, *Bombyx mori*; Bt, *Bombus terrestris*; Dm, *Drosophila melanogaster*; Hs, *Homo sapiens*; Mm, *Mus musculus*; Mr, *Megachile rotundata*; Nv, *Nasonia vitripennis*. (C) Comparison between the genomic structures of *AccGSTO1* and *AccGSTO2*. The translational initiation codons (ATG) and termination codons (TAA) are marked with dots and asterisks, respectively.

MDA content during temperature-induced oxidative stress

To determine the influence of the temperature on oxidative stress, MDA concentration (nmol/mg protein) was measured *in vitro* by the TBA method. As shown in Fig. 3, 4 ℃ (cold) and 43 ℃ (heat) resulted in a higher accumulation of lipid peroxides compared with the other temperatures. In general, the MDA concentration gradually increased from 0.5 to 7 h, suggesting that many harmful secondary products were continuously generated. The high expression of *AccGSTO1* observed after 1 and 3 h of cold or heat stress indicates that *AccGSTO1* may play a role in antioxidant defense when MDA levels are rising.

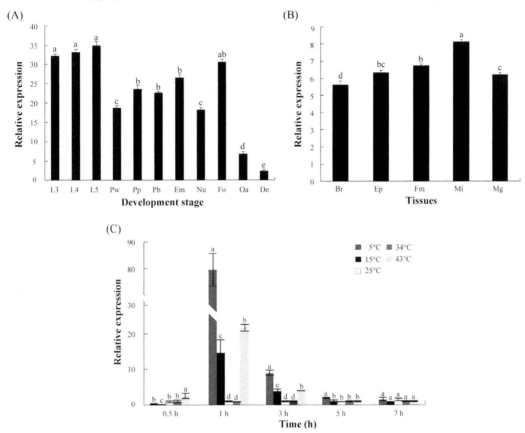

Fig. 2 The expression profile of *AccGSTO1* determined by using qPCR. (A) *AccGSTO1* mRNA expression at different developmental stages. Total RNA was isolated from larvae (L3-the third instars, L4- the fourth instars, and L5- the fifth instars), pupae (Pw-white eyes pupae, Pp-pink eyes pupae, and Pb-brown eyes pupae), and adults (Em-newly emerged bees, Nu-nurse bees, Fo-foragers, Oa-old bees, and De-dead bees). (B) The tissue distribution of *AccGSTO1* expression. Total RNA was extracted from dissected brain (Br), epidermis (Ep), flight muscle (Fm), midgut (Mi), and mandibular gland (Mg) tissues of foragers. (C) Expression of *AccGSTO1* after different temperature treatments (4, 15, 25, 34, and 43℃). Samples were collected at 0.5, 1, 3, 5, and 7 h. Bees at 34℃ were used as a control. The vertical bars represent the mean ± SE (n = 3). The different letters above the columns indicate significant differences ($P < 0.05$) according to Duncan's multiple range test performed using SAS version 9.1 software.

RNAi-mediated silencing of *AccGSTO1* leads to temperature sensitivity

Colder temperatures induced the foragers to enter a sleeping stage, making it impossible to accurately determine their survival. Therefore, the foragers were fed double-stranded *AccGSTO1* RNA and were placed at higher temperatures (50℃) to measure their survival rate; foragers fed a normal diet and ds*GFP* were placed under the same conditions to be used as controls. As shown in Fig. 4A, the survival rate of the *AccGSTO1* RNAi foragers was significantly lower than that of the control foragers, and there were no obvious signs of life in the foragers after 4 h at 50℃. In addition, RNAi-mediated inhibition of *AccGSTO1* also induced a higher MDA concentration after 1 and 2 h at 50℃ (Fig. 4B). These results indicate that the RNAi-mediated silencing of *AccGSTO1* increases the temperature sensitivity of foragers.

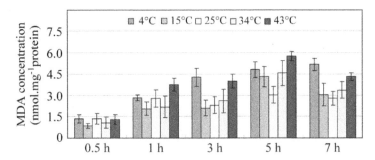

Fig. 3 The detection of the MDA content at different temperatures. In all experiments, foragers at 34℃ were used as controls. The effects of temperature-induced oxidative stress on the MDA concentration (nmol/mg protein) in foragers. The vertical bars represent the mean ± SE (*n* = 3).

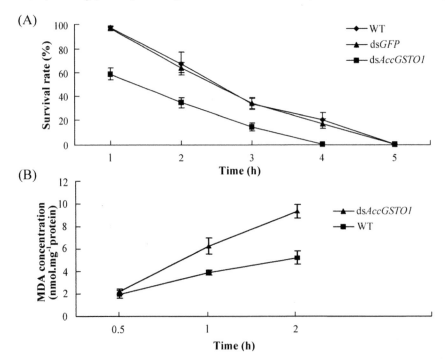

Fig. 4 The effects of the RNAi-mediated silencing of *AccGSTO1*. (A) The survival rate of *AccGSTO1* RNAi-treated foragers compared with controls at 50℃. The foragers were collected and divided into nine groups (*n* = 80/group). Groups 1-3 and Groups 4-6 were used as controls and were fed a normal diet or a normal diet supplemented with ds*GFP*, respectively. Groups 7-9 were fed ds*AccGSTO1*. All the foragers were placed in incubators at a constant temperature (50℃) and humidity (75%) for 1-5 h. (B) The MDA content of RNAi-treated foragers compared with controls at 50℃. All the foragers were placed in incubators at a constant temperature (50℃) and humidity (75%) for 0.5-2 h. WT presents the untreatment honeybees.

Characterization of the recombinant proteins

To determine whether the different Cys residues critically influence the antioxidant functions of AccGSTO1, each residue was mutated to Ala by using site-directed mutagenesis. The three mutants were named C28A, C70A, and C124A and were overexpressed in *E. coli* BL21 (DE3) cells as recombinant proteins. Although the predicted size of the AccGSTO1 monomer and the three mutants was 27.9 kDa, the proteins with a His-tag migrated at approximately 35 kDa during sodium dodecyl sulfate polyacrylamide gel electrophoresis (SDS-PAGE). Further purification was performed by using HisTrap™ FF columns (Fig. S1). These results demonstrate that the wild-type and three mutant versions of AccGSTO1 were successfully overexpressed. Moreover, polyclonal antibodies against AccGSTO1 were raised in white male mice following standard immunization procedures. Immunoblotting showed that AccGSTO1 was obviously detected in both the total proteins of the foragers and the purified AccGSTO1 (Fig. 5A). Moreover, the expression of AccGSTO1 protein in different tissue was consistent with the mRNA expression levels (Fig. 2B, Fig. 5B). These results indicate that the anti-AccGSTO1 antibody is reasonably specific.

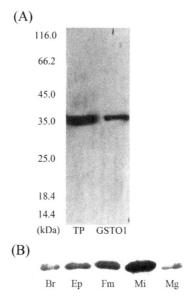

Fig. 5 Immunoblot analysis of AccGSTO1. (A) Immunoblot analysis was performed on total protein from foragers (TP) and purified AccGSTO1. (B) Immunoblot analysis was performed with different tissues from the foragers. Each lane was loaded with an equivalent amount of protein. Br, brain; Ep, epidermis; Fm, flight muscle; Mi, midgut; Mg, mandibular gland.

Effects of the C28A, C70A, and C124A mutations on the oxidative stress response

To evaluate the effect of the three Cys residues on the antioxidant activity of AccGSTO1, *E. coli* overexpressing AccGSTO1, C28A, C70A, and C124A were exposed to oxidative stressors, such as paraquat (internal inducer), cumene hydroperoxide, and tert-butyl hydroperoxide (external inducer). As shown in Fig. S2, the ability of AccGSTO1 to protect the *E. coli* cells from oxidative stress was quite obvious. The killing zones were smaller on the plates with the *E. coli* expressing AccGSTO1 than the plates with the controls, with 43%-55% (paraquat), 36%-42% (cumene hydroperoxide), and 9%-33% (tert-butyl hydroperoxide) halo reduction (Fig. 6, Fig. S2). Interestingly, replacing the three Cys residues with Ala resulted in varied losses of antioxidant activity for each residue. Because Cys-28 is present in the active site, the C28A mutant exhibited almost no antioxidant activity when compared to the control bacteria. The alanine mutations of Cys-70 and Cys-124 also decreased the stress response of AccGSTO1, but the antioxidant ability of the C70A or C124A mutants was observed mostly with cumene hydroperoxide treatment.

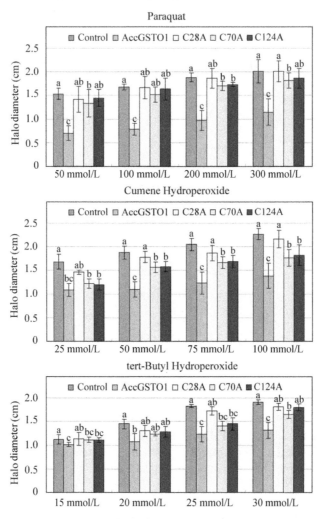

Fig. 6 Disc diffusion assays by using *E. coli*-overexpressed proteins. The halo diameters of the killing zones were detected after an overnight exposure. LB agar plates were flooded with bacteria overexpressing AccGSTO1, C28A, C70A, or C124A. Bacteria transfected with BL21 [pET-30a (+) only] was used as a control. The data presented are the mean ± SE of three independent experiments (*n* = 3). The different letters above the columns indicate significant differences ($P < 0.05$) according to Duncan's multiple range test performed by using SAS version 9.1 software.

Effects of the C28A, C70A, and C124A mutations on DNA damage protection

Oxidative stress destroys DNA integrity and blocks specific DNA repair pathways [42, 43]. In the mixed-function oxidation (MFO) system, the electron donor dithiothreitol (DTT) can react with $FeCl_3$ to produce hydroxyl radicals, which convert supercoiled DNA to nicked DNA forms [44]. When the pUC19 plasmid was incubated alone or together with $FeCl_3$, a completely supercoiled form of DNA was observed; however, when DTT was added, the pUC19 plasmid was nicked, leading to altered migration (Fig. 7). As increasing concentrations of AccGSTO1 were added, the amount of the nicked form gradually decreased, becoming nearly undetectable at 100 μg/ml AccGSTO1 (Fig. 7, lane 10). Notably, regardless of which Cys residue was replaced with Ala, the three mutants were unable to effectively protect the supercoiled DNA from nicking as well as the wild-type AccGSTO1. A control protein (BSA) did not display any nicking protection (Fig. 7). We can conclude that the C28A, C70A, and C124A mutations influence DNA damage protection, which may be conferred by the sulfhydryl moieties of the Cys residues.

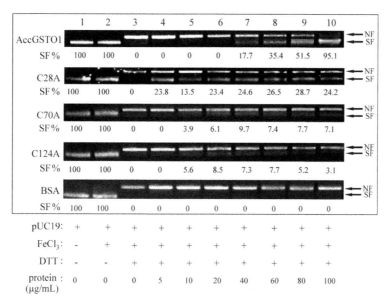

Fig. 7 Protection of DNA from oxidative damage by purified proteins in the MFO system. Lane 1, pUC19 plasmid DNA only; lane 2, pUC19 plasmid DNA + $FeCl_3$; lane 3, pUC19 plasmid DNA + $FeCl_3$ + DTT; lane 4-10, pUC19 plasmid DNA + $FeCl_3$ + DTT + purified proteins (5, 10, 20, 40, 60, 80, and 100 μg/ml, respectively). SF, supercoiled form; NF, nicked form. SF% is SF/(SF + NF) for each lane. The expression level was determined by the signal intensities of the bands.

Effects of the C28A, C70A, and C124A mutations on AccGSTO1 enzyme activity

GSTO1 has unique enzymatic properties, such as undetectable CDNB activity and highly specific DHAR activity [22]. AccGSTO1 also exhibited the lowest CDNB activity and the most specific DHAR activity (Table 1). The maximum DHAR activity for AccGSTO1 was observed at pH 8.0 and an optimal temperature of 25°C (Fig. S3). At a fixed DHA concentration of 1 mmol/L, the K_m and V_{max} values for GSH were (1.63±0.41) mmol/L and (4.17±0.87) μmol/(min·mg protein), respectively. At a fixed GSH concentration of 2.25 mmol/L, the K_m and V_{max} values for DHA were (0.57±0.06) mmol/L and (6.15±1.76) μmol/(min·mg protein), respectively. Additionally, AccGSTO1 exhibited detectable glutathione peroxidase activity against different peroxidase substrates including H_2O_2, cumene hydroperoxide or tert-butyl hydroperoxide, which is similar to Gto1 in *Saccharomyces cerevisiae* [45]. However, mutation of AccGSTO1 Cys-28 or Cys-124 caused an increase in CDNB activity and the C28A, C70A, and C124A mutations showed no activity toward DHA, H_2O_2, cumene hydroperoxide or tert-butyl hydroperoxide. GSTO1 orthologs display different enzymatic activities across species, and their mutants also show differential losses of activity (Table 1). These results indicate that the Cys residues are important for GSTO1 activities.

Table 1 Comparison of specific activities of AccGSTO1 and other GSTO

	CDNB	DHAR	glutathione peroxidase	cumene hydroperoxide	tert-butyl hydroperoxide	Reference
AccGSTO1	0.015±0.002	1.025±0.041	0.193±0.012	0.081±0.008	0.047±0.003	This study
C28A	0.0225±0.003	ND	ND	ND	ND	This study
C70A	0.143±0.006	ND	ND	ND	ND	This study
C124A	0.181±0.003	ND	ND	ND	ND	This study
HsGSTO1	0.18±0.006	0.16±0.005	-	ND	ND	[13]
Gto1	ND	0.23±0.031	ND	ND	ND	[42]
Gto2	ND	0.11±0.009	ND	ND	0.13	[42]
Gto3	ND	0.16±0.004	0.18	0.31	0.99	[42]
bmGSTO	0.67±0.06	0.64±0.08	0.32±0.05	—	—	[24]
C38A	3.8±0.7	ND	ND	—	—	[24]
P39A	ND	ND	ND	—	—	[24]
Y40A	ND	ND	ND	—	—	[24]

Note: specific activity shown in μmol/(min·mg). ND, not detectable. "—" indicated not determined. C38A, P39A and Y40A are mutants for bmGSTO.

Effects of the C28A, C70A, and C124A mutations on the structure of AccGSTO1

By using Swiss-PdbViewer, we predicted the three-dimensional structures of AccGSTO1 and the C28A, C70A, and C124A mutants based on the only established crystal structure of GSTO1 [13]. All the GSTO1s were composed of 241 amino acids and assembled as a dimer that clearly adopted the canonical GST fold. At the N-terminal domain, a central four-stranded β sheet and two α helices established a thioredoxin-like domain (βαβαββ), whereas larger helices (7α) were found in the C-terminal domain. Notably, of the nine helices, two helices (α2, residues 58-59, and α8, residues 211-214) were 3_{10} helices, and two special β sheets (β5, residues 178-180, and β6, residues 185-187) were found in the C-terminal domain of

AccGSTO1 but not in the C28A, C70A, and C124A mutants or in other GST structures.

There were some differences in the predicted three-dimensional structures of the three mutants compared with the structure of AccGSTO1 (Fig. 8). For example, large random coils were distributed differently throughout these structures; β5 and β6 in AccGSTO1 were replaced with other secondary structures, including α6 (residues 167-181) and random coils (residues 185-187) in the C28A, C70A, and C124A proteins. These distinct structures were mainly concentrated in the C-terminal domain. As shown in Fig. 9, we found that the H-bonds in the C-terminal domain were markedly altered between AccGSTO1 and the three mutants and among the C28A, C70A, and C124A mutants. Furthermore, the residue solvent accessibilities among these proteins were also changed, such as Ile-191 and His-193 in AccGSTO1, Met-183 and Phe-189 in C28A and C124A, and Leu-184 and Gln-188 in C70A. These predicted structural differences may be the reason for the functional disparities among the wild-type AccGSTO1 and the C28A, C70A, and C124A mutant proteins.

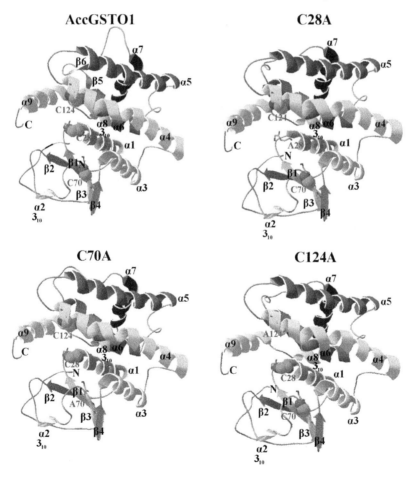

Fig. 8 The predicted tertiary structure of the wild-type AccGSTO1 and C28A, C70A, and C124A mutant proteins. The N-terminus, C-terminus, α helices, and β sheets are marked. Helices α2 and α8 are 3_{10} helices, and the other helices are α helices. The Cys and mutagenic Ala residues are highlighted by a ball and stick representation. These images were generated by using Swiss-PdbViewer.

Discussion

The Omega-class of GSTs (GSTO) is a class of cytosolic GSTs with structural and functional characteristics that differ from those of the other GST groups, such as a specific Cys-32 in the active site, a lower CDNB activity, and unique enzymatic properties (Ttase and DHAR) [13, 20, 22, 23]. In this study, we demonstrated the involvement of GSTO1 from *A. cerana cerana* in the oxidative stress response and investigated the influence of different cysteine residues on this response. The first GSTO1 was identified from human expressed sequence tags [13]. Similar to HsGSTO1, the canonical GST structure and unique N-terminal and C-terminal extensions were also found in AccGSTO1. However, three Cys residues were found in AccGSTO1, and the conserved Cys-28 active site residue of AccGSTO1 was not predicted to be located in helix α1. Additionally, two novel β5 and β6 sheets were found in the predicted AccGSTO1 tertiary structure, which is significantly different from the typical GST.

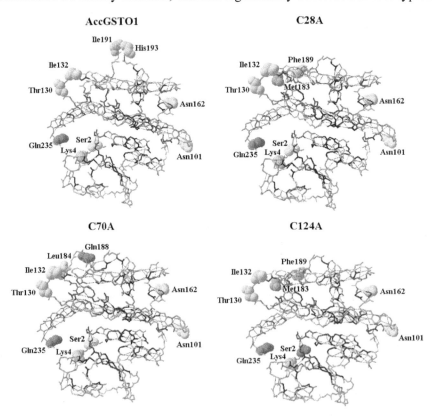

Fig. 9 The distribution of H-bonds and different amino acid residues on the surface of the protein. The H-bonds are highlighted in red. The different colors indicate the different locations of the amino acids within the protein. Blue and green represent interior amino acids. The amino acids on the surface of the protein are shown in ball representation. Orange represents the superficial amino acids that were highly contacted by solvent. These images were drawn by using Swiss-PdbViewer. (For color version, please sweep QR Code in the back cover)

Accumulated evidence suggests that GSTO1 is involved in antioxidant defense [23, 24, 27, 28]. The highest expression of GSTO1 was reported in liver and muscle, suggesting that the encoded protein is responsible for the detoxification of xenobiotics and ROS scavenging [13]. The higher tissue expression of AccGSTO1 was found in the midgut, epidermis, and flight muscle, which differed from the expression of AccGSTO2 in brain tissue [36]. Although *AccGSTO2* may respond to oxidative stress by maintaining ascorbic acid levels in the brain, *AccGSTO1* plays an antioxidant role by detoxification in the muscle or midgut. In fact, this distribution pattern of *AccGSTO1* is similar to the distribution pattern in vertebrates: honeybees do not have a liver, though the midgut is a highly metabolically active organ that can function in the elimination of toxicants. In addition, the protective role of *AccGSTO1* against oxidative stress was also demonstrated in different developmental stages. Foragers are the major pollinators of flowering plants and are often exposed to a variety of environmental stimuli, such as temperature, UV light, and pesticides. As the autologous enzymatic antioxidant systems are more powerful in younger individuals than older individuals [8], it was not surprising that the lowest expression of AccGSTO1 was detected in dead bees. In humans, several reports have confirmed a close association between GSTO1 and age-related diseases [6, 7]. In honeybees, the specific silencing of the *AccGSTO1* gene by using RNAi notably decreased their survival rate at high temperature, which was related to higher oxidative stress. Our findings may provide valuable insight into the correlation between lifespan and antioxidant activity.

The direct evidence of an antioxidant role was derived from the experiments that examined responses to cold and heat shock, resistance to peroxides, and DNA damage protection. Temperature is one of the dominant abiotic factors directly affecting herbivorous insects, and the optimal temperature of honeybee colonies is between 33 and 36 ℃ [46]. Low temperatures decrease metabolic rates and increase oxygen radical formation during flight [47]; conversely, heat stress accelerates mitochondrial respiration and also stimulates the formation of ROS [48]. When honeybees are subjected to cold or heat stress, lipid peroxides begin to accumulate faster, resulting in higher MDA concentrations. The expression of AccGSTO1 was significantly induced at 4, 15, and 43 ℃ in a short time period, specifically by 1 h, suggesting that AccGSTO1 perhaps mediates an immediate antioxidant activity when excessive ROS are generated. While oxidative stress was beyond the scavenging rate of autologous enzymatic antioxidant systems, AccGSTO1 expression was decreased. Furthermore, the RNAi-mediated silencing of AccGSTO1 in foragers increased their sensitivity to heat shock and raised MDA levels. Therefore, when organisms suffer long-term injuries or become older, the antioxidant functions sharply decrease or disappear.

In addition to temperature, cumene hydroperoxide and tert-butyl hydroperoxide can also act as external environmental stressors. Paraquat can act as an intracellular ROS inducer by crossing cell membranes to generate the oxygen radical superoxide anion (O_2^-) during aerobic metabolism [23]. The disc diffusion assay clearly showed that *AccGSTO1* protected against oxidative stress, with the intracellular inducer-mediated oxidative stress being more significantly inhibited than the externally induced oxidative stress. In contrast, the resistance to cumene hydroperoxide and tert-butyl hydroperoxide in *C. elegans* increased 3-fold compared to paraquat resistance [23]. This discrepancy may be due to species differences or the specific structures of AccGSTO1 that are distinct from classic GSTO1. Of the many biological targets of oxidative stress, nucleic acids are often affected [42]; indeed, ROS can inhibit DNA damage repair and uncouple oxidative phosphorylation [42, 43]. In the MFO system, the C28A mutant protected a small amount of supercoiled plasmid at the low concentration, indicating that it retained some antioxidant activity. We speculated that each Cys residue plays a different role in the response to oxidative stress. Cys28 is at the active site and may be required for AccGSTO1 function. Moreover, regardless of which Cys residue was replaced with Ala, the three mutants were unable to effectively protect the supercoiled form of plasmid DNA as well as the wild-type AccGSTO1, especially at protein concentrations greater than 60 μg/ml. AccGSTO1 protected DNA integrity and was involved in DNA damage protection, which is similar to other antioxidant enzymes such as thioredoxin peroxidase [49]. These results demonstrate that AccGSTO1 plays a key role in counteracting oxidative stress.

This protective role against oxidative stress is mediated by glutathione (GSH), which can neutralize a component of toxic xenobiotics. Similar to other GSTs, GSTO1 has a well-defined active site that is adjacent to the GSH-binding site. The H-site is constructed from elements of both the N-terminal and C-terminal domains [13]. However, GSTO1 has unique N-terminal and C-terminal extensions that are not found in other GST classes, and these variations reflect its varied substrate specificity [21]. Although some residues are conservatively replaced in AccGSTO1, these residues also form the walls of the pocket that corresponds to the H-site in the HsGSTO1 structure [13]. The Phe-27 and Pro-29 residues in AccGSTO1 form a wide, deep pocket, and Arg-177 contributes to the bottom of the pocket; residues in the C-terminal α9 helix, particularly His-222, form the top and back of the pocket. The structural similarity suggests that AccGSTO1 has the same substrate specificity as HsGSTO1. In this study, AccGSTO1 exhibited the low CDNB activity, glutathione peroxidase activity and specific DHAR activity, which are similar characteristics to the majority of GSTO1 proteins [13, 22-24, 43].

Unlike HsGSTO1, there are three Cys residues in AccGSTO1. Although the active site residue Cys-28 in AccGSTO1 is not located in helix α1, it has been demonstrated that this Cys residue is crucial for forming a mixed disulphide bond with GSH [50, 51]. Mutations of some other important residues have also been reported. For example, a Cys-32 to Ala mutation in human GSTO1 strongly increases activity toward CDNB [22], whereas mutations at Cys-38 and Pro-39 in silkmoths lead to largely negative changes in the activities of BmGSTO1 [24], and an Asp-140 variant has decreased transferase activity [32]. In the present study, three mutants, C28A, C70A, and C124A, exhibited no enzymatic activity against DHA. Interestingly, the Cys-28 and Cys-124 mutations in AccGSTO1 caused an increase in activity toward CDNB, though C70A had no influence on CDNB activity. An interesting result was also observed in the disc diffusion assay: the C28A protein had almost no antioxidant activity, whereas the antioxidant activity was retained in the C70A and C124A mutants mostly with cumene hydroperoxide. These results may be ascribed to variations in the GSTO1 structure. Cys-28 is located in the active site in the N-terminal domain and Cys-124 is located in the α4 helix in the C-terminal region, which can form a distinct structural unit by attachment of the N-terminal extension [13]. The predicted tertiary structure of the C28A and C124A proteins displayed dramatic alterations in their C-terminal domains compared with AccGSTO1; in particular, the β5 and β6 sheets of AccGSTO1 were replaced with the α6 helix and a random coil. Interestingly, many different random coils were predicted among the three mutants, and the C70A mutant had more differences than the C28A and C124A mutants. Hydrogen bonds play an important role in maintaining protein stability [52], and these distinct structures influenced the location of several H-bonds. Furthermore, the change in protein structure also affected the residue solvent accessibilities that related to hydrophobic interactions. For example, Ile-191 and His-193 in AccGSTO1 were internalized in the three mutants, but Met-183 and Phe-189 became surface residues in C28A and C124A. Moreover, Leu-184 and Gln-188 of C70A had more opportunity to bind to solvents. Cys-replacement mutants changed their solvent accessibilities and hydrophobic interactions, which are the major driving forces for protein structure [53].

Protection against oxidative stress is an eternal theme for the lifespan of living organisms. In conclusion, the GSTO1 gene from *A. cerana cerana* is involved in the response to oxidative stress, and three Cys residues play pivotal roles in its antioxidant activity, particularly Cys-28 and Cys-124. These findings provide valuable insight into the functions of Omega-class GSTs in the oxidative stress response.

References

[1] Valko, M., Izakovic, M., Mazur, M., Rhodes, C. J., Telser, J. Role of oxygen radicals in DNA damage and cancer incidence. Mol Cell Biochem. 2004, 266 (1-2), 37-56.

[2] Touyz, R. M. Reactive oxygen species, vascular oxidative stress, and redox signaling in hypertension: what is the clinical significance? Hypertension. 2004, 44 (3), 248-252.

[3] Dugan, L. L., Quick, K. L. Reactive oxygen species and aging: evolving questions. Sci Aging Knowledge Environ. 2005, 2005 (26), pe20.

[4] Mahadev, K., Wu, X., Zilbering, A., Zhu, L., Lawrence, J. T., Goldstein, B. J. Hydrogen peroxide generated during cellular insulin stimulation is integral to activation of the distal insulin signaling cascade in 3T3-L1 adipocytes. J Biol Chem. 2001, 276 (52), 48662-48669.

[5] Colavitti, R., Pani, G., Bedogni, B., Anzevino, R., Borrello, S., Waltenberger, J., Galeotti, T. Reactive oxygen species as downstream mediators of angiogenic signaling by vascular endothelial growth factor receptor-2/KDR. J Biol Chem. 2002, 277 (5), 3101-3108.

[6] Finkel, T., Holbrook, N. J. Oxidants, oxidative stress and the biology of ageing. Nature. 2000, 408 (6809), 239-247.

[7] Li, Y. J., Oliveira, S. A., Xu, P., et al. Glutathione *S*-transferase omega-1 modifies age-at-onset of Alzheimer disease and Parkinson disease. Hum Mol Genet. 2003, 12 (24), 3259-3267.

[8] Corona, M., Robinson, G. E. Genes of the antioxidant system of the honey bee: annotation and phylogeny. Insect Mol Biol. 2006, 15 (5), 687-701.

[9] Enayati, A. A., Ranson, H., Hemingway, J. Insect glutathione transferases and insecticide resistance. Insect Mol Biol. 2005, 14 (1), 3-8.

[10] Hayes, J. D., Flanagan, J. U., Jowsey, I. R. Glutathione transferases. Ann Rev Pharmacol Toxicol. 2005, 45, 51-88.

[11] Mannervik, B., Board, P. G., Hayes, J. D., Listowsky, I., Pearson, W. R. Nomenclature for mammalian soluble glutathione transferases. Methods Enzymol. 2005, 401, 1-8.

[12] Ding, Y., Ortelli, F., Rossiter, L. C., Hemingway, J., Ranson, H. The *Anopheles gambiae* glutathione transferase supergene family: annotation, phylogeny and expression profiles. BMC Genomics. 2003, 4 (1), 4-35.

[13] Board, P. G., Coggan, M., Chelvanayagam, G., et al. Identification, characterization, and crystal structure of the Omega class glutathione transferases. J Biol Chem. 2000, 275 (32), 24798-24806.

[14] Dixon, D. P., Davis, B. G., Edwards, R. Functional divergence in the glutathione transferase superfamily in plants: identification of two classes with putative functions in redox homeostasis in *Arabidopsis thaliana*. J Biol Chem. 2002, 277 (34), 30859-30869.

[15] Garcera, A., Barreto, L., Piedrafita, L., Tamarit, J., Herrero, E. *Saccharomyces cerevisiae* cells have

[15] three Omega class glutathione S-transferases acting as 1-Cys thiol transferases. Biochem J. 2006, 398 (2),187-196.

[16] Walters, K. B., Grant, P., Johnson, D. L. Evolution of the GST Omega gene family in 12 *Drosophila* species. J Hered. 2009,100 (6), 742-753.

[17] Xun, L., Belchik, S. M., Xun, R., Huang, Y., Zhou, H., et al. S-glutathionyl-(chloro) hydroquinone reductases: a novel class of glutathione transferases. Biochem J. 2010, 428 (3), 419-427.

[18] Whitbread, A. K., Tetlow, N., Eyre, H. J., Sutherland, G. R., Board, P. G. Characterization of the human Omega class glutathione transferase genes and associated polymorphisms. Pharmacogenetics. 2003, 13 (3), 131-144.

[19] Board, P. G. The omega-class glutathione transferases: structure, function, and genetics. Drug Metab Rev. 2011, 43 (2), 226-235.

[20] Caccuri, A. M., Antonini, G., Allocati, N., et al. GSTB1-1 from *Proteus mirabilis*: a snapshot of an enzyme in the evolutionary pathway from a redox enzyme to a conjugating enzyme. J Biol Chem. 2002, 277 (21), 18777-18784.

[21] Ball, L. J., Kuhne, R., Schneider-Mergener, J., Oschkinat, H. Recognition of proline-rich motifs by protein-protein-interaction domains. Angew Chem Int Ed Engl. 2005, 44 (19), 2852-2869.

[22] Whitbread, A. K., Masoumi, A., Tetlow, N., Schmuck, E., Coggan, M., Board, P. G. Characterization of the Omega-class of glutathione transfrases. Methods Enzymol. 2005, 401,78-99.

[23] Burmeister, C., Lüersen, K., Heinick, A., Hussein, A., Domagalski, M., Walter, R. D., Liebau, E. Oxidative stress in *Caenorhabditis elegans*: protective effects of the Omega class glutathione transferase (GSTO-1). FASEB J. 2008, 22 (2), 343-354.

[24] Yamamoto, K., Teshiba, S., Shigeoka, Y., Aso, Y., Banno, Y., Fujiki, T., Katakura, Y. Characterization of an Omega-class glutathione S-transferase in the stress response of the silkmoth. Insect Mol Biol. 2011, 20 (3), 379-386.

[25] Story, M. D., Meyn, R. E. Modulation of apoptosis and enhancement of chemosensitivity by decreasing cellular thiols in a mouse B-cell lymphoma cell line that overexpresses bcl-2. Cancer Chemother Pharmacol. 1999, 44 (5), 362-366.

[26] Chowdhury, U. K., Zakharyan, R. A., Hernandez, A., Avram, M. D., Kopplin, M. J., Aposhian, H. V.. Glutathione-S-transferase-omega [MMA(V) reductase] knockout mice: enzyme and arsenic species concentrations in tissues after arsenate administration. Toxicol Appl Pharmacol. 2006, 216 (3), 446-457.

[27] Yan, X. D., Pan, L. Y., Yuan, Y., Lang, J. H., Mao, N. Identification of platinum resistance associated proteins through proteomic analysis of human ovarian cancer cells and their platinum-resistant sublines. J Proteome Res. 2007, 6 (2), 772-778.

[28] Wan, Q., Whang, I., Lee, J. S., Lee, J. Novel Omega glutathione S-transferases in disk abalone: characterization and protective roles against environmental stress. Comp Biochem Physiol C: Toxicol Pharmacol. 2009, 150 (4), 558-568.

[29] Dulhunty, A., Gage, P., Curtis, S., Chelvanayagam, G., Board, P. The glutathione transferase structural family includes a nuclear chloride channel and a ryanodine receptor calcium release channel modulator. J Biol Chem. 2001, 276 (5), 3319-3323.

[30] Laliberte, R. E., Perregaux, D. G., Hoth, L. R., et al. Glutathione S-transferase Omega 1-1 is a target of cytokine release inhibitory drugs and may be responsible for their effect on interleukin-1beta posttranslational processing. J Biol Chem. 2003, 278 (19), 16567-16578.

[31] Kim, J., Suh, H., Kim, S., Kim, K., Ahn, C., Yim, J. Identification and characteristics of the structural gene for the *Drosophila* eye color mutant sepia, encoding PDA synthase, a member of the Omega class glutathione S-transferases. Biochem J. 2006, 398 (3), 451-460.

[32] Tanaka-Kagawa, T., Jinno, H., Hasegawa, T., Makino, Y., Seko, Y., Hanioka, N., Ando, M. Functional characterization of two variant human GSTO 1-1s (Ala140Asp and Thr217Asn). Biochem Biophys Res Commun. 2003, 301 (2), 516-520.

[33] Meng, F., Lu, W., Yu, F., Kang, M., Guo, X., Xu, B. Ribosomal protein L11 is related to brain maturation during the adult phase in *Apis cerana cerana* (Hymenoptera, Apidae).

Naturwissenschaften. 2012, 99 (5), 343-352.

[34] Scharlaken, B., De Graaf, D. C., Goossens, K., Peelman, L. J., Jacobs, F. J. Differential gene expression in the honeybee head after a bacterial challenge. Dev Comp Immunol. 2008, 32 (8), 883-889.

[35] Elias-Neto, M., Soares, M. P., Simões, Z. L., Hartfelder, K., Bitondi, M. M. Developmental characterization, function and regulation of a Laccase2 encoding gene in the honey bee, *Apis mellifera* (Hymenoptera, Apinae). Insect Biochem Mol Biol. 2010, 40 (3), 241-251.

[36] Zhang, Y., Yan, H., Lu, W., Li, Y., Guo, X., Xu, B. A novel Omega-class glutathione *S*-transferase gene in *Apis cerana cerana*: molecular characterisation of GSTO2 and its protective effects in oxidative stress. Cell Stress Chaperones. 2013, 18 (4), 503-516.

[37] Meng, F., Zhang, L., Kang, M., Guo, X., Xu, B. Molecular characterization, immunohistochemical localization and expression of a ribosomal protein L17 gene from *Apis cerana cerana*. Arch Insect Biochem Physiol. 2010, 75 (2), 121-138.

[38] Habig, W. H., Pabst, M. J., Jakoby, W. B. Glutathione *S*-transferases. The first enzymatic step in mercapturic acid formation. J Biol Chem. 1974, 249 (22), 7130-7139.

[39] Wells, W. W., Xu, D. P., Washburn, M. P. Glutathione: dehydroascorbate oxidoreductases. Methods Enzymol. 1995, 252, 30-38.

[40] Chu, F. F., Doroshow, J. H., Esworthy, R. S. Expression, characterization, and tissue distribution of a new cellular selenium-dependent glutathione peroxidase, GSHPx-GI. J Biol Chem. 1993, 268 (4), 2571-2576.

[41] Lineweaver, H., Burk, D. The determination of enzyme dissociation constants. J Am Chem Soc. 1934, 56, 658-666.

[42] Tsuda, M., Ootaka, R., Ohkura, C., Kishita, Y., Seong, K. H., Matsuo, T., Aigaki, T. Loss of Trx-2 enhances oxidative stress-dependent phenotypes in *Drosophila*. FEBS Lett. 2010, 584 (15), 3398-3401.

[43] Cavallo, D., Ursini, C. L., Setini, A., Chianese, C., Piegari, P., Perniconi, B., Iavicoli, S. Evaluation of oxidative damage and inhibition of DNA repair in an *in vitro* study of nickel exposure. Toxicol In Vitro. 2003, 17 (5-6), 603-607.

[44] Suttiprapa, S., Loukas, A., Laha, T., Wongkham, S., Kaewkes, S., Gaze, S., Brindley, P. J., Sripa, B. Characterization of the antioxidant enzyme, thioredoxin peroxidase, from the carcinogenic human liver fluke, *Opisthorchis viverrini*. Mol Biochem Parasitol. 2008, 160 (2), 116-122.

[45] Garcerá, A., Barreto, L., Piedrafita, L., Tamarit, J., Herrero, E. *Saccharomyces cerevisiae* cells have three Omega class glutathione *S*-transferases acting as 1-Cys thiol transferases. Biochem J. 2006, 398 (2), 187-196.

[46] Kleinhenz, M., Bujok, B., Fuchs, S., Tautz, J. Hot bees in empty broodnest cells: heating from within. J Exp Biol. 2003, 206 (23), 4217-4231.

[47] Harrison, J. F., Fewell, J. H. Environmental and genetic influences on flight metabolic rate in the honey bee, *Apis mellifera*. Comp Biochem Physiol A: Mol Integr Physiol. 2002, 133 (2), 323-333.

[48] Flanagan, S. W., Moseley, P. L., Buettner, G. R. Increased flux of free radicals in cells subjected to hyperthermia: detection by electron paramagnetic resonance spin trapping. FEBS Lett. 1998, 431 (2), 285-286.

[49] Yu, F., Kang, M., Meng, F., Guo, X., Xu, B. Molecular cloning and characterization of a thioredoxin peroxidase gene from *Apis cerana cerana*. Insect Mol Biol. 2011, 20 (3), 367-378.

[50] Gustafsson, A., Pettersson, P. L., Grehn, L., Jemth, P., Mannervik, B. Role of the glutamyl α-carboxylate of the substrate glutathione in the catalytic mechanism of human glutathione transferase A1-1. Biochemistry. 2001, 40 (51), 15835-15845.

[51] Winayanuwattikun, P., Ketterman, A. J. An electronsharing network involved in the catalytic mechanism is functionally conserved in different glutathione transferase classes. J Biol Chem. 2005, 280 (36), 31776-31782.

[52] Vogt, G., Argos, P. Protein thermal stability: hydrogen bonds or internal packing? Fold Des. 1997, 2 (4), S40-S46.

[53] Nicholls, A., Sharp, K. A., Honig, B. Protein folding and association: insights from the interfacial and thermodynamic properties of hydrocarbons. Proteins. 1991, 11 (4), 281-296.

2.3 A Novel Omega-class Glutathione *S*-transferase Gene in *Apis cerana cerana*: Molecular Characterisation of *GSTO2* and Its Protective Effects in Oxidative Stress

Yuanying Zhang, Huiru Yan, Wenjing Lu, Yuzhen Li, Xingqi Guo, Baohua Xu

Abstract: Oxidative stress may be the most significant threat to the survival of living organisms. Glutathione *S*-transferases (GSTs) serve as the primary defences against xenobiotic and peroxidative-induced oxidative damage. In contrast to other well-defined GST classes, the Omega-class members are poorly understood, particularly in insects. Here, we isolated and characterized the *GSTO2* gene from *Apis cerana cerana* (*AccGSTO2*). The predicted transcription factor binding sites in the *AccGSTO2* promoter suggested possible functions in early development and antioxidant defence. Real-time quantitative PCR (qPCR) and Western blotting analysis indicated that *AccGSTO2* gene was highly expressed in larvae and was predominantly localized to the brain tissue in adults. Moreover, *AccGSTO2* transcription was induced by various abiotic stresses. The purified recombinant AccGSTO2 protein exhibited glutathione-dependent dehydroascorbate reductase and peroxidase activity. Furthermore, it could prevent DNA damage. In addition, *Escherichia coli* overexpressing *AccGSTO2* displayed resistance to long-term oxidative stress exposure in disc diffusion assays. Taken together, these results suggest that *AccGSTO2* plays a protective role in counteracting oxidative stress.

Keywords: *Apis cerana cerana*; *GSTO2*; Gene expression pattern; Oxidative stress; Biochemical properties

Introduction

As normal by-products of oxygen metabolism, reactive oxygen species (ROS) are both harmful and beneficial to living systems [1]. In low intracellular concentrations, ROS act as secondary messengers in signal transduction pathways regulating cell growth, while high ROS concentrations induce cellular senescence and apoptosis by oxidising nucleic acids, proteins, and lipids [2]. In normal conditions, a dynamic balance between ROS generation and scavenging exists, resulting in a relatively low level of ROS maintained within cells. However, endogenous or exogenous insults can break this balance and lead to the excessive production or accumulation of ROS (i.e., oxidative stress). To defend against oxidative damage by ROS, organisms have evolved complex antioxidant defence systems to maintain normal cell structures and functions.

The glutathione *S*-transferases (GSTs, EC 2.5.1.18) family is a member of antioxidant defence systems that have dominant roles in regulating the intracellular ROS balance. GSTs catalyse the conjugation of reduced glutathione with compounds containing an electrophilic centre, which forms more soluble, nontoxic peptide derivatives that can be excreted from cells [3]. Therefore, GSTs play central roles in the detoxification of endogenous and xenobiotic compounds, including drugs, herbicides and insecticides [4]. GSTs also participate in the biosynthesis and intracellular transport of hormones and in the protection against oxidative stress [5, 6]. GSTs are ubiquitous in both prokaryotes and eukaryotes. On the basis of insect GSTs sequence similarity, chromosomal location and immunological properties, insect cytosolic GSTs are currently divided into six classes, which are named Delta, Sigma, Epsilon, Zeta, Theta and Omega [7].

The Omega-class GST (GSTO) was originally identified by analysing sequence similarities in the human expressed sequence tag (EST) database [8]. To date, GSTOs appear to be widespread in nature and have been identified in plants [9], yeast [10], insects [11], bacteria [12], and mammals [13]. GSOTs have unique structural and functional characteristics. Within the canonical GST fold, GSTOs have a cysteine residue in their active site rather than the serine or tyrosine found in the other GST classes [14-16]. Consequently, the GSTOs catalyse a range of thiol transfer and reduction reactions that are rarely catalysed by members of the other classes [8]. Recently, GSTOs were shown to scavenge free radicals and ROS by modulating dehydroascorbate reduction and recycling [17, 18]. GSTOs also participate in the biotransformation pathway of inorganic arsenic metabolism, which protects organisms from acute and chronic arsenic toxicosis [19]. There is increasing evidence that the GSTOs are involved in various biological and clinically significant settings, including drug resistance [20], Alzheimer's disease [21], the action of anti-inflammatory drugs [22], and susceptibility to chronic obstructive pulmonary disease (COPD) [23]. By regulating mitochondrial ATP synthase activity, DmGSTO1 has been demonstrated to play a protective role in a *Drosophila* model of Parkinson's disease [24]. However, the role of GSTO in insect have not been fully investigated and should be further determined.

As a pollinator of flowering plants, *A. cerana cerana* is a valuable indigenous species that plays an indispensable role in the balance of regional ecologies and agricultural economic development. Due to environmental deterioration in recent decades, farming this species is extremely difficult. The protective roles of GSTOs during oxidative stress have been invoked [9, 25]. To our knowledge, the molecular identity of Hymenoptera GSTOs has not been investigated in comparison with other GST classes in Hymenoptera species although the *Apis mellifera* genome was annotated. In this study, we isolated and characterized a GSTO gene, named *AccGSTO2*, from *A. cerana cerana*. In addition to assaying the biochemical properties of the purified AccGSTO2, we investigated the antioxidant ability of this protein. On the basis of these results, we infer that *AccGSTO2* might be one of the factors to perform functions in response to oxidative stress.

Materials and Methods

Insects and treatments

A. cerana cerana worker bees were obtained from the experimental apiary of Technology Park of Shandong Agricultural University. The larvae and pupae were identified according to the criteria of Michelette and Soares [26]. The entire bodies of larvae, pupae and 1-day-old adult workers were collected from the hive. 10-day-old adult workers were obtained by marking newly emerged bees with paint and collecting them after 10 days. The brain, epidermis, flight muscle and midgut from adults (10 d) were dissected on ice, frozen immediately in liquid nitrogen, and stored at −80℃. Untreated larvae, pupae and adult workers were also frozen in liquid nitrogen and stored at −80℃ for analysis of developmental expression patterns. Moreover, the 10-day-old adults were fed on the basic adult diet containing water, 70% powdered sugar and 30% honey from the source colonies, and maintained in incubators at a constant temperature (34℃) and humidity (70%) under a 24 h dark regimen [27]. Then, they were divided into ten groups ($n = 40$/group). Groups 1-5 of adult workers were subjected to cold (4, 16, 25℃), heat (42℃), and ultraviolet (UV)-light (30 MJ/cm^2), respectively. Group 6 of adult workers were injected with 20 µl of H_2O_2 (50 µmol/L of H_2O_2/worker) between the first and the second abdominal segments by using a sterile microscale needle. Considering that a bee in the wild is more likely to be exposed to insecticides through the cuticle when it lands on/rubs against a plant, groups 7-10 of adult workers for insecticide treatments, insecticides (cyhalothrin, phoxim, pyridaben, and paraquat) were delivered in 0.5 µl distilled water to the thoracic notum of worker bees and the final concentrations are 20 µg/L, 1 µg/ml, 10 µmol/L, 10 µmol/L, respectively. The adult workers were injected with PBS as H_2O_2 controls in group 6. In insecticide treatments, adult workers were treated with 0.5 µl distilled water as controls. The adult workers were left untreated as controls in group 1-5. The adult workers of groups 1, 3, 7-10 were collected after treatment at 0.5, 1, 1.5, 2, and 2.5 h, while the groups 2, 4-6 at 1, 2, 3, 4, and 5 h. The treated bees were then immersed in liquid nitrogen and stored at −80℃ until used. These experiments were performed in triplicate.

RNA extraction, cDNA synthesis, and DNA isolation

Trizol reagent (Invitrogen, Carlsbad, CA, USA) was used to extract total RNA from the bees according to the protocol of the manufacturer. Total RNA was digested with Rnase-free Dnase I to eliminate any potential DNA contamination. The first-strand cDNA was generated with the EasyScript First-Strand cDNA Synthesis SuperMix (TransGen Biotech, Beijing, China) according to the instructions of the manufacturer. Genomic DNA was isolated from adults by using the EasyPure Genomic DNA Extraction Kit (TransGen Biotech, Beijing, China) according to the instructions of the manufacturer.

Isolation of the cDNA, genomic sequence and the 5'-flanking region of the *AccGSTO2* gene

To determine the cDNA and the genomic DNA sequences of *AccGSTO2*, cloning procedures were performed as previously described [28]. The 5'-flanking region of *AccGSTO2* was amplified as previously described [29]. The primer sequences used were provided in Supplementary Table 1. The MatInspector database (http://www.cbrc.jp/research/db/TFSEARCH.html) was used to predict putative cis-acting element and transcription factor binding sites in the 5'-flanking region of *AccGSTO2*. All of the PCR amplification conditions are shown in Supplementary Table 2.

Bioinformatic analysis

The conserved domains in AccGSTO2 protein were detected with the bioinformatics tools available at the NCBI server (http://blast.ncbi.nlm.nih.gov/Blast.cgi). Tertiary structures were predicted by using SWISS-MODEL (http://swissmodel.expasy.org/). By using the DNAMAN software program version 5.2.2, we identified the *AccGSTO2* ORF and conducted multiple alignments of gene homologues. The Expasy-Sib Bioinformatics Resource Portal (http://web.expasy.org/compute_pi/) was used to predict the theoretical isoelectric point and molecular mass. The phylogenetic tree of the predicted amino acid sequences of GSTs from several insect species was constructed by using the neighbour-joining method and the molecular evolution genetics analysis (MEGA) software program, version 4.1.

Real-time quantitative PCR

cDNA synthesis was performed as described above. *β-actin* (XM640276) gene was selected as a reference gene to normalize qPCR experiments since it was validated among the most stably expressed genes tested in honeybee [30, 31]. The AccGSTO2 primer pairs RPF/RPR and *β-actin* primer pairs β-actin-s/β-actin-x were used for qPCR with the SYBR® PrimeScript™ RT-PCR Kit (TaKaRa, Dalian, China) in a CFX96™ Real-time System (Bio-Rad, Hercules, CA, USA). The qPCR programme was as follows: pre-denaturation at 95℃ for 30 s, 40 cycles of amplification (95℃ for 10 s, 55℃ for 20 s, and 72℃ for 15 s) and a melting cycle from 65℃ to 95℃. mRNA abundance was evaluated in three independent biological replicates for each group with three technical repeats for each primer pair. The data were analyzed with the CFX Manager software programme, version 1.1 by using the $2^{-\Delta\Delta C_t}$ method [32].

Protein expression, purification and antibodies preparation

The expression of the recombinant AccGSTO2 was performed as previously described [29]. The *AccGSTO2* coding region was ligated into pET-30a (+) vector (Novagen, Darmstadt, Germany) after digestion with the restriction endonucleases *Bam*H I and *Kpn* I. The purification of target protein was performed with a method modified from Zhou et al. [33].

Briefly, the cells were centrifuged at 6,000 g at 4℃ for 10 min. The pellet was resuspended in 15 ml of lysis buffer (phosphate-buffered saline containing 15% glycerol, 0.3 mol/L NaCl and 20 mmol/L imidazole, pH 7.4), sonicated and centrifuged again at 4℃ (11,000 g, 15 min). Finally, both the supernatant and the pellet were solubilized in lysis buffer, and expression of the target recombinant protein was analyzed by 12% sodium dodecyl sulphate polyacrylamide gel electrophoresis (SDS-PAGE). The recombinant AccGSTO2 protein was purified on a HisTrap™ FF column (GE Healthcare, Uppsala, Sweden) according to the instructions of the manufacturer. The supernatant was also loaded onto 1 ml of a HisTrap™ FF column equilibrated with lysis buffer for 1 h, and the protein was eluted with lysis buffer containing 0.25 mol/L imidazole, pH 7.4, and 0.5 mol/L arginine to prevent precipitation and encourage refolding. The purified protein was examined by 12% SDS-PAGE. The purified protein was injected subcutaneously into white mice for generation of antibodies as described by Meng et al. [34].

Western blotting analysis

Total protein from four different tissues (brain, epidermis, muscle and midgut) and three different developmental stages (day-3 larvae, day-6 pupae, and day-10 adults) were extracted as described by Li et al. [35]. The total proteins were quantified with the BCA Protein Assay Kit (Thermo Scientific Pierce, IL, USA). Equal amounts of protein from four tissues or three developmental stages were subjected to 12% SDS-PAGE and subsequently electrotransferred onto a polyvinylidene fluoride (PVDF) membrane (Millipore, Bedford, MA) by using a Semi-dry Transfer Conit. Western blotting was performed according to the procedure of Meng et al. [34]. The anti-AccGSTO2 serum was used as the primary antibody at a dilution of 1 : 500 (V/V). Peroxidase-conjugated goat anti-mouse immunoglobulin G (Dingguo, Beijing, China) was used as the secondary antibody at a dilution of 1 : 2,000 (V/V). The proteins were detected by using the SuperSignal® West Pico Trial Kit (Thermo Scientific Pierce, IL, USA).

Enzymatic activity assays of recombinant AccGSTO2

Purified recombinant AccGSTO2 protein was dialysed in 100 mmol/L KH_2PO_4, pH 7.5, and used for the enzyme assays. The assays were performed in triplicate from three independent enzyme preparations. The 1-chloro-2,4-dinitrobenzene (CDNB)-conjugating activity was measured by using spectrophotometry, as described by Habig et al. [36]. Glutathione peroxidase activity was monitored by using a method adapted from Boldyrev et al. [37] in a reaction mixture containing 50 mmol/L sodium phosphate buffer (pH 7.8), 1 mmol/L EDTA, 0.12 mmol/L NADPH, 0.85 mmol/L GSH as a substrate, 0.5 unit/ml glutathione reductase and 0.2 mmol/L cumene hydroperoxide (or t-butyl hydroperoxide). The reaction was initiated by adding the enzymatic extract, and it was recorded at 340 nm for 10 min at 25℃. Glutathione-dependent dehydroascorbate reductase (DHAR) activity was measured by recording the increase in absorbance at 265 nm as described by Wells et al. [38].

Optimum temperature and pH analysis

Because the recombinant AccGSTO2 protein showed detectable activity towards typical

GST substrates, the effects of temperature and pH on enzymatic activity were determined by the standard DHAR assay. Enzyme activity for optimum temperature was determined by incubating the recombinant AccGSTO2 in the reaction mix contained of 200 mmol/L sodium phosphate (pH 6.85), 1 mmol/L EDTA, 2.25 mmol/L GSH and 1 mmol/L DHA in a total volume of 1 ml. The reaction was initiated by adding DHA and recorded for 2 min over a temperature ranges from 5℃ to 55℃ in 5℃ increments with a 1 cm light path. Citrate buffer (50 mmol/L, pH 4.0-5.0), sodium phosphate buffer (50 mmol/L, pH 6.0-7.0) and Tris-HCl buffer (50 mmol/L, pH 8.0-9.0) were used to measure the optimal pH. All of the assays were performed in triplicate.

Kinetic parameter analysis

The DHAR apparent kinetic parameters of the recombinant GST were determined by using a GSH range of 0.25-4.0 mmol/L and a fixed DHA concentration of 1 mmol/L; similarly, the apparent K_m and V_{max} values for DHA were determined by using a DHA range from 0.1 to 2.0 mmol/L and a fixed GSH concentration of 2.25 mmol/L. All of the reactions were measured at pH 6.85 and 25℃. The apparent K_m and V_{max} values for CDNB (or GSH) were determined by using a CDNB (or GSH) range from 0.1 to 4.0 mmol/L and a fixed GSH (or CDNB) concentration of 1.0 mmol/L. All of the reactions were measured at pH 6.5 and 25℃. The Michaelis constant was calculated by using the Lineweaver-Burk method in the Hyper program.

DNA cleavage assay with the MFO system

AccGSTO2 was characterized in a DNA protection assay modified from Yu et al. [29]. Briefly, a reaction mixture (25 μl) containing 100 mmol/L HEPES buffer, 3 mmol/L $FeCl_3$, 10 mmol/L DTT, and increasing concentrations of AccGSTO2 protein ranging from 10 to 150 μg/ml was incubated at 37℃ for 30 min before the addition of 300 ng of pUC19 supercoiled plasmid DNA. The reaction was incubated for an additional 2.5 h at 37℃ and then subjected to 1% agarose gel electrophoresis for the determination of DNA cleavage. N-ethylmaleimide (NEM, 5 mmol/L) and BSA (150 μg/ml) were used as negative controls to add in this reaction mixture.

Disc diffusion assay under oxidative stress

Disc diffusion assay under oxidative stress was performed with a method modified from Burmeister et al. [18]. Briefly, approximately 5×10^8 bacterial cells were plated on LB-kanamycin agar plates and incubated at 37℃ for 1 h. Filter discs (6 mm diameter) soaked in different concentrations of cumene hydroperoxide (0, 25, 50, 80, or 100 mmol/L), t-butyl hydroperoxide (0, 2, 5, 8, or 10 mmol/L), or paraquat (0, 10, 50, 100, or 200 mmol/L) were placed on the surface of the top agar. The cells were grown for 24 h at 37℃ before the inhibition zones around the paper discs were measured.

Data analysis

Error bars denote standard error of the mean (SEM) from three independent experiments. The significant differences were determined by Duncan's multiple range tests using the Statistical Analysis System (SAS) version 9.1 software programmes (SAS Institute, Cary, NC, USA).

Results

Isolation and characterization of the *AccGSTO2* cDNA sequence

The sequence analysis indicated that the full-length cDNA of *AccGSTO2* (JX434029) was 1,365 bp, including a 321 bp 5' untranslated region (UTR), a 324 bp 3' UTR and a 720 bp complete ORF. The ORF encoded a polypeptide of 239 amino acids with a predicted molecular mass of 27.785 kDa and a theoretical isoelectric point of 6.21.

Multiple sequence alignments showed that AccGSTO2 shares 46.67%-99.16% sequence similarity to GSTO from *Apis florae* (AfGSTO1), *Bombus impatiens* (BiGSTO1), *Bombus terrestris* (BtGSTO1), *A. mellifera* (AmGSTO2), *Nasonia vitripennis* (NvGSTO2), *Harpegnathos saltator* (HsGSTO1), *Acromyrmex echinatior* (AeGSTO1), and *Camponotus floridanus* (CfGSTO1). In addition, the predicted AccGSTO2 contains the conserved features of the cytosolic GST superfamily, which include an N-terminal thioredoxin-like domain (H_5-D_{90}, with $\beta\alpha\beta\alpha\beta\beta\alpha$ topology) and a C-terminal α helixes domain (E102-L225). Moreover, the GSH-binding site (G-site) and putative substrate-binding site (H-site) were predicted by using a bioinformatics tool on the NCBI server (Fig. 1A). This analysis indicated the AccGSTO2 belongs to the typical GSTOs [8].

To determine the evolutionary relationships between AccGSTO2 and other insect GSTs, a neighbour-joining phylogenetic tree was built with 40 GSTs from different insects (Fig. 1B). Of the six classes insect GSTs examined, AccGSTO2 was categorized into the Omega class, where it clustered with AmGSTO2. Thus, we named the putative protein AccGSTO2.

To understand the relationships between the structure and function of AccGSTO2, its three-dimensional structure was predicted and reconstructed by SWISS-MODEL; the template that was used was the X-ray crystal structure of human Omega GSTO2-2 (PDB 3qagA), which was the second Omega GST structure determined. The two proteins share 30.67% sequence identity. As shown in Fig. 1C, the N-terminal and C-terminal regions were located on the surface of the tertiary structure, and the overexpression of the pET-30a (+) vector did not affect the protein folding.

Identification of the genomic structure of the *AccGSTO2*

To further elucidate the properties of the *AccGSTO2*, the genomic DNA sequence of *AccGSTO2* was obtained by PCR amplification. The complete *AccGSTO2* gene (JX456219) is 2,147 bp, which includes six exons that are separated by five introns with high AT contents and canonical 5'-GT splice donor and 3'-AG splice acceptor sites. The *AccGSTO2* exons and introns were aligned with *GSTO* sequences from *A. cerana cerana*, *A. florea*, *B. impatiens*, *B. terrestris*, and *N. vitripennis* (Fig. 2). *AccGSTO2*, *AfGSTO1*, *BiGSTO1*, and *BtGSTO1* contained five introns and shared the highest length similarity, whereas *NvGSTO2* contained six introns. Moreover, because the first intron of the *NvGSTO2* was located in the 5' UTR, this gene was longer than the others.

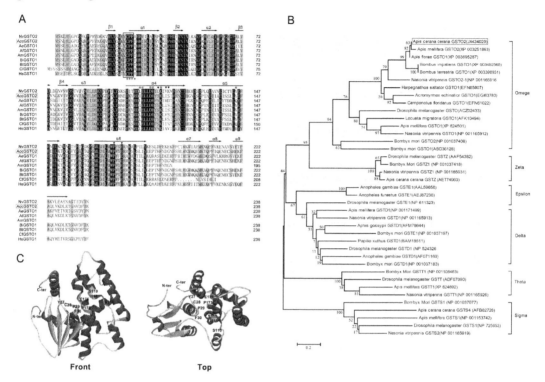

Fig. 1 Molecular properties of AccGSTO2. A. The amino acid sequence alignment of AccGSTO2 and other GSTOs. The putative secondary structure of AccGSTO2 is shown. Identical amino acids are shaded in black. The conserved functional domains are boxed. The putative G-site, H-site and dimer interface of AccGSTO2 are denoted by (❖), (inverted filled triangle) and (※), respectively. B. Phylogenetic relationships between glutathione transferases (GSTs) from different insect species. The six primary classes are shown, and AccGSTO2 is boxed. C. The tertiary structure of AccGSTO2. The conserved G-site residues (Y27, C28, P29, and F30), H-site residues (S115, I118, S119, P173, and E176), N-terminal, and C-terminal are shown.

Identification of partial putative *cis*-acting elements in the 5'-flanking region of *AccGSTO2*

To determine the organisation of the regulatory region of the *AccGSTO2*, a 1,591 bp fragment (JX456219) located upstream of the transcription start site was isolated. Many *cis*-acting elements were predicted by using the TFSEARCH software program (Fig. 3). Sequences involved in tissue development and growth in early stages, including cell factor 2-II (CF2-II), Hunchback (Hb), Dfd, NIT2, Broad-Complex (BR-C), Pbx1 and caudal-related homeobox (CdxA) protein [29, 39-41], were identified. In addition, several important transcription factors required for regulating various environmental stresses, such as heat shock factors (HSF) and activating protein-1 (AP-1), were predicted [42]. The cAMP-responsive element binding protein (CREB) and the CCAAT/enhancer binding protein (C/EBP) were also found; and these elements regulate gene expression during cell differentiation, proliferation, and apoptosis [43, 44]. We speculated that the *AccGSTO2* gene may be involved in organismal development and environmental stress responses.

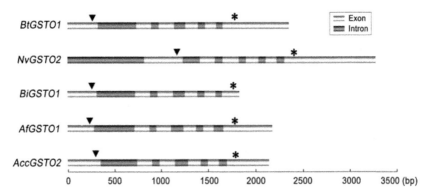

Fig. 2 Genomic structure of the Omega-class glutathione transferase genes (*GSTOs*). The lengths of the exons and introns of genomic DNA from *A. cerana cerana, A. florea, B. impatiens, B. terrestris,* and *N. vitripennis* are shown according to the scale below. Light grey and grey are used to highlighted the exons and introns separately. The translational initiation codons (ATG) and termination codons (TAA) are marked by inverted filled triangle and asterisk, respectively.

```
                                                                                          
GTAGATTAGCAGTTTTTCTCGTAGGAAATAAGGCAAAGTTGCTGTCAAAATAATGGCACTGCCACCTTGTAAAGCGTGGTAATGTAA   -1505
                                                              CREB
TAGGTTTTCGAGTAATTTTACCATGTAACTGTCATAAGCATGAACTTCGTCTAATATCAGTAATTTA TGACGTAA TCCTAATAAACGT  -1417
AACGATTGATGTTTGAAAGGCATAATTGCCATTAACGCTTGATCTAATGTGCCAACTCCTACTTCAGCTAATAACGATTTTTTTCGAG  -1329
   NIT2
 TATCAA CAAACCATTCATTACGTTCTCCACTCACTGATTGTTCACTCTTCGGGTATTTTTGCTGGCTAATTTCATCGTGATTTAAAAT  -1241
    AP-1                                          NIT2       CREB
TGT TTGAGTCAA CTGCGATGTACATAAATAGTTAATAAGTA TATCAA TTACATG TGACGTAA CTTATTTAAGTGTACAAATACATAT  -1154
                                   CdxA                           CF2-II      HSF
ATAGACAAAATTCTATGTGTAGTAATAATTTCTTTTT ATTTATA TATTTCGAAAACTAAACTGAGCAATAAT TATATATA TT AGAAA A  -1066
           CdxA
AGTTATTCAAAATATTG ATTTATA TACAATTTAAAATCTATTCAAAAATAATCAAAATTAGTAAACGAAATTTTTTTTAGATATTAAA  -977
                                HSF
AAAAATATTTTCATTTCATTATATAA AGAAA GCAAAAGATTAGCAAATTACAAATAATGTATAAAATAATAATATTTAGGATTTTTAG  -889
      BR-C
ATC AAAATAAATAAAA GTTAGTATAAAATGTCAGTTAAAGCACGATACACATGAGCGATTTTTTATTGCGCAATATCACTATATATT  -801
TTTTTTTGTTTTCTCAATTAAAAAAACGAAAACAGATATATTATCATAGTGATATCGTAAGGAAAAATGTTGTTCATGAGCATAAAGC  -713
                Hb           Dfd
TAAAAGCATAAAG CATAAAAATT ACTTAAA TTATAAATAATTGAAGTTTGTGTTAGATAAAGTAAAAAAAAGTTATTTTACCTAAATT  -625
TAAATAATTTGTTTACCGATAATTTAATATTGTCACTGATAATACTGATAATTGTTTTTATATATTAATTTTTAGTATTTGATTTGAAG  -536
    HSF                                            CdxA
ATTTGATA AGAAA TCAAATCACTAAATACATATATTAATATCCTGTAT ATTTATA ATAAATTTGATTTCATAATATAAAAATATTATT  -448
    NIT2
GTATTG TATCAA ATGAGTAAAATGAATAAAATATTATTTGTATTGATATTTAATTTTTGTTTCTATTTCATTGATTATATATTTCTCTG  -359
                                       CdxA                         NIT2
CGCAGATTCTATTCGATTTCTACCCTTTAATTTGCAATT ATTTATA TGAATAAAATTTGGTATATTTTTGT TATCAA TCTATTACTAAAT  -270
                                                               C/EBP
ATAATAGAATATAACAGATCTTATTTCAAATTTAATCTTTTTGATGATAAATATTTGCAACA AAATTTTGAAAAAA CTATTTTCAAGT  -182
             TATA box
TTGACTAAGA TATAAA ATATTTTTAAAACAATATTTTATTGAACGATTTATTATCTTAATTATACATATTTTTTTTTATAATTTCTTTTAT  -92
                                                       Pbx1
TATTTAAAAAAATAAAATATTTAAATATCATAAAATAACGCTATTTCTAAGTTCAACTCG ATCAATCAA CATTTGATCATGCGCAGAT   -4
       +1
        transcription start site
TTGACCTATTAATTTCTTTGAAAAAGAAATACATTAGAAAGAAACTATCCGTCTATGAAATTGTTATTGGTACCATAATATCATTATG  +85
GCTTAGTTTTGAAAATGTGAAAATTGAGAAATTGCTTCTATTAATTTTTTAGAATTTCAATCACGCATCATTTACAAAAACATTAATA  +173
TATATATATATATATACATACATAATACGTTACAAATAATCATTGATATTTTATTAATTAAAATACAATATCATTTCAAACATTT  +262
CGAAATATATTGTTTATTTTCTTGTTAGACACAAGTCTACTAAAGAAATACTAACTATT ATG AGTAACTTACATCTTGGACCTGG  +347
                                                         translation start site
```

Fig. 3 The nucleotide sequence and putative *cis*-acting elements of the *AccGSTO2* regulatory region. The translation and transcription start sites are marked with arrows. The *cis*-acting elements are boxed, with the exception of Hb, which is shaded in black.

Temporal and spatial expression patterns of the *AccGSTO2*

qPCR was performed to examine the expression patterns of *AccGSTO2* during different developmental stages. As shown in Fig. 4A, the expression level of *AccGSTO2* gradually decreased from the 3-day larvae to the 1-day adults, and increased rapidly in 10-day adults. The highest expression level was detected in 3-day larvae, and almost no significant differences in *AccGSTO2* expression were observed during pupal stages.

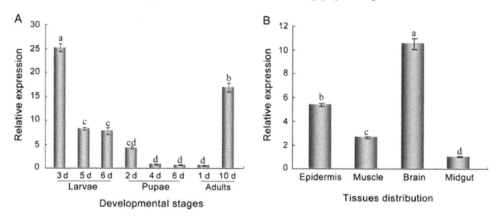

Fig. 4 Expression profile of *AccGSTO2* as determined by qPCR. The relative expression of *AccGSTO2* at different developmental stages (A) and in different tissues (B) is shown. The data are the means ± SE of three independent experiments. The different letters above the columns Indicate significant differences ($P < 0.01$) according to Duncan's multiple range tests.

Among the adult tissues studied, the most pronounced expression appeared in the brain, and decreasing mRNA levels of *AccGSTO2* were observed in the epidermis, muscle and midgut (Fig. 4B). The brain is a highly sensitive tissue for xenobiotics and peroxidative damage, implying that *AccGSTO2* may play a protective role in the brain tissue.

Expression patterns of *AccGSTO2* under abiotic stress treatments

Knowledge on the expression patterns of *AccGSTO2* under environmental stressors could be helpful to better understand the biological functions. Previous studies have indicated that the expression of *GSTO* could be induced by environmental stressors such as heat, heavy metal and endocrine-disrupting chemical in disk abalone [45], insecticides and UV radiation in the silkmoth [46]. To determine whether *AccGSTO2* is involved in different abiotic stress responses, relative expressions of *AccGSTO2* under temperature (4, 16, 25, 42℃), UV, H_2O_2 and insecticides (cyhalothrin, phoxim, pyridaben, paraquat) exposure were detected by qPCR. As shown in Fig. 5, the expression of *AccGSTO2* was upregulated under all treatments tested, although there were different expression patterns and increased degrees. These results indicate that *AccGSTO2* may be involved in abiotic stress responses.

Western blotting analysis

To further verify the temporal and spatial expression patterns of AccGSTO2, total protein was detected from four different tissues (epidermis, muscle, brain and midgut) of the 10-day adults or three different developmental stages (3-day larvae, 6-day pupae, and 10-day adults) and probed with anti-AccGSTO2 serum (Fig. 6). The expression of AccGSTO2 was consistent with its transcript expression in different tissues or developmental stages.

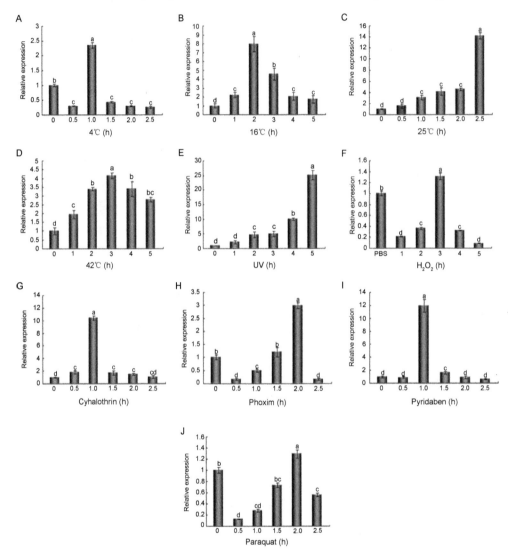

Fig. 5 Expression profile of *AccGSTO2* under different stress conditions. These conditions included 4℃ (A), 16℃ (B), 25℃ (C), 42℃ (D), UV (30 MJ/cm²) (E), H_2O_2 (2 mmol/L) (F), cyhalothrin (20 μg/L) (G), phoxim (1 μg/ml) (H), pyridaben (10 μmol/L) (I), and paraquat (10 μmol/L) (J). Untreated adult worker bees (Lane 0) were used as controls, and adult worker bees injected with PBS for 5 h were used as injection controls. The data are the means ± SE of three independent experiments. The different letters above the columns indicate significant differences ($P < 0.01$) according to Duncan's multiple range tests.

Purification and enzymatic features of recombinant AccGSTO2 protein

To further characterize the recombinant AccGSTO2 protein, the complete *AccGSTO2* ORF lacking a stop codon was cloned into the pET-30a (+) vector, and it was expressed in *E. coli* BL21 (DE3) as a histidine-fusion protein via IPTG induction. A sodium dodecyl sulphate polyacrylamide gel electrophoresis (SDS-PAGE) analysis showed that the recombinant protein was soluble and had a molecular mass of approximately 34 kDa, which is consistent with the predicted molecular mass of 34.804 kDa (Fig. 7). The soluble recombinant protein was further purified by HisTrap™ FF columns, and the concentration of the purified AccGSTO2 was approximately 1.89 mg/ml.

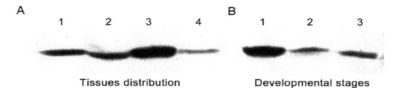

Fig. 6 Western blotting analysis of AccGSTO2 in different tissues and developmental stages. A. Lanes 1-4 were loaded with an equivalent amount of protein: Lane 1, epidermis, Lane 2, muscle, Lane 3, brain, Lane 4, midgut. B. Lanes 1-3 were loaded with an equivalent amount of protein: Lane 1, 3-day larvae, Lane 2, 6-day pupae, and Lane 3, 10-day adults.

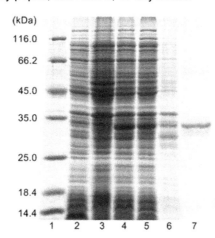

Fig. 7 Expression and purification of AccGSTO2. An SDS-PAGE analysis was used to separate recombinant AccGSTO2 expressed in *E. coli* BL21 cells. Lane 1, low molecular weight protein marker; Lane 2, induced overexpression of pET-30a (+) in BL21. Lanes 3 and 4, non-induced and induced overexpression of pET-30a (+) -AccGSTO2 in BL21, respectively; Lanes 5 and 6, suspension and pellet of sonicated recombinant AccGSTO2, respectively; Lane 7, purified recombinant AccGSTO2.

To identify the catalytic activities and potential biological functions of AccGSTO2, its substrate specificity with several typical GST substrates was determined. The purified enzyme showed detectable GSH-conjugating activity towards CDNB, the K_m and V_{max} values for CDNB were (2.32 ± 0.25) mmol/L and (8.21 ± 1.67) μmol/(min·mg protein), respectively, and the K_m and V_{max} values for GSH were (1.14 ± 0.16) mmol/L and (9.77 ± 0.11) μmol/(min·mg protein), respectively and measurable GSH peroxidase activity towards cumene hydroperoxide [$V_{max} = (3.24 \pm 0.23)$ μmol/(min·mg protein)] and *t*-butyl hydroperoxide [$V_{max} = (1.17 \pm 0.28)$ μmol/(min·protein)]. Moreover, AccGSTO2 could use GSH as an electron donor to reduce dehydroascorbate [the K_m and V_{max} values for DHA were (0.46 ± 0.09) mmol/L and (8.11 ± 1.56) μmol/(min·mg protein), respectively, and the K_m and V_{max} values for GSH were (3.55 ± 0.73) mmol/L and (6.35 ± 1.37) μmol/(min·mg protein), respectively], exhibiting high-affinity specificity towards DHA. Furthermore, the recombinant AccGSTO2 showed a maximum DHAR activity at pH 7.0 and an optimum temperature of 25 ℃ (Fig. 8).

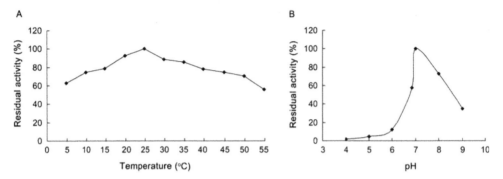

Fig. 8 Temperature (A) and pH (B) effects on the catalytic activity of AccGSTO2. The DHAR activities of recombinant AccGSTO2 were tested at different temperatures (5-55 ℃) and pH values (4.0-9.0). The values are the means of three replicates.

Protective effects of recombinant AccGSTO2 protein in oxidative stress

To provide a direct evidence that AccGSTO2 is responsible for antioxidant defence, the mixed-function oxidation (MFO) system, which produces hydroxyl radicals, was chosen to test whether AccGSTO2 protects DNA from ROS damage [29]. As shown in Fig. 9, with increasing concentrations of recombinant AccGSTO2, the amount of the nicked form of the plasmid declined gradually. When the concentration of AccGSTO2 was 150 μg/ml, the nicked form was nearly undetectable. N-ethylmaleimide (NEM) contains a reactive double bond and is used to modify cysteine residues. When added to the MFO reaction, NEM inhibited the AccGSTO2 protection of DNA. Therefore, we concluded that the sulfhydryl moieties of cysteine residues in AccGSTO2 may play important roles in DNA protection.

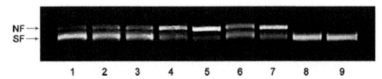

Fig. 9 *AccGSTO2* protected DNA from oxidative damage in the mixed-function oxidation system. Lanes 1-4, pUC19 plasmid DNA + FeCl$_3$ + DTT + purified AccGSTO2 (150, 100, 50 and 10 μg/ml, respectively); Lane 5, pUC19 plasmid DNA + FeCl$_3$ + DTT; Lane 6, pUC19 plasmid DNA + FeCl$_3$ + DTT + purified AccGSTO2 (50 μg/ml) + NEM (5 mmol/L); Lane 7, pUC19 plasmid DNA + FeCl$_3$ + DTT + BSA (150 μg/ml); Lane 8, pUC19 plasmid DNA + FeCl$_3$; Lane 9, pUC19 plasmid DNA only. SF, supercoiled form; NF, nicked form.

Disc diffusion assay provides a further evidence for its protective effects in oxidative stress. After exposed overnight to various stressors, the killing zones around the drug-soaked filters were smaller on the plates containing *E. coli* overexpressing AccGSTO2 than in the control bacteria; 21% (cumene hydroperoxide), 31% (*t*-butyl hydroperoxide), and 31% (paraquat) halo reductions were observed (Fig. 10). The model external pro-oxidants cumene hydroperoxide and *t*-butyl hydroperoxide were used because they are more stable than H_2O_2 under the applied incubation conditions. Because paraquat are redox-active and can cross cell membranes, they were used as intracellular ROS inducers to generate superoxide anions from molecular oxygen during metabolism [18].

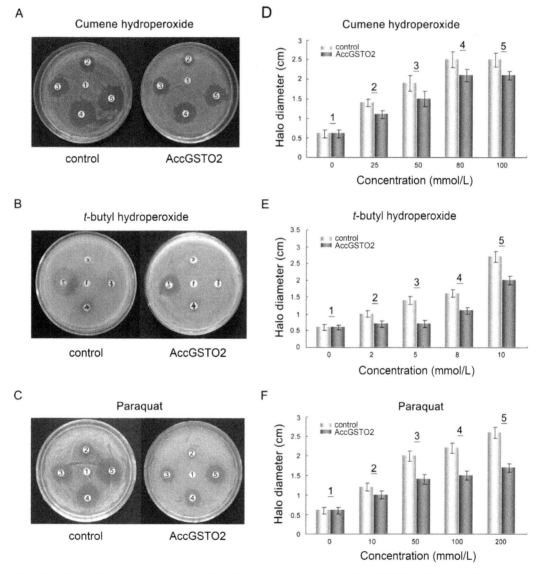

Fig. 10 Disc diffusion assays by using *E. coli* overexpressing AccGSTO2. LB agar plates were inoculated with 5×10^8 cells. AccGSTO2 was overexpressed in *E. coli* and bacteria transfected with pET-30a (+) were used as negative controls. Filter discs soaked with different concentrations of cumene hydroperoxide (A, D), *t*-butyl hydroperoxide (B, E) or paraquat (C, F) were placed on the agar plates. After an overnight exposure, the killing zones around the drug-soaked filters were measured.

Discussion

The GSTOs play central roles in detoxifying endogenous and exogenous agents. This gene family is particularly interesting because it is involved in human disease and exists in a wide range of species [47]. However, research concerning GSTOs has mainly focused on mammals, and there is limited information on GSTO in insects [48]. In this study, we described the cloning and characterization of a novel GSTO from *A. cerana cerana*. Unlike most GSTOs, the recombinant AccGSTO2 had DHAR activity and GSH-dependent peroxidase activity, and it also catalysed the conjugation of CDNB. These results were consistent with two recent studies by Burmeister et al. [18] and Garcera et al. [10]. The precise mechanism involved in the distinct enzymatic features has not yet been elucidated. Wan et al. [45] postulated that a non-traditional GST active site contributed to these differences in activity. The compositions of the cysteine-containing tetramers are as divergent as F-C-P-Y, F-C-P-F, F-C-P-W, Y-C-P-F, Y-C-P-Y, etc. AccGSTO2 was unique because its tetramer sequence was Y-C-P-F; unlike most GSTOs, tyrosine (Y) replaced phenylalanine (F) in the first residue. This substitution likely reflects a stronger interaction between the hydroxyl groups of the tyrosine and glutathione molecules. Although this explanation hints at distinct enzymatic features for different GSTOs, it remains to be tested experimentally.

The 5′-flanking region of *AccGSTO2* contains many predicted transcription factor binding sites involved in early development. To confirm whether *AccGSTO2* is involved in early development, we performed a stage-specific expression analysis of *AccGSTO2*. The qPCR analysis indicated that *AccGSTO2* is highly expressed in larval stage (Fig. 4A). Krishnan and Sehnal [49] reported that larvae and adults suffer higher oxidative stress than pupae because during the larval stages and the first two weeks of adulthood, their diet is restricted to honey, pollen and glandular secretions provided by other colony members. Because the pupal stages and the first-instar adults are at a quiescent non-feeding stage, the active intake and digestion of foods that either contain or produce potentially toxic oxidative radicals may cause higher oxidative stress levels in the larval and adult stages. Moreover, microsomal oxidases have been shown to be active in honeybee larvae [50]. An incomplete antioxidant or immune defence system in larvae may explain this pattern because they are more susceptible to environmental stress. Thus, the higher transcriptional levels of *AccGSTO2* in larval stage suggest a potential role in the detoxification of xenobiotics in their food in early development.

Knowledge on the tissue distributions of *AccGSTO2* mRNA could be helpful to better understand the physiology. A tissue-specific expression analysis revealed that *AccGSTO2* mRNA was expressed at its highest levels in the brain (Fig. 4B), which is very sensitive to oxidative stress [51, 52]. This observation is inconsistent with the low expression of *GSTO* in the mammalian brain [8]. However, CG6781/se from *Drosophila melanogaster* was only found in significant levels in brain regions involved in pteridine metabolism and the biosynthesis of red eye pigments [11]. Furthermore, we found that AccGSTO2 had significant DHAR activity. This activity may be critical for maintaining ascorbic acid (AsA) levels in the brain because AsA is dependent on the uptake and subsequent enzymatic reduction of DHA. AsA plays a major role in scavenging free radicals and specific ROS [53], and due to the significant consumption of oxygen in the brain, the scavenging and detoxification of these reactive species is imperative. Taken together, the abundant expression of *AccGSTO2* in the brain tissue implied that it may play important roles in protection against oxidative stress by maintaining AsA levels.

Studies on the insect GSTs mainly focused on their roles in insecticide resistance [7] and oxidative stress responses [54, 55]. The various insect GSTs play protective roles through different defence mechanisms. The insect-specific Delta-class and Epsilon-class GSTs are thought to be major contributors to detoxification and insecticide resistance [48]. In addition, some Epsilon-class GSTs in mosquito have peroxidase activity and may be important in protection against oxidative stress [56, 57]. The insect Sigma-class and Theta-class GSTs are implicated in the detoxification of lipid peroxidation products, suggesting a protective role against oxidative stress [46, 58]. The Zeta-class GSTs may have common house-keeping functions in the tyrosine degradation pathway, including cellular defence against oxidative stress [59]. For the Omega-class GSTs, this protective role against oxidative stress is mediated by DHA reductase and thiol transferase activities [46], whereas other insect GSTs do not exhibit these activities. Currently, lack of knowledge of endogenous insect GST substrates makes it difficult to elucidate the precise roles of different insect GST classes, and not all insect GSTs are involved in detoxification and antioxidative stress [4]. In this study, several antioxidant response elements, such as HSF and AP-1, were predicted in the 5′-flanking region of *AccGSTO2*. Furthermore, the qPCR analysis revealed that *AccGSTO2* was induced by all abiotic stressors examined, such as temperature, H_2O_2, insecticides and UV radiation can induce oxidative stress [60, 61]. Consistent with these results, *GSTO* from *Bombyx mori* is activated by a variety of environmental stimuli, including bacteria, ultraviolet-B (UV-B) and three commonly used chemical insecticides [46]. These results suggest that the function of GSTOs in insects is conserved. In general, the elevated mRNA levels of *GSTO* are always correlated with their known roles [4, 46, 45]. This appears to the case in *C. elegans*, where specific silencing of the GSTO1-1 by RNAi created worms with an increased sensitivity to several pro-oxidants, arsenite, and heat shock [18]. This observation provides clue for the role of *AccGSTO2* in the defence response. Therefore, an increase in amount of *AccGSTO2* expression under various abiotic stressors may be related to increased tolerance of oxidative stress; however, it should be further explored.

GSH-dependent peroxidase is a well-known enzyme that functions as an antioxidant by reducing organic hydroperoxides to the less toxic monohydroxy alcohols and protecting the lipid membranes and other cellular components against oxidative stress [62]. However, the lack of Se-dependent glutathione peroxidases in insects increases the potential importance of the putative Se-independent peroxidase function of GSTs in antioxidant defence [54]. DHAR activity may maintain AsA levels, which plays a major role in scavenging free radicals and specific ROS [53]. Yamamoto et al. [46] also demonstrated that *bmGSTO* plays a role in scavenging ROS by peroxidase activity and induces resistance to oxidative stress by DHAR activity. In our study, *AccGSTO2* exhibited glutathione-dependent DHAR and peroxidase activities. Moreover, the transcription of *AccGSTO2* is induced by H_2O_2 and paraquat treatment. The disc diffusion assay also provides direct evidence that *E. coli* cells overexpressing AccGSTO2 were protected from environmental stressors (paraquat, cumene hydroperoxide, and *t*-butyl hydroperoxide). In addition, the recombinant AccGSTO2 can protect supercoiled DNA from nicking by hydroxyl radicals in the MFO system. Taken together, these observations indicate that *AccGSTO2* could be associated with the scavenging of ROS and contribute to the cells resistance to oxidative stress.

Protection against oxidative stress is an eternal theme for the survival of living organisms. In conclusion, the unique biochemical features, expression patterns, functional characteristics and potential physiological roles of *AccGSTO2* that were demonstrated in this study offer the basic knowledge for further studies about functions of Omega-class GST.

Reference

[1] Valko, M., Izakovic, M., Mazur, M., Rhodes, C. J., Telser, J. Role of oxygen radicals in DNA damage and cancer incidence. Mol Cell Biochem. 2004, 266, 37-56.

[2] Poli, G., Leonarduzzi, G., Biasi, F., Chiarpotto, E. Oxidative stress and cell signaling. Curr Med Chem. 2004, 11, 1163-1182.

[3] Salinas, A. E., Wong, M. G. Glutathione S-transferases–a review. Cur Med Chem. 1999, 6, 279-309.

[4] Huang, Y. F., Xu, Z. B., Lin, X. Y., Feng, Q. L., Zheng, S. C. Structure and expression of glutathione S-transferase genes from the midgut of the Common cutworm, *Spodoptera litura* (Noctuidae) and their response to xenobiotic compounds and bacteria. J Insect Physiol. 2011, 57, 1033-1044.

[5] Hayes, J. D., Flanagan, J. U., Jowsey, I. R. Glutathione transferases. Ann Rev Pharmacol Toxicol. 2005, 45, 51-88.

[6] Cnubben, N. H., Rietjens, I. M., Wortelboer, H., Van Zanden, J., Van Bladeren, P. J. The interplay of glutathione-related processes in antioxidant defense. Environ Toxicol Pharmacol. 2001, 10, 141-152.

[7] Enayati, A. A., Ranson, H., Hemingway, J. Insect glutathione transferases and insecticide resistance. Insect Mol Biol. 2005, 14, 3-8.

[8] Board, P. G., Coggan, M., Chelvanayagam, G., et al. Identification, characterization, and crystal structure of the Omega class glutathione transferases. J Biol Chem. 2000, 275, 24798-24806.

[9] Dixon, D. P., Davis, B. G., Edwards, R. Functional divergence in the glutathione transferase superfamily in plants: identification of two classes with putative functions in redox homeostasis in *Arabidopsis thaliana*. J Biol Chem. 2002, 277, 30859-30869.

[10] Garcera, A., Barreto, L., Piedrafita, L., Tamarit, J., Herrero, E. *Saccharomyces cerevisiae* cells have three Omega class glutathione S-transferases acting as 1-Cys thiol transferases. Biochem J. 2006, 398, 187-196.

[11] Walters, K. B., Grant, P., Johnson, D. L. Evolution of the GST Omega gene family in 12 *Drosophila* species. J Hered. 2009, 100, 742-753.

[12] Xun, L., Belchik, S. M., Xun, R., et al. S-glutathionyl-(chloro) hydroquinone reductases: a novel class of glutathione transferases. Biochem J. 2010, 428, 419-427.

[13] Rouimi, P., Anglade, P., Benzekri, A., Costet, P., Debrauwer, L., Pineau, T., Tulliez, J. Purification and characterization of a glutathione S-transferase Omega in pig: evidence for two distinct organ-specific transcripts. Biochem J. 2001, 358, 257–262.

[14] Board, P. G., Coggan, M., Wilce, M. C., Parker, M. W. Evidence for an essential serine residue in the active site of the Theta class glutathione transferases. Biochem J. 1995, 311, 247-250.

[15] Stenberg, G., Board, P. G., Mannervik, B. Mutation of an evolutionarily conserved tyrosine residue in the active site of a human class Alpha glutathione transferase. FEBS Lett. 1991, 293, 153-155.

[16] Wilce, M. C., Parker, M. W. Structure and function of glutathione S-transferases. Biochim Biophys Acta. 1994, 1205, 1-1817.

[17] Rice, M. E. Ascorbate regulation and its neuroprotective role in the brain. Trends Neurosci. 2000, 23, 209-216.

[18] Burmeister, C., Luersen, K., Heinick, A., Hussein, A., Domagalski, M., Walter, R. D., Liebau, E. Oxidative stress in *Caenorhabditis elegans*: protective effects of the Omega class glutathione transferase (GSTO-1). FASEB J. 2008, 22, 343-354.

[19] Chowdhury, U. K., Zakharyan, R. A., Hernandez, A., Avram, M. D., Kopplin, M. J., Aposhian, H. V. Glutathione-*S*-transferase-Omega [MMA(V) reductase] knockout mice: enzyme and arsenic species concentrations in tissues after arsenate administration. Toxicol Appl Pharmacol. 2006, 216, 446-457.

[20] Yin, S., Li, X., Meng, Y., Finley Jr, R. L., Sakr, W., Yang, H., Reddy, N., Sheng, S. Tumor-suppressive maspin regulates cell response to oxidative stress by direct interaction with glutathione *S*-transferase. J Biol Chem. 2005, 280, 34985-34996.

[21] Li, Y. J., Oliveira, S. A., Xu, P., et al. Glutathione *S*-transferase Omega-1 modifies age-at-onset of Alzheimer disease and Parkinson disease. Hum Mol Genet. 2003, 12, 3259-3267.

[22] Laliberte, R. E., Perregaux, D. G., Hoth, L. R., et al. Glutathione *S*-transferase Omega 1-1 is a target of cytokine release inhibitory drugs and may be responsible for their effect on interleukin-1β posttranslational processing. J Biol Chem. 2003, 278, 16567-16578.

[23] Dulhunty, A., Gage, P., Curtis, S., Chelvanayagam, G., Board, P. The glutathione transferase structural family includes a nuclear chloride channel and a ryanodine receptor calcium release channel modulator. J Biol Chem. 2001, 276, 3319-3323.

[24] Kim, K., Kim, S. H., Kim, J., Kim, H., Yim, J. Glutathione *S*-transferase Omega 1 activity is sufficient to suppress neurodegeneration in a *Drosophila* model of Parkinson disease. J Biol Chem. 2012, 287, 6628-6641.

[25] Kampkötter, A., Volkmann, T. E., De Castro, S. H., Leiers, B., Klotz, L. O., Johnson, T. E., Link, C. D., Henkle-Dührsen, K. Functional analysis of the glutathione *S*-transferase 3 from *Onchocerca volvulus* (*Ov*-GST-3): a parasite GST confers increased resistance to oxidative stress in *Caenorhabditis elegans*. J Mol Biol. 2003, 325, 25-37.

[26] Michelette, E. R. F., Soares, A. E. E. Characterization of preimaginal developmental stages in Africanized honey bee workers (*Apis mellifera* L). Apidologie. 1993, 24, 431-440.

[27] Alaux, C., Ducloz, F., Crauser, D., Le Conte, Y. Diet effects on honeybee immunocompetence. Biol Lett. 2010, 6, 562-565.

[28] Meng, F., Kang, M., Liu, L., Luo, L., Xu, B., Guo, X. Characterization of the TAK1 gene in *Apis cerana cerana* (*AccTAK1*) and its involvement in the regulation of tissue-specific development. BMB Rep. 2011, 44, 187-192.

[29] Yu, F., Kang, M., Meng, F., Guo, X., Xu, B. Molecular cloning and characterization of a thioredoxin peroxidase gene from *Apis cerana cerana*. Insect Mol Biol. 2011, 20, 367-378.

[30] Lourenco, A. P., Mackert, A., Cristino, A. S., Simoes, Z. L. P. Validation of reference genes for gene expression studies in the honey bee, *Apis mellifera*, by quantitative real-time RT-PCR. Apidologie. 2008, 39, 372-385.

[31] Scharlaken, B., De Graaf, D. C., Goossens, K., Brunain, M., Peelman, L. J., Jacobs, F. J. Reference gene selection for insect expression studies using quantitative real-time PCR: The honeybee, *Apis mellifera*, head after a bacterial challenge. J Insect Sci. 2008, 8, 1-10.

[32] Livak, K. J., Schmittgen, T. D. Analysis of relative gene expression data using real-time quantitative PCR and the 2-(Delta Delta C(T)) method. Methods. 2001, 25, 402-408.

[33] Zhou, H., Brock, J., Liu, D., Board, P. G., Oakley, A. J. Structural insights into the dehydroascorbate reductase activity of human Omega-class glutathione transferases. J Mol Biol. 2012, 420, 190-203.

[34] Meng, F., Zhang, L., Kang, M., Guo, X., Xu, B. Molecular characterization, immunohistochemical localization and expression of a ribosomal protein L17 gene from *Apis cerana cerana*. Arch Insect Biochem Physiol. 2010, 75, 121-138.

[35] Li, X., Zhang, X., Zhang, J., Zhang, X., Starkey, S. R., Zhu, K. Y. Identification and characterization of eleven glutathione *S*-transferase genes from the aquatic midge *Chironomus tentans* (Diptera: Chironomidae). Insect Biochem Mol Biol. 2009, 39, 745-754.

[36] Habig, W. H., Pabst, M. J., Jakoby, W. B. Glutathione *S*-transferases. The first enzymatic step in

[37] mercapturic acid formation. J Biol Chem. 1974, 249, 7130-7139.

[37] Boldyrev, A. A., Yuneva, M. O., Sorokina, E. V., Kramarenko, G. G., Fedorova, T. N., Konovalova, G. G., Lankin, V. Z. Antioxidant systems in tissues of senescence accelerated mice. Biochemistry (Mosc). 2001, 66, 1157-1163.

[38] Wells, W. W., Xu, D. P., Washburn, M. P. Glutathione: dehydroascorbate oxidoreductases. Methods Enzymol. 1995, 252, 30-38.

[39] Ericsson, A., Kotarsky, K., Svensson, M., Sigvardsson, M., Agace, W. Functional characterization of the CCL25 promoter in small intestinal epithelial cells suggests a regulatory role for Caudal-Related Homeobox (Cdx) transcription factors. J Immunol. 2006, 176, 3642-3651.

[40] Rørth, P. Specification of C/EBP function during *Drosophila* development by the bZIP basic region. Science. 1994, 266, 1878-1881.

[41] Stanojević, D., Hoey, T., Levine, M. Sequence specific DNA-binding activities of the gap proteins encoded by hunchback and Krüppel in *Drosophila*. Nature. 1989, 341, 331-335.

[42] Ding, Y. C., Hawkes, N., Meredith, J., Eggleston, P., Hemingway, J., Ranson, H. Characterization of the promoters of Epsilon glutathione transferases in the mosquito *Anopheles gambiae* and their response to oxidative stress. Biochem J. 2005, 387, 879-888.

[43] Ishida, M., Mitsui, T., Yamakawa, K., Sugiyama, N., Takahashi, W., Shimura, H., Endo, T., Kobayashi, T., Arita, J. Involvement of cAMP response element-binding protein in the regulation of cell proliferation and the prolactin promoter of lactotrophs in primary culture. Am J Physiol Endocrinol Metab. 2007, 293, E1529-E1537.

[44] Carlezon Jr, W. A., Duman, R. S., Nestler, E. J. The many faces of CREB. Trends Neurosci. 2005, 28, 436-445.

[45] Wan, Q., Whang, I., Lee, J. S., Lee, J. Novel Omega glutathione S-transferases in disk abalone: characterization and protective roles against environmental stress. Comp Biochem Physiol C: Toxicol Pharmacol. 2009, 150, 558-568.

[46] Yamamoto, K., Teshiba, S., Shigeoka, Y., Aso, Y., Banno, Y., Fujiki, T., Katakura, Y. Characterization of an Omega-class glutathione S-transferase in the stress response of the silkmoth. Insect Mol Biol. 2011, 20, 379-386.

[47] Board, P. G. The Omega-class glutathione transferases: structure, function, and genetics. Drug Metab Rev. 2011, 43, 226–235.

[48] Ketterman, A. J., Saisawang, C., Wongsantichon, J. Insect glutathione transferases. Drug Metab Rev. 2011, 43, 253–265.

[49] Krishnan, N., Sehnal, F. Compartmentalization of oxidative stress and antioxidant defense in the larval gut of *Spodoptera littoralis*. Arch Insect Biochem Physiol. 2006, 63, 1-10.

[50] Gilbert, M. D., Wilkinson, C. F. An inhibitor of microsomal oxidation from gut tissues of the honey bee (*Apis mellifera*). Comp Biochem Physiol B: Biochem Mol Biol. 1975, 50, 613-619.

[51] Ament, S. A., Corona, M., Pollock, H. S., Robinson, G. E. Insulin signaling is involved in the regulation of worker division of labor in honey bee colonies. Proc Natl Acad Sci USA. 2008, 105, 4226-4231.

[52] Rival, T., Soustelle, L., Strambi, C., Besson, M. T., Iche, M., Birman, S. Decreasing glutamate buffering capacity triggers oxidative stress and neuropil degeneration in *Drosophila* brain. Curr Biol. 2004, 17, 599-605.

[53] Frei, B., England, L., Ames, B. N. Ascorbate is an outstanding antioxidant in human blood plasma. Proc Natl Acad Sci USA. 1989, 86, 6377-6381.

[54] Corona, M., Robinson, G. E. Genes of the antioxidant system of the honey bee: annotation and phylogeny. Insect Mol Biol. 2006, 15, 687-701.

[55] Li, H. M., Buczkowski, G., Mittapalli, O., Xie, J., Wu, J., Westerman, R. Transcriptomic profiles of *Drosophila melanogaster* third instar larval midgut and responses to oxidative stress. Insect Mol Biol. 2008, 17, 325-339.

[56] Lumjuan, N., McCarroll, L., Prapanthadara, L., Hemingway, J., Ranson, H. Elevated activity of an

Epsilon class glutathione transferase confers DDT resistance in the dengue vector, *Aedes aegypti*. Insect Biochem Mol Biol. 2005, 35, 861-871.

[57] Ortelli, F., Rossiter, L. C., Vontas, J., Ranson, H., Hemingway, J. Heterologous expression of four glutathione transferase genes genetically linked to a major insecticide-resistance locus from the malaria vector *Anopheles gambiae*. Biochem J. 2003, 373, 957-963.

[58] Singh, S. P., Coronella, J. A., Benes, H., Cochrane, B. J., Zimniak, P. Catalytic function of *Drosophila melanogaster* glutathione S-transferase *DmGSTS1-1* (GST-2) in conjugation of lipid peroxidation end products. Eur J Biochem. 2001, 268, 2912-2923.

[59] Claudianos, C., Ranson, H., Johnson, R. M., Biswas, S., Schuler, M. A., Berenbaum, M. R. A deficit of detoxification enzymes: pesticide sensitivity and environmental response in the honey bee. Insect Mol Biol. 2006, 15, 615-636.

[60] Lushchak, V. I. Environmentally induced oxidative stress in aquatic animals. Aquat Toxicol. 2011, 101, 13-30.

[61] Kottuparambil, S., Shinb, W., Brownc, M. T., Han, T. UV-B affects photosynthesis, ROS production and motility of the freshwater flagellate, *Euglena agilis* Carter. Aquat Toxicol. 2012, 122-123, 206-213.

[62] Mahmoud, K. Z., Edens, F. W. Influence of selenium sources on age-related and mild heat stress-related changes of blood and liver glutathione redox cycle in broiler chicken (*Gallus domesticus*). Comp Biochem Physiol B: Biochem Mol Biol. 2003, 136, 921-934.

2.4 Functional and Mutational Analyses of an Omega-class Glutathione S-transferase (*GSTO2*) That Is Required for Reducing Oxidative Damage in *Apis cerana cerana*

Yuanying Zhang, Xulei Guo, Yaling Liu, Feng Liu, Hongfang Wang, Xingqi Guo, Baohua Xu

Abstract: Glutathione S-transferases perform a variety of vital functions, particularly in reducing oxidative damage. Here, we investigated the expression patterns of *Apis cerana cerana* Omega-class glutathione S-transferase 2 (*AccGSTO2*) under various stresses and explored its connection with antioxidant defences. We found that *AccGSTO2* knockdown by RNA interference triggered increased mortality in *Apis cerana cerana*, and immunohistochemistry revealed significantly decreased *AccGSTO2* expression, particularly in the midgut and fat body. Further analyses indicated that *AccGSTO2* knockdown resulted in decreases in catalase and glutathione reductase activities, ascorbate content and the ratio of reduced to oxidized glutathione, and increases in H_2O_2, malondialdehyde and carbonyl contents. We also analyzed the transcripts of other antioxidant genes and found that many genes were downregulated in the *AccGSTO2* knockdown samples, revealing that *AccGSTO2* may be indispensable for attaining a normal lifespan by enhancing cellular oxidative resistance. In addition, the roles of cysteine residues in AccGSTO2 were explored by using site-directed mutagenesis. Mutants of Cys^{28} and Cys^{124} significantly affected the enzyme and antioxidant activities of AccGSTO2, which may be attributed to the changes in the spatial structures of mutants as determined by homology modeling. In summary, these observations provide novel insight into the structural and functional characteristics of GSTOs.

Keywords: *GSTO*; Antioxidant enzyme; RNA interference; Site-directed mutagenesis; *Apis cerana cerana*

Introduction

Increased levels of reactive oxygen species (ROS) in cells can cause oxidative stress and affect cellular signal transduction, which controls diverse cellular processes, resulting in cell damage and triggering apoptosis [1]. Therefore, ROS detoxification is critical for organism survival. In general, organisms are dependent on various enzymatic and non-enzymatic mechanisms to eliminate or inhibit ROS generation or reactivity. Moreover, organisms have developed systems to repair oxidative damage once it has occurred. One common molecular response is altering the expression of related genes, including the upregulation of antioxidant genes [2]. Cellular antioxidant systems consist of multifunctional proteins and enzymes, such as superoxide dismutase (SOD), catalase (CAT), glutathione reductase (GR), thioredoxin peroxidase (Tpx), thioredoxin (Trx), and glutaredoxin (Grx), to resist oxidative damage [3]. Glutathione S-transferases (GSTs) are major detoxifying enzymes in antioxidant systems [4].

GSTs are found nearly universally in eukaryotes and prokaryotes. Based on their different subcellular localizations, GSTs have been classified into three major groups: cytosolic GSTs, mitochondrial GSTs and microsomal GSTs [5-7]. Cytosolic GSTs are the most diverse group and are divided into at least 13 classes in mammals, plants, insects, parasites, bacteria, and fungi [5, 7-9]. In insects, cytosolic GSTs are divided into the following 6 classes: Sigma, Theta, Omega, Zeta, and insect-specific Delta and Epsilon [7, 10]. The vast substrate diversity of the GST superfamily, which recognizes at least 100 different xenobiotic chemicals [2], allows these proteins to play various enzymatic and non-enzymatic roles in cellular detoxification. GSTs can catalyze the conjugation of reduced glutathione (GSH) with various electrophiles and therefore detoxify both exogenous and endogenous toxic compounds [4]. Some GSTs also have glutathione peroxidase activities and can promote the detoxification of organic hydroperoxides, making these GSTs critical in the defense against oxidative stress [11, 12].

Recently, a role for insect GSTs in mediating the oxidative stress response has been suggested by a number of studies [13-15]. For example, in venom glands of *Apis mellifera*, proteomic analysis indicated that the presence of GSTs seems to defend against oxidative stress [16]. Increased GST levels in *Helicoverpa armigera* adults were also detected under ultraviolet light-induced oxidative stress [17]. In addition, Nair and Choi found upregulated GST expression in *Chironomus riparius* upon exposure to the pro-oxidative stress inducer paraquat, suggesting its participation in the oxidative stress defense response [18]. Although an increasing number of GST genes has been identified and characterized from insect species [10, 13, 14, 19], the properties of hymenopteran GSTs, particularly the Omega-class GSTs (GSTOs), have not been well investigated in comparison with those of dipteran and lepidopteran species.

The GSTOs have unique structural and functional properties that are different from those of the other GST classes. In GSTOs, a novel cysteine residue (Cys) is located in their active site, which differs from the canonical serine and tyrosine residues in the other GST classes [20, 21]. Moreover, GSTO has unique N-terminal and C-terminal extensions that form a distinct domain that is important for substrate specificity [21]. Cys residues play a pivotal role in stabilizing the protein structure by forming intramolecular or intermolecular disulphide bonds or by coordinating with metal ions. The number of Cys residues that each GSTO protein contains varies within species. For instance, human GSTO2 has 11 Cys residues compared with five in GSTO1 [22], whereas *Apis cerana cerana* GSTO1 and GSTO2 both have three Cys residues [12]. Several studies have demonstrated that the site-directed mutagenesis of active site Cys residues can result in undetected or decreased enzymatic activity toward GSTOs [11, 19]. Zhou et al. also adopted a mutagenesis strategy in which six non-catalytic

Cys residues were all mutated to serine residues to improve the solubility of the human GSTO2-2 protein to determine its crystal structure [22]. However, the effects of Cys replacement on GSTO2 folding, structural stability and function remain largely unknown.

Compared to the western honeybee *A. mellifera*, as a pollinator, the Chinese honeybee *A. cerana cerana* has a greater ability to resist cold temperatures and to defend against the ectoparasitic mite *Varroa destructor* [23]. Thus, the Chinese honeybee plays critical roles in the pollination of crops and flora in mountainous areas and in the conservation of biodiversity in China [24]. However, population sizes of Chinese honeybee have rapidly decreased during the past few years, seriously disturbing the symbiotic ecosystem equilibrium of many plants. Apart from competition with western honeybees, this has been caused by an epidemic of several honeybee diseases that caused a rapid loss of honeybees in the 1960s [25, 26]. In general, the antioxidant capacity of honeybee may be positive correlation with its disease resistance. Consequently, study on the antioxidant molecular mechanism for improving honeybee health is urgently needed.

Previously, we identified and characterized an Omega-class GST, *AccGSTO2*, from *A. cerana cerana* [27]. To acquire novel insight into its structure and antioxidant defence capabilities, we further investigated the roles of three Cys residues (positions 28, 124 and 217) in protein folding, catalysis and oxidative stress resistance. Furthermore, AccGSTO2 knockdown by using RNA interference (RNAi) technology was also performed to explore its mechanism of antioxidant defence.

Experimental Procedures

Honeybees and various treatments

Chinese honeybees (*A. cerana cerana*) were maintained at the experimental apiary of the Technology Park of Shandong Agricultural University. The larvae were collected from the hive and were reared collectively in 24-well plates (Corning Incorporated Costar®, NY, USA) as described by Yu et al. [31]. Once the larvae reached the third instar, they were divided into three groups (n = 48/group). The larvae in group 1 were randomly transferred to artificial diets treated with different concentrations of ecdysone (0.001, 0.01, 0.1, and 1 μg/ml, respectively). The larvae in group 2 were inoculated with various microbes (*Aspergillus flavus*, *Bacillus thuringiensis*, *Beauveria bassiana*, and *Bacillus bombysepticus*). The control larvae in group 3 were fed only untreated artificial diets. At three days after treatment, the larvae were immersed in liquid nitrogen and stored at −80°C until used. For the metal treatments, two-week-old adult workers were collected from the hive by marking newly emerged bees with paint and collecting them after two weeks. Then, these adult workers were divided into five groups (n = 40/group). The control adult workers in group 1 were fed only the basic adult diet. The adult workers in groups 2-5, were first fed an artificial diet containing $HgCl_2$ (3 mg/ml) or $CdCl_2$ (0.5, 5, or 10 mg/L) with a micropipette. Then, the workers were transferred to incubators with the basic adult diets. All groups were maintained in an incubator at a constant temperature (34°C) and humidity (70%) under a 24 h dark regimen [53]. All samples were collected at the indicated time points and stored at −80 ℃. These experiments were repeated three times.

Real-time quantitative PCR and Western blotting analysis

RNA extraction, cDNA synthesis, RT-qPCR, and Western blotting were performed as described previously [27]. The primers used are listed in Supplementary Table 1. In brief, the CFX96™ Real-time System (Bio-Rad, Hercules, CA, USA) and the SYBR® PrimeScript™ RT-PCR Kit (TaKaRa, Dalian, China) were used for the qPCR analysis. The *A. cerana cerana β-actin* gene (GenBank accession number: HM_640276) was selected as an internal control. The gene transcriptional level was evaluated in three independent biological replicates for each group with three technical repeats for each primer pair. For Western blotting analysis, four different $CdCl_2$ treatment concentrations after 1 day (0, 0.5, 5, and 10 mg/L $CdCl_2$-fed adults) and five different microbial infection larvae (no infection, Af infection, Bt infection, Bba infection, and Bbo infection) were selected to extract their total protein, respectively. The anti-AccGSTO2 serum was served as the primary antibody at a dilution of 1∶500 (V/V). Peroxidase-conjugated goat anti-mouse IgG (Dingguo, Beijing, China) was served as the secondary antibody at a dilution of 1∶2,000 (V/V). The SuperSignal® West Pico Trial Kit (Thermo Scientific Pierce, IL, USA) was used to detect the proteins.

RNAi of AccGSTO2

dsRNA-*AccGSTO2* and dsRNA-*GFP* for RNAi were performed as described in an established protocol [54]. The primer sequences (GIF and GIR) are provided in Supplementary Table 1. The third-instar larvae were employed for the RNAi experiments. The larvae were divided into three groups (n = 48/group) and were fed normal diets, additional dsRNA-*AccGSTO2* (10 μg), or additional dsRNA-*GFP* (10 μg). Meanwhile, the newly emerged honeybee adults were divided into 4 groups (n = 50/group) and injected with 0.5 μl (5 μg) of dsRNA-*AccGSTO2*, or with an equivalent concentration of a dsRNA-*GFP* into the thoracic muscle. Two of the four groups of newly emerged adults were injected with 0.5 μl of nuclease-free water, or were kept untreated. During this period, their survival rates were recorded, and healthy bees were sampled each day. This experiment was repeated three times.

Metabolite content determination and enzymatic activity assays

The whole-bee homogenates were analyzed by using the test kits. Total proteins were extracted from whole bee larvae inoculated with different microbes after 3 days, from adults fed with $CdCl_2$ after 1 day and from adults injected with dsRNA-*AccGSTO2* after 3 days. Total proteins were quantified by using a BCA Protein Assay Kit (Nanjing Jiancheng Bioengineering Institute, Nanjing, China). CAT and GR activities; H_2O_2, MDA, protein carbonyl and ascorbate contents; and the GSH/GSSG ratio were then tested by using commercially available assay kits (Nanjing Jiancheng Bioengineering Institute, Nanjing, China). These assay kits have been successfully applied in various organisms, including bees

[54-56]. All procedures for each assay were performed according to the instructions of the manufacturer. The whole-bee homogenates were analyzed by using the test kits.

Immunohistochemistry (IHC)

IHC was performed as described in an established protocol [31], with some modifications. Briefly, sections from 4% paraformaldehyde-fixed, paraffin-embedded tissues were deparaffinized and rehydrated. Then, IHC staining of these slices was performed by using commercially available assay kits (Zhongshan Golden Bridge, SP-9002, Beijing, China) according to the instructions of the manufacturer. In addition, the sections were incubated with the polyclonal primary antibody anti-AccGSTO2 (1 : 50 dilution) over night at 4℃. Last, after counterstaining with hematoxylin, the sections were visualized by using a DAB Color Development Kit (Zhongshan Golden Bridge, Beijing, China) until a brown reaction product was obtained. The negative control group was processed simultaneously by using the same steps, except that phosphate-buffered saline (PBS) was substituted for the primary antibody. All slides were observed under a confocal laser microscope (LSM510 META, Carl Zeiss AG, Germany).

Site-directed mutagenesis and mutant AccGSTO2 expression

A site-directed mutagenesis kit (TransGen Biotech, Beijing, China) was used to generate three mutations (C28A, C124A, and C217A). The recombinant plasmid pET-30a (+)-*AccGSTO2* was used as a template for mutagenesis. The primers used for these mutations are listed in Supplementary Table 1, and mutated sequences and the introduced restriction sites in the primers are underlined. Finally, these mutations were verified by nucleotide sequencing, and these constructs were transformed into *Escherichia coli* BL21 (DE3) cells. The expression and purification of these mutants were performed as described previously [27].

Enzyme assay and kinetic parameter determination

For the enzyme assays, the purified AccGSTO2 variants were dialyzed in 100 mmol/L KH_2PO_4 (pH 7.5). Glutathione-dependent dehydroascorbate reductase (DHAR) activity and the apparent kinetic parameters of the GST mutants were determined as described previously [27]. All reactions were performed at least four times.

DNA cleavage assay and disc diffusion assay

DNA cleavage assay of the protein mutant was performed as described by Yu et al. [57]. Specifically, the reaction mixture (25 μl) contained 150 μg/ml wild-type (WT) AccGSTO2 protein or an equal concentration of one of the three AccGSTO2 mutants. Disc diffusion assays were performed as described by Burmeister et al. [11].

Fluorescence measurements

The intrinsic fluorescence emission spectra were monitored with a Hitachi F-4500 spectrofluorimeter by using a 1 cm path-length quartz cuvette. Four tryptophan (Trp) residues (positions 58, 172, 174, and 202) are present in AccGSTO2. For ANS fluorescence measurements, a 10-fold molar excess of ANS was added to the samples. The samples were equilibrated for 30 min in the dark, and then the extrinsic fluorescence was measured. The intrinsic fluorescence emission spectra were recorded from 300 to 400 nm, with the excitation wavelength set at 280 nm, whereas the ANS fluorescence emission spectra were recorded from 400 to 600 nm, with the excitation wavelength set at 380 nm. The final protein concentration for spectroscopic experiments was 3.6 μmol/L. All spectroscopic experiments were performed at 25°C and repeated three times.

Results and Discussion

The expression profiles of *AccGSTO2* and different oxidative statuses under various stressors

Previously, we reported the identification of *AccGSTO2* and its protective effects against oxidative stress [27]; however, whether *AccGSTO2* is involved in heavy metals ($CdCl_2$, $HgCl_2$), ecdysone and microbe (*Aspergillus flavus*, *Bacillus thuringiensis*, *Beauveria bassiana* and *Bacillus bombysepticus*) responses remain unknown. As a pollinator, the polylectic foraging behavior of the honeybee and its habit of concentrating nectar to become honey indicate that honeybees encounter phytochemicals at higher levels than do the nectar-feeding pollinators that do not handle their food [28]. Thus, the over-accumulation of cadmium and mercury due to environmental pollution is detrimental to honeybees because these metals are non-essential heavy metals that are toxic even at extremely low concentrations. In this study, opposite responses were observed for individual metals: $CdCl_2$ elevated *AccGSTO2* expression by varying degrees depending on the treatment concentrations and times (Fig. 1A, C), which is consistent with the observed GST expression induced by cadmium in *Drosophila* [29] and in *Oncorhynchus kisutch* [30]. Interestingly, the induced level of *AccGSTO2* was highest in the 0.5 mg/L $CdCl_2$ treatment compared with those in the other $CdCl_2$ treatment concentrations, but its induced level was higher in the 10 mg/L $CdCl_2$ treatment than in the 5 mg/L $CdCl_2$ treatment. This result revealed that the degree of $CdCl_2$-induced *AccGSTO2* expression was not concentration-dependent. By contrast, the transcription of this gene was inhibited gradually by $HgCl_2$ treatment (Fig. 1E), which is similar to the result for *AccGST sigma 4* [31] but is inconsistent with the finding for *AccGSTS1* [13, 14]. In the present study, the adult workers were fed with 3 mg/ml $HgCl_2$, and treated for 24 h. Yu et al. chose 0.5, 5 and 10 mg/L $HgCl_2$ to test the response of the larvae, and Yan et al. adopted 3 μg $HgCl_2$ to feed the adult workers for 2.5 h [13, 14, 31]. When the concentration of $HgCl_2$ was higher, the inhibition of *GST* expression was more obvious. Cabassi has reported that low doses of sever heavy metals such as Cd and Hg demonstrated immune-potentiating effects, while higher doses are suppressive[32]. Furthermore, these metals may have different mechanisms of toxicity other that ROS generation. Hg, for example, can cause toxicity by binding to metalloproteins (metal-substitution). Cd^{2+} can displace Fenton metal ions like Cu^{2+} and Fe^{2+} in the Fenton reaction to produce toxicity [33]. Therefore, this discrepancy of GST responses may be attributed to the differences in the duration and intensity of the exposure, and GST classes.

It is well known that ecdysone can trigger a complex cascade that modulates the transcription of a broad variety of genes that perform different functions in central physiological events in the life of insect. Broad-Complex (BR-C, −873 to −885), which is an

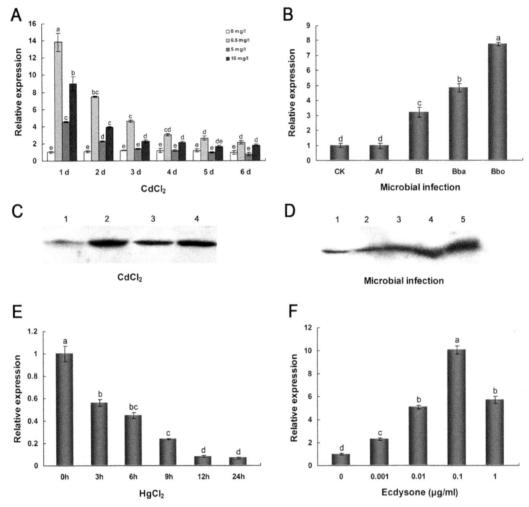

Fig. 1 Expression profiles (A, B, E, F) and Western blotting analyses (C, D) of *Apis cerana cerana* Omega-class glutathione *S*-transferase 2 (*AccGSTO2*) under different stress conditions. A. Expression patterns of *AccGSTO2* under various concentrations of $CdCl_2$ for different treatment times. B. Microbial infection. Af, *Aspergillus flavus*; Bt, *Bacillus thuringiensis*; Bba, *Beauveria bassiana*; Bbo, *Bacillus bombysepticus*. Untreated bees were used as controls (CK). C. Effects of different $CdCl_2$ treatment concentrations after 1 day. Lanes 1-4, 0, 0.5, 5 and 10 mg/L $CdCl_2$ fed, respectively. D. Microbial infection. Lanes 1-5 represent no infection, Af infection, Bt infection, Bba infection, and Bbo infection, respectively. All lanes were loaded with equivalent amounts of extracted protein. E. *AccGSTO2* expression with $HgCl_2$ fed (3 mg/ml) at different treatment times. F. The expression profile of *AccGSTO2* when exposed to different concentrations of ecdysone, as determined by real-time quantitative (qPCR). The data are the means ± SE of three independent experiments. Different letters above the columns indicate significant differences ($P < 0.01$) according to Duncan's multiple range tests.

ecdysone-responsive key regulator of metamorphosis during the third-instar stage and early prepupal development [34], has been predicted in the regulatory region of *AccGSTO2* [27]. To confirm whether *AccGSTO2* is involved in ecdysone-responsive in early development, we performed an ecdysone-treated expression analysis of *AccGSTO2*. Compared to the control group, different ecdysone-supplemented diets significantly induced *AccGSTO2* expression (Fig. 1F). This result further indicated that *AccGSTO2* is related to early development and ecdysone-responses. In addition, upregulated expression of *AccGSTO2* was also observed under the application of various microbes (Fig. 1B, D); however, immune-related promoter elements (e.g., NF-κB) were not found and the antimicrobial activity of recombinant *AccGSTO2* was not detected in the agar dilution assay, revealing that the upregulated expression of *AccGSTO2* may relate to oxidative stress.

In general, intracellular cadmium does not gain or lose electrons; nonetheless, this metal exhausts antioxidant components, particularly (GSH), leading to increased ROS formation [35, 36]. Yamamoto et al. also found an increased level of H_2O_2 in fat bodies of silkmoth larvae after exposure to *Escherichia coli*, but no increase in CAT activity was observed [19]. To determine whether the increases in *AccGSTO2* expression were triggered by oxidative stress under $CdCl_2$ and microbe treatments, we evaluated the activity of CAT and the levels of H_2O_2 and malondialdehyde (MDA), a toxic marker of membrane lipid peroxidation. As demonstrated in Fig. 2A-C, compared with the control group, the honeybees treated with $CdCl_2$ displayed a significant decrease in CAT activity and significant increases in H_2O_2 and MDA levels. Additionally, the honeybees exposed to microbes for 3 d displayed weakened CAT activity and elevated H_2O_2 and MDA levels compared with the control (Fig. 2D, F). Moreover, we found that CAT activity negatively correlated with the H_2O_2 and MDA levels. Consistent with our result, Chandran et al. found that CAT, glutathione peroxidase (GPX) and SOD activities were decreased in both the kidneys and the digestive gland during *Achatina fulica* exposure to $CdCl_2$ and $ZnSO_4$ [37]. Radwan et al. thought that such an inhibition could be a result of the toxic effect of $CdCl_2$ on the enzymes or of the repression of CAT expression via different mechanisms [38]. Orbea et al. reported that CAT enhancement by pollutants may be transient as a result of either an adaptive or compensatory mechanism or a toxic action after extreme or prolonged exposure [39]. Considering these results, we speculated that the induced expression of *AccGSTO2* under $CdCl_2$ and microbe stressors may be directly related to the accumulation of H_2O_2 and MDA and that *AccGSTO2* may be essential for protection against oxidative damage.

AccGSTO2 knockdown

RNAi mediated by dsRNA has served as one of the most promising strategies to study gene function, particularly in insects for which successful transgenesis is not available [40]. To date, two methods have been used to deliver exogenous dsRNA into insects: artificial feeding and microinjection [41, 42]. The former is a non-invasive technique, but individual intake is difficult to control precisely. In contrast, a controllable dose and minimal invasion make the latter widely used in diverse insects [42-44].

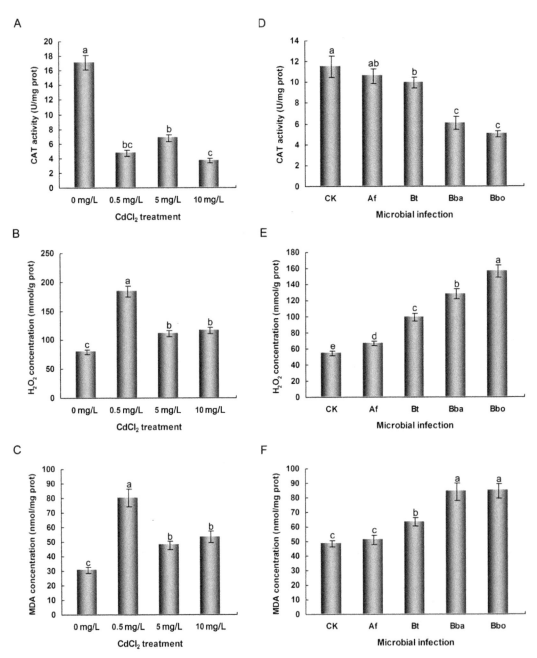

Fig. 2 Detection of oxidative status in *Apis cerana cerana in vivo* under $CdCl_2$ (A-C) and microbe stress (D, E). The effects of different $CdCl_2$ treatment concentrations on catalase (CAT) activity (A), H_2O_2 concentration (B) and malondialdehyde (MDA) concentration (C). Effects of microbe stress on CAT activity (D), H_2O_2 concentration (E), and MDA concentration (F). CK, no infection; Af, *Aspergillus flavus*; Bt, *Bacillus thuringiensis*; Bba, *Beauveria bassiana*; Bbo, *Bacillus bombysepticus* infection. The vertical bars represent the means ± SE of three independent experiments. Different letters above the bars indicate significant differences ($P < 0.05$) according to Duncan's multiple range tests.

Here, *AccGSTO2* was successfully silenced by using artificial feeding and microinjection in larvae and adults, respectively. In the control groups [no injection, water injection and dsRNA-*GFP* injection, for adults], the level of *AccGSTO2* mRNA gradually increased (no injection group) or significantly increased and then went back to normal levels (water injection and dsRNA-*GFP* injection groups) after injection. By contrast, significant transcriptional suppression was observed in the dsRNA-*AccGSTO2*-injected group, particularly at 1 day after injection (Fig. 3A). Similar to these results, a reduced mRNA level of *AccGSTO2* was also apparent in the *AccGSTO2*-knockdown larvae group (Fig. 3B). Histological sections were also used to detect the effect of *AccGSTO2* knockdown further (Fig. 4). As expected, the sections from *AccGSTO2*-knockdown bees were more lightly stained than those from the controls, although no visible morphological alterations were observed, specifically in the midgut and fat body, which play important roles in detoxification and in protection from oxidative stress [10]. Additionally, we found that the survival rate of *AccGSTO2*-knockdown bees was significantly lower than that of the control bees during treatment (Fig. 5A, B), indicating that the knockdown bees were more sensitive to changes in the inner and outer environments.

AccGSTO2 gene knockdown caused oxidative damage

To further elucidate the effects of *AccGSTO2* knockdown, the transcriptional levels of other antioxidant genes, ie *Apis cerana cerana GST* Delta (*AccGSTD*) (GenBank ID: JF798572), *AccGSTs4* (GenBank ID: JN008721), *AccGSTO1* (GenBank ID: KF496073), *AccSOD1* (GenBank ID: JN700517), *AccSOD2* (GenBank ID: JN637476), *AccCAT* (GenBank ID: KF765424), *Apis cerana cerana* methionine sulphoxide reductase A (*AccMsrA*) (GenBank ID: HQ219724), *Apis cerana cerana* cytochrome P450 (*AccCYP4G11*) (GenBank ID: KC243984), *Apis cerana cerana* thioredoxin peroxidase (*AccTpx1*) (GenBank ID: HM641254), *AccTpx3*

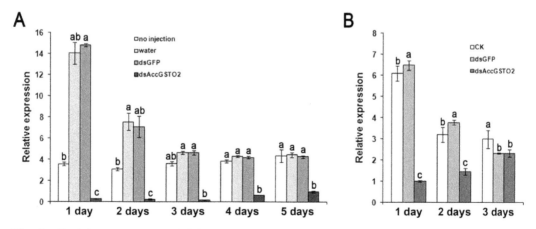

Fig. 3 Real-time quantitative (qPCR) analysis of *Apis cerana cerana* Omega-class glutathione S-transferase 2 (*AccGSTO2*) expression following RNA interference. A. double-stranded RNA (dsRNA)-injected adults. No injection, control; water, water-injected; dsGFP, dsRNA-green fluorescent protein injected; dsAccGSTO2, dsRNA-*AccGSTO2* injected. B. dsRNA-fed larvae. CK, control; dsGFP, dsRNA-*GFP* fed; dsAccGSTO2, dsRNA-*AccGSTO2* fed. The data are shown as the means ± SE based on three independent experiments. Significant differences ($P < 0.01$) were determined by using Duncan's multiple range tests.

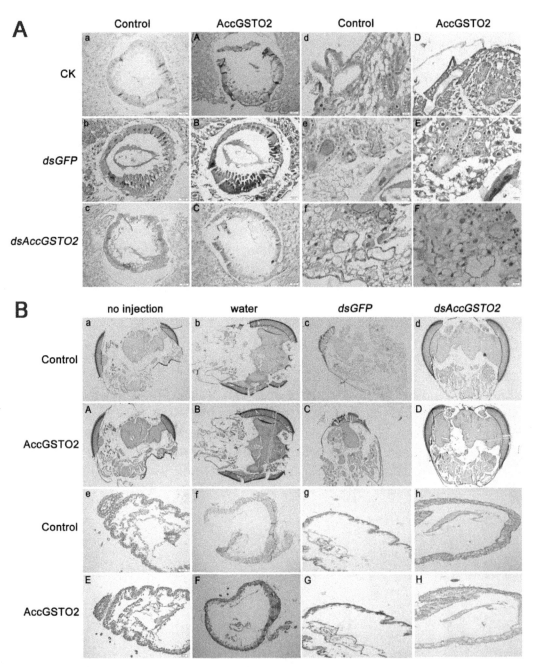

Fig. 4 Immunohistochemical analysis of the effects of *Apis cerana cerana* Omega-class glutathione *S*-transferase 2 (*AccGSTO2*) knockdown in larvae (A) and adult bees (B). a-h are negative controls for A-H, respectively. A. a-c and A-C, midgut; scale bar, 100 μm. d-f and D-F, Malpighian tubule; scale bar, 20 μm. CK, control; dsGFP, double-stranded RNA-green fluorescent protein fed; dsAccGSTO2, dsRNA-*AccGSTO2* fed. B. a-d and A-D, head; e-h and E-H, midgut; scale bar, 100 μm. No injection, control; water, water-injected; *dsGFP*, dsRNA-*GFP* injected; dsAccGSTO2, dsRNA-*AccGSTO2* injected. (For color version, please sweep QR Code in the back cover)

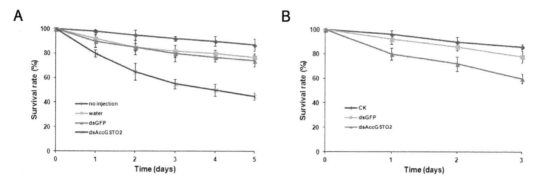

Fig. 5 Survival rates of double-stranded RNA (dsRNA)-treated larvae and adult bees. A. dsRNA-injected adults. No injection, control; water, water-injected; dsGFP, dsRNA-green fluorescent protein injected; dsAccGSTO2, dsRNA-*Apis cerana cerana* Omega-class glutathione S-transferase 2 injected. B. dsRNA-fed larvae. CK, control; dsGFP, dsRNA-GFP fed; dsAccGSTO2, dsRNA-*AccGSTO2* fed. The data are shown as the means ± SE based on three independent experiments. Significant differences ($P < 0.05$) were determined by using Duncan's multiple range tests.

(GenBank ID: JX456217), *AccTpx5* (GenBank ID: KF745893), *Apis cerana cerana* thioredoxin (*AccTrx1*) (GenBank ID: JX844651), *AccTrx2* (GenBank ID: JX844649), *Apis cerana cerana* glutaredoxin (*AccGrx1*) (GenBank ID: JX844656), *AccGrx2* (GenBank ID: JX844655) and *A. cerana cerana* small heat shock protein 22.6 (*AccsHsp22.6*; GenBank ID: KF150016), were also investigated by real-time quantitative (qPCR) in dsRNA-injected 3-day-old adults. As shown in Fig. 6A, *AccGSTD*, *AccGSTs4*, *AccCYP4G11* and *AccsHsp22.6* were induced when *AccGSTO2* was knocked down, suggesting that these induced genes may participate in compensating for *AccGSTO2* knockdown in *A. cerana cerana*. Moreover, the ascorbate content, GSH/GSSG ratio, CAT activity and GR activity of the samples after *AccGSTO2* knockdown were all lower than in the control groups. The H_2O_2, MDA, and protein carbonyl contents in the silenced samples were all higher than in the control groups (Fig. 6B). All of the above results revealed that the downregulation of the antioxidant genes *AccTpx1,3,5*; *AccTrx1,2*; *AccGrx1,2*; *AccSOD1,2*; and *AccCAT*, as well as CAT activity, might result in elevated levels of H_2O_2 when *AccGSTO2* was knocked down. H_2O_2 could cause oxidative damage to proteins and lipids [45] thus, the protein carbonyl and MDA contents were elevated in *AccGSTO2* knockdown samples compared to control groups in *A. cerana cerana*. The methionine sulfoxide reductase A (MrxA) system may mediate methionine sulfoxide modification; thus, the downregulation of *AccMrxA* may lead to elevated levels of protein carbonyls in *AccGSTO2* knockdown samples, consistent with previous reports in yeast and mice [46]. In general, GSTs act as phase II detoxifying enzymes, playing major roles in the protection against oxidative damage. However, different classes of GSTs have non-identical substrates and operational mechanisms due to their distinct molecular components and structures. Of the GSTs superfamily, the Omega class GSTs are thought to be major contributors to the defenses against various environmental stresses by DHAR and thiol transferase activities [19]. Ascorbate and GSH can also eliminate H_2O_2 [47, 48], and GR plays an important role in maintaining a constant GSH level *in vivo* [49]. The downregulation of *AccGSTO1* and GR activity may reduce the ascorbate contents and the GSH/GSSG ratio in *AccGSTO2* knockdown samples compared to the control groups. Considering these findings, we speculate that *AccGSTO2* may be indispensable for attaining a normal lifespan by enhancing cellular oxidative resistance, and few genes could compensate for the loss of *AccGSTO2* function. The possible mechanism for the results discussed above is summarized in Fig. 6C.

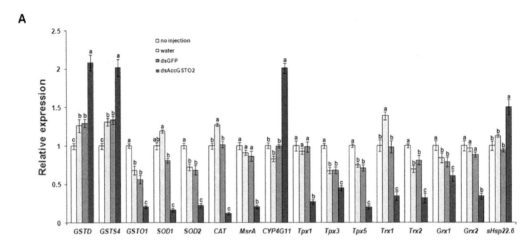

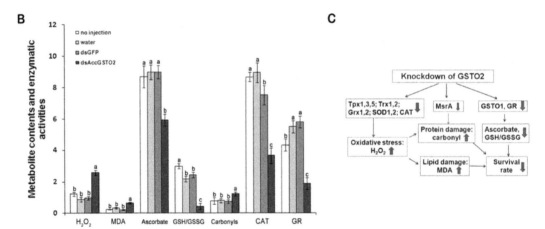

Fig. 6 Effects of *Apis cerana cerana* Omega-class glutathione S-transferase 2 (*AccGSTO2*) knockdown in adult bees. A. Expression patterns of other antioxidant genes in double-stranded RNA (dsRNA)-injected 3-day-old adults as detected by real-time quantitative (qPCR). B. Effects of *AccGSTO2* knockdown on the metabolite contents of H_2O_2 [mmol/g protein (prot)], malondialdehyde (MDA; nmol/mg prot), ascorbate (μg/mg prot), and protein carbonyls (nmol/mg prot), and on the enzymatic activities of catalase (CAT; U/mg prot) and glutathione reductase (GR; U/g prot). C. Possible mechanism of *AccGSTO2* knockdown effects on *A. cerana cerana*. The data are shown as the means ± SE based on three independent experiments. Significant differences ($P < 0.01$) were determined by using Duncan's multiple range tests. *Grx*, glutaredoxin; GSH/GSSG, reduced to oxidized glutathione ratio; *MsrA*, methionine sulphoxide reductase A; *sHsp*, small heat shock protein; *SOD*, superoxide dismutase; *Tpx*, thioredoxin peroxidase; *Trx*, thioredoxin; *GSTD*, GST delta; *CYP4G11*, cytochrome P450.

Expression of the mutant AccGSTO2 and analyses of kinetic parameters

The AccGSTO2 protein shares 30.67%-99.16% sequence identity with the GSTO proteins of the species shown in Supplementary Figure, and three Cys residues (positions 28, 124, and 217) present in AccGSTO2 are highly conserved. Generally, Cys residues play vital roles in sustaining the normal protein structure and function through their sulfhydryl groups. To investigate whether the three Cys residues are indeed involved in the catalytic mechanism of AccGSTO2, these residues were mutated to alanine (Ala) residues by site-directed mutagenesis and named C28A, C124A, and C217A. WT and mutant AccGSTO2 proteins were successfully expressed and purified. The kinetic parameters of WT and mutant AccGSTO2 proteins, which were determined by using dehydroascorbate (DHA) and GSH as substrates, are listed in Table 1. These parameters contributed to elucidating the roles of the three Cys residues. The kinetic parameters were not determined for C28A because Cys^{28} is a G-site (GSH-binding site) residue and its replacement significantly affected GSH binding; the V_{max} values of the mutants C124A and C217A decreased to varying degrees, which ranged from 4.71%-7.56% compared with that of WT-AccGSTO2. In addition, C124A had significantly higher K_m^{DHA} (but not K_m^{GSH}) and lower k_{cat}^{DHA} and k_{cat}^{GSH} than WT-AccGSTO2, indicating that the Cys-to-Ala mutation decreased the affinity for substrate DHA and the reaction turnover rate. However, C217A only showed a slight difference in kinetic parameters compared to WT-AccGSTO2. Therefore, we speculate that Cys^{124} may be more important for the catalytic activities of AccGSTO2 by stabilizing the binding of substrate DHA.

Table 1 Comparison of kinetic parameters for the recombinant wild type and mutant of *Apis cerana cerana* Omega-class glutathione *S*-transferase 2 (AccGSTO2)

WT or Mutants	V_{max}^{DHA} [μmol product/(min·mg)]	K_m^{DHA} (mmol/L)	k_{cat}^{DHA} (s^{-1})	V_{max}^{GSH} [μmol product/(min·mg)]	K_m^{GSH} (mmol/L)	k_{cat}^{GSH} (s^{-1})
WT-AccGSTO2	8.11 ± 1.56*	0.46 ± 0.09*	5.41 ± 1.03	6.35 ± 1.37*	3.55 ± 0.73*	4.23 ± 0.91
C28A	ND	ND	ND	ND	ND	ND
C124A	4.71 ± 0.94	0.75 ± 0.06	3.14 ± 0.86	4.70 ± 0.89	3.63 ± 0.78	3.13 ± 0.61
C217A	7.56 ± 1.03	0.47 ± 0.05	5.04 ± 0.97	6.05 ± 0.83	3.61 ± 0.66	4.03 ± 0.96

Note: (1) Kinetic parameters were obtained from at least three reaction runs. (2) *Data from Zhang et al. [27]. (3) ND: not determined. (4) The superscript of V_{max}, K_m, and k_{cat} with "DHA" and "GSH" in the first line indicate the relevant "saturating" GSH and DHA concentrations that were used.

Effects of the mutations on the antioxidant defense functions of AccGSTO2

In a previous study, we found that the sulfhydryl moieties of Cys residues in AccGSTO2 may be involved in protecting DNA against damage from oxidative stress [27]. To elucidate whether all Cys residues in AccGSTO2 affect antioxidant activity, C28A, C124A, and C217A mutants were employed. As shown in Fig. 7G, in the mixed-function oxidation (MFO) system, WT-AccGSTO2 protected supercoiled pUC19 from nicked by the hydroxyl radical. Compared with WT-AccGSTO2, the substitution of Cys^{217} only slightly impaired this protective effect, whereas

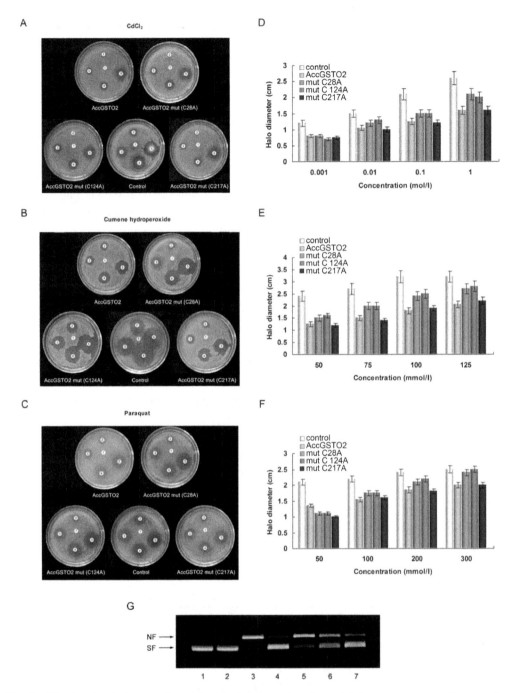

Fig. 7 Effects of mutations on the antioxidant defence function of *Apis cerana cerana* Omega-class glutathione *S*-transferase 2 (AccGSTO2). A-F. Disc diffusion assays. Filter discs that had been immersed in different concentrations of $CdCl_2$ (A, D), cumene hydroperoxide (B, E) or paraquat (C, F) were positioned on the plates. The killing zones around the drug-immersed filters were measured after an overnight exposure. G. Mixed-function oxidation system. Lane 1, pUC19 plasmid DNA only; lane 2, pUC19 plasmid DNA + $FeCl_3$; lane 3, pUC19 plasmid DNA + $FeCl_3$ + dithiothreitol (DTT); lanes 4-7, pUC19 plasmid DNA + $FeCl_3$ + DTT + 150 µg/ml purified proteins (wild-type-AccGSTO2, C28A, C124A and C217A, respectively). SF, supercoiled form; NF, nicked form.

C28A and C124A mutants exhibited significant losses of antioxidant capacity, particularly the C28A mutant, in which the supercoiled pUC19 was nearly undetectable. In addition, the disc diffusion assay provides further evidence that AccGSTO2 can effectively protect cells against oxidative stress [27]. Therefore, we employed this method to investigate the influence of Cys residues on AccGSTO2 activity. After overnight exposure to $CdCl_2$, cumene hydroperoxide, or paraquat, the killing zones of *E. coli* overexpressing WT and mutant *AccGSTO2* were smaller than those of control bacteria transfected with the vector only, although varying degrees of effects were observed (Fig. 7A-F). The substitution of Cys with Ala did not result in a complete loss of protection, particularly the replacement of the active site Cys^{28}. In addition, we found that the protection provided by WT-AccGSTO2 and the C217A mutant was roughly identical; however, C28A and C124A mutants generally had consistent halo size reduction. These observations suggest that Cys^{217} is less responsible for the antioxidant activity of AccGSTO2 and that only Cys^{28} and Cys^{124} may play important roles in maintaining enzyme activity.

Fluorescence spectra analysis

Spectroscopic measurements were employed to ascertain whether the effect of the mutation on enzyme activities might be ascribed only to a local effect on the AccGSTO2 conformation. In Fig. 8A, the emission spectra of the intrinsic fluorescence for the WT and the mutated AccGSTO2 are shown, and different red shifts of the emission maximum are observed at approximately 340 nm for C28A, 338 nm for C124A and 336 nm for C217A compared to the WT-AccGSTO2 (335 nm). This result suggests that the Trp residues in mutant AccGSTO2s were more exposed to solvent environment and that the conformation of the mutant AccGSTO2s were more relaxed than that of WT-AccGSTO2. Furthermore, Fig. 8B shows that the ANS fluorescence values of C28A, C124A, and C217A were enhanced by 43.6%,

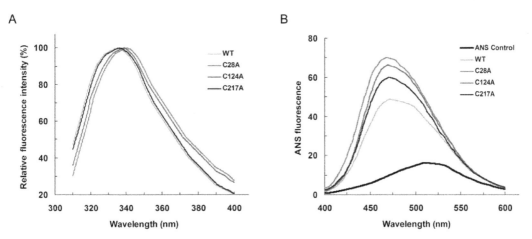

Fig. 8 Effects of mutations on *Apis cerana cerana* Omega-class glutathione S-transferase 2 (AccGSTO2) tertiary structures as determined by spectral analysis. A. Intrinsic fluorescence spectral analysis. B 8-anilino-1-naphthalene-sulfonic acid (ANS) fluorescence spectral analysis. Except for the darkest and thickest curve (ANS control), the curves from light to dark represent the spectra of wild-type (WT)-AccGSTO2, C28A, C124A and C217A, respectively. A, mutagenic Ala residue; C, cysteine residue.

35.9%, and 22.9% and that their emission maximum wavelengths were blue-shifted in comparison with that of WT-AccGSTO2 at the same concentration, suggesting that the mutant AccGSTO2s had more hydrophobic exposure to allow the occupation of the ANS probes than that of WT-AccGSTO2. These spectroscopic experiments indicated that the substitution of Cys residues had different degrees of influence on the AccGSTO2 conformation, which may be interpreted as the effect of the mutations on enzyme activities.

Effect of the mutations on the spatial structure of AccGSTO2

To provide further evidence for the fluorescence spectra analysis results, the three-dimensional structures of WT-AccGSTO2 and its mutants were predicted by using I-TASSER [50]. Based on the predicted results, the X-ray crystal structure of human GSTO1 (PDB 1eemA), which shared the highest sequence identity (32%) with AccGSTO2, was used as the template. These tertiary structures shown in ribbon form (Fig. 9) displayed the canonical GST fold: a thioredoxin-like domain (βαβαββ) in the N-terminal region and a C-terminal α-helixes domain [20]. Interestingly, the spatial structure and hydrophobic patches showed notable changes between WT-AccGSTO2 and its mutants. The predominant structural changes were observed in the α7 and β4 strands of the protein. In C28A and C124A mutants, the α7 strand was divided

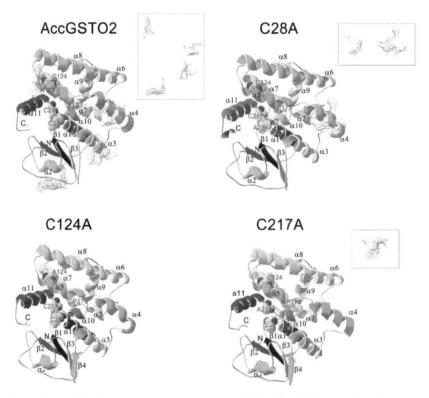

Fig. 9 Comparison of the structures and surface hydrophobic patches in the structure models of AccGSTO2, C28A, C124A, and C217A. The Cys residues and mutagenic Ala residues are shown in ball form. The α-helices, β-sheets, Phe[173], N-terminal, and C-terminal are marked. Gray is used to highlight the hydrophobic patches. These images were colored by secondary structure succession by using Swiss-PdbViewer software, version 4.1.0. (For color version, please sweep QR Code in the back cover)

into two sections, and phenylalanine (Phe^{173}), which is a conserved H-site residue, was between those two sections. The β4 strand was observed in the C124A and C217A mutants but could not be found in WT-AccGSTO2 or in the C28A mutant. Furthermore, many differences in the surface hydrophobic patches were found, particularly in the C124A mutant, in which no patches were found. Moreover, the distribution of some residues on the protein surface was also varied in the backbone and side structure, which are colored by accessibility to solvent molecules in Fig. 10. For instance, Trp^{58} in the C28A and C124A mutants and Trp^{174} in the C127A mutant were more exposed to the protein surface and could further contact solvent molecules, which is consistent with the intrinsic fluorescence analysis. In addition, Collin et al. indicated that the surface charge distribution of the G-site

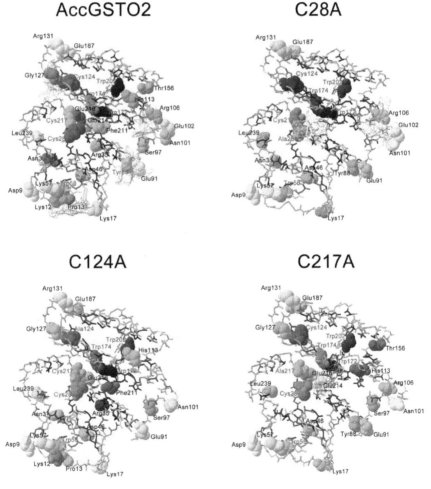

Fig. 10 Comparison of the distribution of some residues in the structural models of *Apis cerana cerana* Omega-class glutathione S-transferase 2 (AccGSTO2), C28A, C124A and C217A. The colour of each residue is determined by its accessibility to solvent molecules. Blue represents the inside amino acids that are the least accessible to solvents, red represents the superficial amino acids that are completely in contact with solvents and the other amino acids are indicated by a range of colours between the two. These images were developed by using Swiss-PdbViewer. A, mutagenic Ala residue; C, cysteine residue. (For color version, please sweep QR Code in the back cover)

was related to the substrate specificity of thioredoxin isoforms in *Arabidopsis* [51]. As shown in Fig. 11, some differences in surface electrostatic potential distributions were found in WT-AccGSTO2 and its mutants. Negatively charged regions surrounding Cys^{28} (or Ala^{28}) decreased, whereas some positively charged regions surrounding Phe^{173} were enlarged in the mutants. In summary, the Cys^{28} and Cys^{124} replacement mutants changed their surface hydrophobic patches and electrostatic potential distributions, which are the major forces that maintain protein spatial structure [52]. The reason for the negligible functional effect of C217A mutants may be attributed to the structural changes in the α7 strand where a conserved H-site residue, Phe^{173}, is located.

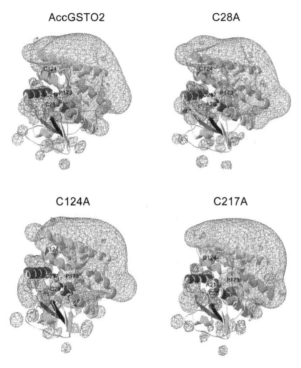

Fig. 11 Comparison of the surface electrostatic potential distributions in the structural models of *Apis cerana cerana* Omega-class glutathione *S*-transferase 2 (AccGSTO2), C28A, C124A and C217A. Negative and positive charges are shown in blue and red, respectively. These images were colour coded by secondary structure succession by using Swiss-PdbViewer. A, mutagenic Ala residue; C, cysteine residue. (For color version, please sweep QR Code in the back cover)

Conclusion

In the present study, we investigated the response of *AccGSTO2* to heavy metal, ecdysone and microbe stresses. Its induction under $CdCl_2$ and microbes may contribute to the reduction of oxidative damage *in vivo*. We also adopted RNAi to further explore the effect of *AccGSTO2* deficiency on honeybees, and a site-directed mutagenesis strategy was used to investigate the roles of Cys residues in the structure and function of the AccGSTO2 protein. Our results indicate that *AccGSTO2* may play a vital role in attaining a normal lifespan and that Cys^{28} and Cys^{124} were predominantly responsible for implementing enzyme activity. In conclusion, these findings may be useful for understanding the structural and functional characteristics of different GST classes.

References

[1] Simon, H. U., Haj-Yehia, A., Levi-Schaffer, F. Role of reactive oxygen species (ROS) in apoptosis induction. Apoptosis. 2000, 5, 415-418.

[2] Hayes, J. D., Pulford, D. J. The glutathione S-transferase supergene family: Regulation of GST and the contribution of the isoenzymes to cancer chemoprotection and drug resistance. Crit Rev Biochem Mol Biol. 1995, 30, 445-600.

[3] Holmgren, A. Thioredoxin and glutaredoxin systems. J Biol Chem. 1989, 264, 13963-13966.

[4] Board, P. G., Menon, D. Glutathione transferases, regulators of cellular metabolism and physiology. Biochim Biophys Acta. 2013, 1830, 3267-3288.

[5] Sheehan, D., Meade, G., Foley, V. M., Dowd, C. A. Structure, function and evolution of glutathione transferases: implications for classification of non-mammalian members of an ancient enzyme superfamily. Biochem J. 2001 360, 1-16.

[6] Robinson, A., Huttley, G. A., Booth, H. S., Board, P. G. Modelling and bioinformatics studies of the human Kappa-class glutathione transferase predict a novel third transferase family with homology to prokaryotic 2-hydroxychromene-2-carboxylate isomerases. Biochem J. 2004, 379, 541-552.

[7] Hayes, J. D., Flanagan, J. U., Jowsey, I. R. Glutathione transferases. Annu Rev Pharmacol Toxicol. 2005, 45, 51-88.

[8] Allocati, N., Federici, L., Masulli, M., Di Ilio, C. Glutathione transferases in bacteria. FEBS J. 2009, 276, 58-75.

[9] Flanagan, J. U., Smythe, M. L. Sigma-class glutathione transferases. Drug Metab Rev. 2011, 43, 194-214.

[10] Ketterman, A. J., Saisawang, C., Wongsantichon, J. Insect glutathione transferases. Drug Metab Rev. 2011, 43, 253-265.

[11] Burmeister, C., Luërsen, K., Heinick, A., Hussein, A., Domagalski, M., Walter, R. D., Liebau, E. Oxidative stress in *Caenorhabditis elegans*: protective effects of the Omega class glutathione transferase (GSTO-1). FASEB J. 2008, 22, 343-354.

[12] Singh, S. P., Coronella, J. A., Benes, H., Cochrane, B. J., Zimniak, P. Catalytic function of *Drosophila melanogaster* glutathione S-transferase DmGSTS1-1 (GST-2) in conjugation of lipid peroxidation end products. Eur J Biochem. 2001, 268, 2912-2923.

[13] Yan, H., Jia, H., Gao, H., Guo, X., Xu, B. Identification, genomic organization, and oxidative stress response of a sigma class glutathione S-transferase gene (*AccGSTS1*) in the honey bee, *Apis cerana cerana*. Cell Stress Chaperones. 2013, 18: 415-426.

[14] Yan, H., Jia, H., Wang, X., Gao, H., Guo, X., Xu, B. Identification and characterization of an *Apis cerana cerana* Delta class glutathione S-transferase gene (*AccGSTD*) in response to thermal stress. Naturwissenschaften. 2013, 100: 153-163.

[15] Meng, F., Zhang, Y., Liu, F., Guo, X., Xu, B. Characterization and mutational analysis of Omega-class GST (GSTO1) from *Apis cerana cerana*, a gene involved in response to oxidative stress. PLoS One. 2014, 9, e93100.

[16] Peiren, N., De Graaf, D. C., Vanrobaeys, F., Danneels, E. L., Devreese, B., Van Beeumen,. J., et al. Proteomic analysis of the honey bee worker venom gland focusing on the mechanisms of protection against tissue damage. Toxicon. 2008, 52, 72-83.

[17] Meng, J. Y., Zhang, C. Y., Zhu, F., Wang, X. P., Lei, C. L. Ultraviolet light-induced oxidative stress: effects on antioxidant response of *Helicoverpa* armigera adults. J Insect Physio. 2009, 55, 588-592.

[18] Nair, P. M. G., Choi, J. Identification characterization and expression profiles of *Chironomus riparius* glutathione S-transferase (GST) genes in response to cadmium and silver nanoparticles exposure.

Aquat Toxicol. 2011, 101, 550-560.

[19] Yamamoto, K., Teshiba, S., Shigeoka, Y., Aso, Y., Banno, Y., Fujiki, T., Katakura, Y. Characterization of an Omega-class glutathione S-transferase in the stress response of the silkmoth. Insect Mol Biol. 2011, 20: 379-386.

[20] Board, P. G., Coggan, M., Chelvanayagam, G., et al. Identification, characterization, and crystal structure of the Omega class glutathione transferases. J Biol Chem. 2000, 27, 24798-24806.

[21] Caccuri, A. M., Antonini, G., Allocati, N., et al. GSTB1-1 from *Proteus mirabilis*: a snapshot of an enzyme in the evolutionary pathway from a redox enzyme to a conjugating enzyme. J Biol Chem. 2000, 277, 18777-18784.

[22] Zhou, H., Brock, J., Liu, D., Board, P. G., Oakley, A. J. Structural Insights into the dehydroascorbate reductase activity of human Omega-class glutathione transferases. J Mol Biol. 2012, 420, 190-203.

[23] Anderson, D. L., Trueman, J. W. H. *Varroa jacobsoni* (Acari: Varroidae) is more than one species. Exp Appl Acarol. 2000, 24, 165-189.

[24] Li, J., Qin, H., Wu, J., Sadd, B. M., Wang, X., Evans, J. D., Peng, W., Chen, Y. The prevalence of parasites and pathogens in Asian honeybees *Apis cerana* in China. PLoS One. 2012, 7, e47955.

[25] Yang, G. H. Harm of introducing the western honey bee *Apis mellifera* L. to the Chinese honey bee *Apis cerana* F. and its ecological impact. Acta Entomologica Sinica. 2005, 48, 401-406.

[26] Gegear, R. J., Otterstatter, M. C., Thomson, J. D. Bumble-bee foragers infected by a gut parasite have an impaired ability to utilize floral information. Proc Biol Sci. 2006, 273, 1073-1078.

[27] Zhang, Y., Yan, H., Lu, W., Li, Y., Guo, X., Xu, B. A novel Omega-class glutathione S-transferase gene in *Apis cerana cerana*: molecular characterisation of GSTO2 and its protective effects in oxidative stress. Cell Stress Chaperones. 2013, 18, 503-516.

[28] Adler, L. S. The ecological significance of toxic nectar. Oikos. 2000, 91, 409-420.

[29] Yepiskoposyan, H., Egli, D., Fergestad, T., Selvaraj, A., Treiber, C., Multhaup, G., Georgiev, O., Schaffner, W. Transcriptome response to heavy metal stress in *Drosophila* reveals a new zinc transporter that confers resistance to zinc. Nucleic Acids Res. 2006, 34, 4866-4877.

[30] Espinoza, H. M., Williams, C. R., Gallagher, E. P. Effect of cadmium on glutathione S-transferase and metallothionein gene expression in coho salmon liver, gill and olfactory tissues. Aquat Toxicol. 2012, 110-111, 37-44.

[31] Yu, X., Sun, R., Yan, H., Guo, X., Xu, B. Characterization of a sigma class glutathione S-transferase gene in the larvae of the honeybee (*Apis cerana cerana*) on exposure to mercury. Comp Biochem Physiol B: Biochem Mol Biol. 2012, 161, 356-364.

[32] Cabassi, E. The immune system and exposure to xenobiotics in animals. Vet Res Commun. 2007, 31: 115-120.

[33] Kozlowski, H., Kolkowska, P., Watly, J., Krzywoszynska, K., Potocki, S. General aspects of metal toxicity. Curr Med Chem. 2014, 21, 3721-3740.

[34] Von Kalm, L., Crossgrove, K., Von Seggern, D., Guild, G. M., Beckendorf, S. K. The Broad-Complex directly controls a tissue-specific response to the steroid hormone ecdysone at the onset of *Drosophila* metamorphosis. EMBO J. 1994, 13, 3505-3516.

[35] Stohs, S. J., Bagchi, D., Hassoun, E., Bagchi, M. Oxidative mechanisms in the toxicity of chromium and cadmium ions. J Environ Pathol Toxicol Oncol. 2001, 20, 77-88.

[36] Filipič, M. Mechanisms of cadmium induced genomic instability. Mutat Res. 2012, 733, 69-77.

[37] Chandran, R., Sivakumar, A., Mohandass, S., Aruchami, M. Effect of cadmium and zinc on antioxidant enzyme activity in the gastropod, *Achatina fulica*. Comp Biochem Physiol. 2005, 140, 422-426.

[38] Radwan, M. A., El-Gendy, K. S., Gad, A. F. Biomarkers of oxidative stress in the land snail, *Theba pisana* for assessing ecotoxicological effects of urban metal pollution. Chemosphere. 2010, 79, 40-46.

[39] Orbea, A., Ortiz-Zarragoitia, M., Cajaraville, M. P. Interactive effects of benzo (a) pyrene and cadmium and effects of di (2-ethylhexyl) phthalate on antioxidant and peroxisomal enzymes and peroxisomal volume density in the digestive gland of mussel *Mytilus galloprovincialis* Lmk.

Biomarkers. 2002, 7, 33-48.

[40] Huvenne, H., Smagghe, G. Mechanisms of dsRNA uptake in insects and potential of RNAi for pest control: a review. J Insect Physiol. 2010, 56, 227-235.

[41] Walshe, D. P., Lehane, S. M., Lehane, M. J., Haines, L. R. Prolonged gene knockdown in the tsetse fly *Glossina* by feeding double stranded RNA. Insect Mol Biol. 2009, 18, 11-19.

[42] Futahashi, R., Tanaka, K., Matsuura, Y., Tanahashi, M., Kikuchi, Y., Fukatsu, T. Laccase 2 is required for cuticular pigmentation in stinkbugs. Insect Biochem Mol Biol. 2001, 41, 191-196.

[43] Elias-Neto, M., Soares, M. P., Simões, Z. L., Hartfelder, K., Bitondi, M. M. Developmental characterization, function and regulation of a Laccase 2 encoding gene in the honey bee, *Apis mellifera* (Hymenoptera, Apinae). Insect Biochem Mol Biol. 2010, 40, 241-251.

[44] Liu, S., Ding, Z., Zhang, C., Yang, B., Liu, Z. Gene knockdown by intro-thoracic injection of double-stranded RNA in the brown planthopper, *Nilaparvata lugens*. Insect Biochem Mol Biol. 2010, 40, 666-671.

[45] Lee, K. S., Kim, S. R., Park, N. S., et al. Characterisation of silkworm thioredoxin peroxidase that is induced by external temperature stimulus and viral infection. Insect Biochem Mol Biol. 2005, 35, 73-84.

[46] Moskovitz, J., Oien, D. B. Protein carbonyl and the methionine sulfoxide reductase system. Antioxid Redox Signaling. 2010, 12, 405-415.

[47] Meister, A. Glutathione–ascorbic acid antioxidant system in animals. J Biol Chem. 1994, 269, 9397-9400.

[48] Chatterjee, I. B., Majumder, A. K., Nandi, B. K., Subramanian, N. Synthesis and some major functions of vitamin C in animals. Ann N Y Acad Sci. 1975, 258, 24-47.

[49] Pannala, V. R., Bazil, J. N., Camara, A. K. S., Dash, R. K. A biophysically based mathematical model for the catalytic mechanism of glutathione reductase. Free Radic Biol Med. 2013, 65, 1385-1397.

[50] Kelley, L. A., Sternberg, M. J. Protein structure prediction on the Web: a case study using the Phyre server. Nat Protoc. 2009, 4, 363-371.

[51] Collin, V., Issakidis-Bourguet, E., Marchand, C., Hirasawa, M., Lancelin, J. M., Knaff, D. B., Miginiac-Maslow, M. The *Arabidopsis* plastidial thioredoxins: new functions and new insights into specificity. J Biol Chem. 2003, 278, 23747-23752.

[52] Nicholls, A., Sharp, K. A., Honig, B. Protein folding and association: insights from the interfacial and thermodynamic properties of hydrocarbons. Proteins. 1991, 11, 281-296.

[53] Alaux, C., Ducloz, F., Crauser, D., Le Conte, Y. Diet effects on honeybee immunocompetence. Biol Lett. 2010, 6, 562-565.

[54] Yao, P., Chen, X., Yan, Y., Liu, F., Zhang, Y., Guo, X., Xu, B. Glutaredoxin 1, glutaredoxin 2, thioredoxin 1, and thioredoxin peroxidase 3 play important roles in antioxidant defense in *Apis cerana cerana*. Free Radic Biol Med. 2014, 68, 335-346.

[55] Lu, R., Wang, S., Xing, G., Ren, C., Han, F., Chen, S., Zhang, Z., Zhu, Q., Michael, A. Effects of acrylonitrile on antioxidant status of different brain regions in rats. Neurochem Int. 2009, 55, 552-557.

[56] Lu, W., Kang, M., Liu, X., Zhao, X., Guo, X. Identification and antioxidant characterisation of thioredoxin-like1 from *Apis cerana cerana*. Apidologie. 2012, 43, 737-752.

[57] Yu, F., Kang, M., Meng, F., Guo, X., Xu, B. Molecular cloning and characterization of a thioredoxin peroxidase gene from *Apis cerana cerana*. Insect Mol Biol. 2011, 20, 367-378.

2.5 Characterization of a Sigma Class Glutathione S-Transferase Gene in the Larvae of the Honeybee (*Apis cerana cerana*) on Exposure to Mercury

Xiaoli Yu, Rujiang Sun, Huiru Yan, Xingqi Guo, Baohua Xu

Abstract: Glutathione S-transferases (GSTs) are multifunctional enzymes that are mainly involved in detoxification of endogenous and xenobiotic compounds and oxidative stress resistance in insects. In this study, we identified a novel Sigma class GST from *Apis cerana cerana* (*AccGSTs4*). The open reading frame of cDNA was 612 bp and encoded a 203-amino acid polypeptide, which exhibited the structural motif and domain organization characteristic of GST. Homology and evolutionary analysis indicated that the induced amino acid sequence of AccGSTs4 belonged to an insect Sigma class group. Expression analysis indicated that *AccGSTs4* was presented in all stages of development with high level in the fourth-instar larvae. Immunolocalization further revealed the distribution of AccGSTs4 in the fourth-instar larvae. RT-qPCR showed that the transcripts of *AccGSTs4* from the larvae were upregulated under dietary $HgCl_2$. The GST activity under stress was higher than the controls fed on a $HgCl_2$-free diet. Disc diffusion assay provided evidence of recAccGSTs4 resistance to long-term exposure of $HgCl_2$ stress. Additionally, analysis of 5′-flanking region further clarified the probable expression patterns of *AccGSTs4*. Taken together, our findings indicate that the larvae *AccGSTs4* may play a role in mercury stress response, and it will help to protect honeybees from heavy metal.

Keywords: GST; *Apis cerana cerana*; RT-qPCR; Immunolocalization; Mercury

Introduction

Environment such as spring water, mining and agriculture has sometimes resulted in the local accumulation of heavy metal to levels toxic to living organisms [1, 2]. Studies have shown that the insects living in polluted areas accumulate heavy metal [3]. Heavy metal affect not only the growth rates and mortality of insects but also some innate insect resistance machanisms such as the immune and detoxification system [4-6]. As their capacity to lose electrons, heavy metals are primarily thought to be toxic by virtue of the generation of reactive oxygen species (ROS) which causes significant damage to cellular macromolecules [7]. Moreover, mercury is a ubiquitous environmental toxin, and the mercury-induced ROS formation and GSH depletion can result in oxidative stress [8]. Aerobic organisms have evolved various enzymatic antioxidant systems to avoid oxidative damage. One of the main enzymes of the detoxification system is glutathione *S*-transferase (GST).

GSTs are a group of multifunctional enzymes whose major functions involve the detoxification of both endogenous and xenobiotic compounds such as ROS, intracellular transport, biosynthesis of hormones, and protection against oxidative stress [9, 10]. GSTs are widespread in both prokaryotes and eukaryotes. There are seven classes of GSTs (Alpha, Mu, Pi, Omega, Sigma, Theta and Zeta) in mammals [11] and six classes of GSTs (Delta, Epsilon, Omega, Sigma, Theta and Zeta) in insects such as *Anopheles gambiae*, *Drosophila melanogaster* and *Aedes aegypti* [12]. The Delta and Epsilon classes, which are specific to insects, are mainly implicated in xenobiotic metabolism [13]. The Omega, Sigma, Theta and Zeta classes are distributed more widely and play essential house-keeping roles [14].

In insects, GSTs have been studied for their role in mediating oxidative stress responses [15, 16]. The Epsilon class GSTs play an important role in DDT resistance in both *A. gambiae* and *A. aegypti* [17, 18]. The Sigma class GSTs from *D. melanogaster* are implicated in the detoxification of lipid peroxidation products, suggesting a protective role against oxidative stress [19]. The expression of GSTs from common cutworm, *Spodoptera litura* increases in response to xenobiotic compounds and bacteria [20]. The response of GSTs from *D. melanogaster* midgut to dietary H_2O_2 shows a dosage-based expression pattern [16]. In addition, the proteomics analysis of the venom glands from *Apis mellifera* indicated that the presence of a Sigma class GSTs appears to protect against oxidative stress [21].

More recently, insect GSTs are studied mainly from the viewpoint of insecticide metabolism [22]. However, the role in the detoxification system which is involved in the defense of insects against stress such as heavy metal, to our knowledge, is limited. The Chinese honeybee (*Apis cerana cerana*), which is one kind of Hymenoptera, is widely farmed in China and benefits for agriculture and ecological balance. Honeybees, which are in the lack of detoxification enzymes, are highly susceptible to various environmental stresses. As bees are the principal insects in agriculture, knowledge of bees GSTs is of great importance. Thus, we choose *A. cerana cerana* as the experimental material to detect the potential function of GSTs in honeybees. The study of the GST gene in Chinese honeybees suggested that it might play a role in regulation of heavy metal response, and it would help the Chinese honeybees defend against injury.

Material and Methods

Insects and mercury treatments

The larvae, pupae and adults of Chinese honeybees (*A. cerana cerana*) which were provided by Shandong Agricultural University, China, were collected according to the criteria of Michelette and Soares [23]. They were used from the first- to the fifth-instar larvae (L1-L5), successive pupal stages including earlier pupal phase which show white (Pw), pink (Pp) and brown (Pb) eyes pupal phase, and then the light- (Pbl), dark- (Pbd) pigmented pupal phase, and 2- and 30-day-old adult workers. The larvae, pupae and 10-day-old adult workers were taken from the hive. The 30-day-old adult workers were collected at the entrance of the hive when they returned to the colony after foraging. The adult workers were captured by marking newly emerged bees with paint and then collected them after 2 and 30 days, respectively.

A. cerana cerana larvae were collectively reared in 24-well plates (Costar, Corning Incorporated, USA) and placed into a desiccator with a constant temperature (34℃) and humidity (70%), by using a 10% glycerol solution inside the desiccator. The larvae were initially fed an artificial diet containing royal jelly (49.0%), D-fructose (6.8%), D-glucose (6.8%), yeast powder (1.1%) and water (36.3%) [24]. The sugars and yeast powder were dissolved in water, filtered through a Millipore membrane (0.22 μm), and then added to the royal jelly. Once the larvae reached their third instars they were randomly transferred to artificial diets treated with different and proper concentration of $HgCl_2$: 0, 0.5, 5 and 10 mg/L. One 24-well plate was filled with 48 larvae (two larvae/well) and three plates were treated with one concentration of $HgCl_2$. Moreover, this experiment was repeated three times. Three days after treatment, the number of living and dead larvae was recorded. The larvae that were still alive after 72 h feeding on the treatment diets have been separated on groups for RNA extraction and GST activity, respectively. These different developmental bees and treated larvae were immediately stored at −80℃ until used.

Isolation of the full-length cDNA of *AccGSTs4*

Total RNA was extracted by using Trizol reagent (Invitrogen, USA) from the larvae, and then was reversely transcribed by using reverse transcriptase system (TransGen Biotech, China) to prepare the cDNA which was used as template. To obtain the internal fragment of *AccGSTs4* cDNA, primers G1 and G2 (Supplementary Table 1) were designed and synthesized (Shanghai Sangon Biotechnological Company, China) by the conserved regions of the GSTs from *A. mellifera*, *Nasonia vitripennis* and *Locusta migratoria*. On the basis of the cloned internal fragment, specific primers (Supplementary Table 1) were designed and synthesized for 5′ and 3′ rapid amplification of cDNA end (5′ and 3′ RACE) to obtain the full-length cDNA. PCR conditions are presented in Supplementary Table 2. All cloned sequences in this study were performed as follows: the PCR products were purified by a gel extraction kit (TaKaRa, Japan), ligated into pEASY-T3 vector (TransGen Biotech, China), transformed into *Escherichia coli* strain DH5α and then sequenced.

Cloning of the 5′-flanking region of *AccGSTs4*

The genomic DNA of the fourth-instar larvae which was isolated by the EasyPure Genomic DNA Extraction Kit (TransGen Biotechnology, China) and pairs GM1/GM2 which were designed and synthesized according to the full-length cDNA of *AccGSTs4* were used to amplify the DNA sequence of *AccGSTs4* as template and primers, respectively. Then we designed specific primers GPR1/GPR2 and GPO1/GPO2 (Supplementary Table 1) according to 5′ end of the indentified genomic DNA to isolate the 5′-flanking region of *AccGSTs4* by the inverse polymerase chain reaction (I-PCR). Total genomic DNA (4.66 μg) which was digested completely with *Taq* I at 37°C overnight and ligated to the circles by using T4 DNA ligase (TaKaRa, Japan) was used as template. The first PCR was performed by using the primers GPR1 and GPR2 and the nested was carried out by using primers GPO1 and GPO2. To further verify the accuracy of this sequence, PCR was carried out to amplify the full-length sequence by using the specific primers GPF1 and GPF2 which were designed based on the obtained promoter sequence. The cloning and sequencing strategies were as the same as described above. The primers and reaction conditions are provided in Supplementary Table 1 and Supplementary Table 2, respectively.

Sequence analysis of AccGSTs4

Homologous GSTs sequences from other insects were retrieved from GenBank at NCBI and aligned by using DNAMAN version 5.2.2 software (Lynnon Biosoft Company, USA). The isoelectric point (pI) and molecular weight (MW) were computed by PeptideMass (http://us.expasy.org/tools/peptide-mass.html). Three-dimensional (3D) homology modeling was carried out by using an online protein 3D structure prediction tool (http://swissmodel.expasy.org/). Phylogenetic analysis was performed to check the evolutionary relationships between the sequences of different origins by using neighbor-joining (NJ) in MEGA version 4.1 software. The MatInspector database (http://www.cbrc.jp/research/db/TFSEARCH.html) was used to search for transcription factor-binding sites in the 5'-flanking region.

RT-qPCR analysis

The method of extraction of total RNA and reverse transcription were as the same as described above. One set of primers (GRT1/GRT2) for *AccGSTs4* and one set (AG1/AG2) for *A. cerana cerana β-actin* gene (GenBank accession number: XM640276) were designed and synthesized. RT-qPCR was performed by using SYBR Premix Ex *Taq* (TaKaRa, Japan) on a CFX 96^{TM} Real-time System (Bio-Rad, USA). The samples were analyzed in triplicate, and the expression levels of *AccGSTs4* were normalized against the corresponding control *β-actin* levels. The $2^{-\Delta\Delta C_t}$ comparative CT method was used to calculate the different expression levels of *AccGSTs4* [25]. Overall differences in abundances of *AccGSTs4* transcripts were determined by one-way ANOVA analysis, based on a post hoc Tukey test using Statistical Analysis System (SAS) version 9.1 (Version 8e, SAS Institute, USA). Significance was set at $P<0.05$. Three independent experiments were replicated.

Overexpression and purification of recombinant protein

The *AccGSTs4* clone, integrated in the pEASY-T3 vector, was digested with *Sac* I and *Hin*dIII and subcloned into the expression vector pET-30a (+). Then, the prepared wide-type and mutant plasmid recAccGSTs4 were transformed into competent *Escherichia coli* BL21 (DE3), which were grown at 37°C in Luria-Bertani (LB) medium containing 30 mg/ml kanamycin. After the cell density reached 0.7 at optical density (OD)$_{600}$, isopropylthio-β-D-galactoside (IPTG) was added at a final concentration of 1 mmol/L to induce the production of recombinant protein. After induction for 4 h, the cells were harvested from liquid culture by centrifugation at 10,000 g for 5 min. The soluble recombinant protein was purified by MagneHis™ Protein Purification System (Promega, USA). Sodium dodecylsulfate-polyacrylamide gel electrophoresis (SDS-PAGE) was performed by using a 12% polyacrylamide gel slab containing 0.1% SDS according to the method of Laemmli [26] with Unstained Protein Molecular Weight Marker (Fermentas, Canada). Protein bands were visualized by staining with CBB.

Antibody production and Western blotting analysis

Six-week-old and specific pathogen free female white mice were immunized with the recAccGSTs4 (50 μg) in Freund's adjuvants (Sigma, USA) three times at 3 weeks intervals. The final boost was injected with 10 μg protein. One week later, the blood was collected by eyeball puncture. The blood was centrifuged at 3,000 g for 5 min and the anti-serum was prepared (anti-recAccGSTs4). The recombinant protein was resolved by SDS-PAGE under the same condition described above. The protein was then transferred to a polyvinylidene difluoride (PVDF) membrane (Whatman, British). After blocking for 1 h in 5% (w/V) skim milk diluted in PBS buffer, the membrane was incubated with anti-recAccGSTs4 polyclonal antibody (primary antibody) for 1 h, followed by HRP-conjugated secondary antibody (Beijing Dingguo Changsheng Biotech Co., Ltd., China) diluted in blocking buffer for 2 h at 37℃. After washing, the membrane was subjected to 4-chloro-1-naphthol (Solarbio, China) to detection.

Immunolocalization

The fresh whole bodies of the fourth-instar larvae were fixed in PBS containing 4% paraformaldehyde overnight at 4℃. The larvae were dehydrated with graded alcohol series, embedded in paraffin blocks and stored in a desiccator until used. Tissue sections (10 μm thick) were mounted on poly-L-lysine-coated glass slides (Beijing Dingguo Changsheng Biotech Co., Ltd., China), deparaffinised and rehydrated. The slides were washed with PBS (150 mmol/L, pH 7.4). The sections were treated with 3% hydrogen peroxide for 30 min in the dark, after which they were blocked with 1% BSA for 30 min at room temperature. The sections were incubated overnight with the anti-recAccGSTs4 antibody (1 ∶ 200 dilution) at 4℃. Normal mouse serum (1 ∶ 200 dilution) was employed as a negative control. Next day, the sections were incubated with HRP-conjugated goat anti-rat IgG (Wuhan Boster Biological Technology Co., Ltd., China) diluted in PBS at 37℃ for 1 h, then incubated with HRP-streptavidin (Solarbio, China) diluted in PBS for 40 min. Color development was performed with DAB Horseradish Peroxidase Color Development Kit (Wuhan Boster Biological Technology Co., Ltd., China). The slides were observed under a laser confocal microscope (LSM 510 META; Carl Zeiss AG, Germany).

Assay of GST activity

The activity of GST in the third-, the fourth- and the fifth-instar larvae was determined spectrophotometrically by using commercially available assay kits (Nanjing Jiancheng Bioengineering Institute, China) following the protocols of the manufacturer, with some modifications. GST activity was determined by using 1-chloro-2,4-dinitrobenzene (CDNB) as a substrate, and the conjugation of GSH (glutathione) to CDNB was determined by recording the change in absorbance at 412 nm. One unit of GST activity was defined as the amount that catalyzes the conjugation of 1 mmol/L GSH with CDNB per minute per mg protein. GST activity was expressed as U/mg protein. Nonenzymatic controls were performed in parallel in order to correct for nonenzymatic conjugation of GSH to the substrates.

Disc diffusion assay

The recAccGSTs4 which was heterologously expressed in *E. coli* BL21 (DE3) cells on the sensitivity to heavy metal $HgCl_2$ was detected. Approximately 5×10^8 cells which were cultured in lipid LB medium, were plated on LB-kanamycin agar plates and incubated at 37℃ for 1 h. Filter discs (8 mm diameter) soaked with suitable concentrations (0, 40, 60, 80, 100 mg/L) of $HgCl_2$, which were demonstrated as the appropriate concentration gradient in our experiment, were put on the surface of the top agar. The *E. coli* BL21 cells with pET-30a (+) only were as the control and dealt with the same method. Cells were grown for 24 h at 37℃.

Results

Identification and molecular properties of a gene encoding *AccGSTs4*

A *GSTs4* gene, referred to *AccGSTs4*, was identified from *A. cerana cerana* larvae by RT-PCR and RACE-PCR methods. The full-length cDNA (GenBank accession number: JN008721), which was 740 bp in length, contained an open reading frame (ORF) of 612 bp encoding a 203-amino acid polypeptide with a predicted molecular weight of 23.3 kDa and a pI of 8.41.

The deduced amino acid sequence of AccGSTs4 was most similar to GST from *A. mellifera*, followed by *L. migratoria* (52.43% identity), *N. vitripennis* (53.17% identity) and *Bombus ignitus* (42.93% identity). These amino acid sequences were aligned in Fig. 1A. Moreover, it contained two conserved domains: N terminal domain spanned amino acids 4-75 and C terminal domain spanned amino acids 88-189. The N terminal domain possessed several conserved residues especially the Tyr residue which was important for the catalytic mechanism. Meanwhile, the three-dimensional structure of AccGSTs4 was predicted and reconstructed by SWISS-MODEL (Fig. 1B). The template named 1m0uA was used to be the homologous modeling of the AccGSTs4 protein and the sequence identity between the two proteins is 43.35%. A phylogenetic tree, generated from the amino acid sequences of the various insect GSTs, indicated that the current protein grouped into the Sigma class (Fig. 1C), thus we named it AccGSTs4.

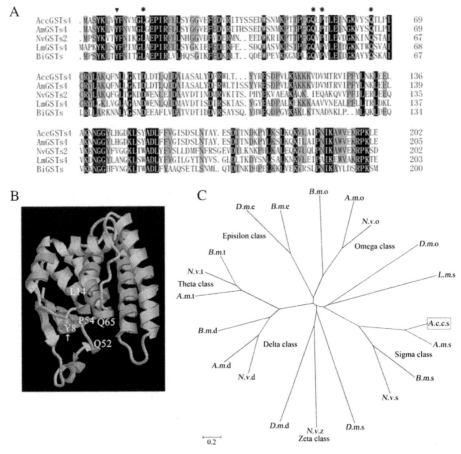

Fig. 1 Molecular properties of AccGSTs4. A. Sequence alignment of AccGSTs4 with other insect GSTs: AmGSTs4 (*Apis mellifera*, NP 001136128), NvGSTs2 (*Nasonia vitripennis*, NP 001165918) and LmGSTs4 (*Locusta migratoria*, AEB91976). Identical or 80% conserved amino acids are shaded in black or gray, respectively. The conserved G-site residues are represented by asterisks. The closed triangle indicates the conserved tyrosine residue at position 8. B. The tertiary structure of AccGSTs4. The locations of the conserved G-site residues are shown. C. A phylogenetic analysis showing the relationship of AccGSTs4 with other class members including: *A.c.c*.s (*Apis cerana cerana*, JN008721), *A.m*.s (*Apis mellifera*, NP 001136128), *B.m*.s (*Bombyx mori*, NP 001036994), *N.v*.s (*Nasonia vitripennis*, NP 001165918), *D.m*.s (*Drosophila melanogaster*, NP001037077), *N.v*.z (*Nasonia vitripennis*, NP 001165931), *D.m*.d (*Drosophila melanogaster*, NP 001034042), *N.v*.d (*Nasonia vitripennis*, NP 001165913), *A.m*.d (*Apis mellifera*, NP 001171499), *B.m*.d (*Bombyx mori*, NP 001037546), *A.m*.t (*Apis mellifera*, XP 624692), *N.v*.t (*Nasonia vitripennis*, NP 00116592), *B.m*.t (*Bombyx mori*, NP 001108463), *D.m*.e (*Drosophila melanogaster*, NP 611325), *B.m*.e (*Bombyx mori*, NP 001108460), *B.m*.o (*Bombyx mori*, NP 001040131), *A.m*.o (*Apis mellifera*, XP 003251893), *N.v*.o (*Nasonia vitripennis*, NP 001165916), *D.m*.o (*Drosophila melanogaster*, NP 648237), *L.m*.s (*Locusta migratoria*, AEB91976).

Expression of *AccGSTs4* by worker bees during development

To determine the expression levels of *AccGSTs4* at an appropriate developmental stage, we firstly performed a general overview on the expression pattern of *AccGSTs4* in various developmental stages of worker bees by RT-qPCR. As shown in Fig. 2, the transcripts of *AccGSTs4* were detected at all developmental stages. It was interesting to observe that there were lower expression levels in early larval stages but a significant difference ($P < 0.05$) of the levels of *AccGSTs4* mRNA in the late larval stages especially the fourth-instar stage. Given these results, we presumed that *AccGSTs4* might be involved in the regulation of development, especially in the larval stages. Thus, we chose the larvae as the appropriate spot for study of the gene expression in this study.

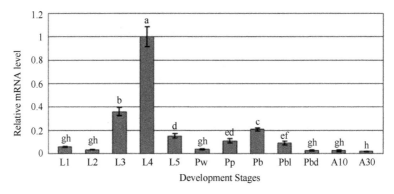

Fig. 2 RT-qPCR analysis of *AccGSTs4* in different developmental stages: the first- to the fifth-instar larvae (L1-L5), successive pupal stages including earlier pupal phase which show white (Pw), pink (Pp) and brown (Pb) eyes pupal phase, and then the light- (Pbl), dark- (Pbd) pigmented pupal phase, and 2- and 30-day-old adult workers. The *A. cerana cerana β-actin* gene was used as reference house-keeping gene to normalize the expression level. Bars represent the triplicate mean±SD from three individuals ($n = 3$). Different capital letters indicate significant difference ($P<0.05$).

Immunological properties of AccGSTs4 in larvae

The mRNA level of *AccGSTs4* gene has been proved to be expressed predominantly in the fourth-instar larvae, so the expression pattern of the native AccGSTs4 was further evaluated by immunolocalization at the protein level. The recAccGSTs4 purified by Ni-NTA spin columns was approximately 23 kDa by SDS-PAGE analysis (Fig. 3A). This recombinant protein was used to generate a specific mouse antibody (anti-recAccGSTs4). Western blotting analysis was performed to show the specificity of anti-recAccGSTs4 (Fig. 3B). Moreover, a highly-ordered distributional pattern of positive signal was observed in the intestinal wall (IW) of midgut, Malpighian tube and salivary gland (Fig. 3a-c). Positive immunoreaction was also found in the secretory cell nucleus (SCN) of salivary gland (Fig. 3c).

Effect of dietary HgCl$_2$ GST in larvae

It has been shown that heavy metals are toxic to insects and affect growth rates and mortality of insects. So we chose HgCl$_2$ to test the expression of *AccGSTs4* mRNA and the response of larvae with HgCl$_2$ exposure. The third-instar larvae (L3) were fed with HgCl$_2$ of 0, 0.5, 5 and 10 mg/L, and treated for 72 h. In phenotype, there was non-significant difference between the HgCl$_2$ treated larvae and the untreated larvae. And we only chose the 10 mg/L

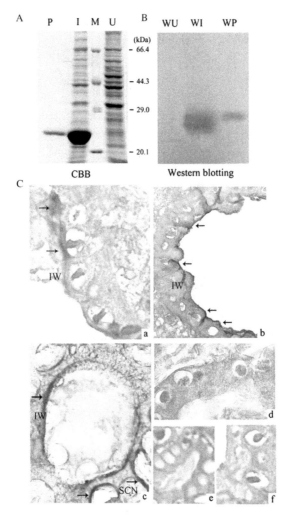

Fig. 3 Expression of the recAccGSTs4 and localization of native AccGSTs4. A. *Escherichia coli* cells were induced with isopropylthio-β-D-galactoside (IPTG) and the recAccGSTs4 was purified by MagneHis™ Protein Purification System. The purified protein was analyzed by 12% SDS-PAGE under reducing conditions. Lane U, uninduced cell; Lane M, Marker in kDa; Lane I, induced fraction; Lane P, purified recAccGSTs4. B. Western blotting analysis of recAccGSTs4. Proteins separated by 12% SDS-PAGE were transferred onto a PVDF membrane and probed with the anti-recAccGSTs4 antibody. Lane WU, uninduced cell; Lane WI, induced fraction; Lane WP, purified recAccGSTs4. C. Immunohistochemical staining of AccGSTs4 in the fourth-instar larvae sections with anti-recAccGSTs4 (a-c) and normal mouse serum (d-f). a, d, midgut; b, e, Malpighian tube; c, f, salivary gland. IW, intestinal wall; SCN, secretory cell nucleus. a, b, d, e and f, ×10; and c, ×20. (For color version, please sweep QR Code in the back cover)

$HgCl_2$ treated larvae to be compared with the untreated larvae in Fig. 4A. However, at the mRNA level, the transcripts of *AccGSTs4* were upregulated with the gradual increase of concentration and time, and peaked at 72 h with treatment of 0.5 mg/ml (Fig. 4B). At the protein level, the activity of GST was also progressively increased compared with controls (Fig. 4C). Meanwhile, the $HgCl_2$ treatment larvae had significantly higher mortality rates than larvae given the control diet (Fig. 4D).

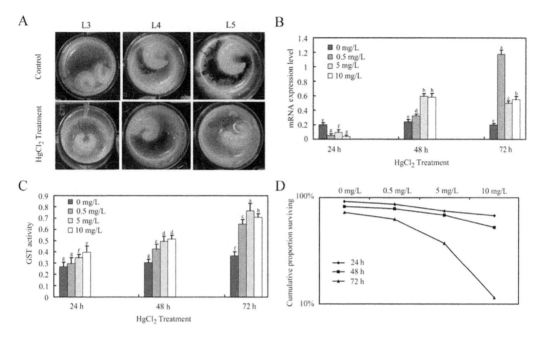

Fig. 4 Effect of dietary HgCl$_2$ on AccGSTs4 in larvae. A. The third-, the fourth- and the fifth-instar larvae which were fed on a HgCl$_2$ diet (10 mg/L) and a HgCl$_2$-free diet (control). B. The expression pattern of larvae *AccGSTs4* at the mRNA level under dietary HgCl$_2$ treatment with different concentrations: 0, 0.5, 5 and 10 mg/L. C. The GST activity of the third-, the fourth- and the fifth-instar larvae under dietary HgCl$_2$ treatment with four different concentrations. D. Mortality rates of third-, fourth- and fifth-instar larvae reared on four different treatments of dietary HgCl$_2$.

Disc diffusion assay under HgCl$_2$ stress

To evaluate the ability of recAccGSTs4 to defend against heavy metal stress, *E. coli* overexpressing AccGSTs4 with IPTG induction was exposed to HgCl$_2$. After overnight exposure, the killing zones of *E. coli* overexpressing AccGSTs4 around the HgCl$_2$-soaked filters on plates were smaller than that of the control bacteria (Fig. 5).

Characterization of the 5'-flanking region of *AccGSTs4*

To clarify the mechanism underlying *AccGSTs4* expression patterns, a 1,123 bp fragment of the 5'-flanking region was generated from the *A. cerana cerana* genome by I-PCR. Then a number of putative transcription factor-binding sites in the 5'-flanking region were predicted by using the TFSEARCH software (Fig. 6). Of these sites, cell factor 2-II (CF2-II), Hunchback (Hb) and Fushi Tarazu (Ftz), which were related to development and growth, were identified. In addition, several important transcription factors such as heat shock factors (HSF) and activating protein-1 (AP-1) required for regulating various environmental stresses were predicted.

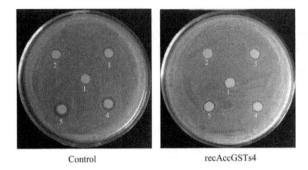

Control   recAccGSTs4

Fig. 5 Disc diffusion assay of recAccGSTs4 under HgCl₂ stress. AccGSTs4 was overexpressed in *E. coli*, and the bacteria transfected with pET-30a (+) only was used as controls. Filter discs soaked with different concentrations of HgCl₂ were placed on the agar plates with 24 h exposure at 37℃. 1, 0 mg/L; 2, 40 mg/L; 3, 60 mg/L; 4, 80 mg/L; 5, 100 mg/L.

```
                                                            HSF
                                     TCGATTCTTTAAAA AGAAA AAAAA      -1106
AATTCAAATTTTTACAAATAATATTTTACGTATAATCACATTTTGTTTATAATATCAAAATTACTAACATTTAAT   -1032
AATTCGAGATCTATATGAACGTTACCATGAAATTCATTTTTATCAAAATCAGGCGAAATCGTAATCACATATTTC  -957
TTGGGAATTACATCTTCCGGGAGACGTTTCTCGATTTCCATAGACTCATTGTAATGTGGACTTAAATTATTGAGA  -882
                                             Ftz
GGAACCTCGGCGTTGATCGCACGAAGATCTTTCGTCTTCATTAAGACT ATTGTAATTAAG ATAAATGGCCAGCTG -807
      HSF
GTCATTTCC AGAAC AAAGATTCTTTGTTTCAACCACACGATTTCTTTATCGTGTTACGAAGTTTAAATTATTGAA -732
AAACGAATCTTTTTTTTTTTTCTGCTCACTGAAGTCTCAACTATGCAAATCACTTAATCTCTTCTTCATAAGATT  -657
GAAAGATTATTGATTCGAAATATACGTTGAATATTATTTTCATTCAAAGGAAACTGGATGATTTATAACCATAAG  -582
GATGGGTTAGATATCTTCACGTCAAGAGATATACTCTGCTTACTTCACTTTGTCCGCACAAGTCGTCTTGATATT  -507
                            Hb
GGAAACAAAATTTTAAAAGTTGATTA GAAAAAAAAAA TTGTTGTCGTTATTTTGTACACATTGTATAAGTTAATA -432
                            CF2-II
AAAAGATAACTCGTGCTGAAATAATTGGTAGCTATACTAT GGTAAATATA TTAAAGTGAAAGTAATCAGAAGAGT -357
GGTCATGACTTCGTAGTAGGCCAGTGAGAGCATTCACAGTATTTTATAGACAGAGAGAATGTGTTCAATGATGAG  -282
          HSF
AAGGACAAG AGAAA TCCAGACTATGTGGAGAGGGAAATGTATAATTCTTATATAACGTTGGGAACCGGTTAATTC -207
     AP-1                                             HSF
AC AGTGACTTACA GTGTCTTATACTCGACTGTTTGTGTCTACTCCATTCATTTTC AGAAC AATAGTGTCAATAAA -132
GGAATAATTTATCATCAAGTTTCATAAAATTTTGTGTAAAATTGAAATTAGTGATTATTAACTCAAAAAAGAAAA  -57
AAGAAAAAAAAGGAAAAAGGTTTAAAACCATTACATTTCACAGATTAACAATA ATG
                    ↧                           ↧
                 Transcription start site     +1 Translation start site
```

Fig. 6 The nucleotide sequence and putative transcription factor-binding sites of the 5′-flanking region of *AccGSTs4*. The translation and transcription start sites are marked with arrows. The transcription factor-binding sites are boxed.

Discussion

Features of the AccGSTs4 deduced protein

The majority of studies on insect GSTs have focused on their role in detoxification of exogenous compounds, especially insecticides and plant allelochemicals and, more recently, their role in mediating oxidative stress responses [9]. In order to understand the role of honeybee GSTs in defending against oxidative stress, we identified a Sigma class GST, referred to *AccGSTs4*, from *A. cerana cerana*. The deduced amino acid sequence was homologous to Sigma class GSTs from other insects. The phylogenetic tree showed that AccGSTs4 was closely related to an insect Sigma class GST already reported. In the N terminal region of amino acid sequence, there were several conserved G-site residues, in particular a tyrosine residue at position 8, which is highly conserved and catalytically essential [27]. Earlier studies have reported that this residue is also present in the mammalian Sigma, Alpha, Mu and Pi class GSTs [28]. The predicted tertiary structure further revealed the conserved G-site residues of AccGSTs4. It is well known that the residues, which contribute to binding glutathione, comprise of a network of specific polar interactions between G-site residues and GSH that are either conserved or conservatively replaced between various classes [29].

Levels of *AccGSTs4* transcripts and its role in detoxification

Although age-dependent alteration of GST activity has been demonstrated in both vertebrates and invertebrates [30], there is a paucity of information on the expression of GSTs during the developmental lifespan of the insect. In the present study, the transcripts of *AccGSTs4* were all detected in the development from the larvae to the adult stages with the high expression in the fourth-instar larvae. The transcription factor-binding sites for CF2-II, Hb and Ftz which are respectively related to developmental regulation and the formation of body segments in the 5′ promoter sequence, were also identified [31-33]. It is generally presumed that the larvae are always exposed to external oxidative stress before sealed period, and the development of larvae can be influenced. Moreover, a study on microsomal oxidation in honeybees revealed that microsomal oxidases were active in larvae [34]. At the protein level, AccGSTs4 was highly present in the midgut, Malpighian tube and salivary gland of the fourth-instar larvae. Insect midgut and Malpighian tube are vital organs involved in digestion, nutrient absorption and interactions with various xenobiotics. The salivary gland, also known as the silk gland, plays important roles in silk production and pupae protection. Interestingly, the silk formation stopped at the end of the fifth-instar in Hymenoptera [35, 36]. Previously, Li et al [37] have studied that the most identified GSTs are expressed at higher levels in the larval stages than at other development stages in *Chironomus tentans* [37]. A Sigma class GST from *Choristoneura fumiferana* is expressed at low levels during the early larval stages but reached high levels in diapausing the second-instar and feeding the sixth-instar larvae [38]. Taken together, these results indicate that *AccGSTs4* may play a crucial role in the detoxification during the larval feeding stages, among other unidentified functions.

Regulation of *AccGSTs4* gene by the dietary HgCl$_2$

In this study, the expression levels of *AccGSTs4* mRNA were upregulated with treatment of dietary HgCl$_2$ compared with that in untreated honeybee larvae. In the 5′-flanking region of *AccGSTs4*, transcription factor-binding site for HSF is involved in regulation of environmental stress [39], while AP-1 takes part in mediating cellular responses to environmental stress through their ability to upregulate genes related to oxidative stress responses [7]. Furthermore, the enzyme activity of honeybee larvae GSTs increased along with a gradient of heavy metal pollution. It is currently believed that the metal with variable valency can enhance oxidative stress and lipid peroxidation in insects, especially when prooxidant components are present in their diet [40]. The enhanced level of GSTs in larvae probably results from adaptation to the metal contamination through inactivation and elimination of endogenous metabolites produced under metal treatment [41]. Similarly, a study noted that an increase in GST activity in the haemolymph of *Galleria mellonella* larvae treated with heavy metal [6]. Notably, when the concentration of $HgCl_2$ was higher, more larvae were died. Cabassi [42] has reported that low doses of some heavy metals such as Hg can improve immune system function, whereas higher doses are suppressive. The activity of recAccGSTs4 *in vitro* under $HgCl_2$ stress further demonstrated its role in the inactivation of accumulated toxic products during prooxidant activity of the metal. In outer sphere, when the larvae are fed with royal jelly, honey or pollen, which are produced by worker bees and may be polluted with heavy metal, their development will be affected and may be die under serious condition. Although the detoxification enzymes such as GSTs take part in these stress responses, the larvae still face death. So it is essential to protect the bees from heavy metal in rearing. Consequently, on the basis of these observations, we conclude that this Sigma class GST in honeybee larvae may play an important role in the detoxification system, and it will benefit for the breeding of honeybees.

References

[1] Sharma, R. K., Agrawal, M. Biological effects of heavy metals: an overview. J Environ Biol. 2005, 26, 301-313.

[2] Boyd, R. S. Heavy metal pollutants and chemical ecology: exploring new frontiers. J Chem Ecol. 2010, 36, 46-58.

[3] Zvereva, E., Serebrov, V., Glupov, V., Dubovskiy, I. Activity and heavy metal resistance of non-specific esterases in leaf beetle *Chrysomela lapponica* from polluted and unpolluted habitats. Comp Biochem Physiol C: Toxicol Pharmacol.2003, 135, 383-391.

[4] Van Ooik, T., Rantala, M. J., Saloniemi, I. Diet-mediated effects of heavy metal pollution on growth and immune response in the geometrid moth *Epirrita autumnata*. Environmental Pollution. 2007, 145, 348-354.

[5] Van Ooik, T., Pausio, S., Rantala, M. J. Direct effects of heavy metal pollution on the immune function of a geometrid moth, *Epirrita autumnata*. Chemosphere. 2008, 71, 1840-1844.

[6] Dubovskiy, L. M., Grizanova, E. V., Ershova, N. S., Rantala, M. J., Glipov, V. V. The effects of dietary nickel on the detoxification enzymes, innate immunity and resistence to the fungus *Beauveria bassiana* in the larvae of the greater wax moth *Galleria mellonella*. Chemosphere. 2011, 85 (1), 92-96.

[7] Limón-Pacheco, J., Gonsebatt, M. E. The role of antioxidants and antioxidant-related enzymes in protective responses to environmentally induced oxidative stress. Mutation Research. 2009, 674, 137-147.

[8] Park, E. J., Park, K. Induction of reactive oxygen species and apoptosis in BEAS-2B cells by mercuric chloride. Toxicol In Vitro. 2007, 21, 789-794.

[9] Enayati, A. A., Ranson, H., Hemingway, J. Insect glutathione transferases and insecticide resistance. Insect Mol Biol. 2005, 14, 3-8.

[10] Hayes, J. D., Flanagan, J. U., Jowsey, I. R. Glutathione transferase. Annu Rev Pharmacol Toxicol. 2005, 45, 51-88.

[11] Mannervik, B., Board, P. G., Hayes, J. D., Listowsky, I., Pearson, W. R. Nomenclature for

mammalian soluble glutathione transferases. Meth Enzymol. 2005, 401, 1-8.
[12] Lumjuan, N., Stevenson, B. J., Prapanthadara, L., Somboon, P., Brophy, P. M., Loftus, B. J., Severson, D. W., Ranson, H. The *Aedes aegypti* glutathione transferase family. Insect. Biochem Mol Biol. 2007, 37, 1026-1035.
[13] Ranson, H., Claudianos, C., Ortelli, F., Abgrall, C., Hemingway, J., Sharakhova, M. V., Unger, M. F., Collins, F. H., Feyereisen, R. Evolution of supergene families associated with insecticide resistance. Science. 2002, 298, 179-181.
[14] Board, P. G., Coggan, M., Chelvanayagam, G., et al. Identification, characterization, and crystal structure of the Omega class glutathione transferases. J Biol Chem. 2000, 75, 24798-24806.
[15] Corona, M., Robinson, G. E. Genes of the antioxidant system of the honey bee: annotation and phylogeny. Insect Mol Biol. 2006, 15, 687-701.
[16] Li, H. M., Buczkowski, G., Mittapalli, O., Xie, J., Wu, J., Westerman, R., Schemerhorn, B. J., Murdock, L. L., Pittendrigh, B. R. Transcriptomic profiles of *Drosophila melanogaster* third instar larval midgut and responses to oxidative stress. Insect Mol Biol. 2008, 17, 325-339.
[17] Lumjuan, N., McCarroll, L., Prapanthadara, L., Hemingway, J., Ranson, H. Elevated activity of an Epsilon class glutathione transferase confers DDT resistance in the dengue vector, *Aedes aegypti*. Insect Biochem Mol Biol. 2005, 35, 861-871.
[18] Ortelli, F., Rossiter, L. C., Vontas, J., Ranson, H., Hemingway, J. Heterologous expression of four glutathione transferase genes genetically linked to a major insecticide-resistance locus from the malaria vector *Anopheles gambiae*. Biochem J. 2003, 373, 957-963.
[19] Singh, S. P., Coronella, J. A., Benes, H., Cochrane, B. J., Zimniak, P. Catalytic function of *Drosophila melanogaster* glutathione *S*-transferase DmGSTS1-1 (GST-2) in conjugation of lipid peroxidation end products. Eur J Biochem. 2001, 268, 2912-2923.
[20] Huang, Y. F., Xu, Z. B., Lin, X. Y., Feng, Q. L., Zheng, S. C. Structure and expression of glutathione *S*-transferase genes from the midgut of the common cutworm, *Spodoptera litura* (Noctuidae) and their response to xenobiotic compounds and bacteria. J Insect Physiol. 2011, 57, 1033-1044.
[21] Peiren, N., De Graaf, D. C., Vanrobaeys, F., Danneels, E. L., Devreese, B., Beeumen, J. V., Jacobs, F. J. Proteomic analysis of the honey bee worker venom gland focusing on the mechanisms of protection against tissue damage. Toxicon. 2008, 52, 72-83.
[22] Li, X., Schuler, M. A., Berenbaum, M. R. Molecular mechanisms of metabolic resistance to synthetic and natural xenobiotics. Annu Rev Entomol. 2007, 52, 231-253.
[23] Michelette, E. R, Soares, A. E. Characterization of preimaginal developmental stages in Africanized honey bee workers (*Apis mellifera* L.). Apidologie. 1993, 24, 431-440.
[24] Silva, I. C., Message, D., Cruz, C. D., Campos, L. A. O., Sousa-Majer, M. J. Rearing Africanized honey bee (*Apis mellifera* L.) brood under laboratory conditions. Genet Mol Res. 2008, 8, 623-629.
[25] Livak, K. J, Schmittgen, T. D. Analysis of relative gene expression data using real-time quantitative PCR and the 2 (−Delta C (T)) method. Methods. 2001, 25, 402-408.
[26] Laemmli, U. K. Cleavage of structural proteins during the assembly of the head of bacteriophage T4. Nature. 1970, 227, 680-685.
[27] Yamamoto, K., Ichinose, H., Aso, Y., Banno, Y., Kimura, M., Nakashima, T. Molecular characterization of an insecticide-induced novel glutathione transferase in silkworm. Biochimica et Biophysica Acta. 2011, 1810, 420-426.
[28] Yamamoto, K., Zhang, P. B., Banno, Y., Fujii, H. Identification of a sigma-class glutathione *S*-transferase from the silkworm, *Bombyx mori*. J Appl Entomol. 2006, 130, 515-522.
[29] Winayanuwattikun, P., Ketterman, A. J. Catalytic and structural contributions for glutathione-binding residues in a Delta class glutathione *S*-transferase. Biochem J. 2004, 382, 751-757.
[30] Papadopoulos, A. I., Polemitou, I., Laifi, P., Yiangou, A., Tananaki, C. Glutathione *S*-transferase in the developmental stages of the insect *Apis mellifera macedonica*. Comp Biochem Physiol C: Toxicol Pharmacol. 2004, 139, 87-92.
[31] Hsu, T., Gogos, J. A., Kirsh, S. A., Kafatos, F. C. Multiple zinc finger forms resulting from

developmentally regulated alternative splicing of a transcription factor gene. Science. 1992, 257, 1946-1950.

[32] Stanojević, D., Hoey, T., Levine, M. Sequence-specific DNA-binding activities of the gap proteins encoded by hunchback and Krüppel in *Drosophila*. Nature. 1989, 341, 331-335.

[33] Florence, B., Handrow, R., Laughon, A. DNA-binding specificity of the Fushi Tarazu homeodomain. Mol Cell Biol. 1991, 11, 3613-3623.

[34] Gilbert, M. D., Wilkinson, C. F. Microsomal oxidases in the honey bee *Apis mellifera* L. Pestic Biochem Physiol. 1974, 4, 56.

[35] Silva-Zacarin, E. C., Silva De Moraes, R. L., Taboga, S. R. Silk formation mechanisms in the larval salivary glands of *Apis mellifera* (Hymenoptera: Apidae). J Biosci. 2003, 28, 753-764.

[36] Silva-Zacarin, E. C., Taboga, S. R., Silva De Moraes, R. L. Nuclear alterations associated to programmed cell death in larval salivary glands of *Apis mellifera* (Hymenoptera: Apidae). Micron. 2008, 39, 117-127.

[37] Li, X., Zhang, X., Zhang, J., Zhang, X., Starkey, S. R., Zhu, K. Y. Identification and characterization of eleven glutathione *S*-transferase genes from the aquatic midge *Chironomus tentans* (Diptera: Chironomidae). Insect Biochem Mol Biol. 2009, 39, 745-754.

[38] Feng, Q., Davey, K. G., Pang, A., Ladd, T. R., Retnakaran, A., Tomkins, B. L., Zheng, S., Palli, S. R. Developmental expression and stress induction of glutathione *S*-transferase in the spruce budworm, *Choristoneura fumiferana*. J Insect Physiol. 2001, 47, 1-10.

[39] Fernandes, M., Xiao, H., Lis, J. T. Fine structure analyses of the *Drosophila* and *Saccharomyces* heat shock factor-heat shock element interactions. Nucleic Acids Res. 1994, 22, 167-173.

[40] Felton, G. W., Summers, C. B. Antioxidant systems in insects. Arch Insect Biochem Physiol. 1995, 2, 187-197.

[41] Ahmad, S. Oxidative stress from environmental pollutants. Arch Insect Biochem Physiol. 1995, 29, 135-157.

[42] Cabassi, E. The immune system and exposure to xenobiotics in animals. Vet Res Commun. 2007, 31, 115-120.

2.6 A Glutathione *S*-Transferase Gene Associated with Antioxidant Properties Isolated from *Apis cerana cerana*

Shuchang Liu, Feng Liu, Haihong Jia, Yan Yan, Hongfang Wang, Xingqi Guo, Baohua Xu

Abstract: Glutathione *S*-transferases (GSTs) are an important family of multifunctional enzymes in aerobic organisms. They play a crucial role in the detoxification of exogenous compounds, especially insecticides, and protection against oxidative stress. Most previous studies of GSTs in insects have largely focused on their role in insecticide resistance. Here, we isolated a Theta class GST gene designated *AccGSTT1* from *Apis cerana cerana* and aimed to explore its antioxidant and antibacterial attributes. Analyses of homology and phylogenetic relationships suggested that the predicted amino acid sequence of *AccGSTT1* shares a high level of identity with the other hymenopteran GSTs and that it was conserved during evolution. Quantitative real-time PCR showed that *AccGSTT1* is most highly expressed in adult stages and that the expression profile of this gene is significantly altered in response to various abiotic stresses. These results were confirmed by using Western blotting analysis. Additionally, a disc diffusion assay showed that a recombinant AccGSTT1 protein may be roughly capable of inhibiting bacterial growth and that it reduces the resistance of *Escherichia coli* cells to multiple adverse stresses. Taken together, these data indicate that *AccGSTT1* may play an important role in antioxidant processes under adverse stress conditions.

Keywords: Glutathione *S*-transferase; *Apis cerana cerana*; Adverse stress; Antibacterial property; Antioxidant property

Introduction

Animals are constantly exposed to a variety of environmental stresses, including global warming, excessive pesticides use and heavy metal, which induce the production of reactive oxygen species (ROS). ROS, such as superoxide radicals, hydrogen peroxide and hydroxyl radicals, can then cause oxidative stress [1-3]. ROS are generally produced via metabolic processes in aerobic cells, which strive to maintain a dynamic balance under normal conditions. However, this balance can be disrupted by biotic and abiotic stressors, which can cause significant damage to cellular macromolecules [4, 5]. Thus, organisms have evolved a variety of complex antioxidant systems to avoid the oxidative damage that can be caused by ROS. One of the major lines of defence against ROS is the family of glutathione S-transferases (GSTs), in addition to other enzymes, such as thioredoxin peroxidases (Tpxs), catalases (CATs), glutathione reductase (GR) and superoxide dismutases (SODs) [6, 7].

Among antioxidant enzymes, the GSTs have been found in many plants and animals. As important members of antioxidant systems, GSTs are generally involved in the biosynthesis of leukotrienes, prostaglandins, testosterone, and progesterone and the degradation of tyrosine [8]. The enzymatic activity of GSTs is correlated with the function and metabolic activity of different tissues [9]. GSTs also have glutathione peroxidase activity, through which they perform an oxygen-detoxifying function [10] that results in the large-scale detoxification of xenobiotics [11, 12]. This activity is particularly important in invertebrates because they lack other glutathione peroxidases [13]. GSTs also play a significant role in decreasing ROS, detoxifying both endogenous and xenobiotic compounds, catalysing the combination of GSH and electrophilic compounds and protecting against oxidative stress [14-17]. Because GSTs perform these vital functions, researchers around the world who work in various fields are interested in the impact of GSTs. Many studies have now been performed in humans and mammals [18, 19]. and other common model organisms, but only limited research has been performed in honeybees. GSTs are likely to be a particularly important type of antioxidant enzyme in queen honeybees because they have much longer lifespans than worker bees [20-22]. It has been shown that the expression of GSTs can be induced in *Apis mellifera* by numerous chemicals, in response to which GSTs act as enzyme biomarkers for animals that are suffering from exposure to xenobiotics (such as thiamethoxam) [23, 24].

Similar to the mammalian GSTs, insect GSTs have been separated into six classes: Delta, Epsilon, Omega, Theta, Sigma and Zeta [17]. A large number of studies have reported on their role in mediating oxidative stress responses in insects [25-29]. In *Anopheles gambiae*, the expression of Epsilon GSTs increased in response to oxidative stress [30]. Kim et al. showed that Delta and Sigma class GSTs may play roles in oxidative stress responses as oxidative stress-inducible antioxidant enzymes in *Bombus ignitus* (*BiGSTD* and *BiGSTS*) [1]. Previous studies have reported that in *Apis cerana cerana*, a Zeta GST (GenBank ID: JN637474) had high antioxidant activity in response to exposure to oxidative stress [31] and that Omega GST (GenBank ID: JX456219) protected the supercoiled form of plasmid DNA from mutative damage caused by surplus ROS [5]. These reports fully support the notion that GSTs may play important roles in protecting insects from oxidative stress. In the context of immune responses, GST-Theta expression was significantly increased in *Macrobrachium rosenbergii* following exposure to viruses and bacteria, suggesting that this marker potentially plays a role in conferring immune protection [32]. Because GSTs have been shown to play important roles in protecting tissues against damage from oxidative stress, we decided to study another GST gene in insects. No previous study has described molecular characterization of the GSTT1 gene in insects or explored its differential influence on responses to oxidative stress or its potential novel functions in comparison to other GST genes.

Although the honeybee population in China has increased in recent years, Chinese native honeybees (*Apis cerana cerana*) have been under threat because of a worsening environment and competition with Western bee species. In addition, it is extremely difficult to farm this species. As a pollinator of flowering plants and an indigenous species in China, it is vital to the balance of regional ecologies and agricultural enterprise [33]. Thus, protecting the Chinese honeybee and gaining an understanding of its antioxidant systems and ROS defence mechanisms have become an extremely urgent issue. Therefore, we cloned and characterized a Theta class GST from the Chinese honeybees to gain further insight into the role that this gene plays. We analyzed responses to oxidative stress at both the messenger RNA (mRNA) and the protein level.

Materials and Methods

Insects and treatments

Chinese honeybee workers were obtained from the experimental apiary of Shandong Agricultural University (Taian, China). Bees were classified as larvae, pupae and adults according to their age, eye colour and shape after the queen had laid eggs [34, 35]. The larvae were divided into the first- to the fifth-instar larvae and pupae, which included prepupae (Po), white eyes (Pw), pink eyes (Pp), brown eyes (Pb) and dark eyes (Pd). Adult workers were collected after 10 days (A10) by marking the newly emerged bees (A1) with paint. These bees were selected for experimental treatments [35]. Untreated larvae, pupae and adult workers were frozen in liquid nitrogen as soon as they were collected to explore differences in developmental expression patterns. The adult samples that were used for the treatment experiments were reared under artificial conditions (34℃ and 70% relative humidity) [36], fed water and pollen and caged in eight treatment groups (each group contained 30 individuals). Samples were obtained from groups 1 and 2 (4℃ and 44℃, respectively) once per hour. Group 3 and its control group were injected with 20μl of hydrogen peroxide (50 μmol/L H_2O_2/worker) or PBS via an abdominal injection by using a sterile microscale needle. Group 4 was treated with $HgCl_2$ (3 mg/ml added to food) for 0, 3, 6, 9, 12 and 24 h after they were starved for approximately 2 days. One of three types of pesticides, including methomyl, pyridaben and cyhalothrin, was applied to groups 5-7, respectively, via fumigation at a final concentration of 10 μmol/L, 10 μmol/L and 20 μg/L, respectively [5]. Samples were collected from the groups that were treated with methomyl and pyridaben after 0, 0.5, 1, 2, 3 and 4 h. Samples from the other insecticide-treated group were collected at 0, 1, 2, 3 and 4 h. Group 8 was exposed to ultraviolet radiation (30 MJ/cm^2 UV) for 0, 1, 2, 3 and 4 h. All of the honeybees that were used in this research were collected at the indicated time points, and each sample contained tissues from five individuals. The samples were immediately flash-frozen in liquid nitrogen and stored at −80℃. Each treatment was repeated in triplicate (all bee materials were collected from bees that originated in the same colony).

RNA isolation and cDNA synthesis

Total RNA was extracted from whole tissues of Chinese honeybee workers by using Trizol reagent (Invitrogen, Carlsbad, CA, USA) and then rapidly stored at −80℃ until used. The potential influence of genomic DNA was eliminated by using RNase-Free Dnase I (Promega).

The first-strand complementary DNA (cDNA) was synthesized by using reverse transcriptase (TransGen Biotech, Beijing, China) according to the protocol of the manufacturer, and the cDNA was used as the template for PCR in subsequent gene cloning procedures and for quantitative real-time PCR.

Cloning of the ORF of *AccGSTT1* and data analysis

The internal region of *AccGSTT1* was isolated by using reverse transcription-PCR (RT-PCR) with a pair of primers (GZ1 and GZ2) (Table 1) that was designed based on regions of *GSTT1* that were found to be conserved in the sequences of *Apis mellifera*, *Apis dorsata*, *Bombus terrestris* and *Bombus impatiens*. The primers were synthesized by Biosune Biotechnological Company, Shanghai, China. All PCR products were purified and cloned into the pEASY-T3 vector (TransGen Biotech, Beijing, China) and then transformed into competent Escherichia coli cells (DH5α) for later sequencing.

Table 1 The primers in this study

Abbreviation	Primer sequence (5'—3')	Description
GZ1	CATGAGTCTTAAATTGTACTATG	cDNA sequence primer, forward
GZ2	CGATACAATGCGCACGTAAGA	cDNA sequence primer, reverse
DL1	GGACAACCAGTGACACAAGAA	Real-time PCR primer, forward
DL2	TGCTCCAATTCACATGCTCCA	Real-time PCR primer, reverse
β-s	TTATATGCCAACACTGTCCTTT	Standard control primer, forward
β-x	AGAATTGATCCACCAATCCA	Standard control primer, reverse
YH1	GGTACCATGAGTCTTAAATTGTAC	Protein expression primer, forward
YH2	GAGCTCTTAAATCTTGCTCTTTACATC	Protein expression primer, reverse

Conserved domains in AccGSTT1 and homologous AccGSTT1 protein sequences were compared to the homologous GSTT1 protein sequences from various species, which were obtained by using the Basic Local Alignment Search Tool from NCBI (http://blast.ncbi.nlm.nih.gov/Blast.cgi) [37]. These sequences were then aligned by using DNAMAN version 5.2.2 (Lynnon Biosoft, Quebec, Canada). The molecular weight and isoelectric point were also predicted by using this software. A phylogenetic analysis was constructed based on the sequences of homologous GSTT1 proteins of multiple species by using Molecular Evolutionary Genetic Analysis (MEGA) version 4.0 with the neighbour-joining (NJ) method and a Poisson correction model. Statistical analyses were performed by using the bootstrap method on 500 replicates.

Quantitative real-time PCR

Quantitative real-time PCR (qRT-PCR) was performed by using a SYBR® PrimeScript™ RT-PCR Kit (TaKaRa, Dalian, China) and a CFX96™ Real-Time PCR Detection System (Bio-Rad, Hercules, CA, USA) to quantify the expression of *AccGSTT1* following exposure to various adverse abiotic stresses and at different developmental stages. The specific primers

(DL1 and DL2) (Table 1) were designed based on the cDNA sequence of *AccGSTT1* and used to amplify a PCR product of 114 bp. The *β-actin* (no. HM640276) gene was selected as a reference gene to normalize the results of qRT-PCR experiments. The β-s and β-x primers were designed for the *β-actin* gene (Table 1). Before the experiments were performed, we first validated the primers of *β-actin* and *AccGSTT1*. The results showed that the efficiency of *β-actin* and *AccGSTT1* were 101.3% and 104.5%, respectively. The melting curves of the *AccGSTT1* gene showed a single peak, and the correlation coefficient (R^2) was 0.996. The PCR amplifications were optimized according to Kim et al. and mixed in a total volume of 20 μl that contained 10 μl of SYBR Premix Ex *Taq*, 1.6 μl of 1∶5 diluted cDNA, 0.4 μl each of the forward and reverse primers and 7.6 μl of PCR-grade water [38]. qRT-PCR was performed by using the following program: 40 cycles at 95 ℃ for 30 s, 55 ℃ for 25 s and 72 ℃ for 15 s followed by a melting curve analysis from 65 ℃ to 95 ℃ [34]. To guarantee the accuracy of the qPCR assays, three candidate reference genes (*ribosomal protein 49, elongation factor 1-Alpha* and *tbp-association factor*) were selected for repeat experiments [39]. Meanwhile, gene transcription levels were evaluated in three independent biological replicates, and three technical repeats were performed for each of the treatment groups. Relative expression data were automatically analyzed by using CFX Manager Software version 1.1 after the PCR program was run, and differences between the samples were determined by using one-way ANOVA followed by Duncan's multiple range tests with SPSS software version 17.0.

Protein expression, purification and polyacrylamide gel electrophoresis

Sense primers that introduced *Bam*H I restriction sites (YH1) and an antisense primer (YH2) harbouring *Sac* I restriction sites (Table 1) were designed based on the cDNA sequence of *AccGSTT1* and used to apply the open reading frame (ORF) of *AccGSTT1*. The obtained PCR products were purified and inserted into the transfer vector pET-30a (+), which was predigested by using the same enzymes.

We transformed the recombinant plasmid pET-30a (+)-*AccGSTT1* into *E. coli* BL21 (DE3) cells, and 4 ml of the produced bacterial solution was added to 400 ml of Luria-Bertani broth (LB) that contained 50 mg/ml kanamycin, and the cells were grown at 37 ℃. Approximately 1 h later (at 0.4-0.6 OD_{600}), the expression of the recombinant protein was induced by using 0.4 mmol/L isopropylthio-β-D-galactoside (IPTG) while shaking at 28 ℃ for approximately 6 h. The AccGSTT1 protein was purified according to the methods described in Zhang et al. [5]. Cells were harvested by centrifugation at 5,000 *g* for 15 min and then suspended in native lysis buffer (20 mmol/L Tris-HCl and 0.5 mol/L NaCl; pH 7.5). After the solution was sonicated and centrifuged again, the supernatant was added to a HisTrap^{TM} FF column (GE Healthcare, Uppsala, Sweden) that had been equilibrated by using lysis buffer for 15 min. The purified protein with the His-tag was eluted by using elution buffer (20 mmol/L Tris-HCl, 0.5 mol/L NaCl and 500 mmol/L imidazole; pH 7.5). Protein samples analyzed by using 12% sodium dodecyl sulphate polyacrylamide gel electrophoresis (SDS-PAGE) and the gels were then stained by using Coomassie brilliant blue and stored at −80 ℃ until processed by using Western blotting analysis.

Disc diffusion assay

A disc diffusion assay was performed in accordance with Chen et al. [40]. First, AccGSTT1 was overexpressed in *E. coli* cells that were plated on LB-kanamycin agar plates, and the cells were cultured at 37℃ for 1 h. Filter discs (6 mm diameter) were then placed on the surface of the agar, and the filters were soaked with different concentrations of cumene hydroperoxide (0, 30, 50, 80 and 100 mmol/L) or *t*-butyl hydroperoxide (0, 20, 40, 60 and 80 mmol/L). Then, the bacteria were cultured for another 24 h at 37℃. Afterwards, the inhibition zones around the discs were measured. Cells containing the pET-30a (+) vector were used as the controls and were treated with the same conditions.

Preparation of anti-AccGSTT1 serum and Western blotting analysis

The purified recombinant AccGSTT1 protein was mixed with an equal volume of sodium chloride (1 ml), and this solution was subcutaneously injected into 6-week-old pathogen-free female white mice to obtain the antiAccGSTT1 serum, as described in Chen et al. [40]. The mice were injected a total of four times at 1-week intervals plus a final booster injection. Blood was collected on the fourth day after the last injection (antigen only) via extraction through the eyeball. The blood was centrifuged at 3,000 g for 5 min and then allowed to clot at 4℃ overnight to generate the AccGSTT1 antibody, which was then stored at −80℃. α-tubulin was used as the reference gene for Western blotting analysis. Total protein was obtained from specimens at four different developmental stages (L3, PP, PD and A10) for each of the four treatment groups. The protein was extracted according to the methods described in Yan et al. [31]. Protein concentration was quantified by using a BCA Protein Assay Kit (Nanjing Jiancheng Bioengineering Institute, Nanjing, China) to equalize results across different samples. Protein samples were mixed with sample buffer, boiled for 5 min, separated by using 12% SDS-PAGE and then blotted onto polyvinylidine difluoride (PVDF) membranes. After the membranes were blotted, they were soaked in 5% (*w/V*) skim milk diluted in TBST for 1 h and then incubated overnight at 4℃ with primary antibodies (anti-AccGSTT1 serum) at a dilution of 1 ∶ 500 (*V/V*). The next day, the membranes were incubated at room temperature for 2 h with peroxidase-conjugated goat anti-mouse IgG (Dingguo, Beijing, China) diluted to 1 ∶ 2,000 (*V/V*) as the secondary antibody. The membranes were then washed in TBST. The results were detected by using a SuperSignal® West Pico Trial Kit (Thermo Scientific Pierce, IL, USA). The method used to perform Western blotting analysis was adopted from the methods described in Meng et al. [41].

Results

Cloning and characterization of *AccGSTT1*

To explore how GSTs respond to oxidative stress in honeybees, we identified the ORF of a gene encoding a Theta class GST (AccGSTT1) in *Apis cerana cerana*. The ORF of *AccGSTT1* (GenBank accession number: KR058858) contains 675 bp that encodes 225 amino acids (Fig. S1) with a predicted molecular mass of 26.614 kDa and a theoretical isoelectric point (pI) of 8.19. The amino acid sequence of AccGSTT1 has a high homology with the sequences of other insects, sharing 90.43%, 93.89%, 90.83%, 67.36% and 68.62% similarity to the GSTT1 proteins of *Apis mellifera*, *Apis dorsata*, *Apis florea*, *Bombus impatiens* and *Bombus terrestris*, respectively. The predicted AccGSTT1 protein contains features that are consistent with Theta class GSTs, specifically including two conserved domains at the N-terminus (domain Ⅰ) and C-terminus (domain Ⅱ) that are similar to those in other cytosolic GSTs. This suggests that Theta class GSTs may have been conserved over a considerable evolutionary period. Moreover, domain Ⅰ, which contains the majority of the residues that are involved in binding glutathione (e.g. the G-site) [42], is located between amino acids L_3 and F_{78}, while a putative substrate-binding site (H-site) in domain Ⅱ is positioned between Q_{92} and K_{216}. Additionally, a characteristic motif with the sequence IHPFQKVPAIEHND was also identified (Fig. 1).

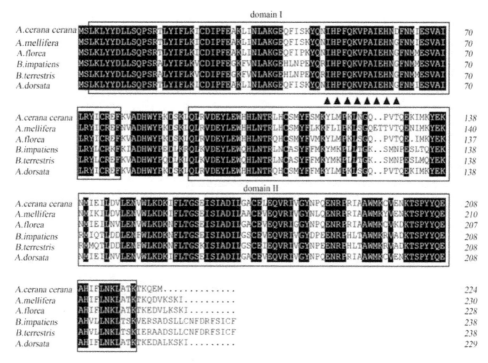

Fig. 1 Amino acid sequence alignment of AccGSTT1 with homologous genes in various species. Alignment of the amino acid sequence of AccGSTT1 with the sequences of known insect Theta class GSTs. The amino acid sequences used in the analysis included the sequences for *Apis mellifera* (GenBank accession no. XP_624692.1), *Apis dorsata* (GenBank accession no. XP_006620387.1), *Apis florea* (GenBank accession no. XP_003693859.1), *Bombus impatiens* (GenBank accession no. XP_003485831.1) and *Bombus terrestris* (GenBank accession no. XP_003397510.1). Identical amino acid residues are shaded in black, and the conserved N-terminal domain (domain I) and C-terminal domain (domain II) are boxed. The characteristic motif (IHPFQKVPAIEHND) is indicated by a black triangle.

The evolutionary distance of AccGSTT1 to homologous genes in closely related species was determined by using the amino acid sequences of the other insect GST gene members.

The results indicated that of the six classes of GSTs that have been identified in insects, the cloned protein could be classified into the Theta class and is closely related to genes in *Apis mellifera*, *Nasonia vitripennis* and *Bombyx mori*. The AccGSTT1 is therefore a divergent member of the GSTT class (Fig. 2).

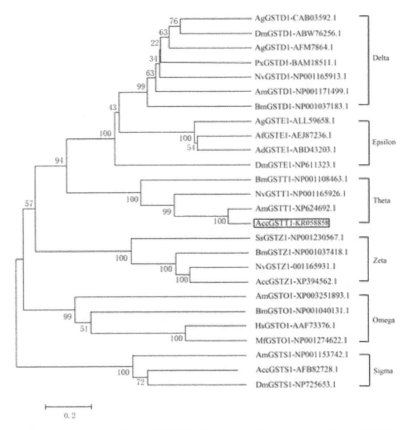

Fig. 2 Phylogenetic analysis of *AccGSTT1*. Phylogenetic analysis of AccGSTT1 amino acid sequences from various species based on the neighbour-joining (NJ) method with bootstrap values of 500 replicates. Six main groups of the GST superfamily are shown, and AccGSTT1 is boxed. The sequences were obtained from the NCBI database. The scale bar is equal to 0.02.

Quantitative RT-PCR of *AccGSTT1* at different developmental stages

The transcripts of *AccGSTT1* were detected at different honeybee life cycle stages, including L1, L3-L5, P0, PW, PP, PD, A1 and A10. As shown in Fig. 3a, the *AccGSTT1* gene is expressed at all developmental stages. Compared to other larval stages, the expression levels of *AccGSTT1* reached a significant peak during the third larval stage. In pupal stages, expression clearly trended downward between the P0 and PP stages, while expression levels in dark pigmented phase pupae were somewhat higher than in pupae examined during other phases. The second highest level of expression was seen during the A10 stage, when expression was higher than during the pupal and A1 stages.

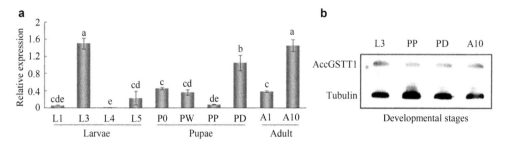

Fig. 3 The relative expression of *AccGSTT1* during different developmental stages. a. mRNA expression levels at different developmental stages. The expression profile of *AccGSTT1* was measured by using qPCR at different developmental stages, including larval (L1, L3-L5) and pupal (P0, PW, PP and PD) stages and adult workers (A1 and A10). The data are shown as the mean ± SE of three independent experiments (*n* = 5 for each experiment). The letters were determined by using Duncan's multiple range tests in SPSS software (version 17.0). b. Western blotting analysis of AccGSTT1 were aligned with the mRNA expression results. Total protein samples were obtained from L3, PP, PD, and A10 stage bees and immunoblotted by using anti-AccGSTT1 antibodies. α-Tubulin expression was used as an internal control for Western blotting analysis.

The expression profile of *AccGSTT1* in response to abiotic stress

To determine the transcriptional expression profile of AccGSTT1, we used qRT-PCR with one normalization step. We examined the relative results after exposing the honeybees to extreme temperatures, and the results showed that the expression of AccGSTT1 was higher in the treated honeybees than in the control group. In the treated group, expression reached a peak after 3 h at both 4 and 44 ℃ (Fig. 4a, b). H_2O_2 also caused an obvious decrease in gene expression that peaked at 3 h (Fig. 4c). The expression of *AccGSTT1* following treatment with $HgCl_2$ was strikingly upregulated but then slowly decreased beginning at 9 h (Fig. 4d). After bees were treated with different pesticides, The expression levels of *AccGSTT1* were significantly increased (Fig. 4e-g). In particular, its transcriptional level returned to normal following 4 h of exposure to cyhalothrin. Of all the treatments that were tested, UV treatment was the only one that resulted in decrease in expression (Fig. 4h). In addition, the results from three reference genes are shown in Fig. S2. These results all indicate that AccGSTT1 is likely to be involved in abiotic stress responses.

Overexpression and purification of AccGSTT1

The recombinant plasmid pET-30a (+)-*AccGSTT1* was transformed into *E. coli* BL21 (DE3) cells in which it was expressed as a histidine-fusion protein to further characterize the behaviour of the recombinant AccGSTT1 protein. By using 12% SDS-PAGE, the expression of the recombinant protein was induced by different IPTG concentrations (Fig. 5). After ultrasonic decomposition was induced by using an Ultrasonic Cell Disruptor, we found that the protein appeared both in inclusion bodies and cell supernatants. The fusion protein was then purified with HisTrap™ FF columns and reexamined by using 12% SDS-PAGE (Fig. S3).

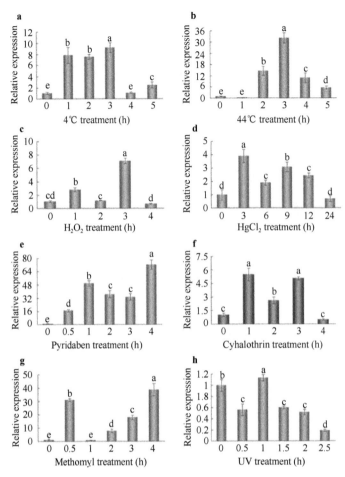

Fig. 4 Expression profile of AccGSTT1 under different abiotic stress conditions. The transcript patterns of AccGSTT1 were analyzed by using qRT-PCR of total RNA that was extracted from *Apis cerana cerana* that had been treated with various stressors. For the stress treatments, 10-day post-emergence worker bees were subjected to the following treatments: 4℃ (a), 44℃ (b), H_2O_2 (c), $HgCl_2$ (d), pyridaben (e), cyhalothrin (f), methomyl (g) or UV (h). The β-actin gene was used as an internal control. The data are shown as the mean ± SE of three independent experiments. The letters above the columns represent significant differences (P < 0.05) based on one-way ANOVA and Duncan's multiple range tests.

Western blotting analysis

To further verify the expression profile of *AccGSTT1* at different stages and in response to different abiotic stressors, we attempted to use Western blotting analysis. Proteins that were extracted from bees in the L3, PP, PD and A10 stages were detected. As shown in Fig. 3b, there was a higher level of AccGSTT1 expression in both the L3 and A10 stage bees than in the PP and PD stage bees. Samples obtained following treatment with 4℃, 44℃, pyridaben and UV were also processed for protein detection. The results showed that the level of protein expression of AccGSTT1 was consistent with its level of expression at the transcription level (Fig. 6).

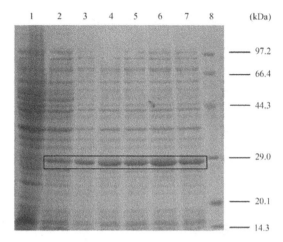

Fig. 5 Expression of the recombinant AccGSTT1 protein. The following are shown on an SDS-PAGE gel: lane 1, expression of AccGSTT1 without IPTG induction; lanes 2-7, expression of AccGSTT1 after induction with different IPTG concentrations (0.2, 0.4, 0.5, 0.6, 0.8, 1.0 mmol/L, in that order); and lane 8, protein molecular weight marker.

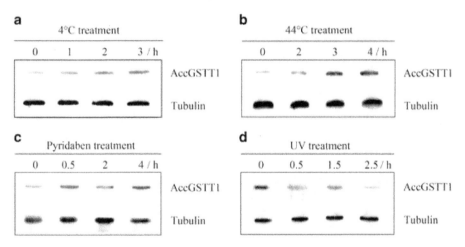

Fig. 6 Western blotting analysis of AccGSTT1. 10-day post-emergence worker bees were subjected to the following treatments: 4℃ (a), 44℃ (b), pyridaben (c) or UV (d). α-Tubulin was used as a reference gene for Western blotting analysis. Total protein samples were immunoblotted with anti-AccGSTT1 antibodies. The signals from the binding reactions were visualized by using HRP substrates.

Disc diffusion assays performed under various conditions

The results showed that after overnight exposure to various stressors, including t-butyl hydroperoxide and cumene hydroperoxide, bacteria that overexpressed AccGSTT1 were more susceptible to adverse stresses than control bacteria and specifically that the killing zones of the bacteria were larger than those of the control group (Fig. 7). Interestingly, this phenomenon is not the same as what has been reported for other GSTs. To verify our experimental results, we explored the mechanism underlying the inhibitory effects that

AccGSTT1 exerted on bacterial growth. Putative antimicrobial motifs were predicted by using the website http://aps.unmc.edu/AP/. The number of amino acid residues that were entered at a time was less than 50, and some positively charged amino acids were considered. We determined that there may be more than one antimicrobial motif in *AccGSTT1* and that the motif may have a neutral charge and antimicrobial activity (Fig. S1). The reason for this phenomenon can therefore be easily explained.

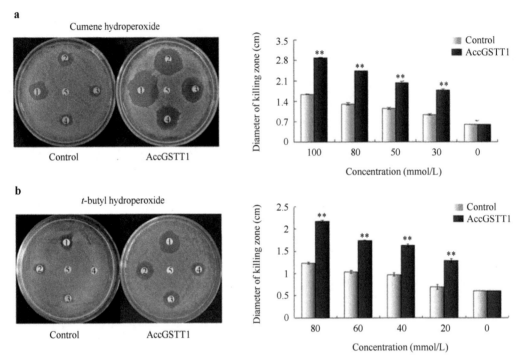

Fig. 7 Disc diffusion assays by using *E. coli* cells that overexpressed AccGSTT1. *E. coli* cells were transfected with a plasmid to overexpress AccGSTT1. Bacteria that were transfected with the empty pET-30a (+) vector were used as controls. Resistance to *t*-butyl hydroperoxide and cumene hydroperoxide was analyzed in bacteria that overexpressed AccGSTT1. The halo diameters of the killing zones were compared by using histograms. a. Cumene hydroperoxide was applied at a concentration of 0, 30, 50, 80 or 100 mmol/L, which are respectively shown from 5 to 1. b. *t*-butyl hydroperoxide was applied at a concentration of 0, 20, 40, 60 or 80 mmol/L. Independent-sample *t* tests were used for comparisons of baseline characteristics between the treatment and control groups. Significant differences are indicated with two asterisks at $P < 0.01$.

Discussion

The GST superfamily contains a subset of enzymes that are active in xenobiotic detoxification. This implies that they are important for helping insects to adapt to various environmental stresses [43]. In this study, we cloned a Theta class GST (*AccGSTT1*) in *Apis cerana cerana* and explored changes in its expression profile in response to oxidative stress. We also explored resistance in bacteria that overexpressed this gene to further analyze its function.

An analysis of the protein sequence of AccGSTT1 indicated that it shares a high level of identity with other typically conserved Theta class GSTs in insects such as *Apis dorsata* and *Apis mellifera*. Most cystolic GSTs contain two binding sites [37, 44]. In *Apis cerana cerana*, *GSTT1* also possesses two conserved domains, one of which contains a G-site, while the other contains an H-site. It has been noted that *AccGSTT1* may possess the same basic properties or functions of other GST family members. In addition, a phylogenetic tree indicated that its amino acid sequence underwent very conservative evolution and had the highest homology to GST1 in *A. mellifera*. Thus, AccGSTT1 is a typical Theta class GST gene, and it is quite appropriate for it to be named AccGSTT1.

To explore the putative roles that *AccGSTT1* might play, we analyzed *AccGSTT1* expression at different developmental stages and in response to various abiotic stressors. Its expression profile was stage-specific at every stage (Fig. 3a), indicating that it may be involved in processes during growth and development in *Apis cerana cerana*. Regarding developmental stages, previous studies have reported that some GSTs exhibit changes in transcription levels with age and that inducing their expression can increase longevity in flies [45]. In comparison, higher levels of expression were detected in the third larval stage and A10 adult stage bees. According to the available literature, some GST genes exhibit the highest level of expression during the larval or late stages [5], which is consistent with the above-mentioned expression levels that we observed for *AccGSTT1* during development. This phenomenon is most likely the result of an increase in sensitivity of larval and adult stage bees to xenobiotic components that are contained in food. It is also possible that they have a higher metabolic rate than pupae. Bees in pupal stages go through a quiescent stage (diapause) during which they actively intake and digest very little food. In comparison, larvae and adults may suffer more adversely from stress than pupae because their diet is restricted to the honey, pollen and glandular secretions that are provided by other colony members, and these sources may contain or result in the production of potentially toxic oxidative radicals, causing an increase in oxidative stress levels [46]. The adult bees can leave the hive to forage for food [47], and they therefore consume more energy and are exposed to worse environmental stresses [40]. The results of Western blotting analysis showed that protein expression levels were roughly equivalent to mRNA expression levels. Taken together, these data indicate that *AccGSTT1* may play an important role in preventing adverse damage during different developmental stages.

With regard for *AccGSTT1* expression levels under various abiotic and biotic stress conditions, GSTs are known to be an important component of oxidative stress response systems [48]. GSTs have primarily been studied because of their roles in insecticide resistance [17] and oxidative stress responses [21, 49]. The dehydrochlorination of the organochlorine insecticide DDT can be catalysed by Epsilon and Delta class insect GSTs [28, 50, 51]. Sigma and Theta class GSTs are associated with the metabolism and detoxification of lipid peroxidation, which are activities that have been suggested to be beneficial for protecting tissues from the products of oxidative stress [15, 52]. Zeta class GSTs may play key roles in catalysing the degradation of tyrosines and defending against oxidative stress [5, 53]. An Omega class GST in *Bombyx mori* has been shown to be induced by exposure to various environmental stresses [54]. There is also evidence that environmental conditions can induce oxidative stress [55, 56]. Various environmental conditions (e.g. temperature, H_2O_2 and pesticides) may induce insect GSTs to perform protective roles through a variety of defensive mechanisms. Extreme environmental conditions independently increase the expression of GSTs in the insect *Apis mellifera* [57]. To explore the function of *AccGSTT1*, the expression of *AccGSTT1* following exposure to adverse environments was analyzed. When samples were incubated at 4 and 44 ℃ or exposed to $HgCl_2$, methomyl, pyridaben or cyhalothrin, the expression of AccGSTT1 at the mRNA level was induced, as expected. The induction of *AccGSTT1* in response to excessive levels of pesticides activates the generation of ROS. Moreover, in *Apis mellifera*, challenging the samples with pesticides (e.g. deltamethrin or fipronil) promoted a dramatic increase in the enzymatic activity of GSTs [58]. We therefore suggest that *AccGSTT1* plays roles in the detoxification of xenobiotics and the defensive responses against oxidative stress [1]. It is worth mentioning that when bees were exposed to 4, 44 ℃, or treated with $HgCl_2$, the expression of *AccGSTT1* first increased and then decreased. It is possible that the *AccGSTT1* gene was initially activated to protect host cells, whereas its decline may reflect other stress-related genes playing important roles in later responses to cellular stress. The role of *AccGSTT1* might also be weakened as the level of stress increases [59]. Another possible cause for the rapid downregulation of *AccGSTT1* is that we applied only one pulse of stress, which was followed by a long lag phase. The subsequent decrease in stress may cause some unknown signalling pathway to directly downregulate the expression of *AccGSTT1*. Contrarily, we found that *AccGSTT1* was first downregulated and then upregulated following exposure to H_2O_2. GSTO2, in Chinese honeybees, displayed an expression profile that was consistent with this phenomenon [5, 60], and it has been verified that $GSTO_2$ plays a protective role by counteracting oxidative stress. Kim et al. showed that some GSTs in the bumblebee *Bombus ignitus* were significantly induced in all tissues, including the fat body, midgut, muscle and epidermis, after overload with H_2O_2 [38]. UV is understood to be a harmful environmental stressor that can produce oxidative stress in Chinese honeybees [34] and cause oxidative stress through the production of ROS, leading to damage to membrane lipids and proteins [61, 62]. The results of qRT-PCR showed that the transcription of AccGSTT1 was slightly induced by UV, which is in agreement with the results obtained in studies of other antioxidant-related genes. This finding suggests that treatment with UV might not produce enough ROS to induce gene expression in host cells [63]. It has also been claimed that *AccGSTT1* might play different roles in response to UV treatments or be involved in different signal transduction processes [64]. The observed changes in protein expression levels also indicate that *AccGSTT1* might be activated by some types of adverse stress. Considering the consistency of the expression profiles between mRNA and protein levels, the above results indicate that *AccGSTT1* plays a significant role in responses to adverse environmental situations in *Apis cerana cerana*. Coincidentally, in previous studies, Theta class GSTs have been implicated in the detoxification of lipid peroxidation products, suggesting that they play a protective role against oxidative stress [15, 52].

In this study, we transformed *AccGSTT1* into *E. coli* cells to perform a disc diffusion assay [65]. We cultured the cells in the experimental group (bacteria with *AccGSTT1*) and the control group until they reached the same cell density. It was noted that the bacteria that overexpressed *AccGSTT1* took a longer time to achieve the same quantity of cells as the bacteria that did not overexpress *AccGSTT1*. A reasonable explanation for this phenomenon is that *AccGSTT1* may be an inhibitor of bacterial growth. When cells were exposed to both cumene hydroperoxide and *t*-butyl hydroperoxide, the cells that overexpressed *AccGSTT1* showed reduced resistance to these treatments. However, these results were contrary to those of similar experiments performed by using other GST genes, such as *AccGSTO2* [5], *AccGSTZ1* [31] and *AccGSTD* [37]. Hence, we next sought to explore this question from an amino acid sequence perspective. We used a website to predict antimicrobial activity based on amino acid sequences [40]. A putative antimicrobial peptide was predicted in the middle of the full-length protein. Antimicrobial peptides are relatively small molecules that perform a broad spectrum of antimicrobial activities [66] that can be activated by allothigenes (e.g. bacteria or viruses) [67, 68]. They function via various mechanisms according to differences in the type of invading allothigene. It may therefore be assumed that the putative antimicrobial peptide in AccGSTT1 is activated after exposure to certain types of damage. This result explains this reversal phenomenon and suggests that AccGSTT1 may possess antibacterial properties that protect honeybees against bacterial damage. This is the first time that a putative antimicrobial peptide has been discovered and reported in an insect GST gene.

Taken together, the results of this study show that AccGSTT1 is a typical member of the GST superfamily and that it is potentially associated with the elimination of ROS and cellular resistance to oxidative stress. Our research, which includes analyses of structure and characteristics of the gene and its expression patterns and putative antioxidative and antibacterial functions, increases our understanding of the detailed mechanisms that underlie Theta class GST responses to adverse stressors. These data increase the probability that *AccGSTT1* may play an important role during this process. However, due to the limitations of these types of experiments, we could not determine the entire set of signalling pathways that *AccGSTT1* may participate in. To further understand the characteristics of *AccGSTT1*, further studies that can confirm the functional significance of this gene and its connections with other genes should be performed.

References

[1] Kim, B. Y., Hui, W. L., Lee, K. S., Wan, H., Yoon, H. J., Gui, Z. Z., Chen, S., Jin, B. R. Molecular cloning and oxidative stress response of a Sigma-class glutathione *S*-transferase of the bumblebee *Bombus ignitus*. Comp Biochem Physiol B: Biochem Mol Biol. 2011, 158 (1), 83-89.

[2] Mount, D. W. Reprogramming transcription. Nature. 1996, 383, 763-764.

[3] Dalton, T. P., Shertzer, H. G., Puga, A. Regulation of gene expression by reactive oxygen. Annu Rev Pharmacol Toxicol. 1999, 39, 67-101.

[4] Brennan, L. J., Keddie, B. A., Braig, H. R., Harris, H. L. The endosymbiont *Wolbachia pipientis* induces the expression of host antioxidant proteins in a *Aedes albopictus* cell line. PLoS One. 2008, 3 (5), e2083.

[5] Zhang, Y., Yan, H., Lu, W., Li, Y., Guo, X., Xu, B. A novel Omega-class glutathione *S*-transferase gene in *Apis cerana cerana*: molecular characterisation of *GSTO2* and its protective effects in oxidative stress. Cell Stress Chaperones. 2013, 18 (4), 503-516.

[6] Asayama, K., Hayashibe, H., Dobashi, K., Niitsu, T., Miyao, A., Kato, K. Antioxidant enzyme status and lipid peroxidation in various tissues of diabetic and starved rats. Diabetes Res. 1989, 12 (2), 85-91.

[7] Dubovskiy, I. M., Martemyanov, V. V., Vorontsova, Y. L., Rantala, M. J., Gryzanova, E. V., Glupov, V.

V. Effect of bacterial infection on antioxidant activity and lipid peroxidation in the midgut of *Galleria mellonella* L. larvae (Lepidoptera, Pyralidae). Comp Biochem Physiol C: Toxicol Pharmacol. 2008, 148 (1), 1-5.

[8] Hayes, J. D., Flanagan, J. U., Jowsey, I. R. Glutathione transferases. Annu Rev Pharmacol Toxicol. 2005, 45, 51-88.

[9] Aucoin, R. R., Philogène, B. J. R., Arnason, J. T. Antioxidant enzymes as biochemical defenses against phototoxin-induced oxidative stress in three species of herbivorous Lepidoptera. Arch Insect Biochem Physiol. 1991, 16, 139-152.

[10] Mannervik, B. Glutathione peroxidase. Methods Enzymol. 1985, 113, 490-495.

[11] Maxwell, D. M. The specificity of carboxylesterase protection against the toxicity of organophosphorus compounds. Toxicol Appl Pharmacol. 1992, 114 (2), 306-312.

[12] Salinas, A. E., Wong, M. G. Glutathione *S*-transferases. Current Med Chem. 1999, 6, 279-309.

[13] Ahmad, S., Duval, D. L., Weinhold, L. C., Pardini, R. S. Cabbage looper antioxidant enzymes: tissue specificity. Insect Biochem. 1991, 21 (5), 563-572.

[14] Grant, D. F., Matsumura, F. Glutathione *S*-transferase 1 and 2 in susceptible and insecticide resistant *Aedes aegypti*. Pestic Biochem Phys. 1989, 33, 132-143.

[15] Singh, S. P., Coronella, J. A., Benes, H., Cochrane, B. J., Zimniak, P. Catalytic function of *Drosophila melanogaster* glutathione *S*-transferase DmGSTS1-1 (GST-2) in conjugation of lipid peroxidation end products. Eur J Biochem. 2001, 268 (10), 2912-2923.

[16] Huang, Y. F., Xu, Z. B., Lin, X. Y., Feng, Q. L., Zheng, S. C. Structure and expression of glutathione *S*-transferase genes from the midgut of the *Common cutworm*, *Spodoptera litura* (Noctuidae) and their response to xenobiotic compounds and bacteria. J Insect Physiol. 2011, 57 (7), 1033-1044.

[17] Enayati, A. A., Ranson, H., Hemingway, J. Insect glutathione transferases and insecticide resistance. Insect Mol Biol. 2005, 14 (1), 3-8.

[18] Tew, K. D. Glutathione-associated enzymes in anticancer drug resistance. Cancer Res. 1994, 54, 4313-4320.

[19] Hayes, J. D., Pulford, D. J. The glutathione *S*-transferase supergene family: regulation of GST and the contribution of the isoenzymes to cancer chemoprotection and drug resistance. Crit Rev Biochem Mol Biol. 1995, 30 (6), 445-600.

[20] Page Jr, R. E., Peng, C. Y. Aging and development in social insects with emphasis on the honey bee, *Apis mellifera* L. Exp Gerontol, 2001, 36, 695-711.

[21] Corona, M., Robinson, G. E. Genes of the antioxidant system of the honey bee: annotation and phylogeny. Insect Mol Biol. 2006, 15 (5), 687-701.

[22] Aurori, C. M., Buttstedt, A., Dezmirean, D. S., Mărghitaş, L. A., Moritz, R. F., Erler, S. What is the main driver of ageing in long-lived winter honeybees: antioxidant enzymes, innate immunity, or vitellogenin J Gerontol A Biol Sci Med Sci. 2013, 69 (6), 633-639.

[23] Stone, D., Jepson, P., Laskowski, R. Trends in detoxification enzymes and heavy metal accumulation in ground beetles (Coleoptera: Carabidae) inhabiting a gradient of pollution. Comp Biochem Physiol C: Toxicol Pharmacol. 2002, 132 (1), 105-112.

[24] Badiou-Bénéteau, A., Carvalho, S. M., Brunet, J. L., Carvalho, G. A., Buleté, A., Giroud, B., Belzunces, L. P. Development of biomarkers of exposure to xenobiotics in the honey bee *Apis mellifera*: application to the systemic insecticide thiamethoxam. Ecotoxicol Environ Saf. 2012, 82, 22-31.

[25] Clark, A. G., Shamaan, N. A., Sinclair, M. D., Dauterman, W. C. Insecticide metabolism by multiple glutathione *S*-transferases in two strains of the house fly *Musca domestica* (L). Pest Biochem Physiol. 1986, 25 (2), 169-175.

[26] Wang, J. Y., McCommas, S., Syvanen, M. Molecular cloning of a glutathione *S*-transferase overproduced in an insecticide-resistant strain of the housfly (*Musca domestica*). Mol Gen Genet. 1991, 227 (2), 260-266.

[27] Fournier, D., Bride, J. M., Poirie, M., Bergé, J. B., Plapp Jr, F. W. Insect glutathione *S*-transferases.

[28] Biochemical characteristics of the major forms from houseflies susceptible and resistant to insecticides. J Biol Chem. 1992, 267 (3), 1840-1845.

[28] Ranson, H., Rossiter, L., Ortelli, F., Jensen, B., Wang, X., Roth, C. W., Collins, F. H., Hemingway, J. Identification of a novel class of insect glutathione S-transferases involved in resistance to DDT in the malaria vector *Anopheles gambiae*. Biochem J. 2001, 359 (pt2), 295-304.

[29] Sawicki, R., Singh, S. P., Mondal, A. K., Benes, H., Zimniak, P. Cloning, expression and biochemical characterization of one Epsilon-class (*GST-3*) and ten Delta-class (*GST-1*) glutathione S-transferases from *Drosophila melanogaster*, and identification of additional nine members of the Epsilon class. Biochem J. 2003, 370 (Pt 2), 661-669.

[30] Ding, Y. C., Hawkes, N., Meredith, J., Eggleston, P., Hemingway, J., Ranson, H. Characterization of the promoters of Epsilon glutathione transferases in the mosquito *Anopheles gambiae* and their response to oxidative stress. Biochem J. 2005, 387 (pt3), 879-888.

[31] Yan, H., Meng, F., Jia, H., Guo, X., Xu, B. The identification and oxidative stress response of a Zeta class glutathione S-transferase (GSTZ1) gene from *Apis cerana cerana*. J Insect Physiol. 2012, 58 (6), 782-791.

[32] Arockiaraj, J., Gnanam, A. J., Palanisamy, R., et al. A cytosolic glutathione S-transferase, GST-Theta from freshwater prawn *Macrobrachium rosenbergii*: molecular and biochemical properties. Gene. 2014, 546 (2), 437-442.

[33] Yang, G. Harm of introducing the western honeybee *Apis mellifera* L. to the Chinese honeybee *Apis cerana* F. and its ecological impact. Acta Entomol Sin. 2005, 3, 401-406.

[34] Jia, H., Sun, R., Shi, W., Yan, Y., Li, H., Guo, X., Xu, B. Characterization of a mitochondrial manganese superoxide dismutase gene from *Apis cerana cerana* and its role in oxidative stress. J Insect Physiol. 2014, 60:68-79.

[35] Meng, F., Zhang, Y., Liu, F., Guo, X., Xu, B. Characterization and mutational analysis of Omega-class GST (*GSTO1*) from *Apis cerana cerana*, a gene involved in response to oxidative stress. PLoS One. 2014, 9 (3), e93100.

[36] Alaux, C., Ducloz, F., Crauser, D., Le Conte Y. Diet effects on honeybee immunocompetence. Biol Lett. 2010, 6 (4), 562-565.

[37] Yan, H., Jia, H., Wang, X., Gao, H., Guo, X., Xu, B. Identification and characterization of an *Apis cerana cerana* Delta class glutathione S-transferase gene (*AccGSTD*) in response to thermal stress. Naturwissenschaften. 2013, 100 (2), 153-163.

[38] Kim, Y. I., Kim, H. J., Kwon, Y. M., et al. Modulation of MnSOD protein in response to different experimental stimulation in *Hyphantria cunea*. Comp Biochem Physiol B: Biochem Mol Biol. 2010, 157 (4), 343-350.

[39] Lourenco, A. P., Mackert, A., Cristino, A. S., Simoes, Z. L. P. Validation of reference genes for gene expression studies in the honey bee, *Apis mellifera*, by quantitative real-time RT-PCR. Apidologie. 2008, 39 (3), 372-385.

[40] Chen, X., Yao, P., Chu, H. L., Guo, X., Xu, B. Isolation of arginine kinase from *Apis cerana cerana* and its possible involvement in response to adverse stress. Cell Stress Chaperones. 2015, 20 (1), 169-183.

[41] Meng, F., Kang, M., Liu, L., Luo, L., Xu, B., Guo, X. Characterization of the TAK1 gene in *Apis cerana cerana* (*AccTAK1*) and its involvement in the regulation of tissue specific development. BMB Rep. 2011, 44, 187-192.

[42] Wilce, M. C. J., Parker, M. W. Structure and function of glutathione S-transferases. Biochem Biophys Acta. 1994, 1205, 1-18.

[43] Ranson, H., Claudianos, C., Ortelli, F., Abgrall, C., Hemingway, J., Sharakhova, M. V., Unger, M. F., Collins, F. H., Feyereisen, R. Evolution of supergene families associated with insecticide resistance. Science. 2002, 298 (5591), 179-181.

[44] Dirr, H., Reinemer, P., Huber R. X-ray structures of cytosolic glutathione S-transferases, implications for protein architecture, substrate recognition and catalytic function. Eur J Biochem. 1994, 220 (3),

645-661.

[45] Zou, S., Meadows, S., Sharp, L., Jan, L. Y., Jan, Y. N. Genome-wide study of aging and oxidative stress response in *Drosophila melanogaster*. Proc Natl Acad Sci USA. 2000, 97 (25), 13726-13731.

[46] Krishnan, N., Sehnal, F. Compartmentalization of oxidative stress and antioxidant defense in the larval gut of *Spodoptera littoralis*. Arch Insect Biochem Physiol. 2006, 63 (1), 1-10.

[47] Ament, S. A., Corona, M., Pollock, H. S., Robinson, G. E. Insulin signaling is involved in the regulation of worker division of labor in honey bee colonies. Proc Natl Acad Sci USA. 2008, 105 (11), 4226-4231.

[48] Yan, H., Jia, H., Gao, H., Guo, X., Xu, B. Identification, genomic organization, and oxidative stress response of a Sigma class glutathione S-transferase gene (*AccGSTS1*) in the honey bee, *Apis cerana cerana*. Cell Stress Chaperones. 2012, 18 (4), 415-426.

[49] Li, H. M., Buczkowski, G., Mittapalli, O., Xie, J., Wu, J., Westerman, R., Schemerhorn, B. J., Murdock, L. L., Pittendrigh, B. R. Transcriptomic profiles of *Drosophila melanogaster* third instar larval midgut and responses to oxidative stress. Insect Biochem Mol Biol. 2008, 17 (4), 325-339.

[50] Tang, A. H., Tu, C. P. Biochemical characterization of *Drosophila* glutathione S-transferases D1 and D21. J Biol Chem. 1994, 269 (45), 27876-27884.

[51] Lumjuan, N., McCarroll, L., Prapanthadara, L. A., Hemingway, J., Ranson, H. Elevated activity of an Epsilon class glutathione transferase confers DDT resistance in the dengue vector, *Aedes aegypti*. Insect Biochem Mol Biol. 2005, 35 (8), 861-871.

[52] Yamamoto, K., Zhang, P., Miake, F., Kashige, N., Aso, Y., Banno, Y., Fujii, H. Cloning, expression and characterization of theta-class glutathione S-transferase from the silkworm, *Bombyx mori*. Comp Biochem Physiol B: Biochem Mol Biol. 2005, 141 (3), 340-346.

[53] Claudianos, C., Ranson, H., Johnson, R. M., Biswas, S., Schuler, M. A., Berenbaum, M. R., Feyereisen, R., Oakeshott, J. G. A deficit of detoxification enzymes: pesticide sensitivity and environmental response in the honeybee. Insect Mol Biol. 2006, 15 (5), 615-636.

[54] Yamamoto, K., Teshiba, S., Shigeoka, Y., Aso, Y., Banno, Y., Fujiki, T., Katakura, Y. Characterization of an Omega-class glutathione S-transferase in the stress response of the silkmoth. Insect Mol Biol. 2011, 20 (3), 379-386.

[55] Lushchak, V. I. Environmentally induced oxidative stress in aquatic animals. Apidologie. 2011, 101 (1), 13-30.

[56] Kottuparambila, S., Shinb, W., Brownc, M. T., Han, T. UV-B affects photosynthesis, ROS production and motility of the freshwater flagellate, *Euglena agilis* Carter. Aquat Toxicol. 2012, 122-123, 206-213.

[57] Papadopoulos, A. I., Polemitou, I., Laifi, P., Yiangou, A., Tananaki, C. Glutathione S-transferase in the insect *Apis mellifera macedonica* kinetic characteristics and effect of stress on the expression of GST isoenzymes in the adult worker bee. Comp Biochem Physiol C: Toxicol Pharmacol. 2004, 139 (1-3), 93-97

[58] Carvalho, S. M., Belzunces, L. P., Carvalho, G. A., Brunet, J. L., Badiou-Bénéteau, A. Enzymatic biomarkers as tools to assess environmental quality: a case study of exposure of the honeybee *Apis mellifera* to insecticides. Environ Toxicol Chem. 2013, 32 (9), 2117-2124.

[59] Yan, Y., Zhang, Y., Huaxia, Y., Wang, X., Yao, P., Guo, X., Xu, B. Identification and characterisation of a novel 1-Cys thioredoxin peroxidase gene (*AccTpx5*) from *Apis cerana cerana*. Comp Biochem Physiol B: Biochem Mol Biol. 2014, 172-173, 39-48.

[60] Yao, P., Chen, X., Yan, Y., Liu, F., Zhang,Y., Guo, X., Xu, B. Glutaredoxin 1, glutaredoxin 2, thioredoxin 1, and thioredoxin peroxidase 3 play important roles in antioxidant defense in *Apis cerana cerana*. Free Radic Biol Med. 2014, 68, 335-346.

[61] Ravanat, J. L., Douki, T., Cader, J. Direct and indirect effects of UV radiation on DNA and its components. J Photochem Photobiol B. 2001, 63 (1-3), 88-102.

[62] Cadet, J., Sage, E., Douki, T. Ultraviolet radiation-mediated damage to cellular DNA. Mutat Res. 2005, 571 (1-2), 3-17.

[63] Yao, P., Hao, L., Wang, F., Chen, X., Yan, Y., Guo, X., Xu, B. Molecular cloning, expression and antioxidant characterisation of a typical thioredoxin gene (*AccTrx2*) in *Apis cerana cerana*. Gene. 2013, 527 (1), 33-41.

[64] Yao, P., Lu, W., Meng, F., Wang, X., Xu, B., Guo, X. Molecular cloning, expression and oxidative stress response of a mitochondrial thioredoxin peroxidase gene (*AccTpx-3*) from from *Apis cerana cerana*. J Insect Physiol. 2013, 59 (3), 273-282.

[65] Burmeister, C., Lüersen, K., Heinick, A., Hussein, A., Domagalski, M., Walter, R. D., Liebau, E. Oxidative stress in *Caenorhabditis elegans*: protective effects of the Omega class glutathione transferase (GSTO-1). FASEB J. 2008, 22 (2), 343-354.

[66] Nappi, A. J., Ottaviani, E. Cytotoxicity and cytotoxic molecules in invertebrates. Bioessays. 2000, 22 (5), 469-480.

[67] An, C., Ishibashi, J., Ragan, E. J., Jiang, H., Kanost, M. R. Functions of *Manduca sexta* hemolymph proteinases HP6 and HP8 in two innate immune pathways. J Biol Chem. 2009, 284 (29), 19716-19726.

[68] Jiang, H., Vilcinskas, A., Kanost, M. R. Immunity in lepidopteran insects. Adv Exp Med Biol. 2010, 708, 181-204.

2.7 The Identification and Oxidative Stress Response of a Zeta Class Glutathione *S*-transferase (GSTZ1) Gene from *Apis cerana cerana*

Huiru Yan, Fei Meng, Haihong Jia, Xingqi Guo, Baohua Xu

Abstract: Glutathione-*S*-transferases (GSTs) play an important role in protecting organisms against the toxicity of reactive oxygen species (ROS). However, no information is available for GSTs in the Chinese honeybee (*Apis cerana cerana*). In this study, we isolated and characterized a Zeta class GST gene (*AccGSTZ1*) from the Chinese honeybee. This gene is present in a single copy and harbors five exons. The deduced amino acid sequence of AccGSTZ1 shared high sequence identity with homologous proteins and contained the highly conserved features of this gene family. The temporal and spatial expression profiles of *AccGSTZ1* showed that *AccGSTZ1* was highly expressed in the fourth-instar larvae during development, and the mRNA level of *AccGSTZ1* was higher in the epidermis than that in other tissues. The expression pattern under oxidative stress revealed that the transcription of *AccGSTZ1* was significantly upregulated by external factors, such as temperature challenges and H_2O_2 treatment. The characterization of the purified protein revealed that AccGSTZ1 had low glutathione-conjugating activity, but the recombinant AccGSTZ1 protein displayed high antioxidant activity under oxidative stress. These data suggest that AccGSTZ1 is an oxidative stress-inducible antioxidant enzyme that plays an important role in the protection against oxidative stress and may be of critical importance for the survival of the honeybees.

Keywords: Glutathione-*S*-transferase; Antioxidant activity; Oxidative stress; *Apis cerana cerana*

Introduction

The Chinese honeybee, *Apis cerana cerana*, is an important species that plays a critical role in the balance of regional ecologies and agricultural economics as a pollinator of flowering plants. However, farming this species is extremely difficult due to the environmental temperature challenge. Yang et al. recently demonstrated that induced thermal stress was associated with the generation of increased reactive oxygen species (ROS) and the production of oxidative damage [1]. Thus, understanding the antioxidant system of the honeybee and its

ROS defense mechanisms has become a primary issue for this industry.

ROS, including superoxide anions ($O_2^{·-}$), hydrogen peroxide (H_2O_2) and hydroxyl radicals (·OH), are generated naturally during aerobic metabolism. In normal conditions, there is a balance between the generation of ROS and antioxidant processes, but exogenous stressors, such as pro-oxidants, heavy metal, pesticides and biotic infections, can break this balance and cause excessive production or accumulation of ROS [2, 3]. The excess ROS can result in lipid peroxidation which disrupts cell membrane fluidity and can lead to apoptosis [4]. The sperm storage of *Apis mellifera* is also affected by ROS [5]. Oxidative damage to proteins can range from specific amino acid modifications and peptide scission to loss of enzyme activity [6, 7]. Surplus ROS can also lead to DNA damage in the form of mutations, base deletions, degradation, single-strand scission and rearrangements [8]. To prevent ROS-mediated damage, complex enzymatic and non-enzymatic defense systems have evolved. The dominant antioxidant enzymes include glutathione *S*-transferase (GST), glutathione peroxidase (GPX), catalase (CAT), superoxide dismutase (SOD) and glutathione reductase (GR) [9, 10].

Among the antioxidant enzymes, GSTs are the major detoxifying enzymes that are primarily localized to the cytosol of the cells in all living organisms [11], and play pivotal roles in oxidative stress resistance and in detoxifying endogenous and xenobiotic compounds in cells [12]. Considering the significant functions in which they are involved in, GSTs have been predominantly studied in mammals and humans. Insect GSTs, which have been grouped into six GST classes, Delta, Epsilon, Omega, Sigma, Theta and Zeta [13], are of particular interest because of their role in insecticide resistance [14, 15]. All of the GSTs directly implicated in insecticide metabolism to date belong to the Delta and Epsilon classes, which are believed to be insect specific [16, 17]. Most recently, insect GSTs have been studied for their role in mediating oxidative stress responses [18]. Kampkotter et al. (2003) showed that overexpressing an *Onchocerca volvulus* Omega class GST (OvGST3) in *Caenorhabditis elegans* increased resistance to oxidative stress in the transgenic worm lines, and the authors speculated that OvGST3 plays an important role in the protection of the parasite against reactive oxygen species derived from the immune system of the host [19]. Putative binding sites and regulatory/response elements involved in the induction of GST expression in response to oxidative stress have been found in Epsilon and Delta GST gene promoters from anophelines, supporting the antioxidant physiological role of some GSTs [20, 21]. Increased GST activity was also found in *Helicoverpa armigera* adults under oxidative stress resulted from ultraviolet light [22].

In *A. mellifera*, proteins involved in oxidative stress, such as vitellogenin and juvenile hormone, have been reported by Corona et al. (2007), but few are available in *A. cerana cerana* [23]. Given the important roles of GSTs in protecting tissues against oxidative damage and oxidative stress [16, 18], we decided to study the oxidative stress response of GSTs in *A. cerana cerana*. Our studies were undertaken to identify the *A. cerana cerana* GST supergene family in response to oxidative stress through molecular cloning studies and analysis of the responses to oxidative stress at both the mRNA and protein levels. Our work provides a better understanding of the role that a Zeta class GST plays in the defense against oxidative stress in *A. cerana cerana* and contributes to our knowledge of this versatile superfamily.

Materials and Methods

Insects and treatments

The Chinese honeybees (*A. cerana cerana*) used in this study were maintained at the experimental apiary of Shandong Agricultural University (Taian, China). Larval and pupal worker honeybees were generally classified according to age, eye color and shape. The whole bodies of the first to the fifth larval instars (L1-L5) and pupae including prepupae (PP), white eyes (Pw), pink eyes (Pp), brown eyes (Pb) and dark eyes (Pd) were obtained from the hive, while the adult worker honeybees (A10-ten days post-emergence and A30-thirty days post-emergence) were collected at the entrance of the hive. The brain, epidermis, muscle and midgut of adults (15 days) were dissected on ice, freshly collected and used to examine tissue specific expression. Two-week-old adult worker honeybees, collected randomly from combs in outdoor beehives, were caged in eight groups of 40 individuals and reared on an artificial diet in an incubator with 60% relative humidity at 34℃ under a 24 h dark regimen.

Groups 1-5 were placed at 4, 15, 25, 34 or 43℃, respectively. The bees kept at 34℃ served as a control. Worker bees in group 6 were injected with 20 μl (50 μmol/L of H_2O_2/worker) H_2O_2 between the first and the second abdominal segments by using a sterile microscale needle. Bees in group 7 injected with phosphate buffered saline (PBS) (20 μl/worker) were the injection controls. Prior to injection, the worker bees were placed on ice for 5 min. Control bees in group 8 were fed a pollen-and-sucrose solution only. The whole bodies of the bees were flash-frozen in liquid nitrogen at the appropriate time and stored at −80℃ until used.

Primers and PCR amplification conditions

Primers and PCR amplification conditions used are listed in Table 1 and Table 2, respectively.

Table 1 Details of the primers used in this study

Abbreviation	Primer sequence (5'—3')	Description
GZ1	GCGGAGTTCTTGTTCGTGGAG	cDNA sequence primer, forward
GZ2	CAGGCTGATTATTTGGATGAGCAG	cDNA sequence primer, reverse
5W	CGATGTGGTCTGGTTTCTTCC	5' RACE reverse primer, outer
5N	GTACCTGCTCCATTGGATTAATCTC	5' RACE reverse primer, inner
3W	GAAAAGAGCACGCGTCAGAG	3' RACE forward primer, outer
3N	CGTAAAAAGGAATGGGCACAAC	3' RACE forward primer, inner
AAP	GGCCACGCGTCGACTAGTAC(G)$_{14}$	Abridged anchor primer
AUAP	GGCCACGCGTCGACTAGTAC	Abridged universal amplification primer
B26	GACTCTAGACGACATCGA(T)$_{18}$	3' RACE universal adaptor primer
B25	GACTCTAGACGACATCGA	3' RACE universal primer
QC1	GTCTGGCAATCGAATAAAAGG	Full-length cDNA sequence primer, forward
QC2	TAAATATATGACAAATGTTTCTACAAAAT	Full-length cDNA sequence primer, reverse

Continued

Abbreviation	Primer sequence (5'—3')	Description
N1	GCGGAGTTCTTGTTCGTGGAG	Genomic sequence primer, forward
N2	CTCTAGTAATCCAATGTTGTGCCC	Genomic sequence primer, reverse
N3	GCACTTCATATTGACAATCACAC	Genomic sequence primer, forward
N4	CACCTTTTGCCACTATAGAACTAC	Genomic sequence primer, reverse
BDL1	CGAATCCGTCCATCAAAGCC	Semi-quantitative RT-PCR primer, forward
BDL2	GTTGTGCCCATTCCTTTTTACG	Semi-quantitative RT-PCR primer, reverse
β-actin-s	GTTTTCCCATCTATCGTCGG	Standard control primer, forward
β-actin-x	TTTTCTCCATATCATCCCAG	Standard control primer, reverse
YH1	GGTACCATGTCCGTTATGGGAAAGCC	Protein expression primer, forward
YH2	GAGCTCCTTGGTTGCCTCTGGAGGAC	Protein expression primer, reverse
TZ1	CGAATAAAAGGGGAGGGAGGAG	Primer of the probe used in Southern blotting, forward
TZ2	CTCCACGAACAAGAACTCCGC	Primer of the probe used in Southern blotting, reverse

Table 2 PCR amplification conditions in this study

Primer pair	Amplification conditions
GZ1/GZ2	10 min at 94℃; 40 s at 94℃, 40 s at 52℃, and 50 s at 72℃ for 35 cycles; 5 min at 72℃
5W/AAP	10 min at 94℃; 40 s at 94℃, 40 s at 49℃, and 40 s at 72℃ for 28 cycles; 5 min at 72℃
5N/AUAP	10 min at 94℃; 40 s at 94℃, 40 s at 51℃, and 40 s at 72℃ for 35 cycles; 5 min at 72℃
3W/B26	10 min at 94℃; 40 s at 94℃, 40 s at 51℃, and 40 s at 72℃ for 28 cycles; 5 min at 72℃
3N/B25	10 min at 94℃; 40 s at 94℃, 40 s at 51℃, and 40 s at 72℃ for 35 cycles; 5 min at 72℃
QC1/QC2	10 min at 94℃; 40 s at 94℃, 40 s at 49℃, and 1 min at 72℃ for 35 cycles; 5 min at 72℃
N1/N2	10 min at 94℃; 40 s at 94℃, 40 s at 53℃, and 1 min at 72℃ for 35 cycles; 5 min at 72℃
N3/N4	10 min at 94℃; 40 s at 94℃, 40 s at 48℃, and 1 min at 72℃ for 35 cycles; 5 min at 72℃
BDL1/BDL2	10 min at 94℃; 40 s at 94℃, 40 s at 50℃, and 25 s at 72℃ for 35 cycles; 5 min at 72℃
β-actin-s/β-actin-x	10 min at 94℃; 40 s at 94℃, 40 s at 50℃, and 20 s at 72℃ for 28 cycles; 5 min at 72℃
YH1/YH2	10 min at 94℃; 40 s at 94℃, 40 s at 53℃, and 20 s at 72℃ for 32 cycles; 5 min at 72℃
TZ1/TZ2	10 min at 94℃; 40 s at 94℃, 40 s at 54℃, and 20 s at 72℃ for 32 cycles; 5 min at 72℃

Isolation of the full-length cDNA of *AccGSTZ1*

Total RNA was extracted by using the Trizol reagent (Invitrogen, USA) according to the instructions of the manufacturer. To eliminate potential DNA contamination, total RNA was digested with RNase-free DNase I, and then the first-strand cDNA was synthesized by using reverse transcriptase (TransGen Biotech, Beijing, China) according to the protocol of the manufacturer. An internal fragment was obtained by reverse transcription-PCR (RT-PCR) by using the primers GZ1 and GZ2, which were designed based on the *GSTZ1* gene consensus sequences from *Nasonia vitripennis*, *Drosophila melanogaster*, *Homo sapiens* and *Bombyx mori*. According to the obtained internal *AccGSTZ1* sequence, two pairs of specific primers (5W/5N and 3W/3N) were designed and synthesized for 5' and 3' rapid amplification of cDNA end (RACE), respectively. For 5' RACE, the first-strand cDNA was purified, and a poly(C) tail was added to provide the primary round template; then, PCR was performed by using 5W and the abridged anchor primer (AAP). The product was diluted and used as the template for the nested PCR. The specific primer 5N and abridged universal amplification primer (AUAP) were used for this round. For 3' RACE, two rounds of PCR were also carried out: primer 3W and the universal primer B26 were used for the first round PCR, and then 3N and B25 were used for the second round. Finally, the full-length cDNA of *AccGSTZ1* was obtained by PCR using primers QC1 and QC2, which were designed based on the full-length cDNA of *AccGSTZ1* deduced by overlapping the 5' and 3' UTR fragments with the internally conserved region. All of the PCR products were purified, cloned into the pEASY-T3 vector (TransGen Biotech, Beijing, China) and then transformed into the *Escherichia coli* strain DH5α for sequencing.

Bioinformatic analysis

The nucleotide and amino acid sequences of AccGSTZ1 were analyzed and compared by using the BLAST algorithm available at the NCBI server (http://blast.ncbi.nlm.nih.gov/Blast.cgi). The deduced amino acid sequence was aligned with homologous GSTZ1 protein sequences from various species by using DNAMAN version 5.2.2 (Lynnon Biosoft, Quebec, Canada), and a neighbor-joining phylogenetic tree was constructed by using Molecular Evolutionary Genetic Analysis (MEGA version 4.0). Finally, the tertiary structure was predicted by using the online protein structure prediction tool SWISS-MODEL (http://swissmodel.xpasy.org/).

Isolation of the genomic sequence of *AccGSTZ1*

Genomic DNA was extracted from whole adult bodies by using the EasyPure Genomic DNA Extraction Kit (TransGen Biotech, Beijing, China) in accordance with the instructions of the manufacturer. Two pairs of primers (N1/N2 and N3/N4) were designed and synthesized according to the full-length cDNA of *AccGSTZ1* by using *A. cerana cerana* genomic DNA as the template. Two fragments were obtained and spliced to form the *AccGSTZ1* genomic DNA.

Southern blotting analysis

To identify the potential complexity of *AccGSTZ1*, we performed a Southern blotting. Thirty micrograms/sample of genomic DNA from *A. cerana cerana* worker adults was completely digested by two restriction endonucleases, *Eco*R I and *Alu* I (TakaRa, Dalian, China), for 48 h at 37℃. After precipitating with ethanol, the DNA was separated on 1.5% agarose gels and then transferred onto Hybond-N+ Nylon membranes (Amersham, Pharmacia, UK) by capillary blotting. The probe was amplified with primers TZ1 and TZ2 and labeled with [α^{32}P] dCTP by using the Primer-a-Gene Labeling System (Promega, Madison, WI, USA) according to the instructions of the manufacturer. The hybridization was performed as described in Wang et al. (2010) [24].

Expression analysis of *AccGSTZ1* by semi-quantitative RT-PCR

Two gene-specific primers, BDL1 and BDL2, were used to amplify a fragment of *AccGSTZ1*. As an internal control, a semi-quantitative RT-PCR reaction for the house-keeping gene *β-actin* was performed under the same conditions. The primers of *β-actin* gene were designed according to Wang et al. (2010) [24]. These experiments were repeated three times. Quantitated data from the stained PCR bands were normalized against that of the *β-actin* by using the Quantity-One™ image analysis software implemented for the VersaDoc 4000 (Bio-Rad, Hercules, CA, USA). The ratios of *AccGSTZ1* to *β-actin* were calculated.

Protein expression and purification of recombinant AccGSTZ1

To obtain the AccGSTZ1 coding region for expression in *E. coli* BL21(DE3) cells, a 654 bp cDNA fragment of *AccGSTZ1* encoding the entire open reading frame (ORF) was amplified with two specific primers (YH1/YH2). After digestion of the PCR product with *Bam*H I and *Sac* I, the obtained fragment was subcloned between the *Bam*H I and *Sac* I sites of the bacterial expression vector pET-30a (+). The recombinant plasmid pET-30a (+)- AccGSTZ1 was transformed into *E. coli*, which were grown at 37℃ in 10 ml of Luria-Bertani (LB) broth supplemented with 30 μg/ml kanamycin. Expression of the recombinant protein was induced by the addition of 1 mmol/L final concentration isopropylthio-β-D-galactoside (IPTG) when the cell density reached 0.4-0.6 OD_{600}. Following incubation for an additional 6 h, cells were harvested and centrifuged at 5,000 *g* at 4℃ for 10 min. Then, the pellet was resuspended in lysis buffer, sonicated and centrifuged again. Finally, the resuspended pellet was subjected to 15% SDS-polyacrylamide gel electrophoresis (SDS-PAGE) to determine the expression level of the target recombinant protein. In addition, the recombinant protein was purified by using the MagneHis™ Protein Purification System (Promega, Madison, WI, USA). The purified recombinant protein concentration was measured by using the Bradford method (1976) [25]. Bovine serum albumin served as a standard.

Measurements of enzyme activity

GST activity was measured spectrophotometrically by using 1-chloro-2,4-dinitrobenzene (CDNB) and GSH as standard substrates under the previously specified conditions [26], with some modifications. The reaction system (1 ml) contained 0.1 mol/L sodium phosphate buffer (pH 6.5) in the presence of 5 mmol/L CDNB, 5 mmol/L GSH and 30 μl purified recombinant protein. The changes in absorbance at 340 nm/min were monitored. Enzymatic activity was expressed as moles CDNB conjugated with GSH per min per mg protein. AccGSTZ1 activity towards 1,2-dichloro-4-nitrobenzene (DCNB) was assayed in a similar manner. All assays were performed at room temperature and were run in triplicate.

The antioxidant activity of the recombinant AccGSTZ1 protein

Recombinant AccGSTZ1 was overexpressed in *E. coli* BL21(DE3). Approximately 5×10^8

cells of the culture were overlaid onto LB-kanamycin agar and incubated at 37℃ for 1 h. Bacteria transfected with the pET-30a (+) vector only were used as control. Filter discs soaked with different concentrations (10, 20, 30, 40 and 50 μmol/L) of *t*-butyl hydroperoxide were placed on the surface of the agarose medium plates. *E. coli* cells were grown for another 24 h at 37℃ before the killing zones around the discs were measured.

Preparation of anti-AccGSTZ1 serum and Western blotting

The purified protein (100 μg) diluted by an equal volume of Freund's complete adjuvant (Sigma, USA) was injected subcutaneously into the six-week-old and pathogen free female white mice. The mice were boosted two times with 50 μg/ml of a mixture of purified protein and Freund's incomplete adjuvant at 1 week interval. Four days later, the blood was collected and centrifuged at 3,000 *g* for 5 min for the generation of AccGSTZ1 antibody. Total protein from different tissues including the brain, epidermis, muscle and midgut were extracted as described by Li et al. (2009) [27]. Then the protein was separated by SDS-PAGE and subsequently electrotransferred onto a polyvinylidine difluoride (PVDF) membrane (Whatman, British). After blocking in 5% (*w/V*) skim milk diluted in PBS for 30 min, the membrane was incubated with primary antibodies at appropriate dilutions at 4℃ overnight. After being rinsed in PBS, the membrane was probed with HRP-conjugated goat anti-rat secondary antibody (Beijing Dingguo Changsheng Biotech Co., Ltd., China) which was diluted by 1∶4,000 at 37℃ for 2 h. After washed by PBS, the membrane was treated with 4-chloro-1-naphthol (Solarbio, China) to detect the reaction. The protein band intensity was quantified by the mean ± SD of three experiments as determined from densitometry relative to β-actin.

Statistical analysis

The results of gene expression and killing zones are present as means ± standard deviation (SD) with $n = 3$. Data were subjected to multiple comparisons by analysis of variance (ANOVA), and means were separated by the Duncan's Multiple Range test with significant difference level $P < 0.05$. ANOVA was performed by using Statistical Analysis System (SAS) version 9.1 software.

Results

Isolation and sequence analysis of the full-length cDNA of *AccGSTZ1*

The full-length cDNA sequence of *AccGSTZ1* (GenBank ID: JN637474) was 1,087 bp including 153 bp in the 5' untranslated region (UTR), 654 bp in the open reading frame (ORF) and 280 bp in the 3' UTR. The ORF encodes a polypeptide of 217 amino acids with a calculated molecular mass of 24.759 kDa and a theoretical isoelectric point (pI) of 7.86.

A multiple sequence alignment revealed that the deduced protein had significant identity with GSTZs from other species, including *D. melanogaster*, *Anopheles gambiae*, *B. mori* and *N. vitripennis*. The identity ranged from 76.82% to 94.93% (Fig. 1A), which suggests that

Zeta class glutathione S-transferases are conserved over a considerable evolutionary period. In addition, the predicted AccGSTZ1 possessed two conserved domains that existed in all cytosolic GSTs, called the N-terminal (domain I) and the C-terminal domains (domain II). Domain I, which provided the GSH-binding site (G-site) was positioned between amino acids P_7 and E_{81}, while its co-substrate binding site was between P_{94} and Q_{208}. The characteristic motif, with the sequence SSCXWRVRIAL, was also identified.

A phylogenetic protein sequence analysis was undertaken to ascertain whether the AccGSTZ1 was a divergent member of the GSTZ class. The deduced AccGSTZ1 and representative sequences from previously described GST classes in insects were used to construct an evolutionary tree by using the neighbor-joining method in MEGA 4.0 software (Fig. 1B). The most likely tree revealed six classes of GSTs in insects, and categorized the newly cloned protein into the Zeta class for it exhibited high relatedness to GSTZs from *N. vitripennis*, *B. mori* and other insects.

To elucidate the function of the conserved residues and determine the catalytic mechanism of AccGSTZ1, the tertiary structure of AccGSTZ1 was predicted by using SWISS-MODEL (Fig. 1C). The model showed that AccGSTZ1 folded to form a typical GST dimeric structure, known as the thioredoxin fold. A *cis*-proline loop located at the N-terminus of β-strand 3 is conserved in most members of the superfamily and is important for the overall structure and structural stability of the active site and for the catalytic activities of these proteins [28].

Genomic structure and Southern blotting analysis of *AccGSTZ1*

To further elucidate the characteristics of the *AccGSTZ1*, the genomic sequence was amplified by using genomic DNA as the template for PCR amplification. The full-length *AccGSTZ1* genomic sequence (GenBank ID: JN637475) was 1,783 bp and comprised five exons and four introns, and all introns were within the ORF. Moreover, the nucleotide sequence at the splice junctions was consistent with the canonical GT-AG rule.

A comparison of the genomic sequence of *AccGSTZ1* and other insect *GSTZ1* sequences from *Apis mellifera*, *Anopheles gambiae* and *D. melanogaster* revealed variations in the intron numbers and gene lengths (Fig. 2A). Notably, contrasting with the two introns in the *A. gambiae GSTZ1* and the three introns in the orthologue of this gene in *D. melanogaster*, the two bees have a total of four introns, which shows the diversity that arose between Hymenoptera and Diptera during evolution. In addition, a comparison of the intron positions within these four Zeta genes shows that a common feature of the honeybees and the fly is the presence of a first exon that encodes a small polypeptide, 18-24 amino acids in length, which is absent in the mosquito.

To determine the potential complexity of the *AccGSTZ1* locus in the *A. cerana cerana* genome, adult honeybee genomic DNA was digested with *Alu* I or *Eco*R I, followed by hybridization with a partial sequence from the 5' end of the *AccGSTZ1* cDNA as the probe. The blotting was performed under stringent conditions to avoid cross-hybridization. As shown in Fig. 2B, only one band was observed in the two different enzyme-digested DNA lanes, which indicates that *AccGSTZ1* is present in a single copy in the *A. cerana cerana* genome.

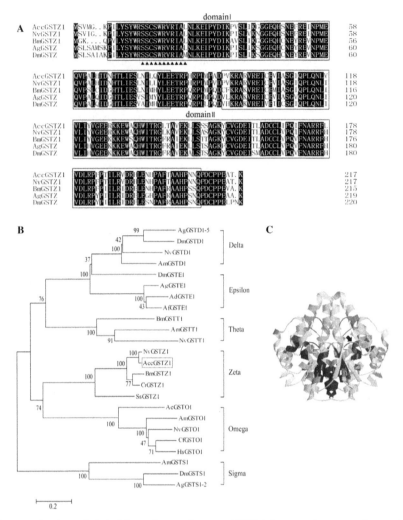

Fig. 1 Characterization of glutathione S-transferases (GSTs) from various species and the tertiary structure of AccGSTZ1. A. An alignment of the deduced AccGSTZ1 protein sequence (GenBank ID: JN637474) from *Apis cerana cerana* with other known insect GSTZs. The amino acid sequences used in the analysis were from *N. vitripennis* (NvGSTZ1, GenBank ID: NP_001165931), *B. mori* (BmGSTZ1, NP_001037418), *Anopheles gambiae* (AgGSTZ, XP_003436290) and *D. melanogaster* (DmGSTZ, NP_731358). Identical amino acid residues are shaded in black. The conserved domains are boxed. The characteristic motifs are marked by ▲. B. A phylogenetic tree of GSTs was constructed from different species using the neighbor-joining method and MEGA 4.0 software. Six main groups are shown and AccGSTZ1 is boxed. The amino acid sequences of GSTs were all obtained from NCBI as follows: AgGSTD1 (*A. gambiae*, CAB03592), DmGSTD1 (*D. mojavensis*, ABW76195), NvGSTD1 (*N. vitripennis*, NP_001165913), AmGSTD1 (*A. mellifera*, NP_001171499), DmGSTE1 (*D. melanogaster*, NP_611323), AgGSTE1 (*A. gambiae*, AAL59658), AdGSTE1 (*Anopheles dirus*, ABD43203), AfGSTE1 (*Anopheles funestus*, AEJ87236), BmGSTT1 (*B. mori*, NP_001108463), AmGSTT1 (*A. mellifera*, XP_624692), NvGSTT1 (*N. vitripennis*, NP_001165926), NvGSTZ1 (*N. vitripennis*, NP_001165931), BmGSTZ1 (*B. mori*, NP_001037418), CrGSTZ1 (*Chironomus riparius*, ADK66969), SsGSTZ1 (*Sus scrofa*, NP_001230567), AcGSTO1 (*Anopheles cracens*, ACY95464), AmGSTO1 (*A. mellifera*, XP_624501), NvGSTO1 (*N. vitripennis*, NP_001165912), CfGSTO1 (*Camponotus floridanus*, EFN62827), HsGSTO1 (*Harpegnathos saltator*, EFN81737), AmGSTS1 (*A. mellifera*, NP_001153742), DmGSTS1 (*D. melanogaster*, NP_725653), AgGSTS1-2 (*A. gambiae*, AAM53611). C. The tertiary structure of AccGSTZ1. The structure was built using homology modeling in the SWISS-MODEL modeling environment. The conserved proline in the loop between β-strand 3 and α-helix 2 is marked by an arrowhead.

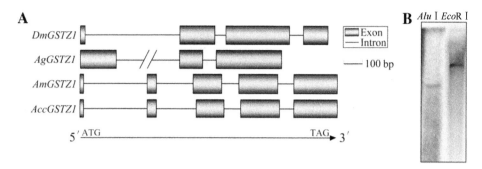

Fig. 2 Structure and Southern blotting analyses of *AccGSTZ1*. A. Exon-intron organization of *AccGSTZ1* from *A. cerana cerana* and its homologs, including *AmGSTZ1* from *A. mellifera* (XP_394562), *AgGSTZ1* from *A. gambiae* (AAAB01008859.1) and *DmGSTZ1* from *D. melanogaster* (NT_033777). The exons are indicated with grey boxes and the introns are shown as lines. The initiation (ATG) and termination (TAG) codons are indicated. B. Southern blotting analysis of *AccGSTZ1*. Genomic DNA was extracted from adult bees, digested with two restriction enzymes, separated on a 1.5% agarose gel and hybridized with a partial sequence.

Temporal and spatial expression of *AccGSTZ1*

To examine the expression level of *AccGSTZ1* during different developmental stages, semi-quantitative RT-PCR was performed by using RNA extracted from larvae, pupae, and adults. As shown in Fig. 3A, *AccGSTZ1* was present at all of the developmental stages ($F = 141.58$; df = 11; $P < 0.0001$). In addition, *AccGSTZ1* transcripts were significantly increased from the first-instar larvae to the fourth-instar larval stage and then increased in a wave until white-eyed pupae developed. From the late pupal stages to adults, the mRNA level decreased. This result suggested that the *AccGSTZ1* may play a fundamental role in the early developmental stages of the honeybee.

In addition, we also analyzed the transcripts induced by H_2O_2 injection and found that H_2O_2 dramatically induced the expression of *AccGSTZ1* (Fig. 4E). The mRNA expression of the *AccGSTZ1* gene peaked at 45 min after H_2O_2 injection, and the mRNA level was as much as

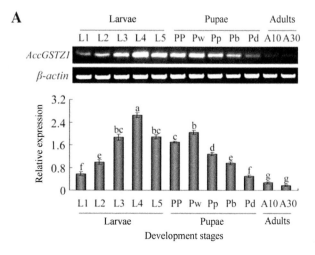

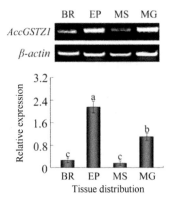

Fig. 3 The expression patterns of *AccGSTZ1*. A. *AccGSTZ1* mRNA expression at different developmental stages: larvae from the first to the fifth instars (L1-L5), pupae (PP-prepupae, Pw-white eyes, Pp-pink eyes, Pb-brown eyes and Pd-dark eyes) and adults (A10-ten days post-emergence and A30-thirty days post-emergence adults). B. The distribution of *AccGSTZ1* in brain (BR), epidermis (EP), muscle (MS) and midgut (MG). The *β-actin* gene was used as an internal control. Histograms (means based on triplicate assays) show the relative expression levels of *AccGSTZ1*. Signal intensities (INT/mm^2) of the bands were assigned using the Quantity-One^{TM} image analysis software. Each value is the mean (SD) of three replicates. Different letters topped on the bar designated significant difference at $P < 0.0001$ in SAS.

20-fold higher than in control bees, but gradually decreased to the normal level in the following hours ($F = 164.22$; df = 7; $P < 0.0001$). Taken together, these results suggest that *AccGSTZ1* participates in the response to oxidative stresses.

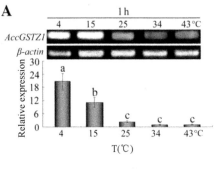
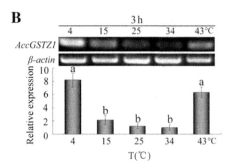
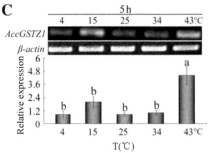
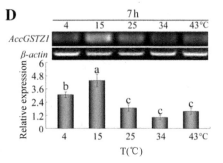

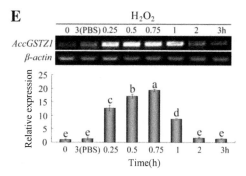

Fig. 4 The expression profiles of *AccGSTZ1* under oxidant stresses. A-D. The expression of glutathione *S*-transferases (GST) under temperature stress treatments for 1, 3, 5 and 7 h, respectively. E. The expression profiles of *AccGSTZ1* induced by H_2O_2 injection. Untreated adult worker bees (Lane 0) were used as controls, and adult worker bees injected with PBS for 3 h were used as injection controls. The *β-actin* gene was used as an internal control. Histograms (means based on triplicate assays) show the relative expression levels of *AccGSTZ1*. Signal intensities (INT/mm^2) of the bands were assigned using the Quantity-One^{TM} image analysis software. Different letters topped on the bar designated significant difference at $P < 0.01$ in SAS.

Protein purification and characterization of recombinant AccGSTZ1 protein

To eliminate the possibility that the enzymatic activity of the AccGSTZ1 protein isolated directly from *A. cerana cerana* may not have the intrinsic properties of the protein due to a putative minor impurity, the enzyme was expressed in *E. coli* BL21(DE3). SDS-PAGE demonstrated that a recombinant protein with a molecular mass of approximately 27 kDa was produced (Fig. 5A). The target recombinant protein was further purified by using the MagneHis^{TM} Protein Purification System, and the concentration of purified AccGSTZ1 protein was approximately 0.5 mg/ml.

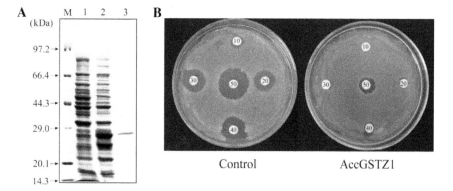

Fig. 5 Characterization of recombinant AccGSTZ1. A. Overexpression and purification of AccGSTZ1 protein in *E. coli* BL21. SDS-PAGE analysis was used to separate the recombinant AccGSTZ1 protein expressed in *E. coli*. Lanes 1 and 2, uninduced or IPTG-induced *E. coli* BL21 containing the AccGSTZ1 protein, respectively; Lane 3, purified AccGSTZ1 protein; M, indicates the molecular mass standard. B. AccGSTZ1 was overexpressed in *E. coli*, and the LB agar plates were flooded with 5 × 10^8 cells. Bacteria transfected with pET-30a (+) (vector only) were used as controls. 10, 20, 30, 40 and 50 on filter discs represent the different concentrations (μmol/L) of *t*-butyl hydroperoxide.

The purified AccGSTZ1 was tested for enzymatic activity with 1-chloro-2, 4-dinitrobenzene (CDNB) and 1,2-dichloro-4-nitrobenzene (DCNB) as substrates. The catalytic efficiency of AccGSTZ1 was found to be very low, 42.3 µmol/(min mg) for CDNB and no measurable conjugation activity with DCNB.

To evaluate the ability of AccGSTZ1 to protect against oxidative stress, $E.\ coli$ overexpressing the recombinant AccGSTZ1 was exposed to t-butyl hydroperoxide, which acted as an external environmental stressor. After a 24 h exposure, the killing zones around the drug-soaked filters were much smaller on plates with $E.\ coli$ overexpressing AccGSTZ1 than in control bacteria (Fig. 5B). Significant differences in halo diameters for killing zones were recorded (Table 3). The halo reductions among different concentration (10, 20, 30, 40, 50 µmol/L) were 7% ($F = 3.37$; df = 1; $P = 0.1401$), 44% ($F = 139.64$; df = 1; $P = 0.003$), 48% ($F = 79.57$; df = 1; $P = 0.0009$), 46% ($F = 57.0$; df = 1; $P = 0.0016$) and 46% ($F = 79.53$; df = 1; $P = 0.0009$), respectively. The results obtained clearly proved that AccGSTZ1 could play a vital role in protecting Chinese honeybees from oxidative stresses.

Table 3 Details of statistical analysis of killing zones

Concentration (µmol/L)	Diameters of the killing zones (cm)	
	Control	AccGSTZ1
10	0.84 ± 0.031a	0.78 ± 0.012a
20	1.44 ± 0.053b	0.8 ± 0.012a
30	1.62 ± 0.085b	0.84 ± 0.023a
40	1.66 ± 0.087b	0.9 ± 0.052a
50	1.95 ± 0.098b	1.04 ± 0.029a

Data are the means of three replications ±standard deviation. Values in the same row with different letters are significantly different at $P < 0.05$.

Western blotting

To confirm that the recombinant AccGSTZ1 could actually reflect the intrinsic property of the AccGSTZ1 protein in the $A.\ cerana\ cerana$ and investigative the tissue distribution of AccGSTZ1 at the translational level, proteins extracted from the brain, epidermis, muscle and midgut were subjected to Western blotting (Fig. 6). The result revealed that the expression of AccGSTZ1 in different tissues was similar to that at the transcriptional level ($F = 37.78$; df = 3; $P < 0.0001$).

Discussion

In this study, we identified a Zeta class GST (*AccGSTZ1*) from the Chinese honeybee $A.\ cerana\ cerana$ and investigated the response of *AccGSTZ1* under oxidative stresses. Sequence analysis indicated that AccGSTZ1 not only had the classical conserved domains but also contained the characteristic motif, SSCXWRVRIAL, in the N-terminal region. The first three

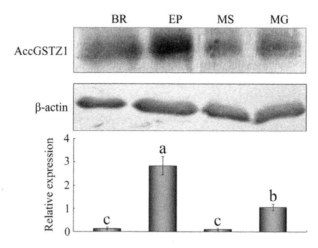

Fig. 6 Western blotting analysis of AccGSTZ1 in different tissues. Proteins extracted from brain, epidermis, muscle and midgut were separated by SDS-PAGE, transferred onto a PVDF membrane and probed with the AccGSTZ1 antibody. Line BR, brain; Line EP, epidermis; Line MS, muscle; Line MG, midgut. β-actin expression is used as an internal control. Different letters topped on the bar designated significant difference at $P < 0.0001$.

residues of this motif line the active-site pocket, and the first serine aligns with the active-site serine found in the mammalian Theta and insect Delta class sequences, which suggests that this serine may be catalytically active [29, 30]. Southern blotting demonstrated that the *AccGSTZ1* gene was present as a single copy in the genome of *A. cerana cerana* (Fig. 2B), which was in agreement with *GSTZ1* from *Apis mellifera*, *Anopheles gambiae* and *Pediculus humanus humanus*. This relatively similar distribution across different insect species suggests that the Zeta class GSTs may have common house-keeping functions in conserved physiological processes, including cellular defense against oxidative stress [31]. In addition, the variations in the intron numbers and gene lengths between the genomic sequence of *AccGSTZ1* and other insect *GSTZ1* sequences from *Apis mellifera*, *Anopheles gambiae* and *D. melanogaster* showed a probable alternate regulatory mRNA splicing mechanism that may result in different molecular functions.

Previous studies have shown that some GSTs are upregulated with aging [32], and their overexpression causes a lifespan extension in flies [33]. Alterations of glutathione *S*-transferase characteristics in the course of development of *Apis mellifera macedonica* have also been demonstrated [34]. In this study, *AccGSTZ1* was constitutively expressed at all stages in the Chinese honeybee (Fig. 3A). The broad spectrum of expression indicates that it may participate in the growth of *A. cerana cerana*. Moreover, the most robust expression of *AccGSTZ1* was observed in the larval stages rather than in the emerged adult stages. There is evidence that the GSTZ1 cDNA species in many libraries were derived from fetal tissues in mammals [29]. These results further demonstrate that *AccGSTZ1* may play a role in developmental regulation. The higher level in the larval stages was probably due to the oxidative stresses caused by feeding on xenobiotic compounds and bacterial infections [12].

The fat body and midgut are believed to be the main metabolic center and are involved in the detoxification of xenobiotics and protection from oxidative stress [15]. Krishnan and

Kodrik (2006) measured the activities of antioxidant enzymes in *Spodoptera littoralis* to examine whether they are enhanced to protect gut tissues during oxidative stress [35]. However, in our study, *GSTZ1* from *A. cerana cerana* is primarily expressed in the epidermis (Fig. 3B), which contributes to the protection of honeybees from environmental stresses and mechanical injuries. Konno and Shishido (1992) detected a GST in the Malpighian tubules [36]. Rogers et al. (1999) discovered an olfactory-specific GST expressed in antennae of the sphinx moth *Manduca sexta* [37]. We speculate that GSTs prior expressed in tissues other than midgut may have additional physiological functions in addition to detoxifying xenobiotics. Taken together, the tissue-specific and developmental expression patterns of *AccGSTZ1* exhibit an interesting GST model for *A. cerana cerana* and, to the best of our knowledge, are unique.

Temperature is one of the most important abiotic environmental factors that cause physiological changes in organisms [38]. During evolutionary history, organisms have evolved many behavioral and physiological strategies to avoid temperature impairments, such as seeking shelters; changing the fluidity of cell membranes; and accumulating sugars, polyols, antifreeze proteins and amino acids [1]. Recently, stress induced by temperature changes has been associated with enhanced reactive oxygen species (ROS) generation and oxidative stress. Relatively low and high environmental temperatures can yield oxidative stress accompanied by a high level of lipid peroxidation, which can be metabolized by GSTs [38], but not all insect GSTs are involved in detoxification and antioxidative stress [12]. To confirm whether the *AccGSTZ1* gene has a role in oxidative stress resistance, the *AccGSTZ1* mRNA level of worker bees exposed to relatively low or high environmental temperature was measured. The results show that changes in the temperature could significantly increase mRNA expression of *AccGSTZ1* in a time-dependent manner (Fig. 4A-D). Notably, the expression profile in honeybees exposed to low or high temperatures were different. We hypothesized that cold and heat cause oxidative stress in cells and induce the expression of *AccGSTZ1* through distinct pathways.

Under the high temperature treatments, H_2O_2 concentrations increased significantly in the ark shell, *Scapharca broughtonii* [38]. Li et al. (2008) also investigated the response of GSTs in the midgut to dietary H_2O_2, and the GSTs showed a dose-dependent response [16]. In this study, the high induction level of *AccGSTZ1* in response to excessive H_2O_2 (Fig. 4E), which is an active ROS, suggests a role for *AccGSTZ1* in the enzymatic protection against oxidative stress. The expression of *AccGSTZ1* was nearly 20-fold higher than in the control at 45 min after H_2O_2 treatment, whereas two GSTs from the fat body of *Bombus ignitus* worker bees were highly upregulated 9 h after H_2O_2 exposure [18]. The different expression patterns imply that *AccGSTZ1* may act more positively than GST genes in *B. ignitus*.

GSTs catalyze a broad range of reactions because different members of the family exhibit distinct substrate specificity [39]. To identify the biological function of AccGSTZ1, the catalytic activity of the purified protein was measured. The results revealed that AccGSTZ1 has low glutathione transferase activity towards CDNB and no measurable activity was observed with DCNB, which was consistent with human GSTZ [29]. Although showing minimal glutathione-conjugating activity, GSTZs have high glutathione peroxidase activity with *t*-butyl and cumene hydroperoxides [29, 40]. As an external environmental stressor, *t*-butyl hydroperoxide is widely used as a model substance for studying the mechanism of cell injury resulting from oxidative stress [41]. In this study, *E. coli* overexpressing AccGSTZ1 were exposed to different concentrations of *t*-butyl hydroperoxide. The obvious distinction of the killing zones between the AccGSTZ1-overexpressing and the control bacteria definitively demonstrated that AccGSTZ1 contributes to the cells resistance to oxidative stress (Fig. 5B).

In conclusion, we have identified a Zeta class GST (*AccGSTZ1*) from *A. cerana cerana*. *AccGSTZ1* was primarily expressed in the early stages during the development, and mainly distributed in the epidermis. The mRNA accumulation of *AccGSTZ1* was significantly induced after temperature challenges and H_2O_2 treatment. The increased expression level may be partially associated with the increasing oxidative stress caused by these external factors. In addition, *E. coli* overexpressing AccGSTZ1 exhibited less sensitivity to *t*-butyl hydroperoxide, providing another evidence that *AccGSTZ1* may play a role in response to oxidative stress. Taken together, our present data improve the understanding of the expression, regulation and function of GST proteins in *A. cerana cerana*.

References

[1] Yang, L. H., Huang, H., Wang, J. J. Antioxidant responses of citrus red mite, *Panonychus citri* (McGregor) (Acari: Tetranychidae), exposed to thermal stress. Journal of Insect Physiology. 2010, 56, 1871-1876.

[2] Narendra, M., Bhatracharyulu, N. C., Padmavathi, P., Varadacharyulu, N. C. Prallethrin induced biochemical changes in erythrocyte membrane and red cell osmotic haemolysis in human volunteers. Chemosphere. 2007, 67, 1065-1071.

[3] Brennan, L. J., Keddie, B. A., Braig, H. R., Harris, H. L. The endosymbiont *Wolbachia pipientis* induces the expression of host antioxidant proteins in an *Aedes albopictus* cell line. PLoS One. 2008, 3, e2083.

[4] Green, D. R., Reed, J. C. Mitochondria and apoptosis. Science. 1998, 281, 1309-1312.

[5] Collins, A. M., Williams, V., Evans, J. D. Sperm storage and antioxidative enzyme expression in the honey bee, *Apis mellifera*. Insect Molecular Biology. 2004, 13, 141-146.

[6] Stadtman, E. R. Oxidation of proteins by mixed-function oxidation systems: implication in protein turnover, aging and neutrophil function. Trends in Biochemistry Sciences. 1986, 11, 11-12.

[7] Stadtman, E. R., Levine, R. L. Free radical-mediated oxidation of free amino acids and amino acid residues in proteins. Amino acids. 2003. 25, 207-218.

[8] Imlay, J. A., Linn, S. DNA damage and oxygen radical toxicity. Science. 1988, 240, 1302-1309.

[9] Rameshthangam, P., Ramasamy, P. Antioxidant and membrane bound enzymes activity in WSSV-infected *Penaeus monodon* Fabricius. Aquaculture. 2006, 254, 32-39.

[10] Dubovskiy, I. M., Martemyanov, V. V., Vorontsova, Y. L., Rantala, M. J., Gryzanova, E. V., Glupov, V. V. Effect of bacterial infection on antioxidant activity and lipid peroxidation in the midgut of *Galleria mellonella* L. larvae (Lepidoptera, Pyralidae). Comparative Biochemistry and Physiology C: Pharmacology Toxicology and Endocrinology. 2008, 148, 1-5.

[11] Lantum, H. B., Baggs, R. B., Krenitsky, D. M., Board, P. G., Anders, M. W. Immunohistochemical localization and activity of glutathione transferase Zeta (*GSTZ1-1*) in rat tissues. Drug Metabolism Disposition. 2002, 30, 616-625.

[12] Huang, Y. F., Xu, Z. B., Lin, X. Y., Feng, Q. L., Zheng, S. C. Structure and expression of glutathione *S*-transferase genes from the midgut of the common cutworm, *Spodoptera litura* (Noctuidae) and their response to xenobiotic compounds and bacteria. Journal of Insect Physiology. 2011, 57, 1033-1044.

[13] Tu, C. P., Akgül, B. *Drosophila* glutathione *S*-transferases. Methods in Enzymology. 2005, 401,

204-226.

[14] Vontas, J. G., Small, G. J., Nikou, D. C., Ranson, H., Hemingway, J. Purification, molecular cloning and heterologous expression of a glutathione S-transferase involved in insecticide resistance from the rice brown planthopper, *Nilaparvata lugens*. Biochemical Journal. 2002, 362, 329-337.

[15] Enayati, A. A., Ranson, H., Hemingway, J. Insect glutathione transferases and insecticide resistance. Insect Molecular Biology. 2005, 14, 3-8.

[16] Li, H. -M., Buczkowski, G., Mittapalli, O., Xie, J., Wu, J., Westerman, R., Schemerhorn, B. J., Murdock, L. L., Pittendrigh, B. R. Transcriptomic profiles of *Drosophila melanogaster* third instar larval midgut and responses to oxidative stress. Insect Molecular Biology. 2008, 17, 325-339.

[17] Yu, Q., Lu, C., Li, B., Fang, S., Zuo, W., Dai, F., Zhang, Z., Xiang, Z. Identification, genomic organization and expression pattern of glutathione S-transferase in the silkworm, *Bombyx mori*. Insect Biochemistry and Molecular Biology. 2008, 38, 158-1164.

[18] Kim, B. Y., Hui, W. L., Lee, K. S., Wan, H., Yoon, H. J., Gui, Z. Z., Chen, S., Jin, B. R. Molecular cloning and oxidative stress response of a Sigma-class glutathione S-transferase of the bumblebee *Bombus ignitus*. Comparative Biochemistry and Physiology, Part B: Biochemistry and Molecular Biology. 2011, 158, 83-89.

[19] Kampkotter, A., Volkmann, T. E., De Castro, S. H., Leiers, B., Klotz, L. O., Johnson, T. E., Link, C. D., Henkle-Dührsen, K. Functional analysis of the glutathione S-transferase 3 from *Onchocerca volvulus* (*OvGST-3*): a parasite GST confers increased resistance to oxidative stress in *Caenorhabditis elegans*. Journal of Molecular Biology. 2003, 325, 25-37.

[20] Ding, Y., Hawkes, N., Meredith, J., Eggleston, P., Hemingway, J., Ranson, H. Characterisation of the promoters of Epsilon glutathione transferases in the mosquito *Anopheles gambiae* and their response to oxidative stress. Biochemical Journal. 2005, 387, 879-888.

[21] Udomsinprasert, R., Pongjaroenkit, S., Wongsantichon, J., Oakley, A. J., Prapanthadara, L., Wilce, M. C. J., Ketterman, A. J. Identification, characterization and structure of a new Delta class glutathione transferase isoenzyme. Biochemical Journal. 2005, 388, 763-771.

[22] Meng, J. Y., Zhang, C. Y., Zhu, F., Wang, X. P., Lei, C. L. Ultraviolet light-induced oxidative stress: effects on antioxidant response of *Helicoverpa armigera* adults. Journal of Insect Physiology. 2009, 55, 588-592.

[23] Corona, M., Velarde, R. A., Remolina, S., Moran-Lauter, A., Wang, Y., Hughes, K. A., Robinson, G. E. Vitellogenin, juvenile hormone, insulin signaling, and queen honey bee longevity. PNAS. 2007, 104, 7128-7133.

[24] Wang, M., Kang, M. J., Guo, X. Q., Xu, B. H. Identification and characterization of two phospholipid hydroperoxide glutathione peroxidase genes from *Apis cerana cerana*. Comparative Biochemistry and Physiology C: Pharmacology Toxicology and Endocrinology. 2010, 152, 75-83.

[25] Bradford, M. M. A rapid and sensitive method for the quantitation of microgram quantities of protein utilizing the principle of protein-dye binding. Analytical Biochemistry. 1976, 72, 248-254.

[26] Habig, W. H., Pabst, M. J., Jakoby, W. B. Glutathione S-transferases. The first enzymatic step in mercapturic acid formation. Journal of Biological Chemistry. 1974, 249, 7130-7139.

[27] Li, J., Zhang, L., Feng, M., Zhang, Z., Pan, Y. Identification of the proteome composition occurring during the course of embryonic development of bees (*Apis mellifera*). Insect Molecular Biology. 2009, 18, 1-9.

[28] Su, D., Berndt, C., Fomenko, D. E., Holmgren, A., Gladyshev, V. N. A conserved *cis*-proline precludes metal binding by the active site thiolates in members of the thioredoxin family of proteins. Biochemistry. 2007, 46, 6903-6910.

[29] Board, P. G., Baker, R. T., Chelvanayagam, G., Jermiin, L. S. Zeta, a novel class of glutathione transferases in a range of species from plants to humans. Biochemical Journal. 1997, 328, 929-935.

[30] Board, P. G., Taylor, M. C., Coggan, M., Parker, M. W., Lantum, H. B., Anders, M. W. Clarification of the role of key active site residues of glutathione transferase Zeta/maleylacetoacetate isomerase by a new spectrophotometric technique. Biochemical Journal. 2003, 374, 731-737.

[31] Claudianos, C., Ranson, H., Johnson, R. M., Biswas, S., Schuler, M. A., Berenbaum, M. R., Feyereisen, R., Oakeshott, J. G.. A deficit of detoxification enzymes: pesticide sensitivity and environmental response in the honey bee. Insect Molecular Biology. 2006, 15, 615-636.

[32] Zou, S., Meadows, S., Sharp, L., Jan, L. Y., Jan, Y. N. Genome-wide study of aging and oxidative stress response in *Drosophila melanogaster*. Proceedings of the National Academy of Sciences. 2000, 97, 13726-13731.

[33] Che-Mendoza, A., Penilla, R. P., Rodríguez, D. A. Insecticide resistance and glutathione S-transferases in mosquitoes: a review. African Journal of Biotechnology. 2009, 8, 1386-1397.

[34] Papadopoulos, A. I., Polemitoua, I., Laifi, P., Yiangoua, A., Tananaki, C. Glutathione S-transferase in the insect *Apis mellifera* macedonica kinetic characteristics and effect of stress on the expression of GST isoenzymes in the adult worker bee. Comparative Biochemistry and Physiology C: Pharmacology Toxicology and Endocrinology. 2004, 139, 93-97.

[35] Krishnan, N., Kodrik, D. Antioxidant enzymes in *Spodoptera littoralis* (Boisduval): are they enhanced to protect gut tissues during oxidative stress? Journal of Insect Physiology. 2006, 52, 11-20.

[36] Konno, Y., Shishido, T. Distribution of glutathione S-transferase activity in insect tissues. Applied Entomology and Zoology. 1992, 27, 391-397.

[37] Rogers, M. E., Jani, M. K., Vogt, R. G. An olfactory-specific glutathione S- transferase in the sphinx moth *Manduca sexta*. Journal of Experimental Biology. 1999, 202, 1625-1637.

[38] An, M. I., Choi, C. Y. Activity of antioxidant enzymes and physiological responses in ark shell, *Scapharca broughtonii*, exposed to thermal and osmotic stress: effects on hemolymph and biochemical parameters. Comparative Biochemistry and Physiology, Part B: Biochemistry and Molecular Biology. 2010, 155, 34-42.

[39] Yamamoto, K., Teshiba, S., Shigeoka, Y., Aso, Y., Banno, Y., Fujiki, T., Katakura, Y. Characterization of an Omega-class glutathione S-transferase in the stress response of the silkmoth. Insect Molecular Biology. 2011, 20, 379-386.

[40] Blackburn, A. C., Matthaei, K. I., Lim, C., Taylor, M. C., Cappello, J. Y., Hayes, J. D., Anders, M. W., Board, P. G. Deficiency of glutathione transferase Zeta causes oxidative stress and activation of antioxidant response pathways. Molecular Pharmacology. 2006, 69, 650-657.

[41] Jung, J. S., Lee, J. Y., Oh, S. O., Jang, P. G., Bae, H. R., Kim, Y. K., Lee, S. H. Effect of *t*-butyl hydroperoxide on chloride secretion in rat tracheal epithelia. Pharmacology and Toxicology. 1998, 82, 236-242.

2.8 Molecular Cloning, Expression and Oxidative Stress Response of a Mitochondrial Thioredoxin Peroxidase Gene (*AccTpx-3*) from *Apis cerana cerana*

Pengbo Yao, Wenjing Lu, Fei Meng, Xiuling Wang, Baohua Xu, Xingqi Guo

Abstract: Thioredoxin peroxidase (Tpx) plays an important role in maintaining redox homeostasis and in protecting organisms from the accumulation of toxic reactive oxygen species (ROS). Here, we isolated a mitochondrial thioredoxin peroxidase gene from *Apis cerana cerana*, *AccTpx-3*. The open reading frame (ORF) of *AccTpx-3* is 729 bp in length and encodes a predicted protein of 242 amino acids, 27.084 kDa and an isoelectric point of 8.70. Furthermore, the 980 bp 5′-flanking region was cloned, and the transcription factor binding sites were predicted. A quantitative RT-PCR (qRT-PCR) analysis indicated that *AccTpx-3* was expressed higher in muscle than other tissues, with its the highest expression occurring on the fourth day of the larval stage, followed by the fifteenth day of the adult stage. Moreover, the expression of the *AccTpx-3* transcript was upregulated by such abiotic stresses as 4℃, 42℃, H_2O_2, cyhalothrin, acaricide and phoxim treatments. In contrast, *AccTpx-3* transcription was downregulated by other abiotic stresses, including 16℃, 25℃, ultraviolet light and $HgCl_2$. Recombinant AccTpx-3 protein acted as a potent antioxidant that resisted paraquat-induced oxidative stress and protected DNA from oxidative damage. Taken together, these results suggest that the AccTpx-3 protein is an antioxidant enzyme that may protect organisms from oxidative stress.

Keywords: *Apis cerana cerana*; *AccTpx-3*; qPCR; Oxidative stress; Antioxidant enzyme

Introduction

Although reactive oxygen species (ROS), including superoxide anions, hydrogen peroxide and hydroxyl radicals, can be generated by endogenous metabolism, there is generally a balance between the generation of ROS and antioxidant processes [1]. High ROS concentrations cause serious oxidative damage to DNA, proteins and lipids [2] and are produced by exogenous stressors, such as pro-oxidants, heavy metal, pesticides [1] and biotic infections [3]. There are three types of antioxidant peroxidases in insects: thioredoxin peroxidases (Tpxs), also known as peroxiredoxins (Prxs) [4]; phospholipid-hydroperoxide glutathione peroxidases (GPX) homologues with thioredoxin peroxidase activity (GTPX) [5] and glutathione S-transferases (GSTs) [6, 7].

Tpxs are ubiquitous antioxidants found in all organisms [8] and protect enzymatic systems against toxic ROS by eliminating H_2O_2 and alkyl hydroperoxidases by using electrons provided by thioredoxin (Trx) [8-11]. H_2O_2 is mainly reduced by GPX and Prxs in the mitochondria [12-14]. To compensate for the lack of GPX in insects, Tpxs play a critical role in the enzymatic removal of ROS [15]. An *Anopheles stephensi* 2-Cys Prx orthologue of *Drosophila melanogaster* Prx-4783 (AsPrx-4783) protects *A. stephensi* cells against the nitrosative and oxidative stresses associated with malaria parasite infection [16]. A thioredoxin peroxidase of *Bombyx mori* (BmTpx) protects against the oxidative stress caused by temperature stimuli and viral infection [17]. Thioredoxin peroxidase-1 of *Plasmodium vivax* (PvTpx-1) has been shown to reduce and detoxify hydrogen peroxides to maintain redox homeostasis and proliferation in the host [18]. Peroxiredoxin 4 of *B. mori* (BmPrx4) was found to protect *B. mori* from oxidative damage [19]. Thioredoxin peroxidase 1 of *Apis cerana cerana* (AccTpx1) plays an important role in protecting honeybees from oxidative injury, potentially extending their lifespan [20].

Based on the number and position of conserved cysteines, Tpxs are classified into two subfamilies, 1-Cys and 2-Cys, with the 2-Cys subfamily having a second conserved Cys in the C-terminus [21].

Tpx-3 is a mitochondrial 2-Cys thioredoxin peroxidase. *Drosophila* peroxiredoxin 3 (dPrx3), found exclusively within mitochondria [22], catalytically removes hydrogen and organic peroxides to maintain the redox state and to ensure the survival of the adult flies [23]. Because Prx3 is essential for the regulation of adipocyte oxidative stress, defects in Prx3 alter the mitochondrial redox state, mitochondrial functions, and adipokine expression in adipocytes, leading to metabolic alterations [24]. Depletion of *Prx3* leads to H_2O_2 accumulation in mammalian mitochondria and increases the apoptosis induced by staurosporine or by tumour necrosis factorα and cycloheximide [25]. Moreover, *Prx3* overexpression in a mammalian cell line could prevent the cells from undergoing apoptosis caused by *t*-butyl hydroperoxide (*t*-BOOH) and H_2O_2 [26]. All of these observations suggest that *Prx3* plays a crucial role in eliminating H_2O_2 from mitochondria.

Although the antioxidant properties of Tpxs proteins have been studied in other insects, there is limited knowledge of Tpxs proteins in honeybees and, particularly, in Chinese honeybees. Because the Chinese honeybee (*A. cerana cerana*) plays an important role in the balance of regional ecology and agricultural development [27], the identification of Tpxs and the elucidation of the ROS defence mechanisms in *A. cerana cerana* is critical. Based on previous observations, we predicted that Tpx-3 protects *A. cerana cerana* from ROS damage. In this study, we characterized a novel *Tpx* gene from *A. cerana cerana* (*AccTpx-3*) and evaluated its expression in different developmental stages and tissues. We investigated the changes in the *AccTpx-3* expression patterns after various oxidative stresses, including temperature changes and the exposure to H_2O_2, cyhalothrin, acaricide, phoxim, ultraviolet light and $HgCl_2$. We also characterized the antioxidant activity of the recombinant AccTpx-3 protein and the purified AccTpx-3 protein which were both expressed in *Escherichia coli* (*E.*

coli). Our results revealed that *AccTpx-3* might play a crucial role in resisting external oxidative pressures.

Materials and Methods

Animal specimens and treatments

The *A. cerana cerana* used in this study were obtained from the experimental apiary of Shandong Agricultural University (Taian, China). The fourth (L4)-, the fifth (L5)-, and the sixth (L6)-instar larvae, white-eyed (Pw), pink-eyed (Pp), brown-eyed (Pb) and dark-eyed (Pd) pupae and 1 day post-emergence (A1), 15 days post-emergence (A15) and 30 days post-emergence adults (A30) were collected from the hive. The 15 days post-emergence adult worker Chinese honeybees were divided into groups ($n = 30$), fed a diet of water and pollen and maintained in an incubator with 60% relative humidity at 34℃ under a 24 h dark regimen. Groups 1, 2, 3, and 4 were placed at 4, 16, 25, and 42℃, respectively. Groups 5-10 were treated with the following conditions: UV (30 MJ/cm^2); H_2O_2 (0.5 μl of a 2 mmol/L dilution was applied to the thoracic notum of worker bees); $HgCl_2$ (3 mg/ml added to food); or pesticides (cyhalothrin, acaricide and phoxim, which were diluted to 20 mg/L and added to the food for the experimental group). The control group fed on normal food. At appropriate times after treatment, the bees were flash-frozen in liquid nitrogen and stored at −80℃.

The brain, epidermis, muscle and midgut of 15 days post-emergence adult workers were used to examine the tissue-specific expression profiles. These tissues were dissected on ice, washed twice with PBS, frozen in liquid nitrogen and stored at −80℃.

RNA extraction, cDNA synthesis, and DNA isolation

Total RNA was extracted with Trizol reagent (TransGen Biotech, Beijing, China) according to the instructions of the manufacturer. To remove the genomic DNA, the RNA samples were treated with RNase-free DNase Ⅰ, and the cDNA was synthesised by using a reverse transcription system (TransGen Biotech, Beijing, China). The genomic DNA was isolated by using an EasyPure Genomic DNA Extraction Kit in accordance with the instructions of the manufacturer (TransGen Biotech, Beijing, China).

2.3. Primers and PCR amplification conditions

The primers and PCR amplification conditions are listed in Table 1, Table 2, respectively.

Table 1　PCR primers in this study

Abbreviation	Primer sequence (5′—3′)	Description
TP1	TTAATAATGCTTCGATTCTTAG	cDNA sequence primer, forward
TP2	CTGATTCAAAATATTGTTTACTATCC	cDNA sequence primer, reverse
3P1	GGGTGGATTAGGTGGTAATTTAG	3′ RACE forward primer, outer
3P2	GAAGCGTAGATGAAACATTGAGAC	3′ RACE forward primer, inner

Continued

Abbreviation	Primer sequence (5'—3')	Description
5P1	GTCGGTTGTTTAGTTTTCACC	5' RACE reverse primer, outer
5P2	CAAGTCTGAGAATGTAAAGATGC	5' RACE reverse primer, inner
AAP	GGCCACGCGTCGACTAGTAC(G)$_{14}$	Abridged anchor primer
AUAP	GGCCACGCGTCGACTAGTAC	Abridged universal amplification primer
B26	GACTCTAGACGACATCGA(T)$_{18}$	3' RACE universal primer, outer
B25	GACTCTAGACGACATCGA	3' RACE universal primer, inner
QP1	TTAATAATGCTTCGATTCTTAG	Full-length cDNA primer, forward
QP2	AATACAATTCGTTATAGTTTAG	Full-length cDNA primer, reverse
β-s	TTATATGCCAACACTGTCCTTT	Standard control primer, forward
β-x	AGAATTGATCCACCAATCCA	Standard control primer, reverse
N1	AATTGTCGCCCTCTAACATTTTACTAG	Genomic sequence primer, forward
N2	CCGGAAAATTCAGGTGCAGG	Genomic sequence primer, reverse
N3	CCTGCACCTGAATTTTCCGG	Genomic sequence primer, forward
N4	AATACAATTCGTTATAGTTTAGTTATAC	Genomic sequence primer, reverse
YH1	GGTACCTTAATAATGCTTCGATTCTTAG	Protein expression primer, forward
YH2	GAGCTCATTAACTGATTCAAAATA	Protein expression primer, reverse
TPr1	AAATCGCCATCCACTACAGCAG	Inverse PCR forward primer, outer
TPr2	GTCTGCCCTACTGAGTTGA	Inverse PCR reverse primer, outer
TPr3	GCAAGTCTGAGAATGTAAAGATG	Inverse PCR forward primer, inner
TPr4	GCATGGACTAATACACCGAG	Inverse PCR reverse primer, inner
TS1	CCGGATGGTACAACAGTTCCC	Promoter special primer, forward
TS2	GCAAGTCTGAGAATGTAAAGATG	Promoter special primer, reverse
TY1	CCTGCACCTGAATTTTCCGG	Real-time PCR primer, forward
TY2	CTCGGTGTATTAGTCCATGC	Real-time PCR primer, reverse

Table 2 PCR amplification conditions

Primer pair	Amplification conditions
TP1/TP2	10 min at 94℃, 40 s at 94℃, 40 s at 50℃, 50 s at 72℃ for 35 cycles, 10 min at 72℃
3P1/B26	10 min at 94℃, 40 s at 94℃, 40 s at 53℃, 40 s at 72℃ for 28 cycles, 10 min at 72℃
3P2/B25	10 min at 94℃, 40 s at 94℃, 40 s at 53℃, 40 s at 72℃ for 35 cycles, 10 min at 72℃
5P1/AAP	10 min at 94℃, 40 s at 94℃, 40 s at 51℃, 40 s at 72℃ for 28 cycles, 10 min at 72℃
5P2/AUAP	10 min at 94℃, 40 s at 94℃, 40 s at 51℃, 40 s at 72℃ for 35 cycles, 10 min at 72℃
QP1/QP2	10 min at 94℃, 40 s at 94℃, 40 s at 51℃, 1 min at 72℃ for 35 cycles, 10 min at 72℃
N1/N2	10 min at 94℃, 40 s at 94℃, 40 s at 51℃, 50 s at 72℃ for 35 cycles, 10 min at 72℃
N3/N4	10 min at 94℃, 40 s at 94℃, 40 s at 52℃, 50 s at 72℃ for 35 cycles, 10 min at 72℃
TPr1/TPr1	10 min at 94℃, 50 s at 94℃, 50 s at 52℃, 2 min at 72℃ for 30 cycles, 10 min at 72℃
TPr3/TPr4	10 min at 94℃, 50 s at 94℃, 50 s at 47℃, 2 min at 72℃ for 30 cycles, 10 min at 72℃
TS1/TS2	10 min at 94℃, 40 s at 94℃, 40 s at 50℃, 2 min at 72℃ for 35 cycles, 10 min at 72℃
YH1/YH2	10 min at 94℃, 40 s at 94℃, 40 s at 54℃, 1 min at 72℃ for 30 cycles, 10 min at 72℃

Cloning of *AccTpx-3* cDNA

The total RNA was extracted with Trizol (TransGen Biotech, Beijing, China), and the first-strand cDNA was synthesised by using reverse transcriptase (TransGen Biotech, Beijing, China) according to the protocol of the manufacturer. To obtain the internal fragment of the *AccTpx-3* cDNA, primers TP1 and TP2 were designed based on the conserved regions of the *Tpxs* from *Apis mellifera*, *Nasonia vitripennis* and *Drosophila melanogaster* and were then synthesised (Sangon Biotechnological Company, Shanghai, China). Subsequently, specific primers (5P1/5P2 and 3P1/3P2) were generated for the 5' rapid amplification of cDNA end (RACE) and 3' RACE, respectively. Following rapid amplification of cDNA end PCR (RACE-PCR), the full-length cDNA sequence was derived by using specific primers (QP1 and QP2) designed according to the assembled full-length cDNA sequence. The PCR product was cloned into the pEASY-T3 vector (TransGen Biotech, Beijing, China), which was transformed into competent *E. coli* DH5α cells for sequencing. All of the PCR primers and amplification conditions are shown in Table 1, Table 2, respectively.

Cloning of the genomic sequence and promoter of *AccTpx-3*

Genomic DNA was extracted from adult bees by using the EasyPure Genomic DNA Extraction Kit according to the instructions of the manufacturer (TransGen Biotech, Beijing, China). The genomic DNA sequence of *AccTpx-3* was cloned by using the N1/N2 and N3/N4 primer pairs with *A. cerana cerana* genomic DNA as the template. Two genomic sequence fragments of 703 bp and 694 bp were cloned into the pEASY-T3 vector (TransGen Biotech, Beijing, China) and transformed into competent *E. coli* DH5α cells for propagation and sequencing.

The *AccTpx-3* promoter region was cloned via inverse PCR (I-PCR). Genomic DNA was digested with *Hpa*II at 37℃ overnight and was ligated into the plasmid vector by using T4 DNA ligase (TaKaRa, Dalian, China). Based on the genomic sequence of *AccTpx-3*, two pairs of primers (TPr1/TPr2 and TPr3/TPr4) were designed to obtain the promoter, and two specific sequencing primers (TS1 and TS2) were used to verify the promoter sequence. The PCR products were cloned into the pEASY-T3 vector and sequenced. The TFSEARCH database (http://www.cbrc.jp/research/db/TFSEARCH.html) was used to search for transcription factor binding sites in the promoter region of *AccTpx-3*. All of the PCR primers and amplification conditions are shown in Table 1, Table 2, respectively.

Bioinformatic analysis and phylogenetic tree construction

NCBI bioinformatics tools (http://blast.ncbi.nlm.nih.gov/Blast.cgi) were used to detect conserved domains in *AccTpx-3*. DNAMAN version 5.2.2 (Lynnon Biosoft, Quebec, Canada) was used to search for the ORF of *AccTpx-3*. The tertiary structure was predicted by using SWISS-MODEL (http://swiss-model.expasy.org/). The theoretical isoelectric point and molecular weight were predicted by using DNAMAN software. Homologous Tpx-3 sequences in other species were retrieved from the NCBI server and aligned by using DNAMAN software. The

qRT-PCR transcriptional analysis

Quantitative RT-PCR (qRT-PCR) was used to identify the expression patterns of *AccTpx-3* at different developmental stages, in different tissues and with different environmental stresses. A 242 bp fragment of *AccTpx-3* was obtained by using primers TY1 and TY2. The *β-actin* gene (GenBank accession number: XM640276) was used to normalize the RNA input between samples, and the same conditions were used to amplify the *AccTpx-3* fragment. qPCR was performed with the SYBR® PrimeScript™ RT-PCR Kit (TaKaRa, Dalian, China) in a 25 μl volume by using a CFX96™ Real-time System (Bio-Rad). The qPCR amplification conditions were as follows: initial denaturation at 95℃ for 30 s; 41 cycles at 95℃ for 5 s, 55℃ for 15 s and 72℃ for 15 s; and a single melting cycle from 65℃ to 95℃. At least three individual samples were prepared for each sample, and each sample was analyzed three times. The linear relationship, amplification efficiency and data analysis were conducted by using CFX Manager Software version 1.1. The analysis of the significant differences was performed by using Duncan's multiple range tests with Statistical Analysis System (SAS) software version 9.1.

Construction of expression plasmids, recombinant protein expression, and purification

The plasmid containing the *AccTpx-3* cDNA sequence was used as the template for the amplification of *AccTpx-3*. The cDNA fragment was subcloned into the plasmid vector pEASY-T1-simple (TransGen Biotech, Beijing, China) for expression in the *E. coli* BL21 strain. The insert was excised by using the corresponding restriction enzymes, and the digested fragment was separated, recovered and cloned into the expression vector pET-30a (+) to generate a His-tag at the N-terminus. The expression plasmid pET-30a (+)-*AccTpx-3* was then transformed into BL21 (DE3) *E. coli*, and *AccTpx-3* expression was induced with 1.0 mmol/L isopropylthio-β-D-galactoside (IPTG) at 28℃ for 8 h. The cells were shaken at 200 r/min. The bacterial cells were the harvested by centrifugation at 5,000 g for 10 min at 4℃, resuspended in loading buffer, boiled for 10 min, and centrifuged again to remove the insoluble debris. Each sample (20 μl) was subjected to 12% SDS-PAGE.

As described above, *AccTpx-3* was expressed in *E. coli* BL21 (DE3) and then purified. The bacteria were lysed by using an ultrasonic disrupter, and the suspension was purified under native conditions with Ni^{2+}-NTA (nitrilotriacetic acid)-binding resin (Qiagen, Shanghai, China) according to the instructions of the manufacturer. The purified protein was examined by 12% SDS-PAGE.

Characterisation of recombinant AccTpx-3 protein by disc diffusion assay

The recombinant AccTpx-3 was expressed in *E. coli* BL21 (DE3). An initial bacterial culture with a density of 10^6 cells/ml was plated on LB-kanamycin agar plates and incubated at 37℃ for 1 h. Subsequently, filter discs (8 mm diameter) soaked with different concentrations (0,

10, 50, 100, and 200 mmol/L) of paraquat were placed on the surface of the agarose medium plates, and the *E. coli* were grown for another 24 h at 37℃. Bacteria transfected with the pET-30a (+) vector were only used as a negative control.

DNA cleavage assay in the MFO system

The ability of *AccTpx-3* to protect DNA was determined with a method modified from Yu et al. (2011) [20]. Reaction mixtures containing 3 μmol/L $FeCl_3$, 10 mmol/L dithiothreitol (DTT), 100 mmol/L 4-(2-hydroxyethyl)-1-piperazineethanesulfonic acid (HEPES) buffer, and increasing concentrations of *AccTpx-3* (ranging from 2.4 μg/ml to 24.0 μg/ml) were incubated in a total volume of 25 μl at 37℃ for 30 min. Supercoiled pUC19 plasmid DNA (300 ng) was added to the mixture for another 2.5 h at 37℃.

Results

Characterisation of *AccTpx-3* and sequence analysis

Reverse-transcription PCR (RT-PCR) and RACE-PCR were used to obtain the full-length 1,205 bp cDNA sequence of *AccTpx-3* (GenBank accession number: JX456217), which included 370 bp in the 5′ untranslated region (UTR), 729 bp in the ORF and 106 bp in the 3′ UTR. The open reading frame (ORF) of *AccTpx-3* encodes a polypeptide of 242 amino acids with a calculated molecular mass of 27.084 kDa and a theoretical isoelectric point (pI) of 8.70. An alignment of the AccTpx-3 amino acid sequence with Tpx reference sequences from other species revealed that the predicted protein is highly homologous to AmTpx-3v1 and AmTpx-3v2 (thioredoxin peroxidase 3 transcript variant 1, NP_001171495.1, and transcript variant 2, NP_001171494.1) from *A. mellifera*, DmTpx-3 (NP_524387.1) from *D. melanogaster* and NvTpx-2like (XP_003428081.1) from *N. vitripennis*. The identity ranged from 56.97% to 98.76%, suggesting that the Tpx family is highly conserved across species. As shown in Fig. 1A, the essential catalytic cysteines Cys^{94} and Cys^{216} are present in the *AccTpx-3* amino acid sequence, indicating that *AccTpx-3* is a 2-Cys Tpx. The mitochondrial-targeting peptide of AccTpx-3 and AmTpx-3v1 predicted by MitoProtII (http://www.mybiosoftware.com/mitoprot-ii-1-101-prediction-of-mitochondrial-targeting-sequences.html) is found in region Ⅰ, and the conserved regions involved in catalysis are present in region Ⅱ and region Ⅳ. As shown in Fig. 1A, the residues (GGLG and YF) in regions Ⅲ and Ⅴ involved in increasing the susceptibility of Tpxs to hyperoxidation are boxed, and the residues involved in stabilising the active site (Pro^{88} and Thr^{91}) are marked with arrows.

To analyze the evolutionary relationships among the Tpxs from different insect species, a phylogenetic tree of the 2-Cys Tpxs was constructed (Fig. 1B). In this analysis, AccTpx-3 and AmTpx-3v1 were more closely related to each other than to homologues in other species, and this result confirmed the relationship predicted from the amino acid alignment.

To understand the conserved residues and clarify the catalytic mechanism of *AccTpx-3* further, the tertiary structure of AccTpx-3 was predicted, and the conserved redox-active cysteines were identified (Fig. 1C).

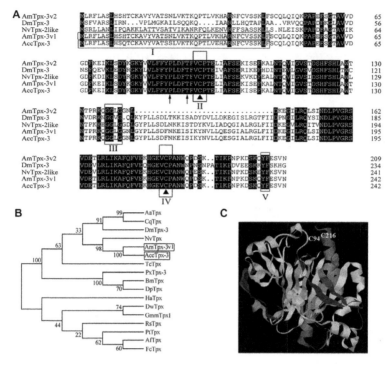

Fig. 1 Characterisation of thioredoxin peroxidases (Tpxs) from various species. A. Multiple amino acid sequence alignment of AccTpx-3 (thioredoxin peroxidase 3, GenBank accession number: JX456217) from *A. cerana cerana*, AmTpx-3v1, AmTpx-3v2 (thioredoxin peroxidase 3 transcript variant 1, NP_001171495.1, and transcript variant 2, NP_001171494.1) from *Apis mellifera*, DmTpx-3 (thioredoxin peroxidase 3, NP_524387.1) from *Drosophila melanogaster* and NvTpx-2like (thioredoxin peroxidase-2-like, XP_003428081.1) from *Nasonia vitripennis*. Identical amino acid residues are shaded in black. The amino acids conserved in the catalytic regions are marked with ▲. As shown in region I, the mitochondrial-targeting peptide of AmTpx-3v1 and *AccTpx-3* is boxed. Conserved regions (II and IV) are boxed. Residues (GGLG and YF) are shown in region III and region V. Pro^{88} and Thr^{91} are marked with arrows. B. Phylogenetic analysis of Tpx amino acid sequences from different insect species. The amino acid sequences of Tpxs were obtained from NCBI as follows: AaTpx (*Aedes aegypti*, XP_001663718.1), CqTpx (*Culex quinquefasciatus*, XP_001852614.1), DmTpx-3 (*Drosophila melanogaster*, NP_524387.1), NvTpx-2like (*Nasonia vitripennis*, XP_003428081.1), AmTpx-3v1 (*Apis mellifera*, NP_001171495.1), AccTpx-3 (*Apis cerana cerana*, JX456217), TcTpx (*Tribolium castaneum*, XP_967356.1), PxTpx-3 (*Papilio xuthus*, BAM18129.1), BmTpx (*Bombyx mori*, NP_001040464.1), DpTpx (*Danaus plexippus*, EHJ73839.1), HaTpx (*Helicoverpa armigera*, ABW96360.1), DwTpx (*Drosophila willistoni*, XP_002071309.1), GmmTpx1 (*Glossina morsitans morsitans*, AAT85824.1), RsTpx (*Rhipicephalus sanguineus*, ACX54025.1), PtTpx (*Portunus trituberculatus*, ACI46625.1), AfTpx (*Artemia franciscana*, ABY62745.1), and FcTpx (*Fenneropenaeus chinensis*, ABB05538.1). C. The tertiary structure of AccTpx-3. The structure was built by using homology modelling in the SWISS-MODEL modelling environment. Cys^{94} and Cys^{216}, which are found in the predicted catalytic active site, are shown.

Structure analysis of the *AccTpx-3* genomic sequence

PCR was used to amplify the genomic DNA sequence of *AccTpx-3*. The full-length genomic DNA sequence (GenBank accession number: JX456218) consisted of 1,433 bp, and three exons and two introns were found. The splice junctions had the canonical nucleotide sequences at the 5′-GT splice donor and 3′-AG splice acceptor sites.

To examine further the relationship of *AccTpx-3* with the Tpxs from *A. mellifera*, *D. melanogaster*, and *N. vitripennis*, the sequences of the exons and introns from these genes were aligned (Fig. 2). *AccTpx-3*, *NvTpx-2like*, and *AmTpx-3v1* contained two introns, whereas *DmTpx-3* and *AmTpx-3v2* have three introns. In accordance with the phylogenetic analysis, the genomic structures of *AccTpx-3* and *AmTpx-3v1* were very similar.

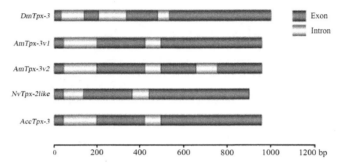

Fig. 2 Schematic representation of the DNA structures of Tpxs. Exon-intron organisation of *AccTpx-3* from *Apis cerana cerana* and its homologues, including *DmTpx-3* from *Drosophila melanogaster* (NP_524387.1), *AmTpx-3v1* from *Apis mellifera* (NP_001171495.1), *AmTpx-3v2* from *Apis mellifera* (NP_001171494.1), and *NvTpx-2like* from *Nasonia vitripennis* (XP_003428081.1). The exons are indicated with grey boxes, and the introns with white boxes.

Identification of the 5′-flanking region of *AccTpx-3*

To identify the regulatory sequences involved in *AccTpx-3* expression, I-PCR was used to amplify the 5′-flanking region of *AccTpx-3*. We isolated a 980 bp DNA fragment upstream of the transcription start site. Transcription factor binding sites were predicted by using the TFSEARCH software. As shown in Fig. 3, three heat shock factor (HSF), two Broad-Complex (BR-C), two Deformed (Dfd), two CF2-II and one AP-1 transcription factor binding sites were identified in the 5′-flanking region of *AccTpx-3*. Dfd determines the specificity of homeotic gene action [28], and BR-C is an ecdysone-responsive key regulator of metamorphosis [29]. CF2-II is a DNA-binding zinc-finger domain that regulates distinct sets of genes during the development of embryonic [30].

Developmental and tissue-specific expression patterns of *AccTpx-3*

qPCR was used to determine the *AccTpx-3* expression patterns at different developmental stages and in various tissues. As shown in Fig. 4A, *AccTpx-3* expression was the highest in the fourth-instar larvae, and the next highest expression level was found in the 15 days post-emergence adult workers. In all of the other developmental stages, the *AccTpx-3* mRNA was expressed in a similar pattern but was lower than in the 15 day s post-emergence adult workers. The spatial expression profiles showed that *AccTpx-3* transcripts were abundant in muscle; although the expression levels in the brain, epidermis and midgut were nearly equal, they were all lower than that in muscle (Fig. 4B).

```
                                              HSF      AP-1
GGGTAATATAGTCACAGTAATGCTCTCGGAAAATAT AGAAGAA GTGACTCAA CCGGATGGTACAA        -916

                                                                  BR-C
CAGTTCCCATGTTGGTTGAGACACATTTCCAAATGAATTATACAAATGGCGAATGGAAG AGGATA         -851

 AAGAAAA ATCGAAGAATTATTACAGAAGAATCAACGTCCACTACGACTCCTACTCCCAGTGTGAC        -786

                                    HSF
GGCAACAGCTTCCAATTGAAAAAAAATAAA AGAAA TTTTTATCGAGTCGCAATTATATAGGTCAA        -721

TAATTCTGTCATTTTTTGGATGACAGTGTTTTGAAAAACTCTGTTCTATAGATAGCAAAAGAATT         -656

TGAATACAGAAGGCAAAACATGGCCTCTGTGTTCCATATGGTGTAAGCAGACTTGTGTATCATGT         -591

TTTAGTAGTAAAACTTTAATGTATGATGTACATGTTACATCAAAAAAGATGTTATTCGAAAAAT          -526

ACAATGTAATAATGAAGATATTCTTGTTTCCATTTTTTTTAAAAATGGAAGAATAGTTAAACGGA         -461

                                                                 BR-C
CACATTTTTAATTGAAAATAACGCTTAGACATTATTTTCAACAGATTTCAAATATT ATTAGAAA         -396

                                                                 Dfd
 TAAAA CTTCTATTTCTATGCGCGCATCTGGTTTCATTGACAATGAATAAAAA TCATTGATTAAT        -331

         Dfd
 TAAA AAATTAAATG CTATTTATTAATAGAT AGTATGATATTTTCATTGTAGAGGAAACCGGATA       -266

                                 CF2-II
TTATTAATTTAAATGCGATAGTTGGTAATTTGAAT TTATATGTA AATATATACATAAAAACAAGT       -201

TTCTTCATTCAATAGAAACGAATATAAAACTTATAATATTGTTATTTGAATAAAATATGGTTATA         -136

                            CF2-II                              HSF
TTGATTAAATTTGATTTAATTAAAAAA TTATATTTAT TATTATAATGCAATATATACTCG AGAAT       -71

GATTTGAAATGTTGAGATGTTAAGATATTGATGTATTGGCAATTCCAAAGATCAATTTTCTTACT         -6

TCGAC AATTGTCGCCCTCTAACATTTTACTAGAAGGTTATTTCGTTTTTCGAAAGAATATTTTAT        +60
      ↳ +1 Transcription start

AAGAATCTAAACAATTATTTTCTCGATAGTAATTAATATTTGATTAACAAAATTAATCAAAATTA         +125

AAATATATTTAATAAATTCTAAGATATTTTCTATTATTTTTTATTAAAATGTGTCATTCATAAGA         +190

AAAGTGAATTATATAACCTCTTTGAAAAGTTCAAGATAAACTTTCAACAAGAAAATAAGTTTTAA         +255

ATTAATATTTAAATTTAATATTTATTGTTTTATGAAGTAATTTTGTATTTCTCTGATTATAAGA         +320

TTTATTATCGTTGCTTTGTCCTTGCATTGAGAATAATTTATATATTAATA ATG                    +373
                                                    ↳ Translation start
```

Fig. 3 The partial nucleotide sequence and putative transcription factor binding sites of the 5'-flanking region of *AccTpx-3*. The transcription factor binding sites are boxed. The transcription and translation start sites are indicated with arrowheads.

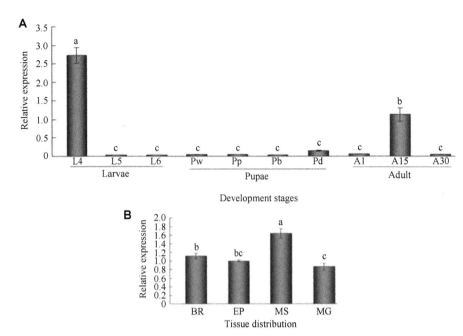

Fig. 4 Expression profiles of *AccTpx-3* as determined by Q-PCR. A. Transcriptional expression level of *AccTpx-3* at different developmental stages: larvae (L4 [4 d fourth instars], L5 [5 d fifth instars], L6 [6 d sixth instars]), pupae (Pw, white-eyed pupae; Pp, pink-eyed pupae; Pb, brown-eyed pupae and Pd, dark-eyed pupae), and adults (A1 [1 day post-emergence], A15 [15 days post-emergence] and A30 [30 days post-emergence]). B. Expression analysis of *AccTpx-3* in various tissues revealed the distribution of AccTpx-3 transcripts in the brain (BR), epidermis (EP), muscle (MS) and midgut (MG). The *β-actin* gene was used as an internal control. Histograms reveal the relative expression profiles of *AccTpx-3*. Each value is given as the mean (SD) of three replicates. The letters on the bar represent a significant difference at $P < 0.001$, as determined by Duncan's multiple range tests using SAS software version 9.1.

Transcriptional expression profiles of *AccTpx-3* in response to various oxidative stresses

qPCR was performed to characterize the transcriptional expression profiles of *AccTpx-3* at 4, 16, 25, and 42°C, with the transcriptional expression level normalized to control bees. Interestingly, *AccTpx-3* gene expression was different at the various temperatures. As shown in Fig. 5A-D, transcription was upregulated at 4 and 42°C. At 4°C, the *AccTpx-3* transcripts reached their maximum level at 1.0 h and then declined to half of the control level at 1.5 h. Under stress at 42°C, the expression level of *AccTpx-3* increased slightly compared to the control and peaked at 1 h. Conversely, mRNA expression was downregulated at 16 and 25°C. We also used Q-PCR to analyze the expression level of *AccTpx-3* after ultraviolet (UV), H_2O_2, mercury ($HgCl_2$) and pesticides (cyhalothrin, acaricide, and phoxim) treatments. As shown in Fig. 5E-J, the ultraviolet and $HgCl_2$ conditions downregulated the expression of *AccTpx-3*, whereas *AccTpx-3* expression was upregulated by the application of H_2O_2 and pesticides. Taken together, these results showed that *AccTpx-3* may play an important role in response to various oxidative stresses.

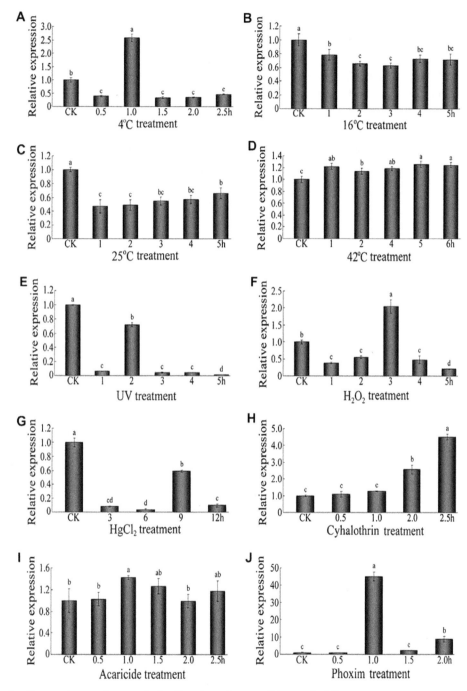

Fig. 5 Expression patterns of *AccTpx-3* under conditions of different environmental stresses. *AccTpx-3* expression was normalized to untreated adult worker bees housed at 34℃. The *β-actin* gene was used as an internal control. Panels A-D show *AccTpx-3* expression in 15 days post-emergence adult workers housed at 4, 16, 25, and 42℃. Panels E-J show *AccTpx-3* expression in 15 days post-emergence adult workers exposed to ultraviolet (UV), H_2O_2, $HgCl_2$, cyhalothrin, acaricide, and phoxim treatments. Each value is given as the mean (SD) of three replicates. The letters on the bar represent a significant difference at $P < 0.001$, as determined by Duncan's multiple range tests using SAS software version 9.1.

Expression, purification and characterisation of recombinant AccTpx-3 protein

The ORF of *AccTpx-3* was cloned into the pET-30a (+) vector, which was digested with *Kpn* I and *Sac* I, and expressed in the *Escherichia coli* (*E. coli*) strain BL21 as a 6×His-tagged fusion protein. SDS-PAGE analysis showed that a target protein was induced by IPTG at 28℃ for 8 h (Fig. 6, lane 2). The recombinant AccTpx-3 protein was purified by using affinity chromatography (Fig. 6, lane 3), and the concentration of the soluble recombinant protein was approximately 0.25 mg/ml.

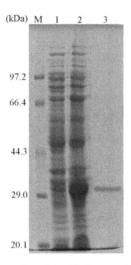

Fig. 6 SDS-PAGE analysis of expressed and purified recombinant AccTpx-3 protein. M, molecular mass standard; lanes 1 and 2, non-induced and induced overexpression of pET-30a (+)-*AccTpx-3* in BL21; lane 3, purified recombinant AccTpx-3 protein.

We characterized the peroxidase activity of the recombinant AccTpx-3 protein by using the disc diffusion method. *E. coli* (BL21) overexpressing recombinant AccTpx-3 was exposed to paraquat, which acted as an external environmental stressor. As shown in Fig. 7A-C, the bacteria expressing the recombinant protein had a smaller killing zone than the pET-30a (+) empty vector control (data not shown), indicating that the recombinant protein had antioxidant activity.

The thiol mixed-function oxidation (MFO) system was employed to verify that the AccTpx-3 protein protects DNA from the oxidative damage induced by ROS. In this experiment, DTT in the presence of Fe^{3+} catalyses the conversion of O_2 to H_2O_2, and the H_2O_2 is then converted to hydroxyl radicals by the Fenton reaction [9, 31]. Hydroxyl radicals induce DNA strand breaks [32], but Tpxs can protect DNA from this ROS-induced damage. As shown in Fig. 8, supercoiled pUC19 plasmid was converted to the nicked form (lane 8) when exposed to MFO-induced stress. However, in the negative control lacking DTT, the plasmid was not converted to the nicked form (lane 10), suggesting that DTT acts as an electron donor [33]. The nicked form of the pUC19 plasmid decreased gradually when the concentration of recombinant AccTpx-3 increased (lanes 1 to 7). These data suggest that AccTpx-3 acts as an antioxidant that protects against the oxidative damage induced by ROS.

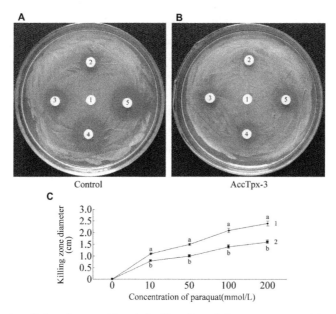

Fig. 7 Antioxidant activity of recombinant AccTpx-3 protein, as measured by the disc diffusion method. A. *E. coli* BL21 bacteria transfected with pET-30a (+) (vector only) were used as the control. B. *E. coli* BL21 bacteria transfected with pET-30a (+)-AccTpx-3. The filter discs labelled "1" represent the negative control. Filter discs 2, 3, 4 and 5 represent 10, 50, 100, and 200 mmol/L paraquat, respectively. C. Detailed analysis of killing zones. The paraquat concentrations include 0, 10, 50, 100 and 200 mmol/L. The diamonds (1) and squares (2) represent the control and AccTpx-3, respectively. Each value is given as the mean (SD) of three replicates. The letters on the bar represent a significant difference at $P < 0.001$, as determined by Duncan's multiple range tests using SAS software version 9.1.

Fig. 8 Antioxidant activity of AccTpx-3, as measured by the mixed-function oxidation system. The reaction conditions were as follows: lanes 1-7, pUC19 plasmid DNA + $FeCl_3$ + dithiothreitol (DTT) + purified AccTpx-3 (2.4, 6.0, 9.6, 13.2, 16.8, 20.4 and 24.0 μg/ml, respectively); lane 8, pUC19 plasmid DNA + $FeCl_3$ +DTT ; lane 9, pUC19 plasmid DNA only; and lane 10, pUC19 plasmid DNA + $FeCl_3$. SF indicates the supercoiled form; NF indicates the nicked form.

Discussion

Research on Tpxs is mainly concentrated in model insect species, such as *D. melanogaster* and *Anopheles stephensi*, whereas the expression of *AccTpx-3* in *A. cerana cerana* due to the exposure to oxidative stress has not been previously reported. In this study, we identified a thioredoxin peroxidase gene (*AccTpx-3*) from *A. cerana cerana* and investigated the response of *AccTpx-3* to various abiotic stresses. Our results indicate that *AccTpx-3* plays a crucial role in protecting *A. cerana cerana* from oxidative stress.

Sequence analysis showed that *AccTpx-3* contains two copies of the Tpx essential catalytic motif, valine-cysteine-proline (VCP) [34], and also contains residues involved in increasing the susceptibility of Tpxs to hyperoxidation (GGLG and YF) and in stabilising its active site (Pro^{88} and Thr^{91}) [35]. These observations indicated that *AccTpx-3* is a typical 2-Cys Tpx. Prx3 is a thioredoxin peroxidase localized to mitochondria; similarly, as predicted by MitoProtII, the presence of a mitochondrial-targeting peptide in *AccTpx-3* indicated that it belongs to the mitochondrial family of Tpxs. The overexpression of *Prx3* has been reported to decrease ROS production and lipid peroxidation [36]. In addition, the phylogenetic tree constructed in the present study was in accordance with the amino acid sequence alignment, indicating that AccTpx-3 and AmTpx-3v1 are very closely related.

The tissue-specific expression analysis of the *AccTpx-3* gene showed a greater mRNA accumulation in muscle compared with the corresponding Tpxs in the epidermis of *Gryllotalpa orientalis* and *B. mori* [37, 38]. In addition, our expression analysis in various developmental stages of *A. cerana cerana* showed that *AccTpx-3* transcription is the highest in the fourth-instar larvae and then followed by 15 days post-emergence adult workers when compared to 14 days post-emergence adult workers [20], suggesting that the *AccTpx-3* might play an important role in preventing damage due to ROS during the growth of the fourth-instar larvae and 15 days post-emergence adult workers.

Temperature is an abiotic environmental factor that causes physiological changes in organisms [39]. In general, heat shock stress can result in polyamine oxidation to generate H_2O_2 [40], and ROS act as key mediators of cold-induced apoptosis [41]. A yeast mutant lacking *Tpx* is thermosensitive, which suggests that *Tpx* acts in the cellular defence against heat shock [17]. In the present study, *AccTpx-3* was upregulated at 4 and 42℃, and this result was similar to that observed for *BmTpx* during the exposure to 4 and 37℃ [42]. Depending on the type of stressors, Tpxs have both beneficial and detrimental effects on cell viability. Cells overexpressing Tpx have an increased susceptibility to the oxidative stress caused by elevated temperatures or copper [15]. Dietz (2011) [43] demonstrated that *Prx* transcriptional regulation varied according to the type of *Prx*, stress intensity and developmental stage in plants. We also found that *AccTpx-3* was downregulated at 16 and 25℃, suggesting that there might not be enough ROS produced by 16 and 25℃ treatments to be stimuli to the host. *AccTpx-3* might also be involved in different signal transduction processes compared to other Tpxs. All of these results indicated that *AccTpx-3* may be involved in preventing the oxidative damage caused by ROS.

BmTpx mRNA expression was upregulated in fatty tissues after H_2O_2 injection [42], and *AccTpx-3* was also upregulated by H_2O_2 injection in the present study. Pesticides destroy the biochemical and physiological functions of erythrocytes and lymphocytes by causing lipid biomembrane oxidation [1], and *AccTpx-3* was also induced during cyhalothrin, acaricide and phoxim treatments, which was consistent with the above results. The Tpx in *Helicoverpa armigera* adults is significantly upregulated at specific time points (from 30 min to 90 min) following UV-A radiation [44]. It was also demonstrated that heavy metal could enhanced endogenous ROS [1]. However, we found that *AccTpx-3* was repressed during the exposure to ultraviolet light and $HgCl_2$, indicating that *AccTpx-3* might play different roles in resistance to ultraviolet light and $HgCl_2$ treatments or might be involved in different signal transduction processes. All of these results suggest that *AccTpx-3* is involved in resisting oxidative stress.

In the presence of DTT, the recombinant thioredoxin peroxidase 1 protein of *Bombus ignitus* (BiTpx1) was able to eliminate H_2O_2 when it was expressed in baculovirus-infected insect Sf9 cell [45]. Consistent with our previous AccTpx1 study [20], recombinant *Cryptosporidium parvum* Tpx (CpTPx) protects plasmid DNA from damage by metal-catalysed oxidation *in vitro* [46], and a mitochondrial peroxiredoxin (LdmPrx) of *Leishmania donovani* can detoxify organic and inorganic peroxides and prevent hydroxyl radical-induced DNA damage [47]. In this study, we demonstrated that recombinant AccTpx-3 protected plasmid DNA from hydroxyl radical damage. Moreover, the ability of the recombinant AccTpx-3 protein to protect plasmid DNA is obviously higher than that of AccTpx1 because the demand for recombinant AccTpx-3 protein is less than that of AccTpx1. In addition, the killing zones of *E. coli* overexpressing AccTpx-3 exposed to paraquat were smaller than those of the control bacteria, which definitively demonstrated that AccTpx-3 contributes to the cellular resistance to oxidative stress.

In conclusion, we have identified an antioxidant enzyme gene (*AccTpx-3*) in *A. cerana cerana* and examined the expression profiles of *AccTpx-3* in different developmental stages and tissues. We also investigated *AccTpx-3* mRNA expression with various oxidative stresses and, interestingly, found that *AccTpx-3* transcription varied in response to the oxidative stresses: *AccTpx-3* expression was induced by some stresses and repressed by others. In addition, the purified recombinant AccTpx-3 was able to protect plasmid DNA from ROS damage better than AccTpx1, and *E. coli* overexpressing AccTpx-3 exhibited less sensitivity to paraquat than the control bacteria. These results provided evidence that *AccTpx-3* protects against oxidative stress. In summary, the cloning, characterization, expression analysis and functional testing of *AccTpx-3* enrich our understanding of Tpx proteins in *A. cerana cerana*.

References

[1] Narendra, M., Bhatracharyulu, N. C., Padmavathi, P., Varadacharyulu, N. C. Prallethrin induced biochemical changes in erythrocyte membrane and red cell osmotic haemolysis in human volunteers. Chemosphere. 2007, 67, 1065-1071.

[2] Halliwell, B., Gutterridge, J. M. C. Production against Radical Damage: Systems with Problems, second ed. Free Radical Biology and Medicine Clarendon Press, Oxford. 1989.

[3] Brennan, L. J., Keddie, B. A., Braig, H. R., Harris, H. L. The endosymbiont *Wolbachia pipientis* induces the expression of host antioxidant proteins in an *Aedes albopictus* cell line. PLoS One. 2008, 3, e2083.

[4] Radyuk, S. N., Klichko, V. I., Spinola, B., Sohal, R. S., Orr, W. C. The peroxiredoxin gene family in *Drosophila melanogaster*. Free Radical Biology and Medicine. 2001, 31, 1090-1100.

[5] Missirlis, F., Rahlfs, S., Dimopoulos, N., Bauer, H., Becker, K., Hilliker, A., Phillips, J. P., Jäckle, H. A putative glutath-ione peroxidase of *Drosophila* encodes a thioredoxin peroxidase that provides resistance against oxidative stress but fails to complement a lack of catalase activity. Biological Chemistry. 2003, 384, 463-472.

[6] Tang, A. H., Tu, C. P. Biochemical characterization of *Drosophila* glutathione *S*-transferases D1 and D21. Journal of Biological Chemistry. 1994, 269, 27876-27884.

[7] Toba, G., Aigaki, T. Disruption of the microsomal glutathione *S*-transferase-like gene reduces lifespan of *Drosophila melanogaster*. Gene. 2000, 253, 179-187.

[8] Chae, H. Z., Robison, K., Poole, L. B., Church, G., Storz, G., Rhee, S. G. Cloning and sequencing of thiol-specific antioxidant from mammalian brain: alkyl hydroperoxide reductase and thiol-specific antioxidant define a large family of antioxidant enzymes. Proceedings of the National Academy of Sciences of the United States of America. 1994, 91, 7017-7021.

[9] Lim, Y. S., Cha, M. K., Uhm, T. B., Park, J. W., Kim, K., Kim, I. H. Removals of hydrogen peroxide and hydroxyl radical by thiol-specific antioxidant protein as a possible role *in vivo*. Biochemical and Biophysical Research Communications. 1993, 192, 273-280.

[10] Chae, H. Z., Chung, S. J., Rhee, S. G. Thioredoxin-dependent peroxide reductase from yeast. Journal of Biological Chemistry. 1994, 269, 27670-27678.

[11] Kang, S. W., Baines, I. C., Rhee, S. G. Characterization of a mammalian peroxiredoxin that contains one conserved cysteine. Journal of Biological Chemistry. 1998, 273, 6303-6311.

[12] Panfili, E., Sandri, G., Ernster, L. Distribution of glutathione peroxidases and glutathione reductase in rat brain mitochondria. FEBS Letters. 1991, 290, 35-37.

[13] Esworthy, R. S., Ho, Y. S., Chu, F. F. The *Gpx1* gene encodes mitochondrial glutathione peroxidase in the mouse liver. Archives of Biochemistry and Biophysics. 1997, 340, 59-63.

[14] Ho, Y. S., Magnenat, J. L., Bronson, R. T., Cao, J., Gargano, M., Sugawara, M., Funk, C. D. Mice deficient in cellular glutathione peroxidase develop normally and show no increased sensitivity to hyperoxia. Journal of Biological Chemistry. 1997, 272, 16644-16651.

[15] Radyuk, S. N., Sohal, R. S., Orr, W. C. Thioredoxin peroxidases can foster cytoprotection or cell death in response to different stressors: over- and under-expression of thioredoxin peroxidase in *Drosophila* cells. Biochemical Journal. 2003, 371, 743-752.

[16] Peterson, T. M. L., Luckhart, S. A mosquito 2-Cys peroxiredoxin protects against nitrosative and oxidative stresses associated with malaria parasite infection. Free Radical Biology and Medicine. 2006, 40, 1067-1082.

[17] Lee, S. M., Park, J. W. Thermosensitive phenotype of yeast mutant lacking thioredoxin peroxidase. Archives of Biochemistry and Biophysics. 1998, 359, 99-106.

[18] Hakimi, H., Asada, M., Angeles, J. M. M., Inoue, N., Kawazu, S. I. Cloning and characterization of *Plasmodium vivax* thioredoxin peroxidase-1. Parasitology Research. 2012, 111, 525-529.

[19] Shi, G. Q., Yu, Q. Y., Shi, L., Zhang, Z. Molecular cloning and characterization of peroxiredoxin 4 involved in protection against oxidative stress in the silkworm *Bombyx mori*. Insect Molecular Biology. 2012, 21, 581-592.

[20] Yu, F., Kang, M., Meng, M., Guo, X., Xu, B. Molecular cloning and characterization of a thioredoxin peroxidase gene from *Apis cerana cerana*. Insect Molecular Biology. 2011, 20, 367-378.

[21] Trivelli, X., Krimm, I., Ebel, C., Verdoucq, L., Prouzet-Mauleon, V., Chartier, Y., Tsan, P., Lauquin, G., Meyer, Y., Lancelin, J. M. Characterization of the yeast peroxiredoxin Ahp1 in its reduced active and overoxidized inactive forms using NMR. Biochemistry. 2003, 42, 14139-14149.

[22] Chae, H. Z., Kim, H., Kang, S. W., Rhee, S. G. Characterization of three isoforms of mammalian peroxiredoxin that reduce peroxides in the presence of thioredoxin. Diabetes Research and Clinical Practice. 1999, 45, 101-112.

[23] Radyuk, S. N., Rebrin, I., Klichko, V. I., Sohal, B. H., Michalak, K., Benes, J., Sohal, R. S., Orr, W. C. Mitochondrial peroxiredoxins are critical for the maintenance of redox state and the survival of adult *Drosophila*. Free Radical Biology and Medicine. 2010, 49, 1892-1902.

[24] Huh, J. Y., Kim, Y., Jeong, J., et al. Peroxiredoxin 3 is a key molecule regulating adipocyte oxidative stress, mitochondrial biogenesis, and adipokine expression. Antioxidants & Redox Signaling. 2012, 16, 229-243.

[25] Chang, T. S., Cho, C. S., Park, S., Yu, S., Kang, S. W., Rhee, S. G. Peroxiredoxin III, a mitochondrion-specific peroxidase, regulates apoptotic signaling by mitochondria. Journal of Biological Chemistry. 2004, 279, 41975-41984.

[26] Nonn, L., Berggren, M., Powis, G. Increased expression of mitochondrial peroxiredoxin-3 (thioredoxin peroxidase-2) protects cancer cells against hypoxia and drug-induced hydrogen peroxide-dependent apoptosis. Molecular Cancer Research. 2003, 1, 682-689.

[27] Yang, G. Harm of introducing the western honeybee *Apis mellifera* L. to the Chinese honeybee *Apis cerana* F. and its ecological impact. Acta Entomologica Sinica. 2005, 3, 401-406.

[28] Ekker, S. C., Von Kessler, D. P., Beachy, P. A. Differential DNA sequence recognition is a determinant of

specificity in homeotic gene action. The EMBO Journal. 1992, 11, 4059-4072.
[29] von Kalm, L., Crossgrove, K., Von Seggern, D., Guild, G. M., Beckendorf, S. K. The Broad-Complex directly controls a tissue-specific response to the steroid hormone ecdysone at the onset of *Drosophila metamorphosis*. The EMBO Journal. 1994, 13, 3505-3516.
[30] Gogos, J. A., Hsu, T., Bolton, J., Kafatos, F. C. Sequence discrimination by alternatively spliced isoforms of a DNA binding zinc finger domain. Science. 1992, 257, 1951-1955.
[31] Lim, Y. S., Cha, M. K., Yun, C. H., Kim, H. K., Kim, K., Kim, I. H. Purification and characterization of thiol-specific antioxidant protein from human red blood cell: a new type of antioxidant protein. Biochemical and Biophysical Research Communications. 1994, 199, 199-206.
[32] Cochrane, C. G. Mechanisms of oxidant injury of cells. Molecular Aspects of Medicine. 1991, 12, 137-147.
[33] Suttiprapa, S., Loukas, A., Laha, T., Wongkham, S., Kaewkes, S., Gaze, S., Brindley, P. J., Sripa, B. Characterization of the antioxidant enzyme, thioredoxin peroxidase, from the carcinogenic human liver fluke, *Opisthorchis viverrini*. Molecular and Biochemical Parasitology. 2008, 160, 116-122.
[34] Wood, Z. A., Schroder, E., Robin Harris, J., Poole, L. B. Structure, mechanism and regulation of peroxiredoxins. Trends in Biochemical Sciences. 2003, 28, 32-40.
[35] Cox, A. G., Winterbourn, C. C., Hampton, M. B. Mitochondrial peroxiredoxin involvement in antioxidant defence and redox signaling. Biochemical Journal. 2010, 425, 313-325.
[36] Chen, L., Na, R., Gu, M., et al. Reduction of mitochondrial H_2O_2 by overexpressing peroxiredoxin 3 improves glucose tolerance in mice. Aging Cell. 2008, 7, 866-878.
[37] Kim, I., Lee, K. S., Hwang, J. S., Ahn, M. Y., Li, J., Sohn, H. D., Jin, B. R. Molecular cloning and characterization of a peroxiredoxin gene from the mole cricket, *Gryllotalpa orientalis*. Comparative Biochemistry and Physiology Part B: Biochemistry and Molecular Biology. 2005, 140, 579-587.
[38] Wang, Q., Chen, K., Yao, Q., Zhao, Y., Li, Y., Shen, H., Mu, R. Identification and characterization of a novel 1-Cys peroxiredoxin from silkworm, *Bombyx mori*. Comparative Biochemistry and Physiology Part B: Biochemistry and Molecular Biology. 2008, 149, 176-182.
[39] An, M. I., Choi, C. Y. Activity of antioxidant enzymes and physiological responses in ark shell, *Scapharca broughtonii*, exposed to thermal and osmotic stress: effects on hemolymph and biochemical parameters. Comparative Biochemistry and Physiology Part B: Biochemistry and Molecular Biology. 2010, 155, 34-42.
[40] Hariari, P. M., Fuller, D. J., Gerner, E. W. Heat shock stimulate polyamine oxidation by two distinct mechanisms in mammalia cell cultures. International Journal of Radiation Oncology Biology Physics. 1989, 16, 451-457.
[41] Rauen, U., Polzar, B., Stephan, H., Mannherz, H. G., de Groot, H. Cold-induced apoptosis in cultured hepatocytes and liver endothelial cells: mediation by reactive oxygen species. FASEB Journal. 1999, 13, 155-168.
[42] Lee, K. S., Kim, S. R., Park, N. S., et al. Characterization of a silkworm thioredoxin peroxidase that is induced by external temperature stimulus and viral infection. Insect Biochemistry and Molecular Biology. 2005, 35, 73-84.
[43] Dietz, K. J. Peroxiredoxins in plants and cyanobacteria. Antioxidants & Redox Signaling. 2011, 15, 1129-1159.
[44] Wang, Y., Wang, L. J., Zhu, Z. H., Ma, W. H., Lei, C. L. The molecular characterization of antioxidant enzyme genes in *Helicoverpa armigera* adults and their involvement in response to ultraviolet-A stress. Journal of Insect Physiology. 2012, 58, 1250-1258.
[45] Hu, Z., Lee, K. S., Choo, Y. M., Yoon, H. J., Lee, S. M., Lee, J. H., Kim, D. H., Sohn, H. D., Jin, B. R. Molecular cloning and characterization of 1-Cys and 2-Cys peroxiredoxins from the bumblebee *Bombus ignitus*. Comparative Biochemistry and Physiology Part B: Biochemistry and Molecular Biology. 2010, 155, 272-280.
[46] Joung, M., Yoon, S., Choi, K., Kim, J. Y., Park, W. Y., Yu, J. R. Characterization of the thioredoxin peroxidase from *Cryptosporidium parvum*. Experimental Parasitology. 2011, 129, 331-336.
[47] Harder, S., Bente, M., Isermann, K., Bruchhaus, I. Expression of a mitochondrial peroxiredoxin prevents programmed cell death in *Leishmania donovani*. Eukaryotic Cell. 2006, 5, 861-870.

2.9 A Novel 1-Cys Thioredoxin Peroxidase Gene in *Apis cerana cerana*: Characterization of *AccTpx4* and Its Role in Oxidative Stresses

Yifeng Huaxia, Fang Wang, Yan Yan, Feng Liu, Hongfang Wang, Xingqi Guo, Baohua Xu

Abstract: Thioredoxin peroxidase (Tpx), also named peroxiredoxin (Prx), is an important peroxidase that can protect organisms against stressful environments. *AccTpx4*, a 1-Cys thioredoxin peroxidase gene from the Chinese honeybee *Apis cerana cerana*, was cloned and characterized. The *AccTpx4* gene encodes a protein that is predicted to contain the conserved PVCTTE motif from 1-Cys peroxiredoxin. Quantitative real-time PCR (Q-PCR) and Western blotting revealed that *AccTpx4* was induced by various oxidative stresses, such as cold, heat, insecticides, H_2O_2, and $HgCl_2$. The *in vivo* peroxidase activity assay showed that recombinant AccTpx4 protein could efficiently degrade H_2O_2 in the presence of DL-dithiothreitol (DTT). In addition, disc fusion assays revealed that AccTpx4 could function to protect cells against oxidative stresses. These results indicate that *AccTpx4* plays an important role in oxidative stress responses and may contribute to the conservation of honeybees.

Keywords: *Apis cerana cerana*; Thioredoxin peroxidase; Oxidative stress

Introduction

Oxidative stress arises from a significant increase in the amount of reactive oxygen species (ROS) and reactive nitrogen species (RNS) or a decrease in their detoxification mechanisms. Living organisms are constantly exposed to many natural sources of oxidative stress that are caused by environmental oxidants, toxic heavy metal, ionizing irradiation, heat shock, and inflammation [1]. To protect against excessive ROS, organisms have developed various antioxidative enzymes; among the family of cysteine-based peroxidases, thioredoxin peroxidase (Tpx, also named peroxiredoxin) is an important thiol-specific antioxidant protein [2-5]. Tpxs are antioxidant enzymes that are present ubiquitously in mammals, plants, yeast, and bacteria [6-8]. The antioxidative activity of peroxiredoxin is the first to be studied and is known as the most general biological function of the compound. However, recent research has shown that peroxiredoxins also function in signal transduction [9-11] and the regulation of phospholipid metabolism [12-15]. However, the inactivation of peroxiredoxins at high H_2O_2 levels can turn off peroxide defenses and preserve the pool of reduced thioredoxin; thus, peroxiredoxin can be used to repair proteins vital to survival [16].

Peroxiredoxins are responsible for antioxidant defense in bacteria (AhpC), yeast (thioredoxin peroxidase), and trypanosomatids (tryparedoxin peroxidase). They also appear to be fairly promiscuous with respect to the hydroperoxide substrates, and the specificities for the donor substrate—e.g., glutathione (GSH), thioredoxin, tryparedoxin, and the analogous CXXC motifs of bacterial AhpF proteins, vary considerably between subfamilies [17]. Peroxiredoxins are a family of cysteine-based peroxidases that are classified into 1-Cys Prx, typical 2-Cys Prx, and atypical 2-Cys Prx based on the number of conserved cysteine residues. The second conserved Cys that is present near the C terminus distinguishes the "2-Cys" from "1-Cys" groups of Prx enzymes. However, Prxs share a common fold and active site as well as a catalytic cycle that uses a conserved Cys residue, called Cys47 [10, 18, 19]. In the typical catalytic mechanism for Prxs, sulfenic acid is captured by the active site, and H_2O_2 is reduced to H_2O while the active site cysteine is oxidized to sulfenic acid [10, 16].

Among the types of peroxiredoxins, 1-Cys Prx, a homodimer with a subunit molecular weight of 25 kDa, has been studied the least. Recently, TgPrx2 of *Toxoplasma gondii* [20], 1-Cys-Prx of *Bombyx mori* (BmPrx) [1], BiPrx1 of *Bombus ignitus* [21], and 1-Cys-Prx of *Plamodium vivax* (Pv1-Cys-Prx) and *Plamodium knowlesi* (Pk1-Cys-Prx) [22] have been characterized. These proteins share high homology and play important roles in oxidation resistance; however, the roles of 1-Cys Prx in bees have not been investigated extensively.

As an important indigenous species of the Chinese honeybee, *Apis cerana cerana* (*A. cerana cerana*) not only provides rich nutrition via bee products but also plays an important role in crop pollination. It has been reported that bees are one of the animals that function as fresh power, promoting the rapid development of the agricultural economy due to their pollination function. Furthermore, Prxs play an important role in the antioxidant system of the honeybee [5]. In this study, we cloned a 1-Cys Prx and tested its antioxidant activity against H_2O_2. The results suggest that *AccTpx4* may function in the oxidative stress response of *A. cerana cerana*.

Materials and Methods

Apis cerana cerana and treatments

The Chinese honeybee (*A. cerana cerana*) was collected in mimetic hives at experimental apiary of Shandong Agricultural University (Taian, China). The adult worker honeybees were caught at the 50th day after emergence. These honeybees were housed 30 per group and kept in an incubator with a constant temperature 34℃, 60% relative humidity under a 24 h dark condition. They were fed with basic adult food containing 30% honey from the source colonies, 70% powdered sugar, and water [23] for 1 day before treatments. Group 1 was fed only the pollen-and-sucrose solution during the treatments as control. Groups 2-4 were incubated under 4, 25, and 42℃ conditions. Groups 5-8 were treated as follows: H_2O_2 (0.5 μl of a 2 mmol/Ldilution was injected on the thoracic notum of each worker bee),

$HgCl_2$ (3 mg/mL), and pesticides (cyhalothrin, acetamiprid, and phoxim which were diluted to the final concentration 20 mg/L) were added to food for stress. The worker honeybees also collected based on the age and shape named as L4-L6 (the fourth- to the sixth-instar larvae), Pw (white-eyed pupae), Pp (pink-eyed pupae), Pb (brown-eyed pupae), Pd (dark-eyed pupae), A1 (1-day post-emergence adult), A15 (15-day post-emergence adult), and A30 (30-day post-emergence adult). At appropriate time after inoculation, the bees were flash-frozen in liquid nitrogen and stored at −80℃.

Primers and PCR amplification conditions

The sequences of primers and PCR amplification conditions used in the study are listed in Table 1 and Table 2, respectively.

Table 1 The primers used in this study

Abbreviation	Primer sequence (5′—3′)	Description
TP1	GATTGGCAAGGAGATTCATGGGTTG	cDNA sequence primer, forward
TP2	CAGAAGGCATTGACACTTTATCCACTC	cDNA sequence primer, reverse
3P1	TCTTGCGAGTTATAGATTCCTTGC	3′ RACE forward primer, outer
3P2	GATGATCCAGAACAAGCGTT	3′ RACE forward primer, inner
5P1	CAACCCATGAATCTCCTTGCCAAT	5′ RACE reverse primer, outer
5P2	TCATGATCAGCAATTATCGGGT	5′ RACE reverse primer, inner
AAP	GGCCACGCGTCGACTAGTAC(G)14	Abridged anchor primer
AUAP	GGCCACGCGTCGACTAGTAC	Abridged universal amplification primer
B26	GACTCTAGACGACATCGA(T)18	3′ RACE universal primer, outer
B25	GACTCTAGACGACATCGA	3′ RACE universal primer, inner
QP1	ATGAGAATTAATTCTATTGTGCCG	Full-length cDNA primer, forward
QP2	TTAATAATTTGTGGTTGTGCGGAC	Full-length cDNA primer, reverse
β-s	TTATATGCCAACACTGTCCTTT	Standard control primer, forward
β-x	AGAATTGATCCACCAATCCA	Standard control primer, reverse
YH1	GGTACCATGAGAATTAATTCTATTGTG	Protein expression primer, forward
YH2	GGATCCATAATTTGTGGTTGTGCGGAC	Protein expression primer, reverse
TY1	GCGTTAACTGTTCGTGCTTT	Real-time PCR primer, forward
TY2	CAAGGAATCTATAACTCGCAAG	Real-time PCR primer, reverse

Table 2 PCR amplification conditions

Primer pair	Amplification conditions
TP1/TP2	10 min at 94℃; 40 s at 94℃; 40 s at 54℃; 40 s at 72℃ for 32 cycles; 10 min at 72℃
3P1/B26	10 min at 94℃; 40 s at 94℃; 40 s at 50℃; 40 s at 72℃ for 28 cycles; 10 min at 72℃
3P2/B25	10 min at 94℃; 40 s at 94℃; 40 s at 53℃; 30 s at 72℃ for 35 cycles; 10 min at 72℃
5P1/AAP	10 min at 94℃; 40 s at 94℃; 40 s at 50℃; 40 s at 72℃ for 28 cycles; 10 min at 72°C
5P2/AUAP	10 min at 94℃; 40 s at 94℃; 40 s at 54℃; 30 s at 72℃ for 35 cycles; 10 min at 72℃
QP1/QP2	10 min at 94℃; 40 s at 94℃; 40 s at 50℃; 50 s at 72℃ for 32 cycles; 10 min at 72℃
YH1/YH2	10 min at 94℃; 40 s at 94℃; 40 s at 50℃; 50 s at 72℃ for 30 cycles; 10 min at 72℃

RNA extraction, cDNA synthesis

The total RNA was extracted with Trizol reagent (TransGen Biotech, Beijing) based on the instructions of the manufacturer. The RNA samples for different treatments were digested with RNase-free DNase I and used for complementary DNA (cDNA) synthesis with the reverse transcriptase system (TransGen Biotech, Beijing).

Isolation of the full-length cDNA of *AccTpx4*

The full-length cDNA of *AccTpx4* was amplified using reverse transcription PCR (RT-PCR) and rapid amplification of cDNA end PCR (RACE-PCR) as described previously [24].

Bioinformation and phylogenetic analysis

Sequence identity of AccTpx4 1-Cys Prxs was analyzed using NCBI bioinformatics tools (http://blast.ncbi.nlm.nih.gov/Blast.cgi). Multiple alignment analysis, the theoretical isoelectric point, and molecular weight prediction were conducted by using DNAMAN software 7.0.2 (Lynnon Biosoft). Tertiary structure was predicted by using SWISS-MODEL (http://swiss-model.expasy.org/). The phylogenetic tree was constructed by the neighbor-joining method by using MEGA software 4.0.

Protein expression and purification

The coding region of *AccTpx4* was cloned into the pET-30a (+) expression vector by using primers YH1 and YH2. The recombinant plasmid pET-30a (+)-*AccTpx4* was then transformed into *Escherichia coli* BL21 (DE3) cells, and expression in *E. coli* was induced by 1.0 mmol/L isopropylthio-β--D-galactoside (IPTG) at 37℃. Purification of the fusion protein was performed as described previously [24].

Western blotting

The purified protein was injected subcutaneously into white mice for the generation of antibodies as described by Yan et al. (2014) [25]. Total proteins were extracted from whole adult bees by using the Tissue Protein Extraction Kit (CoWin Bioscience Co., Beijing, China). These total proteins were quantified by using the BCA Protein Assay Kit (Nanjing Jiancheng Bioengineering Institute, Nanjing, China). Western blotting was performed according to the procedure by Zhang et al. (2013) [26] by using wet transfer method for preparing the replica.

Disc diffusion assay

Disc diffusion assays under oxidative stress were performed according to Zhang et al. (2013) [26]. Approximately, 5×10^8 bacterial cells were coated plates on LB-kanamycin agar plates and incubated at 37℃ for 1 h. Next, filter discs (8 mm diameter) soaked in different concentrations of cumene hydroperoxide (0, 10, 20, 50, or 100 mmol/L) or *t*-butyl hydroperoxide (0, 10, 30, 50, or 100 mmol/L) were placed on the surface of the culture medium. The cells were cultivated at 37℃ for 24 h before the inhibition zones around the paper discs were measured.

In vitro peroxidase activity

Reduction of H_2O_2 by purified recombinant AccTpx4 protein was estimated as described previously [24]. Increasing concentrations (0, 20, 40, 60, 80, and 100 μg/ml) of the purified recombinant AccTpx4 proteins were added into the reaction system, respectively, and then added H_2O_2 (final concentration of 200 mmol in reaction system) to initiate the reactions and incubated for 10 min at 37℃. The decrease in absorbance at 475 nm was used to measure the peroxidase activity.

Quantitative real-time PCR

To identify the expression pattern of *AccTpx4*, the *AccTpx4* primers TY1/TY2 and the *β-actin* gene (XM640276) primers β-s/β-x were used for Q-PCR with a modified volume containing 1.6 μl of diluted cDNA from different samples, 10.0 μl TaKaRa SYBR® premix Ex Taq^{TM}, 0.4 μl of each primer (10 pmol/ml), and 7.6 μl ddH$_2$O from Yao et al. 2013 [27]. The qPCR amplification condition was as follows: initial denaturation at 95℃ for 30 s; 40 cycles (95℃ for 5 s, 55℃ for 15 s, and 72℃ for 15 s) and a single melting cycle from 65℃ to 95℃. The linear relationship, amplification efficiency, and data analysis were conducted by using CFX Manager Software version 1.1. The analysis of significant difference was determined by Duncan's multiple range tests using Statistical Analysis System (SAS) software version 9.1.

Results

Cloning and molecular characterization of *AccTpx4*

The full-length cDNA sequence of *AccTpx4* (GenBank accession number: KJ551847), containing a 73-base pair (bp) 5' untranslated region (UTR) and a 196 bp 3' UTR, was cloned by RT-PCR and RACE-PCR. The 663 bp open reading frame (ORF) encoded a protein of 220 amino acids that was predicted to be localized in the cytoplasm and to have a molecular mass of approximately 25.087 kDa and an isoelectric point of 6.30. Analysis of the deduced amino acid sequence indicated that the PVCTTE motif was conserved in 1-Cys peroxiredoxin [19]. Fig. 1 indicated the peroxidatic cysteine and nearby histidine. Comparisons of AccTpx4 with other 1-Cys peroxiredoxins showed that AccTpx4 had the highest similarity to AmPrx (*Apis mellifera*, NP_001164444.1), BiPrx (*Bombus ignitus*, ACP44068.1), AePrx6 (*Acromyrmex echinatior*, EGI64430.1), and GoPrx (*Gryllotalpa orientalis*, AAX18657.1), ranging from 100% to 76.82%. To investigate the evolutionary relationship of the 1-Cys peroxidases in insects, a phylogenetic tree (Fig. 2) obtained by using a neighbor-joining method showed that

AccTpx4 was closely related to AmPrx through evolution.

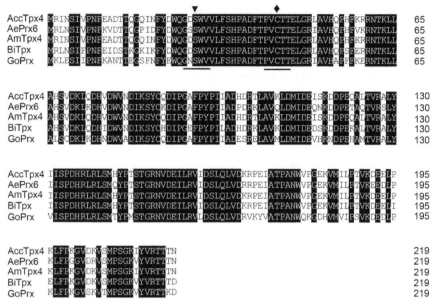

Fig. 1 Multiple amino acid alignments. A comparison of the amino acid sequences of AccTpx4 (GenBank accession number: KJ551847) from *A. cerana cerana*, AmTpx4 (NP_001164444.1) from *Apis mellifera*, AePrx6 (EGI64430.1) from *Acromyrmex echinatior*, BiPrx (ACP44068.1) from *Bombus ignitus*, and GoPrx (AAX18657.1) from *Gryllotalpa orientalis*. Identical amino acid residues are shaded in dark color. The conserved 1-Cys Prx signature (FTPVCTTE) is underlined. The peroxidatic cysteine (C43) is marked with a black quadrangle, and the serine at position 28 (S28) of motif GXSXG is marked with a black triangle.

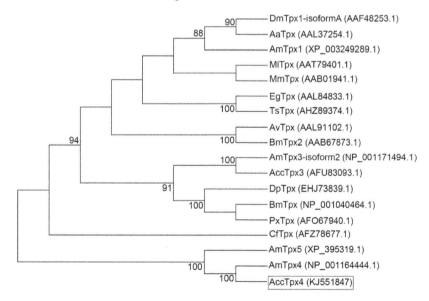

Fig. 2 A phylogenetic tree generated by MEGA version 4.0 with Tpxs from various species. The numbers above and below the branches are the bootstrap values from 500 replicates. AccTpx4 is boxed. The GenBank accession numbers or NCBI references are indicated in parentheses.

Furthermore, to understand the conserved peroxiredoxin active site and catalytic mechanism of AccTpx4, the tertiary structure of AccTpx4 (Fig. 3) was predicted by the SWISS-MODEL. The catalytic site is surrounded by a loop-helix structural motif that forms a narrow pocket, contributing to peroxide substrate attack by reaching across the dimer interface to interact with the other subunit of the domain-swapped homodimer. In addition, the sulfur atom of C47, which binds to arginine in the monomer, forms a new hydrogen bond with the nearby histidine, forming a dimer, and this interaction is conserved for 1-Cys Prx [19, 28].

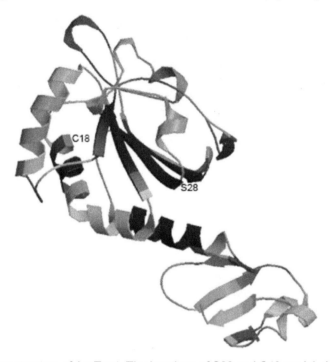

Fig. 3 The tertiary structure of AccTpx4. The locations of S28 and C43 are labeled respectively.

Temporal and spatial expression patterns of *AccTpx4*

Peroxiredoxins are a widespread and highly expressed family of cysteine-based peroxidases that react very rapidly with H_2O_2, organic peroxides, and peroxynitrite [29]. This large protein family, which has diverged from thioredoxin-like redox proteins, is widespread across phylogeny, and one or more members are typically expressed at high levels in many cell types [19]. To determine the expression profile of AccTpx4 during different developmental stages, the transcription levels from larvae, pupae, and adults were measured by using qPCR. Fig. 4a showed that *AccTpx4* was expressed at high levels in larvae, pupae, and adult bees. In addition, among the adult tissues studied, the messenger RNA (mRNA) level of *AccTpx4* in the head and thorax was higher than that in the abdomen. However, the mRNA level of *AccTpx4* in the muscle and midgut was higher than that in the epidermis (Fig. 4b). These results showed that *AccTpx4* was expressed at a higher level in the head and muscle than that in other tissues.

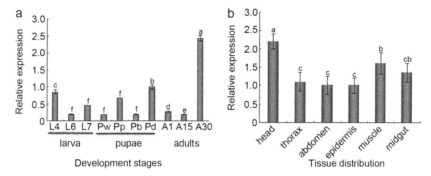

Fig. 4 Expression patterns of *AccTpx4*. a. The transcription level of *AccTpx4* at different developmental stages. L4-L6 (the fourth- to the sixth-instar larvae), Pw (white-eyed pupae), Pp (pink-eyed pupae), Pb (brown-eyed pupae), Pd (dark-eyed pupa), A1 (1-day post-emergence adult), A15 (15-day post-emergence adult), and A30 (30-day post-emergence adult). b. The relative expression of *AccTpx4* in different body regions and tissues. Each value is given as the mean (SD) of three replicates. Letters on the bar represent significant difference at $P < 0.001$ as determined by Duncan's multiple range tests using SAS version 9.1 software.

Expression patterns of *AccTpx4* under environmental stresses

Prxs are induced by various stimuli such as temperature stress [1, 30] and multiple other environmental stresses [24]. To characterize the transcription levels of *AccTpx4* under various environmental stresses, adult bees were exposed to an artificial hostile environment or were fed reagents that cause oxidative stress and then qPCR was performed. As shown in Fig. 5a, the transcription levels of *AccTpx4* were increased 2.22-fold when the bees were exposed to 4℃ for 1 h and were 5.38-fold higher when the bees were exposed to 25℃ treatments for 1.5 h (Fig. 5b). However, heat treatment (42℃) reached a peak after 1.5 h with an approximately 2.07-fold relative mRNA increase, followed by a sharp decrease (Fig. 5c), which is possibly due to the bees being nearly dead. When the bees were fed insecticide, the expression patterns of *AccTpx4* were different. The mRNA level of *AccTpx4* changed slightly under acetamiprid treatment (Fig. 5d). For cyhalothrin and phoxim treatments, the mRNA levels of *AccTpx4* were increased 3.06- and 1.74-fold at 2 and 1 h (Fig. 5e, f), respectively. When the bees treated with H_2O_2 or $HgCl_2$, the expression of *AccTpx4* was increased 1.69-fold after 1.5 h of H_2O_2 treatment (Fig. 5h). In addition, the transcription level of *AccTpx4* was induced within 3 h under $HgCl_2$ treatment (Fig. 5g).

Western blotting analysis

To further understand the expression pattern of the AccTpx4 protein in response to environmental stresses, the recombinant AccTpx4 protein containing a 6× His-tag was produced. After induction with IPTG, the proteins were separated by SDS-PAGE, which showed that the AccTpx4 fusion protein had a molecular mass of approximately 31 kDa (Fig. 7a), and the prediction was confirmed by using DNAMAN. The recombinant protein was then purified by using Ni-nitrilotriacetic acid (NTA) spin columns to test the antioxidant activity, and then, the purified protein used to make the antibody for AccTpx4. Proteins from bees that exposed to acetamiprid, cyhalothrin, phoxim, H_2O_2, $HgCl_2$, and heat at 42℃ were extracted and subjected to Western blotting analysis. These results revealed that changes in the AccTpx4 protein levels were consistent with the changes in the transcription levels (Fig. 6).

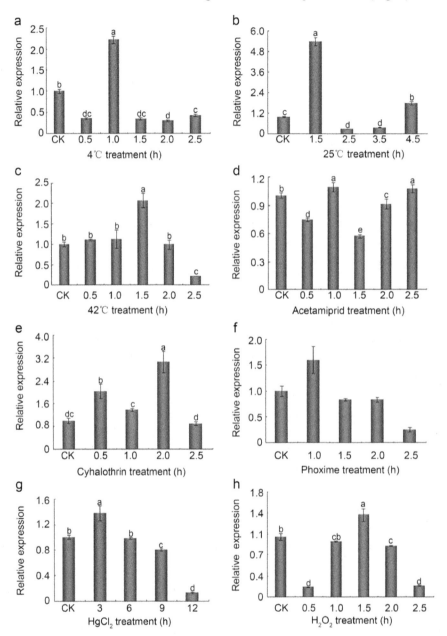

Fig. 5 Expression patterns of *AccTpx4* under different environmental stresses. The relative expression of *AccTpx4* under different stress conditions, such as cold (4, 25℃) (a, b), heat (42℃) (c), acetamiprid (d), cyhalothrin (e), phoxim (f), $HgCl_2$ (g) (3 mg/ml), and H_2O_2 (h) is shown. The expression levels were normalized to that of untreated bees (0 h), and each value is given as the mean (SD) of three replicates. Letters on the bar represent significant differences at $P < 0.001$ as determined by Duncan's multiple range tests using SAS version 9.1 software.

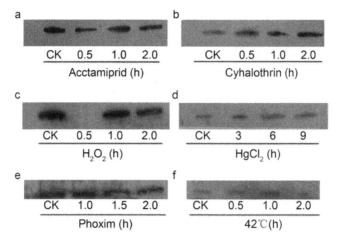

Fig. 6 The Western blotting assays of recombinant AccTpx4 protein exposed to acetamiprid, cyhalothrin, H$_2$O$_2$, HgCl$_2$, phoxim, and 42℃ treatments (a-f). The total protein was immunoblotted with anti-AccTpx4. The signal of the binding reactions was visualized with HRP substrates

Peroxidase activity of recombinant AccTpx4 protein

Prx can maintain proper cellular levels of H$_2$O$_2$ by removing unnecessary H$_2$O$_2$ [4, 10, 19, 31]. To examine the H$_2$O$_2$ peroxidase activity of the AccTpx4 fusion protein, the ability to remove H$_2$O$_2$ was measured *in vitro*. Purified recombinant AccTpx4 protein degraded H$_2$O$_2$ in a concentration-dependent manner in the presence of DTT, and the H$_2$O$_2$ content barely changed in the absence of DTT (Fig. 7b). Thus, AccTpx4 possesses the thiol-dependent antioxidant activity of removing H$_2$O$_2$.

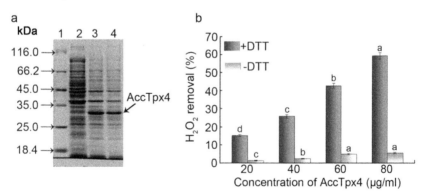

Fig. 7 a. Expression of recombinant AccTpx4 protein. Recombinant AccTpx4 protein was separated by SDS-PAGE analysis, lane 1, molecular weight marker; lane 2: non-induced overexpression of pET-30a (+)-*AccTpx4* in BL21; lane 3 and lane 4, induced overexpression of pET-30a (+)-*AccTpx4* in BL21. The recombinant AccTpx4 protein was marked by an arrow. b. The peroxidase activity of the recombinant AccTpx4 protein. The H$_2$O$_2$ degradation activity with or without DTT for different concentrations of AccTpx4 protein. Letters on the bar represent significant differences at $P<0.001$ as determined by Duncan's multiple range tests using SAS version 9.1 software.

Disc fusion assays for oxidative stresses

To further investigate the protective function of AccTpx4 toward oxidative stresses, disc fusion assays were performed in the presence of oxidative stresses. Cumene hydroperoxide and t-butyl hydroperoxide were chosen as intracellular ROS because they are more stable than H_2O_2 when applied under cultivation conditions [32]. After overnight cultivation, the zone diameters were measured. Fig. 8 showed that the killing zones around the drug-soaked filters were smaller on the plates containing control bacteria than on those containing bacteria overexpressing AccTpx4.

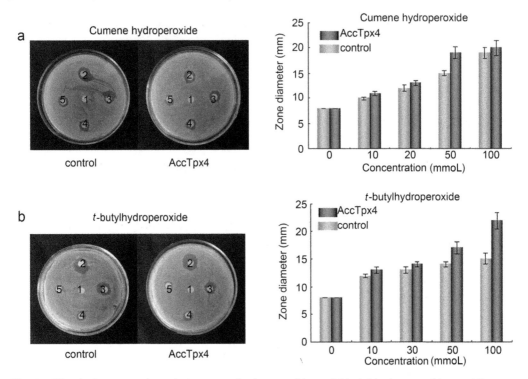

Fig. 8 Disc fusion assay by using cumene hydroperoxide and t-butyl hydroperoxide, and the zone diameters of the killing zones were showed in the histograms. a. The cumene hydroperoxide concentrations of discs 1-5 are 0, 10, 20, 50, and 100 mmol/L, respectively. b. The t-butyl hydroperoxide concentrations of discs 1-5 are 0, 10, 30, 50, and 100 mmol/L, respectively.

Discussion

Peroxiredoxin (also called thioredoxin peroxidase) is a ubiquitous peroxidase that plays important roles in the antioxidative response [10], signal transduction [9, 11], and the regulation of phospholipid metabolism [12, 13]. 1-Cys Prx, also classified as Prx6 [28, 29], is a special type of peroxidase in which the resolving Cys is missing and the sulfenic acid of the peroxidatic Cys can be reduced by a heterologous thiol-containing reductant [33]. In this study, we isolated a 1-Cys Prx (*AccTpx4*) from *A. cerana cerana*. As Fig. 1 showed, AccTpx4 was predicted to have the PVCTTE motif, similar to TgPrx2 of *T. gondi* [20], AmTpx4 [5], and 1-Cys peroxiredoxin of yeast [31]. The PVCTTE motif may act as a peroxidase to form a specific conformation similar to that of hORF6 [28] and is conserved in 1-Cys Prx [19, 34]. In addition, the tertiary structure of AccTpx4 (Fig. 3) indicated that it was a 1-Cys Prx that might have common functions and might participate in similar physiological processes, including cellular stress and antioxidative processes. However, it was noted that a GXSXG motif was located 14 amino acids before the PVCTTE motif and could be responsible for the phospholipase A_2 activities of 1-cysteine peroxiredoxin [15].

Peroxiredoxins are a fascinating group of thiol-dependent peroxidases that are ubiquitously expressed, with multiple isoforms present in most organisms [10]. Each of these isoforms is unique in that they exhibit different expression patterns during the development, distribute differently in organelles, and produce different reaction intermediates during catalysis [4]. Moreover, Prxs can scavenge reactive oxygen species to protect organisms against oxidative damage. Fig. 4b showed that *AccTpx4* was expressed in the head than in other tissues, a finding that may be attributed to the head sensing a hostile environment early. This result was different from that of *AccTpx1*, which was expressed at a higher level in the thorax [24]. It has been reported that Prxs may act as a house-keeping gene to regulate the intracellular ROS level because its expression can be found in all stages of the bee life cycle [1]. In addition, the accumulation of oxidized Prxs may indicate the disruption of cellular redox homeostasis [35]. qPCR analysis showed that *AccTpx4* was expressed throughout the life cycle of bees (Fig. 4a), in keeping with the results of previous studies. All of these results indicated that *AccTpx4* might play a positive role in the cellular redox homeostasis of bees.

The role of Prx enzymes as peroxidases has been demonstrated by increasing or decreasing their expression levels in various cell lines and by evaluating the sensitivity of the cells to oxidative insults [36]. Some members of the Prx family are upregulated in cells under stress conditions. The upregulation of the Prxs in cells and tissues under oxidative stress conditions is one of the cellular recovery responses that occur after oxidative damage [37]. The mRNA level of *AccTpx4* was induced by various stress conditions, such as cold and heat treatment (Fig. 5a-c). In addition, *AccTpx4* was more sensitive to cyhalothrin and phoxim treatment to acetamiprid treatment (Fig. 5d-f). Western blotting results also demonstrated the response of AccTpx4 to various oxidative stresses at the protein level (Fig. 6). These results provide insight to the role of *AccTpx4* in the oxidative stress responses. Additionally, the increased expression of *AccTpx4* exposure to various environmental stresses may be involved in the increased protection of oxidative stress. However, this finding needs to be further explored.

To further test the peroxidase activity of the AccTpx4 protein, its ability to remove H_2O_2 was measured *in vitro*. The recombinant protein existed mostly as a soluble protein (Fig. 7a), the same as BmPrx4 [30]. Furthermore, the purified recombinant AccTpx4 protein could efficiently degrade H_2O_2 in the presence of DTT (Fig. 7b), similar to *Arenicola marina* PRDX6 [34] and AbTPx2 [3] but different from the *AccTpx4* induced by H_2O_2 (Fig. 5h). In addition, the disc diffusion assay also proved that overexpressing AccTpx4 in *E. coli* cells could protect cells from oxidative stresses (cumene hydroperoxide and *t*-butyl hydroperoxide) (Fig. 8). These results suggest that AccTpx4 can function as a peroxidase to remove H_2O_2. However, the regulation of the *AccTpx4* mRNA was more complex and requires more research.

Peroxiredoxins are a family of multifunctional antioxidant, thioredoxin-dependent peroxidases that have been identified in numerous organisms. The major functions of Prxs comprise cellular protection against oxidative stress, modulation of intracellular signaling cascades that apply H_2O_2 as a second messenger molecule, and regulation of cell proliferation [38]. In conclusion, we have isolated and characterized a 1-Cys Prx from *A. cerana cerana* that is abundantly expressed in body regions and tissues. Additionally, *AccTpx4* responds to temperature change as well as insecticide, H_2O_2, and $HgCl_2$ treatments. Furthermore, the AccTpx4 protein expressed in *E. coli* could efficiently degrade H_2O_2, indicating that *AccTpx4* might play important roles in responses to environmental stress.

References

[1] Wang, Q., Chen, K., Yao, Q., Zhao, Y., Li, Y. J., Shen, H. X., Mu, R. H. Identification and characterization of a novel 1-Cys peroxiredoxin from silkworm. Bombyx Mori Comp Biochem Physiol B: Biochem Mol Biol. 2008, 149, 176-182.

[2] McGonigle, S., Dalton, J. P., James, E. R. Peroxidoxins a new antioxidant family. Parasitol Today. 1998, 14, 139-145.

[3] Pushpamali, W. A., De Zoysa, M., Kang, H. S., Oh, C. H., Whang, I., Kim, S. J., Lee, J. Comparative study of two thioredoxin peroxidases from disk abalone (*Haliotis discus discus*): cloning, recombinant protein purification, characterization of antioxidant activities and expression analysis. Fish Shellfish Immunol. 2008, 24, 294-307.

[4] Rhee, S. G., Kang, S. G., Chang, T. S., Jeong, W., Kim, K. Peroxiredoxin, a novel family of peroxidases. IUBMB Life. 2001, 52, 35-41.

[5] Corona, M., Robinson, G. E. Genes of the antioxidant system of the honey bee: annotation and phylogeny. Insect Mol Biol. 2006, 15, 687-701.

[6] Pedrajas, J. R., Carreras, A., Valderrama, R., Barroso, J. B. Mitochondrial 1-Cys-peroxiredoxin/thioredoxin system protects manganese-containing superoxide dismutase (Mn-SOD) against inactivation by peroxynitrite in *Saccharomyces cerevisiae*. Nitric Oxide. 2010, 23, 206-213.

[7] Perez-Sanchez, J., Bermejo-Nogales, A., Calduch-Giner, J. A., Kaushik, S., Sitja-Bobadilla, A. Molecular characterization and expression analysis of six peroxiredoxin paralogous genes in gilthead sea bream (*Sparus aurata*): insights from fish exposed to dietary, pathogen and confinement stressors. Fish Shellfish Immunol. 2011, 31, 294-302.

[8] Srinivasa, K., Kim, N. R., Kim, J., Kim, M., Bae, J. Y., Jeong, W., Kim, W., Choi, W. Characterization of a putative thioredoxin peroxidase prx1 of *Candida albicans*. Mol Cells. 2012, 33, 301-307.

[9] Jarvis, R. M., Hughes, S. M., Ledgerwood, E. C. Peroxiredoxin 1 functions as a signal peroxidase to receive, transduce, and transmit peroxide signals in mammalian cells. Free Radic Biol Med. 2012, 53, 1522-1530.

[10] Poole, L. B., Hall, A., Nelson, K. J. Overview of peroxiredoxins in oxidant defense and redox regulation. Curr Protoc Toxicol. 2011, Chapter 7, Unit7-9.

[11] Wood, Z. A., Poole, L. B., Karplus, P. A. Peroxiredoxin evolution and the regulation of hydrogen peroxide signaling. Science. 2003, 300, 650-653.

[12] Chen, J. W., Dodia, C., Feinstein, S. I., Jain, M. K., Fisher, A. B. 1-Cys peroxiredoxin, a bifunctional enzyme with glutathione peroxidase and phospholipase A2 activities. J Biol Chem. 2000, 275, 28421-28427.

[13] Manevich, Y., Fisher, A. B. Peroxiredoxin 6, a 1-Cys peroxiredoxin, functions in antioxidant defense and lung phospholipid metabolism. Free Radic Biol Med. 2005, 38, 1422-1432.

[14] Manevich, Y., Hutchens, S., Tew, K. D., Townsend, D. M. Allelic variants of glutathione S-transferase P1-1 differentially mediate the peroxidase function of peroxiredoxin VI and alter membrane lipid peroxidation. Free Radic Biol Med. 2013, 54, 62-70.

[15] Nevalainen, T. J. 1-Cysteine peroxiredoxin, a dual-function enzyme with peroxidase and acidic Ca^{2+}-independent phospholipase A2 activities. Biochimie. 2010, 92, 638-644.

[16] Karplus, P. A., Poole, L. B. Peroxiredoxins as molecular triage agents, sacrificing themselves to enhance cell survival during a peroxide attack. Mol Cell. 2012, 45, 275-278.

[17] Hofmann, B., Hecht, H. J., Flohé, L. Peroxiredoxins. Biol Chem. 2002, 383(3-4), 347-364.

[18] Hall, A., Sankaran, B., Poole, L. B., Karplus, P. A. Structural changes common to catalysis in the Tpx peroxiredoxin subfamily. J Mol Biol. 2009, 393, 867-881.

[19] Wood, Z. A., SchrÖde, E., Harris, J. R., Poole, L. B. Structure, mechanism and regulation of peroxiredoxins. Trends Biochem Sci. 2003, 28, 32-40.

[20] Deponte, M., Becker, K. Biochemical characterization of Toxoplasma gondii 1-Cys peroxiredoxin 2 with mechanistic similarities to typical 2-Cys Prx. Mol Biochem Parasitol. 2005, 140, 87-96.

[21] Hu, Z., Lee, K. S., Choo, Y. M., Yoon, H. J., Lee, S. M., Lee, J. H., Kim, D. H., Sohn, H. D., Jin, B. R. Molecular cloning and characterization of 1-Cys and 2-Cys peroxiredoxins from the bumblebee *Bombus ignitus*. Comp Biochem Physiol B: Biochem Mol Biol. 2010, 155, 272-280.

[22] Hakimi, H., Asada, M., Angeles, J. M., Kawai, S., Inoue, N., Kawazu, S. *Plasmodium vivax* and *Plasmodium knowlesi*: cloning, expression and functional analysis of 1-Cys peroxiredoxin. Exp Parasitol. 2013, 133, 101-105.

[23] Alaux, C., Ducloz, F., Crauser, D., Le Conte, Y. Diet effects on honeybee immunocompetence. Biol Lett. 2010, 6, 562-565.

[24] Yu, F., Kang, M., Meng, F., Guo, X., Xu, B. Molecular cloning and characterization of a thioredoxin peroxidase gene from *Apis cerana cerana*. Insect Mol Biol. 2011, 20, 367-378.

[25] Yan, Y., Zhang, Y., Huaxia, Y., Wang, X., Yao, P., Guo, X., Xu, B. Identification and characterisation of a novel 1-cys thioredoxin per-oxidase gene (*Acctpx5*) from *Apis cerana cerana*. Comp Biochem Physiol Part B: Biochem Mol Biol. 2014, 39-48.

[26] Zhang, Y., Yan, H., Lu, W., Li, Y., Guo, X., Xu, B. A novel Omega-class glutathione S-transferase gene in *Apis cerana cerana*: molecular characterisation of GSTO2 and its protective effects in oxidative stress. Cell Stress Chaperones. 2013, 18, 503-516.

[27] Yao, P., Hao, L., Wang, F., Chen, X., Yan, Y., Guo, X., Xu, B. Molecular cloning, expression and antioxidant characterisation of a typical thioredoxin gene (*AccTrx2*) in *Apis cerana cerana*. Gene. 2013, 527 (1), 33-41.

[28] Mizohata, E., Sakai, H., Fusatomi, E., Terada, T., Murayama, K., Shirouzu, M., Yokoyama, S. Crystal structure of an archaeal peroxiredoxin from the aerobic hyperthermophilic crenarchaeon *Aeropyrum pernix* K1. J Mol Biol. 2005, 354, 317-329.

[29] Nelson, K. J., Knutson, S. T., Soito, L., Klomsiri, C., Poole, L. B., Fetrow, J. S. Analysis of the peroxiredoxin family: using active-site structure and sequence information for global classification and residue analysis. Proteins. 2011, 79, 947-964.

[30] Shi, G. Q., Yu, Q. Y., Shi, L., Zhang, Z. Molecular cloning and characterization of peroxiredoxin 4 involved in protection against oxidative stress in the silkworm *Bombyx mori*. Insect Mol Biol. 2012, 21, 581-592.

[31] Greetham, D., Grant, C. M. Antioxidant activity of the yeast mitochondrial one-Cys peroxiredoxin is dependent on thioredoxin reduc-tase and glutathione in vivo. Mol Cell Biol. 2009, 29, 3229-3240.

[32] Burmeister, C., Luersen, K., Heinick, A., Hussein, A., Domagalski, M., Walter, R. D., Liebau, E. Oxidative stress in *Caenorhabditis elegans*: protective effects of the Omega class glutathione transferase (*GSTO-1*). FASEB J. 2008, 22, 343-354.

[33] Loumaye, E., Andersen, A. C., Clippe, A., Degand, H., Dubuisson, M., Zal, F., Morsomme, P., Rees, J. F., Knoops, B. Cloning and characterization of *Arenicola marina* peroxiredoxin 6, an annelid two-cysteine peroxiredoxin highly homologous to mammalian one-cysteine peroxiredoxins. Free Radic Biol Med. 2008, 45, 482-493.

[34] Trujillo, M., Mauri, P., Benazzi, L., et al. The mycobacterial thioredoxin peroxidase can act as a one-cysteine peroxiredoxin. J Biol Chem. 2006, 281, 20555-20566.

[35] Poynton, R. A., Hampton, M. B. Peroxiredoxins as biomarkers of oxidative stress. Biochim Biophys Acta. 2014, 1840 (2), 906-912.

[36] Rhee, S. G., Woo, H. A., Kil, I. S., Bae, S. H. Peroxiredoxin functions as a peroxidase and a regulator and sensor of local peroxides. J Biol Chem. 2012, 287 (7), 4403-4410.

[37] Ishii, T., Yanagawa, T. Stress-induced peroxiredoxins. Subcell Biochem. 2007, 44, 375-384.

[38] Immenschuh, S., Baumgart-Vogt, E. Peroxiredoxins, oxidative stress, and cell proliferation. Antioxid Redox Signal. 2005, 7 (5-6), 768-777.

2.10 Identification and Characterisation of a Novel 1-Cys Thioredoxin Peroxidase Gene (*AccTpx5*) from *Apis cerana cerana*

Yan Yan, Yuanying Zhang, Yifeng Huaxia, Xiuling Wang, Pengbo Yao, Xingqi Guo, Baohua Xu

Abstract: Thioredoxin peroxidases (Tpxs), members of the antioxidant protein family, play critical roles in resisting oxidative stress. In this work, a novel 1-Cys thioredoxin peroxidase gene was isolated from *Apis cerana cerana* and was named *AccTpx5*. The open reading frame (ORF) of *AccTpx5* is 663 bp in length and encodes a 220-amino acid protein with a predicted molecular mass and isoelectric point of 24.921 kDa and 5.45, respectively. Promoter sequence analysis of *AccTpx5* revealed the presence of putative transcription factor binding sites related to early development and stress responses. Additionally, real-time quantitative PCR (qPCR) analysis indicated that *AccTpx5* was primarily present in some developmental stages, with the highest expression levels in the first-instar larvae. The expression level of *AccTpx5* was upregulated under various abiotic stresses, including 4, 42℃, $HgCl_2$, H_2O_2, phoxim and acaricide treatments. Conversely, it was downregulated by UV and pyriproxyfen treatments. Moreover, H_2O_2 concentration dramatically increased under a variety of stressful conditions. Finally, the purified recombinant AccTpx5 protein protected the supercoiled form of plasmid DNA from damage in the thiol-dependent mixed-function oxidation (MFO) system. These results suggest that *AccTpx5* most likely plays an essential role in antioxidant defence.

Keywords: Abiotic stress; Antioxidant defence; *Apis cerana cerana*; qPCR; Thioredoxin peroxidases 5

Introduction

As a pollinator of flowering plants, the Chinese honeybee (*Apis cerana cerana*) is critical in maintaining the balance of regional ecologies and agricultural enterprise [1]. However, recent colony losses raise concerns. Multiple factors have been investigated as potential causes of or factors contributing to the losses, including honeybee pathogens and exogenous stressors associated with the generation of increased reaction oxygen species (ROS) and subsequent oxidative damage [2, 3]. Therefore, there is a need to understand the antioxidant system of *A. cerana cerana* and its ROS defence mechanisms and address the new challenges in antioxidant stress.

Naturally, organisms living in aerobic environments generate excessive ROS that could damage all of the major classes of biological macromolecules, causing lipid peroxidation, protein oxidation, DNA base modifications and strand breaks [4, 5]. To protect themselves from the damage caused by ROS, insects have evolved a complex network of enzymatic antioxidant systems including enzymatic and non-enzymatic components [6]. The dominant components of the antioxidant system in insects include thioredoxin peroxidases (Tpxs, which are also named peroxiredoxins), catalases (CATs), glutathione *S*-transferases (GSTs) and superoxide dismutases (SODs) [7-10].

Tpxs are a ubiquitous family of antioxidant enzymes that play an important role in the detoxification of ROS and balancing the redox environment in the insect system [11-14]. All Tpxs contain one conserved cysteine residue that catalyses peroxide reduction during the reaction with the hydroperoxide. According to the number and location of the cysteine residues directly involved in catalysis, Tpxs are divided into two categories, namely 1-Cys Tpxs and 2-Cys Tpxs [15]. In contrast to the 2-Cys Tpxs, the 1-Cys Tpxs lack a second conserved resolving cysteine residue.

To date, 1-Cys Tpxs have been identified in several organisms, and these proteins show unique functional characteristics. The human 1-Cys Tpx protein, hORF06, is a bifunctional enzyme because it has phospholipase A_2 (PLA_2) activity in addition to its peroxidase function [16]. Mutational studies in the mouse indicated that 1-Cys PRDX6 plays important roles in resisting certain exogenous sources of oxidative stress [17]. In insects, the two 1-Cys Tpxs in *Drosophila melanogaster* possess distinctive temporal patterns of expression, and both function to protect against ROS resulting from airborne oxidants and xenobiotics [11]. It was also shown that a 1-Cys Tpx in *Bombyx mori* was induced by an external temperature stimulus [18].

However, the molecular characterisation of the 1-Cys Tpxs gene in honeybees, particularly in Chinese honeybees, has not been performed. In this paper, we cloned a novel Tpx gene (*AccTpx5*) encoding a 1-Cys thioredoxin peroxidases from *A. cerana cerana*. We evaluated its expression patterns in different developmental stages and under oxidative stresses including cold, heat, ultraviolet light, $HgCl_2$, phoxim, acaricide and pyriproxyfen. We evaluated the antioxidant activity of the purified recombinant AccTpx5 protein by expressing the recombinant enzyme in *Escherichia coli*. Based on these results, we speculate that *AccTpx5* might play an important role in alleviating oxidative stress.

Material and Methods

Animals and various treatments

The Chinese honeybees (*A. cerana cerana*) used in this study were reared routinely in the experimental apiary at Shandong Agricultural University(Taian, China). The bees were staged according to the criteria established by Thompson [19], Michelette and Soares [20] and

Elias-Neto et al. [21]; the stages included egg, larvae, pupae and adults. The second or the third day eggs; the first- to the fifth-instar larvae (L1-L5); pupae, including prepupae (P0), white-eyed (Pw), pink-eyed (Pp), brown-eyed (Pb) and dark-eyed (Pd); newly emerged adults (A1, 1 day post-emergence); nurses (A7, usually 7-10 days post-emergence); and foragers (A15, older than 15 days post-emergence) were collected from the hive, frozen in liquid nitrogen and stored at $-70℃$ for the analysis of developmental expression patterns. Ten-day-old adult bees were collected randomly from the hive and reared under artificial conditions (34℃, 60% relative humidity and constant darkness). The bees were randomly divided into 8 groups, and each group contained 30 individuals. Groups 1, 2 and 3 were exposed to 4, 42℃ and UV (254 nm, 30 MJ/cm^2), respectively. Groups 4-8 were treated with H_2O_2 (0.5 μl of 2 mmol/L H_2O_2 was injected to the thoracic notum of bees), $HgCl_2$ (3 mg/ml was added to the basic adult diet), pesticides (acaricide, phoxim and pyriproxyfen, 2.0 mg/L was added to the basic adult diet), respectively. The control group fed on a normal diet containing water, 70% powdered sugar and 30% honey from the source colonies. The whole bodies of the bees were flash-frozen in liquid nitrogen at the appropriate time and stored at $-70℃$.

RNA extraction, cDNA synthesis, and DNA preparation

Total RNA was isolated from adult bees by using Trizol reagent (Invitrogen, Carlsbad, CA, USA). The extracted RNA was incubated with RNase-free DNase I (Promega, Madison, WI, USA). Then, the RNA was used to obtain the first-strand cDNA by using the EasyScript First-strand cDNA Synthesis SuperMix (TransGen Biotech, Beijing, China) following the instructions of the manufacturer. Genomic DNA was extracted from the adult bees by using the EasyPure Genomic DNA Extraction Kit (TransGen Biotech) according to the instructions of the manufacturer.

Primers and PCR amplification conditions

The primers and PCR amplification conditions are listed in Table 1 and Table 2, respectively.

Table 1 The primers in this study

Abbreviation	Primer sequence (5'—3')	Description
PF	CGCTACTATGGTTTTGTTAGGTG	cDNA sequence primer, forward
PR	CGATACAATGCGCACGTAAGA	cDNA sequence primer, reverse
5PW	TGGCAGTCATAGGTATACC	5' RACE reverse primer, outer
5PN	TAGTGCAATAACTTTCACTCCT	5' RACE reverse primer, inner
3PW	TCTTATATCCTGCTACTACT	3' RACE forward primer, outer
3PN	TCACTACAATTAACTGAGAAG	3' RACE forward primer, inner
AAP	GGCCACGCGTCGACTAGTAC(G)$_{14}$	Abridged anchor primer
AUAP	GGCCACGCGTCGACTAGTAC	Abridged universal amplification primer
B25	GACTCTAGACGACATCGA	3' RACE universal primer, outer

Abbreviation	Primer sequence (5'—3')	Description
B26	GACTCTAGACGACATCGA(T)$_{18}$	3' RACE universal primer, inner
CSF	CGCTACTATGGTTTTGTTAGGTG	Full-length cDNA sequence primer, forward
CSR	TTACAGCGGTTGCGATACAATG	Full-length cDNA sequence primer, reverse
DLF	GGGGTATTCTATTTTCGCATCCA	Real-time PCR primer, forward
DLR	CCATTTACGATGAGAATCGACTGA	Real-time PCR primer, reverse
β-s	TTATATGCCAACACTGTCCTTT	Standard control primer, forward
β-x	AGAATTGATCCACCAATCCA	Standard control primer, reverse
G1	CGCTACTATGGTTTTGTTAGGTG	Genomic sequence primer, forward
G2	TCAACTTCTAAAGGATCTAAC	Genomic sequence primer, reverse
G3	CTTATATCCTGCTACTACTGG	Genomic sequence primer, forward
G4	TTACAGCGGTTGCGATACAATG	Genomic sequence primer, reverse
YHF	GATATCATGGTTTTGTTAGGTGAA	Protein expression primer, forward
YHR	GGATCCCAGCGGTTGCGATACAAT	Protein expression primer, reverse
QS1	GTACAACTGAGTTAGCTCGAG	I-PCR forward primer, outer
QS2	GTCTTACGAATTATCTAGCCAATC	I-PCR reverse primer, outer
QS3	CAGTCGATTCTCATCGTAAA	I-PCR forward primer, inner
QS4	CACCTAACAAAACCATAGTAGCG	I-PCR reverse primer, inner
QYZ1	TCAGCCTGCTATTATAAAACTCAT	Promoter special primer, forward
QYZ2	CACCTAACAAAACCATAGTAGCG	Promoter special primer, reverse

Table 2 PCR amplification conditions

Primer pair	Amplification conditions
PF/PR	10 min at 94℃, 40 s at 94℃, 40 s at 51℃, 60 s at 72℃ for 35 cycles, 10 min at 72℃
5PW/AAP	10 min at 94℃, 40 s at 94℃, 40 s at 45℃, 60 s at 72℃ for 35 cycles, 10 min at 72℃
5PN/AUAP	10 min at 94℃, 40 s at 94℃, 40 s at 49℃, 60 s at 72℃ for 35 cycles, 10 min at 72℃
3PW/B26	10 min at 94℃, 40 s at 94℃, 40 s at 43℃, 40 s at 72℃ for 35 cycles, 10 min at 72℃
3PN/B25	10 min at 94℃, 40 s at 94℃, 40 s at 47℃, 40 s at 72℃ for 35 cycles, 10 min at 72℃
CSF/CSR	10 min at 94℃, 40 s at 94℃, 40 s at 50℃, 60 s at 72℃ for 35 cycles, 10 min at 72℃
G1/G2	10 min at 94℃, 40 s at 94℃, 40 s at 45℃, 90 s at 72℃ for 35 cycles, 10 min at 72℃
G3/G4	10 min at 94℃, 40 s at 94℃, 40 s at 48℃, 90 s at 72℃ for 35 cycles, 10 min at 72℃
QS1/QS2	10 min at 94℃, 40 s at 94℃, 40 s at 47℃, 2 min at 72℃ for 35 cycles, 10 min at 72℃
QS3/QS4	10 min at 94℃, 40 s at 94℃, 40 s at 48℃, 2 min at 72℃ for 35 cycles, 10 min at 72℃
QYZ1/QYZ2	10 min at 94℃, 40 s at 94℃, 40 s at 48℃, 90 s at 72℃ for 35 cycles, 10 min at 72℃
YHF/YHR	10 min at 94℃, 40 s at 94℃, 40 s at 55℃, 40 s at 72℃ for 33 cycles, 10 min at 72℃

Cloning of the full-length cDNA of *AccTpx5*

To obtain internal conserved gene fragments, two PCR primers PF and PR were designed and synthesised based on the conserved regions of Tpxs in several insects. Then, rapid amplification of cDNA end PCR (RACE-PCR) was used to obtain the 5′ and 3′ cDNA fragments using specific primers (5PW/5PN and 3PW/3PN, respectively) as described by Yan et al. [9]. After assembly, specific primers for CSF and CSR were designed to clone the complete encoding sequence of AccTpx5. All of the PCR products were purified, cloned into the pEASY-T3 vectors (TransGen Biotech), and then transformed into *E. coli* competent cells for sequencing. The PCR primers and amplification conditions are shown in Table 1 and Table 2, respectively.

Amplification of genomic sequence and the 5′-flanking region of *AccTpx5*

For amplification of the genomic DNA sequence of *AccTpx5*, two pairs of specific primers (G1/G2 and G3/G4) were designed according to the full-length cDNA of *AccTpx5* by using *A. cerana cerana* genomic DNA as the template. The PCR products were purified, cloned into the pEASY-T3 vectors, and then transformed into *E. coli* competent cells for sequencing.

To better understand the organisation of the regulatory elements, primers (QS1/QS2 and QS3/QS4) were designed to obtain the 5′-flanking region of *AccTpx5* by using inverse-PCR (IPCR). The total genomic DNA was completely digested with the restriction endonuclease *Hin*dIII at 37℃ overnight, self-ligated to form circles by using T4 DNA ligase (TaKaRa, Dalian, China), and used as the template. Finally, specific primers (QYZ1 and QYZ2) were used to examine the whole promoter sequence. The purified PCR product was cloned into the pEASY-T3 vector and then sequenced. Furthermore, the TFSEARCH database (http://www.cbrc.jp/research/db/TFSEARCH.html) was used to analyze the putative *cis*-acting elements in the promoter region of *AccTpx5*.

Bioinformatic analysis and phylogenetic tree construction

The nucleotide sequences of *AccTpx5* were assembled and analyzed by using DNAMAN software 6.0.3. Multiple protein sequence alignments among homologues were conducted by using DNAMAN software 6.0.3 and NCBI bioinformatics tools (http://blast.ncbi.nlm.nih.gov/Blast.cgi). Tertiary structures were predicted by using SWISS-MODEL software (http://swissmodel.expasy.org/). The theoretical isoelectric point and molecular weight were predicted with ExPASy (http://web.expasy.org/compute_pi/). The phylogenetic analysis was performed by using the Molecular Evolutionary Genetics Analysis (MEGA version 5.1) software with the neighbour-joining method.

Real-time quantitative PCR

To investigate the expression pattern of *AccTpx5* in different developmental stages and abiotic stresses, real-time quantitative PCR was carried out by using the SYBR® PrimeScript™ RT-PCR Kit (TaKaRa) with two specific primers (DLF and DLR) based on the cDNA of *AccTpx5*. The PCR reaction was performed by using a CFX96™ Real-time System (Bio-Rad) following the procedures as described by Yao et al. [14]. The *β-actin* gene (GenBank accession number: HM640276) was used as an internal control. All of the samples were analyzed in triplicate. The $2^{-\Delta\Delta C_t}$ comparative CT method was used to calculate the different expression levels of *AccTpx5* [22]. Significant differences were identified by using Duncan's multiple range tests with Statistical Analysis System (SAS) software version 9.1.

Expression and purification of recombinant AccTpx5

The entire coding region of *AccTpx5*, flanked by an *Eco*R V restriction site and a *Bam*H I restriction site, was amplified and subcloned into the expression vector pET-30a (+) with a His-tag. The recombinant plasmid pET-30a (+)-*AccTpx5* was then transfected into *E. coli* BL21 (DE3), and a transformed colony was cultured in Luria-Bertani (LB) broth with 50 μg/ml kanamycin at 37℃ until the optical density at 600 nm reached 0.4-0.6. Expression of the recombinant AccTpx5 was induced by adding a final concentration of 75 μg/ml isopropylthio-β-D-galactoside (IPTG) at 22℃ overnight. The bacterial cells were then centrifuged at 8,000 g for 5 min at 4℃ and resuspended in Tris-HCl buffer (pH 7.4). After the suspension was sonicated for 40 min on ice, the supernatant was the harvested by centrifugation at 16,000 g for 15 min at 4℃. Subsequently, the supernatant was used for purifying the recombinant AccTpx5 protein under native conditions using a HisTrap™ FF column (GE Healthcare, Uppsala, Sweden) following the instructions of the manufacturer. After elution, the purified protein was evaluated by using 12% sodium dodecyl sulphate polyacrylamide gel electrophoresis (SDS-PAGE) and then Coomassie Brilliant Blue staining.

Western blotting and determination of H_2O_2 contents

The purified protein was injected subcutaneously into white mice for generation of antibodies as described by Yan et al. [9]. Total proteins were extracted from whole adult bees by using the Tissue Protein Extraction Kit (CoWin Bioscience Co., Beijing, China). These total proteins were quantified with the BCA Protein Assay Kit (Nanjing Jiancheng Bioengineering Institute, Nanjing, China). Western blotting was performed according to the procedure of Zhang et al. [23] with some modifications. In addition, H_2O_2 concentration was tested by using the hydrogen peroxide test kit (Nanjing Jiancheng Bioengineering Institute) according to the instructions of the manufacturer.

Antioxidant activity assay

Supercoiled pUC19 plasmid DNA (TaKaRa) was used as a substrate for detecting DNA damage by using the thiol-dependent mixed-function oxidation (MFO) system. The reaction mixture containing HEPES buffer, $FeCl_3$, dithiothreitol (DTT), was pre-incubated without or with the purified AccTpx5 protein (0.8, 1.6, 3.2, 9.6, 19.2, 32 μg/ml) in a total volume of 50 μl at 37℃ for 30 min. After the pre-incubation period, 300 ng of supercoiled pUC19 plasmid DNA was added to each reaction and developed at 37℃ for another 2.5 h. DNA degradation by the MFO was assessed by electrophoresis on a 1.5% agarose gel with ethidium bromide.

Results

Cloning of *AccTpx5* gene

A portion of the cDNA sequence of Tpx showing high homology with reported 1-Cys Tpx genes was amplified by reverse transcription PCR (RT-PCR) and rapid amplification of cDNA end PCR (RACE-PCR). After sequencing, blasting and matching, the full-length cDNA of *AccTpx5* gene (GenBank accession no. KF745893) was obtained. The *AccTpx5* gene is 1,003 bp in length, containing a 263 bp 5′ untranslated region (UTR), a 77 bp 3′ UTR and a 663 bp open reading frame (ORF). The ORF codes for a protein composed of 220 amino acid residues with a predicted molecular mass and isoelectric point of 24.921 kDa and 5.45, respectively. As shown in Fig. 1, one conserved 1-Cys Tpx signature motif (PVCTTE) and cysteine residue were found.

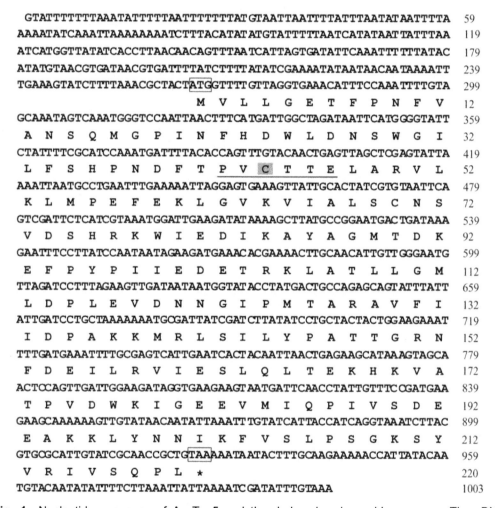

Fig. 1 Nucleotide sequence of AccTpx5 and the deduced amino acid sequence. The cDNA sequence is shown on the top line, and the deduced amino acid sequence is indicated on the second line. The asterisk marks the stop codon. The start codon (ATG) and stop codon (TAG) are boxed. The conserved PVCTTE motifs are underlined, and the conserved cysteine residue is indicated by a shadow. The sequence was deposited in GenBank (GenBank accession no. KF745893).

Sequence analysis of *AccTpx5*

The deduced amino acid sequence of AccTpx5 was closely related to those of *Apis mellifera* AmTpx5 (XP 395319.2, 99% protein sequence identity), *Anopheles gambiae* AgTpx5 (XP 308753.2, 58% protein sequence identity), *Drosophila melanogaster* DmTpx5 (NP 523463.2, 59% protein sequence identity), and *Bombus impatiens* BiPrx6-like (XP 003493744.1, 90% protein sequence identity). The highly conserved region and sites belonging to the 1-Cys Tpx signature were found, indicating that AccTpx5 belongs to the 1-Cys Tpx family of proteins (Fig. 2A).

To investigate the evolutionary relationship amongst Tpxs in insects, a phylogenetic tree of various species was constructed by using the neighbour-joining method. As shown in Fig. 2B, AccTpx5 was classified as a 1-Cys Tpx and had a closer relationship with AmTpx5, which is in agreement with the results of the amino acid alignment.

To understand the conserved residues and determine the catalytic mechanism of AccTpx5, the tertiary structure of AccTpx5 was predicted by using SWISS-MODEL software [24]. The conserved redox-active cysteines are shown in Fig. 2C.

Structure analysis of the genomic sequence of *AccTpx5*

To further elucidate the properties of *AccTpx5*, the genomic DNA sequence of *AccTpx5* (KF745894) consisting of 2,581 bp (five exons and four introns) was obtained. All of the introns have typical eukaryotic characteristics with high AT content and conserved 5'-GT splice donor and 3'-AG splice acceptor sites. The exons and introns of *AccTpx5* were aligned with 1-Cys Tpxs from *A. mellifera*, *D. melanogaster*, and *B. impatiens* (Fig. 3). This analysis showed that *AccTpx5*, *AmTpx5*, *BiPrx6-like* contained four introns and possessed the highest length similarity, whereas *DmTpx5* contained only two introns.

Identification of partial putative *cis*-acting elements in the 5'-flanking region of *AccTpx5*

To further clarify the molecular mechanism of transcriptional regulation of *AccTpx5*, a 1,062 bp 5'-flanking region of *AccTpx5* was isolated (KF745894). By using the TFSEARCH software programme, we analyzed the promoter sequence of *AccTpx5*, and a series of putative *cis*-acting elements were predicted. Some important putative *cis*-acting elements involved in development were found, including cell factor 2-II (CF2-II), Hunchback (Hb), Dfd, Broad-Complex (BR-C) and Croc [25-28]. Additionally, three heat shock elements, which are the binding sites for heat shock factor (HSF), were also found [29]. Parts of the obtained sequence and several putative transcription factors are shown in Fig. 4. These data imply that the *AccTpx5* may be involved in early development and abiotic stress responses.

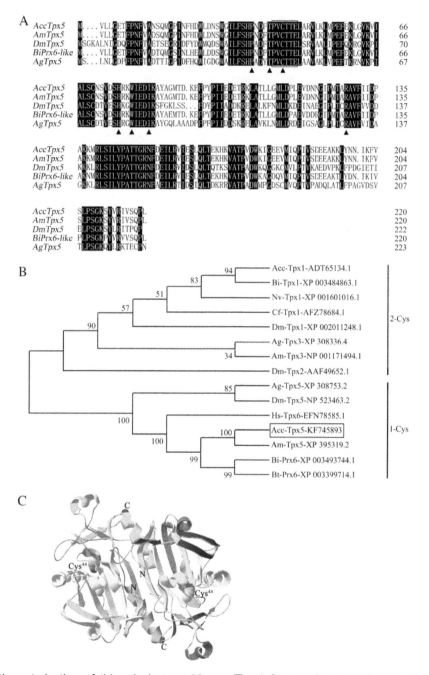

Fig. 2 Characterisation of thioredoxin peroxidases (Tpxs) from various species and the tertiary structure of AccTpx5. A. Alignment of the deduced AccTpx5 protein sequence with other known insect 1-Cys Tpxs. The amino acid sequences used in the analysis are from *Apis mellifera* (AmTpx5, GenBank accession no. XP_395319.2), *Anopheles gambiae* (AgTpx5, XP_308753.2), *Drosophila melanogaster* (DmTpx5, NP_523463.2), and *Bombus impatiens* (BiPrx6-like, XP_003493744.1). Identical amino acid residues are shaded in black, and 1-Cys Tpx signature regions (PVCTTE) are marked by a frame. The conserved sites are marked by (▲). B. Phylogenetic analysis of Tpxs from different species. The amino acid sequences of Tpxs were all obtained from NCBI. C. The tertiary structure of AccTpx5. The conserved sites (Cys^{44}), N-terminus, and C-terminus are shown.

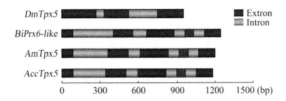

Fig. 3 Schematic representation of the DNA structures. Length of the exons and introns of genomic DNA of *Apis cerana cerana* (AccTpx5, GenBank accession no. KF745894), *Apis mellifera* (AmTpx5, NC_007081.3), *Drosophila melanogaster* (DmTpx5, NT_033779.4), and *Bombus impatiens* (BiPrx6-like, NT_177694.1) are shown according to the scale below.

```
                                                         Hb
       TCAGCC TGCTAT TATAAAAC TCATTC ATTG TT TTGT CACG TAT GCGAAAAAAT AAT GAAT TGC CT    -998
       AT TCCA TATCGA TCAT CCGT TTT TAAC TTCT TTT TTCT TTTTT CAT TTT GTAA TAA CTT TTTT TCT AG    -933
                                                  HSF
       TC GATT TTTT TG TTTC ATTT CT ATTT TCCA AAGAAA TTCT ATA TAA GATC TAAATA TTTA TGT AA    -868
              BR-C                                           Dfd
       AT TAGAATT AAAATAAAAATAA ATTT TTAA AT AATT TTTA T TTT TTAAATT AAT ATGA AAT CCAAC    -803
       CT TCTT TGTC CT TTTG TTGA TGATTT TAGT AAATGT AACC AAT TAG TTGA TGT TAA TGTT TTAGG    -738
       TACTTT TCCT GT ATTA TACT TT ATGT ATTC CAATTC TAAA TTA TAT CTAT CGT AAA TAGC TTT AT    -673
       AT TTTG GATC TTT CTAG AAGT GT TTTT AATT CT TCTT CATT TAA TTT GTTA TAA TGT TCAC ATA TC    -608
       AC TTTA CTTG GA TTTT CTTT AC AATA ATTT CG TACA TGTA CTT GCA TAGG AAT TTT AAAT TGT GA    -543
                         HSF
       AA TTTT AACT TTA TGT ATAT GT TGAA TAGAAA CATT AGAA TAA AAT TTAG TTA TAA AAGT AAC AA    -478
       AA TTTA AATA AAAT CGATTA TACATT TTTA AT TATA TATG GAA TTT AAC AAAT AAA ATAT TTG CT    -413
                                                                BR-C
       TT ATAT ATCT TAAGAT TGCT AC TATT GCAA CAGATA TA AT GTC ATA GACA AAA TAC AAGA TAG AT    -348
       AGATCT AAAT TC ATGT TAAA TT TGAA TTT GAT CTCA TTT GAGT TTG ATCT AAA TTA TAGT ACAAT    -283
       TT GAAT TTTA TAAT CAAAAA AT AATT TGAA TTT TGT TTTA ATA TTT CTAT TAT TCA AAAAA AC TG    -218
       AT GATA TCAA TA TAAAATAT TA TTTA TGTT AAATTG CGAAATT TTA CATG CAT ATA TAAT TTT TG    -153
             CF2-II
       CA TTTT GTATATATAA TTTA TA TTAA TTTT AAAACT ATTT TTAAAT TTTAATT ATT AAAATCTAT    -88
                 Croc                HSF
       TAC AAAAAATAAAAATATTC AGAAA TTTT TT TTAAATGAAAGATATAAT TTT TTA TTAT TTTGC    -23
       AA TTTT TACAAA TATT TTTT TAGTAT TTTT TT AAAT ATTT TTAATT TTTT TAT GTA ATT AATT TT    +43
                                  └→ +1 transcription start
       AT TTAA TATAAT TTTA AAAA TA TCAA ATTA AAAAAAATCT TTACAT ATATGTATTT TTAATCATA    +108
       TAATTA TTTAAA TCATGGTT ATATCA CCTT AACAAC AGTT TAATCA TTAGTGA TAT TCAAATT TT    +173
       TTATAC ATAT GT AACG TGATAA CGTG ATTT TA TCTT TTAT ATC GAA AATA TAA TAA CAAT AAA AT    +237
       TT GAAA GTATCT TTTAAACG CT ACTA TG                                            +266
                                  └→ translation start
```

Fig. 4 Partial nucleotide sequences and putative cis-acting elements in the 5'-flanking region of AccTpx5. The transcription and translation start sites are marked with arrows. The putative transcription factor binding sites mentioned in the text are boxed.

Developmental expression patterns of *AccTpx5*

To determine the expression pattern of the *AccTpx5* gene at different developmental stages, the RNA extracted from egg, larvae, pupae, and adult bees were assessed by qPCR. As showed in Fig. 5, mRNA expression of *AccTpx5* was the highest in the first-instar larvae, followed by the fifth-instar larvae. However, the mRNA expression was not detected in other developmental stages including the egg and larvae. Interestingly, a clear increase in the transcription levels was identified after pupal cuticle apolysis, occurring at the onset of pharate-adult development. The qPCR analyses also showed that *AccTpx5* mRNA in adult stage is plentiful in the newly emerged adult but then decreases in 7- to 10-day post-emergence adults, which generally nourish the brood. This result revealed that *AccTpx5* might play an essential role in the early developmental stages of the honeybee.

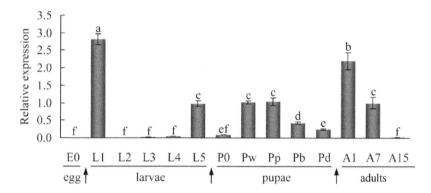

Fig. 5 Expression profile of AccTpx5 in different developmental stages. The expression profile of AccTpx5 was measured by qPCR in different developmental stages including egg, larvae (L1-L5, from the first to the fifth instars), pupae (P0, prepupae; Pw, white-eyed pupae; Pp, pink-eyed pupae; Pb, brown-eyed pupae and Pd, dark-eyed pupae), newly emerged adult workers (A1, 1 day post-emergence), nurses (A7, 7-10 days post-emergence), and foragers (A15, > 15 days post-emergence). The data are the mean ± SE of three independent experiments. The letters above the columns indicate significant differences ($P < 0.01$) according to Duncan's multiple range tests.

Expression patterns of *AccTpx5* under abiotic stresses

A growing body of evidence has shown that 2-Cys AccTpxs play an important role in resistance to environmental stresses in *A. cerana cerana* [7, 14]. To determine whether the 1-Cys *AccTpx5* gene is also involved in different environmental stress responses, the transcriptional expression profiles of *AccTpx5* were determined by qPCR after exposure to stressful temperatures (4, 42 ℃), $HgCl_2$, ultraviolet light (UV), H_2O_2 and insecticides (phoxim, acaricide, pyriproxyfen). As shown in Fig. 6A, B, expression of *AccTpx5* mRNA was enhanced by 4 and 42 ℃ treatments and reached the highest level at 2.5 h and 2.0 h of exposure, respectively. $HgCl_2$ and H_2O_2 treatments caused the expression of *AccTpx5* to increase significantly (Fig. 6C, D). However, the expression of *AccTpx5* was downregulated after ultraviolet (UV) treatment (Fig. 6E). For the insecticide (phoxim, acaricide, pyriproxyfen) treatments, the phoxim and acaricide treatments upregulated the expression of *AccTpx5*, whereas expression of *AccTpx5* was downregulated by the pyriproxyfen treatment in Fig. 6F-H. These results suggested that *AccTpx5* may have various roles in the responses to different abiotic stresses.

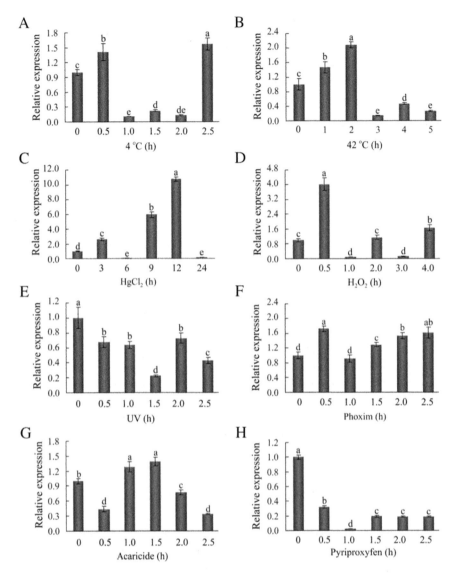

Fig. 6 Expression profile of AccTpx5 under different abiotic stresses. The expression of AccTpx5 was analyzed via qPCR by using total RNA extracted from the *A. cerana cerana*, which were treated by various stresses. For the stress treatments, 10-day-old adult bees were subjected to treatments with 4℃ (A), 42℃ (B), $HgCl_2$ (C), H_2O_2 (D), UV (E), phoxim (F), acaricide (G), and pyriproxyfen (H). The *β-actin* gene was employed as an internal control. The data are the mean ± SE of three independent experiments. The letters above the columns represent significant differences ($P < 0.01$) based on Duncan's multiple range tests.

Expression, purification and characterisation of recombinant AccTpx5 protein

AccTpx5 was successfully expressed in transformed *E. coli* BL21 (DE3) with the pET-30a (+) vector as a histidine-fusion protein. After induction with IPTG overnight, the *E. coli* cells were the harvested from the LB medium. Sodium dodecyl sulphate polyacrylamide gel electrophoresis (SDS-PAGE) analysis showed that the recombinant protein was overexpressed in a soluble form and had a molecular mass of approximately 31 kDa (Fig. 7A, lane 4). Subsequently, the fusion protein was further purified by HisTrap™ FF columns, and the concentration of the purified AccTpx5 was approximately 80 μg/ml.

To verify the antioxidant activity of AccTpx5, a DNA nicking assay was carried out to evaluate protection of the DNA from damage by using the thiol-dependent mixed-function oxidation (MFO) system. In the reaction mixture, the hydroxyl radicals produced in the presence of DTT and Fe^{3+} can disrupt the supercoiled form of pUC19 plasmid DNA; however, this damage was abrogated by Tpx [30, 31]. As shown in Fig. 7B, in the absence of the recombinant AccTpx5 protein, the hydroxyl radical caused conversion of the supercoiled form of pUC19 plasmid to the nicked form. However, the addition of the purified Tpx5 prevented the nicking of the supercoiled form of pUC19 by hydroxyl radicals. Additionally, with increasing concentrations of recombinant AccTpx5, the amount of plasmid nicking declined gradually. Therefore, we concluded that 1-Cys AccTpx5 may also play important roles in counteracting oxidative stress.

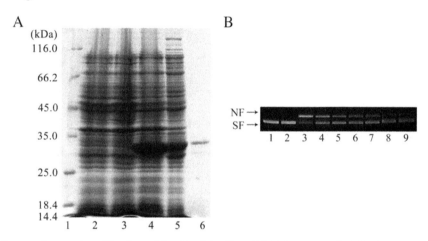

Fig. 7 Expression and purification of AccTpx5 in *E. coli* BL21 cells and DNA cleavage assay. A. Cell extracts are separated by SDS-PAGE and stained with Coomassie Brilliant blue. Lane 1, protein molecular weight markers. Lane 2, induced overexpression of pET-30a (+) in BL21. Lane 3 and Lane 4, non-induced and induced overexpression of pET-30a (+)-AccTpx5 in BL21. Lane 5, suspension of sonicated recombinant AccTpx5. Lane 6, purified recombinant AccTpx5. B. Lane 1, pUC19 plasmid DNA only. Lane 2, pUC19 plasmid DNA and $FeCl_3$. Lane 3, pUC19 plasmid DNA, $FeCl_3$ and dithiothreitol (DTT). Lanes 4-9, pUC19 plasmid DNA, $FeCl_3$, DTT and varying concentrations of AccTpx5: 0.8, 1.6, 3.2, 9.6, 19.2, and 32 μg/ml. SF, supercoiled form; NF, nicked form.

Assay of oxidant status *in vivo* under 42℃, UV, phoxim, and pyriproxyfen stressors

Heat, UV, phoxim, and pyriproxyfen are the four most common stresses *A. cerana cerana* always suffer. To further determine the correlation between the expression of *AccTpx5* and the oxidant stresses, we detected the levels of H_2O_2 under the four stresses. The treatments of sample are same with that used in qPCR. As showed in Fig. 8, following 42℃, UV, phoxim, and pyriproxyfen treatments, H_2O_2 concentration dramatically increased. Therefore, we concluded the expression of *AccTpx5* may be associated with the oxidative damage.

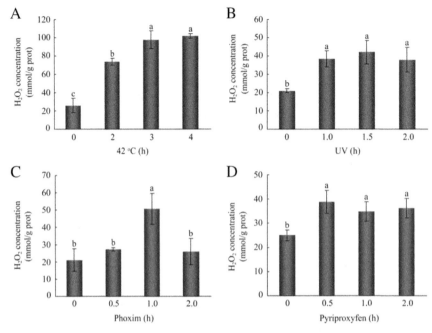

Fig. 8 The detection of H_2O_2 concentration *in vivo* under 42℃ (A), UV (B), phoxim (C), and pyriproxyfen (D) treatments. The whole-bee homogenates were analyzed by using a test kit according to the protocol of the manufacturer. Each value is given as the mean (SD) of three replicates. The letters above the columns represent significant differences ($P < 0.01$) based on Duncan's multiple range tests.

Western blotting analysis

To further verify the expression patterns of *AccTpx5* in response to various oxidative stresses, samples with 42℃, UV, phoxim, and pyriproxyfen treatments were subjected to Western blotting (Fig. 9). The result revealed that the expression of AccTpx5 was consistent with that at the transcriptional level.

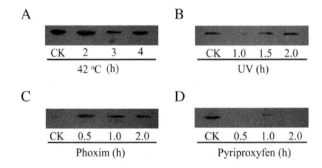

Fig. 9 Western blotting analysis of AccTpx5. 10-day-old adult bees were subjected to treatments with 42℃ (A), UV (B), phoxim (C), and pyriproxyfen (D). All lanes were loaded with an equivalent amount of extracted protein in every group.

Discussion

Tpxs are a newly discovered class of thiol-dependent antioxidant enzymes that protect organisms against cellular damages caused by reactive oxygen species produced from oxidation processes [32]. We previously reported that two 2-Cys Tpxs (AccTpx1 and AccTpx-3) play important roles in protecting honeybees from oxidative injury and may extend the lifespan of honeybees [7, 14]. However, the roles of the 1-Cys Tpxs remain poorly understood in *A. cerana cerana*. In this study, a novel 1-Cys Tpx gene was isolated from *A. cerana cerana*. Based on the deduced amino acid sequence, sequence analysis showed that both the number of amino acid residues and the predicted molecular size of the AccTpx polypeptide coincide well with other 1-Cys Tpxs from insects. Like other members of Tpx family, AccTpx5 had a conserved Cys residue (Cys^{44}) near the N-terminus. By using cysteine mutants, Kang et al. [33] demonstrated that the conserved Cys residue is absolutely essential for the function of the protein, forming the site of oxidation. The alignment constructed for the present study revealed the conspicuous presence of the motif (PVCTTE) in nearly all insect 1-Cys Tpx subunits. Taken together, these data suggest an important conserved role for these motifs.

Some putative *cis*-acting elements involved in development were found in the 5'-flanking region of *AccTpx5*, prompting us to investigate the temporal expression patterns of *AccTpx5* in *A. cerana cerana*. qPCR indicated that transcripts of *AccTpx5* are primarily present in the larvae, pupal-adult metamorphosis and adult stages, with the highest expression levels in the first-instar larvae (Fig. 5). In general, Tpx family members have great biological relevance in terms of antioxidant function. Li et al. [34] also reported that the overexpression of peroxiredoxin 2540 and *Tpx1* at 72 h in worker larvae suggests a key role for the antioxidant system in the developing larvae by changing oxidation products into harmless chemicals. Moreover, Corona and Robinson [35] demonstrated that ROS can cause oxidant damage, especially in fast growing organisms, which demand high oxygen levels. Thus, the specific temporal expression patterns of the *AccTpx5* suggested a potential role in reducing the oxidative damage in early development.

ROS could be generated during many "routine" metabolic processes occurring in living cells. Increased respiratory metabolism was observed in insects exposed for long time periods to pollution with metal [36, 37], suggesting production of highly reactive hydroxyl radicals. Previous studies have demonstrated that oxidative stress could be induced by temperature, UV radiation, and insecticides [38, 39]. In this study, we found that 42°C, UV, phoxim, and pyriproxyfen treatments increased H_2O_2 concentration, indicating that oxidative stress might be associated with alterations in oxidative stress-related gene expression. When the adults were exposed to 4, 42°C, $HgCl_2$, and H_2O_2 treatments, the transcripts of *AccTpx5* increased before decreasing. A possible explanation for this was that with the increase of stress, other stress-related genes might play important roles in cellular stress responses and the role of *AccTpx5* might be weakened. In addition, the transcript of *AccTpx5* was very low compared to the control group at 6 h when exposed to $HgCl_2$ treatment. This could be because some other antioxidant genes might reduce the transcripts of *AccTpx5* in the antioxidant process [40].

There is increasing evidence that UV is one of the most ubiquitous environmental hazards for most animals, which can cause oxidative stress through the generation of ROS and in turn leads to damage to membrane lipids and proteins [41, 42]. However, expression of *AccTpx5* was repressed by UV treatment in present study, indicating that *AccTpx5* might play different roles in resistance to UV radiation or might be involved in different signal transduction processes [14]. For the different insecticide treatments, the Tpxs may play protective roles through different defence mechanisms. Li et al. [43] found that following the addition of cerium, increased expression of Tpx may be closely associated with decreased oxidative damage of *B. mori* caused by exposure to phoxim. The increased expression of *AccTpx5* was in agreement with previous works in which researchers have found that enhanced activity of metabolic detoxification was a major mechanism for phoxim resistance in oriental fruit fly [44]. Wang et al. [44] found that mixed function oxidase enzymes were the most important enzymes in acaricide detoxification and resistance. Our data also showed that the expression of *AccTpx5* was upregualted after exposure to acaricide treatment. Pyriproxyfen is a juvenile hormone analogue effective against some arthropod pests. However, the mode of action of pyriproxyfen is not fully understood due to the lack of a known signaling pathway and/or a receptor [45]. The repression of *AccTpx5* transcripts under pytiproxyfen treatment implied that *AccTpx5* might possess different roles in confronting with various adverse conditions. Interestingly, a high fluctuation was showed at 0.5 h under phoxim and acaricide treatments. This could be because the response time of *AccTpx5* was different in various signaling transduction pathway. All of these results suggest that the expression profile of *AccTpx5* correlated with the occurrence and elimination of oxidative stress. However, these results should be further confirmed.

Because numerous environmental factors increase the levels of ROS, which exerts their destructive effects on various biomolecules within the cell including DNA, the cell is equipped with various enzymatic and non-enzymatic responses to counteract the damage [46-48]. The number of DNA strand breaks will depend upon the efficiency of the enzymatic antioxidant system. The functional studies by using recombinant AccTpx5 in the MFO system showed that, in the presence of DTT, the recombinant AccTpx5 protein was able to more efficiently prevent strand breaks caused by hydroxyl radicals in supercoiled plasmid DNA, as demonstrated in *B. mori* [18], *Plasmodium vivax* and *Plasmodium knowlesi* [49], and *Bombus ignitus* [50]. These results further suggest that AccTpx5 may be involved in the DNA protection mechanism after under ROS damage.

In conclusion, a novel 1-Cys *Tpx* gene was identified from *A. cerana cerana*, and the expression pattern in different developmental stages was examined. Moreover, expression of the *AccTpx5* transcript under various environmental stresses indicated that *AccTpx5* may play a potential role in the responses to various environmental stresses. The purified recombinant AccTpx5 protein showed a capacity to protect DNA from oxidative damage. Taken together, our data provide very useful information for further understanding the mechanisms of *A. cerana cerana* in response to environmental pressures.

References

[1] Weinstock, G. M., Robinson, G. E., Gibbs, R. A., et al. Insights into social insects from the genome of the honeybee *Apis mellifera*. Nature. 2006, 443, 931-949.

[2] Narendra, M., Bhatracharyulu, N. C., Padmavathi, P., Varadacharyulu, N. C. Prallethrin induced biochemical changes in erythrocyte membrane and red cell osmotic haemolysis in human volunteers. Chemosphere. 2007, 67, 1065-1071.

[3] Brennan, L. J., Keddie, B. A., Braig, H. R., Harris, H. L. The endosymbiont *Wolbachia pipientis* induces the expression of host antioxidant proteins in an *Aedes albopictus* cell line. PLoS One. 2008, 3, e2083.

[4] Halliwell, B., Gutterridge, J. M. C. Production against Radical Damage: Systems with Problems, second ed. Free Radical Bio Med. Oxford. 1989.

[5] Lee, K. S., Kim, S. R., Park, N. S., et al. Characterization of a silkworm thioredoxin peroxidase that is induced by external temperature stimulus and viral infection. Insect Biochem and Molec. 2005, 35, 73-84.

[6] Felton, G. W., Summers, C. B. Antioxidant systems in insects. Arch Insect Biochem. 1995, 29, 187-197.

[7] Yu, F., Kang, M., Meng, M., Guo, X., Xu, B. Molecular cloning and characterization of a thioredoxin peroxidase gene from *Apis cerana cerana*. Insect Mol Biol. 2011, 20, 367-378.

[8] Wang, Y., Wang, L. J., Zhu, Z. H., Ma, W. H., Lei, C. L. The molecular characterization of antioxidant enzyme genes in *Helicoverpa armigera* adults and their involvement in response to ultraviolet-A stress. J Insect Physiol. 2012, 58, 1250-1258.

[9] Yan, H., Meng, F., Jia, H., Guo, X., Xu, B. The identification and oxidative stress response of a Zeta class glutathione S-transferase (GSTZ1) gene from *Apis cerana cerana*. J Insect Physiol. 2012, 58, 782-791.

[10] Yan, H., Jia, H., Wang, X., Gao, H., Guo, X., Xu, B. Identification and characterization of an *Apis cerana cerana* Delta class glutathione S-transferase gene (*AccGSTD*) in response to thermal stress. Naturwissenschaften. 2013, 100, 153-163.

[11] Radyuk, S. N., Klichko, V. I., Spinola, B., Sohal, R. S., Orr, W. C. The peroxiredoxin gene family in *Drosophila melanogaster*. Free Radic Biol Med. 2001, 31, 1090-1100.

[12] Radyuk, S. N., Sohal, R. S., Orr, W. C. Thioredoxin peroxidases can foster cytoprotection or cell death in response to different stressors: over- and under- expression of thioredoxin peroxidase in *Drosophila* cells. Biochem J. 2003, 371, 743-752.

[13] Tsuda, M., Ootaka, R., Ohkura, C., Kishita, Y., Seong, K. H., Matsuo, T. Aigaki, T. Loss of Trx-2 enhances oxidative stress-dependent phenotypes in *Drosophila*. FEBS Lett. 2010, 584, 3398-3401.

[14] Yao, P., Lu, W., Meng, F., Wang, X., Xu, B., Guo, X. Molecular cloning, expression and oxidative stress response of a mitochondrial thioredoxin peroxidase gene (*AccTpx-3*) from *Apis cerana cerana*. J Insect Physiol. 2013, 59, 273-282.

[15] Wood, Z. A., Schröder, E., Harris, J. R.,., Poole, L. B. Structure, mechanism and regulation of peroxiredoxins. Trends Biochem Sci. 2003, 28, 32-40.

[16] Fisher, A. B. Peroxiredoxin 6: a bifunctional enzyme with glutathione peroxidase and phospholipase A_2 activities. Antioxid Redox Signal. 2011, 15, 831-844.

[17] Muller, F. L., Lustgarten, M. S., Jang, Y., Richardson, A., Van Remmen, H. Trends in oxidative aging theories. Free Radic Biol Med. 2007, 43, 477-503.

[18] Wang, Q., Chen, K., Yao, Q., Zhao, Y., Li, Y., Shen, H., Mu, R. Identification and characterization of a novel 1-Cys peroxiredoxin from silkworm, *Bombyx mori*. Comp Biochem Physiol B: Biochem Mol Biol. 2008, 149, 176-182.

[19] Thompson, P. R. Histological development of cuticle in the worker honeybee, *Apis mellifera adansonii*. J Apicult Res. 1978, 17, 32-40.

[20] Michelette, E. R. F., Soares, A. E. E. Characterization of preimaginal developmental stages in Africanized honey bee workers (*Apis mellifera* L.). Apidologie. 1993, 24, 431-440.

[21] Elias-Neto, M., Soares, M. P., Simões, Z. L., Hartfelder, K., Bitondi, M. M. Developmental characterization, function and regulation of a Laccase2 encoding gene in the honey bee, *Apis mellifera* (Hymenoptera, Apinae). Insect Biochem Mol Biol. 2010, 40, 231-241.

[22] Livak, K. J., Schmittgen, T. D. Analysis of relative gene expression data using real-time quantitative PCR and the 2-(Delta Delta C (T)) method. Methods. 2001, 25, 402-408.

[23] Zhang, Y., Yan, H., Lu, W., Li, Y., Guo, X., Xu, B. A novel Omega-class glutathione S-transferase gene in *Apis cerana cerana*: molecular characterisation of *GSTO2* and its protective effects in oxidative stress. Cell Stress Chaperones. 2013, 18, 503-516.

[24] Smeets, A., Loumaye, E., Clippe, A., Rees, J. F., Knoops, B., Declercq, J. P. The crystal structure of the C45S mutant of annelid *Arenicola marina* peroxiredoxin 6 supports its assignment to the mechanistically typical 2-Cys subfamily without any formation of toroid-shaped decamers. Protein Sci. 2008, 17, 700-710.

[25] Gogos, J. A., Hsu, T., Bolton, J., Kafatos, F. C. Sequence discrimination by alternatively spliced isoforms of a DNA binding zinc finger domain. Science. 1992, 257, 1951-1955.

[26] Rørth, P. Specification of C/EBP function during *Drosophila* development by the bZIP basic region. Science. 1994, 266, 1878-1881.

[27] Von Kalm, L., Crossgrove, K., Von Seggern, D., Guild, G. M., Beckendorf, S. K. The Broad-Complex directly controls a tissue-specific response to the steroid hormone ecdysone at the onset of *Drosophila* metamorphosis. EMBO J. 1994, 13, 3505-3516.

[28] Stanojević, D., Hoey, T., Levine, M. Sequence specific DNA-binding activities of the gap proteins encoded by hunchback and Krüppel in *Drosophila*. Nature. 1989, 341, 331-335.

[29] Ding, Y. C., Hawkes, N., Meredith, J., Eggleston, P., Hemingway, J., Ranson, H. Characterization of the promoters of Epsilon glutathione transferases in the mosquito *Anopheles gambiae* and their response to oxidative stress. Biochem J. 2005, 387, 879-888.

[30] Sauri, H., Butterfield, L., Kim, A., Shau, H. Antioxidant function of recombinant human natural killer enhancing factor. Biochem Biophys Res Commun. 1995, 208, 964-969.

[31] Li, J., Zhang, W. B., Loukas, A., Lin, R. Y., Ito, A., Zhang, L. H., Jones, M., McManus, D. P. Functional expression and characterization of *Echinococcus granulosus* thioredoxin peroxidase suggests a role in protection against oxidative damage. Gene. 2004, 326, 157-165.

[32] Rhee, S., Chae, H., Kim, K. Peroxiredoxins: a historical overview and speculative preview of novel mechanisms and emerging concepts in cell signaling. Free Radic Biol Med. 2005, 38, 1543-1552.

[33] Kang, S. W., Baines, I. C., Rhee, S. G. Characterization of a mammalian peroxiredoxin that contains one conserved cysteine. J Biol Chem. 1998, 273, 6303-6311.

[34] Li, J., Wu, J., Begna Rundassa, D., Song, F., Zheng, A., Fang, Y. Differential protein expression in honeybee (*Apis mellifera* L.) larvae: underlying caste differentiation. PLoS One. 2010, 20, e13455.

[35] Corona, M., Robinson, G. E. Genes of the antioxidant system of the honey bee: annotation and phylogeny. Insect Mol Biol. 2006, 15, 687-701.

[36] Kramarz, P., Kafel, A. The respiration rate of the beet armyworm pupae (*Spodoptera exigua*) after multi-generation intoxication with cadmium and zinc. Environ Pollut. 2003, 126, 1-3.

[37] Bednarska, A. J., Stachowicz, I. Costs of living in metal polluted areas: respiration rate of the ground beetle *Pterostichus oblongopunctatus* from two gradients of metal pollution. Ecotoxicology. 2013, 22, 118-124.

[38] Lushchak, V. I. Environmentally induced oxidative stress in aquatic animals. Aquat Toxicol. 2011, 101, 13-30.

[39] Kottuparambila, S., Shinb, W., Brownc, M. T., Han, T. UV-B affects photosynthesis, ROS production and motility of the freshwater flagellate, *Euglena agilis* Carter. Aquat Toxicol. 2012, 122-123, 206-213.

[40] Yao, P., Hao, L., Wang, F., Chen, X., Yan, Y., Guo, X. Xu, B. Molecular cloning, expression and antioxidant characterisation of a typical thioredoxin gene (*AccTrx2*) in *Apis cerana cerana*. Gene. 2013, 527, 33-41.

[41] Ravanat, J. L., Douki, T., Cader, J. Direct and indirect effects of UV radiation on DNA and its components. J Photochem Photobiol B. 2001, 63, 88-102.

[42] Cadet, J., Sage, E., Douki, T. Ultraviolet radiation-mediated damage to cellular DNA. Mutat Res.

2005, 571, 3-17.

[43] Li, B., Sun, Q., Yu, X., Xie, Y., Hong, J., Zhao, X. Sang, X., Shen, W., Hong, F. Molecular mechanisms of silk gland damage caused by phoxim exposure and protection of phoxim-induced damage by cerium chloride in *Bombyx mori*. Environ Toxicol. 2015, 30, 1102-1111.

[44] Wang, J. J., Wei, D., Dou, W., Hu, F., Liu, W. F., Wang, J. J. Toxicities and synergistic effects of several insecticides against the oriental fruit fly (Diptera: Tephritidae). J Econ Entomol. 2013, 106, 970-978.

[45] Karatolos, N., Williamson, M. S., Denholm, I., Gorman, K., Ffrench-Constant, R. H., Bass, C. Over-expression of a cytochrome P450 is associated with resistance to pyriproxyfen in the greenhouse whitefly *Trialeurodes vaporariorum*. PLoS One. 2012, 7, e31077.

[46] Ahmad, S. Oxidative stress from environmental pollutants. Arch Insect Biochem Physiol. 1995, 29, 135-157.

[47] Boesch, P., Weber-Lotfi, F., Ibrahim, N., Tarasenko, V., Cosset, A., Paulus, F. Lightowlers, R. N., Dietrich, A. DNA repair in organelles: pathways, organization, regulation, relevance in disease and aging. Biochim Biophys Acta. 2011, 1813, 186-200.

[48] Augustyniak, M., Orzechowska, H., Kędziorski, A., Sawczyn, T., Doleżych, B. DNA damage in grasshoppers' larvae—Comet assay in environmental approach. Chemosphere. 2014, 96, 180-187.

[49] Hakimi, H., Asada, M., Angeles, J. M. M., Kawai, S., Inoue, N., Kawazu, S. *Plasmodium vivax* and *Plasmodium knowlesi*: cloning, expression and functional analysis of 1-Cys peroxiredoxin. Exp Parasitol. 2013, 133, 101-105.

[50] Hu, Z., Lee, K. S., Choo, Y. M., Yoon, H. J., Lee, S. M., Lee, J. H. Kim, D. H., Sohn, H. D., Jin, B. R. Molecular cloning and characterization of 1-Cys and 2-Cys peroxiredoxins from the bumblebee *Bombus ignitus*. Comp Biochem and Physiol B: Biochem Mol Biol. 2010, 155, 272-280.

2.11 Glutaredoxin 1, Glutaredoxin 2, Thioredoxin 1, and Thioredoxin Peroxidase 3 Play Important Roles in Antioxidant Defense in *Apis cerana cerana*

Pengbo Yao, Xiaobo Chen, Yan Yan, Feng Liu, Yuanying Zhang, Xingqi Guo, Baohua Xu

Abstract: Glutaredoxins (Grxs) and thioredoxins (Trxs) play important roles in maintaining intracellular thiol redox homeostasis by scavenging reactive oxygen species. However, few Grxs and Trxs have been functionally characterized in *Apis cerana cerana*. In this study, we identified three genes, *AccGrx1*, *AccGrx2*, and *AccTrx1*, and investigated their connection to antioxidant defense. *AccGrx1* and *AccGrx2* were mainly detected in dark-eyed pupae, whereas *AccTrx1* was highly concentrated in 15-day post-emergence adults. The expression levels of *AccGrx1* and *AccTrx1* were the highest in fat body and epidermis, respectively. However, the expression level of *AccGrx2* was the highest in muscle, followed by the epidermis. *AccGrx1*, *AccGrx2*, and *AccTrx1* were induced by 4, 16, and 42℃; H_2O_2; and pesticides (acaricide, paraquat, cyhalothrin, and phoxim) treatments and repressed by UV light. *AccGrx1* and *AccGrx2* were upregulated by $HgCl_2$ treatment, whereas *AccTrx1* was downregulated. We investigated the knockdown of *AccGrx1*, *AccGrx2*, *AccTpx-3*, and *AccTrx1* in *A. cerana cerana* and surprisingly found that knockdown of these four genes enhanced the enzymatic activities of CAT and POD; the metabolite contents of hydrogen peroxide, carbonyls, and ascorbate; and the ratios of GSH/GSSG and $NADP^+$/NADPH. In addition, we also analyzed the transcripts of other antioxidant genes and found that some were upregulated and others were downregulated, revealing that the upregulated genes may be involved in compensating for the knockdown of *AccGrx1*, *AccGrx2*, *AccTpx-3*, and *AccTrx1*. Taken together, these results suggest that *AccGrx1*, *AccGrx2*, *AccTpx-3*, and *AccTrx1* may play critical roles in antioxidant defense.

Keywords: *Apis cerana cerana*; RNA interference; qRT-PCR; Oxidative stress; Antioxidant enzyme; Free radicals

Introduction

Reactive oxygen species (ROS) can be generated as by-products in all oxygenic organisms during aerobic metabolism [1, 2]. External or internal oxidants such as superoxide, hydrogen peroxide, and hydroxyl radicals, referred to as ROS, can result in oxidative damage to proteins, lipids, and nucleic acids [3]. Cells possess a battery of antioxidant enzymes, including catalase (CAT), superoxide dismutase (SOD), glutathione peroxidase (GPX), thioredoxin peroxidase (Tpx), glutaredoxin (Grx), and thioredoxin (Trx) systems, to resist oxidative damage [4]. The Grx and Trx superfamilies actively maintain intracellular thiol redox homeostasis by scavenging ROS [5].

Grxs are small heat-stable disulfide oxidoreductases, which are ubiquitously detected in both prokaryotes and eukaryotes [6]. Grxs generally bear a highly conserved active site, Cys-X-X-Cys/Ser [7]. Grxs appear to participate in various cellular processes, such as metabolism, actin polymerization, signal transduction, gene transcription [8-10], immune defense, cardiac hypertrophy, hypoxia-reoxygenation insult, neurodegeneration, and cancer development, progression, and treatment, and play important roles in protecting cells from oxidative damage [11]. In mammals, there are two isoforms of Grxs characterized, cytosolic glutaredoxin 1 (Grx1) and mitochondrial glutaredoxin 2 (Grx2) [3]. Grx1 plays an important role in protecting cells from H_2O_2-induced apoptosis by regulating the redox state of protein kinase B (Akt) [12, 13]. Overexpression of Grx1 enhances the resistance of MCF7 breast cancer cells to doxorubicin, which is widely used as an anticancer agent [14]. In human lens epithelial cells, Grx2 could prevent H_2O_2-induced cell apoptosis by protecting complex I activity in the mitochondria [15]. Overexpression of Grx2 in HeLa cells attenuates apoptosis by preventing cytochrome c release [16]. There is little research concerning Grx in insects. Glutaredoxin 2 of *Ostrinia furnacalis* plays an important role in insects when exposed to adverse environments [17].

Trxs are small, evolutionarily conserved, and ubiquitous proteins that are involved in antioxidant defense and maintaining cellular redox homeostasis [18-20]. There are two isoform of Trxs in mammals, cytosolic thioredoxin 1 (Trx1) and mitochondrial thioredoxin 2 (Trx2) [21]. In humans, overexpression of mitochondrial thioredoxin could protect against diamide-induced oxidation and cytotoxicity [22]. However, there are three Trxs in *Apis mellifera*: AmTrx1, with predicted mitochondrial localization; AmTrx2, a putative ortholog of *Drosophila* Trx2 with predicted cytoplasm localization; and AmTrx3 [23]. The mitochondrion is an organelle responsible for aerobic respiration and is also the most common source of ROS produced in eukaryotes [24], suggesting AmTrx1 might be involved in scavenging ROS in mitochondria. There are also some other studies of Trxs in *Apis cerana cerana*. Thioredoxin-like 1 and thioredoxin 2 of *A. cerana cerana* (AccTrx-like1 and AccTrx2) have been shown to participate in the antioxidant response [25, 26]. All the above observations reveal that Grxs and Trxs play important roles in protecting organisms from ROS damage.

The Chinese honeybee (*A. cerana cerana*) is an important pollinator species, playing an essential role in the balance of regional ecology and agricultural development [27]. *A. cerana cerana* has acquired some incomparable advantages in the history of evolution compared with *A. mellifera*. It can adapt to extreme fluctuations in temperature (e.g., $-0.1°C$) and long periods of rainfall, whereas *A. mellifera* cannot survive at $-0.1°C$. However, colonies of this species have declined severely because of various environmental stressors including indiscriminate use of pesticides, infectious diseases, and global warming [28]. To our knowledge, there is little research on Grx and Trx in insects, particularly in the Chinese honeybee. Because Grx and Trx play critical roles in the response to oxidative stress in mammals and other species and the characteristic active-site motifs of Grx and Trx are highly conserved, we predicted that Grx and Trx may also participate in antioxidant defense. So it is essential to investigate the function of Grx and Trx in *A. cerana cerana*. In this study, we isolated and characterized three antioxidant genes, AccGrx1, AccGrx2, and AccTrx1, from *A. cerana cerana* and investigated the expression profiles during various developmental stages and tissues. We also used quantitative real-time PCR (qRT-PCR) to evaluate the transcripts of these three genes after various oxidative stresses, including 4, 16, and 42℃; H_2O_2; mercury ($HgCl_2$); UV; and pesticides (acaricide, paraquat, cyhalothrin, and phoxim). In addition, we also used RNA interference (RNAi) technology to knock down AccGrx1, AccGrx2, AccTpx-3, and AccTrx1 and then examined the antioxidant enzyme activities and metabolite contents of downregulated samples and also investigated the expression levels of other antioxidant genes. Our results provide a new perspective on the resistance mechanisms of AccGrx1, AccGrx2, AccTpx-3, and AccTrx1 to oxidative stress in *A. cerana cerana* and help to develop strategies for protecting the honeybee from oxidative damage.

Materials and Methods

Experimental insects and treatments

Chinese honeybees (*A. cerana cerana*) used in this study were obtained from the experimental apiary of Shandong Agricultural University (Taian, China). Honeybees of various development stages, including the 4th-, the 5th-, and the 6th-instar larvae; white-eyed (Pw), pink-eyed (Pp), brown-eyed (Pb), and dark-eyed (Pd) pupae; and 1-day post- emergence adults, were collected from the hive, whereas 15-day post-emergence and 30-day post-emergence adults were collected at the entrance of the hive. The worker honeybees of 15-day post-emergence adults were treated according to Yao et al [26] with some modifications.

RNA extraction, cDNA synthesis, and DNA isolation

Total RNA was extracted with Trizol reagent (TransGen Biotech, Beijing, China) from worker honeybees according to the instructions of the manufacturer. The RNA samples were treated with RNase-free DNase I and were reverse transcribed by using TranScript First-Strand cDNA Synthesis SuperMix (TransGen Biotech). The genomic DNA was isolated from worker honeybees by using an EasyPure Genomic DNA extraction kit according to the instructions of the manufacturer (TransGen Biotech).

Primers and PCR amplification conditions

The primers and PCR amplification conditions are listed in Supplementary Materials 1 and 2, respectively.

Cloning of the full-length cDNA sequences of *AccGrx1*, *AccGrx2*, and *AccTrx1*

To obtain the internal regions of *AccGrx1*, *AccGrx2*, and *AccTrx1*, primers G1R1/G1R2, G2R1/G2R2, and T1R1/T1R2 were designed and synthesized based on the conserved regions of glutaredoxins and thioredoxins from other insects. Based on the internal fragment sequences of *AccGrx1* and *AccGrx2*, primers 5G1R1/5G1R2, 5G2R1/5G2R2, 3G1R1/3G1R2, and 3G2R1/3G2R2 were synthesized and used for 5′ rapid amplification of cDNA end (RACE) and 3′ RACE, respectively. The 3′ RACE of *AccTrx1* was performed by using primers 3T1R1/3T1R2. Subsequently, primers QG1R1/QG1R2, QG2R1/QG2R2, and QT1R1/QT1R2 were synthesized to amplify the complete cDNA sequences of *AccGrx1* and *AccGrx2* and the partial cDNA sequence of *AccTrx1*. All the PCR primers and amplification conditions are shown in Supplementary Materials 1 and 2, respectively. All the PCR products separated by electrophoresis were purified by using a gel extraction kit (Solarbio, Beijing, China), ligated into pEASY-T3 vectors (TransGen Biotech), and transformed into competent *Escherichia coli* DH5α cells for sequencing.

Cloning of the genomic DNA sequences of *AccGrx1*, *AccGrx2*, and *AccTrx1*

Primers G1N1/G1N2, G2N1/G2N2, and T1N1/T1N2 were used to amplify the genomic DNA sequences of *AccGrx1*, *AccGrx2*, and *AccTrx1* by using genomic DNA of *A. cerana cerana* as the template. The PCR products were separated, purified, and cloned into the pEASY-T3 vectors (TransGen Biotech) and then transformed into competent *E. coli* DH5α cells for sequencing.

Bioinformatic analysis of AccGrx1, AccGrx2, and AccTrx1

Homologous AccGrx1, AccGrx2, and AccTrx1 protein sequences were retrieved by using the Basic Local Alignment Search Tool program from the NCBI (http://blast.ncbi.nlm.nih.gov/Blast.cgi) and aligned by using DNAMAN version 5.2.2 (Lynnon Biosoft, Pointe-Claire, QC, Canada). Conserved domains in AccGrx1, AccGrx2, and AccTrx1 were detected by using bioinformatics tools available at the NCBI server (http://www.ncbi.nlm.nih.gov/Structure/cdd/cdd.shtml). The tertiary structures were predicted by using Swiss-Model (http://swiss-model.expasy.org/). Molecular Evolutionary Genetic Analysis version 4.0 with the neighbor-joining method was used to perform the phylogenetic analysis. MitoProtII (http://ihg.gsf.de/ihg/mitoprot.html) was used to predict the mitochondrial-targeting peptides of AccGrx2 and AccTrx1.

qRT-PCR analysis

qRT-PCR was used to analyze the gene transcriptional level by using the SYBR Premix Ex *Taq* (TaKaRa, Dalian, China) and the CFX96 real-time PCR detection system (Bio-Rad, Hercules, CA, USA) according to the procedure of Yao et al. [29] with some modifications. The *A. cerana cerana β-actin* gene (GenBank Accession No. HM_640276) was used as an internal control.

dsRNA synthesis of *AccGrx1*, *AccGrx2*, *AccTrx1*, and *AccTpx-3*

To synthesize dsRNA of *AccGrx1*, *AccGrx2*, *AccTrx1*, and *AccTpx-3*, primers G1RA1/G1RA2, G2RA1/G2RA2, T1RA1/T1RA2, and T3RA1/T3RA2, containing a T7 polymerase promoter sequence at their 5′ end, were used to amplify the target sequences of the four genes by RT-PCR. PCR amplification was performed according to the following conditions: 10 min at 95℃, 40 cycles (40 s at 94℃, 40 s at 60℃, and 50 s at 72℃), and a final extension of 10 min at 72℃. The PCR products were purified by using the Gel Extraction Kit (Solarbio). Then RiboMAX T7 large-scale RNA production systems (Promega, Madison, WI, USA) were used to synthesize the dsRNA of *AccGrx1*, *AccGrx2*, *AccTrx1*, and *AccTpx-3* by using the purified PCR products as the template. The synthesized dsRNA was digested by using DNase I, precipitated with isopropanol, redissolved in RNase-free water, heated to 95℃ for 1 min, and cooled to room temperature. As a control, dsRNA of the green fluorescent protein (*GFP*) gene [30] was also synthesized.

Knockdown of *AccGrx1*, *AccGrx2*, *AccTrx1*, and *AccTpx-3* gene expression

Fifteen-day post-emergence adult workers were fed with dsRNA of *AccGrx1*, *AccGrx2*, *AccTrx1*, and *AccTpx-3* (8 μg/individual). As controls, 15-day post-emergence adult workers were also fed the equivalent amount of *GFP* dsRNA. The experimental groups and control groups were maintained in an incubator with 60% relative humidity at 34℃ under a 24 h dark regimen. At the appropriate time after treatment, the bees were flash-frozen in liquid nitrogen and stored at −80℃.

Enzymatic activities and metabolite contents of RNAi-mediated silencing samples of *AccGrx1*, *AccGrx2*, *AccTpx-3*, and *AccTrx1*

Total proteins were extracted from whole bee adults fed with dsRNA-*AccGrx1* at 3 and 4 days post-feeding, adults fed with dsRNA-*AccGrx2* at 2 and 3 days post-feeding, adults fed with dsRNA-*AccTpx-3* at 1 and 2 days post-feeding, and adults fed with dsRNA-*AccTrx1* at 2 and 4 days post-feeding. The total proteins were quantified with the BCA Protein Assay Kit (Nanjing Jiancheng Bioengineering Institute, Nanjing, China). Then, a catalase test kit and a peroxidase assay kit (Nanjing Jiancheng Bioengineering Institute) were used to assay catalase capacity and peroxidase capacity, respectively. In addition, a hydrogen peroxide test kit, protein carbonyls test kit, ascorbate test kit, glutathione test kit (Nanjing Jiancheng Bioengineering Institute), and Amplite colorimetric NADP/NADPH ratio assay kit (AAT Bioquest, Sunnyvale, CA, USA) were used to assay the hydrogen peroxide, carbonyl, and ascorbate contents and the ratios of reduced to oxidized glutathione (GSH/GSSG) and $NADP^+$/NADPH, respectively, according to the protocols of the manufacturer. The whole-bee homogenates were analyzed by using the test kits.

Transcriptional levels of other antioxidant genes after knockdown of *AccGrx1*, *AccGrx2*, *AccTpx-3*, and *AccTrx1*

To analyze the expression level of CAT, the *AccCAT* gene was isolated by using CR1/CR2, 3CR1/3CR2, 5CR1/5CR2, and QCR1/QCR2 from *A. cerana cerana*. qRT-PCR was used to analyze the expression levels of *AccCAT* (GenBank Accession No. KF765424), *AccGSTD* (GenBank Accession No. JF798573), *AccGSTO2* (GenBank Accession No. JX434029), *AccGSTs4* (GenBank Accession No. JN008721), *AccMsrA* (GenBank Accession No. HQ219724), *AccSOD1* (GenBank Accession No. JN700517), *AccSOD2* (GenBank Accession No. JN637476), *AccTpx1* (GenBank Accession No. HM641254), *AccTpx5* (GenBank Accession No. KF745893), *AccTrx2* (GenBank Accession No. JX844649), *AccGrx2* (GenBank Accession No. JX844655), *AccTpx-3* (GenBank Accession No. JX456217), and *AccTrx1* (GenBank Accession No. JX844651) when *AccGrx1* (GenBank Accession No. JX844656) was inhibited; *AccCAT*, *AccGSTD*, *AccGSTO2*, *AccGSTs4*, *AccMsrA*, *AccSOD1*, *AccSOD2*, *AccTpx1*, *AccTpx5*, *AccTrx2*, *AccGrx1*, *AccTpx-3*, and *AccTrx1* when *AccGrx2* was knocked down; *AccCAT*, *AccGSTD*, *AccGSTO2*, *AccGSTs4*, *AccMsrA*, *AccSOD1*, *AccSOD2*, *AccTpx1*, *AccTpx5*, *AccTrx2*, *AccGrx1*, *AccGrx2*, and *AccTrx1* when *AccTpx-3* was inhibited; and *AccCAT*, *AccGSTD*, *AccGSTO2*, *AccGSTs4*, *AccMsrA*, *AccSOD1*, *AccSOD2*, *AccTpx1*, *AccTpx5*, *AccTrx2*, *AccGrx1*, *AccGrx2*, and *AccTpx-3* when *AccTrx1* was knocked down.

Statistical analysis

The results of gene expression and enzymatic activity assays are presented as means ± standard deviation (SD) with $n = 3$. Data were subjected to multiple comparisons by analysis of variance (ANOVA), and means were separated by the Duncan multiple range test with significant difference level at $P < 0.05$. ANOVA was performed by using Statistical Analysis System version 9.1 software.

Results

Isolation of *AccGrx1*, *AccGrx2*, and *AccTrx1*

The internal fragment sequences of *AccGrx1*, *AccGrx2*, and *AccTrx1* obtained from Chinese honeybees by RT-PCR showed significant similarity to *Grx* and *Trx* from other insect species. Subsequently, 5' and 3' RACE PCR was performed to amplify the full-length cDNA sequence. The full-length cDNA sequences of *AccGrx1* and *AccGrx2* were 436 bp and 627 bp long, containing a 297 bp and a 402 bp open reading frame (ORF), respectively. The partial cDNA sequence of *AccTrx1* was 649 bp long, containing a 411 bp ORF. Multiple sequence alignment revealed high amino acid identity (77.55%-91.59%) between *AccGrx1* and Grx sequences from other species, including *Apis mellifera*, *A. florae*, *Bombus impatiens*, and *Megachile rotundata*; high amino acid identity (74.10%-99.25%) between *AccGrx2* and other Grx2 sequences including *Apis. mellifera*, *Acromyrmex echinatior*, *Apis florae*, and *B. impatiens*; and high amino acid identity (88.24%-99.26%) between *AccTrx1* and other Trx sequences including *Apis mellifera*, *Apis florae*, *B. impatiens*, *Bombus terrestris*, and *M. rotundata*, suggesting that the Grx and Trx families are highly conserved across species. As shown in Fig. 1, the characteristic active-site motifs CPYC or CPFC, CGFS, and CNPC of AccGrx1, AccGrx2, and AccTrx1 were highly conserved among all selected species. In addition, AccGrx2 and AccTrx1 have the mitochondrial-targeting peptide in regions I and II, respectively (Fig. 1).

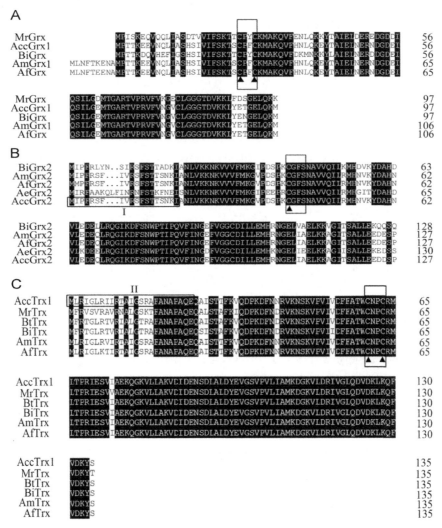

Fig. 1 Molecular characterization of AccGrx1, AccGrx2, and AccTrx1. A-C. Multiple amino acid sequence alignments of AccGrx1, AccGrx2, and AccTrx1 (GenBank Accession No. JX844656, JX844655, and JX844652) from *Apis cerana cerana*; AmGrx1 (XP_003250703.1), AmGrx2 (XP_625213.1), and AmTrx (XP_393603.1) from *Apis mellifera*; BiGrx (XP_003491196.1), BiGrx2 (XP_003487837.1), and BiTrx (XP_003485122.1) from *Bombus impatiens*; AfGrx (XP_003694808.1), AfGrx2 (XP_003691919.1), and AfTrx (XP_003695035.1) from *Apis florae*; MrGrx (XP_003699811.1), MrTrx (XP_003708266.1), from *Megachile rotundata*; AeGrx2 (EGI69671.1) from *Acromyrmex echinatior*; and BtTrx (XP_003400766.1) from *Bombus terrestris*. Identical amino acid residues are shaded in blue. The conserved catalytic motif is boxed and the active sites are marked by ▲. The mitochondrial-targeting peptides of AccGrx2 and AccTrx1 are boxed.

Phylogenetic analysis was also performed to investigate the evolutionary relationships among AccGrx1, AccGrx2, AccTrx1, and their homologs in insects. As shown in Fig. 2, phylogenetic analysis revealed that AccGrx1, AccGrx2, and AccTrx1 were more closely related to AmGrx, AmGrx2 and AmTrx, respectively, than to homologs in other species ,and this result was consistent with the relationship predicted from the multiple sequence alignment.

The tertiary structures of AccGrx1, AccGrx2, and AccTrx1 were predicted by Swiss-Model and the conserved redox-active cysteines were also identified (Fig. 3).

Genomic structures of AccGrx1, AccGrx2, and AccTrx1

To investigate the genomic structures of *AccGrx1*, *AccGrx2*, and *AccTrx1*, the genomic sequences were amplified by using genomic DNA as the template for PCR amplification. The genomic sequences of *AccGrx1*, *AccGrx2*, and *AccTrx1* (GenBank ID: JX844657, JX844658, and JX844652) were 495, 959, and 796 bp in length and contained one, two, and two introns, respectively. The nucleotide sequences at the splice junctions are in line with the canonical GT-AG rule. The genomic structures of *AccGrx1*, *AccGrx2*, and *AccTrx1* are very similar to those of *BiGrx*, *AmGrx2*, and *AmTrx* (Fig. 4), respectively.

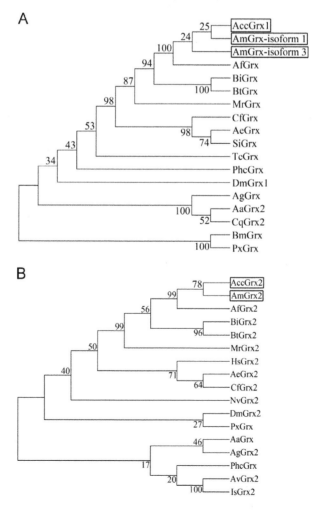

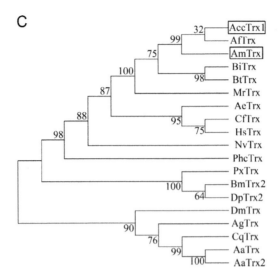

Fig. 2 Phylogenetic analysis of AccGrx1 (A), AccGrx2 (B), and AccTrx1 (C) with other known Grx and Trx protein sequences. The amino acid sequences of various species are as follows: AeGrx (*Acromyrmex echinatior*, EGI70869.1), AaGrx2 (*Aedes aegypti*, ABF18388.1), AgGrx (*Anopheles gambiae*, XP_309539.2), AfGrx (*Apis florae*, XP_003694808.1), AmGrx1 isoforms 1 and 3 (*Apis mellifera*, XP_001123018.1 and XP_003250703.1), BiGrx (*Bombus impatiens*, XP_003491196.1), BtGrx (*Bombus terrestris*, XP_003400925.1), BmGrx (*Bombyx mori*, NP_001040246.1), CfGrx (*Camponotus floridanus*, EFN72746.1), CqGrx2 (*Culex quinquefasciatus*, XP_001850698.1), DmGrx1 (*Drosophila melanogaster*, NP_611609.1), MrGrx (*Megachile rotundata*, XP_003699811.1), PxGrx (*Papilio xuthus*, BAM18349.1), PhcGrx (*Pediculus humanus corporis*, XP_002430689.1), SiGrx (*Solenopsis invicta*, EFZ19571.1), TcGrx (*Tribolium castaneum*, XP_975253.1), AeGrx2 (*Ac. echinatior*, EGI69671.1), AaGrx (*Ae. aegypti*, XP_001652385.1), AvGrx2 (*Amblyomma variegatum*, DAA34722.1), AgGrx2 (*An. gambiae*, XP_312440.3), AfGrx2 (*A. florae*, XP_003691919.1), AmGrx2 (*A. mellifera*, XP_625213.1), BiGrx2 (*B. impatiens*, XP_003487837.1), BtGrx2 (*B. terrestris*, XP_003399257.1), CfGrx2 (*C. floridanus*, EFN62391.1), DmGrx2 (*D. melanogaster*, NP_572974.1), HsGrx2 (*Harpegnathos saltator*, EFN75709.1), IsGrx2 (*Ixodes scapularis*, XP_002416216.1), MrGrx2 (*M. rotundata*, XP_003700609.1), NvGrx2 (*Nasonia vitripennis*, XP_001605234.1), PxGrx (*Pa. xuthus*, BAM18344.1), PhcGrx (*Pe. humanus corporis*, XP_002432885.1), AeTrx (*Ac. echinatior*, EGI58159.1), AaTrx (*Ae. aegypti*, XP_001659976.1), AaTrx2 (*Ae. albopictus*, AAV90706.1), AgTrx (*An. gambiae*, XP_308583.2), AfTrx (*A. florae*, XP_003695035.1), AmTrx (*A. mellifera*, XP_393603.1), BiTrx (*B. impatiens*, XP_003485122.1), BtTrx (*B. terrestris*, XP_003400766.1), BmTrx2 (*Bombyx mori*, NP_001040283.1), CfTrx (*C. floridanus*, EFN69069.1), CqTrx (*Cu. quinquefasciatus*, XP_001845536.1), DpTrx2 (*Da. plexippus*, EHJ64981.1), DmTrx (*D. melanogaster*, NP_647716.1), HsTrx (*H. saltator*, EFN79238.1), MrTrx (*M. rotundata*, XP_003708266.1), NvTrx (*N. vitripennis*, XP_001601189.1), PxTrx (*Pa. xuthus*, BAM18395.1), and PhcTrx (*Pe. humanus corporis*, XP_002430252.1).

Developmental regulation and tissue-specific distribution of *AccGrx1*, *AccGrx2*, and *AccTrx1*

To investigate the expression profiles of *AccGrx1*, *AccGrx2*, and *AccTrx1* during various developmental stages and in different tissues in the honeybee, qRT-PCR was performed by using total RNA extracted from larvae, pupae, adults, and various tissues of adults. As shown

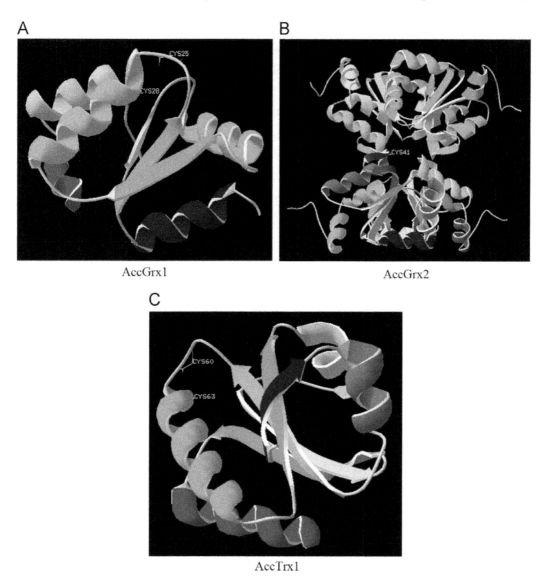

Fig. 3 The tertiary structures of AccGrx1 (A), AccGrx2 (B), and AccTrx1 (C). Swiss-Model was used to build the tertiary structure and the conserved active sites in the catalytic motif are marked.

in Fig. 5A, *AccGrx1* was mainly expressed in the 4th-instar larvae, pink-eyed pupae, dark-eyed pupae, 1-day post-emergence adults, and 15-day post-emergence adults; *AccGrx2* was highly expressed in the 4th- instar larvae, dark-eyed pupae, 1-day post-emergence adults, and 15-day post-emergence adults. However, *AccTrx1* was predominantly detected in the 4th-instar larvae, dark-eyed pupae, and 15-day post-emergence adults. The expression of *AccGrx1*, *AccGrx2*, and *AccTrx1* in various tissues is presented in Fig. 5B. Among the adult tissues studied, *AccGrx1* was expressed the highest in fat body, followed by epidermis, and *AccTrx1* was expressed the highest in epidermis, whereas *AccGrx2* was expressed the highest in muscle, followed by epidermis.

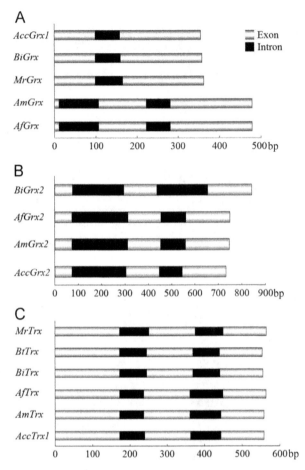

Fig. 4 Schematic representations of the intron positions of Grx and Trx family genes. The lengths of the genomic DNA sequences (JX844657, JX844658, and JX844652) from *Apis cerana cerana*; AmGrx (XP_003250703.1), AmGrx2 (XP_625213.1), and AmTrx (XP_393603.1) from *Apis mellifera*; BiGrx (XP_003491196.1), BiGrx2 (XP_003487837.1), and BiTrx (XP_003485122.1) from *Bombus impatiens*; AfGrx (XP_003694808.1), AfGrx2 (XP_003691919.1), and AfTrx (XP_003695035.1) from *Apis florae*; MrGrx (XP_003699811.1) and MrTrx (XP_003708266.1) from *Megachile rotundata*; and BtTrx (XP_003400766.1) from *Bombus terrestris*, are shown. The exons and introns are represented by white boxes and black boxes, respectively.

Expression profiles of *AccGrx1*, *AccGrx2*, and *AccTrx1* under various oxidative stresses

Fifteen-day post-emergence adults were subjected to 4, 16, and 42°C; H_2O_2; $HgCl_2$; UV light; and pesticides (acaricide, paraquat, cyhalothrin, and phoxim) treatments. As shown in Fig. 6A, *AccGrx1*, *AccGrx2*, and *AccTrx1* were all induced and reached their maximum levels at 1.0 h under 4°C treatment. *AccGrx2* was induced obviously, whereas *AccGrx1* and *AccTrx1* were hardly induced when exposed to 16°C treatment (Fig. 6B). At 42°C, *AccGrx1*, *AccGrx2*, and *AccTrx1* were all induced and reached their highest levels at 6, 6, and 5 h, respectively (Fig. 6C). All three genes were upregulated during H_2O_2 treatment and reached their peak at 3 h

(Fig. 6D). Under stress of acaricide (Fig. 6E), *AccGrx1*, *AccGrx2*, and *AccTrx1* were all upregulated and reached their highest transcripts at 1.0, 0.5, and 1.0 h, respectively. As shown in Fig. 3F, all three genes were repressed under UV treatment. *AccGrx1*, *AccGrx2*, and *AccTrx1* were all induced during paraquat stress. *AccGrx1* and *AccGrx2* reached their peak at 2.0 h, whereas *AccTrx1* reached it at 2.5 h (Fig. 6G). *AccGrx1* and *AccGrx2* were induced, whereas *AccTrx1* was repressed, in response to $HgCl_2$ treatment (Fig. 6H). As shown in Fig. 6I, all three genes were upregulated by cyhalothrin. *AccGrx1* and *AccGrx2* reached their peak at 0.5 h, whereas *AccTrx1* reached the highest level at 2.0 h. All the genes were induced during phoxim treatment and reached their maximum levels at 1.0 h (Fig. 6J).

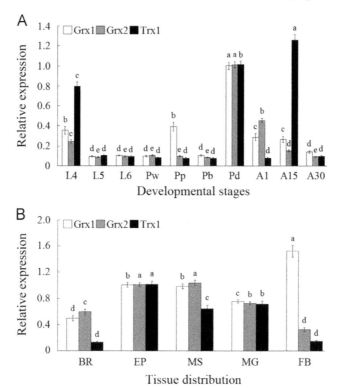

Fig. 5 Expression profiles of *AccGrx1*, *AccGrx2*, and *AccTrx1* during different developmental stages and in different tissues. A. The mRNA transcripts of *AccGrx1*, *AccGrx2*, and *AccTrx1* at the following developmental stages: larvae from the fourth to the sixth instars (L4-L6), pupae (Pw, white-eyed pupae; Pp, pink-eyed pupae; Pb, brown-eyed pupae; and Pd, dark-eyed pupae), and adults (A1, 1 day post-emergence; A15, 15 days post-emergence; and A30, 30 days post-emergence). B. The expression profiles of *AccGrx1*, *AccGrx2*, and *AccTrx1* in the brain (BR), epidermis (EP), muscle (MS), midgut (MG), and fat body (FB), compared to that of the *β-actin* gene. Vertical bars represent the means ± SE (n =3). Different letters above the bars indicate significant differences ($P <$ 0.01), as determined by Duncan's multiple range tests using SAS software version 9.1.

Knockdown of *AccGrx1*, *AccGrx2*, *AccTpx-3*, and *AccTrx1*

To investigate the functions of *AccGrx1*, *AccGrx2*, *AccTpx-3*, and *AccTrx1* in adult antioxidant defense, RNAi experiments were performed to ascertain their roles. Fifteen-day

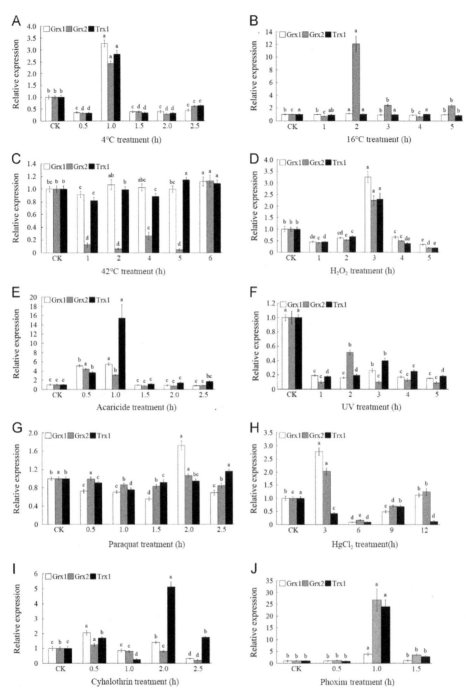

Fig. 6 qRT-PCR analysis of *AccGrx1*, *AccGrx2*, and *AccTrx1* expression levels after treatment with various abiotic stresses. Total RNA was extracted from adult bees (15 days post-emergence) treated at the indicated times with 4℃(A), 16℃(B), 42℃(C), H_2O_2 (D), acaricide (E), UV (F), paraquat (G), $HgCl_2$ (H), cyhalothrin (I), and phoxim (J) and qRT-PCR was then performed. Each value is given as the mean (SD) of three replicates. Different letters above the bars indicate significant differences [$P < 0.01$, except for 16℃ treatment of *AccTrx1* ($P < 0.05$) and 42℃ treatment of *AccGrx1* ($P < 0.05$)], as determined by Duncan's multiple range tests using SAS software version 9.1.

post-emergence adult workers were fed with dsRNAs of *GFP*, *AccGrx1*, *AccGrx2*, *AccTpx-3*, and *AccTrx1*. qRT-PCR results showed that the fewest transcripts of *AccGrx1*, *AccGrx2*, *AccTpx-3*, and *AccTrx1* were observed in dsRNA-fed adults at 3, 3, 2, and 2 days post-feeding, respectively. The transcripts of control groups and dsRNA of *GFP* groups were almost equal, showing that feeding with ds*GFP* did not influence the expression of the above four genes in *A. cerana cerana* (Fig. 7).

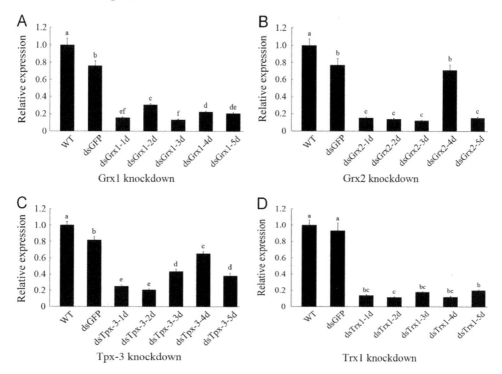

Fig. 7 Effects of RNA interference on mRNA levels of 15-day post-emergence *A. cerana cerana* adults, as induced by feeding 8 μg dsRNAs. The mRNA levels of *AccGrx1* (A), *AccGrx2* (B), *AccTpx-3* (C), and (D) *AccTrx1* are shown. The *β-actin* gene was used as an internal control. Each value is given as the mean (SD) of three replicates. Different letters above the bars indicate significant differences ($P < 0.01$), as determined by Duncan's multiple range tests using SAS software version 9.1.

Determination of enzymatic activities and metabolite contents after knockdown of *AccGrx1*, *AccGrx2*, *AccTpx-3*, and *AccTrx1*

As shown in Fig. 8, catalase capacity and peroxidase capacity of the honeybees after silencing of *AccGrx1*, *AccGrx2*, *AccTpx-3*, and *AccTrx1* were all higher than in the control groups. The contents of hydrogen peroxide, carbonyl, and ascorbate and the ratios of GSH/GSSG and $NADP^+$/NADPH in silenced samples were all higher than in the control groups (Fig. 9). All the above results indicate that some other antioxidant genes may compensate for the knockdown of *AccGrx1*, *AccGrx2*, *AccTpx-3*, and *AccTrx1*.

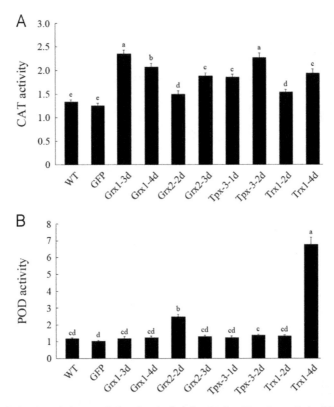

Fig. 8 Effects of the knockdown of *AccGrx1*, *AccGrx2*, *AccTpx-3*, and *AccTrx1* on antioxidant enzymatic activities of CAT (A) and POD (B). The whole-bee homogenates were analyzed by using a test kit according to the protocol of the manufacturer. Each value is given as the mean (SD) of three replicates. Different letters above the bars indicate significant differences ($P < 0.01$), according to SAS software.

Expression profiles of other antioxidant genes after knockdown of *AccGrx1*, *AccGrx2*, *AccTpx-3*, and *AccTrx1*

The cDNA sequence of *AccCAT* was 1,905 bp and consisted of a 109 bp 5′ UTR, a 254 bp 3″ UTR, and an ORF of 1,542 bp. qRT-PCR results showed that *AccTrx1*, *AccCAT*, *AccGSTO2*, *AccGSTs4*, and *AccMsrA* were induced when *AccGrx1* was knocked down; *AccCAT*, *AccGSTD*, *AccGSTs4*, *AccMsrA*, and *AccTrx2* were induced when *AccGrx2* was knocked down; *AccCAT*, *AccGSTD*, *AccGSTs4*, and *AccMsrA* were upregulated when *AccTpx-3* was knocked down; and *AccCAT*, *AccGSTs4*, *AccMsrA*, and *AccTrx2* were upregulated when *AccTrx1* was knocked down, suggesting that all the above induced genes participate in the compensation for knockdown of *AccGrx1*, *AccGrx2*, *AccTpx-3*, and *AccTrx1* in *A. cerana cerana* (Fig. 10).

Discussion

Grxs are involved in maintaining and regulating the cellular redox state and redox-dependent signaling pathways [11]. Trxs participate in resisting oxidative damage caused by ROS and maintaining cellular redox homeostasis [18-20]. A majority of research

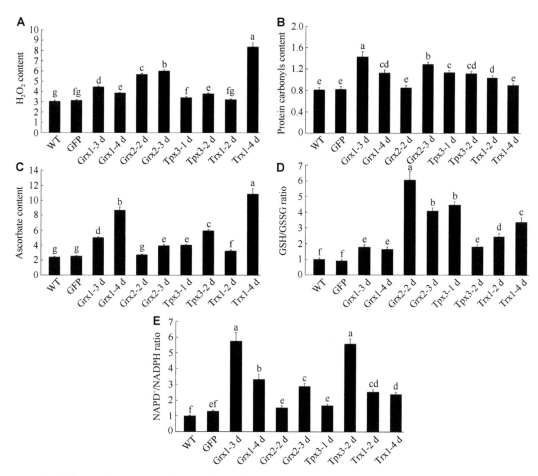

Fig. 9 Effects of the knockdown of *AccGrx1*, *AccGrx2*, *AccTpx-3*, and *AccTrx1* on metabolite contents of hydrogen peroxide (A), carbonyl (B), and ascorbate (C) and ratios of GSH/GSSG (D) and $NADP^+$/NADPH (E). The whole-bee homogenates were analyzed by using a test kit according to the protocol of the manufacturer. Each value is given as the mean (SD) of three replicates. Different letters above the bars indicate significant differences ($P < 0.01$), according to SAS software.

concerning the roles of Grxs and Trxs has been completed in mammalian systems. However, little study of Grxs and Trxs has been conducted in insects, particularly in the Chinese honeybee. To gain new insight into the functions of Grxs and Trxs in other important species, we isolated three antioxidant genes (*AccGrx1*, *AccGrx2*, and *AccTrx1*) from *A. cerana cerana* and investigated their response to various oxidative stresses. In addition, we also used RNAi technology to knock down *AccGrx1*, *AccGrx2*, *AccTpx-3*, and *AccTrx1* and then examined the activities of antioxidant enzymes, metabolites contents and expression levels of other antioxidant genes. This study demonstrated that AccGrx1, AccGrx2, AccTpx-3, and AccTrx1 play important roles in antioxidant defense when honeybees are subjected to oxidative stress.

There are two main groups of Grxs characterized, dithiol Grxs containing a CXXC active-site sequence (mostly CPYC) and monothiol Grxs with a CXXS motif (mostly CGFS) [11]. Trxs contain a highly conserved redox-active dithiol group in the form of CXXC [31]. The cysteine residues of conserved active-site motifs are essential for their catalytic activity [32]. Grx1 and Grx3 of *Chlamydomonas reinhardtii* belong to the classic CPYC type and the CGFS,

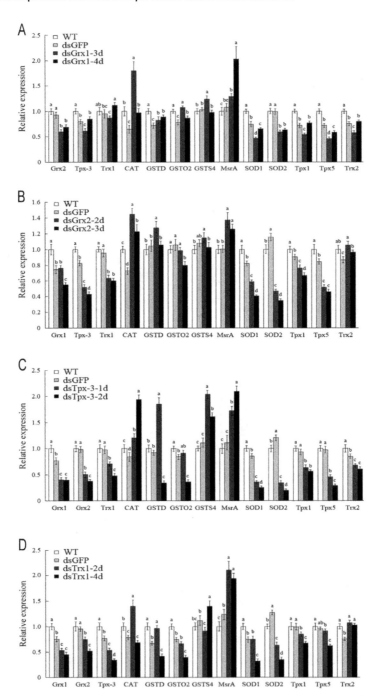

Fig. 10 Expression profiles of other antioxidant genes performed by using qRT-PCR when *AccGrx1* (A), *AccGrx2* (B), *AccTpx-3* (C), and *AccTrx1* (D) were knocked down. The *β-actin* gene was used as an internal control. Each value is given as the mean (SD) of three replicates. Different letters above the bars indicate significant differences [$P < 0.01$, except for *AccTrx1* ($P < 0.05$) and *AccGSTs4* ($P < 0.05$) in (A) and (B), respectively], according to SAS software.

respectively [33]. Thioredoxin 1 of *Cynoglossus semilaevis* (CsTrx1) has a highly conserved active-site motif in the form of CQPC [34]. AccGrx1, AccGrx2, and AccTrx1 have active-site motifs in the form of CPFC, CGFS, and CNPC, respectively, which are in agreement with the above studies.

The fat body is the central metabolic organ that plays an important role in detoxification/degradation of xenobiotics and resisting oxidative stress [35-37]. *AccGrx1*, *AccGrx2*, and *AccTrx1* were all detected in the fat body and *AccGrx1* was expressed higher than *AccGrx2* and *AccTrx1*, suggesting that *AccGrx1* may play a more important role in antioxidant defense than the other two genes in the fat body. The epidermis may be important in resistance to oxidative stresses because it bears the brunt of external attacks [38]. The tissue-specific expression analysis of *AccGrx1* and *AccTrx1* revealed a higher expression in epidermis than in other tissues, which is consistent with *AccTrx-like1* [25], suggesting that it may play an important role in epidermis. However, the expression of *AccGrx2* mRNA was mainly detected in muscle, followed by the epidermis, revealing that it functions in muscle and epidermis. In addition, our transcriptional analysis of *AccTrx1* during various developmental stages in *A. cerana cerana* revealed that it was expressed highest in the 15-day post-emergence adults, followed by the dark-eyed pupae, which is similar to *AccTrx2* [26]. However, *AccGrx1* and *AccGrx2* were mainly detected in dark-eyed pupae, suggesting they may play important roles in this stage.

Environmental conditions such as temperature, heavy metal, pesticides, and ultraviolet radiation can induce oxidative stress [39, 40]. Trx is a stress-inducible protein, which is stimulated by a variety of stresses, including heat, paraquat, H_2O_2, UV radiation, and microorganisms such as bacteria and viruses [41-43]. Mitochondrial thioredoxin 2 from disk abalone (*Haliotis discus discus*; AbTrx2) was upregulated after H_2O_2 oxidative stress, revealing that AbTrx2 functions in reducing harmful ROS [44]. In previous studies, *AccTrx-like1* was induced by 4, 15, and 25 ℃ and H_2O_2 treatments and *AccTrx2* was upregulated by 4, 16, and 25 ℃; H_2O_2; cyhalothrin; acaricide; paraquat; phoxim; and $HgCl_2$ treatments [25, 26]. *AccTrx1* was stimulated by 4, 16, and 42 ℃; H_2O_2; and pesticides (acaricide, paraquat, cyhalothrin, and phoxim) treatments in this study, suggesting that *AccTrx1* may play a critical role in resisting oxidative stress.

Grxs belong to the large oxidoreductase family and participate in maintaining redox homeostasis [3, 11]. The knockdown of Grx1 increases oxidant-induced apoptosis in H9c2 cells [45]. Glutaredoxin 1 and glutaredoxin 2 of *Venerupis philippinarum*(*VpGrx1* and *VpGrx2*) were significantly upregulated at 24 h after *Vibrio* and benzo [a] pyrene challenge [46]. Overexpression of Grx2 could protect cells from H_2O_2-induced damage, whereas knockdown of Grx2 showed the opposite effect, suggesting Grx2 plays an important role in resisting H_2O_2-induced oxidative stress [15]. Glutaredoxin 2 from *Ostrinia furnacalis* (*OfurGrx2*) could be dramatically stimulated by UV radiation, mechanical injury, *E. coli* exposure, and high and low temperatures [17]. In this study, *AccGrx1* and *AccGrx2* were induced by 4, 16, and 42 ℃; H_2O_2; $HgCl_2$; acaricide; paraquat; cyhalothrin; and phoxim treatments, indicating that *AccGrx1* and *AccGrx2* may play important roles in scavenging ROS.

It was demonstrated that the *Drosophila*, which lose the function of *Trx2*, are hypersusceptible to paraquat and express high levels of antioxidant genes, such as SOD, CAT, and glutathione synthetase [47]. Hydrogen peroxide could cause oxidant damage to DNA, proteins, and lipids [42]. CAT and POD, which are antioxidant enzymes, play important roles in scavenging hydrogen peroxide [23]. GSH and ascorbate could also eliminate hydrogen peroxide [48, 49]. In this study, we found that the enzymatic activities of CAT and POD (Fig. 8), the contents of hydrogen peroxide and ascorbate, and the ratio of GSH/GSSG in *A. cerana cerana* were enhanced when *AccGrx1*, *AccGrx2*, *AccTpx-3*, and *AccTrx1* were knocked down, revealing that *A. cerana cerana* was exposed to a high level of oxidative stress when the above four genes were suppressed (Fig. 9A, C, and D), and CAT, POD, GSH, and ascorbate may be involved in scavenging hydrogen peroxide. As shown in Fig. 9B, the elevated levels of protein carbonyls were observed in silenced samples compared to control groups. Hydrogen peroxide could result in formation and accumulation of protein carbonyls, and methionine sulfoxide modifications to protein mediated by the methionine sulfoxide reductase (Msr) system may precede the formation of protein carbonyls. Lacking the MsrA enzyme could cause elevated levels of protein carbonyls in yeast and mice [50]. Interestingly, elevated expression levels of *AccMsrA* were observed when *AccGrx1*, *AccGrx2*, *AccTpx-3*, and *AccTrx1* were downregulated (Fig. 10), suggesting that AccMsrA may play an important role in reducing the production of protein carbonyls in *A. cerana cerana*. All the above results are summarized in Fig. 11. GSH plays an important role in scavenging ROS and serving as a cofactor in enzymatic redox reactions [51, 52]. NADPH plays a crucial role as a donor of electrons that provide reducing equivalents and maintain thiol-disulfide protein activities [53]. Glutathione reductase catalyzes the reduction of oxidized glutathione to GSH with the help of electrons from NADPH and thereby maintains a constant GSH level in the system [54]. As shown in Fig. 9D and E, we found that the GSH/GSSG ratio and the $NADP^+$/NADPH ratio of silenced samples were both higher than those of control groups, which was consistent with previous studies. Altering the expression of one antioxidant enzyme may influence the expression of other antioxidant enzymes as compensatory mechanisms to sustain a reduced redox environment [55]. An increase in the expression level of *AccTrx2* was observed when *AccTrx1* and *AccGrx2* were inhibited and an increase in *AccTrx1* was detected when *AccGrx1*

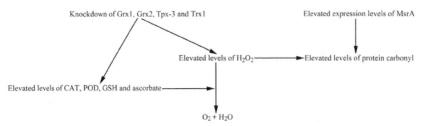

Fig. 11 Knockdown of *Grx1*, *Grx2*, *Tpx-3*, and *Trx1* in *Apis cerana cerana* resulted in elevated levels of hydrogen peroxide. Hydrogen peroxide could cause oxidant damage to DNA, proteins, and lipids and could result in formation and accumulation of protein carbonyls. So the contents of protein carbonyls were elevated when *Grx1*, *Grx2*, *Tpx-3*, and *Trx1* were downregulated compared to control groups in *A. cerana cerana*. Methionine sulfoxide modification to protein mediated by the methionine sulfoxide reductase system may precede the formation of protein carbonyls. Lack of the methionine sulfoxide reductase A (MsrA) enzyme could cause elevated levels of protein carbonyls in yeast and mice. Interestingly, *AccMsrA* was induced in silenced samples compared to control groups. So AccMsrA may play an important role in reducing the generation of protein carbonyls. Simultaneously, the activities of CAT and POD and the contents of ascorbate and GSH in silenced samples were higher than those in control groups. CAT, POD, ascorbate, and GSH could eliminate hydrogen peroxides.

was knocked down, which are inconsistent with loss of *Trx2* in *Drosophila* [47]. In addition, as shown in Fig. 10, we also found that some other antioxidant genes, such as *AccCAT*, *AccGSTs4*, *AccMsrA*, etc., were induced when *AccGrx1*, *AccGrx2*, *AccTpx-3*, and *AccTrx1* were suppressed, revealing that all the above induced genes are involved in compensating for the knockdown of *AccGrx1*, *AccGrx2*, *AccTpx-3*, and *AccTrx1* to resist oxidative damage.

In conclusion, we have identified three antioxidant genes (*AccGrx1*, *AccGrx2*, and *AccTrx1*) from *A. cerana cerana*. These three antioxidant enzymes possess conserved functional domains of the Grx and Trx superfamilies. The expression levels of *AccGrx1*, *AccGrx2*, and *AccTrx1* in response to various environmental stresses indicate that *AccGrx1*, *AccGrx2*, and *AccTrx1* may play critical roles in oxidative stress-related defense mechanisms. Very surprisingly, knockdown of *AccGrx1*, *AccGrx2*, *AccTpx-3* and *AccTrx1* resulted in elevated enzymatic activities of CAT and POD; increased contents of hydrogen peroxide, carbonyls, and ascorbate; and elevated ratios of GSH/GSSG and $NADP^+$/NADPH in *A. cerana cerana* and other antioxidant genes, such as *AccCAT*, *AccGSTs4*, *AccMsrA* etc., were also induced. This further verifies the antioxidant roles of AccGrx1, AccGrx2, AccTpx-3, and AccTrx1 in *A. cerana cerana* These findings may be useful for studying the mechanisms of the antioxidant defense of the proteins.

Acknowledgments This work was financially supported by an Earmarked Fund of the China Agriculture Research System (No. CARS-45), the Special Fund for Agro-scientific Research in the Public Interest (No. 200903006), and the National Natural Science Foundation of China (No. 31172275).

References

[1] Droge, W. Free radicals in the physiological control of cell function. Physiol Rev. 2002, 82, 47-95.
[2] Balaban, R. S., Nemoto, S., Finkel, T. Mitochondria, oxidants, and aging. Cell. 2005, 120, 483-495.
[3] Wu, H. L., Lin, L. R., Giblin, F., Ho, Y. S., Lou, M. F. Glutaredoxin 2 knockout increases sensitivity to oxidative stress in mouse lens epithelial cells. Free Radic Biol Med. 2011, 51, 2108-2117.
[4] Holmgren, A. Thioredoxin and glutaredoxin systems. J Biol Chem. 1989, 264, 13963-13966.
[5] Adluri, R. S., Thirunavukkarasu, M., Zhan, L., et al. Glutaredoxin-1 overexpression enhances neovascularization and diminishes ventricular remodeling in chronic myocardial infarction. PLoS One. 2012, 7, e34790.
[6] Fernandes, A. P., Holmgren, A. Glutaredoxins: glutathione-dependent redox enzymes with functions far beyond a simple thioredoxin backup system. Antioxid Redox Signaling. 2004, 6, 63-74.
[7] Bushweller, J. H., Aslund, F., Wuthrich., K., Holmgren. A Structural and functional characterization of the mutant *Escherichia coli* glutaredoxin (C14-S) and its mixed disulfide with glutathione. Biochemistry. 1992, 31, 9288-9293.
[8] Christensen, C. A., Starke, D. W., Mieyal, J. J. Acute cadmium exposure inactivates thioltransferase (glutaredoxin), inhibits intracellular reduction of protein glutathionyl mixed disulfides and initiates apoptosis. J Biol Chem. 2000, 275, 26556-26565.
[9] Rodriguez-Manzaneque, M. T., Tamarit, J., Belli, G., Ros, J., Herrero, E. Grx5 is a mitochondrial glutaredoxin required for the activity of iron/sulfur enzymes. Mol Biol Cell. 2002, 13, 1109-1121.
[10] Muhlenhoff, U., Gerber, J., Richhardt, N., Lill, R. Components involved in assembly and dislocation of iron-sulfur clusters on the scaffold protein Isu1p. EMBO J. 2003, 22, 4815-4825.
[11] Lillig, C. H., Berndt, C., Holmgren, A. Glutaredoxin systems. Biochim Biophys Acta. 2008, 1780,

1304-1317.

[12] Murata, H., Ihara, Y., Nakamura, H., Yodoi, J., Sumikawa, K., Kondo, T. Glutaredoxin exerts an antiapoptotic effect by regulating the redox state of Akt. J Biol Chem. 2003, 278, 50226-50233.

[13] Wang, J., Pan, S., Berk, B. C. Glutaredoxin mediates Akt and eNOS activation by flow in a glutathione reductase-dependent manner. Arterioscler Thromb Vasc Biol. 2007, 27, 283-288.

[14] Meyer, E. B., Wells, W. W. Thioltransferase overexpression increases resistance of MCF-7 cells to adriamycin. Free Radic Biol Med. 1999, 26, 770-776.

[15] Wu, H., Xing, K., Lou., M. F. Glutaredoxin 2 prevents H_2O_2-induced cell apoptosis by protecting complex I activity in the mitochondria. Biochim Biophys Acta. 2010, 1797, 1705-1715.

[16] Enoksson, M., Fernandes, A. P., Prast, S., Lillig, C. H., Holmgren, A., Orrenius, S. Overexpression of glutaredoxin 2 attenuates apoptosis by preventing cytochrome c release. Biochem Biophys Res Commun. 2005, 327, 774-779.

[17] An, S., Zhang, Y., Wang, T., Luo, M., Li, C. Molecular characterization of glutaredoxin 2 from *Ostrinia furnacalis*. Integr Zool. 2013, Sl, 30-38.

[18] Holmgren, A. Thioredoxin. Annu Rev Biochem. 1985, 54, 237-271.

[19] Arnér, E. S. J., Holmgren, A. Physiological functions of thioredoxin and thioredoxin reductase. Eur J Biochem. 2000, 267, 6102-6109.

[20] Myers, J. M., Antholine, W. E., Myers, C. R. Hexavalent chromium causes the oxidation of thioredoxin in human bronchial epithelial cells. Toxicology. 2008, 246, 222-233.

[21] Lee, S., Kim, S. M., Lee, R. T. Thioredoxin and thioredoxin target proteins: from molecular mechanisms to functional significance. Antioxid Redox Signaling. 2013, 18, 1165-1207.

[22] Chen, Y., Cai, J., Jones, D. P. Mitochondrial thioredoxin in regulation of oxidantinduced cell death. FEBS Lett. 2006, 580, 6596-6602.

[23] Corona, M., Robinson, G. E. Genes of the antioxidant system of the honey bee: annotation and phylogeny. Insect Mol Biol. 2006, 15, 687-701.

[24] Green, D. R., Reed, J. C. Mitochondria and apoptosis. Science. 1998, 281, 1309-1312.

[25] Lu, W., Kang, M., Liu, X., Zhao, X., Guo, X., Xu, B. Identification and antioxidant characterisation of thioredoxin-like1 from *Apis cerana cerana*. Apidologie. 2012, 43, 737-752.

[26] Yao, P., Hao, L., Wang, F., Chen, X., Guo, X., Xu, B. Molecular cloning, expression and antioxidant characterisation of a typical thioredoxin gene (*AccTrx2*) in *Apis cerana cerana*. Gene. 2013, 527, 33-41.

[27] Yang, G. Harm of introducing the western honeybee *Apis mellifera* L. to the Chinese honeybee *Apis cerana* F. and its ecological impact. Acta Entomol Sin. 2005, 3, 401-406.

[28] Xu, P., Shi, M., Chen, X. Antimicrobial peptide evolution in the Asiatic honey bee *Apis cerana*. PLoS One. 2009, 4, e4239.

[29] Yao, P., Lu, W., Meng, F., Wang, X., Xu, B., Guo, X. Molecular cloning, expression and oxidative stress response of a mitochondrial thioredoxin peroxidase gene (*AccTpx-3*) from *Apis cerana cerana*. J Insect Physiol. 2013, 59, 273-282.

[30] Elias-Neto, M., Soares, M. P., Simões, Z. L., Hartfelder, K., Beyond, M. M. Developmental characterization, function and regulation of a Laccase 2 encoding gene in the honey bee, *Apis mellifera* (Hymenoptera, Paine). Insect Biochem Mol Biol. 2010, 40, 241-251.

[31] Powis, G., Montfort, W. R. Properties and biological activities of thioredoxins. Annu Rev Pharmacol Toxicol. 2001, 41, 261-295.

[32] Kallis, G. B., Holmgren, A. Differential reactivity of the functional sulfhydryl groups of cysteine-32 and cysteine-35 present in the reduced form of thioredoxin from *Escherichia coli*. J Biol Chem. 1980, 255, 10261-10265.

[33] Zaffagnini, M., Michelet, L., Massot, V., Trost, P., Lemaire, S. D. Biochemical characterization of glutaredoxins from *Chlamydomonas reinhardtii* reveals the unique properties of a chloroplastic CGFS-type glutaredoxin. J Biol Chem. 2008, 283, 8868-8876.

[34] Sun, J., Li, Y., Sun, L. *Cynoglossus semilaevis* thioredoxin: a reductase and an antioxidant with

immunostimulatory property. Cell Stress Chaperones. 2012, 17, 445-455.

[35] Arrese, E. L., Soulages, J. L. Insect fat body: energy, metabolism, and regulation. Annu Rev Entomol. 2010, 55, 207-225.

[36] Enayati, A. A., Ranson, H., Hemingway, J. Insect glutathione transferases and insecticide resistance. Insect Mol Biol. 2005, 14, 3-8.

[37] Sawicki, R., Singh, S. P., Mondal, A. K., Benes, H., Zimniak, P. Cloning, expression and biochemical characterization of one Epsilon class (GST-3) and ten Delta-class (GST-1) glutathione S-transferases from Drosophila melanogaster, and identification of additional nine members of the Epsilon class. Biochem J. 2003, 370, 661-669.

[38] Marionnet, C., Bernerd, F., Dumas, A., et al. Modulation of gene expression induced in human epidermis by environmental stress in vivo. J Invest Dermatol. 2003, 121, 1447-1458.

[39] Lushchak, V. I. Environmentally induced oxidative stress in aquatic animals. Aquat Toxicol. 2011, 101, 13-30.

[40] Kottuparambila, S., Shinb, W., Brownc, M. T., Han, T. UV-B affects photosynthesis, ROS production and motility of the freshwater flagellate, Euglena agilis Carter. Aquat Toxicol. 122-2012, 123, 206-213.

[41] Kim, I., Lee, K. S., Hwang, J. S., Ahn, M. Y., Yun, E. Y., Li, J. H., Sohn, H. D., Jin, B. R. R. Molecular cloning and characterization of ATX1 cDNA from the mole cricket, Gryllotalpa orientalis. Arch Insect Biochem Physiol. 2006, 61, 231-238.

[42] Lee, K. S., Kim, S. R., Park, N. S., et al. Characterisation of silkworm thioredoxin peroxidase that is induced by external temperature stimulus and viral infection. Insect Biochem Mol Biol. 2005, 35, 73-84.

[43] Nakamura, H., Nakamura, K., Yodoi, J. Redox regulation of cellular activation. Annu Rev Immunol. 1997, 15, 351-369.

[44] Zoysa, M. D., Pushpamali, W. A., Whang, I., Kim, S. J., Lee, J. Mitochondrial thioredoxin-2 from disk abalone (Haliotis discus discus): molecular characterization, tissue expression and DNA protection activity of its recombinant protein. Comp Biochem Physiol B: Biochem Mol Biol. 2008, 149, 630-639.

[45] Gallogly, M. M., Shelton, M. D., Qanungo, S., Pai, H. V., Starke, D. W., Hoppel, C. L., Lesnefsky, E. J., Mieyal, J. J. Glutaredoxin regulates apoptosis in cardiomyocytes via NFκB targets Bcl-2 and Bcl-xL: implications for cardiac aging. Antioxid Redox Signaling. 2010, 12, 1339-1353.

[46] Mu, C., Wang, Q., Yuan, Z., Zhang, Z., Wang, C. Identification of glutaredoxin 1 and glutaredoxin 2 genes from Venerupis philippinarum and their responses to benzo[a]pyrene and bacterial challenge. Fish Shellfish Immun. 2012, 32, 482-488.

[47] Tsuda, M., Ootaka, R., Ohkura, C., Kishita, Y., Seong, K. H., Matsuo, T., Aigaki, T. Loss of Trx-2 enhances oxidative stress-dependent phenotypes in Drosophila. FEBS Lett. 2010, 584, 3398-3401.

[48] Meister, A. Glutathione–ascorbic acid antioxidant system in animals. J Biol Chem. 1994, 269, 9397-9400.

[49] Chatterjee, I. B., Majumder, A. K., Nandi, B. K., Subramanian, N. Synthesis and some major functions of vitamin C in animals. Ann N Y Acad Sci. 1975, 258, 24-47.

[50] Moskovitz, J., Oien, D. B. Protein carbonyl and the methionine sulfoxide reductase system. Antioxid Redox Signaling. 2010, 12, 405-415.

[51] Schafer, F. Q., Buettner, G. R. Redox environment of the cell as viewed through the redox state of the glutathione disulfide/glutathione couple. Free Radic Biol Med. 2001, 30, 1191-1212.

[52] Townsend, D. M., Tew, K. D., Tapiero, H. The importance of glutathione in human disease. Biomed Pharmacother. 2003, 57, 145-155.

[53] Liang, Y., Harris, F. L., Jones, D. P., Brown, L. A. S. Alcohol induces mitochondrial redox imbalance in alveolar macrophages. Free Radic Biol Med. 2013, 65, 1427-1434.

[54] Pannala, V. R., Bazil, J. N., Camara, A. K. S., Dash, R. K. A biophysically based mathematical model for the catalytic mechanism of glutathione reductase. Free Radic Biol Med. 2013, 65, 1385-1397.

[55] Mates, J. M., Segura, J. A., Alonso, F. J., Marquez, J. Oxidative stress in apoptosis and cancer: an update. Arch Toxicol. 2012, 86, 1649-1665.

2.12 Molecular Cloning, Expression and Antioxidant Characterisation of a Typical Thioredoxin Gene (*AccTrx2*) in *Apis cerana cerana*

Pengbo Yao, Lili Hao, Fang Wang, Xiaobo Chen,
Yan Yan, Xingqi Guo, Baohua Xu

Abstract: Thioredoxins (Trxs) are a family of small, highly conserved and ubiquitous proteins that are involved in protecting organisms against toxic reactive oxygen species (ROS). In this study, a typical thioredoxin 2 gene was isolated from *Apis cerana cerana*, *AccTrx2*. The full-length cDNA sequence of *AccTrx2* was composed of 407 bp containing a 318 bp open reading frame (ORF) that encodes a predicted protein of 105 amino acids, 11.974 kDa and an isoelectric point of 4.45. Expression profile of *AccTrx2* as determined by a quantitative real-time PCR (qRT-PCR) analysis was higher in brain than in other tissues, with its highest transcript occurring on the 15 days post-emergence adults and upregulated by such abiotic stresses as 4, 16, 25℃, H_2O_2, cyhalothrin, acaricide, paraquat, phoxim and mercury ($HgCl_2$) treatments. However, *AccTrx2* was slightly repressed when exposed to 42℃ treatment. Characterisation of the recombinant protein showed that the purified AccTrx2 had insulin disulfide reductase activity and could protect DNA from ROS damage. These results indicate that AccTrx2 functions as an antioxidant that plays an important role in response to oxidative stress.

Keywords: *AccTrx2*; *Apis cerana cerana*; qRT-PCR; Oxidative damage; Antioxidant capacity

Introduction

Reactive oxygen species (ROS) including superoxide anions, hydrogen peroxide and hydroxyl radicals can be induced by external or internal factors, such as pro-oxidants, heavy metal, pesticides or the respiratory system in mitochondria [1]. Excessive ROS could cause oxidative damage to DNA, lipid membranes, and proteins [2]. Furthermore, high concentrations of ROS could result in certain diseases and ageing [3]. To control oxidative stress caused by ROS, aerobic cells possess a set of antioxidant enzymes such as catalase (CAT), superoxide dismutase (SOD), glutathione peroxidase (GPX), thioredoxin peroxidase (TPx), and thioredoxin (Trx) [4] systems, which maintain the intracellular ROS at proper levels [5].

Trx is approximately 12 kDa in size and has a highly conserved active site, Cys-Gly-Pro-Cys (CGPC) [6]. Trx is a family of small, evolutionarily conserved and ubiquitous proteins that act as protein-disulfide reductase [7] and participate in essential antioxidant and redox-regulatory processes via a pair of conserved cysteine residues [8]. Trx plays a crucial role in the maintenance of cellular thiol redox balance [9, 10], and is related to the cellular antioxidant network [11]. Trx, which acts as an antioxidant, functions as a catalytic cycle in which peroxiredoxin (Prx) oxidizes the Trx and then Trx reductase (TrxR) reduces it by using NADPH [12].

In mammals, there are two isoforms of thioredoxin characterized, Trx1 and Trx2, which are located in the cytoplasm and the mitochondria, respectively. However, Trx1 can be translocated into the nucleus and secreted out of the cell under certain circumstances [13]. In mammalian cells, Trxs are major disulfide reductases of the cells participating in redox regulation of different metabolic pathways and oxidative stress defense [14, 15]. Overexpression *Trx* in mice could prevent miscellaneous disease phenotypes associated with oxidative stress [16, 17] and improve resistance to a variety of oxidative stresses including infection and inflammation [18].

In *Drosophila*, three *Trx* genes have been identified: *Trx1* [19, 20], *Trx2* [21] and *TrxT* [22]. *Trx1* and *TrxT* are located in the nucleus whereas *Trx2* is located in the cytoplasm suggesting that *Trx2* plays an important role in redox homeostasis [21]. The antioxidant genes such as SOD, CAT, and glutathione synthetase (GSS) of *Trx2* mutants are expressed higher than the wild type in *Drosophila* [23]. The *Apis mellifera* genome also has a *Trx2* gene, *AmTrx2* which is a putative ortholog of *Drosophila Trx2* [24] suggesting that *AmTrx2* may also participate in redox homeostasis.

There are also some other studies about Trx proteins in other insects. Thioredoxin from the silkworm, *Bombyx mori* (*BmTrx*) is found to protect from oxidative stress caused by extreme temperatures and microbial infection [25]. Thioredoxin-like1 of *Apis cerana cerana* (AccTrx-like1), which acts as an antioxidant, has been shown to play an important role in preventing oxidative stress caused by ROS [26]. All the above observations indicate that Trxs play a crucial role in protecting organisms from ROS damage. In addition, Trxs are also involved in multiple biological processes including regulation of growth [27], transcription factors [28], protein folding [29] and scavenging radicals [30].

Although the functions of Trx proteins have been investigated in other species, there is limited knowledge of Trx proteins in honeybees, particularly, in Chinese honeybees. As a pollinator of flowering plants, *A. cerana cerana* plays an important role in the balance of regional ecology and agricultural development [31]; moreover, the survival environment of *A. cerana cerana* is becoming more and more deteriorated due to various environmental stressors including indiscriminate use of pesticides, infectious diseases and global warming [32]. So it is crucial to identify the Trxs and elucidate the ROS defence mechanisms in *A. cerana cerana*. According to previous observations, we predicted that Trx2 protects *A. cerana cerana* from ROS damage. To gain insight into the role of *AccTrx2*, we isolated a typical *Trx2* gene which is highly similar to *AmTrx2* from *A. cerana cerana* in this study and examined its expression profiles in different developmental stages and tissues. We also investigated the transcripts of *AccTrx2* when exposed to various oxidative stresses, including different temperatures, H_2O_2, cyhalothrin, acaricide, paraquat, phoxim, and $HgCl_2$. In addition, we characterized the antioxidant activity and the insulin disulfide reduction ability of the purified AccTrx2 protein. Our results suggested that *AccTrx2* might play an important role in resisting oxidative damage by ROS.

Materials and Methods

Biological specimens and treatments

The Chinese honeybee (*A. cerana cerana*) used in this study was obtained from the experimental apiary of Shandong Agricultural University (Taian, China). The worker honeybees were generally classified into larvae, pupae, and adults according to age [33], eye colour [34, 35] and cuticle pigmentation [36]. Honeybees at various stages of development including the fourth (L4)-, the fifth (L5)-, and the sixth (L6)-instar larvae, white-eyed (Pw), pink-eyed (Pp), brown-eyed (Pb), dark-eyed (Pd) pupae and 1 day post-emergence adults (A1) were collected from the hive, while 15 days post-emergence (A15) and 30 days post-emergence adults (A30) were collected at the entrance of the hive. The adult workers were captured by marking 1 day post-emergence adults (A1) with paint and then collected them after 15 and 30 days, respectively. The worker honeybees of 15 days post-emergence adults were divided into several groups ($n = 30$), fed a diet of water and pollen and maintained in an incubator with 60% relative humidity at 34℃ under a 24 h dark regimen. The bees incubated at 34℃ served as a control. Groups 1, 2, 3, and 4 were placed at 4, 16, 25, and 42℃, respectively. Group 5 was treated with H_2O_2 at a final concentration of 2 mmol/L H_2O_2 by diluting a 30% stock in distilled water. Group 6 was treated with $HgCl_2$ (3 mg/ml added to food). Groups 7-10 were treated with pesticides (cyhalothrin, acaricide, paraquat and phoxim, which were diluted to 20 mg/L and added to the food for the experimental group). The control group fed on normal food. The adult workers of groups 1 and 7-10 were collected after treatments at 0.5 h intervals while the adult workers of groups 2-5 and group 6 at 1.0 h and 3.0 h intervals, respectively. The treated bees were flash-frozen in liquid nitrogen and stored at −80℃ until used

The brain, epidermis, muscle and midgut of adults (15 days) were dissected on ice, freshly collected and used to examine tissue specific expression. At least three biological replicates were performed in all the treatments.

Total RNA extraction, cDNA synthesis, and DNA isolation

RNA was extracted with Trizol reagent (TransGen Biotech, Beijing, China) according to the instructions of the manufacturer. To remove the genomic DNA, the RNA samples were treated with RNase-free DNase I, and the cDNA was synthesised by using a reverse transcription system (TransGen Biotech, Beijing, China). The quality of total RNA was assessed by electrophoresis on 1% agarose gel and analysed by a UV-visible protein nucleic acid analyzer (Thermo Scientific). The genomic DNA was isolated by using an EasyPure Genomic DNA Extraction Kit in accordance with the instructions of the manufacturer (TransGen Bio-tech, Beijing, China).

Primers and PCR amplification conditions

The primers and PCR amplification conditions are listed in Table 1 and Table 2, respectively.

Table 1 PCR primers in this study

Abbreviation	Primer sequence (5'—3')	Description
TR1	CGGTAGTACTTGTGGACAAGCTTAGAA	cDNA sequence primer, forward
TR2	ATTTATTTATTTGTGCTTCTGAATTGTGC	cDNA sequence primer, reverse
3R1	CTTTGCTATGTGGTGTGGTCC	3' RACE forward primer, outer
3R2	GAAAGTTGATGTAGATGAATGCGAAG	3' RACE forward primer, inner
5R1	GGACCACACCACATAGCAAAG	5' RACE reverse primer, outer
5R2	TCACTGGCATTCTTAATTTGGTAAAC	5' RACE reverse primer, inner
AAP	GGCCACGCGTCGACTAGTAC(G)$_{14}$	Abridged anchor primer
AUAP	GGCCACGCGTCGACTAGTAC	Abridged universal amplification primer
B26	GACTCTAGACGACATCGA(T)$_{18}$	3' RACE universal primer, outer
B25	GACTCTAGACGACATCGA	3' RACE universal primer, inner
QR1	TCAGTCGTTAACGAAGCATCGAAATTGT	Full-length cDNA primer, forward
QR2	ATTTATTTATTTGTGCTTCTGAATTGTGC	Full-length cDNA primer, reverse
β-s	TTATATGCCAACACTGTCCTTT	Standard control primer, forward
β-x	AGAATTGATCCACCAATCCA	Standard control primer, reverse
N1	TCAGTCGTTAACGAAGCATCGAAATTGT	Genomic sequence primer, forward
N2	CAAGGACCACACCACATAGCAAAG	Genomic sequence primer, reverse
N3	CTTTGCTATGTGGTGTGGTCCTTG	Genomic sequence primer, forward
N4	ATTTATTTATTTGTGCTTCTGAATTGTGC	Genomic sequence primer, reverse
YH1	GGTACCATGGTTTACCAAATTAAGAAT	Protein expression primer, forward
YH2	GAGCTCTTTGTGCTTCTGAATTGTGC	Protein expression primer, reverse
TYR1	GGTTCGGTAGTACTTGTGGAC	Real-time PCR primer, forward
TYR2	GGACCACACCACATAGCAAAG	Real-time PCR primer, reverse

Table 2 PCR amplification conditions

Primer pair	Amplification conditions
TR1/TR2	10 min at 94℃, 40 s at 94℃, 40 s at 52℃, 40 s at 72℃ for 35 cycles, 10 min at 72℃
3R1/B26	10 min at 94℃, 40 s at 94℃, 40 s at 51℃, 40 s at 72℃ for 28 cycles, 10 min at 72℃
3R2/B25	10 min at 94℃, 40 s at 94℃, 40 s at 50℃, 40 s at 72℃ for 35 cycles, 10 min at 72℃
5R1/AAP	10 min at 94℃, 40 s at 94℃, 40 s at 51℃, 40 s at 72℃ for 28 cycles, 10 min at 72℃
5R2/AUAP	10 min at 94℃, 40 s at 94℃, 40 s at 51℃, 40 s at 72℃ for 35 cycles, 10 min at 72℃
QR1/QR2	10 min at 94℃, 40 s at 94℃, 40 s at 51℃, 60 s at 72℃ for 35 cycles, 10 min at 72℃
N1/N2	10 min at 94℃, 40 s at 94℃, 40 s at 52℃, 90 s at 72℃ for 35 cycles, 10 min at 72℃
N3/N4	10 min at 94℃, 40 s at 94℃, 40 s at 52℃, 50 s at 72℃ for 35 cycles, 10 min at 72℃
YH1/YH2	10 min at 94℃, 40 s at 94℃, 40 s at 52℃, 50 s at 72℃ for 35 cycles, 10 min at 72℃

Isolation of the cDNA of *AccTrx2*

The internal conserved fragment of *AccTrx2* was obtained by reverse transcription PCR (RT-PCR) by using the primers TR1 and TR2, which were designed based on the conserved regions of the Trxs from *Apis mellifera*, *Bombus impatiens*, *A. florae* and *Bombus terrestris* and were then synthesised (Biosune Biotechnological Company, Shanghai, China). Based on the internal fragment sequence of *AccTrx2*, primers 3R1/3R2, B26/B25, 5R1/5R2, and AAP/AUAP were designed and synthesised for the 5' rapid amplification of cDNA end (RACE) and 3' RACE. Subsequently, the full-length cDNA sequence was derived by using specific primers (QR1 and QR2) designed based on the assembled full-length cDNA sequence. All the PCR products were purified, cloned into the pEASY-T3 vector (TransGen Biotech, Beijing, China) and then transformed into competent *Escherichia coli* (*E. coli*) DH5α cells for sequencing. All of the PCR primers and amplification conditions are shown in Table 1 and Table 2, respectively.

Amplification of the genomic DNA sequence of *AccTrx2*

To acquire the genomic DNA sequence of AccTrx2, two pairs of primers (N1/N2 and N3/N4) were designed and synthesised according to the full-length cDNA of AccTrx2 by using *A. cerana cerana* genomic DNA as the template. Two fragments were obtained, cloned into the pEASY-T3 vector (TransGen Biotech, Beijing, China) and then transformed into competent *E. coli* DH5α cells for sequencing.

Sequence alignment, phylogenetic tree and tertiary structure analysis of AccTrx2

Homologous Trx protein sequences from various species were retrieved from the NCBI database and aligned by using DNAMAN software version 5.2.2 (Lynnon Biosoft, Quebec, Canada). Conserved domains in AccTrx2 were detected by using bioinformatics tools available at the NCBI server (http://blast.ncbi.nlm.nih.gov/Blast.cgi). DNAMAN software was also used to predict the theoretical isoelectric point and molecular weight of AccTrx2. Phylogenetic analysis was performed by using the neighbour-joining method with the Molecular Evolutionary Genetics Analysis (MEGA version 4.1) software. Finally, the tertiary structure was predicted by using the online protein structure prediction tool I-TASSER server (http://zhanglab.ccmb.med.umich.edu/I-TASSER) [37-39].

AccTrx2 transcriptional profiling by quantitative real-time PCR (qRT-PCR)

Two gene-specific primers, TYR1 and TYR2, were used to amplify a 155 bp fragment of *AccTrx2*. The gene coding for clam *β-actin* was amplified by using Acc-β-actin (GenBank accession number: XM_640276) gene-specific primers (β-s and β-x) and used as an invariant control. qRT-PCR was performed with the SYBR® PrimeScript™ RT-PCR Kit (TaKaRa, Dalian, China) in a 25 μl volume containing 2 μl of diluted cDNA from different samples, 12.5 μl of TaKaRa SYBR® premix Ex *Taq*™, 0.5 μl of each primer (10 pmol/ml), and 9.5 μl ddH$_2$O by using a CFX96™ Real-time System (Bio-Rad). The qRT-PCR amplification conditions were as follows: initial denaturation at 95℃ for 30 s; 41 cycles at 95℃ for 5 s, 55℃ for 15 s and 72℃ for 15 s; and a single melting cycle from 65℃ to 95℃. The relative *AccTrx2* gene expression was analyzed by the comparative CT method ($2^{-\Delta\Delta C_t}$ method). At least three individual samples were prepared for each sample and all samples were analyzed in three duplications.

Expression plasmid construction of AccTrx2

To construct AccTrx2-pET-30a (+), which expresses a His-tagged fusion protein, the open reading frame (ORF) sequence of *AccTrx2* was amplified by PCR using the plasmid containing the *AccTrx2* cDNA with primers YH1 and YH2. To express in the *E. coli* BL21 (DE3) strain, the PCR product was subcloned into the plasmid vector pEASY-T1-simple (TransGen Biotech, Beijing, China). The recombinant plasmid and pET-30a (+) empty vector were digested with the restriction enzymes *Kpn* I and *Sac* I, then the digested fragments were ligased with T4 DNA ligase and transformed into *E. coli* DH5α cells for sequencing.

Recombinant protein expression and purification

To overexpress the recombinant AccTrx2 protein, *E. coli* BL21 (DE3) was transformed with AccTrx2-pET-30a (+). Cells at log phase were induced by 1.0 mmol/L of isopropylthio-β-D-galactoside (IPTG) at 28℃ for 8 h, and then harvested by centrifugation at 5,000 *g* for 10 min at 4℃, resuspended in loading buffer, boiled for 10 min, and centrifuged again to remove the insoluble debris. Each sample (20 µl) was subjected to 12% SDS-PAGE. To purify the recombinant protein, cells were lysed by using an ultrasonic disrupter, and the suspension was purified under native conditions with Ni^{2+}-NTA (nitrilotriacetic acid)-binding resin (Qiagen, Shanghai, China) according to the instructions of the manufacturer. The purified protein was also examined by 12% SDS-PAGE.

Reducing activity of AccTrx2

The enzymatic activity of AccTrx2 to catalyse reduction of insulin was determined with a method modified from Umasuthanetal [40]. Reaction mixtures (500 µl) contain 100 mmol/L potassium phosphate buffer (pH 7.0), 2 mmol/L EDTA, 130 µmol/L insulin from bovine pancreas (Sigma-Aldrich), 0.6 mmol/L dithiothreitol (DTT) and increasing concentrations (ranging from 10.0 to 40.0 µg/ml) of purified recombinant protein AccTrx2. Insulin was added to initiate the reaction and then allowed to occur at 25℃ for 10 min, generating reduced-insulin polypeptides, which formed a white precipitate and hence increased the turbidity of the solution, and was monitored by measuring absorbance at 650 nm. A non-enzymatic reduction of insulin by DTT was concurrently monitored in a solution in which the purified protein was absent. The assay was performed in triplicate.

Protective effect of AccTrx2 against oxidative damage by MFO system

The ability of AccTrx2 to protect DNA against oxidative damage was determined with a method modified from Yao et al. [41]. In brief, the reaction was carried out in a 25 µl system, containing 3 µmol/L $FeCl_3$, 10 mmol/L dithiothreitol (DTT), 100 mmol/L 4-(2-hydroxyethyl)-1-piperazineethanesulfonic acid (HEPES) buffer, and increasing concentrations of purified recombinant protein (ranging from 2.4 to 24.0 µg/ml). The mixture was incubated at 37℃ for 30 min and then supercoiled pUC19 plasmid DNA (300 ng) was added to the mixture for another 2.5 h at 37℃.

Statistical analysis

Error bars denote standard error of the mean (SEM) from three independent experiments. The significant differences were determined by Duncan's multiple range tests using the Statistical Analysis System (SAS) version 9.1 software programmes (SAS Institute, Cary, NC, USA).

Results

Characterization of the sequence of *AccTrx2* and analysis

The full-length cDNA of *AccTrx2* (GenBank accession number: JX844649) is composed of 407 bp containing a 318 bp open reading frame (ORF), which encodes a polypeptide of 105 amino acids with the predicted molecular weight of 11.974 kDa and theoretical isoelectric point of 4.45. Multiple sequence alignment revealed high amino acid identity (81.9%-98.1%) between AccTrx2 and Trx sequences from other species including *B. impatiens*, *B. terrestris*, *A. florae* and *A. mellifera*. The mitochondrial targeting peptide was not identified at the N-terminal amino acid sequence of AccTrx2 by MitoProtII (http://ihg.gsf.de/ihg/mitoprot.html). In addition, the characteristic Trx active site motif CGPC was highly conserved among all selected species (Fig. 1A).

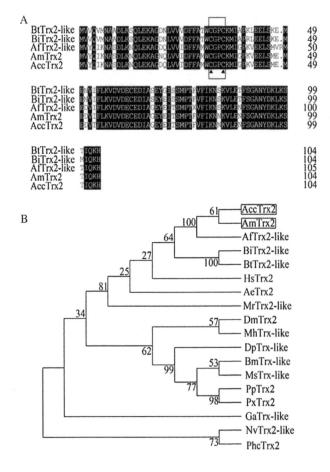

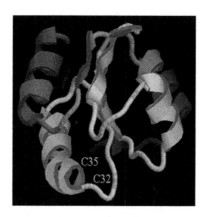

Fig. 1 Molecular characterization, phylogenetic tree and tertiary structure analysis of Trxs from various species. A. Comparison of the deduced amino acid sequences of AccTrx2 (thioredoxin 2, GenBank accession number: JX844649) from *A. cerana cerana*, AmTrx2 (thioredoxin 2, XP_392963.1) from *Apis mellifera*, AfTrx2-like (thioredoxin 2-like, XP_003693957.1), BtTrx2-like (thioredoxin 2-like, XP_003396074.1) and BiTrx2-like (thioredoxin 2-like, XP_003490398.1). Identical amino acid residues are shaded in blue. The conserved CGPC motif is boxed and the active sites are marked by ▲. B. Phylogenetic analysis of AccTrx2 with other known Trx protein sequences. The protein sequences of various species are as follows: AeTrx2 (*Acromyrmex echinatior*, EGI64147.1), AfTrx2-like (*Apis florae*, XP_003693957.1), AmTrx2 (*Apis mellifera*, XP_392963.1), BiTrx2-like (*B. impatiens*, XP_003490398.1), BtTrx2-like (*B terrestris*, XP_003396074.1), BmTrx-like (*B. mori*, NP_001091804.1), DpTrx-like (*Danaus plexippus*, EHJ64037.1), DmTrx2 (*D. melanogaster*, NP_523526.1), GaTrx-like (*Graphocephala atropunctata*, ABD98743.1), HsTrx2 (*Harpegnathos saltator*, EFN85525.1), MhTrx-like (*Maconellicoccus hirsutus*, ABM55528.1), MsTrx-like (*Manduca sexta*, AAF16695.1), MrTrx2-like (*Megachile rotundata*, XP_003705823.1), NvTrx2-like (*Nasonia vitripennis*, XP_001602650.2), PpTrx2 (*Papilio polytes*, BAM19091.1), PxTrx2 (*Papilio xuthus*, BAM17831.1), and PhcTrx2 (*Pediculus humanus corporis*, XP_002424855.1). C. Predicted tertiary structure of AccTrx2. I-TASSER server was employed to build the tertiary structure and the active site (Cys^{32} and Cys^{35}) was marked. (For color version, please sweep QR Code in the back cover)

A phylogenetic tree of Trxs from various species was constructed by using a neighbour-joining method implemented in MEGA 4.1 to analyze the possible evolutionary relationships among Trxs. As shown in Fig. 1B, phylogenetic protein sequence analysis revealed that AccTrx2 was more closely related to AmTrx2 than to homologues in other species, and this result was in agreement with the relationship predicted from the multiple sequence alignment.

To further clarify the catalytic mechanism of AccTrx2, the potential tertiary structure of AccTrx2 was constructed by using I-TASSER server, and the cysteines (Cys^{32} and Cys^{35}) of active motif were identified (Fig. 1C).

Genomic structure of *AccTrx2*

The full-length genomic DNA sequence (GenBank accession number: JX844650) is 1,680 bp in length and contains four exons and three introns, which is similar to the genomic structure of AmTrx2 (Fig. 2). The first intron of *AccTrx2* is apparently longer than *BtTrx2-like* and *BiTrx2-like* during the evolutionary period.

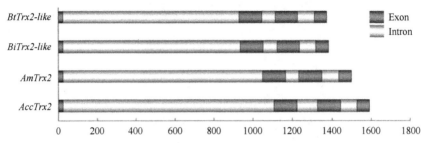

Fig. 2 Schematic representation of the genomic structures of Trx family genes. Lengths of genomic DNA sequences from *A. cerana cerana* (JX844650) and its homologues, including *AmTrx2* from *A. mellifera* (XP_392963.1), *BtTrx2-like* from *B. terrestris* (XP_003396074.1), and *BiTrx2-like* from *B. impatiens* (XP_003490398.1) are shown. The exons are indicated with grey boxes, and the introns with white boxes.

Expression profiles of *AccTrx2*

As shown in Fig. 3A, the expression of *AccTrx2* mRNA was mainly detected in the 15 days post-emergence adult, followed by the fourth-instar larvae and dark-eyed pupae, and the mRNA expressions of which were not apparently detected in all of the other developmental stages. The spatial expression profiles showed that *AccTrx2* transcripts are expressed the highest in the brain, followed by the midgut (Fig. 3B)

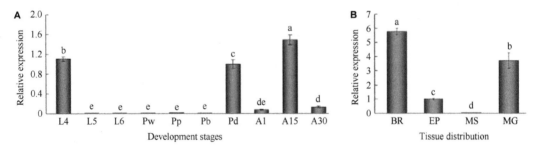

Fig. 3 Transcriptional profiles of AccTrx2 during different developmental stages and in different tissues. A. The transcription level of *AccTrx2* at different developmental stages: larvae from the fourth to the sixth instars (L4-L6), pupae (Pw white-eyed pupae, Pp pink-eyed pupae, Pb brown-eyed pupae and Pd dark-eyed pupae), and adults (A1 1 day post-emergence, A15 15 days post-emergence and A30 30 days post-emergence). B. AccTrx2 expression profiles in brain (BR), epidermis (EP), muscle (MS) and midgut (MG), with the *β-actin* gene shown for comparison. Vertical bars represent the means ± SEM ($n = 3$). Letters above the bars designate significant differences ($P < 0.001$).

To detect the expression profiles of AccTrx2 in response to various oxidative stresses, the 15 days post-emergence adult was chosen and total RNA was extracted. As shown in Fig. 4A-D, AccTrx2 expression was induced by 4, 16 and 25℃ treatments. At 4℃, AccTrx2 transcripts reached its maximum level at 1.0 h and then declined to half during the other treated time compared to the control group. Under stress at 16℃, the mRNA level of AccTrx2 peaked at 2 h. Furthermore, the AccTrx2 expression was all induced in the treated course except after 3 h of 16℃

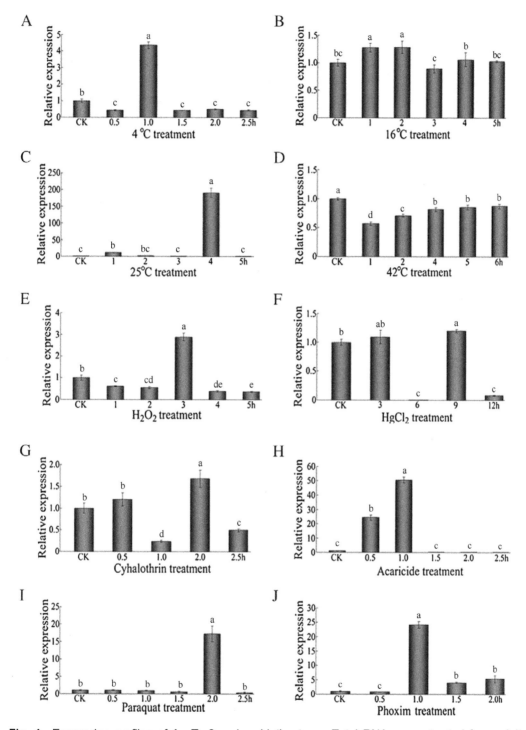

Fig. 4 Expression profiles of AccTrx2 under abiotic stress. Total RNA was extracted from adult bees (15 days post-emergence) treated at the indicated times with 4℃ (A), 16℃ (B), 25℃ (C), 42℃ (D), H_2O_2 (E), $HgCl_2$ (F), cyhalothrin (G), acaricide (H), paraquat (I), and phoxim (J). qRT-PCR was employed to investigate the transcription levels of AccTrx2 under different conditions. Each value is given as the mean (SD) of three replicates. The letters on the bar represent a significant difference at $P <$ 0.001, as determined by Duncan's multiple range tests using SAS software version 9.1.

treatment compared with the unchallenged group. At 25℃, the highest mRNA expression level was detected at 4 h. However, AccTrx2 expression was repressed when exposed to the 42℃ treatment. In addition, we also characterize the transcripts of AccTrx2 when exposed to H_2O_2, mercury ($HgCl_2$) and pesticides (cyhalothrin, acaricide, paraquat and phoxim) treatments. As shown in Fig. 4E-J, the expression profiles of AccTrx2 were all induced and reached their highest level at 3, 9, 2, 1, 2 and 1 h, respectively. Taken together, all the above results showed that AccTrx2 may play an important role in response to various oxidative stresses.

Recombinant expression and purification of AccTrx2

The recombinant AccTrx2 was transformed into *E. coli* BL21 (DE3) cells and was overexpressed as a fusion protein. SDS-PAGE was employed to detect that a target protein was induced by IPTG (Fig. 5). In addition, the recombinant protein of AccTrx2 was purified by using affinity chromatography, and the concentration of the soluble recombinant protein was approximately 0.16 mg/ml.

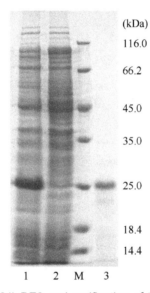

Fig. 5 Expression in *E. coli* (BL-21) DE3 and purification of the recombinant AccTrx2 protein as analyzed by SDS-PAGE. Lane M, molecular mass standard; lanes 1 and 2, induced and non-induced overexpression of recombinant AccTrx2 in *E. coli*; lane 3, purified recombinant AccTrx2 protein.

Bioassay of recombinant AccTrx2

To verify whether AccTrx2 is involved in the clam antioxidant system, insulin disulfide reduction assay was employed to evaluate its ability to reduce insulin disulfides mediated by DTT. As shown in Fig. 6, the insulin disulfide reduction was increased when the concentration of recombinant AccTrx2 increased. When the concentration of AccTrx2 reached 40.0 μg/ml, insulin disulfide reduction was not apparently increased. The negative control without AccTrx2 showed no insulin disulfide reduction.

DNA protection from oxidative cleavage of AccTrx2

The antioxidant role of AccTrx2 was determined by DNA nicking assay by using supercoiled pUC19 plasmid in a mixed-function oxidation (MFO) system. The supercoiled form of pUC19 plasmid was converted to the nicked form of plasmid (Fig. 7, lane 3) when AccTrx2 was absent. The nicked form of the pUC19 plasmid increased gradually when the concentration of recombinant AccTrx2 decreased (Fig. 7, lanes 4 to 10). These results suggest that AccTrx2 acts as an antioxidant that protects against oxidative damage induced by ROS.

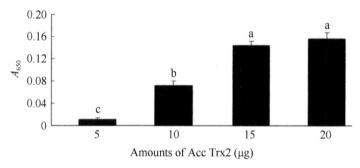

Fig. 6 Reducing insulin disulfide of purified AccTrx2, as measured by the insulin disulfide reduction assay. The assay system without the purified protein was used as the negative control. Reduction activity was determined by measuring the absorbance at 650 nm compared with negative control. The letters on the bar represent a significant difference ($P < 0.001$) among different concentrations of AccTrx2.

Fig. 7 Potential of AccTrx2 to protect supercoiled DNA against oxidative damage in the mixed-function oxidation system. Lane 1, pUC19 plasmid DNA + $FeCl_3$; lane 2, pUC19 plasmid DNA; lane 3, pUC19 plasmid DNA + $FeCl_3$ + dithiothreitol (DTT); lanes 4-10, pUC19 plasmid DNA + $FeCl_3$ + DTT + purified AccTrx2 (24.0, 20.4, 16.8, 13.2, 9.6, 6.0 and 2.4 µg/ml, respectively).

Discussion

Previous studies have demonstrated that Trx ubiquitously exists in all the kingdoms of living organisms [42]. Although the characterization of *Trx* genes in other species has been studied, most studies are performed in model insect species, such as *Drosophila*. However, researches concerning the expression of *AccTrx2* due to exposure to abiotic oxidative stress and the function of Trxs have rarely been previously reported in honeybee. In the present study, we isolated a thioredoxin gene (*AccTrx2*) from *A. cerana cerana* and examined the response of *AccTrx2* to various abiotic stresses. The results of experiments involving the cloning, expression, and functional characterization of an *A. cerana cerana Trx* gene suggest that AccTrx2 could protect against oxidative stresses caused by ROS.

Trxs are characterized by the CGPC motif in their redox active sites [42]. It was demonstrated that the Trx family active motif CGPC is highly conserved in thioredoxin 2 of mitochondria (RpTrx2) from Manila clam (*Ruditapes philippinarum*) [40]. The CGPC active site motif of mitochondrial thioredoxin 2 from disk abalone (*Haliotis discus discus*) (AbTrx-2) is also conserved [43]. Thioredoxin 1 of rock bream, *Oplegnathus fasciatus* (RbTrx1) has a conserved CGPC redox-active motif to control protein function by the redox state of structural or catalytic thiol groups [44]. Sequence analysis revealed that AccTrx2 also contains the conserved CGPC motif, suggesting that AccTrx2 belongs to the typical Trx family and the conserved motif for Trx proteins may play a crucial role of antioxidant.

The brain tissue is very sensitive to oxidative stress [45, 46] and tissue-specific transcriptional analysis of *AccTrx2* showed that it is expressed higher in the brain than in other tissues whereas *AccTrx-like1* had a higher transcript in epidermis, followed by the brain in our previous study [26], suggesting that *AccTrx2* may play functions in the brain. In addition, we also investigated the transcription of *AccTrx2* during different developmental stages and the results showed that *AccTrx2* expression is the highest in the 15 days post-emergence adult and then followed by the fourth-instar larvae. Larvae and adults may suffer higher oxidative stress than pupae because their diet is restricted to honey, pollen and glandular secretions provided by other colony members [47]. These results suggest that *AccTrx2* might play an important role in preventing damage caused by ROS during the development stages of 15 days post-emergence adults and the fourth-instar larvae. The transcripts remained unchanged from the fifth-instar larvae to brown-eyed pupae, indicating that *AccTrx2* may be constitutively expressed. AccTrx2 is expressed higher in dark-eyed pupae than in other pupal stages, suggesting that it may play an important role in the conversion from dark-eyed pupae to 1 day post-emergence adult.

Lushchak [48] and Kottuparambila et al. [49] demonstrated that environmental conditions such as temperature, heavy metal, pesticides, and UV radiation can induce oxidative stress in the previous studies. Expression of a thioredoxin homologue of *Cynoglossus semilaevis* (*CsTrx1*) was upregulated by exposure to iron, copper, and hydrogen peroxide [50]. Previous studies have shown that *AbTrx-2* could be induced in response to H_2O_2-induced oxidative stress [41]. Pesticides could impair the biological and physiological functions of erythrocytes and lymphocytes by causing lipid oxidation of biomembranes [51]. Temperature has also been demonstrated to be the dominant abiotic factor that could cause physiological changes in organisms [52] and affect herbivorous insects [53]. In our previous study, *AccTrx-like1* was upregulated by the 4, 15 and 25°C treatments [26]. In the present study, the expression of *AccTrx2* in response to various abiotic stresses was analyzed by qRT-PCR. The results showed that AccTrx2 could be induced by 4, 16, 25°C, H_2O_2, cyhalothrin, acaricide, paraquat, phoxim and mercury ($HgCl_2$) treatments, which are consistent with previous studies. Interestingly, the transcripts of *AccTrx2* are very low compared to the control group at 6 and 12 h when exposed to $HgCl_2$ treatment. A possible explanation for this is that some other antioxidant genes may reduce the transcripts of AccTrx2 in the antioxidant process. However, *AccTrx2* transcript was slightly repressed by 42°C, suggesting that 42°C treatment might not produce enough ROS to induce gene expression to the host. Lu et al. [26] also demonstrated that *AccTrx-like1* was downregulated when exposed to the 43°C treatment. All of these results indicate that *AccTrx2* may participate in protecting organisms against the oxidative damage caused by ROS.

Shrimp Trx is a potent antioxidant protein that can reduce insulin disulfides [54].

Thioredoxin of *Epinephelus coioides* (EcTrx) was demonstrated to possess the enzymatic redox activity to reduce insulin disulfides [55]. Zoysa et al. [43], Umasuthan et al. [40] and Sun et al. [50] also demonstrated that purified recombinant AbTrx-2, RpTrx2 and CsTrx1 exhibited insulin disulfide reductase activity, respectively. To characterize the function of AccTrx2, insulin disulfide reduction assay was performed in the presence of DTT. In this study, we found that AccTrx2 also has the activity of reducing insulin disulfides, which is consistent with previous studies.

ROS could be generated by the MFO system (DTT/Fe^{3+}/O_2), which is responsible for DNA damage described as nicking of supercoiled DNA form into linear [56]. DNA could be damaged by high concentrations of ROS [57]. Wei et al. [55] and Zoysa et al. [43] demonstrated that recombinant EcTrx and AbTrx-2 have the abilities to protect supercoiled DNA from oxidative damage, respectively. In our study, the results revealed that AccTrx2 also has the capacity to prevent pUC19 plasmid DNA against hydroxyl radical damage. All of the results indicated that AccTrx2 might play an important role in scavenging oxidative injury caused by ROS *in vivo*.

In conclusion, we have isolated and characterized a typical *Trx* gene from *A. cerana cerana* and investigated the transcriptional profiles of *AccTrx2* in different developmental stages and tissues. We analyzed the genomic structure of *AccTrx2*. *AccTrx2* was induced by exposure to various abiotic stresses except for the 42℃ treatment. In addition, the purified recombinant AccTrx2 protein was able to reduce insulin disulfides and protect pUC19 plasmid DNA from hydroxyl radical.

References

[1] Imlay, J. A. Pathways of oxidative damage. Annu Rev Microbiol. 2003, 57, 395-418.
[2] Lee, K. S., Kim, S. R., Park, N. S., et al. Characterization of a silkworm thioredoxin peroxidase that is induced by external temperature stimulus and viral infection. Insect Biochem Mol Biol. 2005, 35, 73-84.
[3] Gertz, M., Fischer, F., Leipelt, M., Wolters, D., Steegborn, C. Identification of peroxiredoxin 1 as a novel interaction partner for the lifespan regulator protein p66Shc. Aging. 2009, 1, 254-265.
[4] Abele, D., Puntarulo, S. Formation of reactive species and induction of antioxidant defense systems in polar and temperate marine invertebrates and fish. Comp Biochem Physiol A: Mol Integr Physiol. 2004, 138, 405-415.
[5] Kobayashi-Miura, M., Shioji, K., Hoshino, Y., Masutani, H., Nakamura, H., Yodoi, J. Oxygen sensing and redox signaling: the role of thioredoxin in embryonic development and cardiac diseases. Am J Physiol. 2007, 292, H2040-H2050.
[6] Powis, G., Montfort, W. R. Properties and biological activities of thioredoxins. Annu Rev Pharmacol Toxicol. 2001, 41, 261-295.
[7] Holmgren, A. Thioredoxin. Annu Rev Biochem. 1985, 54, 237-271.
[8] Wahl, M. C., Irmler, A., Hecker, B., Schirmer, R. H., Becker, K. Comparative structural analysis of oxidized and reduced thioredoxin from *Drosophila melanogaster*. J Mol Biol. 2005, 345, 1119-1130.
[9] Arnér, E. S. J., Holmgren, A. Physiological functions of thioredoxin and thioredoxin reductase. Eur J Biochem. 2000, 267, 6102-6109.
[10] Myers, J. M., Antholine, W. E., Myers, C. R. Hexavalent chromium causes the oxida-tion of thioredoxin in human bronchial epithelial cells. Toxicology. 2008, 246, 222-233.
[11] Gromer, S., Urig, S., Becker, K. The thioredoxin system—from science to clinic. Med Res Rev. 2004, 24, 40-89.

[12] Kalinina, E. V., Chernov, N. N., Saprin, A. N. Involvement of thio-, peroxi-, andglutaredoxins in cellular redox-dependent processes. Biochem Mosc. 2008, 73, 1493-1510.

[13] Lee, S., Kim, S. M., Lee, R. T. Thioredoxin and thioredoxin target proteins: from molecular mechanisms to functional significance Antioxid Redox Signal. 2013, 1165-1207.

[14] Arnér, E. S. Focus on mammalian thioredoxin reductases—important selenoproteins with versatile functions. Biochim Biophys Acta. 2009, 1790, 495-526.

[15] Lillig, C. H., Holmgren, A. Thioredoxin and related molecules—from biology to health and disease. Antioxid Redox Signal. 2007, 9, 25-47.

[16] Widder, J. D., Fraccarollo, D., Galuppo, P., Hansen, J. M., Jones, D. P., Ertl, G., Bauersachs, J. Attenuation of angiotensin II-induced vascular dysfunc-tion and hypertension by overexpression of thioredoxin 2. Hypertension. 2009, 54, 338-344.

[17] Zhou, F., Gomi, M., Fujimoto, M., et al. Attenuation of neuronal degeneration in thioredoxin-1 overexpressing mice after mild focal ischemia. Brain Res. 2009, 1272, 62-70.

[18] Yoshida, T., Nakamura, H., Masutani, H., Yodoi, J. The involvement of thioredoxin and thioredoxin binding protein-2 on cellular proliferation and aging process. Ann N Y Acad Sci. 2005, 1055, 1-12.

[19] Kanzok, S. M., Fechner, A., Bauer; H., Ulschmid, J. K., Müller, H. M., Botella-Munoz, J., Schneuwly, S., Schirmer, R., Becker, K. Substitution of the thioredoxin system for glutathione reductase in *Drosophila melanogaster*. Science. 2001, 291, 643-646.

[20] Pellicena-Palle, A., Stitzinger, S. M., Salz, H. K. The function of the *Drosophila* thioredoxin homolog encoded by the deadhead gene is redox-dependent and blocks the initiation of development but not DNA synthesis. Mech Dev. 1997, 62, 61-65.

[21] Bauer, H., Kanzok, S. M., Schirmer, R. H. Thioredoxin-2 but not thioredoxin-1 is a substrate of thioredoxin peroxidase-1 from *Drosophila melanogaster*: isolation and char-acterization of a second thioredoxin in *Drosophila melanogaster* and evidence for distinct biological functions of Trx-1 and Trx-2. J Biol Chem. 2002, 277, 17457-17463.

[22] Svensson, M. J., Chen, J. D., Pirrotta, V., Larsson, J. The thioredoxinT and deadhead gene pair encode testis- and ovary-specific thioredoxins in *Drosophila melanogaster*. Chromosoma. 2003, 112, 133-143.

[23] Tsuda, M., Ootaka, R., Ohkura, C., Kishita, Y., Seong, K. H., Matsuo, T., Aigaki, T. Loss of Trx-2 enhances oxidative stress-dependent phenotypes in *Drosophila*. FEBS Lett. 2010, 584, 3398-3401.

[24] Corona, M., Robinson, G. E. Genes of the antioxidant system of the honey bee: annotation and phylogeny. Insect Mol Biol. 2006, 15, 687-701.

[25] Kim, Y. J., Lee, K. S., Kim, B. Y., Choo, Y. M., Sohn, H. D., Jin, B. R. Thioredoxin from the silkworm, *Bombyx mori*: cDNA sequence, expression, and functional characterization. Comp Biochem Physiol B: Biochem Mol Biol. 2007, 147, 574-581.

[26] Lu, W. J., Kang, M. J., Liu, X. F., Zhao, X. C., Guo, X. Q., Xu, B. H. Identification and antioxidant characterisation of thioredoxin-like1 from *Apis cerana cerana*. Apidologie. 2012, 43, 737-752.

[27] Rubartelli, A., Bonifaci, N., Sitia, R. High rates of thioredoxin secretion correlate with growth arrest in hepatoma cells. Cancer Res. 1995, 55, 675-680.

[28] Matthews, J. R., Wakasugi, N., Virelizier, J. L., Yodoi, J., Hay, R. T. Thioredoxin regu-lates the DNA binding activity of NF-kappa B by reduction of a disulphide bond involving cysteine 62. Nucleic Acids Res. 1992, 20, 3821-3830.

[29] Berndt, C., Lillig, C. H., Holmgren, A. Thioredoxins and glutaredoxins as facilitators of protein folding. Acta Biochim Biophys Sin. 1992, 1783, 641-650.

[30] Schallreuter, K. U., Wood, J. M. The role of thioredoxin reductase in the reduction of free radicals at the surface of the epidermis. Biochem Biophys Res Commun. 1986, 136, 630-637.

[31] Yang, G. Harm of introducing the western honeybee Apis mellifera L. to the Chinese honeybee *Apis cerana* F. and its ecological impact. Acta Entomol Sin. 2005, 401-406.

[32] Xu, P., Shi, M., Chen, X. X. Antimicrobial peptide evolution in the Asiatic honey bee *Apis cerana*. PLoS One. 2009, 4, e4239.

[33] Robinson, G. E. Regulation of division of labor in insect societies. Annu Rev Entomol. 1992, 37,

637-665.
- [34] Kerr, W. E., Laldlaw, H. H. General genetics of bees. Adv Genet. 1956, 8, 109-153.
- [35] Rothenbuhler, W. C., Gowen, J. W., Park, O. W. Androgenesis with zygogenesis in *gynandromorphic honeybees* (*Apis mellifera* L.). Science. 1952, 115, 637-638.
- [36] Ziegler, I. Genetic aspects of ommochrome and pterin pigments. Adv Genet. 1961, 10, 349-403.
- [37] Roy, A., Kucukural, A., Zhang, Y. I-TASSER: a unified platform for automated protein structure and function prediction. Nat Protoc. 2010, 5, 725-738.
- [38] Roy, A., Yang, J., Zhang, Y. COFACTOR: an accurate comparative algorithm for structure-based protein function annotation. Nucleic Acids Res. 2012, 40, W471-W477.
- [39] Zhang, Y. I-TASSER server for protein 3D structure prediction. BMC Bioinforma. 2008, 9, 40.
- [40] Umasuthan, N., Revathy, K. S., Lee, Y., Whang, I., Lee, J. Mitochondrial thioredoxin-2 from Manila clam (*Ruditapes philippinarum*) is a potent antioxidant enzyme involved in antibacterial response. Fish Shellfish Immunol. 2012, 32, 513-523.
- [41] Yao, P. B., Lu, W. J., Meng, F., Wang, X. L., Xu, B. H., Guo, X. Q.. Molecular cloning, expression and oxidative stress response of a mitochondrial thioredoxin peroxidase gene (*AccTpx-3*) from *Apis cerana cerana*. J Insect Physiol. 2013, 59, 273-282.
- [42] Holmgren, A., Johansson, C., Berndt, C., Lönn, M. E., Hudemann, C., Lillig, C. H. Thiol redox control via thioredoxin and glutaredoxin systems. Biochem Soc Trans. 2005, 33, 1375-1377.
- [43] Zoysa, M. D., Pushpamali, W. A., Whang, I., Kim, S. J., Lee, J. Mitochondrial thioredoxin-2 from disk abalone (*Haliotis discus discus*): molecular characterizationtissue expression and DNA protection activity of its recombinant protein. Comp Biochem Physiol B: Biochem Mol Biol. 2008, 149, 630-639.
- [44] Kim, D. H., Kim, J. W., Jeong, J. M., Park, H. J., Park, C. I. Molecular cloning and expression analysis of a thioredoxin from rock bream. *Oplegnathus fasciatus*, and biological activity of the recombinant protein. Fish Shellfish Immunol. 2011, 31, 22-28.
- [45] Ament, S. A., Corona, M., Pollock, H. S., Robinson, G. E. Insulin signaling is involved in the regulation of worker division of labor in honey bee colonies. Proc Natl Acad Sci USA. 2008, 105, 4226-4231.
- [46] Rival, T., Soustelle, L., Strambi, C., Besson, M. T., Iché, M., Birman, S. Decreasing glutamate buffering capacity triggers oxidative stress and neuropil degeneration in *Drosophila* brain. Curr Biol. 2004, 17, 599-605.
- [47] Krishnan, N., Sehnal, F. Compartmentalization of oxidative stress and antioxidant defense in the larval gut of *Spodoptera littoralis*. Arch Insect Biochem Physiol. 2006, 63, 1-10.
- [48] Lushchak, V. I. Environmentally induced oxidative stress in aquatic animals. Aquat Toxicol. 2011, 101, 13-30.
- [49] Kottuparambila, S., Shinb, W., Brownc, M. T., Han, T. UV-B affects photosynthesis, ROS production and motility of the freshwater flagellate, *Euglena agilis* Carter. Aquat Toxicol. 2012, 122-123, 206-213.
- [50] Sun, J. S., Li, Y. S., Sun, L. *Cynoglossus semilaevis* thioredoxin: a reductase and an antioxidant with immunostimulatory property. Cell Stress Chaperones. 2012, 17, 445-455.
- [51] Narendra, M., Bhatracharyulu, N. C., Padmavathi, P., Varadacharyulu, N. C. Prallethrin induced biochemical changes in erythrocyte membrane and red cell osmotic haemolysis in human volunteers. Chemosphere. 2007, 67, 1065-1071.
- [52] An, M. I., Choi, C. Y. Activity of antioxidant enzymes and physiological responses in ark shell, *Scapharca broughtonii*, exposed to thermal and osmotic stress: effects on hemolymph and biochemical parameters. Comp Biochem Physiol B: Biochem Mol Biol. 2010, 155, 34-42.
- [53] Kleinhenz, M., Bujok, B., Fuchs, S., Tautz, J. Hot bees in empty broodnest cells: heating from within. J Exp Biol. 2003, 206, 4217-4231.
- [54] Aispuro-Hernandez, E., Garcia-Orozco, K. D., Muhlia-Almazan, A., et al. Shrimp thioredoxin is a potent antioxidant protein. Comp Biochem Physiol C: Pharmacol Toxicol Endocrinol. 2008, 148, 94-99.
- [55] Wei, J. G., Guo, M. L., Ji, H. S., Yan, Y., Ouyang, Z. L., Huang, X. H., Hang, Y. H., Qin, Q. W.

Cloning, characterization, and expression analysis of a thioredoxin from orange-spotted grouper (*Epinephelus coioides*). Dev Comp Immunol. 2012, 38, 108-116.

[56] Kwon, O. J., Lee, S. M., Floyd, R. A., Park, J. W. Thiol-dependent metal-catalyzed oxidation of copper, zinc superoxide dismutase. Biochim Biophys Acta. 1998, 1387, 249-256.

[57] Valko, M., Rhodes, C. J., Moncol, J., Izakovic, M., Mazur, M. Free radicals, metals and antioxidants in oxidative stress-induced cancer. Chem Biol Interact. 2006, 160, 1-40.

2.13 Characterization of a Mitochondrial Manganese Superoxide Dismutase Gene from *Apis cerana cerana* and Its Role in Oxidative Stress

Haihong Jia, Rujiang Sun, Weina Shi, Yan Yan, Han Li, Xingqi Guo, Baohua Xu

Abstract: Mitochondrial manganese superoxide dismutase (mMnSOD) plays a vital role in the defense against reactive oxygen species (ROS) in eukaryotic mitochondria. In this study, we isolated and identified a mMnSOD gene from *Apis cerana cerana*, which we named AccSOD2. Several putative transcription factor-binding sites were identified within the 5'-flanking region of AccSOD2, which suggests that AccSOD2 may be involved in organismal development and/or environmental stress responses. Quantitative real-time PCR analysis showed that AccSOD2 is highly expressed in larvae and pupae during different developmental stages. In addition, the expression of AccSOD2 could be induced by cold (4℃), heat (42℃), H_2O_2, ultraviolet light (UV), $HgCl_2$, and pesticide treatments. By using a disc diffusion assay, we provide evidence that recombinant AccSOD2 protein can play a functional role in protecting cells from oxidative stress. Finally, the *in vivo* activities of AccSOD2 were measured under a variety of stressful conditions. Taken together, our results indicate that AccSOD2 plays an important role in cellular stress responses and antioxidative processes and that it may be of critical importance to honeybee survival.

Keywords: *Apis cerana cerana*; Mitochondrial manganese superoxide dismutase; Molecular characterization; Oxidative stress; Antioxidant activity

Introduction

Organisms are exposed to a variety of unfavorable environmental stressors on a nearly constant basis, including temperature swings, heavy metal, pesticides, and ultraviolet (UV) radiation, all of which are believed to induce the formation of reactive oxygen species (ROS) [1, 2]. High ROS concentrations contribute to degenerative diseases and aging through the oxidation of nucleic acids, proteins, and lipid membranes [3, 4]. To survive these stresses, organisms have developed sophisticated antioxidant mechanisms to avoid oxidative damage, and antioxidant enzymes are key components in regulating the intracellular ROS balance.

Superoxide dismutases (SODs) are one class of antioxidant enzymes used to defend against

ROS damage; these enzymes have been identified in both eukaryotic and prokaryotic organisms [5-7]. SODs are a family of metalloenzymes that specifically catalyze the conversion of the superoxide radical (O^{2-}) to H_2O_2 and O_2 to protect oxygen-metabolizing cells against the harmful effects of superoxide free radicals [8]. Based on the metal cofactors within their active sites, SODs can be separated into three classes: copper/zinc SODs (CuZnSOD or SOD1), manganese SODs (MnSOD or SOD2), and iron SODs (FeSOD or SOD3) [7]. Two types of MnSODs have been characterized in eukaryotes: mitochondrial MnSOD (mMnSOD), which contains a peptide for translocation to the mitochondria, and cytosolic MnSOD (cMnSOD), which does not contain a translocation peptide and has been identified in various crustacean species [9, 10].

In eukaryotic cells, mMnSOD is located within the matrix of the mitochondria [11, 12], an organelle that is known to be a major source of ROS production and is a major target of ROS-induced cellular injury [13]. Therefore, mitochondrial MnSOD is thought to play a vital role in the cellular defense against ROS-mediated oxidative damage.

mMnSOD genes have been cloned and characterized from a variety of organisms. Tissue-specific knockout of mMnSOD function has been shown to affect heart and muscle phenotypes in mice [14]. Overexpression of mMnSOD can be used to extend the lifespan of adult *Drosophila melanogaster* [15]. Yeast mMnSOD mutants are hypersensitive to oxygen toxicity [16]. Furthermore, preliminary studies have shown that mMnSOD may be involved in stress responses to stimulators such as heat, cold, starvation, and heavy metal [17, 18]. Recently, Gao et al. [19] reported that the activity of mMnSOD increased under a variety of conditions and that the mMnSOD plays an important role in cellular stress responses and anti-oxidative processes in *Bemisia tabaci*. In addition, mMnSOD has been studied for its role in mediating oxidative stress responses in *Apis mellifera* [20, 21], which is the most studied species in the genus *Apis*. Although mMnSODs have been identified from different species, the studies regarding molecular characterization and function of mMnSOD in *Apis cerana cerana* have not been investigated previously to our knowledge. *A. cerana cerana* – a well-known pollinator of flowering plants that is widespread throughout China – plays an indispensable role in agricultural economic development and in the balance of regional ecologies. However, the Chinese honeybee is highly susceptible to numerous environmental stresses and is currently suffering from significant population decreases [22]. Therefore, we chose to study the function of mMnSOD in *A. cerana cerana* to gain further insight into the characteristics of this enzyme and its role in the cellular defense systems of honeybees. In this study, we isolated and characterized a mMnSOD gene from *A. cerana cerana*, which we named AccSOD2. Next, we examined the expression profile of AccSOD2 during various developmental stages and under different abiotic stress conditions. In addition, we performed a disc diffusion assay to evaluate the capacity of recAccSOD2 to defend against oxidative stress. Finally, we investigated the activity of *A. cerana cerana* mMnSOD *in vivo* under various stress conditions, including cold, heat, UV radiation, insecticides, and heavy metal exposure.

Materials and Methods

Insects and treatments

A. cerana cerana used in the following experiments were maintained at the experimental apiary of the College of Animal Science and Technology, Shandong Agricultural University, Taian, Shandong, China. According to age, eye color and shape, the honeybees were generally classified into larvae, pupae, and adults. The whole bodies of the first- to the fifth-instar larval (L1-L5) and pupae including prepupae (PP), white eyes (Pw), pink eyes (Pp), brown eyes (Pb) and dark eyes (Pd) were identified according to the criteria of Michelette and Soares [23]. The fourth- through the sixth-instar larvae (L4-L6), pupae (Pw, Pp, Pb, and Pd), and adults (A1, one day post-emergence) were collected from the hive randomly. While the adult worker honeybees (A10-ten days post-emergence and A15-fifteen days post-emergence) were collected at the entrance of the hive randomly, and A10 were caged in eight groups of 40 individuals at a constant temperature (34℃) and humidity (70%) and fed on the basic adult diet containing water, 70% powdered sugar and 30% honey for two days before treatments under a 24 h dark regimen [24]. Groups 1-3 were exposed to high heat (treat 0, 1, 2, 3, 4, and 5 h in 42℃), severe cold (treat 0, 0.5, 1, 1.5, 2, and 2.5 h in 4℃), ultraviolet radiation (treat 0, 0.5, 1, 1.5, 2, and 2.5 h with 30 MJ/cm^2 UV). Group 4 were injected with 20 μl (50 mmol/L H_2O_2/worker) of H_2O_2 between the first and the second abdominal segments by using a sterile microscale needle, and treated for 0, 0.5, 1, 1.5, 2, and 2.5 h. Group 5 were treated with $HgCl_2$ (3 mg/ml added to food) for 0, 3, 6, 9, 12, and 24 h. In addition, groups 6-8 were exposed to three different pesticides (pyriproxyfen, phoxim, and acaricide), which were delivered in 0.5 μl distilled water to the final concentrations were 10 μmol/L/L, 1 μg/ml, 10 mmol/L, respectively. The treated workers were collected after treatment at 0, 0.5, 1, 1.5, 2, and 2.5 h. All honeybees in this research were collected at the indicated time points (each sample containing three individuals), flash-frozen in liquid nitrogen, and stored at −80℃.

Total RNA extraction, cDNA preparation, and DNA isolation

Total RNA for cDNA synthesis was extracted from treated honeybees (~100 mg) by using Trizol reagent (Invitrogen, USA) according to the protocol of the manufacturer, resuspended in DEPC-treated water, and stored at −80℃. The first-strand cDNA was synthesized by using EasyScript First-Strand cDNA Synthesis SuperMix (Transgen, Beijing, China) according to the instructions of the manufacturer, and the cDNA was used as the PCR template in the gene cloning procedures. Genomic DNA was isolated by using the EasyPure Genomic DNA Extraction Kit (TransGen, Beijing, China). The total RNA and genomic DNA samples were analyzed by agarose gel electrophoresis and ethidium bromide (EB) staining, and spectrophotometry.

Primers and amplification conditions

Primer sequences and PCR amplification conditions used in this study are listed in Table 1, Table 2, respectively.

Table 1 The primers used in this study

Primer name	Primer sequence (5'—3')	Description
SZ1	GGAACCTATTATTTCTGCCG	Degenerate primer, forward
SZ2	CAATTCACGACATCAAAGATAG	Degenerate primer, reverse
5W1	GAATGATTCAAATGACCACCACC	5' RACE reverse primer, primary
5W2	CAACCAGAACCTTGAATGGCAA	5' RACE reverse primer, nested
AAP	GGCCACGCGTCGACTAGTAC(G)16	Abridged anchor primer
AUAP	GGCCACGCGTCGACTAGTAC	Abridged universal amplification primer
3W1	TTGCCATTCAAGGTTCTGGTTG	3' RACE forward primer, primary
3W2	GTGCTAATCAAGATCCTTTG	3' RACE reverse primer, nested
B26	GACTCTAGACGACATCGA(T)17	Universal primer, primary
B25	GACTCTAGACGACATCGA	Universal primer, nested
QC1	ACGAAGGTGTATAGTGTATCTG	Genomic sequence primer, forward
QC2	CACATTTCGATTTGAGCGC	Genomic sequence primer, reverse
SF1	GCTATCTTTGATGTCGTG	I-PCR outer primer, forward
SF2	GGTATTGATGTTTGGGAACATGC	I-PCR inner primer, forward
SR1	CGGCAGAAATAATAGGTTCC	I-PCR outer primer, reverse
SR2	GGAAAATAGGATACGTCTTACTGC	I-PCR inner primer, reverse
QDZ1	GAACGAAAACCTAATCAAGCGG	Promoter specific primer, forward
QDZ2	GTGCTTTGAATGATGAAGTTGC	Promoter specific primer, reverse
SQ1	TTGCCATTCAAGGTTCTGGTT	Real-time qPCR primer, forward
SQ2	GCATGTTCCCAAACATCAATACC	Real-time qPCR primer, reverse
β-s	GTTTTCCCATCTATCGTCGG	Standard control primer, forward
β-x	TTTTCTCCATATCATCCCAG	Standard control primer, reverse
BD1	GGATCCATGTTTGCAGTAAGACGTATC *Bam*H I	Primers of constructing expression vector, forward
BD2	GAGCTCAGATCCAATAGCTTTC *Sac* I	Primers of constructing expression vector, reverse

Table 2 PCR amplification conditions

Primer pair	Amplification conditions
SZ1, SZ2	10 min at 94℃, 50 s at 94℃, 40 s at 53℃, 40 s at 72℃ for 35 cycles, 5 min at 72℃
3W1, B26	10 min at 94℃, 50 s at 94℃, 40 s at 50℃, 50 s at 72℃ for 31 cycles, 5 min, at 72℃
3W2, B25	10 min at 94℃, 50 s at 94℃, 40 s at 52℃, 50 s at 72℃ for 35 cycles, 5 min at 72℃
5W1, AAP	10 min at 94℃, 50 s at 94℃, 40 s at 49℃, 60 s at 72℃ for 31 cycles, 5 min at 72℃
5W2, AUAP	10 min at 94℃, 50 s at 94℃, 40 s at 50℃, 60 s at 72℃ for 35 cycles, 5 min at 72℃
QC1, QC2	10 min at 94℃, 50 s at 94℃, 40 s at 52℃, 120 s at 72℃ for 35 cycles, 5 min at 72℃
SF1, SR1	10 min at 94℃, 50 s at 94℃, 40 s at 51℃, 120 s at 72℃ for 35 cycles, 5 min at 72℃
SF2, SR2	10 min at 94℃, 50 s at 94℃, 40 s at 53℃, 120 s at 72℃ for 35 cycles, 5 min at 72℃
QDZ1, QDZ2	10 min at 94℃, 50 s at 94℃, 40 s at 54℃, 120 s at 72℃ for 35 cycles, 5 min at 72℃
BD1, BD2	10 min at 94℃, 50 s at 94℃, 40 s at 51℃, 60 s at 72℃ for 35 cycles, 5 min at 72℃

Cloning the full-length cDNA of *AccSOD2*

An internal fragment of the *AccSOD2* gene was isolated by reverse transcription-PCR (RT-PCR) by using a pair of primers (SZ1 and SZ2) designed based on the SOD2 gene consensus sequences LEPIISAE and AIFDVVNW, which were derived from *A. mellifera*, *Harpegnathos saltator*, *Nasonia vitripennis*, *Bombus ignitus*, and *Riptortus pedestris* sequences. Next, to obtain the 5′ and 3′ untranslated region (UTR) sequences, 5′ and 3′ rapid amplification of cDNA end (RACE) procedures were performed by using specific primers (5W1/5W2 and 3W1/3W2) designed based on the sequence of the internal fragment. Finally, the full-length cDNA of *AccSOD2* was PCR-amplified by using the primers QC1 and QC2. All relevant PCR products were purified, cloned into the pEASY-T3 vector (TransGen Biotech, Beijing, China), and transformed into competent *Escherichia coli* cells (DH5α) for later sequencing.

Isolation of genomic DNA and the 5′-flanking region of *AccSOD2*

To isolate the *AccSOD2* genomic DNA, RT-PCR was performed by using two specific primers (QC1 and QC2) that were designed based on the deduced full-length cDNA of *AccSOD2* by using genomic DNA as the template. Next, inverted PCR (I-PCR) was conducted to obtain the *AccSOD2* promoter sequence, as described by Liu et al.; the primer sequences (SF1/SR1 and SF2/SR2) are listed in Table 1. Briefly, genomic DNA was digested with the restriction endonuclease *Sca* I at 37℃, self-ligated with T4 DNA ligase (TaKaRa, Dalian, China), and used as a template for I-PCR. Ultimately, the primers QDZ1 and QDZ2 were used to amplify the 5′-flanking region of *AccSOD2*. The PCR amplification conditions are shown in Table 2. All of the relevant PCR products were cloned into the pEASY-T3 vector (TransGen, Beijing, China) and transformed into competent *E. coli* cells (DH5α) for later sequencing.

Bioinformatic analysis

Conserved amino acid domains in AccSOD2 were analyzed by using the BLAST search program (http://blast.ncbi.nlm.nih.gov/Blast.cgi). The predicted amino acid sequence was aligned with homologous SOD2 proteins from multiple species by using DNAMAN version 6 (http://www.lynnon.com). The molecular weight and theoretical isoelectric point of the predicted protein were determined by using the online tool PeptideMass (http://us.expasy.org/tools/peptide). The signal peptide was predicted by using the Signal P4.0 software program (http://www.cbs.dtu.dk/services/SignalP). MITOPROT (http://ihg.gsf.de/ ihg/mitoprot.html) and TargetP 1.1 (http://www.cbs.dtu.dk/services/TargetP) were used to predict whether the amino acid sequence contains a putative mitochondrial targeting sequence (MTS). A phylogenetic tree constructed from SOD sequences from various species was constructed by using the Molecular Evolutionary Genetic Analysis (MEGA version 4.0) software program. Finally, putative transcription factor-binding sites within the 5′-flanking region were predicted by using the TFSEARCH database (http://www.cbrc.jp/research/ db/TFSEARCH.html).

Fluorescent real-time quantitative PCR

To determine the expression profile of the *AccSOD2* gene, fluorescent real-time quantitative PCR (qRT-PCR) was performed by using SYBR Premix Ex *Taq* (TaKaRa, Dalian, China) and a CFX96™ Real-Time PCR Detection System (Bio-Rad, Hercules, CA, USA). The house-keeping gene *β-actin* (GenBank Accession No. XM640276) was used as a reference gene for normalization purposes. The primer sequences using to amplify AccSOD2 (SQ1/SQ2) and the control gene (β-s/β-x) are listed in Table 1. The PCR mix was composed of 12.5 μl SYBR Premix Ex *Taq*, 2.0 μl cDNA (diluted 1∶10), 0.5 μl each primer (10 mmol each), and 9.5 μl PCR-grade water in a final volume of 25 μl. The PCRs were performed by using the following cycling conditions: 95℃ for 30 s, followed by 40 cycles of 95℃ for 5 s, 55℃ for 15 s, and 72℃ for 15 s. Afterwards, a melting cycle was performed from 65℃ to 95℃. Each sample was analyzed in triplicate, and AccSOD2 expression levels were calculated by using the $2^{-\Delta\Delta C_t}$ comparative CT method [25]. To maximize accuracy, all samples were prepared from at least two individuals, and three parallel experiments were performed.

Protein expression and purification of recombinant AccSOD2

By using the primers BD1 and BD2, which contained *Sac* I and *Hin*dIII restriction sites, respectively, the *AccSOD2* ORF was amplified and inserted into the expression vector pET-30a (+). Next, the recombinant plasmid pET-30a (+)-AccSOD2 was transformed into *E. coli* BL21 (DE3) pLysS cells for protein expression. The bacteria were grown in 400 ml Luria-Bertani (LB) medium containing 50 mg/ml kanamycin at 37℃ until the cells reached an optical density of 0.4-0.6 at 600 nm. Next, the cultures were induced with 0.4 mmol/L isopropylthio-β-D-galactoside (IPTG) and grown at 18℃ for 18 h. The recombinant AccSOD2 protein was then purified by using the HisTrap™ FF column (GE Healthcare, Uppsala, Sweden) and analyzed by using 12% SDS-PAGE (sodium dodecyl sulfate polyacrylamide gel electrophoresis) followed by Coomassie brilliant blue staining.

Analysis of AccSOD2 activity

Total protein was extracted from the treated honeybees by using the Tissue Protein Extraction Kit (ComWin Biotech, Beijing, China) according to the protocol of the manufacturer and was quantified by using the BCA Protein Assay Kit (Thermo Scientific Pierce, IL, USA). AccSOD2 activity was determined by using the CuZn/Mn-SOD Assay Kit (Jiancheng, Nanjing, China) according to the protocol of the manufacturer with certain modifications, as described by Meng et al., [26]. Each treatment was performed in triplicate.

Disc diffusion assay

Disc diffusion assays were performed as described by Yan et al. [27] to evaluate the activity of recombinant AccSOD2. *E. coli* cells overexpressing AccSOD2 were overlaid onto LB-kanamycin agar and incubated at 37℃ for 24 h. Filter discs (6 mm diameter) soaked with different concentrations of cumene hydroperoxide (10, 25, 50, 80, and 100 mmol/L), paraquat

(10, 50, 100, 200, and 300 mmol/L), $HgCl_2$ (20, 40, 60, 80, and 100 mg/L), or $CdCl_2$ (50, 100, 200, 400, and 800 mmol/L) were placed on the agar surface. Afterwards, the inhibition zones around the paper discs were measured as described by Burmeister et al. [28]. *E. coli* BL21 cells transformed with empty pET-30a (+) were used as the control.

Data analysis

The results are expressed as the mean ± SD (standard deviation) from triplicate experiments (n = 3). Statistical significance was determined by using Duncan's multiple range test with analysis of variance (ANOVA), and calculations were performed by using the Statistical Analysis System (SAS) version 9.1 software program (SAS Institute, Cary, NC, USA). Significance was set at $P < 0.05$.

Results

Cloning and characterization of the full-length cDNA of *AccSOD2*

By using RT-PCR and RACE-PCR, a mMnSOD gene was cloned from *A. cerana cerana*, which named *AccSOD2*. The cDNA of *AccSOD2* is 1,003 bp in length (GenBank Accession No: JN637476) and includes a 170 bp 5′ UTR and a 176 bp 3′ UTR containing a putative polyadenylation consensus signal (AATAAA) and an 18 bp poly(A) tail. The open reading frame (ORF) is 657 bp and encodes a 218-amino acid polypeptide with a predicted molecular mass of 24.4 kDa and a theoretical isoelectric point (pI) of 9.00. The full-length nucleotide and deduced amino acid sequences are shown in Fig. 1 Analyses using the online tools MITOPROT and TargetP revealed that the predicted polypeptide contains a putative N-terminal mitochondrial targeting sequence (MTS) consisting of 15 amino acids (Fig. 1), indicating that this enzyme is likely found in the mitochondria. However, the SignalP program identified no putative signal peptides within AccSOD2. Analysis of the deduced amino acid sequence indicated that AccSOD2 contains a completely conserved mMnSOD motif (DVWEHAYY) and four putative manganese-binding sites (His-46, His-94, Asp-178, and His-182) (Fig. 1).

Multiple sequence alignments showed that the predicted amino acid sequence of *AccSOD2* shares 96.79%, 85.32%, 80.91%, 89.45%, and 62.39% similarities with the mMnSOD proteins from *A. mellifera*, *H. saltator*, *N. vitripennis*, *B. ignitus*, and *R. pedestris*, respectively; in addition, three conserved putative manganese superoxide dismutase motifs (FNGGGHLNH, IQGSGWGWLG, and DVWEHAYY) were also identified (Fig. 2). By using the amino acid sequences of select MnSODs (Table 1), a phylogenetic tree was constructed by using the neighbor-joining method and the MEGA version 4 software program (Fig. 3). The phylogenetic analysis revealed the evolutionary relationship of AccSOD2 with other MnSODs from various species, and MnSODs were found to classified into two groups with robust separate branches of mMnSODs and cMnSODs, and that AccSOD2 belongs to the former. In addition, the gene exhibited the strongest similarities to *A. mellifera*-SOD2; therefore, we named this gene *AccSOD2*.

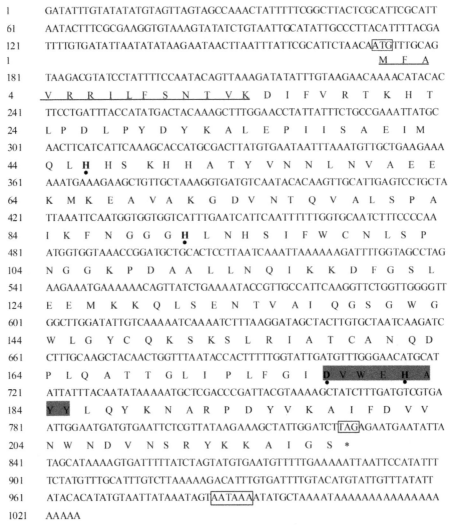

Fig. 1 The nucleotide (top) and deduced amino acid (bottom) sequences of *AccSOD2*. The stop codon is indicated with an asterisk (*). The start codon (ATG), stop codon (TAG), and polyadenylation consensus signal (AATAAA) are indicated with boxes. The putative MTS is underlined. The Mn superoxide dismutase signature motif (DVWEHAYY) is highlighted in gray. The four highly conserved Mn-binding sites (His-46, His-94, Asp-178, and His-182) are indicated with black spots.

Analysis of the genomic structure of *AccSOD2*

By using the specific primer pair SQC1 and SQC2, we amplified the 1,943 bp *AccSOD2* genomic sequence (JN662941). A sequence alignment between the genomic DNA and cDNA sequences indicated that the *AccSOD2* gene contains three exons (38, 290, and 329 bp) separated by two introns (853 and 88 bp). The introns exhibit typical intronic characteristics, possessing high AT contents and canonical 5′-GT splice donors and 3′-AG splice acceptors. A comparison of the genomic sequences from *AccSOD2* and the homologous genes from

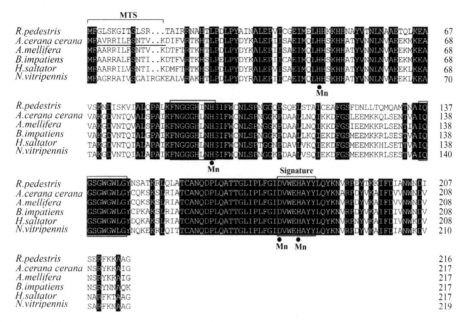

Fig. 2 A multiple alignment analysis of AccSOD2 and other known mMnSODs. Mn-binding sites are indicated with dots. Three putative Mn superoxide dismutase signatures are indicated with black boxes. The MTS is underlined.

A. mellifera (NC_007080.3), *B. impatiens* (NT_176479), *N. vitripennis* (NC_015871), and *A. florae* (NW_003789977) revealed that the number and the sizes of the exons have been conserved, although the lengths of the introns have not (Fig. 4).

Identification of partial putative transcription factor binding sites within the 5'-flanking region of *AccSOD2*

To investigate the mechanisms underlying *AccSOD2* expression and regulation, we determined the sequence of the ~1,638 bp promoter region of *AccSOD2* (JN662941) by using I-PCR and nested PCR. Next, the online software program TFSEARCH was used to predict putative transcription factor-binding sites within this promoter (Fig. 5). We identified binding sites for several heat shock factors (HSFs), which play important roles in heat-induced transcriptional activation in *Drosophila* [29, 30]. Sequences involved in tissue development and early growth stages were identified, including binding sites for caudal-related homeobox (CdxA) protein, cell factor 2-II (CF2-II), Oct-1, ADR1, GATA-3, Sox-5, and the protein encoded by the nitrogen regulatory (Nit-2) gene (NIT2).

Expression of *AccSOD2* during development

To determine the expression profile of *AccSOD2* during different developmental stages, transcript ionlevels from larvae, pupae, and adults were measured by using qPCR. The fourth-through the sixth-instar larvae (L4-L6), pupae (Pw, white eyes; Pp, pink eyes; Pb, brown eyes; and Pd, dark eyes), and adults (A1, one day post-emergence; A15, 15 days post-emergence)

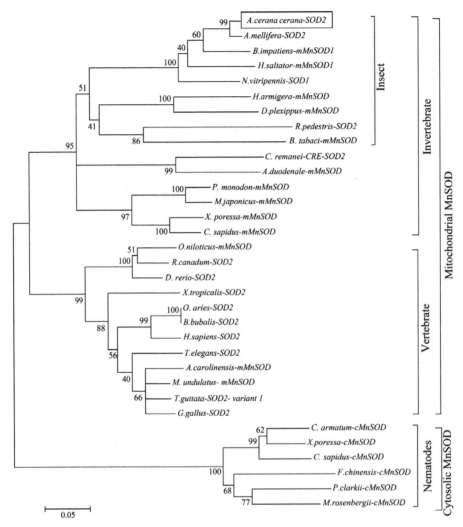

Fig. 3 A phylogenetic tree of the MnSOD amino acid sequences from several species based on the neighbor-joining method. Species abbreviations and GenBank accession numbers are listed in Table 3. The scale bar is 0.05.

Table 3 Sequences of MnSOD used in this study

Species	Abbreviation	GenBank No.
Apis cerana cerana	*A. cerana cerana*-SOD2	JN637476
Apis mellifera	*A. mellifera*-SOD2	NP_00117151.1
Bombus impatiens	*B. impatiens*-mMnSOD1	XP_003485546.1
Danio rerio	*D. rerio*-SOD2	NP_956270.1
Nasonia vitripennis	*N. vitripennis*-SOD1	XP_001607380
Riptortus pedestris	*R. pedestris*-SOD2	BAN20578.1
Rachycentron canadum	*R. canadum*-SOD2	ABC71306.2
Harpegnathos saltator	*H. saltator*-mMnSOD1	EFN87555.1
Penaeus monodon	*P. monodon*-mMnSOD	AGI99530.1
Xantho poressa	*X. poressa*-mMnSOD	CAR82602.1

Species	Abbreviation	GenBank No.
Marsupenaeus japonicus	*M. japonicus*-mMnSOD	ADB90402.1
Helicoverpa armigera	*H. armigera*-mMnSOD	ACY70995.1
Bemisia tabaci	*B. tabaci*-mMnSOD	AFW97646.1
Callinectes sapidus	*C. sapidus*-mMnSOD	AAF74770.1
Danaus plexippus	*D. plexippus*-mMnSOD	EHJ68139.1
Oreochromis niloticus	*O. niloticus*-mMnSOD	XP_003449988.1
Caenorhabditis remanei	*C. remanei*-CRE-SOD2	XP_003115144.1
Taeniopygia guttata	*T. guttata*-SOD2-variant 1	NP_001232398.1
Melopsittacus undulatus	*M. undulatus*- mMnSOD	AAO72712.1
Ancylostoma duodenale	*A. duodenale*-mMnSOD	ACK44110.1
Xenopus tropicalis	*X. tropicalis*-SOD2	NP_001005694.1
Gallus gallus	*G. gallus*-SOD2	NP_989542.1
Anolis carolinensis	*A. carolinensis*-mMnSOD	XP_003215814.1
Thamnophis elegans	*T. elegans*-SOD2	AFS63894.1
Homo sapiens	*H. sapiens*-SOD2	NP_000627.2
Ovis aries	*O. aries*-SOD2	ACV04835.1
Bubalus bubalis	*B. bubalis*-SOD2	AFQ00705.1
Callinectes sapidus	*C. sapidus*-cMnSOD	AAF74771.1
Cardisoma armatum	*C. armatum*-cMnSOD	CAR85672.1
Procambarus clarkii	*P. clarkii*-cMnSOD	ABX44762.3
Fenneropenaeus chinensis	*F. chinensis*-cMnSOD	ABR10072.1
Xantho poressa	*X. poressa*-cMnSOD	CAR85671.1
Macrobrachium rosenbergii	*M. rosenbergii*-cMnSOD	AAY79405.1

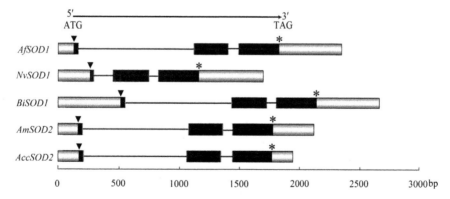

Fig. 4 Structure of the *AccSOD2* gene. Exon-ntron organization of AccSOD2 from *A. cerana cerana* and its homologs, including *AmSOD2* from *A. mellifera* (NC_007080.3), *BiSOD1* from *B. impatiens* (NT_176479), *NvSOD1* from *N. vitripennis* (NC_015871), and *AfSOD1* from *A. florae* (NW_003789977). The lengths of the exons and introns are indicated according to the scale. UTRs and exons are highlighted by gray and black bars, respectively. Lines indicate introns. The start (ATG) and stop (TAG) codons are marked with (▼) and (∗), respectively.

```
                          Oct-1                    Nkx-2
GAACGAAAACCTAATCAAGCGG ATCATAATGATGA AAGC TCAAGTG ATTCAGCTGCTTTGATAATCAAAG    -1638
                  GATA-3                 HSF
TTGAGAATGAGGAAATTAA GTGATAGGG AAGAAGAACATGA AGAAA ATCGAAATAATAAAAGGATATTA    -1568
              Sox-5                                                 ADR1
CGAAGCAAAAGTCATAATAG AATAACAATA AAACTAATGGGGAAGAGATGCTAACTAGAATA ACT AGGG    -1498
GT AGGGATAGGACCAAAAGGGACAAAAGTGCACCAGGTAGTGATGATGAAAGGACACATAAAAAAAG         -1428
AGAAGAGAGTATTCTTTCTCTTGATTATAATGAAGCTGATGATGATGCGTTAGTTTTAACTAATGGTGAG     -1358
ACACTTCAAAGCGATCCAGAATGAACAATTTTCTAGAACAATTTTATACACAAACACTTGTTATATACAAAT  -1288
AAGGGTTAACGTGACTTAATATACAACATAAGTTGAATATAATATTTCATATTTAAAAGTTATAATACACTCAC -1218
ATCACGATAAATATTTCTAATAATTTTATATAATTTTTCGTAATATAAATTTGTAAATATATTTCATATTTGAATA -1148
TTTTGAAGTCACGTTAATGCCGAAAAAATTGCACTTGATTACATATGATAAATATGTTCACAAAATAATGTGGC -1078
         ADR1
GTTGGAATAA AGGGGT AGATAGTTAAAACTATATTTCTTTTTCTTGCAATGAAATCGCATTAAATTCTTCAA -1008
TAAGTATTCGTTAAAAAAATGCTTCAATGTTTTTAATGGCTTTTTTTGTAAATTCGTTGAAACATTGATAGTA  -938
                                  HSF           Sox-5
GCTCATCTTGTTAGATTTTTTGAAACAATTATA AGAAA TATA AATAACAATA AAGTAATAATTAGGGCTATG  -868
                                                                      CdxA
CCTTTTCTTTATTATGTAAATACACAAATGTTAATATCTGGTACACGACGACGCACATACATAG ATTTATAT  -798
ATGATTATGTGCGCAATAATTGTATGTATGTACTATTTTTTTCATTGTGACATGCGTGTCAAAATTCGTTATT  -728
ACATAAAATGTGCTTTTAATTATGTATAATAAAATTGAGGTCTATTTGTTAGATTTATTAAATACCTGTAAAT  -658
TATATGAAATATAAAACAACATTAATTTCAGTAAAACTTGTAGATTAATCTTTTAGATTACAGAATATCACA   -588
                                                         CdxA
ATTATATTTTTTTTTCAAAGATAAATATAAAATATTTACATTATTATT TTATATT ATTTATA TAAAAACATTTTT -518
                                             CdxA
AATTAATTTCTTTGTAAAAATTAATTTTTCATTTTAATCC ATTTATA ATTTTAGTATGTAACATTCAATGAGT  -448
CF2-II
ATATGTA TATTCTATTTAGGATAAAATTTTTAAAAAACTATTCATGATTAAAATATAATTTGTGTAATTCAAT  -378
TTGATATGTCTACATAAAAATTTATTATTATTATTATATAATGTACAATTTATTAATTAAAGATTATTTTCTAAC -308
  HSF
CT AGAAA AAAATGACAATACACTAATATAATAAATAGTTTTAATTCACCAGTTTTTACGATTATAAAATT   -238
AATATATCTGAAGATTATTAAATTAAAAATAAAATTATTTCTGTAATGTTATGTAAAATATCTTTAATTTATT   -168
                                                        Cap
AATTGAACTACAAAAAAATAAATAGTTATTAAATTTATTATTATAT TCAGTTTT TTTTTTAATTATAAGAATA  -98
NIT2    CF2-II                                                       HSF
TCTC AATAA ATATGTA GAGATATTAATTAAAAAAATATTATTTTAGTATTTGTTTAGCCCCAATAGA AGA  -28
    CREB
AA CTGACGTAA TAACTGATATCGCGAAGGTGTAAAGTATATCTGTAATTGCATATTGCCCTTACATTTTAC   +43
                       ◄ +1 transcription start
GATTTTGACTTTGTATATATGTAGTTAGTAGCCAAACTATTTTTCGGCTTACTCGCATTCGCATTAATATG    +113
ATATTAATATATAAGAATAACTTAATTTATTCGCATTCTAACA ATG TTTGCAGTAAGACGTATCCTATTTTC  +183
CAATACAGT                                   ◄ translation start             +192
```

Fig. 5 The 5′-flanking region and putative *cis*-acting elements of the *AccSOD2* gene. The translation and transcription start sites are marked with arrows. The *cis*-acting elements are indicated with boxes.

were selected for total RNA extraction. As shown in Fig. 6, the expression levels of *AccSOD2* rapidly decreased between the L4 and the L6 larval stages and between A1 and A15 during adulthood. However, the transcription of *AccSOD2* significantly increased during the pupal stage and reached a peak during the Pd stage.

Expression of *AccSOD2* under a variety of abiotic stresses

To verify whether abiotic stresses could affect *AccSOD2* expression, the transcriptional expression profile of *AccSOD2* was measured by using qPCR. As shown in Fig. 7A, B, following both cold (4℃) and heat (42℃) treatments, *AccSOD2* transcription was dramatically elevated, peaking after 0.5 and 2 h, respectively. We observed a striking upregulation of *AccSOD2* expression immediately following H_2O_2 treatment; however, this expression sharply decreased after 0.5 h (Fig. 7C). In addition to temperature shifts and H_2O_2 exposure, honeybees are often subjected to other abiotic stresses, such as ultraviolet (UV) radiation, mercury ($HgCl_2$), and

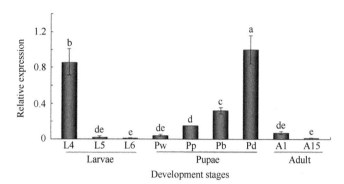

Fig. 6 The expression profile of *AccSOD2* during different developmental stages: larvae from the fourth to the sixth instars (L4-L6), pupae (Pw, white eyes; Pp, pink eyes; Pb, brown eyes; and Pd, dark eyes), and adults (A1, 1 day post-emergence; A15, 15 days post-emergence). Vertical bars represent the means ± SE (n = 3). Letters above the bars designate significant differences (P < 0.05).

pesticides. Following UV light and $HgCl_2$ treatments, *AccSOD2* expression levels increased significantly (Fig. 7D, E). Interestingly, the different pesticides treatments manifested different effects on *AccSOD2* expression (Fig. 7F). All the above-mentioned results indicate that AccSOD2 is likely involved in abiotic stress responses.

Overexpression and purification of AccSOD2

To further characterize the AccSOD2 protein, the complete *AccSOD2* ORF (without the stop codon) was cloned into the pET-30a (+) vector, which contains a cleavable N-terminal His-tag. This plasmid was transformed into *E. coli* BL21 (DE3) cells for *in vitro* overexpression, which was induced with 0.4 mmol/L IPTG at 18 ℃ for 12 h. The expressed protein was found in both inclusion bodies and in soluble form (data not shown). The soluble recombinant protein was then purified by using HisTrap™ FF columns and examined by using SDS-PAGE, which yielded a single 31.4 kDa band (Fig. 8, lane 1) containing the AccSOD2 protein (24.4 kDa) and the His-tag (7 kDa).

Characterization of recombinant AccSOD2

To evaluate the capacity of recombinant AccSOD2 to defend against oxidative stress, a disc diffusion assay was performed. To test the antioxidant properties of the enzyme, *E. coli* cells overexpressing AccSOD2 were exposed to paraquat and cumene hydroperoxide, two known inducers of oxidative stress [28]. $HgCl_2$ and $CdCl_2$ were used to test the effects of the enzyme after heavy metal exposure. Following 12 h of exposure to the various stressors, the death zones around the stressor-soaked filters were smaller in diameter on the plates containing *E. coli* overexpressing AccSOD2 compared with plates containing control bacteria. Furthermore, the halo diameters of the death zones were significantly different for the various treatment concentrations (Fig. 9). These results indicate that AccSOD2 may play an important role in protecting Chinese honeybees from oxidative stress.

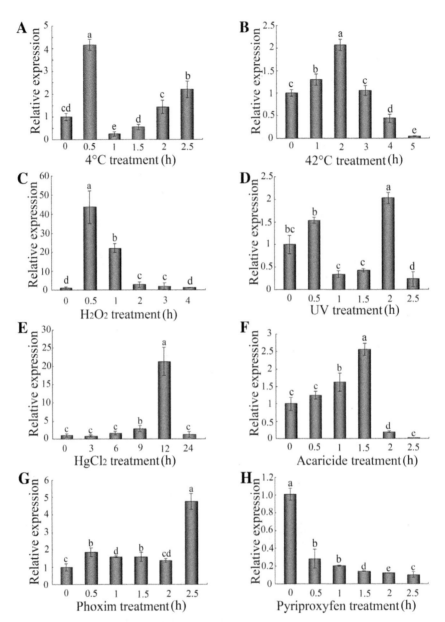

Fig. 7 The expression profiles of *AccSOD2* under different environmental stresses. qPCR was performed on total RNA extracted from adult bees at the indicated times post-treatment with cold (A), heat (B), H_2O_2 (C), UV (30 MJ/cm^2) (D), $HgCl_2$ (2 mmol/L) I, Acaricide (10 μmol/L) (F), Phoxim (1 μg/ml) (G), and Pyriproxyfen (10 μmol/L) (H). Vertical bars represent the means ± SE (n = 3). The expression of AccSOD2 was quantified and normalized to *β-actin*. Letters above the bars indicate significant differences (P < 0.01 in SAS).

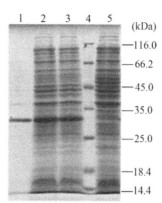

Fig. 8 Expression and purification of the recombinant AccSOD2 protein. The following are shown on an SDS-PAGE gel: lane 1, purified recombinant AccSOD2 protein; lanes 2, 3, overexpression of recombinant AccSOD2 protein after IPTG induction; lane 4, protein molecular weight marker; lane 5, uninduced expression of recombinant AccSOD2 protein.

Effects of temperature, $HgCl_2$, and UV treatments on the activity of AccSOD2

The *in vivo* activity of AccSOD2 was assessed following exposure of bees to high and low temperatures, $HgCl_2$, and UV treatments. Untreated bees were used as the control. For the external temperature stresses, worker bees were exposed to low (4℃) and high (42℃) temperatures for 0, 30, 60, 120, or 180 min. As shown in Fig. 10A, the activity of AccSOD2 first increased and then decreased following incubation at the low temperature (4℃). Following high temperature (42℃) treatment, the activity of AccSOD2 was significantly increased compared with the control (Fig. 10B). To characterize the effects of $HgCl_2$, worker bees were fed $HgCl_2$ for 0, 3, 6, 9, or 12 h. AccSOD2 activity was significantly increased over the first 9 h of feeding compared with the 0 h group and reached a peak level at 9 h; however, by 12 h, the activity of the enzyme had sharply decreased (Fig. 10C). Further UV treatment, AccSOD2 activity was significantly increased by 1 h, after which time it declined gradually (Fig. 10D). No significant differences were observed with respect to AccSOD2 activity in the untreated honeybees at the different time points (control). The observed elevations in enzymatic activity for all of the tested treatments strongly suggest that AccSOD2 activity is essential in defending against oxidative damage.

Discussion

In recent years, much attention has been paid to the function of MnSOD, which is thought to act as an important scavenger of damaging ROS metabolites within the mitochondrial matrix. In the present study, we report, for the first time, the cloning and preliminary characterization of the *AccSOD2* gene from *A. cerana cerana*. Our results suggest that AccSOD2 plays a crucial role in cellular stress responses and antioxidative processes.

Sequence analysis shows that *AccSOD2* encodes four metal-binding residues (His-46, His-94, Asp-178, and His-182) and the highly conserved MnSOD amino acid motif

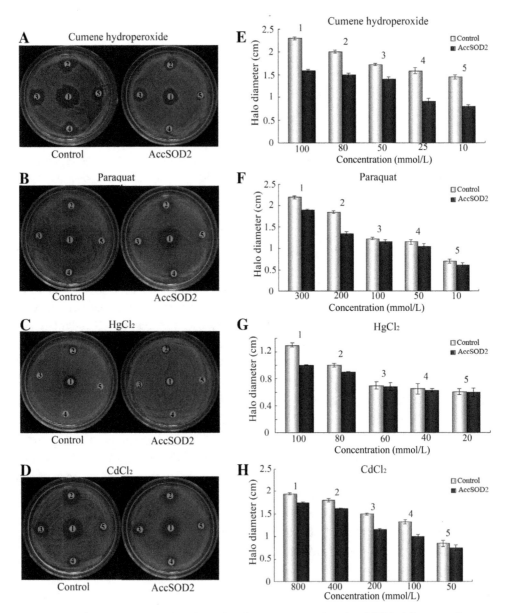

Fig. 9 Disc diffusion assays using *E. coli* cells overexpressing AccSOD2. LB agar plates were inoculated with 5×10^8 cells. *E. coli* cells were transfected with a plasmid to overexpress AccSOD2; bacteria transfected with empty pET-30a (+) vector were used as the negative controls. Filter discs soaked with different concentrations of cumene hydroperoxide (A and E), paraquat (B and F), $HgCl_2$ (C and G), and $CdCl_2$ (D and H) were placed on the agar plates. The numbers in the left Figs. correspond with the numbers in the right ones. After overnight exposure, the diameters of the death zones surrounding the drug-soaked filters were measured. The data are the means ± SE of three independent experiments.

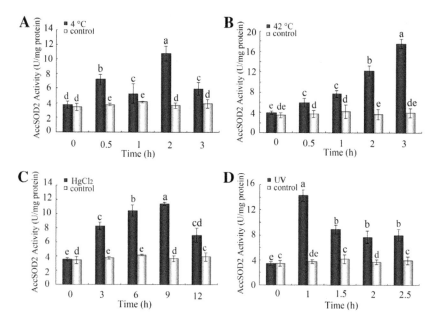

Fig. 10 The effects of thermal stress, HgCl₂, and UV treatment on AccSOD2 activity *in vivo*. A. 4℃ treatments; B. 42℃ treatments; C. HgCl₂ treatment; D. UV treatment.

DVWEHAYY. This signature sequence and the conserved metal-binding residues are essential for MnSOD structure and function, suggesting that *AccSOD2* possesses the essential properties of a functional MnSOD family member [19]. MITOPROT analysis revealed that *AccSOD2* contains a putative MTS consisting of ^{15}N-terminal amino acid residues (Fig. 1). Phylogenetic analysis indicated that *AccSOD2* clustered with the mMnSOD group, suggesting that it is indeed a mMnSOD family member. With the features of mMnSOD and the strongest similarities to *A. mellifera-SOD2*, we named this gene *AccSOD2*. A comparison of the genomic sequence of *AccSOD2* with the sequences of homologous genes from related insect species revealed that both the number and sizes of the exons have been conserved in these genes (Fig. 3). This conservation of exon structure among the different insect species suggests that the encoded mMnSODs may share common functions and may participate in similar physiological processes, such as cellular stress responses and antioxidative processes [19]. However, we also observed differences in intron lengths between the species, which may be indicative of minor functional changes.

A large number of predicted binding sites for transcription factors were identified within the 5′-flanking region of *AccSOD2*. Among these, CdxA is known to contribute to tissue-specific expression [31]. The NIT2-binding site has been described as the major activation site for the nitrogen regulatory gene in *Neurospora crassa* [32, 29]. Nkx-2 is involved in organogenesis [33]. CF2-II determines the specificity of homeotic genes possessing zinc-finger DNA-binding domains in *Drosophila* [34, 35]. Sox-5 is a regulatory molecule involved in the function of the sex-determining factor SRY (sex-determining region Y) [36]. Finally, ADR1 is a nutrient-regulated transcription factor in *Saccharomyces cerevisiae* that activates genes necessary for growth [37]. Importantly, we also identified binding sites for the cAMP-responsive element-binding (CREB) protein, which regulates gene expression during cell differentiation and is involved in the oxidative stress response [38-40]. Therefore, we propose that AccSOD2 may be involved in organismal development, growth, and environmental stress responses.

Previous studies have demonstrated that the *Caenorhabditis elegans* SOD-2 transcript is significantly upregulated during the dauer stage, which is roughly equivalent to the diapause stage in insects [41]. To investigate whether *AccSOD2* is involved in development and growth, the expression profile of *AccSOD2* was measured at different developmental stages by using qPCR. We show that *AccSOD2* is highly expressed during the L4 larval and Pd pupal stages, suggesting that *AccSOD2* may play a role in development, particularly in the larvae and pupae (Fig. 5). Gilbert and Wilkinson [42] reported that microsomal oxidases are active in honeybee larvae and that larvae can suffer from increased oxidative stress compared with other stages [43]. It is possible that the increased levels of *AccSOD2* transcription observed during larval stages. We also observed that *AccSOD2* is highly expressed during the Pd pupal stage, which is a period during which cells become highly differentiated, and many organs tend to mature, for example, enlargement of the individual, formation of the head, thorax, abdomen, and legs, as well as the change of the colour of the body surface. Metabolism will speed up to generate more energy to meet development, and more ROS is generated by β-oxidation of fatty acid and electron transport in peroxisom [44]. Otherwise, during periods of lifecycle transition when bees are exposed to increased environmental stress and are more susceptible to toxic chemicals and oxidative damage [45]. Therefore, it is reasonable to deduce that *AccSOD2* may be involved in differentiation or development. Considering all of these results, we suggest that *AccSOD2* likely plays an important protective role against oxidative damage in the honeybee.

SODs are a family of metalloenzymes that are regarded as the first line of defense against ROS. These enzymes convert superoxide radicals to oxygen and hydrogen peroxide [20], and they play vital roles in reducing oxidative stress in cells [46] and in inhibiting the immune responses to stimulators [17]. Previous studies have indicated that environmental stressors, such as cold temperature or heavy metal, can induce SOD2 expression [18]. Temperature change is a common abiotic environmental stress; it has been described as one of the key mediators of ROS generation. In this study, we found that following both cold (4℃) and heat (42℃) treatments, the transcription levels of *AccSOD2* and enzymatic activity dramatically increased. Consistent with these results, previous studies have shown that temperature stress can induce MnSOD expression in a number of species, including disk abalone and *Chlamys farreri* [47-49]. Harari et al.[50] reported that heat-shock stress can result in polyamine oxidation and the generation of H_2O_2 and that cold stress can induce hepatocytes and liver endothelial cells to undergo apoptosis due to ROS damage. Furthermore, we found that *AccSOD2* expression could be induced by H_2O_2 treatment. These results suggest that *AccSOD2* may function by removing intracellular superoxide anions generated under high- or low-temperature conditions to protect honeybees from ROS damage.

UV radiation is understood to be a powerful and harmful environmental stress for most animals [51]. Previous studies have shown that UV radiation can directly affect insect behavior, developmental physiology, and biochemistry, with UV-B (280-320 nm) radiation being particularly damaging [52, 53]. Furthermore, it has also been reported that UV-A (320-400 nm) radiation can result in damage to photoreceptive cells within the compound eyes of the butterfly species *Papilio xuthus* and *Pieris napi* [54], and UV-A has the potential to generate high levels of oxidative stress in the cotton bollworm *H. armigera* [55, 56]. Here, by using qPCR analyses, we show that *AccSOD2* can be induced by UV radiation. This finding was corroborated by enzymatic activity assays, which demonstrated that the *in vivo* activity of AccSOD2 was significantly increased following UV treatment. These findings support the hypothesis that *AccSOD2* plays an important role in protecting cells against oxidative damage.

Many heavy metals, including Cd and Hg, can enhance the intracellular formation of ROS, thus promoting cellular oxidative stress [57, 58]. In *Musca domestica*, MdSOD2 can be significantly induced by cadmium (Cd) stress [58]. In addition, insecticides are a significant threat to insects, and they can induce the oxidation of lipid biomembranes, leading to impaired biochemical and physiological functions [59]. Therefore, increased *AccSOD2* expression following $HgCl_2$ and insecticide stressors could increase the tolerance of honeybees to oxidative stress. In this study, we used disc diffusion assays (Fig. 8) to show that overexpressing *AccSOD2* in *E. coli* cells protects these cells from oxidative stressors (paraquat, cumene hydroperoxide, $HgCl_2$, and $CdCl_2$). The obvious differences in the diameters of the death zones between the *AccSOD2*-overexpressing bacteria and the control bacteria clearly demonstrate that *AccSOD2* can significantly contribute to cellular resistance to oxidative stress (Fig. 8). Considering our results in the context of previous studies, we conclude that the *AccSOD2* gene product possesses potent antioxidant properties.

In conclusion, we have identified a mMnSOD gene from *A. cerana cerana*, *AccSOD2*, that exhibits molecular characteristics and expression profiles consistent with a physiological role in the amelioration of oxidative stress. Taken together, these findings should be helpful in furthering our understanding of the expression, regulation, and function of *AccSOD2* in honeybees, which may be of critical importance to the survival of this insect species.

References

[1] Lushchak, V. I. Environmentally induced oxidative stress in aquatic animals. Aquat Toxicol. 2011, 101, 13-30.

[2] Kottuparambila, S., Shinb, W., Brownc, M. T., Han, T. UV-B affects photosynthesis, ROS production and motility of the freshwater flagellate, *Euglena agilis*. Carter Aquat Toxicol. 2012, 122-123, 206-213.

[3] Lee, K. S., Kim, S. R., Park, N. S., et al. Characterization of a silkworm thioredoxin peroxidase that is induced by external temperature stimulus and viral infection. Insect Biochem Mol Biol. 2005, 35, 73-84.

[4] Poli, G., Leonarduzzi, G., Biasi, F., Chiarpotto, E. Oxidative stress and cell signaling. Curr Med Chem. 2004, 11, 1163-1182.

[5] Smith, M. W., Doolittle, R. F. A comparison of evolutionary rates of the 2 major kinds of superoxide-dismutase. J Mol Evol. 1992, 34, 175-184.

[6] Bordo, D., Djinovic, K., Bolognesi, M. Conserved patterns in the Cu, Zn superoxide dismutase family. J Mol Biol. 1994, 238, 366-386.

[7] Zelko, I. N., Mariani, T. J., Folz, R. J. Superoxide dismutase multigene family: a comparison of the Cu-Zn-SOD (SOD1), Mn-SOD (SOD2) and EC-SOD (SOD3) gene structures, evolution and expression.

Free Radic Biol Med. 2002, 33, 337-349.

[8] Fridovich, I. Superoxide dismutases: defence against endogenous superoxide radical. Ciba Found Symp. 1978, 65, 77-93.

[9] Cheng, W., Tung, Y. H., Liu, C. H., Chen, J. C. Molecular cloning and characterisation of cytosolic manganese superoxide dismutase (cytMn-SOD) from the giant freshwater prawn *Macrobrachium rosenbergii*. Fish Shellfish Immunol. 2006, 20, 438-449.

[10] Lin, Y. C., Lee, F. F., Wu, C. L., Chen, J. C. Molecular cloning and characterization of a cytosolic manganese superoxide dismutase (cytMnSOD) and mitochondrial manganese superoxide dismutase (mtMnSOD) from the kuruma shrimp *Marsupenaeus japonicus*. Fish Shellfish Immunol. 2010, 28, 143-150.

[11] Ken, C. F., Lee, C. C., Duan, K. J., Lin, C. T. Unusual stability of manganese superoxide dismutase from a new species, *Tatumella ptyseos* ct: its gene structure, expression and enzyme properties. Protein Expres Purif. 2005, 40, 42-50.

[12] Weisiger, R. A., Fridovich, I. Mitochondrial superoxide dismutase. Site of synthesis and intramitochondrial localization. J Biol Chem. 1973, 248, 4793-4796.

[13] Daiber, A. Redox signaling (cross-talk) from and to mitochondria involves mitochondrial pores and reactive oxygen species. Biochim Biophys Acta. 2010, 1797, 897-906.

[14] Koyama, H., Nojiri, H., Kawakami, S., Sunagawa, T., Shirasawa, T., Shimizu, T. Antioxidants improve the phenotypes of dilated cardiomyopathy and muscle fatigue in mitochondrial superoxide dismutase-deficient mice. Molecules. 2013, 18, 1383-1393.

[15] Sun, J., Folk, D., Bradley, T. J., Tower, J. Induced overexpression of mitochondrial Mn-superoxide dismutase extends the life span of adult *Drosophila melanogaster*. Genetics. 2002, 161, 661-672.

[16] Van Loon, A. P., Pesold-Hurt, B., Schatz, G. A yeast mutant lacking mitochondrial manganese-superoxide dismutase is hypersensitive to oxygen. Proc Natl Acad Sci USA. 1986, 83, 3820-3824.

[17] Kim, Y. I., Kim, H. J., Kwon, Y. M., et al. Modulation of MnSOD protein in response to different experimental stimulation in *Hyphantria cunea*. Comp Biochem Phys B: Biochem Mol Biol. 2010, 157, 343-350.

[18] Wang, M. Q., Su, X. R., Li, Y., Jun, Z., Li, T. W. Cloning and expression of the Mn-SOD gene from *Phascolosoma esculenta*. Fish Shellfish Immun. 2010, 29, 759-764.

[19] Gao, X. L., Li J. M., Wang, Y. L., Jiu, M., Yan, G. H., Liu, S. S., Wang, X. W. Cloning, expression and characterization of mitochondrial manganese superoxide dismutase from the whitefly, *Bemisia tabaci*. Int J Mol Sci. 2013, 14, 871-887.

[20] Corona, M., Hughes, K. A., Weaver, D. B., Robinson, G. E. Gene expression patterns associated with queen honey bee longevity. Mech Ageing Dev. 2005, 126, 1230-1238.

[21] Corona, M., Robinson, G. E. Genes of the antioxidant system of the honey bee: annotation and phylogeny. Insect Mol Biol. 2006, 15, 687-701.

[22] Xu, P., Shi, M., Chen, X. X. Antimicrobial peptide evolution in the Asiatic honey bee *Apis cerana*. PLoS One. 2009, 4, e4239.

[23] Michelette, E. R. F., Soares, A. E. E. Characterization of preimaginal developmental stages in Africanized honey bee workers (*Apis mellifera* L.). Apidologie. 1993, 24, 431-440.

[24] Alaux, C., Ducloz, F., Crauser, D., Le Conte, Y. Diet effects on honey bee immunocompetence. Biol Lett. 2010, 6, 562-565.

[25] Livak, K. J., Schmittgen, T. D. Analysis of relative gene expression data using real-time quantitative PCR and the 2 (-Delta C (T)) method. Methods. 2001, 25, 402-408.

[26] Meng, Q. G., Chen, J., Xu, C. C., Huang, Y. Q., Wang, Y., Wang, T. T., Zhai, X. T., Gu, W., Wang, W. The characterization, expression and activity analysis of superoxide dismutases (SODs) from *Procambarus clarkia*. Aquaculture. 2013, 406-407, 131-140.

[27] Yan, H., Jia, H., Gao, H., Guo, X., Xu, B. Identification, genomic organization, and oxidative stress response of a sigma class glutathione *S*-transferase gene (*AccGSTS1*) in the honey bee, *Apis cerana*

cerana. Cell Stress Chaperones. 2013, 18, 415-426.

[28] Burmeister, C., Luërsen, K., Heinick, A., Hussein, A., Domagalski, M., Walter, R. D., Liebau, E. Oxidative stress in *Caenorhabditis elegans*: protective effects of the Omega class glutathione transferase (GSTO-1). FASEB J. 2008, 22, 343-354.

[29] Fernandes, M., Xiao, H., Lis, J. T. Fine structure analyses of the *Drosophila* and *Saccharomyces* heat shock factor-heat shock element interactions. Nucleic Acids Res. 1994, 22, 167-173.

[30] Ding, Y. C., Hawkes, N., Meredith, J., Eggleston, P., Hemingway, J., Ranson, H. Characterization of the promoters of Epsilon glutathione transferases in the mosquito *Anopheles gambiae* and their response to oxidative stress. Biochem J. 2005, 387, 879-888.

[31] Ericsson, A., Kotarsky, K., Svensson, M., Sigvardsson, M., Agace, W. Functional characterization of the CCL25 promoter in small intestinal epithelial cells suggests a regulatory role for Caudal-Related Homeobox (Cdx) transcription factors. J Immunol. 2006, 176, 3642-3651.

[32] Yu, F., Kang, M., Meng, F., Guo, X., Xu, B. Molecular cloning and characterization of a thioredoxin peroxidase gene from *Apis cerana cerana*. Insect Mol Biol. 2011, 20, 367-378.

[33] Durocher, D., Charron, F., Warren, R., Schwartz, R. J., Nemer, M. The cardiac transcription factors Nk2-5 and GATA-4 are mutual cofactors. EMBO J. 1997, 16, 5687-5696.

[34] Ekker, S. C., Von Kessler, D. P., Beachy, P. A. Differential DNA sequence recognition is a determinant of specificity in homeotic gene action. EMBO J. 1992, 11, 4059-4072.

[35] Gogos, J. A., Hsu, T., Bolton, J., Kafatos, F. C. Sequence discrimination by alternatively spliced isoforms of aDNA binding Zinc finger domain. Science. 1992, 257, 1951-1954.

[36] Kyung-II, K., Park, Y., Im, G. Changes in the epigenetic status of the SOX-9 promoter in human osteoarthritic cartilage. J Bone Miner Res. 2013, 28, 1050-1060.

[37] Katherine, A., Braun, P. K., Parua, K. M., Dombek, G. E., Miner, E. T., Young, E. T. Proteins regulate combinatorial transcription following RNA polymerase II recruitment by binding at Adr1-dependent Promoters in *Saccharomyces cerevisiae*. Mol Cell Biol. 2013, 33 (4), 712-724.

[38] Ishida, M., Mitsui, T., Yamakawa, K., Sugiyama, N., Takahashi, W., Shimura, H., Endo, T., Kobayashi, T., Arita, J. Involvement of cAMP response element-binding protein in the regulation of cell proliferation and the prolactin promoter of lactotrophs in primary culture. Am J Physiol Endocrinol Metab. 2007, 293, E1529-E1537.

[39] Carlezon Jr, W. A., Duman, R. S., Nestler, E. J. The many faces of CREB. Trends Neurosci. 2005, 28, 436-445.

[40] Shi, Y. L., Venkataraman, S. L., Dodson, G. E., Mabb, A. M., LeBlanc, S., Tibbetts, R. S. Direct regulation of CREB transcriptional activity by ATM in response to genotoxic stress. Proc Natl Acad Sci USA. 2004, 101, 5898-5903.

[41] Honda, Y., Tanaka, M., Honda, S. Modulation of longevity and diapause by redox regulation mechanisms under the insulin-like signaling control in *Caenorhabditis elegans*. Exp Gerontol. 2008, 43, 520-529.

[42] Gilbert, M. D., Wilkinson, C. F. An inhibitor of microsomal oxidation from gut tissues of the honey bee (*Apis mellifera*). Comp Biochem Physiol B: Biochem Mol Biol. 1975, 50, 613-619.

[43] Krishnan, N., Sehnal, F. Compartmentalization of oxidative stress and antioxidant defense in the larval gut of *Spodoptera littoralis*. Arch Insect Biochem Physiol. 2006, 63, 1-10.

[44] Cadenas, E., Davies, K. J. Mitochondrial free radical generation, oxidative stress, and aging. Free Radical Biol Med. 2000, 29, 222-230.

[45] Alias, Z., Clark, A. G. Studies on the glutathione S-transferase proteome of adult *Drosophila melanogaster*: responsiveness to chemical challenge. Proteomics. 2007, 7, 3618-3628.

[46] Greenberger, J. S., Epperly, M. W. Radioprotective antioxidant gene therapy: potential mechanisms of action. Gene Ther Mol Biol. 2004, 8, 31-44.

[47] Kim, K. Y., Lee, S. Y., Cho, Y. S., Bang, I. C., Kim, K. H., Kim, D. S., Nam, Y. K. Molecular characterization and mRNA expression during metal exposure and thermal stress of copper/zinc- and manganese-superoxide dismutases in disk abalone, *Haliotis discus discus*. Fish Shellfish Immunol.

2007, 23, 1043-1059.

[48] Gao, Q., Song, L. S., Ni, D. J., Wu, L. T., Zhang, H., Chang, Y. Q. cDNA cloning and mRNA expression of heat shock protein 90 gene in the haemocytes of Zhikong scallop *Chlamys farreri*. Comp Biochem Physiol B: Biochem Mol Biol. 2007, 147, 704e15.

[49] Rauen, U., Polzar, B., Stephan, H., Mannherz, H. G., De Groot, H. Cold-induced apoptosis in cultured hepatocytes and liver endothelial cells: mediation by reactive oxygen species. FASEB J. 1999, 13, 155-168.

[50] Harari, P. M., Fuller, D. J., Gerner, E. W. Heat-shock stimulates polyamine oxidation by 2 distinct mechanisms in mammalian-cell cultures. Int J Radiat Oncol. 1989, 16, 451-457.

[51] Schauen, M., Hornig-Do, H. T., Schomberg, S., Herrmann, G., Wiesner, R. J. Mitochondrial electron transport chain activity is not involved in ultraviolet A (UV-A)-induced cell death. Free Radical Biol Med. 2007, 42, 499-509.

[52] An Nguyen, T. T., Michaud, D., Cloutier, C. A proteomic analysis of the aphid *Macrosiphum euphorbiae* under heat and radiation stress. Insect Biochem Mol Biol. 2009, 39, 20-30.

[53] Jung, I., Kim, T. Y., Kim-Ha, J. Identification of *Drosophila* SOD3 and its protective role against phototoxic damage to cells. FEBS Lett. 2011, 585, 1973-1978.

[54] Meyer-Rochow, V. B., Kashiwagi, T., Eguchi, E. Selective photoreceptor damage in four species of insects induced by experimental exposures to UV- irradiation. Micron. 2002, 33, 23-31.

[55] Meng, J. Y., Zhang, C. Y., Zhu, F., Wang, X. P., Lei, C. L. Ultraviolet light-induced oxidative stress: effects on antioxidant response of *Helicoverpa armigera* adults. J Insect Physiol. 2009, 55, 588-592.

[56] Meng, J. Y., Zhang, C. Y., Lei, C. L. A proteomic analysis of *Helicoverpa armigera* adults after exposure to UV light irradiation. J Insect Physiol. 2010, 56, 405-411.

[57] Park, E. J., Park, K. Induction of reactive oxygen species and apoptosis in BEAS-2B cells by mercuric chloride. Toxicol In Vitro. 2007, 21, 789-794.

[58] Tang, T., Huang, D., Zhou, C. Q., Li, X., Xie, Q. J., Li, F. S. Molecular cloning and expression patterns of copper/zinc superoxide dismutase and manganese superoxide dismutase in *Musca domestica*. Gene. 2012, 505, 211-220.

[59] Narendra, M., Bhatracharyulu, N. C., Padmavathi, P., Varadacharyulu, N. C. Prallethrin induced biochemical changes in erythrocyte membrane and red cell osmotic haemolysis in human volunteers. Chemosphere. 2007, 67, 1065-1071.

2.14 Cloning and Molecular Identification of Triosephosphate Isomerase Gene from *Apis cerana cerana* and Its Role in Response to Various Stresses

Yuli Zhou, Fang Wang, Feng Liu, Chen Wang, Yan Yan, Xingqi Guo, Baohua Xu

Abstract: *Apis cerana cerana* triosephosphate isomerase (*AccTPI*), a key regulatory enzyme in glycolysis and gluconeogenesis pathway, which catalyzes the interconversion of glyceraldehyde-3-phosphate to dihydroxyacetone phosphate. However, its role in *A. cerana cerana* has not been completely clarified. In this study, a TPI gene was cloned from *A. cerana cerana* and named *AccTPI*. The open reading frame (ORF) of *AccTPI* is 744 bp, which encodes 247 amino acids with a predicted molecular weight and an isoelectric point of 26.8 kDa and 8.42, respectively. The expression of *AccTPI* was upregulated by some abiotic stresses, while it was downregulated by $HgCl_2$, ultraviolet light (UV), and vitamin C (VC) treatment. Finally, a disc diffusion assay revealed that recombinant AccTPI proteins can protect cells from oxidative stresses. Taken together, all of the results indicate that *AccTPI* may play an important role in the metabolism, antioxidant defense and in many biological functions in the growth and development of *A. cerana cerana*.

Keywords: Triosephosphate isomerase; *Apis cerana cerana*; Abiotic stress; Metabolism

Introduction

As a pollinator, *Apis cerana cerana* is significant in keeping the balance between regional ecologies and agricultural economics [1]. *A. cerana cerana* has been raised in Asian countries for thousands of years and has brought considerable economic benefits to the apicultural industry [2]. The honeybee is an important model organism for neuroethological studies; it exhibits high behavioral plasticity and a remarkable ability to learn [3]. Compared with *A. mellifera*, *A. cerana cerana* has actually been shown to learn better in a controlled laboratory setting [4]. Recent colony losses of *A. cerana cerana* have thus raised concern, and possible explanations for bee decline include nutritional deficiencies and exposures to pesticides and pathogens. Thus, we chose *A. cerana cerana* as the experimental organism for this study. *A. cerana cerana* has unique characteristics including heat hardiness and disease resistance [5]. However, the survival of *A. cerana cerana* is under increasing threat due to various stressors including the indiscriminate use of pesticides and global warming. Moreover, organisms are exposed to a variety of unfavorable environmental stressors which are believed to induce the formation of reactive oxygen species (ROS) [6].

ROS, which include superoxide anions, hydrogen peroxide and hydroxyl radicals, can be induced by external or internal factors, such as pro-oxidants, heavy metal, and pesticides [7]. Generally, there is a balance between the generation of ROS and antioxidant processes, but exogenous stressors can break this balance [8]. High ROS concentrations contribute to degenerative diseases and aging through the oxidation of nucleic acids, proteins, and lipid membranes [9]. To avoid the damage caused by ROS, insects have evolved an intricate mechanism of enzymatic antioxidant systems [10]. Reduced ROS generation, increased antioxidant protection or a combination of both increases the lifespan of social insects [11]. Aerobic organisms have also developed sophisticated antioxidant mechanisms and evolved various enzymatic antioxidant systems to avoid oxidative damage [12].

Metabolic regulation is very important for cellular tolerance to oxidants, providing the reducing power for the antioxidant machinery [13]. Previous studies have demonstrated that triosephosphate isomerase (TPI) catalyzes the isomerization of glyceraldehyde-3-phosphate and dihydroxyacetone phosphate and plays an essential role in glycolysis, gluconeogenesis, and fatty acid synthesis [14]. TPI plays an important function in generating energy and contributing to metabolism. Without TPI, there would be no net production of ATP to promote aerobic metabolism. Furthermore, the deficiency of TPI can result in metabolic diseases [15].

Aerobic metabolism depends on the activities of glycosomal TPI and a mitochondrial glycerophosphate oxidase in *Trypanosoma brucei* [16]. During the life cycle of *Clonorchis sinensis*, adult worms can take in external glucose for their energy supply through the glycolysis pathway [15]. TPI deficiency in *Drosophila melanogaster* exhibits phenotypes such as susceptibility to infection, neurodegeneration, and reduced lifespan [17]. The catalytic activity of TPI is unrestricted, and it is crucial to behavior and longevity [18]. Yeast cells expressing a TPI variant are more resistant against oxidative stress and so TPI cannot tolerate all oxidative stress [19]. Furthermore, several studies demonstrate variations of TPI activity in relation to the levels of antioxidants in lymphocytes [19]. In bees, Wang et al. [20] found that mixed-function oxidative enzymes are the most important enzymes in acaricide detoxification and resistance. Hence, we speculated that the TPI gene in bees might be associated with ROS, so we investigated the TPI antioxidant system in *A. cerana cerana* and its role in ROS defense mechanisms.

As the biochemical characterization of *AccTPI* was still unclear, this is the first study describing the cloning, overexpression and characterization of a recombinant TPI from *A. cerana cerana*. The function of TPI in *A. cerana cerana* was studied to understand its role in cellular defense systems and metabolism. The RNA and protein expression levels of *AccTPI* were calculated. Furthermore, a disc diffusion assay was performed to evaluate the capacity of AccTPI to defend against oxidative stress.

Materials and Methods

Animals and various treatments

A. cerana cerana were obtained from the experimental apiary at Shandong Agricultural

University, Taian, China. Based on age, eye colour and shape, the worker honeybees were generally classified as larvae, pupae and adults [21]. The first (L1), the third (L3), the fifth (L5), and the seventh (L7) day larvae; white-eyed (Pw), pink-eyed (Pp), brown-eyed (Pb) and dark-eyed (Pd) pupae; and newly emerged adults (A1, 1 day post-emergence), 1 week-old bees (A7, usually 7-10 days post-emergence), which are likely nurses, and 2 week-old bees (A15, older than 15 days post-emergence), which can be foragers, were gathered during the spring season from two hives maintained in the experimental apiary. For tissue-specific analyses of *AccTPI* expression levels, the antenna (AN), abdomen (AB), head (HE), hemolymph (HA), epidermis (EP), venom gland (VG), muscle (MU), thoracic (TH), and midgut (MI) of adult workers were used. These were dissected on ice, washed twice with PBS, frozen in liquid nitrogen and stored at $-70\,^\circ\mathrm{C}$.

Fifteen-day-old adult bees were collected randomly and reared under $34\,^\circ\mathrm{C}$ and 60% relative humidity, and fed a diet of water and a combination of pollen and sucrose [22]. The worker honeybees were randomly separated into eight groups, and each group contained 30 individuals. Groups 1, 2 and 3 were exposed to 4, 14, and $44\,^\circ\mathrm{C}$, respectively. Group 4 was treated with ultraviolet light (UV) (254 nm, 30 MJ/cm^2). Groups 5 and 6 were exposed to two pesticides (2.0 mg/L of paraquat and pyridaben was added to the basic adult diet, respectively). The control group fed on a normal diet containing water, 70% powdered sugar and 30% honey from the source colonies. Group 7 was fed $HgCl_2$ (3 mg/ml). Group 8 was subjected to vitamin C (20,000 mg/kg). The honeybees were flash-frozen in liquid nitrogen at a suitable time and stored at $-70\,^\circ\mathrm{C}$ until used.

RNA extraction, cDNA synthesis, and DNA preparation

Total RNA was extracted from adult bees with the Trizol reagent (Invitrogen, Carlsbad, CA, USA), and the RNA was treated with RNase-free DNase I (Promega, Madison, WI, USA). Then, the cDNA was synthesized by using a reverse transcription system (TransGen Biotech, Beijing, China). The genomic DNA was isolated from the honeybees with the EasyPure Genomic DNA Extraction Kit (TransGen Biotech, Beijing, China).

Primers and PCR procedure

The primer sequences and PCR procedure are listed in Table 1 and Table 2, respectively.

Table 1 The primers in this study

Abbreviation	Primer sequence (5'—3')	Description
TF	GGCCATGGGTCGTAAATTTTTG	cDNA sequence primer, forward
TR	CATCTCACTGCTTAGCGTTAACG	cDNA sequence primer, reverse
5TW	TGCAACCTTATATGTATTTTGTCCAGC	5' RACE reverse primer, outer
5TN	TGCAACCTTATATGTATTTTGTCCAGC	5' RACE reverse primer, inner
3TW	GTAGCCCATGCTTTAGATAGTGGAC	3' RACE forward primer, outer
3TN	CACACCACAACAAGCTCAAGAG	3' RACE forward primer, inner

Abbreviation	Primer sequence (5'—3')	Description
AAP	GGCCACGCGTCGACTAGTAC(G)$_{14}$	Abridged anchor primer
AUAP	GGCCACGCGTCGACTAGTAC	Abridged universal amplification primer
B25	GACTCTAGACGACATCGA	3' RACE universal primer, outer
B26	GACTCTAGACGACATCGA(T)$_{18}$	3' RACE universal primer, inner
DLF	TGTTGAGGTTGTCGTTGGAGTAC	Real-time PCR primer, forward
DLR	GGACTTATTTCTCCAGTAAATGCTC	Real-time PCR primer, reverse
β-s	TTATATGCCAACACTGTCCTTT	Standard control primer, forward
β-x	AGAATTGATCCACCAATCCA	Standard control primer, reverse
G1	GGGGATTTCATAAAATCGGGTTGG	Genomic sequence primer, forward
G2	GGTTATGTGATTACTTATGCACTATATTTGTG	Genomic sequence primer, reverse
YHF	GATATCATGGGTCGTAAATTTTTTGTTGGTGG	Protein expression primer, forward
YHR	CGAGCTCTCACTGCTTAGCGTTAACGAT	Protein expression primer, reverse

Table 2 PCR amplification conditions

Primer pair	Amplification conditions
TF/TR	10 min at 94℃, 40 s at 94℃, 40 s at 54℃, 60 s at 72℃ for 35 cycles, 10 min at 72℃
5TW/AAP	10 min at 94℃, 40 s at 94℃, 40 s at 54℃, 60 s at 72℃ for 35 cycles, 10 min at 72℃
5TN/AUAP	10 min at 94℃, 40 s at 94℃, 40 s at 55℃, 60 s at 72℃ for 35 cycles, 10 min at 72℃
3TW/B26	10 min at 94℃, 40 s at 94℃, 40 s at 57℃, 60 s at 72℃ for 35 cycles, 10 min at 72℃
3TN/B25	10 min at 94℃, 40 s at 94℃, 40 s at 55℃, 60 s at 72℃ for 35 cycles, 10 min at 72℃
G1/G2	10 min at 94℃, 40 s at 94℃, 40 s at 55℃, 120 s at 72℃ for 35 cycles, 10 min at 72℃
YHF/YHR	10 min at 94℃, 40 s at 94℃, 40 s at 58℃, 60 s at 72℃ for 35 cycles, 10 min at 72℃

Bioinformatics analysis and phylogenetic tree construction

The conserved nucleotide sequences of *AccTPI* were analyzed by using DNAMAN software 6.0.3 and the National Center for Biotechnology Information (NCBI) bioinformatics tools (http://blast.ncbi.nlm.nih.gov/Blast.cgi). The predicted amino acid sequence was aligned with the homologous sequences of various insects, which were acquired from the NCBI Genbank database. The molecular weight and theoretical isoelectric point were predicted by using the online tool ExPASy (http://web.expasy.org/compute_pi/). A phylogenetic tree was constructed by using the Molecular Evolutionary Genetic Analysis (MEGA version 4.0) software.

Quantitative real-time PCR

To evaluate the expression profile of *AccTPI* in different developmental stages tissues and abiotic stresses, RT-qPCR was performed by using the SYBR® PrimeScript™ RT-PCR Kit (TaKaRa, Dalian, China) with two specific primers (DLF and DLR). The reaction volume and the reaction requirements were as those of Yao et al. [23]. The PCR reactions were performed by using a CFX96™ Real-time System (Bio-Rad), and the *β-actin* gene (GenBank accession no. HM640276; *A. cerana cerana* actin-related protein 1) was used as an internal control. We also obtained experimental data in which we tested and compared additional reference genes (*β-actin, ribosomal protein 49, elongation factor 1-alpha, tbp-association factor*), see Supplemental Fig. S3. All of the samples were analyzed in triplicate. The relative gene expression levels of *AccTPI* were calculated by using the $2^{-\Delta\Delta C_t}$ comparative CT method [24]. The analysis was performed with Duncan's multiple range test by using Statistical Analysis System (SAS) software version 9.1.

Protein expression and purification of recombinant AccTPI

To obtain the AccTPI protein, the recombinant plasmid pET-30a (+)-*AccTPI* was transfected into *Escherichia coli* BL21 (DE3). The bacteria were incubated in Luria-Bertani (LB) broth with kanamycin at 37℃ until the cell density at 600 nm reached 0.4-0.6. Then, the AccTPI expression was induced with 1.0 mmol/L isopropylthio-β-D-galactoside (IPTG) at 18℃ overnight. The protein was purified on a HisTrap™ FF column (GE Healthcare, Uppsala, Sweden). The expression of the target protein was analyzed by 12% sodium dodecyl sulfate polyacrylamide gel electrophoresis (SDS-PAGE).

Western blotting analysis

The white mice for the generation of antibodies came from China Biologic Products, Inc. (Shandong). The protocol number of certificates that they produced and used for animal research were SCXK (LU) 2013-0006 and SYXK (LU) 2013-0011 respectively. The Tissue Protein Extraction Kit (CoWin Bioscience Co., Beijing, China) was used to extract the total *A. cerana cerana* protein, and it was quantified with the BCA Protein Assay Kit (Nanjing Jiancheng Bioengineering Institute, Nanjing, China). Western blotting was performed according to the procedure of Chen et al. [25]. The following treatments were performed to analyze the protein expression levels: (1) 4℃, (2) H_2O_2, (3) methomyl, (4) pyridaben, (5) UV, and (6) vitamin C (VC).

Disc diffusion assay

Disc diffusion assays were performed according to Chen et al. [25]. for the evaluation of the activity of recombinant AccTPI. *E. coli* BL21 (DE3) cells were incubated in LB liquid medium with kanamycin at 37℃ until the cell density reached 0.4-0.6 (OD_{600}), then IPTG was introduced to the LB broth at a final concentration of 1 mmol/L. They were induced by using IPTG and incubated for 8 h at 18℃ with shaking at 180 r/min. Approximately 5×10^8 bacterial cells overexpressing AccTPI were seeded onto LB-kanamycin agar plates and incubated at 37℃ for 1 h. Filter discs (8 mm diameter) soaked with different concentrations of *t*-butyl hydroperoxide (0, 15, 20, 30, or 40 mmol/L) and cumene hydroperoxide (0, 30, 50, 80, or 100 mmol/L) were placed on the surface of the top agar. Eventually, the cells were incubated at 37℃ for 24 h before the inhibition zones around the paper discs were measured as described by Burmeister et al. [26]. *E. coli* BL21 cells transformed with empty pET-30a (+) were used as the control. The greater of the bacteriostatic circle bacteria is more sensitive, rather than vise versa.

Results

Identification of the full-length cDNA and sequence analysis

The full-length cDNA of the *AccTPI* gene (GenBank accession no. KP994676) was cloned. It comprised 1,894 bp containing a 744 bp open reading frame (ORF), a 158 bp 5' untranslated region (UTR) and a 992 bp 3' UTR. The ORF encodes a polypeptide of 247 amino acids with a predicted molecular weight of 26.8 kDa and theoretical isoelectric point of 8.42.

Multiple sequence alignments of several TPI proteins indicates that the putative AccTPI shares high homology with *A. mellifera* AmTPI (NP 001090623.1), *Megachile rotundata* MrTPI (XP 003700534.1), *Harpegnathos saltator* HsTPI (XP 011146748.1), and *Acromyrmex echinatior* AeTPI (EGI64718.1), respectively (Fig. 1A). As shown in Fig. 1, we found one TPI locus and several protein kinase C phosphorylation sites in the ORF. A phylogenetic tree was built to illustrate. the evolutionary relationship of TPI among insects. The exons and introns of *AccTPI* were aligned with *AmTPI* (NM 001097154.2), *MrTPI* (XM 003700486.1), and *HsTPI* (XM 011148446.1). This analysis showed that *AccTPI, AmTPI, MrTPI* and *HsTPI* contain two introns and three exons (Supplemental Fig. S1).

Developmental and tissue-specific expression patterns of *AccTPI*

To investigate the expression patterns at different developmental stages and specific tissues, RT-qPCR assays were performed. As shown in Fig. 2A, the expression levels peaked in the forager stage. The mRNA expression levels of *AccTPI* are higher in the L3 larval stages. Additionally, the expression levels in different *AccTPI* tissues were higher in the abdomen and thorax (Fig. 2B).

Expression patterns of *AccTPI* under abiotic stresses

To verify whether abiotic stresses could affect *AccTPI* expression, a 15-day post-emergence adult stage was chosen and total RNA was extracted. The expression of *AccTPI* was induced by 4℃ treatment and reached the highest level at 2 h after exposure. Moreover, the expression of *AccTPI* was inhibited by 14℃ treatment and the expression level subsequently increased after 44℃ treatment (Fig. 3A-C). The expression of *AccTPI* was upregulated in response to the paraquat and pyridaben treatments (Fig. 3D, E). In contrast, the expression of *AccTPI* was downregulated after UV, $HgCl_2$ and VC treatments (Fig. 3F-H).

Expression and purification of recombinant AccTPI protein

To further characterize the properties of *AccTPI* via protein levels, AccTPI was transformed into *E. coli* BL21 (DE3) with the pET-30a (+) vector. The *E. coli* cells were obtained from the

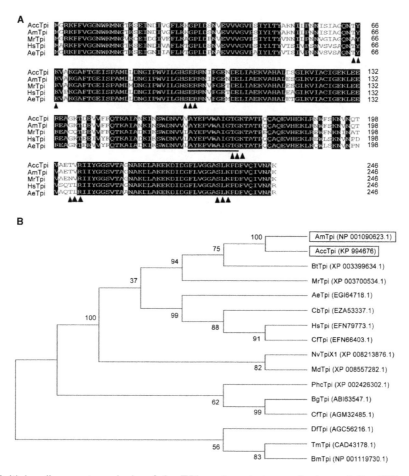

Fig. 1 Multiple alignment analysis of AccTPI and a phylogenetic tree of the TPI amino acid sequences of several species. A. Alignment of the deduced AccTPI protein sequence with other known species of TPI. The amino acid sequences of AmTPI, MrTPI, HsTPI and AeTPI were used in the analysis. The TPI locus is marked by a horizontal line. The protein kinase C phosphorylation sites are marked by a (▲). B. Phylogenetic tree analysis of TPI from different species. The amino acid sequences of TPI were all obtained from NCBI.

LB solution after induction with IPTG overnight. SDS-PAGE analysis indicated that a recombinant protein was expressed and had a molecular mass of approximately 33 kDa (Supplemental Fig. S2, lane 2). It contained the AccTPI protein (26.8 kDa) and the His-tag (7 kDa). The recombinant protein was purified with HisTrap™ FF columns. The result is shown in Supplemental Fig. S2, lane 7.

Western blotting analysis

Western blotting analysis was performed to further detect the expression profiles of AccTPI under various adverse stresses. The results displayed that the expression of AccTPI protein levels were consistent with the transcriptional expression levels. The expression levels of AccTPI were upregulated by 4°C, H_2O_2, methomyl and pyridaben treatments. Furthermore, the expression levels of AccTPI were downregulated by UV and VC treatments (Fig. 4).

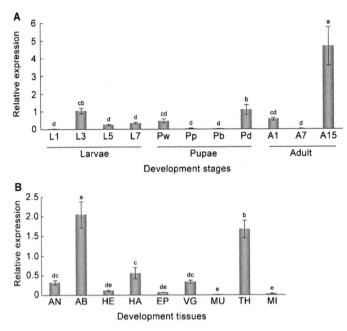

Fig. 2 Expression profile of *AccTPI* in different developmental stages and tissues. A. The expression profile of AccTPI was examined by RT-qPCR in different developmental stages including larvae, pupae, newly emerged adult workers, nurses and foragers. B. The expression profile of AccTPI was studied by RT-qPCR in various tissues. The data are the mean ± SE of three independent experiments. The letters above the columns indicate significant differences ($P < 0.01$) according to Duncan's multiple range test.

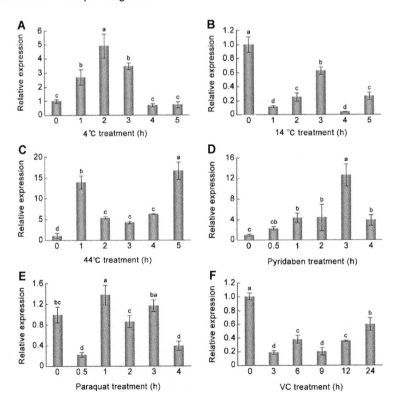

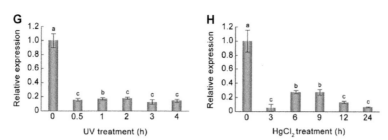

Fig. 3 Expression profile of *AccTPI* under different abiotic stresses. The expression of *AccTPI* was analyzed by RT-qPCR by using total RNA extracted from *A. cerana cerana*, which was treated with different stressors. For the stress treatments, 15-day-old adult bees were subjected to treatments with 4 ℃ (A), 14 ℃ (B), 44 ℃ (C), pyridaben (D), paraquat (E), VC (F), UV (G) and HgCl₂ (H). CK is the abbreviation for control check. The letters above the bars indicate significant differences ($P < 0.01$) according to the SAS software version 9.1. The data are the mean ± SEM of three independent experiments. The letters above the columns indicate significant differences ($P < 0.01$) according to Duncan's multiple range test.

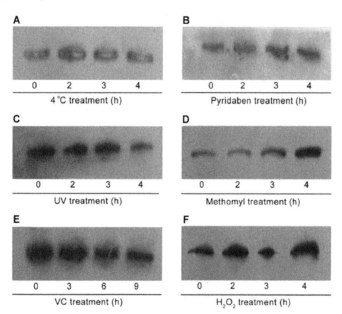

Fig. 4 Western blotting analysis of AccTPI. Western blotting conditions include 15-day-old adult bees that were subjected to treatments with 4 ℃ (A), pyridaben (B), UV (C), methomyl (D), VC (E) and H_2O_2 (F). All lanes were loaded with an equivalent amount of extracted protein for each group.

Disc diffusion assay under various stresses

To provide direct evidence that AccTPI is responsible for antioxidant defenses, a disc diffusion assay was performed according to Zhang et al. [27]. After overnight exposure to various stressors, the halo diameters of the death zones were significantly different for the various treatment concentrations. Compared with plates containing the control bacteria, the inhibition zones around the stressor-soaked filters were smaller in diameter containing *E. coli* overexpressing AccTPI (Fig. 5).

Discussion

Sequence alignment and the phylogenetic tree analysis showed that AccTPI has a high degree of homology with several representative TPI amino acid sequences (Fig. 1). Taken together, these results indicate that the gene we isolated is a bona fide *AccTPI*.

To better understand its physiology, *AccTPI* expression at different development stages and in various tissues was analyzed. The expression of *AccTPI* reached a maximum in the 15-day post-emergence workers, which likely are foragers (Fig. 2A). TPI, a key regulatory enzyme of glycolysis and gluconeogenesis, plays an essential role in metabolism and development of most organisms [15]. These results revealed that *AccTPI* might be essential for the growth and development of bees. Additionally, the abdomen and thorax have higher *AccTPI* expression levels compared with other tissues (Fig. 2B). Therefore, it is reasonable to deduce that *AccTPI* may be involved in differentiation or development relating to the mechanism of maturation, especially in the case of oxidation resistance. Considering all of these results, we propose that *AccTPI* may be involved in organismal development and growth.

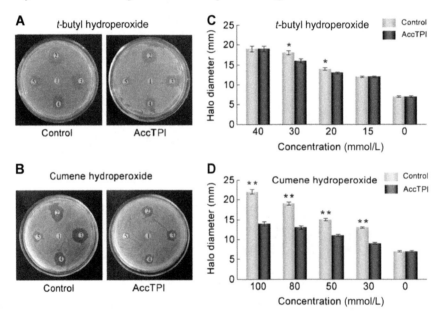

Fig. 5 Disc diffusion assays for AccTPI. Filter discs soaked with different concentrations of *t*-butyl hydroperoxide (A and C) and cumene hydroperoxide (B and D) were placed on agar plates. The picture on the left corresponds with the numbers on the right. The data are the mean ± SE of three independent experiments. Statistical analysis was performed by using Statistical Product and Service Solutions (SPSS) software version 19.0. Statistical significance was determined by unpaired Student's t-test; * $P < 0.05$, ** $P < 0.01$.

Previous studies have indicated that *A. cerana cerana* are stressed due to environmental changes and ROS [28]. As the expression of *AccTPI* in *A. cerana cerana* due to exposure to oxidative stress has not yet been systematically reported, we investigated in this study the expression levels of *AccTPI* in response to various abiotic stresses. The results demonstrate that *AccTPI* may play an important role in protecting *A. cerana cerana* from oxidative stresses.

Temperature is an abiotic environmental factor that causes physiological changes in organisms and affects herbivorous insects [29]. In general, heat shock stress can result in polyamine oxidation to generate H_2O_2, and ROS act as key mediators of cold-induced apoptosis [30]. In the present study, the expression levels of *AccTPI* under 4 and 44 ℃ conditions were upregulated following treatment. Western blotting analysis of AccTPI under 4 ℃ (Fig. 4A) treatments confirmed the result, which demonstrates that AccTPI may play a vital role in response to extreme temperature, low temperature in particular. These results suggest that *AccTPI* may function by removing intracellular superoxide anions generated under high- or low-temperature conditions to protect honeybees from ROS damage. All of these results indicated that *AccTPI* may be associated with the prevention of the oxidative damage.

Oxidative stress can be induced by insecticides, temperature and UV radiation [6]. The role of pesticides in honeybee colony losses, with their sub-lethal and synergistic effects, has recently regained consideration [31]. TPI also plays an important role in responses to adverse environmental conditions [32]. TPI is highly upregulated under stress conditions in mammalian cells [33]. In this study, paraquat and pyridaben treatments upregulated TPI expression levels in *A. cerana cerana* workers. The expression of *AccTPI* was upregulated under some pesticide treatments. All of these results demonstrate that *AccTPI* expression is induced by abiotic stresses and that *AccTPI* may play a crucial role in survival under adverse environmental conditions.

There is much evidence that UV is one of the most ubiquitous environmental hazards for most animals can cause oxidative stress and in turn lead to protein damage [34]. Many heavy metals can induce the formation of endogenous ROS [35]. $HgCl_2$ treatment suppresses the expression of *Acctpx-3* [23]. When honeybees were treated with UV and $HgCl_2$, expression levels of *AccTPI* were significantly downregulated. These results indicated that *AccTPI* may play various roles in dealing with various adverse conditions. The mRNA expression of *AccTPI* is specifically induced by certain stresses and repressed by others, and these results indicate that TPI is a mediator of the stressors. With an increase in the treatment concentration or an extension of the treatment time, the antioxidant system may temporarily be ineffective due to decreased expression. All of these results suggest that *AccTPI* is involved in resisting oxidative stress of *A. cerana cerana* restrictedly.

Western blotting results for AccTPI under H_2O_2 and VC treatments verified that the expression is upregulated and downregulated depending on the stressor. H_2O_2-induced ROS increase results in cell death caused by oxidative stress [36]. H_2O_2-induced ROS increase results in cell death caused by oxidative stress [37]. Vitamin C, a well-known antioxidant, induces the decomposition of lipid hydroperoxides to endogenous genotoxins and results in DNA oxidative damage [38]. The expression levels of *AccTPI* in VC treatments were significantly downregulated. A possible explanation for this was that with the increase of stress, other stress-related genes might play important roles in cellular stress responses and the role of *AccTPI* might be weakened. The expression levels of the *AccTPI* gene in *A. cerana cerana* were varied in response to the oxidative stresses. Thus, *AccTPI* may have a significant influence on the response to environmental stresses.

To understand the role of honeybee TPI in defending against oxidative stresses, disc diffusion assays (Fig. 5) were used to detect the level of protection of AccTPI in *E. coli* cells from oxidative stressors (cumene hydroperoxide and *t*-butyl hydroperoxide). The model external pro-oxidants cumene hydroperoxide and *t*-butyl hydroperoxide were used because they are more stable than H_2O_2 under the applied incubation conditions [27]. There were variations in the diameters of the death zones between the *AccTPI*-overexpressing bacteria and the control bacteria. Hence, we can conclude that *AccTPI* can significantly contribute to cellular resistance under oxidative stresses and that the *AccTPI* gene has antioxygenic properties. The purified recombinant AccTPI protein showed a capacity to protect DNA from oxidative damage. Taken together, our data provide very useful information for further understanding the mechanisms responding to environmental pressures in *A. cerana cerana*.

In conclusion, all of the results indicate that *AccTPI* may play an important role in the metabolism and antioxidant defense of honeybees. The unique functional characteristics, expression patterns and potential physiological roles of *AccTPI* that were demonstrated in this study offer the basic knowledge for further studies. It is an eternal theme for the survival of living organisms to protection against oxidative stress.

References

[1] Weinstock, G. M., Robinson, G. E., Gibbs, R. A., et al. Insights into social insects from the genome of the honeybee *Apis mellifera*. Nature. 2006, (443), 931-949.

[2] Park, D., Jung, J. W., Choi, B. S., et al. Uncovering the novel characteristics of Asian honey bee, *Apis cerana*, by whole genome sequencing. BMC Genomics. 2005, 2, 16, 1.

[3] Zhang, L. Z., Yan, W. Y., Wang, Z. L., Guo, Y. H., Yi, Y., Zhang, S. W., Zeng, Z. J. Differential protein expression analysis following olfactory learning in *Apis cerana*. J Comp Physiol A: Neuroethol Sens Neural Behav Physiol. 2015, (201), 1053-1061.

[4] Qin, Q., He, X., Tian, L., Zhang, S., Zeng, Z. Comparison of learning and memory of *Apis cerana* and *Apis mellifera*. J Comp Physiol A. 2012, (198), 777-786.

[5] Li, H., Zhang, Y., Gao, Q., Cheng, J., Lou, B. Molecular identification of cDNA, immunolocalization, and expression of a putative odorant-binding protein from an Asian honey bee, *Apis cerana cerana*. J Chem Ecol. 2008, (34), 1593-1601.

[6] Kottuparambila, S., Shinb, W., Brownc, M. T., Han, T. UV-B affects photosynthesis, ROS production and motility of the freshwater flagellate, *Euglena agilis* Carter. Aquat Toxicol. 2012, (122-123), 206-213.

[7] Imlay, J. A. Pathways of oxidative damage. Annu Rev Microbiol. 2003, (57), 395-418.

[8] Brennan, L. J., Keddie, B. A., Braig, H. R., Harris, H. L. The endosymbiont *Wolbachia pipientis* induces the expression of host antioxidant proteins in an *Aedes albopictus* cell line. PLoS One. 2008, 3, e2083.

[9] Valko, M., Rhodes, C. J., Moncol, J., Izakovic, M., Mazur, M. Free radicals, metals and antioxidants in oxidative stress-induced cancer. Chem Biol Interact. 2006, (160), 1-40.

[10] Felton, G. W., Summers, C. B. Antioxidant systems in insects. Arch Insect Biochem & Physiol. 1995, (29), 187-197.

[11] Keller, L., Jemiolity, S. Social insects as a model to study the molecular basis of ageing. Exp Gerontol. 2006, (41), 553-556.

[12] Jia, H. H., Sun, R. J., Shi, W. N., Yan, Y., Li, H., Guo, X. Q., Xu, B. H. Characterization of a mitochondrial manganese superoxide dismutase gene from *Apis cerana cerana* and its role in oxidative stress. J Insect Physiol. 2014, (60), 68-79.

[13] Chu, C., Lai, Y., Huang, H., Sun, Y. Kinetic and structural properties of triosephosphate isomerase from *Helicobacter pylori*. Proteins. 2008, (71), 396-406.

[14] Zinsser, V. L., Hoey, E. M., Trudgett, A., Tpison, D. J. Biochemical charaterisation of triosephosphate isomerase from the liver fluke *Fasciola hepatica*. Biochimie. 2013, (95), 2182-2189.

[15] Zhou, J., Liao, H., Li, S., et al. Molecular identification, immunolocalization, and characterization of *Clonorchis sinensis* triosephosphate isomerase. Parasitol Res. 2015, (114), 3117-3124.

[16] Helfert, S., Estevez, A. M., Bakker, B., Michels, P., Clayton, C. Roles of triosephosphate isomerase and aerobic metabolism in *Trypanosoma brucei*. Biochem J. 2001, (357), 117-125.

[17] Hrizo, S. L., Palladino, M. J. Hsp70- and Hsp90-mediated proteasomal degradation underlies TPI^{sugarkill} pathogenesis in *Drosophila*. Neurobiol Dis. 2010, (3), 676-683.

[18] Roland, B. P., Stuchul, K. A., Larsen, S. B., Amrich, C. G., Vandemark, A. P., Celotto, A. M., Palladino, M. J. Evidence of a triosephosphate isomerase non-catalytic function crucial to behavior and longevity. J Cell Sci. 2013, (126), 3151-3158.

[19] Ralser, M., Heeren, G., Breitenbach, M., Lehrach, H., Krobitsch, S. Triosephosphate isomerase deficiency is caused by altered dimerization-not catalytic inactivity-of the mutant enzymes. PLoS One. 2006, 1, e30.

[20] Wang, J., Wei, D., Dou, W., Hu, F., Liu, W., Wang, J. Toxicities and synergistic effects of several insecticides against the oriental fruit fly (Diptera: Tephritidae). J Econ Entomol. 2013, (106), 970-978.

[21] Elias-Neto, M., Soares, M. P., Simoes, Z. L., Hartfelder, K., Bitondi, M. M. Developmental characterization, function and regulation of a Laccase2 encoding gene in the honey bee, *Apis mellifera* (Hymenoptera, Apinae). Insect Biochem Mol Biol. 2010, (40), 231-241.

[22] Alaux, C., Ducloz, F., Crauser, D., Le Conte, Y. Diet effects on honeybee immunocompetence. Biol Lett. 2010, (6), 562-565.

[23] Yao, P., Lu, W., Meng, F., Wang, X., Xu, B., Guo, X. Molecular cloning, expression and oxidative stress response of a mitochondrial thioredoxin peroxidase gene (*AccTpx-3*) from *Apis cerana cerana*. J Insect Physiol. 2013, (59), 273-282.

[24] Livak, K. J., Schmittgen, T. D. Analysis of relative gene expression data using real-time quantitative PCR and the 2 (-Delta C (T)) method. Methods. 2001, (25), 402-440.

[25] Chen, X., Yao, P., Chu, X., Hao, L., Guo, X., Xu, B. Isolation of arginine kinase from *Apis cerana cerana* and its possible involvement in response to adverse stress. Cell Stress Chaperones. 2014, (20), 169-183.

[26] Burmeister, C., Luersen, K., Heinick, A., Hussein, A., Domagalski, M., Walter, R. D., Liebau, E. Oxidative stress in *Caenorhabditis elegans*: protective effects of the Omega class glutathione transferase (GSTO-1). FASEB J. 2008, (22), 343-354.

[27] Zhang, Y., Yan, H., Lu, W., Li, Y., Guo, X., Xu, B. A novel Omega-class glutathione *S*-transferase gene in *Apis cerana cerana*: molecular characterisation of *GSTO2* and its protective effects in oxidative stress. Cell Stress Chaperones. 2013, (18), 503-516.

[28] Yan, H., Meng, F., Jia, H., Guo, X., Xu, B. The identification and oxidative stress response of a Zeta class glutathione *S*-transferase (*GSTZ1*) gene from *Apis cerana cerana*. J Insect Physiol. 2012, (58), 782-791.

[29] An, M. I., Choi, C. Y. Activity of antioxidant enzymes and physiological responses in ark shell, *Scapharca broughtonii*, exposed to thermal and osmotic stress: effects on hemolymph and biochemical parameters. Comp Biochem Physiol B: Biochem Mol Biol. 2010, (155), 34-42.

[30] Rauen, U., Polzar, B., Stephan, H., Mannherz, H. G., De Groot, H. Cold-induced apoptosis in cultured hepatocytes and liver endothelial cells: mediation by reactive oxygen species. FASEB J. 1999, (13), 155-168.

[31] Boncristiani, H., Underwood, R., Schwarz, R., Evans, J. D., Pettis, J., Van Engelsdorp, D. Direct effect of acaricides on pathogen loads and gene expression levels in honey bees *Apis mellifera*. J Insect Physiol. 2012, (5), 613-620.

[32] Hewitson, J. P., Ruckerl, D., Harcus, Y., Murray, J., Webb, L. M., Babayan, S. A., Allen, J. E.,

Kurniawan, A., Maizels, R. M. The secreted triosephosphate isomerase of *Brugia malayi* is required to sustain microfilaria production *in vivo*. PLoS Pathog. 2014, 10 (2), e1003030.

[33] Yamaji, R., Fujita, K., Nakanishi, I., Nagao, K., Naito, M. Hypoxic upregulation of triosephosphate isomerase expression in mouse brain capillary endothelial cells. Arch Biochem Biophys. 2004, (423), 332-342.

[34] Schauen, M., Hornig-Do, H. T., Schomberg, S., Herrmann, G., Wiesner, R. J. Mitochondrial electron transport chain activity is not involved in ultraviolet A (UV-A)-induced cell death. Free Radic Biol Med. 2007, (42), 499-509.

[35] Liu, J., Qu, W., Kadiiska, M. B. Role of oxidative stress in cadmium toxicity and carcinogenesis. Toxicol Appl Pharmacol. 2009, (238), 209-214.

[36] Casini, A. F., Ferrali, M., Pompella, A., Maellaro, E., Comporti, M. Lipid peroxidation and cellular damage in extrahepatic tissues of bromobenzene-intoxicated mice. Am J Pathol. 1986, (123), 520-531.

[37] Goldshmit, Y., Erlich, S., Pinkas-Kramarski, R. Neuregulin rescues PC12-ErbB4 cells from cell death induced by H_2O_2 regulation of reactive oxygen species levels by phosphatidylinositol 3-kinase. J Biol Chem. 2001. (49), 379-385.

[38] Lee, S. H., Oe, T., Blair, I. A. Vitamin C-induced decomposition of lipid hydroperoxides to endogenous genotoxins. Science. 2001, (292), 2083-2086.

2.15 Identification and Characterization of a Novel Methionine Sulfoxide Reductase B Gene (*AccMsrB*) from *Apis cerana cerana* (Hymenoptera: Apidae)

Feng Liu, Zhihong Gong, Weixing Zhang, Ying Wang, Lanting Ma, Hongfang Wang, Xingqi Guo, Baohua Xu

Abstract: Methionine sulfoxide reductase B genes (MsrBs) play a crucial role protecting cells from oxidative damage. In this study, we isolated and characterized an MsrB gene from *Apis cerana cerana*, designated *AccMsrB*. The full cDNA of *AccMsrB* was 757 bp with an open reading frame (ORF) of 414 bp, and the predicted translation product was a 137-amino acid polypeptide with an estimated molecular mass of 15.5 kDa and an isoelectric point of 7.77. Multiple sequence alignment revealed that AccMsrB shares high identity with other known MsrBs (*Apis florea, Apis mellifera, Apis dorsata, and Bombus terrestris,*) and contains conservative Cys residues. Quantitative real-time polymerase chain reaction showed *AccMsrB* to be highly expressed in the epidermis of adult workers, reaching high levels in the first-instar larvae, prepupae, and 15-day-old adults. Furthermore, the expression of *AccMsrB* was upregulated by various oxidative stresses, including 4, 16, 25 and 42℃, ultraviolet light (30 MJ/cm^2), H_2O_2, $CdCl_2$, $HgCl_2$, paraquat, imidacloprid, and cyhalothrin. However, *AccMsrB* was downregulated when exposed to phoxim. These results indicate that *AccMsrB* might respond to various environmental stresses and protect against reactive oxygen species.

Keywords: Methionine sulfoxide reductase B; *Apis cerana cerana*; Gene expression; Oxidative stresses

Introduction

Methionine is readily oxidized to methionine sulfoxide (MetO) by reactive oxygen species (ROS) that are generated in response to stress and that lower the cellular antioxidant capacity, alter protein function, and damage proteins[1]. However, oxidation can be reversed by the enzymatic reduction system, which consists of methionine sulfoxide reductases (Msrs) [2, 3]. Msr proteins are thought to maintain the methionine sulfoxide reduction system in different cellular compartments, ready to repair oxidatively damaged proteins, protect against oxidative stress and regulate protein function [4]. Two Msr families, MsrA and MsrB, reduce the two epimers of Met (O), called Met-*S*-(O) and Met-*R*-(O), respectively [5].

MsrBs have been found in almost all organisms except some bacteria, whereas certain prokaryotes have *MsrA* and only one type of *MsrB* [3, 6, 7], mammals possess three: *MsrB1*, *MsrB2* and *MsrB3* [4, 8]. The number of *MsrB* genes within organisms also varies greatly. Some contain one copy of *MsrB*, while others contain three or more *MsrB*s [9-11]. The first Msr enzyme was identified in *Escherichia coli* and was shown to convert MetO to Met [12]. Subsequent studies have clearly elucidated the structure and characteristics of MsrB family members. For example, most MsrBs belong to a group of zinc-containing proteins based on the presence of two well conserved Zn-bonding CXXC motifs [13]. However, different MsrBs have varying amino acids in their active site, which influences their catalytic efficiency. A signature domain GCGWP was determined that is believed to interact with Trx [14]. Another catalytic Cys was found in the RXCXN motif near the C-terminus [12].

Stresses such as cold, heat, H_2O_2, UV light, peroxynitrite, heavy metal and pesticides always increase ROS [15]. MsrA has been described as playing an important role in defending against oxidative stress, while a previous study of ours found MsrB to act in concert with MsrA [16]. Other prior studies have shown that MsrBs scavenge ROS and protect proteins by reducing the sulfoxide group of R-MetO [17]. The expression level of *MsrB* genes is influenced by ROS, and the MsrB plays an important role in protecting against oxidative damage and the regulation of aging [18-20]. Decreased *MsrB* expression makes cells more sensitive to low or high concentrations of $ONOO^-$ [21].

The honeybee is an excellent model organism for studying the effects of the stress response on a particular gene. While studies of MsrBs in other species have mainly focused on antioxidant defense and the modulation of the aging process, little is known about the role of MsrB in honeybees, particularly Chinese honeybees. When collecting pollen, honeybees encounter such diverse environmental stresses as sunlight (UV radiation), high and low temperatures, infectious diseases, H_2O_2, heavy metal, hydroxyls, and pesticides, all of which can increase ROS levels [22-24], resulting in oxidative damage. Identifying the *MsrB* gene in honeybees would enable us to better understand the ROS defense mechanisms in *Apis cerana cerana*. To this end, we isolated *AccMsrB*, which is highly similar to *AmMsrB*, from *A. cerana cerana*. *AccMsrB* is an *MsrB* gene that encodes a putative methionine sulfoxide-R-reductase. *AccMsrB* mRNA was detected in different developmental stages and tissues, suggesting that *AccMsrB* plays an important role in all stages of honeybee development. Furthermore, accumulation of *AccMsrB* mRNA was induced by different temperatures, $CdCl_2$, $HgCl_2$, UV light, H_2O_2, paraquat, imidacloprid, cyhalothrin and phoxim. *AccMsrB* was upregulated immediately after mild oxidative stress, likely to allow it to respond to the same oxidative stresses.

Materials and Methods

Animal specimens and treatments

The Chinese honeybees, *A. cerana cerana*, used in this study were obtained from the experimental apiary of Shandong Agricultural University in Taian, China. The bees were fed in incubators and were maintained at 33℃, 60% relative humidity, and a 24 h dark.

Honeybees (in various life stages) were classified as larvae, pupae or adults according to age [25], pupal eye color [26] and cuticle pigmentation[27] and divided into the following stages: the first (L1)-, the second (L2)-, the third (L3)-, the fourth (L4)- and the fifth (L5)-instar larvae, prepupal (PP) phase pupae, white-eyed (Pw), pink-eyed (Pp) and dark-eyed (Pd) pupae, 1-day post-emergence adult workers (A1), 15-day post-emergence adults (A15), virgin queen, queen, worker and drone. Individuals from each stage were collected from the hive. Larvae, pupae, A1, virgin queen, queen, worker and drone were frozen in liquid nitrogen and stored at −80℃. Adult workers were captured by marking newly emerged bees with paint on 1 day (A1 workers) and then collecting them either that day or 14 days later (A15 workers). Worker honeybees were divided into several groups of 30 each ($n = 30$), fed a diet of water and pollen and maintained in an incubator under the same conditions as described above. Bees incubated at 33℃ served as a control. Groups 1, 2, 3, and 4 were held at 4, 16, 25 and 42℃, respectively. Groups 5-8 were treated with H_2O_2 (2 mmol/L H_2O_2 added to food), UV (30 MJ/cm^2), $CdCl_2$ (3 mg/ml added to food) and $HgCl_2$ (3 mg/ml added to food), respectively. Groups 9-12 were treated with the pesticides paraquat, imidacloprid, cyhalothrin and phoxim, respectively, which were diluted to 20 mg/L and added to the food. The control group was fed normal food. At appropriate times after treatment, bees were flash-frozen in liquid nitrogen and stored at −80℃.

The brain, epidermis, muscle, haemolymph, hindgut, and midgut tissues of adults (15 days) were removed by dissection on ice, and the freshly collected tissue was then used to determine tissue-specific gene expression. At least three biological replicates were performed for each treatment.

Total RNA and DNA extraction

Total RNA was extracted with Trizol reagent (TransGen Biotech, Beijing, China) according to the instructions of the manufacturer. RNase-free DNase I (Promega, USA) was used to eliminate potential genomic DNA contamination. The quality of total RNA was assessed by electrophoresis on a 1% agarose gel and analyzed by a UV-visible protein nucleic acid analyzer (Thermo Scientific). The First-strand cDNA was synthesized with 2 μg of total RNA by reverse transcriptase (TransGen Biotech, Beijing, China) at 42℃ for 50 min using the oligo-d $(T)_{18}$ adaptor primer. Genomic DNA was isolated from larvae by using the EasyPure Genomic DNA Extraction Kit (TransGen Biotech, Beijing) in accordance with the instructions of the manufacturer.

Primers and PCR amplification conditions

Primers and PCR amplification conditions are listed in Table 1 and Table 2, respectively.

Table 1 Primers used in experiments

Abbreviation	Primer sequence (5'→ 3')	Description
MP1	ATGACTGTGGAAATCGATAAAGAAG	cDNA sequence primer, forward
MP2	TTATGATGTGATCTTCTTCTCTTCTC	cDNA sequence primer, reverse
3P1	CAGTGGTTGTGGTTGGC	3' RACE forward primer, outer
3P2	CAGCAGGAGAAGAGAAGAAGA	3' RACE forward primer, inner

Continued

Abbreviation	Primer sequence (5'→ 3')	Description
B26	GACTCTAGACGACATCGA(T)$_{18}$	Universal primer, primary
B25	GACTCTAGACGACATCGA	Universal primer, nested
5P1	TCTCTCTGTACCCTTTTCCTG	5' RACE reverse primer, outer
5P2	TGACGTGCCATTGTAGCG	5' RACE reverse primer, inner
AAP	GGCCACGCGTCGACTAGTAC(G)$_{16}$	Abridged anchor primer
AUAP	GGCCACGCGTCGACTAGTAC	Abridged universal amplification primer
FP1	GTGCCGCTTCTGTTCTCTT	Full length cDNA primer, forward
FP2	TTATGATGTGATCTTCTTCTCTTCTC	Full length cDNA primer, reverse
β-s	TTATATGCCAACACTGTCCTTT	Standard control primer, forward
β-x	AGAATTGATCCACCAATCCA	Standard control primer, reverse
G1	GTGCCGCTTCTGTTCTCTT	Genomic sequence primer, forward
G2	GCCAACCACAACCACTG	Genomic sequence primer, reverse
G3	CAGTGGTTGTGGTTGGC	Genomic sequence primer, forward
G4	TTATGATGTGATCTTCTTCTCTTCTC	Genomic sequence primer, reverse
QP1	AAGTATTAGATCAGGGACGAG	Real-time PCR primer, forward
QP2	TCTTCTTCTCTTCTCCTGCTGG	Real-time PCR primer, reverse

Table 2 PCR amplification conditions

Primer pair	Amplification conditions
MP1, MP2	5 min at 94℃, 40 s at 94℃, 40 s at 49℃, 40 s at 72℃ for 35 cycles, 5 min at 72℃
3P1, B26	5 min at 94℃, 40 s at 94℃, 40 s at 47℃, 40 s at 72℃ for 31 cycles, 5 min, at 72℃
3P2, B25	5 min at 94℃, 30 s at 94℃, 30 s at 50℃, 30 s at 72℃ for 35 cycles, 5 min at 72℃
5P1, AAP	5 min at 94℃, 40 s at 94℃, 40 s at 49℃, 40 s at 72℃ for 31 cycles, 5 min at 72℃
5P2, AUAP	5 min at 94℃, 40 s at 94℃, 30 s at 50℃, 30 s at 72℃ for 35 cycles, 5 min at 72℃
QP1, QP2	5 min at 94℃, 60 s at 94℃, 60 s at 47℃, 50 s at 72℃ for 35 cycles, 5 min at 72℃
GP1, GP2	5 min at 94℃, 30 s at 94℃, 30 s at 51℃, 30 s at 72℃ for 35 cycles, 5 min at 72℃
GP3, GP4	5 min at 94℃, 60 s at 94℃, 60 s at 47℃, 60 s at 72℃ for 35 cycles, 5 min at 72℃
β-s, β-x	5 min at 94℃, 30 s at 94℃, 30 s at 53℃, 30 s at 72℃ for 35 cycles, 5 min at 72℃

Isolation of the cDNA of *AccMsrB*

To isolate the internal conserved cDNA fragment of *AccMsrB*, primers MP1 and MP2 were designed based on the conserved regions of the MsrBs from *Apis mellifera*, *Apis florae*, *Apis dorsata*, and *Bombus terrestris*, which were obtained from NCBI GenBank, and synthesized for use (Biosune Biotechnological Company, Shanghai, China). PCR amplification was then performed. Based on the sequence of the cloned internal fragment of *AccMsrB*, the primers 5P1/5P2, AAP/AUAP, 3P1/ 3P2 and B26/B25 were designed and synthesized for the 5' rapid amplification of cDNA end (RACE) and 3' RACE. Subsequently, the full-length cDNA sequence was derived by using specific primers (FP1 and FP2) whose design was based on the assembled full-length cDNA sequence. All PCR products were cloned into the pEASY-T3 vector (TransGen Biotech, Beijing, China) and then transformed into competent *Escherichia coli* (*E. coli*) DH5α cells for sequencing. All primers used in this study are shown in Table 1, and all of the PCR amplification conditions are shown in Table 2.

Amplification of the genomic sequence of *AccMsrB*

To clone the genomic DNA of *AccMsrB*, three pairs of gene-specific primers (N1/N2, N3/N4 and N5/N6) were designed based on the full-length cDNA of *AccMsrB*, by using *A. cerana cerana* genomic DNA as the template. The PCR products were purified and cloned into the pEASY-T3 vector (TransGen Biotech, Beijing, China) and then transformed into competent *E. coli* DH5α cells for sequencing.

Bioinformatics analysis and phylogenetic tree construction

NCBI bioinformatics tools (https://blast.ncbi.nlm.nih.gov/Blast.cgihttp://blast.ncbi.nlm.nih.gov/Blas-t.cgi) were used to detect conserved domains in *AccMsrB*. The deduced amino acid sequence of *AccMsrB* was analyzed and predicted by using DNAMAN version 5.2.2 (Lynnon Biosoft, Quebec, Canada). DNAMAN software was also used to predict the theoretical isoelectric point and molecular weight of AccMsrB. The tertiary structure was predicted by using SWISS-MODEL (http://swiss-model.expasy.org/) [28]. A multiple sequence alignment of AccMsrB with three other insect sequences obtained from the NCBI database was also performed with DNAMAN version 5.2.2. A phylogenetic analysis was performed by using the neighbor-joining method with Molecular Evolutionary Genetics Analysis (MEGA version 5.1, free download from http://www.megasoftware.net/, accessed 8 March 2013) software.

AccMsrB transcriptional profiling by quantitative real-time PCR (qRT-PCR)

Specific primer pairs (QP1/QP2) were designed according to *AccMsrB* cDNA for use in quantitative real-time PCR (qRT-PCR) reactions. The gene coding for clam β-actin, which is equally expressed in organisms, was amplified by using *Acc-β-actin* (GenBank accession number: XM_640276) gene-specific primers (β-s and β-x) and was used as an invariant control. qRT-PCR was performed using the SYBR® PrimeScript™ RT-PCR Kit (TaKaRa Dalian, China) in a 25 μl volume by using a CFX96™ Real-time System (Bio-Rad) under the following thermal cycling profile: initial denaturation at 95℃ for 30 s, followed by 41 cycles of amplification (95℃ for 5 s, 55℃ for 15 s and 72℃ for 15 s) and a single melting cycle from 65℃ to 95℃. The C_t of each sample was used to calculate the ΔC_t values. The $2^{-\Delta\Delta C_t}$ method was used to calculate the different expression levels of *AccMsrB* [29]. At least three individual samples were prepared for each sample, and all samples were analyzed in triplicate.

Statistical analysis

Error bars in figures denote the SE of the mean (SEM) from three independent experiments. Significant differences were determined by Tukey's HSD test by using SPSS software (Version 16.0, SPSS Inc, Chicago, Illinois). A P value of 0.05 was used to determine statistical significance.

Results

Cloning and sequence analysis of *AccMsrB*

The full-length cDNA of *AccMsrB* (GenBank accession number: KP317814) was 757 bp with an ORF of 414 bp, a 138 bp 5' untranslated region and a 205 bp 3' untranslated region. The ORF of *AccMsrB* encodes a 137-amino-acid polypeptide with an estimated molecular mass of 15.5 kDa and theoretical isoelectric point of 7.77, which is consistent with other insect MsrBs. A multiple sequence alignment revealed high amino acid identity (74.3%-99.3%) between AccMsrB and MsrB sequences from other species, including *B. terrestris*, *M. rotundata* (F.), *A. dorsata* and *A. mellifera*.

Characterization and phylogenetic analysis of *AccMsrB*

Four conserved Cys residues in domains I and III (Fig. 1A) constituting a zinc-binding motif were identified [17, 30]. Moreover, the conserved Cys was found in the GCGWP domain (box II, Fig. 1A). The catalytic Cys in RXCXN near the C-terminus (box IV, Fig.1A) indicated AccMsrB to have high activity by using thioredoxins [31-33].

A phylogenetic tree of MsrBs from various species was constructed by using a neighbor-joining method implemented in MEGA 5.1 to analyze the possible evolutionary relationships among MsrBs. Phylogenetic protein sequence analysis revealed that AccMsrB was more closely related to AmMsrB than to homologs in other species (Fig. 1B), and this result was in agreement with the relationship predicted from the multiple sequence alignment.

To further clarify the catalytic mechanism of AccMsrB, the potential tertiary structure of AccMsrB was constructed by using SWISS-MODEL server, and the cysteines (Cys34, Cys46, Cys49, Cys64, Cys95, Cys98 and Cys118) of catalytic cores were identified (Fig. 1C).

Genomic structure analysis of AccMsrB

By cloning and assembling cDNA fragments and intron, we obtained the *AccMsrB* genomic sequence, which spans a region of 3,833 bp and contains three exons and two introns. All introns of *AccMsrB* had expected typical characteristics, such as a high A+T content (Table 3). The splice junctions had the canonical nucleotide sequences at the 5'-GT splice donor and 3'-AG splice acceptor sites. In addition, the exon-intron structure was fairly conserved among MsrBs from different species. In accordance with the phylogenetic analysis, the genomic structures of *AccMsrB* and *AfMsrB* were very similar (Fig. 2).

Table 3 The size of exons and introns of *AccMsrB* gene and GC content

No.	Exon		No.	Intron	
	Size (bp)	AT content (%)		Size (bp)	AT content (%)
1	1–86 (86)	58.14	1	87–728 (642)	78.82
2	729–901 (173)	61.85	2	902–3315 (2414)	81.48
3	3316–3470 (155)	58.06			

Expression profiles of *AccMsrB*

The relative expression level of *AccMsrB* mRNA among the bee's life stages (using stage terms defined above) was the highest ($P < 0.05$) in L1, decreased rapidly to L2, and gradually increased from L2 to PP, decreased rapidly at the pupal stage and reached a peak at A15 (Fig. 3A). The highest two expression levels were detected in L1 and PP, suggesting that *AccMsrB* might

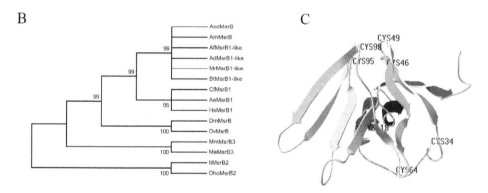

Fig. 1 Homology comparison and phylogenetic analysis of MsrBs and MsrB1-like proteins. A. Alignment of the deduced AccMsrB protein sequence with other known insect MsrBs. Amino acid sequences used in the analysis were from *Apis florea* (AfMsrB, GenBank accession number: XP003692728.1), *Apis mellifera* (AmMsrB, GenBank accession number: XP001120023.2), *Apis dorsata* (AdMsrB1-like, GenBank accession number: XP006612999.1), *Bombus terrestris* (BtMsrB1-like, GenBank accession number: XP002435561). Four putative MsrB signature motifs were boxed. B. Phylogenetic analysis of MsrBs from different species. The amino acid sequences of MsrBs were all obtained from NCBI as follows: *Apis mellifera* (GenBank accession number: XP001120023.2), *Apis florea* (GenBank accession number: XP003692728.1), *Apis dorsata* (GenBank accession number: XP006612999.1), *Megachile rotundata* (GenBank accession number: XP003704590.1), *Bombus terrestris* (GenBank accession number: XP_003398072.1), *Camponotus floridanus* (GenBank accession number: EFN65046.1), *Acromyrmex echinatior* (GenBank accession number: EGI61330.1), *Harpegnathos saltator* (GenBank accession number: EFN85865.1), *Drosophila melanogaster* (GenBank accession number: NP650030.1), *Drosophila virilis* (GenBank accession number: XP002053684.1), *Mus musculus* (GenBank accession number: NP796066.1), *Mesocricetus auratus* (GenBank accession number: XP_005081577.1), *Ictidomys tridecemlineatus* (GenBank accession number: XP005325873.1), *Opisthocomus hoazin* (GenBank accession number: P009936228.1).

play an essential role in the early developmental stages of the honeybee. Spatial expression profiles showed that *AccMsrB* transcripts were expressed at the highest level in the epidermis, followed by the muscle (Fig. 3B), suggesting that *AccMsrB* may play roles in the epidermis and muscle protecting against ROS. To determine whether sex had any effect on *AccMsrB* expression, the different bee type expressions of the *AccMsrB* gene were analyzed, and the

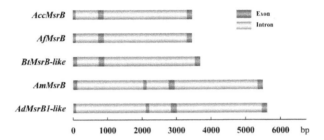

Fig. 2 Schematic representation of the DNA structures. Length of the exons and introns of genomic DNA of *Apis florea* (*AfMsrB*, GenBank accession number: NW003790158.1), *Apis dorsata* (*AdMsrB*, GenBank accession number: NW006263422), *Apis mellifera* (*AmMsrB*, GenBank accession number: NC007079.3), *Bombus terrestris* (*BtMsrB1-like*, GenBank accession number: NC015771.1) were indicated according to the scale below.

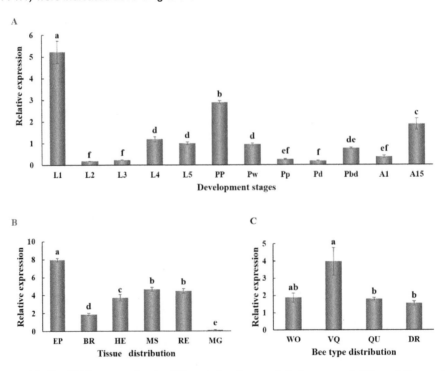

Fig. 3 *AccMsrB* mRNA expression in different developmental phases and different tissues in *Apis cerana cerana*. A. *AccMsrB* mRNA expression in different developmental phases measured by qRT-PCR. Vertical bars represented the mean ± SEM (n = 3). B. *AccMsrB* mRNA expression in different tissues measured by qPCR. Vertical bars represented the mean ± SEM (n = 3). C. *AccMsrB* mRNA expression in different bee types measured by qRT-PCR. Vertical bars represent the mean ± SEM (n = 3). Significant differences across mRNA expression are indicated with letters, etc, a, b, c, d. as determined by the Tukey-HSD test (P < 0.05).

greatest mRNA accumulation was found in females (virgin queens, followed by workers and queens), while the least accumulation was found in males (drones; Fig. 3C).

Expression profiles of *AccMsrB* under diverse environmental stresses

To verify whether the expression level of *AccMsrB* was influenced by abiotic stresses, total RNA was extracted from stage A15 (15-day-old adults) that had been exposed to different environmental stresses. *AccMsrB* expression was induced by 4, 16, 25 and 42 ℃ treatments (Fig. 4A-D), and the expression level was normalized to unstressed bees. At 4, 16 and 25 ℃, *AccMsrB* transcript was dramatically induced and peaked at 1, 1.5 and 0.5 h, respectively. At 42 ℃, *AccMsrB* transcript peaked at 1 and 3 h. *AccMsrB* expression was also clearly induced throughout the entire treatment course at 16 ℃ compared with the control group, in which *AccMsrB* expression decreased. The expression level of *AccMsrB* also increased when exposed to H_2O_2 (Fig. 4E), and the expression level decreased at 0.2 and 0.5 h, then was induced

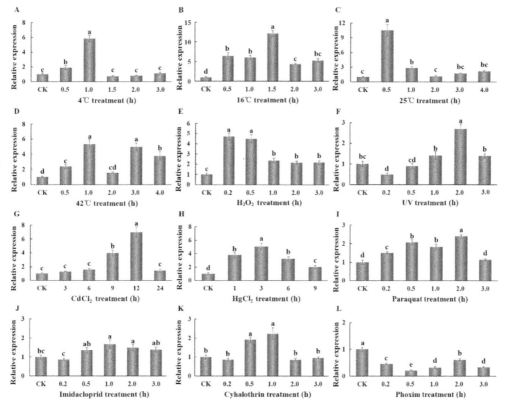

Fig. 4 Expression patterns of *AccMsrB* under conditions of different environmental stresses. *AccMsrB* expression was normalized to untreated adult worker bees housed at 33 ℃. The *β-actin* gene was used as an internal control. Panels A-D show *AccMsrB* expression in 15-day post-emergence adult workers housed at 4, 16, 25, and 42 ℃. Panels E-J show *AccMsrB* expression in 15-day post-emergence adult workers exposed to H_2O_2, ultraviolet (UV), $CdCl_2$, $HgCl_2$, paraquat, imidacloprid, cyhalothrin and phoxim treatments. Each value is given as the mean (SD) of three replicates. The letters on the bar represent a significant difference at $P < 0.05$, as determined by the Tukey-HSD test.

throughout the entire treatment course when exposed to UV (Fig. 4F). When exposed to heavy metals (CdCl$_2$, HgCl$_2$) or pesticides (paraquat, imidacloprid, cyhalothrin, and phoxim), the expression profiles of *AccMsrB* were all induced and reached their highest levels at 12, 3, 2, 1 and 1 h, respectively (Fig. 4G-K). However, *AccMsrB* expression was reduced when exposed to phoxim. Taken together, these results suggest that *AccMsrB* may play an important role in the response to various oxidative stresses.

Discussion

Previous studies have demonstrated that MsrA and MsrB exist ubiquitously in all eukaryotic and most prokaryotic organisms [2, 3, 9, 5, 34]. While MsrBs have recently been described in insects [35], research concerning the expression of *MsrB* due to the exposure to abiotic oxidative stress has rarely been reported in honeybees. Here, we report the cloning and characterization of the *AccMsrB* gene from *A. cerana cerana* for the first time and examine the response of *AccMsrB* to various abiotic stresses. Our results suggest that *AccMsrB* could protect honeybees against oxidative damage caused by ROS.

A multiple sequence alignment indicated that the amino acid sequence of *AccMsrB* shares high sequence similarity with MsrBs from other insects (Fig. 1A). Strictly conserved cysteines (Cys46, Cys49, Cys95, Cys98 Cys64 and Cys118) were found in the CXXC motifs, GCGWP domain and RXCXN fingerprint region, respectively [9, 17, 30, 36-38]. Interestingly, The seventh Cys (Cys34) site in the MsrB amino acid sequences was observed, which is in agreement with a few other MsrBs [39], and suggests that MsrBs in different insects potentially serve different functions. Further study of this phenomenon is required for a comprehensive analysis. The phylogenetic tree constructed in this study was in accordance with the amino acid sequence alignment, indicating that *AccMsrB* and *AmMsrB* are very closely related. All our results confirmed that AccMsrB belongs to the specific B-type MSRs and may play a crucial role as an antioxidant.

Because it is a barrier to external stressors, the epidermis serves an important role in resistance to oxidative stresses. Muscles suffer significant oxidative stress because of their high oxygen consumption, which leads to increased production of ROS. Interestingly, previous studies have found the *MsrA* gene is upregulated in human skin in response to UV radiation and hydrogen peroxide exposure [40]. But in Chinese honeybees the highest mRNA accumulation of *AccMsrA* was found in the head [16] while the *AccMsrB* gene showed the highest mRNA accumulation in the epidermis, followed by muscles, revealing that *AccMsrB*, not *AccMsrA*, is mainly responsible for clearing ROS from the epidermis [16].

Our analysis of *AccMsr* gene expression in the various developmental stages of *A. cerana cerana* showed that these genes were most expressed in L1, followed by PP (a stage on the verge of molting). As bees begin to pupate, their cells become highly differentiated and many organs mature. High expressions of *AccMsrB* at these stages suggests that this gene's production might play an important role in the transition from one development stage to the next. The third highest expression level was found in A15. This is the time that nurse bees change into foraging bees, when they may encounter UV, heat and other sources of oxidative stress [41].

Temperature has been shown to be a dominant abiotic cause of physiological changes in organisms. High and low temperature can increase oxygen consumption and ROS [42-44]. When Chinese honeybees were held at 42°C, H_2O_2 concentration increased dramatically [24], and *MsrBs* were also correlated with H_2O_2 [45]. In our study, the transcription levels of *AccMsrB* in honeybees were upregulated by exposure to different temperatures (4, 16, 25, 42°C), indicating that *AccMsrB* was involved in the ROS response caused by cold and heat, which is similar with other *MsrB* genes [20].

A decrease in *MsrB* gene expression has been found to make cells or conidiospores more sensitive to H_2O_2 or UV [45, 46], and H_2O_2 and UV can, in turn, increase ROS [15, 41]. The methione-rich protein can be readily oxidized by ROS, leading to the upregulation of the expression of *MsrB* to reduce the MetO species [47]. H_2O_2 concentration *in vivo* increased dramatically when Chinese honeybees were exposed to UV radiation [24], and the expression of *AccMsrB* was upregulated by exposure to UV, results similar to those found in *Aspergillus niddulans* [46], which suggests that *AccMsrB* plays a key role in scavenging ROS produced by H_2O_2 and UV.

Metal ions are well-known to induce oxidative stress by stimulating ROS production, and chlorate is a strong oxidizing agent toxic to cells due to its conversion to chlorite [48]. Several studies have found that yeast growth is affected by Cd and the degree of the effect is related to the action of MsrB on ROS [49, 50]. In our study, *AccMsrB* expression was enhanced by cadmium chloride and mercury chloride, similar to results found in *Enterococcus faecalis*, in which the *MsrB* gene was upregulated in cells exposed to cadmium or mercury [51]. This suggests that *AccMsrB* is a major antioxidant gene responsible for honeybee tolerance of $CdCl_2$ and $HgCl_2$.

Several studies have described an increase in oxidative stress from exposure to insecticides [52-54]. Some insecticides increase the concentration of H_2O_2 [24]. The herbicide paraquat, which can generate ROS directly and thus induce oxidative stress, has been found to enhance the expression of the *MsrB* gene in *Aspergillus nidulans* [55, 56]. In our study, all the pesticides tested except phoxim induced the expression of *AccMsrB*, demonstrating that *MsrB* genes play an important role in the response to ROS induced by insecticide or herbicide exposure. Interestingly, the transcripts of *AccMsrB* clearly declined compared to the control group after exposure to phoxim. This might have occurred because some other antioxidant genes responded to the presence of phoxim, thereby reducing the need for *AccMsrB* transcripts in the antioxidant process [23, 57].

The expression of the *AccMsrB* transcript in response to a variety of environmental stresses provides solid evidence for the importance of AccMsrB in protecting *A. cerana cerana* against oxidative stress. All of these results indicate that *AccMsrB* may be involved in the response to environmental stresses and may play a central role in repairing cell damage caused by oxidative stress. However, how MsrB cooperates with other genes, for example *AccMsrA* and *AccTrx*, is unknown, and further examination of the mechanisms involved in honeybee gene regulation is needed.

References

[1] Zhang, C., Jia, P. P., Jia, Y. Y., Weissbach, H., Webster, K. A., Huang, X. P., Lemanski, S. L., Achary, M., Lemanski, L. F. Methionine sulfoxide reductase A (MsrA) protects cultured mouse embryonic stem cells from H_2O_2-mediated oxidative stress. J Cell Biochem. 2010, 111, 94-103.

[2] Moskovitz, J., Poston, J. M., Berlett, B. S., Nosworthy, N. J., Szczepanowski, R., Stadtman, E. R. Identification and characterization of a putative active site for peptide methionine sulfoxide reductase (MsrA) and its substrate stereospecificity. J Biol Chem. 2000, 275, 14167-14172.

[3] Grimaud, R., Ezraty, B., Mitchell, J. K., Lafitte, D., Briand, C., Derrick, P. J., Barras, F. Repair of oxidized proteins identification of a new methionine sulfoxide reductase. J Biol Chem. 2001, 276, 48915-48920.

[4] Kim, H. Y., Gladyshev, V. N. Methionine sulfoxide reduction in mammals: characterization of methionine-R-sulfoxide reductases. Mol Biol Cell. 2004, 15, 1055-1064.

[5] Weissbach, H., Etienne, F., Hoshi, T., Heinemann, S. H., Lowther, W. T., Matthews, B., St John, G., Nathan, C., Brot, N. Peptide methionine sulfoxide reductase: structure, mechanism of action, and biological function. Arch Biochem Biophys. 2002, 397, 172-178.

[6] Olry, A., Boschi-Muller, S., Marraud, M., Sanglier-Cianferani, S., Van Dorsselear, A., Branlant, G. Characterization of the methionine sulfoxide reductase activities of PILB, a probable virulence factor from *Neisseria meningitidis*. J Biol Chem. 2002, 277, 12016-12022.

[7] Etienne, F., Spector, D., Brot, N., Weissbach, H. A methionine sulfoxide reductase in *Escherichia coli* that reduces the R enantiomer of methionine sulfoxide. Biochem Biophys Res Commun. 2003, 300, 378-382.

[8] Kim, H. Y., Gladyshev, V. N. Characterization of mouse endoplasmic reticulum methionine-R-sulfoxide reductase. Biochem Biophys Res Commun. 2004, 320, 1277-1283.

[9] Kryukov, G. V., Kumar, R. A., Koc, A., Sun, Z., Gladyshev, V. N. Selenoprotein R is a zinc-containing stereo-specific methionine sulfoxide reductase. Proc Natl Acad Sci USA. 2002, 99, 4245-4250.

[10] Hansel, A., Heinemann, S. H., Hoshi, T. Heterogeneity and function of mammalian MSRs: enzymes for repair, protection and regulation. Biochim Biophys Acta 2005, 1703, 239-247.

[11] Lee, B. C., Dikiy, A., Kim, H. Y., Gladyshev, V. N. Functions and evolution of selenoprotein methionine sulfoxide reductases. Biochim Biophys Acta 2009, 1790, 1471-1477.

[12] Brot, N., Weissbach, L., Werth, J., Weissbach, H. Enzymatic reduction of protein-bound methionine sulfoxide. Proc Natl Acad Sci USA. 1981, 78, 2155-2158.

[13] Zhang, Y., Gladyshev, V. N. Comparative genomics of trace element dependence in biology. J Biol Chem. 2011, 286, 23623-23629.

[14] Kim, H. Y., Gladyshev, V. N. Different catalytic mechanisms in mammalian selenocysteine-and cysteine-containing methionine-R-sulfoxide reductases. PLoS Biology. 2005, 3, e375.

[15] Lushchak, V. I. Environmentally induced oxidative stress in aquatic animals. Aquat Toxicol. 2011, 101, 13-30.

[16] Gong, Z., Guo, X., Xu, B. Molecular cloning, characterisation and expression of methionine sulfoxide reductase A gene from *Apis cerana cerana*. Apidologie. 2011, 43, 182-194.

[17] Kumar, R. A., Koc, A., Cerny, R. L., Gladyshev, V. N. Reaction mechanism, evolutionary analysis, and role of zinc in *Drosophila* methionine-R-sulfoxide reductase. J Biol Chem. 2002, 277, 37527-37535.

[18] Ruan, H., Tang, X. D., Chen, M. L., et al. High-quality life extension by the enzyme peptide methionine sulfoxide reductase. Proc Natl Acad Sci USA. 2002, 99, 2748-2753.

[19] Koc, A., Gasch, A. P., Rutherford, J. C., Kim, H. Y., Gladyshev, V. N. Methionine sulfoxide reductase regulation of yeast lifespan reveals reactive oxygen species-dependent and -independent components of aging. Proc Natl Acad Sci USA. 2004, 101, 7999-8004.

[20] Lim, D. H., Han, J. Y., Kim, J. R., Lee, Y. S., Kim, H. Y. Methionine sulfoxide reductase B in the endoplasmic reticulum is critical for stress resistance and aging in *Drosophila*. Biochem Biophys Res

Commun. 2012, 419, 20-26.

[21] Jia, Y., Zhou, J., Liu, H., Huang, K. Effect of methionine sulfoxide reductase B1 (SelR) gene silencing on peroxynitrite-induced F-actin disruption in human lens epithelial cells. Biochem Biophys Res Commun. 2014, 443, 876-881.

[22] Xu, P., Shi, M., Chen, X. X. Antimicrobial peptide evolution in the Asiatic honey bee *Apis cerana*. PLoS One. 2009, 4 (1), e4239.

[23] Yao, P. B., Hao, L. L., Wang, F., Chen, X. B., Yan, Y., Guo, X. Q., Xu, B. H. Molecular cloning, expression and antioxidant characterisation of a typical thioredoxin gene (*AccTrx2*) in *Apis cerana cerana*. Gene. 2013, 527, 33-41.

[24] Yan, Y., Zhang, Y. Y., Huaxia, Y. F., Wang, X. L., Yao, P. B., Guo, X. Q., Xu, B. H. Identification and characterisation of a novel 1-Cys thioredoxin peroxidase gene (*AccTpx5*) from *Apis cerana cerana*. Comp Biochem Physiol B: Biochem Mol Biol. 2014, 172, 39-48.

[25] Robinson, G. E. Regulation of division of labor in insect societies. Annu Rev Entomol. 1992, 37, 637-665.

[26] Kerr, W. E., Laidlaw Jr, H. H. General genetics of bees. Adv Genet. 1956, 8, 109-153.

[27] Ziegler, I. Genetic aspects of ommochrome and pterin pigments. Adv Genet. 1961, 10, 349-403.

[28] Benkert, P., Biasini, M., Schwede, T. Toward the estimation of the absolute quality of individual protein structure models. Bioinformatics. 2011, 27, 343-350.

[29] Schmittgen, T. D., Livak, K. J. Analyzing real-time PCR data by the comparative CT method. Nat Protoc. 2008, 3, 1101-1108.

[30] Olry, A., Boschi-Muller, S., Yu, H., Burnel, D., Branlant, G. Insights into the role of the metal binding site in methionine-*R*-sulfoxide reductases B. Protein Sci. 2005, 14, 2828-2837.

[31] Dos Santos, C. V., Cuiné, S., Rouhier, N., Rey, P. The *Arabidopsis plastidic* methionine sulfoxide reductase B proteins. Sequence and activity characteristics, comparison of the expression with plastidic methionine sulfoxide reductase A, and induction by photooxidative stress. Plant Physiol. 2005, 138, 909-922.

[32] Rouhier, N., Dos Santos, C. V., Tarrago, L., Rey, P. Plant methionine sulfoxide reductase A and B multigenic families. Photosynth Res. 2006, 89, 247-262.

[33] Sagher, D., Brunell, D., Hejtmancik, J. F., Kantorow, M., Brot, N., Weissbach, H. Thionein can serve as a reducing agent for the methionine sulfoxide reductases. Proc Nati Acad Sci USA. 2006, 103, 8656-8661.

[34] Ezraty, B., Aussel, L., Barras, F. Methionine sulfoxide reductases in prokaryotes. BBA-Proteins Proteom. 2005, 1703, 221-229.

[35] Lim, D. H., Han, J. Y., Kim, J. R., Lee, Y. S., Kim, H. Y. Methionine sulfoxide reductase B in the endoplasmic reticulum is critical for stress resistance and aging in *Drosophila*. Biochem Biophys Res Commun. 2012, 419, 20-26.

[36] Rouhier, N., Jacquot, J. P. The plant multigenic family of thiol peroxidases. Free Radical Biology and Medicine. 2005, 38, 1413-1421.

[37] Ding, D., Sagher, D., Laugier, E., Rey, P., Weissbach, H., Zhang, X. H. Studies on the reducing systems for plant and animal thioredoxin-independent methionine sulfoxide reductases B. Biochem Biophys Res Commun. 2007, 361, 629-633.

[38] Rouhier, N., Kauffmann, B., Tete-Favier, F., Palladino, P., Gans, P., Branlant, G., Jacquot, J. P., Boschi-Muller, S. Functional and structural aspects of poplar cytosolic and plastidial type a methionine sulfoxide reductases. J Biol Chem. 2007, 282, 3367-3378.

[39] Oke, T. T., Moskovitz, J., Williams, D. L. Characterization of the methionine sulfoxide reductases of *Schistosoma mansoni*. J Parasitol. 2009, 95, 1421-1428.

[40] Ogawa, F., Sander, C. S., Hansel, A., Oehrl, W., Kasperczyk, H., Elsner, P., Shimizu, K., Heinemann, S. H., Thiele, J. J. The repair enzyme peptide methionine-*S*-sulfoxide reductase is expressed in human epidermis and upregulated by UVA radiation. J Invest Dermatol. 2006, 126, 1128-1134.

[41] Kottuparambil, S., Shin, W., Brown, M. T., Han, T. UV-B affects photosynthesis, ROS production

and motility of the freshwater flagellate, *Euglena agilis* Carter. Aquat Toxicol. 2012, 122, 206-213.

[42] Malek, R. L., Sajadi, H., Abraham, J., Grundy, M. A., Gerhard, G. S. The effects of temperature reduction on gene expression and oxidative stress in skeletal muscle from adult zebrafish. Com Biochem Physiol C: Toxico Pharm. 2004, 138, 363-373.

[43] Bagnyukova, T. V., Danyliv, S. I., Zin'ko, O. S., Lushchak, V. I. Heat shock induces oxidative stress in rotan *Perccottus glenii* tissues. J Therm Biol. 2007, 32, 255-260.

[44] Bocchetti, R., Lamberti, C. V., Pisanelli, B., et al. Seasonal variations of exposure biomarkers, oxidative stress responses and cell damage in the clams, *Tapes philippinarum*, and mussels, *Mytilus galloprovincialis*, from Adriatic sea. Mar Environ Res. 2008, 66, 24-26.

[45] Jia, Y., Li, Y., Du, S., Huang, K. Involvement of MsrB1 in the regulation of redox balance and inhibition of peroxynitrite-induced apoptosis in human lens epithelial cells. Exp Eye Res. 2012, 100, 7-16.

[46] Soriani, F. M., Kress, M. R., De Gouvêa, P. F., Malavazi, I., Savoldi, M., Gallmetzer, A., Strauss, J., Goldman, M. H., Goldman, G. H. Functional characterization of the *Aspergillus nidulans* methionine sulfoxide reductases (*msrA* and *msrB*). Fungal Genet and Biol. 2009, 46, 410-417.

[47] Liang, X., Kaya, A., Zhang, Y., Le, D. T., Hua, D., Gladyshev, V. N. Characterization of methionine oxidation and methionine sulfoxide reduction using methionine-rich cysteine-free proteins. BMC Biochem. 2012, 13, 21.

[48] Cove, D. Chlorate toxicity in *Aspergillus nidulans*: the selection and characterization of chlorate resistant mutants. Heredity. 1976, 36, 191-203.

[49] Jo, H., Cho, Y. W., Ji, S. Y., Kang, G. Y., Lim, C. J. Protective roles of methionine-*R*-sulfoxide reductase against stresses in *Schizosaccharomyces pombe*. J Basic Microb. 2014, 54, 72-80.

[50] Lim, C. J., Jo, H., Kim, K. A protective role of methionine-*R*-sulfoxide reductase against cadmium in *Schizosaccharomyces pombe*. J Microbiol. 2014, 52 (11), 976-981.

[51] Laplace, J., Hartke, A., Giard, J., Auffray, Y. Cloning, characterization and expression of an *Enterococcus faecalis* gene responsive to heavy metals. Appl Microbiol Biot. 2000, 53, 685-689.

[52] Piner, P., Sevgiler, Y., Üner, N. *In vivo* effects of fenthion on oxidative processes by the modulation of glutathione metabolism in the brain of *Oreochromis niloticus*. Environ toxicol. 2007, 22, 605-612.

[53] Monteiro, D. A., Rantin, F. T., Kalinin, A. L. The effects of selenium on oxidative stress biomarkers in the freshwater characid fish matrinxã, *Brycon cephalus* (Gunther, 1869) exposed to organophosphate insecticide Folisuper 600 BR® (methyl parathion). Comp Biochem Physiol C: Toxico Pharm. 2009, 149, 40-49.

[54] Thomaz, J. M., Martins, N. D., Monteiro, D. A., Rantin, F. T., Kalinin, A. L. Cardio-respiratory function and oxidative stress biomarkers in Nile tilapia exposed to the organophosphate insecticide trichlorfon (NEGUVON®). Ecotox Environ Safe. 2009, 72, 1413-1424.

[55] Hook, S. E., Skillman, A. D., Small, J. A., Schultz, I. R. Gene expression patterns in rainbow trout, *Oncorhynchus mykiss*, exposed to a suite of model toxicants. Aquat Toxicol. 2006, 77, 372-385.

[56] Parvez, S., Raisuddin, S. Effects of paraquat on the freshwater fish *Channa punctata* (Bloch): non-enzymatic antioxidants as biomarkers of exposure. Arch Environ Con Tox. 2006, 50, 392-397.

[57] Yao, P., Lu, W., Meng, F., Wang, X., Xu, B., Guo, X. Molecular cloning, expression and oxidative stress response of a mitochondrial thioredoxin peroxidase gene (*AccTpx-3*) from *Apis cerana cerana*. J Insect Physiol. 2013, 59, 273-282.

2.16 Molecular Cloning, Characterization and Expression of Methionine Sulfoxide Reductase A Gene from *Apis cerana cerana*

Zhihong Gong, Xingqi Guo, Baohua Xu

Abstract: Methionine sulfoxide reductases (Msrs) catalyse the reduction of methionine sulfoxide to methionine and play key roles in protein repair and reactive oxygen species scavenging. Here, an *MsrA* gene designated *AccMsrA* was isolated from *Apis cerana cerana* for the first time. The full-length cDNA of *AccMsrA* is 1,540 bp, and it encodes a protein with 217 amino acids. Sequence alignment analysis showed that AccMsrA shares high similarity with other insect MsrAs. Analysis of the 5′-flanking region of *AccMsrA* revealed a group of transcription factor-binding sites that are associated with the regulation of development and responses to environmental stresses. Quantitative real-time PCR showed high expression levels of *AccMsrA* mRNA in prepupae and in the head of adult workers. Furthermore, the expression of *AccMsrA* was upregulated by multiple oxidative stresses, including ultraviolet light (30 MJ/cm^2), heat (42℃) and H_2O_2 (2 mmol/L). These results indicate that *AccMsrA* might fulfill an important role in the regulation of insect development and in their responses to various environmental stresses. This report is the first description of the characteristics of the *AccMsrA* gene in the Chinese honeybee, and it establishes a primary foundation for further study.

Keywords: Methionine sulfoxide reductase A; *Apis cerana cerana*; cDNA; Genomic DNA; Quantitative real-time PCR

Introduction

Organisms generate a large amount of reactive oxygen species (ROS) when they are challenged with external environmental stresses. ROS generated by stimulation lowers the cellular antioxidant capacity, alters protein function and damages proteins [1-4]. Methionine is one of the most oxidation-sensitive amino acids. Methionine can be easily oxidized to methionine sulfoxide (MetO) by ROS, and the resulting MetO can be reduced back to methionine by antioxidant enzymes. Methionine sulfoxide-*S*-reductase (MsrA) is one such antioxidant enzyme that can act not only as a repair enzyme, but also as an indirect scavenger of ROS by reducing the amount of MetO [5].

The first MsrA enzyme was identified in *Escherichia coli* and was shown to convert MetO to Met [6]. Subsequent studies have clearly elucidated the importance of MsrAs in protecting cells against oxidative damage. For instance, deletion or silencing of the *MsrA* gene in yeast or animal cells renders the cells more sensitive to oxidative stress [7, 8]. In human epidermal melanocytes, a small interfering RNA (siRNA) specific for *MsrA* was used to suppress its expression, and this decreased *MsrA* expression led to an increased sensitivity to H_2O_2 oxidative stress, resulting in more cell death [8]. *In vivo*, mice lacking MsrA accumulate high levels of oxidized protein under 100% oxygen treatment and exhibit sensitivity to oxidative stress [9]. Moreover, the overexpression of the *MsrA* gene in animal cells improves resistance to oxidative stresses [1, 10]. In PC-12 cells, the overexpression of bovine *MsrA* protected neuronal cells from hypoxia/reoxygenation injury, causing diminished ROS accumulation and facilitating cell survival under hypoxia treatment [10]. Furthermore, fruit flies overexpressing the *MsrA* gene show significant resistance to paraquat-induced oxidative stress and a markedly extended lifespan [11].

Previous studies of MsrAs have mainly focused on antioxidant defence and modulation of the aging process and little information about the role of MsrA in bees has been available until now. Honeybees are pollinating insects that play an important role in maintaining ecological balance and increasing crop yields. When collecting pollen, honeybees encounter diverse environmental stresses, such as sunlight (UV radiation) and high temperatures, that cause the formation of ROS [12, 13], which inevitably results in oxidative damage. In this study, we isolated *AccMsrA*, an *MsrA* gene from *Apis cerana cerana* encoding a putative methionine sulfoxide-*S*-reductase, and we identified a series of transcription factor-binding sites related to tissue development and stress-responses. *AccMsrA* mRNA was detected in different developmental stages and tissues by quantitative real-time PCR (qRT-PCR), and a high level of expression was observed in the fifth-instar larvae and in the head of adult workers. Furthermore, the accumulation of *AccMsrA* mRNA was induced by UV-light, heat and H_2O_2. Together, these results indicate that *AccMsrA* might play an important role in the regulation of insect development and responses to various environmental stresses.

Materials and Methods

Animals and treatments

Chinese honeybees (*A. cerana cerana*) maintained at Shandong Agricultural University, China, were used in the study. The colonies were fed in incubators and were kept at a constant temperature (33°C) and humidity (80%). The entire body of the second-instar larvae (L2), the fourth-instar larvae (L4), prepupal (PP) phase pupae, earlier pupal phase (Pw) pupae, pink (Pp) phase pupae and dark (Pd) phase pupae were taken from the hive, and the adult workers (1-day-old) were collected at the entrance of the hive when returning to the colony after foraging [14]. Larvae, pupae and 1-day-old adults and the head, thorax, abdomen and midgut which were dissected from the adults were frozen in liquid nitrogen and stored at −80°C. The 1-day-old adults were divided into 4 groups (n=10-15). Group 1 was treated with 30 MJ/cm^2 of UV light. Group 2 was subjected to 42°C heat. Group 3 was treated with 2 mmol/L H_2O_2.

The control bees in group 4 were fed a pollen and sucrose solution exclusively. Three bees were harvested at the appropriate times in each condition and stored at −80 °C until used.

RNA and DNA extraction

Total RNA was extracted with Trizol reagent (Invitrogen, USA) according to the instructions of the manufacturer. RNase-free DNase I (Promega, USA) was applied to eliminate potential genomic DNA contamination. The first-strand cDNA was synthesised with 2 μg of total RNA by reverse transcriptase (TransGen biotechnology, Beijing) at 42 °C for 50 min by using the oligo-d(T)$_{18}$-adaptor primer. Genomic DNA was isolated from larvae by using the EasyPure Genomic DNA Extraction Kit (TransGen biotechnology, Beijing) according to the protocol of the manufacturer.

Isolation of the cDNA of *AccMsrA*

To isolate the internal conserved cDNA fragment of *AccMsrA*, primers MP1 and MP2 were designed and synthesised (Shanghai Sangon Biotechnological Company, China) based on amino acid and nucleotide sequences that are conserved among *Aedes aegypti*, *Culex quinquefasciatus*, and *Apis mellifera*, which were obtained from NCBI GenBank. Then PCR was performed. Based on the sequence of the cloned internal fragment, the 5′ and 3′ end of the mRNA were cloned by rapid amplification of cDNA end (RACE) methodology by using gene-specific primers. In the 3′ RACE-PCR, the primers 3P1 and B26 were utilized in the primary PCR reaction, and the nested PCR was performed with the primers 3P2 and B25 by using the primary PCR products as the template. In the 5′ RACE-PCR, the first-strand cDNA was purified with DNA Clean-Up System (Promega, USA) and then polyadenylated at its 5′ end with dGTP by terminal deoxynucleotidyl transferase (TaKaRa, Japan) according to the instructions of the manufacturer. PCR was initially performed with the primers 5P1 and AUAP. Subsequently, the primary PCR products were used as the template and the primers 5P2 and AAP were chosen to amplify the related sequence. All PCR products were cloned into the pMD18-T vector (TaKaRa, Japan) and then sequenced. All primers used in the present study are shown in Table 1, and all of the PCR amplification conditions are shown in Table 2.

Table 1 Primers used in this study

Primer	Primer sequence (5′→3′)	Description
For cloning the full-length cDNA		
MP1	GGAATGGGATGTTTCTGG	conserved fragment primer, forward
MP2	CGAAGTCGATATTTCTGATG	conserved fragment primer, reverse
3P1	CACGTGAACAAGAGCAATACAAAC	3′-RACE forward primer, primary
3P2	GTTTTATCTAGCAGAAGATTATCATC	3′-RACE forward primer, nested
B26	GACTCTAGACGACATCGA(T)$_{18}$	Universal primer, primary
B25	GACTCTAGACGACATCGA	Universal primer, nested
5P1	CAAGTTCTAATTACACCGGG	5′-RACE reverse primer, primary

Continued

Primer	Primer sequence (5'→3')	Description
5P2	CGCCAAATAAACAATCACCGGC	5'-RACE reverse primer, nested
AAP	GGCCACGCGTCGACTAGTAC(G)$_{16}$	Abridged anchor primer
AUAP	GGCCACGCGTCGACTAGTAC	Abridged universal amplification primer
FP1	CTTTCAGTCACCAACGAAATATC	Full-length cDNA primer, forward
FP2	GCTCAAGACTTATTTGCGAGAG	Full-length cDNA primer, reverse
For cloning the genomic and 5'-flanking region		
G1	CTTTCAGTCACCAACGAAATATC	Genomic sequence primer, forward
G2	CGCCAAATAAACAATCACC	Genomic sequence primer, reverse
G3	GGTGATTGTTTATTTGGCG	Genomic sequence primer, forward
G4	TTTGTATTGCTCTTGTTCACG	Genomic sequence primer, reverse
G5	CGTGAACAAGAGCAATACAAA	Genomic sequence primer, forward
G6	GCTCAAGACTTATTTGCGAGAG	Genomic sequence primer, reverse
PP1	TGTTTTTTGATTACTTCGCA	I-PCR forward primer, outer
PP2	CCAGAAACATCCCATTCC	I-PCR reverse primer, outer
PP3	CTGCGTTATTAAATACGCACGC	I-PCR forward primer, inner
PP4	GATATTTCGTTGGTGACTGAAAG	I-PCR reverse primer, inner
Primers used for Quantitative real-time PCR and protein expression assays		
QP1	GGTGATTGTTTATTTGGCG	Real-time PCR primer, forward
QP2	TTTGTATTGCTCTTGTTCACG	Real-time PCR primer, reverse
β-action-u	GTTTTCCCATCTATCGTCGG	Standard control primer, forward
β-action-d	TTTTCTCCATATCATCCCAG	Standard control primer, reverse
Yup	AGTACTCTTTCAGTCACCAACGAAATATC	Protein expression primer, forward
Ydown	GGTACCGCTCAAGACTTATTTGCGAGAG	Protein expression primer, reverse

Table 2 PCR amplification conditions

Primer pair	Amplification conditions
MP1, MP2	5 min at 94℃, 40 s at 94℃, 40 s at 49℃, 40 s at 72℃ for 35 cycles, 5 min at 72℃
3P1, B26	5 min at 94℃, 40 s at 94℃, 40 s at 47℃, 50 s at 72℃ for 31 cycles, 5 min, at 72℃
3P2, B25	5 min at 94℃, 30 s at 94℃, 30 s at 48℃, 50 s at 72℃ for 35 cycles, 5 min at 72℃
5P1, AAP	5 min at 94℃, 40 s at 94℃, 40 s at 49℃, 40 s at 72℃ for 31 cycles, 5 min at 72℃
5P2, AUAP	5 min at 94℃, 30 s at 94℃, 30 s at 49℃, 30 s at 72℃ for 35 cycles, 5 min at 72℃
FP1, FP2	5 min at 94℃, 60 s at 94℃, 60 s at 47℃, 50 s at 72℃ for 35 cycles, 5 min at 72℃
G1, G2	5 min at 94℃, 30 s at 94℃, 30 s at 51℃, 60 s at 72℃ for 35 cycles, 5 min at 72℃
G3, G4	5 min at 94℃, 60 s at 94℃, 60 s at 47℃, 60 s at 72℃ for 35 cycles, 5 min at 72℃
G5, G6	5 min at 94℃, 60 s at 94℃, 60 s at 47℃, 60 s at 72℃ for 35 cycles, 5 min at 72℃
PP1, PP2	10 min at 94℃, 50 s at 94℃, 75 s at 50℃, 2 min at 72℃ for 30 cycles, 5 min at 72℃
PP3, PP4	10 min at 94℃, 50 s at 94℃, 75 s at 51℃, 2 min at 72℃ for 30 cycles, 5 min at 72℃
β-action-u, β-action-d	5 min at 94℃, 30 s at 94℃, 15 s at 50℃, 30 s at 72℃ for 35 cycles, 5 min at 72℃
QP1, QP2	5 min at 94℃, 40 s at 94℃, 15 s at 50℃, 40 s at 72℃ for 35 cycles, 5 min at 72℃
Yup, Ydown	5 min at 94℃, 60 s at 94℃, 60 s at 47℃, 50 s at 72℃ for 35 cycles, 5 min at 72℃

Amplification of the genomic sequence of *AccMsrA* and its 5′-flanking region

To clone the genomic DNA of *AccMsrA*, four gene-specific primers were designed based on the sequence of the coding region of *AccMsrA*. Genomic DNA from *A. cerana cerana* was used as the template, and the primers G1/G2, G3/G4 and G5/G6 were utilized to identify three genomic DNA fragments. The PCR products were purified and cloned into the vector pMD18-T (TaKaRa, Japan) and then sequenced. To amplify the 5′-flanking region of *AccMsrA*, inverse polymerase chain reaction (I-PCR) was used. Total genomic DNA from *A. cerana cerana* was extracted, completely digested with the restriction endonuclease *Eco*R I and then self-ligated to form plasmids by using T4 DNA ligase (TaKaRa, Dalian, China). Based on the *AccMsrA* genomic sequence, the specific primers PP1/PP2 and PP3/PP4 were designed and used in the I-PCR to clone the 5′-flanking region as described by Liu [15]. The PCR products were purified and cloned into pMD18-T and then sequenced. Additionally, the MatInspector database (http://www.cbrc.jp/research/db/ TFSEARCH.html) was used to search for transcription factor-binding sites in the 5′-flanking region.

Sequence analysis and phylogenetic construction

The deduced amino acid sequence of AccMsrA was analyzed and predicted by using DNAMAN version 5.2.2 (Lynnon Biosoft Company). A multiple sequence alignment of AccMsrA with three other insects obtained from the NCBI database was also performed with DNAMAN version 5.2.2. A phylogenetic tree was generated with eleven MsrAs from invertebrates and plants obtained from the NCBI database by MEGA version 4.

Expression analysis of *AccMsrA* by real-time quantitative PCR

Specific primer pairs (QP1/QP2) were designed according to the cDNA of *AccMsrA* for use in quantitative real-time PCR (qRT-PCR) reactions. The house-keeping gene *β-actin*, which is equally expressed in organisms, was used as a control. Reactions were performed by using the SYBR® PrimeScript™ RT-PCR Kit (TaKaRa Code: DRR063A) on a CFX96™ Real-time System (Bio-Rad) under the following thermal cycling profile: 95℃ for 30 s, followed by 40 cycles of amplification (95℃ for 10 s, 50℃ for 15 s and 72℃ for 15 s) and a melting cycle from 65℃ to 95℃. The C_t of each sample was used to calculate the ΔC_t values. The $2^{-\Delta\Delta C_t}$ method was used to calculate the different expression levels of *AccMsrA*. The different transcription levels of all samples and significant differences among samples were verified by Statistical Analysis System (SAS) version 9.1.

Results

Cloning and sequence analysis of *AccMsrA*

Based on the conserved region of insect *MsrA* genes, primers MP1 and MP2 were designed to amplify the middle region of *AccMsrA*. A fragment of about 600 bp was obtained, and then

RACE-PCR was performed according to the internal sequence. The deduced full-length cDNA of *AccMsrA* (GenBank accession no. HQ219724) is 1,540 bp, and it contains a 136 bp 5′ untranslated region (UTR) and a 753 bp 3′ UTR. The cDNA harbours a 651 bp open reading frame (ORF) that encodes a protein of 217 amino acids. The AccMsrA protein has a theoretical molecular mass of 25.0 kDa and a predicted isoelectric point of 6.78. Compared with MsrA sequences from other insects, *AccMsrA* exhibits higher A+T content (72.6%), which is consistent with the genomic characteristics of *A. mellifera* [16].

Characterisation and phylogenetic analysis of AccMsrA

A multiple sequence alignment showed that AccMsrA shares 49.6%, 45.8% and 51.3% identity with the MsrAs from *Aedes aegypti*, *Drosophila erecta* and *Culex quinquefasciatus*, respectively (Fig. 1). The highly conserved motif GCFW, a fingerprint of MsrAs, was found in the N-terminus. A cysteine residue was identified in the conserved GCFW region that is involved in the catalytic mechanism. Moreover, the conserved Cys site in the MsrA amino acid sequences of insects was observed in C-terminus end, and it is distinct from previous studies.

Fig. 1 Alignment of the deduced AccMsrA protein sequence with other known insect MsrAs. The amino acid sequences used in the study are from *Aedes aegypti* (AaMsrA, GenBank accession no. XP001649797), *Culex quinquefasciatus* (CqMsrA, GenBank accession no. XP001868737), *Drosophila erecta* (DeMsrA, GenBank accession no. XP001973088), and *Apis cerana cerana* (AccMsrA, GenBank accession no. HQ219724). Identical amino acid residues in this alignment are shaded in black; gaps are introduced for optimal alignment and maximum similarity among all compared sequences. The short sequence motif characteristic of MsrAs is marked by a box, and conserved MsrA Cys sites are marked by ▼ and *.

To investigate the evolutionary relationships among AccMsrA and other MsrA proteins, a phylogenetic tree was generated based on the MsrA amino acid sequence alignment. The phylogenetic tree displays two distinct clusters: A, metazoans and B, plants and bacteria (Fig. 2). AccMsrA and *Nasonia vitripennis* (NvMsrA) are in the same clade, belonging to Hymenoptera, and they are close to the Diptera-originated *D. erecta* MsrA (DeMsrA) and *Aedes aegypti* MsrA (AaMsrA). Thus, the function of AccMsrA might be similar to those of NvMsrA, DeMsrA and AaMsrA.

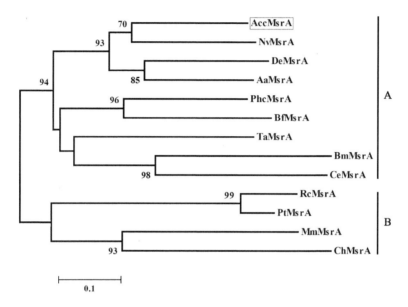

Fig. 2 Phylogenetic analysis of MsrAs from different species. Amino acid sequences of MsrAs were obtained from GenBank as follows: *Nasonia vitripennis* (NvMsrA, GenBank accession no. XP001603389), *Drosophila erecta* (DeMsrA, GenBank accession no. XP001973088), *Aedes aegypti* (GenBank accession no. XP001649797), *Pediculus humanus corporis* (GenBank accession no. XP002426843), *Branchiostoma floridae* (GenBank accession no. XP002594093), *Trichoplax adhaerens* (GenBank accession no. XP002113880), *Brugia malayi* (GenBank accession no. XP001893631), *Caenorhabditis elegans* (GenBank accession no. NP495540), *Ricinus communis* (GenBank accession no. XP002530748), *Populus trichocarpa* (GenBank accession no. XP002310304), *Methanococcus maripaludis* (GenBank accession no. YP001097623), *Cytophaga hutchinsonii* (GenBank accession no. YP678260), and *Apis cerana cerana* (AccMsrA, GenBank accession no. HQ219724).

Genomic structure analysis of *AccMsrA*

To further elucidate the genomic structure of *AccMsrA*, three fragments of about 1,851 bp, 521 bp and 623 bp were obtained by using the specific genomic primers G1+G2, G3+G4 and G5+G6, respectively. By assembling the cDNA fragments and intron sequences, we obtained the *AccMsrA* genomic sequence, which spans a region of 3,690 bp (GenBank accession no. HQ219725) that contains five exons and four introns (Fig. 3). All introns of *AccMsrA* had the expected characteristics typical of introns, such as high A+T content (Table 3) and signals flanked by the 5' splice donor GT and the 3' splice donor AG. Furthermore, the exon-intron structure is fairly conserved among MsrAs from different species. Notably, the first intron of *AccMsrA* is located in the 5' UTR; however, no intron was detected in the UTR of *MsrA* genes from other species. This result implies that the features of introns are specific to different species in the process of evolution.

Table 3 Exon and intron sizes of the *AccMsrA* gene and AT content

Exon			Intron		
Number	Size (bp)	AT content (%)	Number	Size (bp)	AT content (%)
1	1-127(127)	77.95	1	128-1804(1677)	84.14
2	1805-1971(167)	60.48	2	1972-2051(80)	82.5
3	2052-2172(121)	66.94	3	2173-2250(78)	84.62
4	2251-2387(137)	27.01	4	2388-2702(315)	10.48
5	2703-3690(988)	25.40			

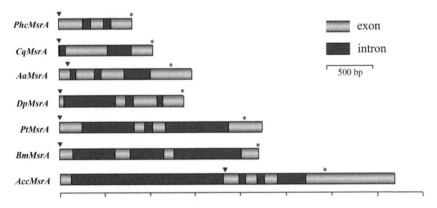

Fig. 3 Schematic representation of the genomic organisation. Lengths of the exons and introns of the genomic DNA of *Pediculus humanus corporis* (GenBank accession no. 8229465), *Culex quinquefasciatus* (CqMsrA, GenBank accession no. 6052472), *Aedes aegypti* (AaMsrA, GenBank accession no. 5565445), *Drosophila persimilis* (DpMsrA, GenBank accession no. 6597072), *Populus trichocarpa* (AmRPL17, GenBank accession no. 7495540), *Bombyx mori* (BmMsrA, GenBank accession no. 732898), and *Apis cerana cerana* (AccMsrA, GenBank accession no. HQ219725) are indicated according to the scale below. The exons are highlighted with gray bars, and the introns are indicated with black bars. The initiation codon (ATG) is marked by ▼. The stop codon (TAA) is marked by *.

Identification of putative *cis*-acting elements in the 5'-flanking region of AccMsrA

To obtain more information on *AccMsrA* gene expression and regulation, a 1,400 bp

5′-flanking region of *AccMsrA* was generated from genomic DNA with the primers PP3 and PP4. By using a comparison with the *AccMsrA* cDNA sequence, the putative transcription start site was defined as +1 at 1,364 bp upstream from the ATG translation start site (Fig. 4). Then several transcription factor-binding sites in the 5′-flanking region of *AccMsrA* were predicted by the web software program MatInspector (http://www.cbrc.jp/research/db/TFSEARCH.html). Of these sites, five heat shock factor (HSF), which are involved in heat stress responsiveness, were identified [17]. In addition, a series of binding sites related to development were found including: (a) tramtrack (Ttk), which is related to the regulation of developmental genes [18]; (b) cell factor 2-II (CF2-II), which is related to developmental regulation [19]; (c) hunchback (Hb), which functions in early embryo segmentation [20]; and (d) Broad-Complex (BR-C), which directly controls a tissue-specific response to the steroid hormone ecdysone at the onset of metamorphosis [21]. The predicted transcription factor-binding sites and their positions are shown in Fig. 4 and Table 4, respectively. Together, these results indicate that *AccMsrA* might be involved in responses to external factors and might also have an effect on development from the early embryo to adulthood.

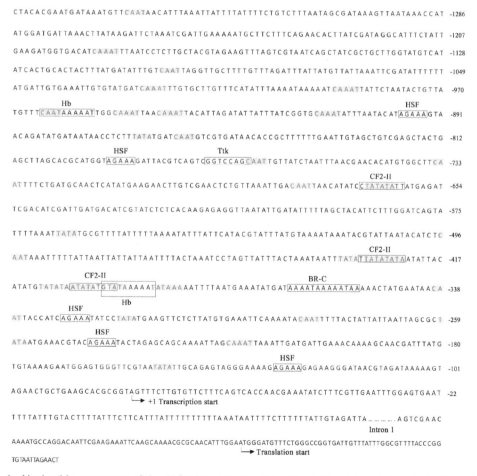

Fig. 4 Nucleotide sequence of the 5′-flanking region of *AccMsrA*. The transcription start site and the translation start site are marked with arrowheads. The putative transcription factors are indicated, their binding sites are boxed, and the putative core promoter consensus sequences are highlighted in grey.

Table 4 Putative transcription factor binding sites in the 5'-flanking region of AccMsrA: positions and possible functions

Transcription factor	Putative binding site and position (bp)	Function
HSF	AGAAA: −52, −165, −249, −715, −819	heat shock factors [17]
Ttk	GGTCCAGC: −697	regulation of developmental genes [18]
CF2-II	CTATATATT: −590, TTATATATA: −353, ATATATGTA: −326	developmentally regulated [19]
Hb	GTATAAAAAT: −320, CAATAAAAAT: −885	function in the segmentation of early embryo [20]
BR-C	AAAATAAAAATAA: −286	controls a tissue-specific response to the steroid hormone ecdysone at the onset of metamorphosis [21]

Expression pattern of *AccMsrA* in different developmental stages and tissues

To determine the expression of *AccMsrA* in different developmental stages, qRT-PCR was carried out by using RNA extracted from larvae, pupae, and 1-day old adults. As shown in Fig. 5A, the expression of *AccMsrA* was detected at all developmental stages. It is interesting to observe that there was a significant high expression ($P<0.05$) of the levels of *AccMsrA* mRNA in the PP_2 stage. Moreover, high expression levels were observed in the PP_1 and Pp stages.

A tissue-specific expression analysis of *AccMsrA* was also performed by using RNA extracted from the head, thorax, abdomen, and midgut of adult *A. cerana cerana* (Fig. 5B). The expression levels in the head and midgut were 3.5-fold and 1.2-fold greater than those in the abdomen ($P < 0.05$).

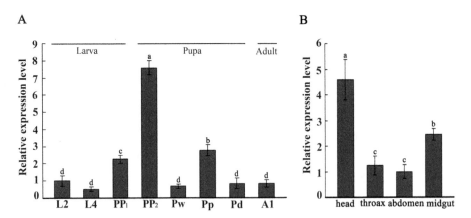

Fig. 5 *AccMsrA* mRNA expression in different developmental phases (A) and tissues (B) determined by qRT-PCR in *A. cerana cerana*. A. Expression of *AccMsrA* in the entire body of the second (L2)-instar larva, the fourth (L4)-instar larva, prepupal (PP) phase pupa, earlier pupal phase (Pw) pupa, pink (Pp) phase pupa and dark (Pd) phase pupa and adult workers (1-day-old). B. Distribution of *AccFABP4* in brain, abdomen, thorax and midgut. Vertical bars represent the mean ± standard error of mean (SEM) (n = 3). Different letters above the bars indicate significant difference as determined using SAS software analysis.

Expression profile of *AccMsrA* under diverse environmental stresses

MsrAs play an important role in the oxidative stress response. Previous studies reported that the expression of MsrA could be regulated by environmental stressors such as UV light, heat and H_2O_2 [22, 23]. To investigate whether *AccMsrA* is involved in the response to these environmental stresses, qRT-PCR was performed. UV light induced the expression of *AccMsrA* rapidly; the expression level of *AccMsrA* peaked at 1 h and then declined (Fig. 6A). When bees were exposed to heat treatment, the expression of *AccMsrA* showed no significant change between 0 and 1 h, and then it increased gradually after 1 h and remained high through 3 h (Fig. 6B). When treated with H_2O_2, the *AccMsrA* expression pattern was similar to that under UV light treatment. However, the highest expression level was observed 3 h after H_2O_2 treatment (Fig. 6C). Thus, these results demonstrate *AccMsrA* might participate in antioxidant systems under environmental stresses in bees.

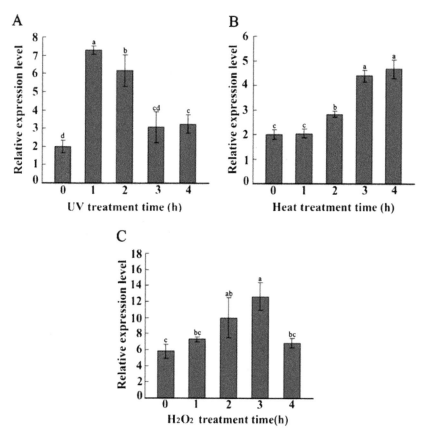

Fig. 6 *AccMsrA* mRNA expression under different environmental stresses in *A. cerana cerana*. A. *AccMsrA* mRNA expression in *A. cerana cerana* after UV light treatment, as measured by qRT-PCR. B. *AccMsrA* mRNA expression in *A. cerana cerana* after heat treatment, as measured by qRT-PCR. C. *AccMsrA* mRNA expression in *A. cerana cerana* after H_2O_2 treatment, as measured by qRT-PCR. Vertical bars represent the mean ± SEM (n = 3). Different letters above the bars indicate significant difference as determined by using SAS software analysis (P < 0.05).

Discussion

In the present study, we report the cloning and characterisation of the *AccMsrA* gene from *Apis cerana cerana* for the first time. Multiple sequence alignment indicated that the deduced amino acid sequence of AccMsrA shares high similarity with MsrAs from other insects. The Cys in the GCFW fingerprint region of the N-terminus and three other Cys sites, located at positions 25, 37 and 216, were observed in the AccMsrA sequence. Previous studies have showed that the catalytic mechanism of MsrAs involves three cysteine residues. The first Cys site, CysA, belongs to the fingerprint GCFW of N-terminus and attacks the sulfoxide group of S-MetO to release the reduced methionine, while the second and the third Cys sites, CysB and CysC, facilitate responsiveness to thioredoxin as the reductant [5, 24]. MsrAs from bovine, human and *Caenorhabditis elegans* have similar structures [25, 26]. Interestingly, a Cys site in the MsrA amino acid sequences of the insects was observed in the C-terminus end, in contrast to other studies. This suggests that MsrAs in insects potentially possess different functions. However, further study would be required for a comprehensive analysis. Phylogenetic analysis showed that AccMsrA belongs to the Hymenoptera insect. In addition, it is interesting to observe that the first intron is located in the 5' UTR, suggesting that the *AccMsrA* gene has evolved differently from other species. However, further studies should be performed.

Previous studies have shown that MsrAs scavenge ROS and protect proteins by reducing the sulfoxide group of S-MetO [27]. The expression level of *MsrA* genes is influenced by ROS [23, 28]. Stresses such as UV light, heat and H_2O_2 always increase ROS [13]. The expression of *MsrA* is upregulated by UVA and H_2O_2 in human keratinocytes [22]. In our multiple stresses study, the transcription levels of *AccMsrA* in honeybees were upregulated by exposure to UV light (UV, 30 MJ/cm^2), heat (42°C) and H_2O_2 (2 mmol/L), with UV light having the most profound influence. This suggests that *AccMsrA* may be involved in insects' response to environmental stresses and that it may play a central role in ROS reduction. In addition, binding sites for putative stress response transcription factors, such as the HSFs, were identified in the 5'-flanking region of *AccMsrA*.

MsrA is considered the major component of the honeybee antioxidant system [29, 30]. Corona (2005) has indicated that mRNA from the MsrA gene is expressed widely in the head, thorax and abdomen of adult workers and that expression levels tend to increase with age. Here, qRT-PCR has demonstrated that the *AccMsrA* gene was highly expressed in the head and midgut. However, transcription factor-binding sites for BR-C, which functions in tissue-specific responses, were observed in the promoter sequence of *AccMsrA*. This indicates that *AccMsrA* may be involved in the clearance of ROS in the head and midgut. High expression profiles of *AccMsrA* were observed in the PP2 and Pp developmental stages, which have histo-differentiation. We hypothesize that the significant change of *AccMsrA* expression was regulated by hormones such as ecdysone. Previous studies have demonstrated that steroid hormones such as ecdysone are systemic signalling molecules that temporally coordinate the juvenile-adult transition in insects and control many aspects of development including moulting and metamorphosis [31-33]. Moreover, the MsrA gene in *Drosophila* Kc cells was induced by ecdysone independently of ROS treatment [34]. Additionally, transcription factor-binding sites related to development such as Ttk, CF2-II and Hb were observed in the 5'-flanking region. This suggests that *AccMsrA* gene was participates in the developmental regulation of bees growth and differentiation.

The identification of the MsrA gene from *A. cerana cerana* represents a critical step in understanding the various roles played by the MsrA gene in response to environmental stresses and in the regulation of the growth and development of bees. However, further elucidation of the mechanisms involved and analysis of transgenic honeybees are necessary.

References

[1] Zhang, C., Jia, P. P., Jia, Y. Y., Weissbach, H., Webster, K., Huang, X. P., Lemanski, S. L., Achary, M., Lemanski, L. F. Methionine sulfoxide reductase A (MsrA) protects cultured mouse embryonic stem cells from H_2O_2-mediated oxidative stress. J Cell Biochem. 2010, 111, 94-103.

[2] Brunell, D., Weissbach, H., Hodder, P., Brot, N. A high-throughput screening compatible assay for activators and inhibitors of methionine sulfoxide reductase A. Assay Drug Dev Technol. 2010, 8 (8), 615-620.

[3] Brot, N., Weissbach, H. Biochemistry of methionine sulfoxide residues in proteins. BioFactors. 1991, 3, 91-96.

[4] Moskovitz, J. Methionine sulfoxide reductases: ubiquitous enzymes involved in antioxidant defense, protein regulation, and prevention of aging-associated diseases. Biochim Biophys Acta. 2005, 1703, 213-219.

[5] Weissbach, H., Etienne, F., Hoshi, T., Heinemann, S. H., Lowther, W. T., Matthews, B., St, John, G., Nathan, C., Brot, N. Peptide methionine sulfoxide reductase: structure, mechanism of action, and biological function. Arch Biochem Biophys. 2002, 397, 172-178.

[6] Brot, N., Weissbach, L., Werth, J., Weissbach, H. Enzymatic reduction of protein-bound methionine sulfoxide. Proc Natl Acad Sci USA. 1981, 78, 2155-2158.

[7] Moskovitz, J., Berlett, B. S., Poston, J. M., Stadtman, E. R. The yeast peptide-methionine sulfoxide reductase functions as an antioxidant *in vivo*. Proc Natl Acad Sci USA. 1997, 94, 9585-9589.

[8] Zhou, Z., Li, C. Y., Li, K., Wang, T., Zhang, B., Gao, T. W. Decreased methionine sulphoxide reductase A expression renders melanocytes more sensitive to oxidative stress: a possible cause for melanocyte loss in vitiligo. Br J Dermatol. 2009, 161, 504-509.

[9] Moskovitz, J., Bar-Noy, S., Williams, W. M., Requena, J., Berlett, B. S., Stadtman, E. R. Methionine sulfoxide reductase (MsrA) is a regulator of antioxidant defense and lifespan in mammals. Proc Natl Acad Sci USA. 2001, 6, 12920-12925.

[10] Yermolaieva, O., Xu, R., Schinstock, C., Brot, N., Weissbach, H., Heinemann, S. H., Hoshi, T. Methionine sulfoxide reductase A protects neuronal cells against brief hypoxia/reoxygenation. Proc Natl Acad Sci USA. 2004, 101, 1159-1164.

[11] Ruan, H., Tang, X. D., Chen, M. L., et al. High-quality life extension by the enzyme peptide methionine sulfoxide reductase. Proc Natl Acad Sci USA. 2002, 99, 2748-2753.

[12] Kielbassa, C., Roza, L., Epe, B. Wavelength dependence of oxidative DNA damage induced by UV and visible light. Carcinogenesis. 1997, 18, 811-816.

[13] Heise, K., Puntarulo, S., Portner, H. O., Abele, D. Production of reactive oxygen species by isolated mitochondria of the Antarctic bivalve *Laternula elliptica* (King and Broderip) under heat stress. Comp

Biochem Physiol C: Toxicol Pharmacol. 2003, 134, 79-90.

[14] Bitondi, M. M. G., Nascimento, A. M., Cunha, A. D., Guidugli, K. R., Nunes, F. M. F., Simões, Z. L. P. Characterization and expression of the hex 110 Gene encoding a glutamine-rich hexamerin in the honey bee, *Apis mellifera*. Arch Insect Biochem. 2006, 63, 57-72.

[15] Liu, Y., Gao. Q., Wu, B., Ai, T., Guo, X. NgRDR1, an RNA-dependent RNA polymerase isolated from *Nicotiana glutinosa*, was involved in biotic and abiotic stresses. Plant Physiol Biochem. 2009, 47, 359-68.

[16] Honeybee Genome Sequencing Consortium. Insights into social insects from the genome of the honeybee *Apis mellifera*. Nature. 2006, 443, 931-949.

[17] Fernandes, M., Xiao, H., Lis, J. T. Fine structure analyses of the *Drosophila* and *Saccharomyces* heat shock factor-heat shock element interactions. Nucleic Acids Res. 1994, 22, 167-173.

[18] Read, D., Manley, J. L. Alternatively spliced transcripts of the Drosophila tramtrack gene encode zinc finger proteins with distinct DNA binding specificities. EMBO J. 1992, 11, 1035-1044.

[19] Hsu, T., Gogos, J. A., Kirsh, S. A., Kafatos, F. C. Multiple zinc finger forms resulting from developmentally regulated alternative splicing of a transcription factor gene. Science. 1992, 257, 1946-1950.

[20] Stanojević, D., Hoey, T., Levine, M. Sequence-specific DNA-binding activities of the gap proteins encoded by hunchback and Krüppel in *Drosophila*. Nature. 1989, 341, 331-335.

[21] Kalm, L.V., Crossgrove, K., Seggern, D.V., Guild, G. M., Beckendorf, S. K. The Broad-Complex directly controls a tissuespecific response to the steroid hormone ecdysone at the onset of *Drosophila* metamorphosis. EMBO J. 1994, 13, 3505-3516.

[22] Ogawa, F., Sander, C. S., Hansel, A., Oehrl, W., Kasperczyk, H., Elsner, P., Shimizu, K., Heinemann, S. H., Thiele, J. J. The repair enzyme peptide methionine-S-sulfoxide reductase is expressed in human epidermis and upregulated by UVA radiation. J Invest Dermatol. 2006, 126, 1128-1134.

[23] Sreekumar, P. G., Kannan, R., Yaung, J., Spee, C. K., Ryan, S. J., Hinton, D. R. Protection from oxidative stress by methionine sulfoxide reductases in RPE cells. Biochem Biophys Res Commun. 2005, 334, 245-53.

[24] Weissbach, H., Resnick, L., Brot, N. Methionine sulfoxide reductases: history and cellular role in protecting against oxidative damage. Biochim Biophys Acta. 2005, 1703, 203-212.

[25] Kauffmann, B., Aubry, A., Favier, F. The three-dimensional structures of peptide methionine sulfoxide reductases: current knowledge and open questions. Biochim Biophys Acta. 2005, 1703, 249-260.

[26] Lee, B. C., Lee, Y. K., Lee, H., Stadtman, E. R., Lee, K. H., Chung, N. Cloning and characterization of antioxidant enzyme methionine sulfoxide-S-reductase from Caenorhabditis elegans. Arch Biochem Biophys. 2005, 434, 275-281.

[27] Levine, R. L., Berlett B. S., Moskovitz, J., Mosoni, L., Stadtman, E. R. Methionine residues may protect proteins from critical oxidative damage. Mech Ageing Dev. 1999, 107, 323-332.

[28] Picot, C. R., Petropoulos, I., Perichon, M., Moreau, M., Nizard, C., Friguet, B. Overexpression of MsrA protects WI-38 SV40 human fibroblasts against H_2O_2-mediated oxidative stress. Free Radic Biol Med. 2005, 39, 1332-1341.

[29] Corona, M., Robinson, G. E. Genes of the antioxidant system of the honey bee: annotation and phylogeny. Insect Mol Biol. 2006, 15, 687-701.

[30] Corona, M., Hughes, K. A., Weaver, D. B., Robinson, G. E. Gene expression patterns associated with queen honey bee longevity. Mech Ageing Dev. 2005, 126, 1230-1238.

[31] McBrayer, Z., Ono, H., Shimell, M., Parvy, J. P., Beckstead, R. B., Warren, J. T., et al. Prothoracicotropic hormone regulates developmental timing and body size in *Drosophila*. Dev Cell. 2007, 13, 857-871.

[32] Rewitz, K. F., Larsen, M. R., Lobner-Olesen, A., Rybczynski, R., O'Connor, M. B., Gilbert, L. I. X. A phosphoproteomics approach to elucidate neuropeptide signal transduction controlling insect

metamorphosis. Insect Biochem Mol Biol. 1992, 39, 475-483.

[33] Gilbert, L. I., Rybczynski, R., Warren, J. T. Control and biochemical nature of the ecdysteroidogenic pathway. Annu Rev Entomol. 2002, 47, 883-916.

[34] Roesijadi, G., Rezvankhah, S., Binninger, D. M., Weissbach, H. Ecdysone induction of MsrA protects against oxidative stress in *Drosophila*. Biochem Biophys Res Commun. 2007, 354, 511-516.

3

The Nonenzymatic Antioxidant Molecular Mechanisms of *Apis cerana cerana*

Apis cerana cerana (*A. cerana cerana*) is a well-known subspecies of oriental bees. Compared with *A. mellifera*, *A. cerana cerana* has strong resistance to mites, has an acute sense of smell, and can forage nectar and pollen from a wide range of flowers, including wild plants. However, recently, the excessive use of pesticides, the existence of pollutants, climate change with extreme heat and cold, exposure to ultraviolet radiation, and the presence of heavy metal in the environment, which can lead to the generation of reactive oxygen species (ROS), have caused serious harm to the survival of honeybees.

Increased levels of reactive oxygen species (ROS) in cells can cause oxidative stress and affect cellular signal transduction, which controls diverse cellular processes, resulting in cell damage and triggering apoptosis. Therefore, ROS detoxification is critical for organism survival. In general, organisms are dependent on various enzymatic and nonenzymatic mechanisms to eliminate or inhibit ROS generation or reactivity.

To explore the antioxidant capacity of Chinese honeybees in coping with oxidative damage, we isolated and characterized a series of antioxidant genes. In the experiment, we used quantitative real-time PCR (qRT-PCR) to evaluate the transcripts of these genes after various oxidative stresses and used RNA interference (RNAi) technology to knock down the genes. In addition, antioxidant enzyme activities and expression levels of other antioxidant genes were detected.

Cytochrome P450 monooxygenases (P450) are widely distributed multifunctional enzymes that play an important role in the oxidative metabolism of endogenous compounds and xenobiotics. The expression level of *AccCYP336A1* was upregulated by cold, heat, ultraviolet (UV) radiation, H_2O_2 and pesticide treatments. These results were confirmed by Western blotting assays. The recombinant AccCYP336A1 protein acted as an antioxidant that resisted paraquat-induced oxidative stress.

Syntheses of cytochrome c oxidase (SCO) proteins are copper-donor chaperones involved in metalation of the CuA redox center of COX. *AccSCO2* expression was induced by cold, $CdCl_2$, $HgCl_2$, ultraviolet (UV) light, and H_2O_2 treatments and was inhibited by different pesticide treatments. A disc diffusion assay of recombinant AccSCO2, AccSCO2-R1, and AccSCO2-R2 proteins showed that these proteins play a functional role in protecting cells from oxidative stress in a copper-dependent manner. The expression levels of most antioxidant genes were significantly decreased in *AccSCO2*-silenced bees. Moreover, the antioxidant enzymatic activities of SOD, POD, CAT and COX were lower in the silenced bees. These results suggest *AccSCO2* plays important roles in cellular stress responses and antioxidative processes.

Cyclin-dependent kinases (CDKs), a family of serine/threonine kinases, form complexes with cyclins, components that are essential for the kinase activity of CDKs. qPCR analysis indicated that *AccCDK5* was induced by several environmental oxidative stresses. The overexpression of the AccCDK5 protein in *E. coli* enhanced the resistance of the bacteria to oxidative stress. Yeast two-hybrid analysis demonstrated the interaction between *AccCDK5* and *AccCDK5r1*. These results suggest that *AccCDK5* plays a pivotal role in the response to oxidative stresses and that *AccCDK5r1* is a potential activator of *AccCDK5*. The mRNA level of *AccCDK6* in adult workers was influenced by H_2O_2, UV, heat, $HgCl_2$, and pyriproxyfen treatments. These results indicated that *AccCDK6* might participate in intracellular reactions of reactive oxygen species (ROS).

Carboxylesterases (CarEs) play vital roles in metabolizing different physiologically important endogenous compounds and in detoxifying various harmful exogenous compounds in insects. AcceFE4 recombinant protein-expressing bacteria had a smaller killing zone than the control group following paraquat, $HgCl_2$ and cumyl hydroperoxide treatments. Multiple oxidant genes were downregulated when *AcceFE4* was knocked down by RNA interference. Enzyme activities of SOD, POD, CAT and carboxylesterase were reduced in the knockdown samples. These data suggest that *AcceFE4* may be involved in the oxidative resistance response during adverse stress.

Dpp (the decapentaplegic gene) is a member of the TGF-β superfamily in insects and is a secreted molecule. The 5′-flanking region of *AccDpp* has many transcription factor binding sites that are relevant to development and stress responses. The transcription of *AccDpp* could be repressed by 4℃ and UV treatments, but induced by other treatments. Recombinant AccDpp restrained the growth of *Escherichia coli*. Vitellogenin (Vg) is a yolk precursor protein in most oviparous females. RT-PCR and Western blotting analysis indicated that the expression of *AccVg* was induced by cold, $CdCl_2$, and pesticide treatments. The transcription levels of *AccCacyBP* in the brains of honeybees were developmentally induced and upregulated by exposure to oxidative stresses, including UV-light, acetamiprid and $HgCl_2$.

3.1 Characterization of an *Apis cerana cerana* Cytochrome P450 Gene (*AccCYP336A1*) and Its Roles in Oxidative Stresses Responses

Ming Zhu, Weixing Zhang, Feng Liu, Xiaobo Chen, Han Li, Baohua Xu

Abstract: Cytochrome P450 monooxygenases (P450), widely distributed multifunctional enzymes, that play an important role in the oxidative metabolism of endogenous compounds and xenobiotics. Studies have found that these enzymes show peroxidase-like activity and may thus be involved in protecting organisms against reactive oxygen species (ROS). In this work, *Apis cerana cerana* was used to investigate the molecular mechanisms of P450 family genes in resisting ROS damage. A cytochrome P450 gene was isolated, AccCYP336A1. The open reading frame (ORF) of AccCYP336A1 is 1,491 bp in length and encodes a predicted protein of 496 amino acids. The obtained amino acid sequence of AccCYP336A1 shared a high sequence identity with homologous proteins and contained the highly conserved features of this protein family. Quantitative real-time PCR (qRT-PCR) analysis showed that AccCYP336A1 was present in some fast developmental stages and had a higher expression in the epidermis than in other tissues. Additionally, the expression levels of AccCYP336A1 were upregulated by cold (4℃), heat (42℃), ultraviolet (UV) radiation, H_2O_2 and pesticides (thiamethoxam, deltamethrin, methomyl and phoxim) treatments. These results were confirmed by the Western blotting assays. Furthermore, the recombinant AccCYP336A1 protein acted as an antioxidant that resisted paraquat-induced oxidative stress. Taken together, these results suggest that AccCYP336A1 may play a very significant role in antioxidant defense against ROS damage.

Keywords: Cytochrome P450 monooxygenases; Reactive oxidative stress; *Apis cerana cerana*; Antioxidant defense

Introduction

The Chinese honeybee (*Apis cerana cerana*) plays a critical role in the balance between agricultural economic development and regional ecology as a pollinator of flowering plants [1]. During their lifespan, honeybees face a variety of adverse environmental stressors, including temperature swings, ultraviolet (UV) radiation, H_2O_2 and pesticides, all of which are considered to produce reactive oxygen species (ROS) [2]. High ROS concentrations may cause serious oxidative damage to DNA, proteins and lipids [3]. High ROS levels can cause DNA damage in the form of base deletions, degradation, single-strand breaks, and rearrangements, giving rise to mutations [4]. High ROS damage to proteins can lead to particular amino acid modifications and peptide rupture, leading in turn to a loss of enzymatic activity [5]. High ROS levels can also lead to lipid peroxidation which destroys cell membrane fluidity and results in apoptosis [6]. Moreover, the sperm storage of *Apis mellifera* may be affected by high ROS concentrations [7]. Due to increasing ROS damage, the bee population is facing serious survival problems. Thus, it is important to understand the antioxidant system and its mechanism of defense against ROS to prevent further damage to the bee population from ROS.

Organisms have developed complex antioxidant mechanisms to avoid oxidative damage. In most cases, organisms protect themselves by varieties of antioxidant enzymes, such as peroxidase (POX), superoxide dismutase (SOD), catalase (CAT), glutathione *S*-transferases (GST) and glutathione peroxidase (GPX) [8-10]. Cytochrome P450 proteins share semblable characteristics with the antioxidant enzymes, as an uncoupled catalytic cycle of cytochrome P450 possesses properties resembling those of a peroxidase [11].

Cytochrome P450s (P450) are a large superfamily of enzymes found in almost all living organisms [12]. Their ancient origin and ubiquity may reflect their physiological importance. These enzymes constitute an extremely important metabolic system because of their involvement in regulating the titers of endogenous compounds such as hormones, fatty acids, and steroids [13, 14]. Additionally, this enzyme system plays a central role in the metabolism of xenobiotics, such as drugs, pesticides, and plant toxins by catalyzing oxidation reactions [15-18]. Besides, those P450s may be involved in the receipt and transmission of semiochemical signals.

In insects, more than 1,000 cytochrome P450 genes have been identified, and this number is rapidly increasing due to the increasing number of insect genome sequences. Most insect cytochrome P450 genes belong to four clades: the mitochondrial P450s and CYP2, CYP3 and CYP4 clades. However, the specific functions of the four clades have not been determined. The CYP3 and CYP4 clades were once thought to be largely responsible for the environmental response and detoxifying functions in insects [14, 19]. Particularly, CYP4 has been implicated in both pesticide metabolism and chemical communication in insects [20]. However, there is emerging evidence that CYP4 is involved in the response to oxidative stress and may have a role in protecting honeybees from oxidative injury in *A. cerana cerana* [21]. Meanwhile, several members of the mitochondrial CYP and CYP2 families have been implicated for essential roles in hormone biosynthesis [22]. Earlier studies on the function of CYP3 from insects focused on their roles on environmental response and detoxifying functions. For example, CYP6Z1 (belongs to CYP3 clade) is overexpressed in pyrethroid resistant in *Anopheles gambiae* [23]. Similarly, CYP321A1 of *Helicoverpa zea*, a member of the CYP3 clade, metabolizes cypermethrin when produced in a baculovirus expression system [24]. However, limited studies have reported the antioxidant functions of CYP3 clades, which are of significant relevance to the organisms.

A. cerana cerana is the main honeybee species in China and has enjoyed exceptional advantages over other species, such as a longer period of collecting honey, an increased disease resistance, and a lower food cost. However, with the increasing trend of ROS damage in China, the honeybee is facing increasing survival problems. Although proteins involved in oxidative stress, such as vitellogenin and juvenile hormone, have been researched in *A. mellifera* [25], little information is available on these proteins in *A. cerana cerana*. Considering the significant roles of antioxidative processes in insects, we isolated the AccCYP336A1 gene from *A. cerana cerana* to further study its role in resisting oxidative stress. We evaluated its expression patterns at different developmental stages and in different tissues. Moreover, the honeybees were exposed to cold (4℃), heat (42℃), ultraviolet (UV) radiation, H_2O_2 and pesticides (thiamethoxam, deltamethrin, methomyl, and phoxim) to evaluate the expression patterns in response to oxidative stress. Moreover, the Western blotting assays confirmed the results. Based on these results, we speculate that AccCYP336A1 might play an important role in the response to oxidative stress.

Material and Methods

Insects and treatments

The Chinese honeybees used in this study were obtained from the experimental apiary of Shandong Agricultural University (Taian, China). Based on their shape, age and eye color, the honeybees were divided into egg, larvae, pupae, and adults, which were collected from the hive. The 15 days post-emergence adult workers were fed a basic adult diet of water and powdered sugar for two days before treatments and were placed in the dark, and divided into 13 groups. Adult workers in groups 1-4 were placed in a cold (4℃) or heat (42℃) environment for 0.25, 0.5, 1, 2, 3 and 4 h. The adult workers of group 5 were treated by ultraviolet (UV)-light (30 MJ/cm^2) for 0.25, 0.5, 1, 2, 3 and 4 h. Groups 6-9 were treated with four types of pesticides (thiamethoxam, deltamethrin, methomyl, and phoxim) that were diluted into 1 μg/ml for 0.25, 0.5, 1, 2, 3 and 4 h. Group 10 honeybees were injected with 20 μl H_2O_2 for 0.25, 0.5, 1, 2, 3 and 4 h. Group 11 honeybees were fed normal food as a control. The tissues of the brain, epidermis, muscle, hemolymph, rectum and midgut were collected from groups 12 and 13, the 15 days post-emergence adult workers Chinese honeybees. All bees and tissues were frozen in liquid nitrogen and stored at −80℃.

Primers

The sequences of the primers used in this study are provided in Table 1.

Table 1 The primers in this study

Abbreviation	Primer sequence (5'—3')	Description
TP1	AATGGGATTCCGTACAGCAA	cDNA sequence primer, forward
TP2	GTAGTTGCATTTTCGGCGAAACCTC	cDNA sequence primer, reverse
5P1	CCGGTTTCATTCCTTTGTACAT	5' RACE reverse primer, outer
5P2	CTTCATTATCAGAGGTAGAAAATG	5' RACE reverse primer, inner
3P1	CCGAATTATTGATACCGATCTGTG	3' RACE forward primer, outer
3P2	CAGCGATGAGAATAAGCAAAGAA	3' RACE forward primer, inner

		Continued
Abbreviation	Primer sequence (5'—3')	Description
AAP	GGCCACGCGTCGACTAGTAC(G)$_{14}$	Abridged anchor primer
AUAP	GGCCACGCGTCGACTAGTAC	Abridged universal amplification primer
B25	GACTCTAGACGACATCGA	3' RACE universal primer, outer
B26	GACTCTAGACGACATCGA(T)$_{18}$	3' RACE universal primer, inner
QP1	GTTCAGTTTTTTTTTCTCGGCG	Full-length cDNA sequence primer, forward
QP2	CAACTCGAACTTTGACTGCAC	Full-length cDNA sequence primer, reverse
SQ1	TGTTCGGTTATTTGCCATTCC	Real-time PCR primer, forward
SQ2	GGTCTGCCAGTAAACTCTTCC	Real-time PCR primer, reverse
β-s	TTATATGCCAACACTGTCCTTT	Standard control primer, forward
β-x	AGAATTGATCCACCAATCCA	Standard control primer, reverse
PRS	CGCCATATGCTGCCTACGAAACACGAC	Protein expression primer, forward
PRX	TTGCGGCCGCTTTCATCAATGTAGCCAAAC	Protein expression primer, reverse

RNA extraction, cDNA synthesis

Total RNA was extracted with Trizol (TransGen Biotench, Beijing, China), and the first-strand cDNA was synthesized by using EasyScript First-Strand cDNA Synthesis SuperMix (TransGen, Beijing, China) according to the instructions of the manufacturer. The cDNA was used as the PCR template in the gene cloning procedures and qRT-PCR transcriptional analysis.

Isolation of the full-length cDNA of *AccCYP336A1*

Reverse transcription-PCR (RT-PCR) was used to amplify full-length of AccCYP336A1. To obtain the internal fragment of the AccCYP336A1 cDNA, primers TP1 and TP2 were designed based on conserved regions of the P450 from *A. mellifera*, *Nasonia vitripennis*, and *Drosophila melanogaster* and synthesized (Sangon Biotechnological Company, Shanghai, China). Specific primers (5P1/5P2 and 3P1/3P2) were generated for the 5' RACE and 3' RACE, respectively. Based on the deduced cDNA, specific primers QP1 and QP2 were designed, and the complete coding sequence of AccCYP336A1 was obtained by RT-PCR.

Bioinformatic analysis and phylogenetic analysis

Conserved domains of *AccCYP336A1* were retrieved by using the BLAST search program (http://blast.ncbi.nlm.nih.gov/Blast.cgi). The ORF of *AccCYP336A1* was retrieved from the DNAMAN version 5.2.2 (Lynnon Biosoft, Quebec, Canada). The online tool PeptideMass (http://web.expasy.org/peptide_mass/) was used to forecast the molecular weight and theoretical isoelectric point of AccCYP336A1. The phylogenetic analysis was conducted by using the neighbor-joining method with the Molecular Evolutionary Genetics Analysis (MEGA version 4.1). Hydrophilicity was analyzed on the web site of ProtScale (http://web.expasy.org/protscale/). The transmembrane segments were analyzed by using TMHMM Server (http://www.cbs.dtu.dk/services/TMHMM/).

Fluorescent real-time quantitative PCR (qRT-PCR) for transcriptional analysis

To confirm the expression pattern of the *AccCYP336A1* gene at different developmental stages and in different tissues under different environmental stresses, real-time quantitative PCR (qRT-PCR) was carried out by using SYBR Premix Ex *Taq* (TaKaRa, Dalian, China) and a CFX96™ Real-time PCR Detection System (Bio-Rad, Hercules, CA, USA). Two specific primers (SQ1/SQ2) were used. The *β-actin* (β-s/β-x) gene (GenBank accession number: HM640276) was used as a reference gene. The primers are listed in Table 1. Three individual samples were prepared for each sample, and each sample was analyzed three times. The linear relationship, amplification efficiency and data analysis were conducted by using CFX Manager Software version 1.1. An analysis of the significant differences was performed by using Duncan's multiple range tests with Statistical Analysis System (SAS) software version 9.1.

Expression and purification of recombinant AccCYP336A1

The 240-460 amino acid sequence of the *AccCYP336A1* ORF, flanked by a *Nde* I site and a *Not* I restriction site, was amplified and subcloned into the expression vector pET-21a (+) with a His-tag. The recombinant plasmid pET-21a (+)-AccCYP336A1 was then transformed into *Escherichia coli* BL21, and a transformed colony was cultured in Luria-Bertani (LB) broth with 40 μg/ml ampicillin (37℃) until the cell density reached 0.4-0.6 (OD_{600}). Expression of the recombinant AccCYP336A1 was induced by adding a final concentration of 0.4 mmol/L isopropylthio-β-D-galactoside (IPTG) for 10 h (24℃). Subsequently, the recombinant AccCYP336A1 protein was purified via the C-term His-tag following the instructions of the manufacturer. The expression of the purified protein was evaluated by 12% sodium dodecyl sulfate polyacrylamide gel electrophoresis (SDS-PAGE) with Coomassie Brilliant Blue staining.

Characterization of recombinant AccCYP336A1 protein by disc diffusion assay

The recombinant AccCYP336A1 was expressed in *E. coli* BL21. Cells transformed with the empty pET-21a (+) vector was used as the control. All the cells were cultivated to the same OD_{600} (OD_{600} = 0.6). Then, all the cells were induced by 0.4 mmol/L isopropylthio-β-D-galactoside (IPTG) for 8 h (24℃). An initial bacterial culture with a density of 10^7 cells/ml was plated on LB-ampicillin agar plates and incubated at 37℃ (1 h). Then, filter discs (8 mm diameter) soaked in different concentrations (0, 10, 20, 40, and 80 mmol/L) of paraquat were placed on the surface of the agar. The cells were then grown for 24 h (37℃).

Anti-AccCYP336A1 preparation and Western blotting analysis

The purified protein was injected subcutaneously into white mice for generation of antibodies as described by Yan et al. [26]. The total proteins were extracted from entire adult bees. A BCA Protein Assay Kit (Nanjing Jiancheng Bioengineering Institute, Nanjing, China) was used to quantify these proteins. Western blotting was performed according to the procedure described by Meng et al. [27] with some modifications. The anti-AccCYP336A1 serum was used as the primary antibody at a 1 : 600 (V/V) dilution. Peroxidase-conjugated goat anti-mouse immunoglobulin G (Dingguo, Beijing, China) was used as the secondary antibody at a 1 : 2,000 (V/V) dilution. A SuperSignal® West Pico Trial Kit (Thermo Scientific Pierce, IL, USA) was used to detect the binding reaction. The house-keeping gene tubulin was used as the control. The tubulin antibody (Beyotime, Jiangsu, China) was used as the house-keeping gene antibody.

Results

Sequence analysis of AccCYP336A1

By using RT-PCR and RACE-PCR, a P450 gene was cloned from *A. cerana cerana* and named *AccCYP336A1*. The open reading frame (ORF) of *AccCYP336A1* is 1,491 bp in length and encodes a predicted protein of 496 amino acids with a predicted molecular mass of 57 kDa and an isoelectric point of 8.8. Genes encoding cytochrome P450 enzymes metabolizing endogenous compounds are generally well conserved [14, 28, 29]. A BLAST search showed that the putative AccCYP336A1 protein is present in many bees. An alignment of the AccCYP336A1 amino acid sequence with the P450 reference sequences from other species demonstrated that the predicted protein is highly homologous to AmCYP336A1 from *A. mellifera*, AfCYP336A1 from *A. florea*, and AdCYP336A1 from *A. dorsata* (Fig. 1A). The identity ranged from 58.52% to 99.26%, suggesting that the P450 family is highly conserved across species.

Phylogenetic analysis

A phylogenetic tree was produced to determine the evolutionary relationship between CYP336A1 in different species. The neighbor-joining method with MEGA 4.0 software was used to build an evolutionary tree containing the predicted amino acid sequence of AccCYP336A1 and other similar sequences from various species (Fig. 1B). In this analysis, AccCYP336A1 and AmCYP336A1 were found to be more closely related to each other than to homologues in other species and they all belong to CYP3 clade. We may safely draw the conclusion that the homologous protein AmCYP336A1 of *A. mellifera* shares a higher similarity with the AccCYP336A1 of *A. cerana cerana* than with those of any other species.

Developmental and tissue-specific expression patterns of AccCYP336A1

We used qRT-PCR to determine the expression patterns of *AccCYP336A1* at different developmental stages and in various tissues. As shown in Fig. 2A, *AccCYP336A1* expression was on average the highest during the egg stage and the lowest during the adult stages. In the larval stage, the highest level of expression appeared at the first larval instar stage; in the pupal stage, the expression levels in the brown eyes pupae were higher than the other phases. The spatial expression patterns showed that *AccCYP336A1* transcripts were most abundant in the epidermis (Fig. 2B). Although the expression levels in the brain, midgut and muscle were nearly the same, none of the expression levels in these tissues were higher than that in the epidermis (Fig. 2B). Total RNA was collected from the head, epidermis, muscle, hemolymph, rectum and midgut.

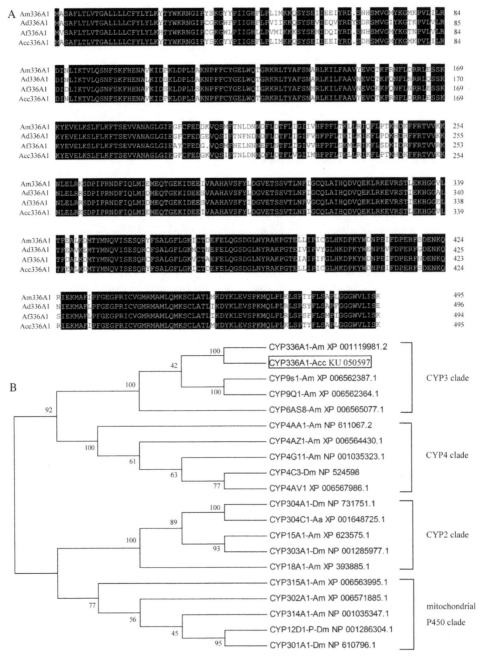

Fig. 1 Characterization of cytochrome P450 monooxygenases (P450) from various species. A. Multiple alignment analysis of cytochrome P450 monooxygenases (P450) from various species. A multiple amino acid sequence alignment of AccCYP336A1 from *A. cerana cerana*, AmCYP336A1 from *A. mellifera*, AfCYP336A1 from *A. florea*, and AdCYP336A1 from *A. dorsata*. The sequences were obtained from the NCBI database. B. Phylogenetic tree created by the neighbor-joining method by using MEGA 4.0 software with sequences from other species.

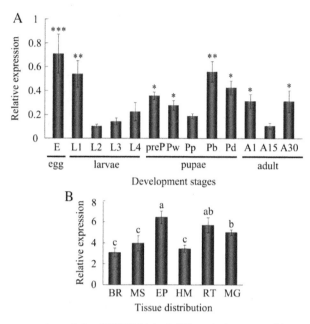

Fig. 2 The relative expression of *AccCYP336A1* at different stages and in several tissues. A. The different developmental stages: egg, larvae (L1-L4, from the first to the fourth instars), pupae (preP prepupae, Pw white-eyed, pupae, Pp pink-eyed pupae, Pb brown-eyed pupae and Pd dark-eyed pupae), and adult workers (A1, 1 day post-emergence; A15, 15 days post-emergence; A30, 30 days post-emergence). B. Expression analysis of *AccCYP336A1* in various tissues: brain (BR), muscle (MS), epidermis (EP), hemolymph (HM), rectum (RT) and midgut (MG). The tissues were collected from 15 days post-emergence adult workers Chinese honeybees. The data are given as the mean ± SE of three replicates. The tissue letters above the bar indicate significant differences ($P < 0.0001$) as determined by Duncan's multiple range tests using SAS software version 9.1 and the one-way analysis of variance (ANOVA) and the least significant difference (LSD) test were used to analyze the differences at developmental stages.

Expression profiles of *AccCYP336A1* under conditions of oxidative stresses

qRT-PCR was used to characterize the transcriptional expression patterns of *AccCYP336A1* after exposure to several types of oxidative stress. Surprisingly, *AccCYP336A1* was induced by all of the tested treatments (Fig. 3). *AccCYP336A1* robustly responded to cold temperatures (4℃) treatment, which dramatically induced the *AccCYP336A1* transcripts to a maximum level at 3 h (Fig. 3A). Similarly, after heat (42℃) treatment, the expression level of *AccCYP336A1* was enhanced slightly compared to the control, reaching its maximum at 0.25 h (Fig. 3B). Furthermore, with the ultraviolet and H_2O_2 treatments, the expression of *AccCYP336A1* was dramatically induced and quickly climbed to a maximum value. Moreover, the results also show that the expression of *AccCYP336A1* was enhanced more drastically after exposure to ultraviolet treatment (Fig. 3C and D). Although *AccCYP336A1* expression changed after treatment with all tested pesticides, it seemed that the induction varied by treatment type, as thiamethoxam and methomyl caused a much more rapid increase in expression than deltamethrin and phoxim (Fig. 3E, F, G and H). The above results suggest that AccCYP336A1 may play an important part in the response to various reactive oxidative stresses.

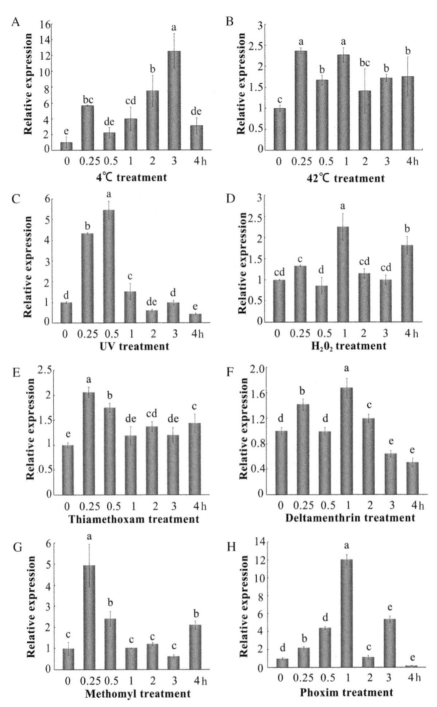

Fig. 3 Expression profiles of *AccCYP336A1* under antioxidant stresses. qPCR was performed on total RNA extracted from 15 days post-emergence adult bees. These stresses are as follows: 4℃ (A), 42℃ (B), UV (C), H_2O_2 (D), thiamethoxam (E), deltamethrin (F), methomyl (G), and phoxim (H). The data are given as the mean ± SE of three replicates. The letters above the bar indicate significant differences ($P < 0.0001$) as determined by Duncan's multiple range tests using SAS software version 9.1.

Expression, purification and characterization of recombinant AccCYP336A1 protein

To further characterize the AccCYP336A1 protein, we obtained a part of the *AccCYP336A1* ORF. Through the analysis of the AccCYP336A1 amino acid sequence, the hydrophobicity is too high to support the expression of a recombinant protein in the N-terminal sequence (Fig. 4A). Through the analysis of the AccCYP336A1 transmembrane segments, potential transmembrane segments were identified in the N-terminal sequence (Fig. 4B). According to the results of the above analysis, the 240-460 amino acid sequence of the *AccCYP336A1* ORF was cloned into the pET-21a (+) vector, which was digested with *Nde* I and *Not* I . The recombinant plasmid was then transformed into *E. coli* BL21 cells. The result of SDS-PAGE showed that the target protein was induced by IPTG at 28℃ for 10 h (Fig. 5). The target protein contained the AccCYP336A1 protein and a C-term His-tag. We characterized the activity of the recombinant AccCYP336A1 protein by using the disc diffusion method. We cultured the bacteria with part of AccCYP336A1 and the control bacteria to achieve the same cell density. The AccSOD2 was proved that may play an important role in protecting Chinese honeybees from oxidative stress by Jia et al. [30]. The death zones around the paraquat filters were

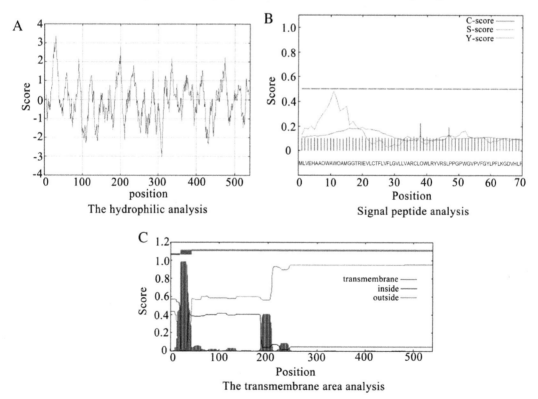

Fig. 4 The hydrophilic analysis, signal peptide analysis and transmembrane analysis. The higher the ordinate value is, the higher the hydrophobicity. It is generally believed that if the value is greater than 2, the hydrophobicity is high to support the expression of a recombinant protein (A); The red area denotes the transmembrane segments and the strong hydrophobic area (B). (For color version, please sweep QR Code in the back cover)

smaller in diameter on the plates containing *E. coli* overexpressing AccSOD2 compared with plates containing control bacteria, which we used as the positive control. The killing zones of *E. coli* overexpressing the target AccCYP336A1 exposed to paraquat were smaller than those of the bacteria with empty pET-21a (+) vector (Fig. 6), which definitively demonstrated that AccCYP336A1 contributes to cellular resistance to oxidative stress.

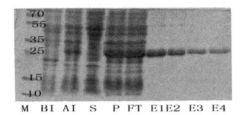

Fig. 5 The expression of the recombinant AccCYP336A1 protein. Lane M, protein molecular weight marker; lane BI, expression of AccCYP336A1 without IPTG induction; lane AI, expression of AccCYP336A1 after IPTG induction. Lane S, supernatant. Lane P-FT, inclusion body of AccCYP336A1 after IPTG induction. E1-E4, purified recombinant AccCYP336A1.

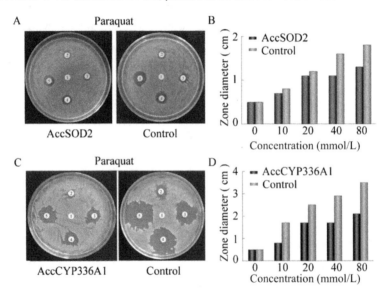

Fig. 6 The resistance of bacterial cells overexpressing AccCYP336A1 to paraquat. A. *E. coli* cells were transformed with a plasmid to overexpress AccCYP336A1 by IPTG inducing. *E. coli* BL21 bacteria transformed with pET-21a (+) (vector only) without IPTG inducing were used as the control. B. The halo diameters of the killing zones are compared in the histograms. The paraquat concentrations of discs 1-5 are 0, 10, 20, 40, and 80 mmol/L, respectively. The data are the mean ± SE of three replicates.

Western blotting analysis

To further understand the expression patterns of AccCYP336A1 in response to various types of ROS damage, Western blotting analysis was used to assess the AccCYP336A1 changes after 4℃ (A), UV (B), phoxim (C) and methomyl (D) treatments (Fig. 6). Anti-AccCYP336A1 was used to detect AccCYP336A1. Following 4℃ treatment for 1 h, the protein level of AccCYP336A1 was clearly induced (Fig. 7A). After exposure to UV for 3 h, the AccCYP336A1 expressions were also induced (Fig. 7B). After exposure to phoxim and methomyl, the AccCYP336A1 was obviously induced rapidly (Fig. 7C and D). Although the Western blotting data for UV treatment and pesticides (phoxim and methomyl) treatments does not correlate with qRT-PCR data (Fig. 3) with respect to time points, the total tendency was induced. The reason for this discrepancy might because that there is a certain hysteresis and intervention by other factors at the protein level.

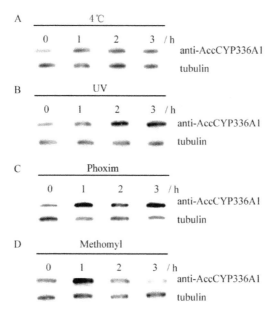

Fig. 7 Western blotting analysis of AccCYP336A1 changes after 4℃ (A), UV (B), phoxim (C) and methomyl treatment (D). A part of the AccCYP336A1 protein was immunoblotted with anti-AccCYP336A1. The signal of the binding reactions was visualized with HRP substrates.

Discussion

The cytochrome P450 proteins form an important body detoxification enzyme system, and are involved in pathways of the biosynthesis and the degradation of endogenous metabolites. They are known to play a vital role in protecting organisms against environmental stress [31, 32]. Studies on the insect P450 have mainly focused on their roles in insecticide resistance [33, 34]. However, a few studies have systematically studied the antioxidative stress that insects experience during their lives.

To achieve this goal, we imitated several types of ROS damage that *A. cerana cerana* may encounter during their lifespan and explored the resulting changes in P450 expression at the mRNA and protein levels to look for evidence of their antioxidant functionality. We first isolated a predicted P450 gene from *A. cerana cerana* and named it *AccCYP336A1*. Sequence analysis revealed that AccCYP336A1 contained highly considerable conservation believed to determine substrate specificity. Meanwhile, the phylogenetic tree indicated that CYP336A1 orthologs are present in many insects and crustaceans and AccCYP336A1 belongs to CYP3 clade, sharing the highest degree of homology with AmCYP336A1 of *A. mellifera*. Taken together, these results indicate that AccCYP336A1 is a member of the P450 family.

Knowledge on the stages and tissue distributions of *AccCYP336A1* mRNA could be useful to better understand their functional mechanisms. To achieve this goal, we investigated the expression patterns of *AccCYP336A1*. The transcription level of *AccCYP336A1* was strikingly higher in the egg and pupal stages than that in the adult stages. Corona and Robinson [35]. demonstrated that ROS can cause oxidant damage, especially in fast-growing organisms, which is caused by high oxygen levels. Thus, we infer that AccCYP336A1 may play an essential role in the early developmental stages and may be essential for the development of the honeybee or a greater exposure of stages to ROS damage. Furthermore, the tissue-specific expression analysis of the AccCYP336A1 gene showed greater mRNA accumulation in the epidermis. It was believed that the influence of AccCYP336A1 might play an important role in preventing damage due to ROS in the epidermis.

Previous research has revealed that ROS can be induced by temperature, H_2O_2, UV radiation, and insecticides [36, 37]. Temperature is an abiotic environmental factor that causes physiological changes in organisms [38]. Indeed, temperature has been described as one of the key mediators of ROS generation. Studies have shown that heat-shock stress can result in polyamine oxidation and the generation of H_2O_2 and that cold stress can induce hepatocytes and liver endothelial cells to undergo apoptosis due to ROS damage [39, 40]. In this study, we found that after cold (4℃) or heat (42℃) treatment, the transcription levels of *AccCYP336A1* increased before decreasing. These results suggest that *AccCYP336A1* may involve in the process that regulates body temperature and the heating ceiling, thus protecting honeybees from ROS damage. UV radiation and H_2O_2 treatment, a typical oxidant and antioxidant, both cause oxidative damage [41-44]. Here, the qRT-PCR results showed that *AccCYP336A1* can be induced by UV radiation and H_2O_2 treatment, and the Western blotting assays further confirmed the results. These findings support the hypothesis that *AccCYP336A1* plays an important role in protecting cells against oxidative damage. Pesticides are the main threat to a honeybee's life. Pesticides destroy the biochemical and physiological functions of erythrocytes and lymphocytes by causing lipid biomembrane oxidation [2]. For many years, the available data on P450 suggested that oxidative metabolism was not a major process in parasitic nematodes and that P450 activity was generally absent, or present only at a low level, in parasite extracts [45, 46]. However, there is now strong evidence that this is not the case. The increased expression of *AccCYP336A1* observed here was in agreement with previous studies in which researchers observed the enhanced activity of metabolic detoxification as a major mechanism for pesticide resistance [47-49]. Our data also showed that the expression of *AccCYP336A1* was upregulated after exposure to pesticide treatment at the mRNA and protein levels. Taken together, all the results indicated that the *AccCYP336A1* would be induced after a series of oxidative stress, which indicates that *AccCYP336A1* quite possibly plays a significant role in the response of *A. cerana cerana* to ROS. Previous studies have identified many genes involved in the response to adverse reactive oxidative stress in *A. cerana cerana*. The several stressors could lead to the transformation of these genes in different expression patterns. For example, cold temperature (4℃) treatment induced the expression of *AccSOD2* [30]. UV treatment inhibited the expression of *AccTpx-3* [50].

Pyriproxyfen treatment inhibited the expression of *AccTpx5* [51]. Although part of these stressors caused an accumulation of gene expressions, the cumulates and patterns are significantly different from those of *AccCYP336A1*. This phenomenon showed that the several stressors would not lead to an accumulation of all other genes and that the *AccCYP336A1* mRNA is specifically upregulated in response to ROS damage in *A. cerana cerana*.

Because several environmental factors lead to an accumulation of ROS, cells react to ROS threat with a variety of enzymatic and non-enzymatic protection mechanisms [52]. In this study, after exposure to paraquat treatment, the killing zones of *E. coli* overexpressing AccCYP336A1 were smaller than those for the control bacteria. This result showed that the overexpression of AccCYP336A1 increased the resistance of the bacterial cells to ROS.

In conclusion, we identified and characterized a P450 gene from *A. cerana cerana* named *AccCYP336A1*. We also imitated several stresses that *A. cerana cerana* may experience during their lifetime and demonstrated corresponding expression changes in AccCYP336A1 at the mRNA and protein levels. In addition, we explored the resistance of bacteria overexpressing AccCYP336A1 to determine the role that AccCYP336A1 may play. The experimental results provided evidence for the functionality of AccCYP336A1 in antioxidation. The results of this study are helpful in further revealing antioxidant mechanisms and physiological responses in insects exposed to reactive oxidative stress and may help facilitate the identification of the functional mechanisms of antioxidation.

References

[1] Weinstock, G. M., Robinson, G. E., Gibbs, R. A., et al. Insights into social insects from the genome of the honeybee *Apis mellifera*. Nature. 2006, 443, 931-949.

[2] Narendra, M., Bhatracharyulu, N., Padmavathi, P., Varadacharyulu, N. Prallethrin induced biochemical changes in erythrocyte membrane and red cell osmotic haemolysis in human volunteers. Chemosphere. 2007, 67, 1065-1071.

[3] Halliwell, B., Gutterridge, J. M. C. Production against Radical Damage: Systems with Problems. Free Radicals in Biology and Medicine. Second ed. Clarendon Press, Oxford. 1989.

[4] Imlay, J. A., Linn, S. DNA damage and oxygen radical toxicity. Science. 1988, 240, 1302-1309.

[5] Stadtman, E., Levine, R. Free radical-mediated oxidation of free amino acids and amino acid residues in proteins. Amino Acids. 2003, 25, 207-218.

[6] Green, D. R., Reed, J. C. Mitochondria and apoptosis. Science. 1998, 281, 1309-1312.

[7] Collins, A., Williams, V., Evans, J. Sperm storage and antioxidative enzyme expression in the honey bee, *Apis mellifera*. Insect Mol Biol. 2004, 13, 141-146.

[8] Dubovskiy, I. M., Martemyanov, V. V., Vorontsova, Y. L., Rantala, M. J., Gryzanova, E. V., Glupov, V. V. Effect of bacterial infection on antioxidant activity and lipid peroxidation in the midgut of *Galleria mellonella* L. larvae (Lepidoptera, Pyralidae). Comp Biochem Physiol C: Pharmacol Toxicol. 2008, 148, 1-5.

[9] Felton, G. W., Summers, C. B. Antioxidant systems in insects. Arch Insect Biochem Physiol. 1995, 29, 187-197.

[10] Wang, Y., Oberley, L. W., Murhammer, D. W. Antioxidant defense systems of two Lepidopteran insect cell lines. Free Radic Biol Med. 2001, 30, 1254-1262.

[11] Matteis, F. D., Ballou, D. P., Coon, M. J., Estabrook, R. W., Haines, D. C. Peroxidase-like activity of uncoupled cytochrome P450 studies with bilirubin and toxicological implications of uncoupling. Biochem Pharmacol. 2012, 84, 374-382.

[12] Nelson, D. R. Metazoan cytochrome P450 evolution. Comp Biochem Physiol C Pharmacol Toxicol

[13] Li, X., Schuler, M., Berenbaum, M. Molecular mechanisms of metabolic resistance to synthetic and natural xenobiotics. Annu Rev Entomol. 2007, 52, 231e253.
[14] Feyereisen, R. Insect cytochrome P450. Compr Mol Insect Sci. 2005, 4, 1-77.
[15] Nelson, D., Kamataki, T., Waxman, D., et al. The P450 superfamily: update on new sequences, gene mapping, accession numbers, early trivial names of enzymes, and nomenclature. DNA Cell Biol. 1993, 12, 1-51.
[16] Anzenbacher, P., Anzenbacherova, E. Cytochromes P450 and metabolism of xenobiotics. Cell Mol Life Sci. 2001, 58, 737-747.
[17] Nebert, D. W., Russell, D. Clinical importance of the cytochromes P450. Lancet. 2002, 360, 1155-1162.
[18] Thomas, J. H. Rapid birth-death evolution specific to xenobiotic cytochrome P450 genes in vertebrates. PLoS Genet. 2007, 3, e67.
[19] Berenbaum, M. Postgenomic chemical ecology: from genetic code to ecological interactions. J Chem Ecol. 2002, 28, 873-896.
[20] Claudianos, C., Ranson, H., Johnson, R. M., Biswas, S., Schuler, M. A., Berenbaum, M. R., Feyereisen, R., Oakeshott, J. G. A deficit of detoxification enzymes: pesticide sensitivity and environmental response in the honeybee. Insect Mol Biol. 2006, 15, 615-636.
[21] Shi, W., Sun, J., Xu, B., Li, H. Molecular characterization and oxidative stress response of a cytochrome P450 gene (*CYP4G11*) from *Apis cerana cerana*. Z. Naturforsch. C. 2013, 68, 509-521.
[22] Gilbert, L. I. Halloween genes encode P450 enzymes that mediate steroid hormone biosynthesis in *Drosophila melanogaster*. Mol Cell Endocrinol. 2004, 215, 1-10.
[23] Nikou, D., Ranson, H., Hemingway, J. An adult-specific CYP6 P450 gene is overexpressed in a pyrethroid-resistant strain of the malaria vector, *Anopheles gambiae*. Gene. 2003, 318, 91-102.
[24] Sasabe, M., Wen, Z., Berenbaum, M. R., Schuler, M. A. Molecular analysis of CYP321A1, a novel cytochrome P450 involved in metabolism of plant allelochemicals (furanocoumarins) and insecticides (cypermethrin) in *Helicoverpa zea*. Gene. 2004, 338, 163-175.
[25] Corona, M., Velarde, R. A., Remolina, S., Moran-Lauter, A., Wang, Y., Hughes, K. A., Robinson, G. E. Vitellogenin, juvenile hormone, insulin signaling, and queen honey bee longevity. Proc Natl Acad Sci USA. 2007, 104, 7128-7133.
[26] Yan, H., Jia, H., Wang, X., Gao, H., Guo, X., Xu, B. Identification and characterization of an *Apis cerana cerana* Delta class glutathione S-transferase gene (*AccGSTD*) in response to thermal stress. Naturwissenschaften. 2013, 100, 153-163.
[27] Meng, F., Zhang, L., Kang, M., Guo, X., Xu, B. Molecular characterization, immunohistochemical localization and expression of a ribosomal protein L17 gene from *Apis cerana cerana*. Arch Insect Biochem Physiol. 2010, 75, 121-138.
[28] Rewitz, K. F., Gilbert, L. I. *Daphnia* Halloween genes that encode cytochrome P450s mediating the synthesis of the arthropod molting hormone: evolutionary implications. BMC Evol Biol. 2008, 8, 60.
[29] Rewitz, K. F., O'Connor, M. B., Gilbert, L. I. Molecular evolution of the insect Halloween family of cytochrome P450s: phylogeny, gene organization and functional conservation. Insect Biochem Mol Biol. 2007, 37, 741-753.
[30] Jia, H., Sun, R., Shi, W., Yan, Y., Li, H., Guo, X., Xu, B. Characterization of a mitochondrial manganese superoxide dismutase gene from *Apis cerana cerana* and its role in oxidative stress. J Insect Physiol. 2014, 60, 68-79.
[31] Gilbert, M., Wilkinson, C. An inhibitor of microsomal oxidation from gut tissues of the honey bee (*Apis mellifera*). Comp. Biochem. Physiol. B. 1975, 50, 613-619.
[32] Yu, S., Robinson, F., Nation, J. Detoxication capacity in the honey bee, *Apis mellifera*. Pestic Biochem Physiol. 1984, 22, 360-368.
[33] Pilling, E., Bromleychallenor, K., Walker, C., Jepson, P. Mechanism of synergism between the pyrethroid insecticide λ-cyhalothrin and the imidazole fungicide prochloraz, in the honeybee (*Apis*

mellifera L.). Pestic Biochem Physiol. 1995, 51, 1-11.

[34] Suchail, S., Debrauwer, L., Belzunces, L. P. Metabolism of imidacloprid in *Apis mellifera*. Pest Manag Sci. 2004, 60, 291-296.

[35] Corona, M., Robinson, G. E. Genes of the antioxidant system of the honey bee: annotation and phylogeny. Insect Mol Biol. 2006, 15, 687-701.

[36] Lushchak, V. Environmentally induced oxidative stress in aquatic animals. Aquat Toxicol. 2011, 101, 13-30.

[37] Kottuparambil, S., Shin, W., Brown, M., Han, T. UV-B affects photosynthesis, ROS production and motility of the freshwater flagellate, *Euglena agilis* Carter. Aquat Toxicol. 2012, 122, 206-213.

[38] An, M. I., Choi, C. Y. Activity of antioxidant enzymes and physiological responses in ark shell, *Scapharca broughtonii*, exposed to thermal and osmotic stress: effects on hemolymph and biochemical parameters. Comp. Biochem. Physiol. B. 2010, 155, 34-42.

[39] Harari, P. M., Fuller, D. J., Gerner, E. W. Heat-shock stimulates polyamine oxidation by 2 distinct mechanisms in mammalian-cell cultures. Int J Radiat Oncol. 1989, 16, 451-457.

[40] Rauen, U., Polzar, B., Stephan, H., Mannherz, H. G., De Groot, H. Cold-induced apoptosis in cultured hepatocytes and liver endothelial cells: mediation by reactive oxygen species. FASEB J. 1999, 13, 155-168.

[41] Goldshmit, Y., Erlich, S., Pinkas-Kramarski, R. Neuregulin rescues PC12-ErbB4 cells from cell death induced by H_2O_2 regulation of reactive oxygen species levels by phosphatidylinositol 3-kinase. J Biol Chem. 2001, 276, 46379-46385.

[42] Casini, A., Ferrali, M., Pompella, A., Maellaro, E., Comporti, M. Lipid peroxidation and cellular damage in extrahepatic tissues of bromobenzene-intoxicated mice. Am J Pathol. 1986, 123, 520.

[43] Nguyen, T. T., Michaud, D., Cloutier, C. A proteomic analysis of the aphid *Macrosiphum euphorbiae* under heat and radiation stress. Insect Biochem Mol Biol. 2009, 39, 20-30.

[44] Schauen, M., Hornig-Do, H. T., Schomberg, S., Herrmann, G., Wiesner, R. J. Mitochondrial electron transport chain activity is not involved in ultraviolet A (UVA)-induced cell death. Free Radic Biol Med. 2007, 42, 499-509.

[45] Pemberton, K., Barrett, J. The detoxification of xenobiotic compounds by *Onchocerca gutturosa* (Nematoda: Filarioidea). Int J Parasitol. 1989, 19, 875-878.

[46] Yadav, M., Singh, A., Rathaur, S., Liebau, E. Structural modeling and simulation studies of *Brugia malayi* glutathione-S-transferase with compounds exhibiting antifilarial activity: implications in drug targeting and designing. J Mol Graph Model. 2010, 28, 435-445.

[47] Eziah, V. Y., Rose, H. A., Wilkes, M., Clift, A. D. Biochemical mechanisms of insecticide resistance in the diamondback moth (DBM), *Plutella xylostella* L. (Lepidopterata: Yponomeutidae), in the Sydney region. Australia Aust J Entomol. 2009, 48, 321-327.

[48] Zhu, F., Sams, S., Moural, T., Haynes, K. F., Potter, M. F., Palli, S. R. RNA interference of NADPH-cytochrome P450 reductase results in reduced insecticide resistance in the bed bug, *Cimex lectularius*. PLoS One. 2012, 7, e31037.

[49] Johnson, R., Dahlgren, L., Siegfried, B., Ellis, M. Acaricide, fungicide and drug interactions in honey bees (*Apis mellifera*). PLoS One. 2013, 8, e54092.

[50] Yao, P., Lu, W., Meng, F., Wang, X., Xu, B., Guo, X. Molecular cloning, expression and oxidative stress response of a mitochondrial thioredoxin peroxidase gene (*AccTpx-3*) from *Apis cerana cerana*. J Insect Physiol. 2013, 59, 273-282.

[51] Yan, Y., Zhang, Y., Huaxia, Y., Wang, X., Yao, P., Guo, X., Xu, B. Identification and characterisation of a novel 1-Cys thioredoxin peroxidase gene (*AccTpx5*) from *Apis cerana cerana*. Comp Biochem Physiol B. 2014, 172-173, 39-48.

[52] Ahmad, S. Oxidative stress from environmental pollutants. Arch Insect Biochem Physiol. 1995, 29, 135-157.

3.2 Roles of a Mitochondrial *AccSCO2* Gene from *Apis cerana cerana* in Oxidative Stress Responses

Haihong Jia, Manli Ma, Na Zhai, Zhenguo Liu, Hongfang Wang, Xingqi Guo, Baohua Xu

Abstract: In eukaryotes, cytochrome c oxidase (COX) is a multimeric protein complex that is the last enzyme in the respiratory electron transport chain of mitochondria. Syntheses of cytochrome c oxidase (SCO) proteins are copper-donor chaperones involved in metalation of the CuA redox center of COX. However, its other precise actions are not yet understood. Here, we report the characterization of *AccSCO2* from *Apis cerana cerana* (*Acc*). Our data showed that *AccSCO2* expression was induced by cold (4℃), $CdCl_2$, $HgCl_2$, ultraviolet (UV) light, and H_2O_2 and was inhibited by different pesticide treatments. In addition, a disc diffusion assay of recombinant AccSCO2, AccSCO2-R1, and AccSCO2-R2 proteins showed that they played a functional role in protecting cells from oxidative stress involved in copper-dependent manner. Further, following knockdown of *AccSCO2* in *A. cerana cerana* by using RNA interference (RNAi), the expression levels of most antioxidant genes (*AccGSTD*, *AccGSTO1*, *AccGSTs4*, *AccSOD1*, *AccSOD2*, etc.) were significantly decreased in the *AccSCO2*-silenced bees compared with the control bees. Moreover, the antioxidant enzymatic activities of superoxide dismutase (SOD), peroxidase (POD) and catalase (CAT) were all lower in the silenced bees than those in the control bees. Finally, the *in vivo* activity of COX was measured after the knockdown of *AccSCO2*, which revealed a strong reduction in COX activity in the silenced bees. Thus, we hypothesize that *AccSCO2* plays important roles in cellular stress responses and anti-oxidative processes, which help to regulate the production of mitochondrial reactive oxygen species and/or the impairment of mitochondrial activity under oxidative stress.

Keywords: *Apis cerana cerana*; *AccSCO2*; Oxidative stress; RNA inference; Cytochrome c oxidase

Introduction

Cytochrome c oxidase (COX; respiratory complex IV) is the last enzyme in the mitochondrial respiratory chain in aerobic organisms. This enzyme catalyzes the transfer of electrons from cytochrome c to molecular oxygen, a biochemical reaction that contributes to the proton gradient essential for aerobic ATP production. In higher eukaryotes, COX is composed of 13 structural subunits and is embedded in the inner mitochondrial membrane [1]. In humans, defects in the assembly and function of COX are frequent causes of oxidative phosphorylation disorders, which primarily affect organs with high energy demands [2, 3]. Mutations in COX subunits or COX assembly factors are among the most common defects that cause mitochondrial diseases, including encephalomyopathies, Leigh syndrome, hypertrophic cardiomyopathies and fatal lactic acidosis [4].

Many genes required for the assembly of mitochondrial COX have been identified in several organisms. COX assembly alone has been reported to require over 20 different genes in yeast [5]. Syntheses of cytochrome c oxidase (SCO) are crucial in metalation of the CuA redox center of COX and are conserved among all kingdoms of life [6, 7]. Eukaryotic SCO proteins generally share several common features and contain 250-300 amino acid residues. According to previous reports, SCO proteins possess a mitochondrial-targeting peptide in their N-terminal domains and a CXXXC motif, with a conserved pair of cysteines separated by three amino acid residues, in their C-terminal domains; this motif has been proposed to bind copper and to be responsible for the main biological functions of these proteins [8, 9].

In many organisms, SCO proteins are required for COX activity, and they have also been suggested to fulfill other essential functions. Mutations in two human SCO-encoding genes, *SCO1* and *SCO2*, result in tissue-specific COX deficiencies associated with distinct clinical phenotypes. However, the precise actions of these proteins in *Apis cerana cerana* remain unclear. Vertebrates possess two paralogous SCO-encoding genes, whereas other eukaryotes apparently only require a single SCO protein, and the reason for this discrepancy is unknown [10].

Organisms are constantly exposed to diverse types of unfavorable environmental stressors, including temperature fluctuations, heavy metal, pesticides, and ultraviolet (UV) radiation, all of which are believed to induce the formation of reactive oxygen species (ROS) [11, 12]. A high level of ROS contributes to aging and disease through the oxidation of nucleic acids, proteins, and lipid membranes[13]. In cancer cells, several functional changes have been observed, such as the increased production of mitochondrial ROS, decreased oxidative phosphorylation, and a corresponding increase in glycolysis [14]. In eukaryotic cells, SCO2 is located within the mitochondrion [10], an organelle that is known to be a major source of ROS production and is a main target of ROS-induced cellular injury [15]. Therefore, mitochondrial *AccSCO2* may play a vital role in cellular defense against ROS-mediated oxidative damage.

The use of animal models (including insects) has enabled the achievement of great advances in understanding of the pathophysiology of diseases related to cellular oxidative stress. In particular, several studies using the powerful genetic tools available in *Drosophila* have provided significant information on genes with human homologs. For example, mutations in several *Drosophila* genes related to mitochondrial function have been demonstrated to be particularly dependent on an adequate energy supply, which is provided by oxidative phosphorylation [10]. *A. cerana cerana*, a well-known pollinator of flowering plants, is highly susceptible to numerous environmental stresses and is currently suffering from significant population decreases [16]. Therefore, we aimed to study the function of *AccSCO2* to gain further insights into the characteristics of this gene and its role in cellular defense systems. Our findings have provided a foundation for further research of human antioxidants.

Although SCOs have been identified in different species, to the best of our knowledge, the functions of these proteins in *A. cerana cerana* have not yet been investigated. In this study, we explored the role of *AccSCO2* by performing gene expression analyses, disc diffusion assay, RNA interference (RNAi) of *AccSCO2*, and analyses of superoxide dismutase (SOD), peroxidase (POD), catalase (CAT) and COX activities *in vivo* after the knockdown of *AccSCO2*. Analysis of physiological parameters revealed that *AccSCO2* played important roles in cellular stress responses and anti-oxidative processes in *A. cerana cerana*. Overall, our work has laid the foundation for further characterization of this important *SCO* gene in humans.

Materials and Methods

Insects and treatments

The insects (*A. cerana cerana*) used in this study were maintained in the experimental apiary of the College of Animal Science and Technology, Shandong Agricultural University (Taian, Shandong, China). Adult worker honeybees (15-30 days post-emergence) were randomly collected at the entrance of the hive and kept under standard conditions, as described by Alaux et al. [17]. The 15-day worker bees were divided into nine groups of 40 individuals each and were subjected to various treatments. The bees in groups 1 and 2 were exposed to severe cold (4℃) and UV radiation (30 MJ/cm^2), respectively, for 0, 0.5, 1, 1.5, or 2 h. In addition, the bees in group 3 were injected with 20 µl (50 mmol/L H_2O_2/worker) H_2O_2 and treated for 0, 0.5, 1, 1.5 or 2 h. Further, the bees in groups 4 and 5 were treated with $HgCl_2$ (by addition of 3 mg/ml to food) and $CdCl_2$ (by addition of 10 mg/ml to food), respectively, for 0, 3, 6, 9, or 12 h. Moreover, the bees in groups 6-9 were each exposed to a different pesticide as previously described [13]. Untreated honeybees were used as controls. All samples were collected at the designated time points (each sample included three individuals) and stored at −80℃ for RNA and protein extractions. Each treatment was repeated at least twice.

RNA extraction and quantitative real-time PCR

For quantitative real-time PCR (qRT-PCR) analysis, total RNA extraction and the first-strand cDNA synthesis were performed by using the method described by Jia et al. [13]. The PCR mixture (10 µl SYBR Premix Ex *Taq*, 1.0 µl cDNA, 0.5 µl of each primer (10 mmol/L each), and 8 µl PCR-grade water in a final volume of 20 µl) and reaction conditions (95℃ for 30 s, followed by 40 cycles at 95℃ for 5 s, 55℃ for 15 s, and 72℃ for 15 s) used have been previously described Jia et al. [13]. The house-keeping gene *β-actin* (No. HM640276) was used as an internal control to quantify relative transcription levels. Relative expression data were automatically analyzed by using CFX Manager Software version 1.1 after the PCR program was run. Three independent experiments were analyzed, and the expression levels were calculated by using the $2^{-\Delta\Delta C_t}$ comparative CT method [18]. The sequences of the primers used for qRT-PCR are listed in Table S1.

Protein expression and disc diffusion assay

To obtain an AccSCO2 recombinant, the open reading frame (ORF) of *AccSCO2* was amplified by using the primers BD1 and BD2 and inserted into an expression vector, pET-30a (+). Deletion mutants of AccSCO2-R1 and AccSCO2-R2 were obtained byk using overlapping PCR. The primers used were shown in Table S1. Then, the recombinant plasmids pET30a (+)-AccSCO2, PET30a (+)-AccSCO2-R1, and PET30a (+)-AccSCO2-R2 were transformed into *E. coli* strain BL21 (DE3) for protein expression. The recombinant AccSCO2, AccSCO2-R1, and AccSCO2-R2 *E. coli* cells were induced with 0.4 mmol/L isopropylthio-β-D-galactoside (IPTG) and grown at 37℃ for 8 h. The recombinant protein was analyzed by 12% sodium dodecyl sulfate-polyacrylamide gel electrophoresis (SDS-PAGE), followed by Coomassie brilliant blue staining. To evaluate the activity of recombinant AccSCO2, AccSCO2-R1, and AccSCO2-R2, disc diffusion assays were performed as described by Yan et al. [19]. Approximately 5×10^8 *E. coli* cells overexpressing *AccSCO2* were overlaid onto LB-kanamycin agar and incubated at 37℃ for 1 h. *E. coli* BL21 cells containing an empty pET-30a (+) vector were used as controls. Filter discs (6 mm diameter) soaked in different concentrations of paraquat (0, 50, 100, 200, and 400 mmol/L), cumene hydroperoxide (0, 10, 20, 40, and 50 mmol/L), or $CdCl_2$ (0, 50, 100, 200, and 400 mmol/L) were placed on the agar surface. Following incubation at 37℃ for 12 h, the killing zones around the paper discs were measured as described by Burmeister et al. [20].

Protein extraction, production of anti-AccSCO2 and immunoblot analysis

Total protein was extracted from honeybees subjected to the above mentioned treatments by using a Tissue Protein Extraction Kit (ComWin Biotech, Beijing, China) according to the protocol of the manufacturer. Protein concentrations were quantified by using a BCA Protein Assay Kit (Nanjing Jiancheng Bioengineering Institute). The purified recombinant AccSCO2 protein was injected subcutaneously into white mice (China Biologic Products, Inc., Shandong) to generate the AccSCO2 antibody, as described by Chen et al. [21]. Anti-α-tubulin antibody (Beyotime Biotechnology, Shanghai, China) was used as the reference for Western blotting analysis. Immunoblot detection was performed as previously described [21].

Subcellular localization analysis of AccSCO2, AccSCO2-R1, and AccSCO2-R2

The ProteoPrep™ Universal Extraction Kit (Sigma) is designed to separate the soluble cytoplasmic and membrane proteins. By using the Universal Extraction Kit (Sigma-Aldrich, USA), the total membrane protein and the soluble fraction were extracted from *E. coli* cells according to the protocol of the manufacturer. Then, the membranes and the soluble fraction proteins were analyzed by 12% sodium dodecyl sulfate-polyacrylamide gel electrophoresis (SDS-PAGE) of the recombinant AccSCO2, AccSCO2-R1, and AccSCO2-R2. Anti-AccSCO2 antibody was used for Western blotting analysis. To evaluate the localization of AccSCO2, AccSCO2-R1, and AccSCO2-R2, immunoblot detection was performed as described above.

RNAi of *AccSCO2*

An RNAi experiment was performed to silence *AccSCO2* expression. The double-stranded RNAs (dsRNAs) dsRNA-*AccSCO2* and dsRNA-*GFP* were prepared by using RiboMAX T7 large-scale RNA production systems (Promega, Madison, WI, USA), as described in an established protocol [22]. No *A. cerana cerana* genes share homology with dsRNA-*GFP*; thus, *GFP* (GenBank accession number: U87974) was used as a control [23]. The primers SA1/SA2 and GA1/GA2, containing T7 polymerase promoter sequences at their ends, were used to amplify dsRNA-*AccSCO2* by RT-PCR. Fifteen-day worker bees were selected for use in the RNAi experiments. Then, dsRNA-Acc*SCO2* (8 μg/individual) or dsRNA-*GFP* (8 μg/individual) was injected into each adult between the first and the second abdominal segments by using a microsyringe. Uninjected bees were used as controls. The silencing efficiency was determined by qRT-PCR. The silenced bees were used for further analyses of ROS-related antioxidant genes/enzymes. This experiment was repeated three times.

Determination of COX activity and antioxidant enzymes activity

Crude total protein was extracted from the silenced honeybees in 1 ml of isolation medium by using the Tissue Protein Extraction Kit (ComWin Biotech, Beijing, China) according to the protocol of the manufacturer and was quantified by using the BCA Protein Assay Kit (Thermo Scientific Pierce, IL, USA). Reduced cytochrome c characteristically absorbs light at 550 nm; mitochondrial complex IV catalytic reduction of cytochrome c generates oxidized cytochrome c, so the decline rate of 550 nm light absorption can reflect mitochondrial complex IV enzyme activity. According to this determination principle, the activity of COX was measured by using kits produced by the Nanjing Jiancheng Institute following the protocol of the manufacturer with certain modifications. Each treatment was performed in triplicate. Enzymes were extracted and quantified by using the abovementioned method. The activity levels of the antioxidant enzymes SOD, POD, and CAT were measured by using kits from the Nanjing Jiancheng Institute. The CAT kit allows for the calculation of the CAT level via the following mechanism: catalase (CAT) can decompose hydrogen peroxide, and the reaction can be quickly suspended by adding ammonium molybdate; the rest of the hydrogen peroxide and ammonium molybdate produces a pale-yellow complex compound, whose change is measured at 405 nm. The SOD kit works via the following mechanism: xanthine and xanthine oxidase react to produce superoxide ion free radicals (O_2^-); these free radicals oxidize hydroxylamine to form nitrite, which, under the action of a chromogenic agent, generates a purple product; the absorbance is then measured at 550 nm with a visible light spectrophotometer. The POD kit works via the following mechanism: peroxide (POD) catalytically degrades hydrogen peroxide, and its change is measured at 420 nm to calculate the level of POD. Three independent experiments were performed.

Bioinformatic analysis

The isoelectric point (pI) and molecular weight (MV of the predicted protein were estimated by using the online tool PeptideMass (http://web.expasy.org/cgi-bin/peptide_mass/peptide-mass.pl). Multiple alignments to homologous SCO2 proteins from multiple species were generated by using DNAMAN version 6 (http://www.lynnon.com). In addition, conserved motifs were analyzed by using BLASTP at NCBI (http://blast.ncbi.nlm.nih.gov/Blast.cgi) and the online tool PROSITE (http://prosite.expasy.org/prosite.html). Further, a phylogenetic tree was constructed by using the neighbor-joining (NJ) method with Molecular Evolutionary Genetic Analysis (MEGA version 4.1) software. Further, the tertiary structure was predicted with SWISS MODEL software (https://swissmodel.expasy.org/). Moreover, the putative mitochondrial targeting sequence (MTS) was predicted by using MITOPROT (http://ihg.gsf.de/ihg/mitoprot.html).

Statistical analysis

The results are presented as the mean ± standard deviation (SD) of triplicate experiments ($n = 3$). Statistical significance was determined by using Duncan's multiple range test with analysis of variance (ANOVA), and calculations were performed by using SPSS Statistics. Significance was set at $P < 0.01$.

Results

Identification and sequence analysis of AccSCO2

By using the specific primers SQ1 and SQ2, the ORF of *AccSCO2* was obtained, which encodes a 291-amino acid protein with a predicted average mass of 33.748 kDa and a theoretical pI of 8.36. This protein contains a putative N-terminal MTS consisting of 15 amino acids (MRTMFGLINLFSLRC), as predicted by using the online tool MITOPROT (Fig. 1A), indicating that it is likely localized to mitochondria. Sequence analysis with other SCO proteins showed that AccSCO2 contains a CXXXC motif and a histidine residue at position 234, both of which are known to be involved in copper binding [7, 8] (Fig. 1A). To elucidate the functions of the conserved residues and determine the catalytic mechanism of AccSCO2, its three-dimensional structure was predicted by using SWISS-MODEL (Fig. 1B). The model showed that AccSCO2 has more β folds (10) than α helices (4), and this finding may be related to the function of this protein.

A phylogenetic tree was constructed by using the amino acid sequences of select SCOs (Table 1). Phylogenetic analysis revealed the evolutionary relationships of AccSCO2 with other SCOs from various species and showed that it belonged to branches of insects. In addition, *A. cerana cerana* SCO2 exhibited the strongest similarity to *A. mellifera* SCO2 (Fig. S1).

Expression of *AccSCO2* under a variety of environmental stresses

Honeybees are exposed to a variety of unfavorable environmental stresses, including cold,

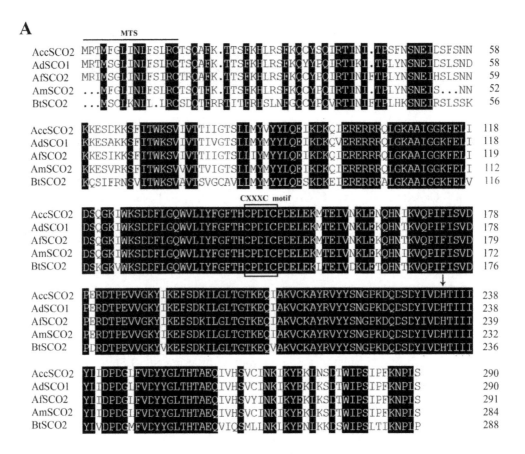

Fig. 1 Multiple alignment analysis and the tertiary structure of AccSCO2. A. Multiple alignment analysis against other known SCO proteins. The copper-binding domain of the CXXXC motif is boxed, and the MTS is underlined. The histidine residue at position 234 is indicated by the arrow. Identical amino acids are shaded in black. B. The tertiary structure of AccSCO2. Helices, sheets, and coils are presented in different colors. The histidine residue at position 234 is highlighted in red. (For color version, please sweep QR Code in the back cover)

Table 1 Sequences of SCO used in this study

Species	Protein	Abbreviation	GenBank no.
Apis cerana cerana	Mitochondrial SCO	A. cerana cerana-SCO2	XP_016909902
Apis mellifera	Mitochondrial SCO	A. mellifera-SCO2	XP_001122061
Apis dorsata	Mitochondrial SCO	A. dorsata-SCO1	XP_006618475
Megachile rotundata	Mitochondrial SCO	M. rotundata-SCO1	XP_003701243
Camponotus floridanus	Mitochondrial SCO	C. floridanus-SCO1	XP_011268485
Nasonia vitripennis	Mitochondrial SCO	N. vitripennis-SCO1	XP_001605752
Poeciliopsis prolifica	Mitochondrial SCO	P. prolifica-SCO1	JAO35691
Macaca mulatta	Mitochondrial SCO	M. mulatta-SCO1	XP_001118271
Capra hircus	Mitochondrial SCO	C. hircus-SCO1	XP_005694282
Danio rerio	Mitochondrial SCO	D. rerio-SCO1	NP_001296727
Oreochromis niloticus	Mitochondrial SCO	O. niloticus-SCO1	XP_013125257
Taeniopygia guttata	Mitochondrial SCO	T. guttata-SCO1	XP_002187064
Serinus canaria	Mitochondrial SCO	S. canaria-SCO1	XP_009097334
Xenopus tropicalis	Mitochondrial SCO	X. tropicalis-SCO1	XP_012810229
Anolis carolinensis	Mitochondrial SCO	A. carolinensis-SCO1	XP_003217223
Thamnophis sirtalis	Mitochondrial SCO	T. sirtalis-SCO1	XP_013914293
Ursus maritimus	Mitochondrial SCO	U. maritimus-SCO1	XP_008686345
Sparus aurata	Mitochondrial SCO	S. aurata-SCO1	AGV76855
Homo sapiens	Mitochondrial SCO	H. sapiens-SCO1	NP_004580

heavy metal, UV radiation, and pesticides, all of which are believed to induce the formation of ROS [11, 12]. Therefore, to determine the role of *AccSCO2* in oxidative stress, its expression profile was examined by qRT-PCR. As shown in Fig. 2A, *AccSCO2* transcription was dramatically elevated by cold (4°C) treatment, with peak expression detected after 0.5 h. We also observed the substantial upregulation of *AccSCO2* expression following treatment with heavy metals, including $CdCl_2$ and $HgCl_2$ (Fig. 2B, C). In addition to temperature fluctuations and heavy metal, honeybees are often exposed to other abiotic stresses, such as UV radiation, H_2O_2, and pesticides. Following the UV light and H_2O_2 treatments, *AccSCO2* expression was significantly increased, with the highest levels observed at 2 h (6.8-fold increase) (Fig. 2D) and 0.5 h (6.5-fold increase) (Fig. 2E). Different types of pesticides, such as organophosphate (OP) pesticides as well as new neonicotinoid, abamectin and pyrethroid insecticides, have been used in agricultural practice. These pesticides are very common in nature, but their mechanisms differ. OP binds with the enzyme cholinesterase (ChE) and inhibits its activity by irreversible phosphorylation; this leads to accumulation of acetylcholine, stimulating the muscarinic and nicotinic receptors and resulting in toxicity. Phoxim belongs to the category of

OP pesticides. Imidacloprid is a type of methylene group insecticide. This pesticide interferes with the motor system of insect pests and leads to the failure of chemical signal transmission. Abamectin is a type of acaricides, which acts on GABA receptors at the insect neuromuscular junction and affects normal nerve conduction. Therefore, we chose these different types of pesticides. Interestingly, the different pesticide treatments inhibited *AccSCO2* expression to different extents (Fig. 2F-I). All of the above-mentioned results indicate that *AccSCO2* might participate in regulating defense responses to various environmental stresses.

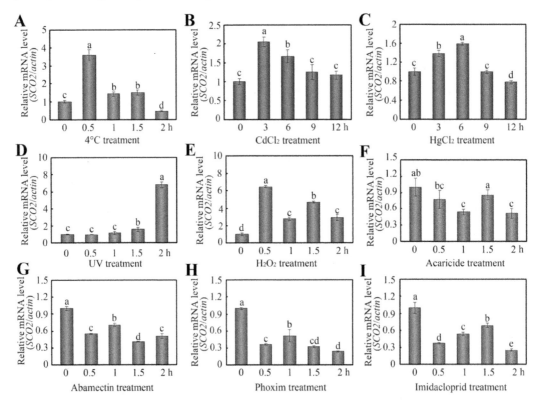

Fig. 2 The expression profiles of *AccSCO2* under different environmental stresses. qPCR was performed on total RNA extracted from adult bees at the indicated time points by following exposure to 4℃ (A), $CdCl_2$ (2 mmol/L) (B), $HgCl_2$ (2 mmol/L) (C), UV (30 MJ/cm^2) (D), H_2O_2 (E), acaricide (10 μmol/L) (F), abamectin (10 μmol/L) (G), phoxim (1 μg/ml) (H), and imidacloprid (10 μmol/L) (I). The vertical bars represent the mean ± SE (*n* = 3). *β-Actin* (GenBank accession number: HM640276) was used as a control. The letters above the bars indicate significant differences ($P < 0.01$ in SPSS) based on Duncan's multiple range test.

To further examine *AccSCO2* expression in response to different abiotic stresses, Western blotting analysis was performed. Following exposure of bees to the treatments described above, samples were obtained to detect AccSCO2 by using anti-AccSCO2. The results showed that *AccSCO2* expression was induced by following exposure to 4℃, $CdCl_2$, $HgCl_2$, UV, or H_2O_2, (Fig. 3A-E). In contrast, its expression was suppressed by following treatment with the different pesticides (acaricide, abamectin, and phoxim) (Fig. 3F-H). These results showed that the protein and mRNA levels of AccSCO2 were consistent.

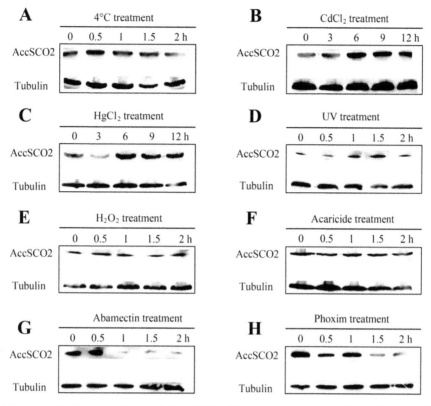

Fig. 3 Western blotting analysis of *AccSCO2*. Honeybees were subjected to the following treatments: 4 ℃ (A), CdCl$_2$ (2 mmol/L) (B), HgCl$_2$ (2 mmol/L) (C), UV (30 MJ/cm^2) (D), H$_2$O$_2$ (E), acaricide (10 μmol/L) (F), abamectin (10 μmol/L) (G), and phoxim (1 μg/ml) (H). Total protein was extracted from samples collected at specific time points. Tubulin was used as a control.

Characterization of recombinant AccSCO2

AccSCO2-R1 lacks the N-terminal MTS (MRTMFGLINLFSLRCT), and AccSCO2-R2 lacks the copper-binding domain of the CXXXC motif (CPDIC). To determine the localization of AccSCO2, AccSCO2-R1, and AccSCO2-R2, we extracted the membranes and the soluble fraction proteins for Western blotting analysis of the recombinant AccSCO2, AccSCO2-R1, and AccSCO2-R2. As shown in Fig. 4, the recombinant AccSCO2, AccSCO2-R1, and AccSCO2-R2 were localized in the cytoplasm of *E. coli* cells. These observations demonstrate that *AccSCO2* may play an important role in cytoplasm.

To evaluate the antioxidant properties of recombinant AccSCO2, AccSCO2-R1, and AccSCO2-R2 that play roles in defense responses against oxidative stress, disc diffusion assay was performed. *E. coli* cells overexpressing *AccSCO2*, *AccSCO2-R1*, and *AccSCO2-R2* were exposed to two known inducers of oxidative stresses (paraquat and cumene hydroperoxide) [20]. A Western blotting showed that AccSCO2 was expressed in *E. coli* (Fig. S2). As shown in Fig. 5, the death zones around the stressor-soaked filters were smaller in diameter on the plates containing *E. coli* overexpressing *AccSCO2* than on the plates containing control bacteria after 12 h of exposure to the various stresses. Furthermore, the halo diameters of the death zones significantly differed

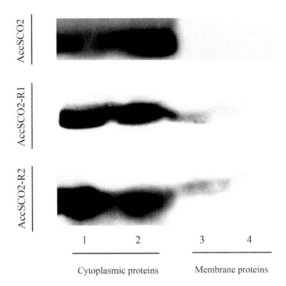

Fig. 4 The localization of AccSCO2, AccSCO2-R1 and AccSCO2-R2. The membranes and the soluble fraction proteins were extracted. Western blotting was used to test the localization of AccSCO2 and the two AccSCO2 mutants by using AccSCO2 antibodies. Lanes 1 and 2, the cytosolic fraction; Lanes 3 and 4, the membrane fraction. The two lines just reflect biological replicates with same protein amounts.

according to the various treatment concentrations (Fig. 5). As shown in Fig. 5, the death zones around the stressor-soaked filters were smaller in diameter on the plates containing *E. coli* overexpressing *AccSCO2-R1* than on the plates containing control bacteria. However, the death zones around the stressor-soaked filters did not change between the plates containing *E. coli* overexpressing *AccSCO2-R2* and the plates containing control bacteria after 12 h of exposure to the various stresses. These results indicate that AccSCO2 play an important role in oxidative stress, and it may perform an antioxidant function involved in copper-dependent manner.

AccSCO2 knockdown

The results of analysis of differential expression patterns (Fig. 2, Fig. 3) and disc diffusion assay (Fig. 5) suggested that *AccSCO2* might play roles in multiple stress defense responses, especially in adult antioxidant defense. To confirm the function of *AccSCO2* in *A. cerana cerana*, we performed loss-of-function experiments on *AccSCO2*-silenced honeybees by using the RNAi technique. dsRNA-mediated RNAi has become one of the most promising strategies for studying the functions of genes, particularly those in insects for which successful transgenesis cannot be performed [24]. Here, 15-day post-emergence adult workers were microinjected with dsRNA of *GFP* or *AccSCO2* for RNAi-mediated silencing. The qRT-PCR results showed that the *AccSCO2* gene was successfully silenced by the dsRNA. The lowest transcription levels of *AccSCO2* were observed at 12, 24, and 36 h after injection. In addition, the *AccSCO2* mRNA level was higher in the control (uninjected and dsRNA-*GFP*-injected) groups than in the dsRNA-*AccSCO2*-injected group (Fig. 6A). A Western blotting analysis showed a lower level of *AccSCO2* expression in the dsRNA-

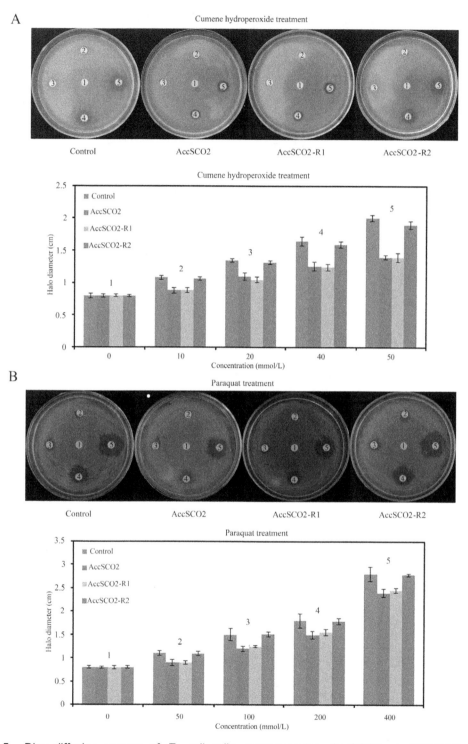

Fig. 5 Disc diffusion assays of *E. coli* cells overexpressing *AccSCO2*, *AccSCO2-R1* and *AccSCO2-R2*. Filter discs soaked with different concentrations of cumene hydroperoxide (A) and paraquat (B) were placed on agar plates. The numbers (1-5) on the filter discs represent the concentrations of the reagents, and the images on the below of the figure correspond to the numbers on the above. The data are presented as the mean ± SE of three independent experiments.

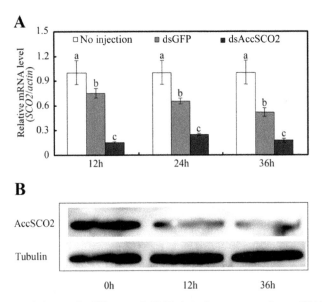

Fig. 6 *AccSCO2* knockdown. A. Effects of RNA interference on the mRNA levels of 15-day post-emergence *Apis cerana cerana* adults. B. The expression level of AccSCO2 protein in control and *AccSCO2*-silenced honeybees measured by Western blotting. The data are presented as the mean ± SE of three independent experiments. Different letters in lower case denote significant differences. Significant differences ($P < 0.01$) were detected by using Duncan's multiple range test.

AccSCO2-injected group 12 and 36 h after injection compared with the control groups (Fig. 6B), indicating that *AccSCO2* was successfully knocked down.

Effects of *AccSCO2* silencing on oxidative stress in *A. cerana cerana*

To further explore the effects of *AccSCO2* knockdown, the transcription levels of other antioxidant genes, including *AccGSTD*, *AccGSTO1*, *AccGSTs4*, *AccSOD1*, *AccSOD2*, *AccCAT*, *AccMsrA*, *AccCYP4G11*, *AccTpx1*, *AccTpx3*, *AccTrx1*, and *AccTrx2*, were also measured by qRT-PCR at 12 and 36 h after injection. In addition, the activities of antioxidant enzymes (including SOD, POD, and CAT) were also analyzed. As shown in Fig. 7, by following *AccSCO2* knockdown, the expression levels of most of the antioxidant genes were significantly decreased in the silenced bees compared with the control bees. Moreover, the SOD, POD and CAT activities were lower in the silenced bees than in the control bees (Fig. 8). These results indicate that *AccSCO2* may be indispensable for attaining a normal lifespan by enhancing cellular oxidative resistance.

COX activity is decreased in bees with *AccSCO2* knockdown

Enzymatic assays were performed to determine the effect of *AccSCO2* knockdown on COX activity. No significant difference in COX activity was observed between the uninjected bees and the dsRNA-*GFP*-injected or dsRNA-*AccSCO2*-injected bees (Fig. 9). However, a strong reduction in COX activity was observed in the silenced bees. In addition, at 36 h after dsRNA-AccSCO2 injection, the specific COX activity was only (12.0 ± 0.91) nmol/(min·g)

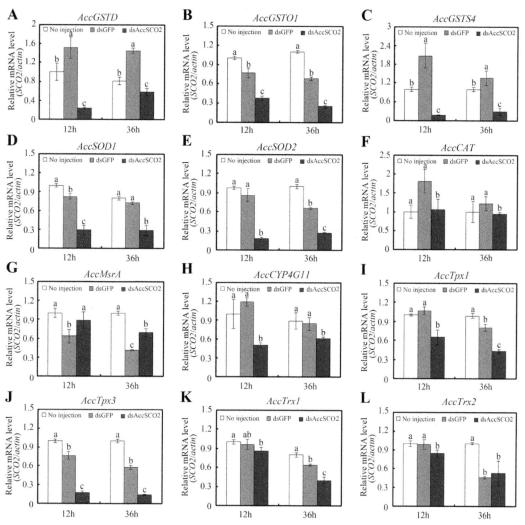

Fig. 7 Expression patterns of other antioxidant genes by following *AccSCO2* knockdown in adult bees: (A) *AccGSTD* (GenBank ID: JF798572), (B) *AccGSTO1* (GenBank ID: KF496073), (C) *AccGSTs4* (GenBank ID: JN008721), (D) *AccSOD1* (GenBank ID: JN700517), (E) *AccSOD2* (GenBank ID: JN637476), (F) *AccCAT* (GenBank ID: KF765424), (G) *AccMsrA* (GenBank ID: HQ219724), (H) *AccCYP4G11* (GenBank ID: KC243984), (I) *AccTpx1* (GenBank ID: HM641254), (J) *AccTpx3* (GenBank ID: JX456217), (K) *AccTrx1* (GenBank ID: JX844651), and (L) *AccTrx2* (GenBank ID: JX844649). The values are presented as the mean ± SE of three independent experiments. The different letters above the columns indicate significant differences ($P < 0.01$) according to Duncan's multiple range test.

(approximately 38.22% of the controls level) in enriched mitochondrial preparations from AccSCO2 knockdown bees, whereas it was (31.4 ± 2.8) nmol/(min·g) in preparations from control (uninjected) bees.

Discussion

The ability of SCO proteins to bind copper is indispensable to their functioning. Previous

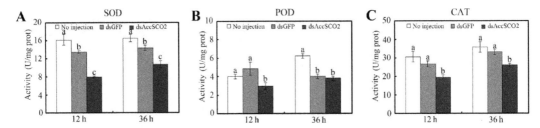

Fig. 8 Effects of *AccSCO2* knockdown on antioxidant enzymatic activities of (A) SOD, (B) POD and (C) CAT. Each value is presented as the mean (SD) of three replicates. The different letters above the bars indicate significant differences ($P < 0.01$) according to SPSS software.

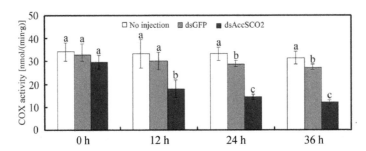

Fig. 9 Cytochrome c oxidase enzymatic activity in *AccSCO2* knockdown bees. The error bar shows the standard deviation determined from three independent experiments. The different letters above the bars indicate significant differences ($P < 0.01$) according to SPSS software.

studies have shown that human SCO1 and SCO2 fulfill unique but partially overlapping functions in COX assembly and cellular copper homeostasis [25, 26]. However, their precise functions remain unclear. To the best of our knowledge, the present study is the first to report the identification of an SCO gene, *AccSCO2*, in *A. cerana cerana*. The Chinese honeybee is highly susceptible to a variety of unfavorable environmental stresses and is currently suffering from significant population decreases [16]. Our results suggest that *AccSCO2* plays crucial roles in cellular stress responses and anti-oxidative processes, and the data presented here provide a foundation for further exploration of the biological functions of the SCO genes in humans.

The identification of a putative N-terminal MTS (MRTMFGLINLFSLRC) (Fig. 1A) in *AccSCO2* indicated that it might play a role in mitochondria, which is known to be a major source of ROS production [15]. In humans, oxidative metabolic processes important for cardiac function are performed in mitochondria [27]. In addition, the conserved histidine and the two cysteines in the conserved CXXXC motif (Fig. 1A) have been proposed to bind copper, suggest an essential role of SCO proteins in metalation of the CuA redox center of COX [6, 7]. Copper is essential for a wide range of metabolic pathways, including mitochondrial respiration, free radical scavenging, and neurotransmitter biosynthesis [28]. The predicted three dimensional structure of AccSCO2 (Fig. 1B) may contribute to a better understanding of the role of AccSCO2. Williams et al. [29] reported that hSCO1 and other authentic SCO family members have similar, highly conserved sequences (SCO loop) including two antiparallel β-strands (β6 and β7) at their base. The CXXXC site of hSCO1 is located at the β3-α2 turn in the heart of the Trx fold. The crystal structure predictions suggest that hSCO1 contains the Trx (thioredoxin) fold, which is most similar to other related oxidoreductases of the peroxiredoxin (Prx) and glutathione peroxidase (GPX) families [30]. The structural similarity of the Sco proteins to the thioredoxin family has led to the proposal that SCO functions as a thiol: disulfide oxidoreductase instead of, or in addition to, a copper trafficking protein, but the specific functions need to be further explored. In this study, a strong reduction in COX activity was observed in the silenced bees (Fig. 9). Previous research has shown that human SCO1 and SCO2 are copper-binding proteins whose cooperative interaction is necessary for the formation of the CuA site of COX II. The decreased Cox activity may be due to disturbed SCO1/SCO2 protein interaction and failure of the two molecules to overlap completely in their function as COX assembly factors [26]. Porcelli et al. [10] reported that a single SCO-encoding gene may be present in *Apis cerana cerana*. A strong reduction in COX activity may be due to incomplete copper loading in the CuA site after knockdown *AccSCO2*. However, the implications remain unclear regarding the role and mechanism of *AccSCO2* in COX assembly. Therefore, we speculate that in *A. cerana cerana*, the SCO protein AccSCO2, which has antioxidant activity, is required for COX activity.

Honeybees are constantly exposed to diverse types of environmental stresses. Temperature change is a common abiotic environmental stress that has been associated with increases in ROS generation and oxidative stress [31]. UV radiation is also a powerful and harmful environmental stress for most animals [32]. Meng et al. have reported that UV-A has the potential to generate a high level of oxidative stress in the cotton bollworm *Helicoverpa armigera* [33]. In addition, the heavy metals Cd and Hg have been reported to enhance intracellular ROS formation [34, 35]. Many studies have demonstrated that pesticides can induce oxidative stress in cells and biological macromolecules, leading to the generation of oxygen and other free radicals, altering the antioxidant enzyme system, and causing lipid peroxidation. For example, OP pesticides have been reported to induce oxidative stress through increased levels of ROS, H_2O_2, nitrate (NO_3^-), and nitrite (NO_2^-) [36]. In our study, *AccSCO2* transcription was induced by cold (4°C), $CdCl_2$, $HgCl_2$, UV, and H_2O_2 (Fig. 2), indicating that this gene might function by removing intracellular superoxide anions generated under oxidative stress conditions to protect honeybees from ROS-induced damage. However, *AccSCO2* transcription was inhibited by pesticide treatment. One explanation for this result is that exposure to pesticides damages biological macromolecules such as RNA, DNA, DNA repair proteins, and other proteins [37, 38]. The biological mechanisms involved in protection against oxidative stress conditions are very complex and require further study. It is known that the expression pattern of a gene is usually an indicator of its functions. For example, mitochondrial *AbTrx2* has been reported to be upregulated in *Haliotis discus discus* following exposure to H_2O_2 oxidative stress, revealing that this protein functions in reducing the levels of harmful ROS [39]. In addition, in *Ostrinia furnacalis*, *OfurGrx2* expression has been shown to be markedly upregulated following exposure to UV radiation, mechanical injury, *E. coli*, and high and low temperatures, indicating that it may play an important role in insects under adverse environmental conditions [40]. These results collectively suggest that *AccSCO2* is a stress-inducible gene that may play roles in responses to various oxidative stresses.

Further, we found that *AccSCO2* overexpression in *E. coli* cells protected them from oxidative stressors (paraquat and cumene hydroperoxide) by performing disc diffusion assays (Fig. 5). The death zones around the stressor-soaked filters were smaller in diameter on the plates containing *AccSCO2*-overexpressing bacteria, than on those containing control bacteria. In addition, the death zones were smaller in diameter on the plates containing *E. coli* overexpressing *AccSCO2-R1* (lacking the N-terminal MTS) than on the plates with control bacteria. However, the death zones around the stressor-soaked filters did not change between the plates containing *E. coli* overexpressing *AccSCO2-R2* (lacking the copper-binding domain of the CXXXC motif) and plates containing control bacteria (Fig. 5). Considering our results in the context of previous studies, we conclude that the *AccSCO2* gene product possesses potent antioxidant properties, and it may perform this antioxidant function involved in copper-dependent manner, suggesting that copper and redox homeostasis are intimately linked. Morgada reported that copper binding to SCO2 is essential to elicit its redox function and to maintain the reduced state of its own cysteine residues in the oxidizing environment of the mitochondrial intermembrane space (IMS). Furthermore, copper acts as a guardian of the thiol groups in various proteins, preventing oxidation [41]. These results provide a detailed molecular mechanism for the antioxidant function of AccSCO2. However, the molecular mechanism of copper and redox homeostasis are entangled and possibly coupled in the mitochondrion; this possibility must be further explored.

High ROS concentrations in cells can cause oxidative stress and contribute to degenerative diseases and aging through the oxidation of nucleic acids, proteins, and lipid membranes [13]. Therefore, ROS detoxification is critical for the survival of organisms. Cells possess a variety of antioxidant systems for resisting oxidative damage, including the glutathione *S*-transferases (GSTs), superoxide dismutases(SODs), catalase (CAT), glutathione peroxidase (GPX), thioredoxin peroxidase (TPx), methionine sulfoxide reductase A (MsrA), glutaredoxin (Grx), and thioredoxin (Trx) systems, which consist of multifunctional proteins and enzymes. Previous reports have suggested that *AccSOD2* plays important roles in cellular stress responses and anti-oxidative processes [13]. The Omega-class glutathione *S*-transferase (GSTO2) is required for reducing oxidative damage in *A. cerana cerana* [42]. In addition, Yao has reported that *AccTpx3* and *AccTrx1* play crucial roles in antioxidant defense [22]. Further, Tsuda has reported that the loss of Trx-2 enhances oxidative stress-dependent phenotypes in *Drosophila*, which are hypersusceptible to paraquat and exhibit high levels of expression of antioxidant genes, such as those encoding SOD, CAT, and glutathione synthetase [43]. Moreover, it has been reported that the MsrA system may mediate methionine sulfoxide modifications and that absence of the MsrA enzyme may result in elevated levels of protein carbonyls in yeast and mice [44]. In mitochondria, such redox reactions modulate the activity of a group of proteins, most of whose functions consist of essential roles in delivering copper to cytochrome c oxidase (COX) during holoenzyme biogenesis. Compelling evidence from studies of SCO1 and SCO2 argues that regulation of the redox state of their cysteines is integral to their metallochaperone function [45, 46] Here, I review our understanding of redox-dependent modulation of copper delivery to COX and of the many anti-oxidant enzymes that are regulated in an SCO-dependent manner. By investigating the transcription levels of most of the antioxidant genes (i.e. *AccGSTD*, *AccGSTO1*, *AccGSTs4*, *AccSOD1*, *AccSOD2*, *AccCAT*, *AccMsrA*, *AccCYP4G11*, *AccTpx1*, *AccTpx3*, *AccTrx1*, and *AccTrx2*), which serve as antioxidant-related marker genes, one can indirectly show that a gene has antioxidant properties. As is common knowledge, the functions of a gene are implemented by its expression products; therefore, to determine the functions and activities of antioxidant enzymes, one must study the mRNA levels of antioxidant genes. Interestingly, after *AccSCO2* knockdown, the transcription levels of most of the antioxidant genes were significantly lower in the silenced bees than in the control bees (Fig. 7), and the SOD, POD and CAT activities were decreased (Fig. 8). Considering our results in the context of previous studies, we conclude that the *AccSCO2* possesses potent antioxidant properties.

In conclusion, our study revealed the presence of an *AccSCO2* gene in *A. cerana cerana* that exhibits molecular characteristics and expression patterns consistent with a physiological role in oxidative stress responses. We propose that *AccSCO2* plays important roles in mitochondrial anti-oxidative stress mechanisms that are important to the survival of this insect species.

References

[1] Kadenbach, B., Hüttemann, M. The subunit composition and function of mammalian cytochrome c oxidase. Mitochondrion. 2015, 24, 64-76.
[2] Bourens, M., Boulet, A., Leary, S. C., Barrientos, A. Human COX20 cooperates with SCO1 and SCO2 to mature COX2 and promote the assembly of cytochrome c oxidase. Hum Mol Genet. 2014, 23 (11), 2901-2913.
[3] Zeviani, M., Tiranti, V. C. Piantadosi, C. Mitochondrial disorders. Medicine (Baltimore). 1998, 77, 59-72.
[4] Diaz, F. Cytochrome c oxidase deficiency: patients and animal models. Biochim Biophys Acta. 2010, 1802 (1), 100-110.
[5] Sotoontanesi, I. C., Horn, D., Barrientos, A. Assembly of mitochondrial cytochrome c-oxidase, a complicated and highly regulated cellular process. Am J Phys Cell Physiol. 2006, 291, C1129-C1147.
[6] Beers, J., Glerum, D. M., Tzagoloff, A. Purification and characterization of yeast Sco1p, a mitochondrial copper protein. J Biol Chem. 2002, 277, 22185-22190.
[7] Horng, Y. C., Leary, S. C., Cobine, P. A., Young, F. B., George, G. N., Shoubridge, E. A., Winge, D. R. Human Sco1 and Sco2 function as copper-binding proteins. J Biol Chem. 2005, 280, 34113-34122.
[8] Leary, S. C., Kaufman, B. A., Pellecchia, G., Guercin, G. H., Mattman, A., Jaksch, M., Shoubridge, E. A. Human SCO1 and SCO2 have independent, cooperative functions in copper delivery to cytochrome c oxidase. Hum Mol Genet. 2004, 13, 1839-1848.
[9] Nittis, T., George, G. N., Winge, D. R. Yeast Sco1, a protein essential for cytochrome c oxidase function is a Cu(I)-binding protein. J Biol Chem. 2001, 276, 42520-42526.
[10] Porcelli, D., Oliva, M., Duchi, S., Latorre, D., Cavaliere, V., Barsanti, P., Villani, G., Gargiulo, G., Caggese, C. Genetic, functional and evolutionary characterization of *scox*, the *Drosophila melanogaster* ortholog of the human SCO1 gene. Mitochondrion. 2010, 10 (5), 433-448.
[11] Lushchak, V. I. Environmentally induced oxidative stress in aquatic animals. Aquat Toxicol. 2011, 101, 13-30.
[12] Kottuparambila, S., Shinb, W., Brownc, M. T., Han, T. UV-B affects photosynthesis, ROS production and motility of the freshwater flagellate, *Euglena agilis* Carter. Aquat Toxicol. 2012, 122-123, 206-213.
[13] Jia, H. H., Sun, R. J., Shi, W. N., Yan, Y., Li, H., Guo, X. Q., Xu, B. H. Characterization of a mitochondrial manganese superoxide dismutase gene from *Apis cerana cerana* and its role in oxidative stress. J Insect Physiol. 2014, 60, 68-79.
[14] Chen, R., Cairns, I., Papandreou, A., Koong, N. C. Oxygen consumption can regulate the growth of

[15] Daiber, A. Redox signaling (cross-talk) from and to mitochondria involves mitochondrial pores and reactive oxygen species. Biochim Biophys Acta. 2010, 1797, 897-906.

[16] Xu, P., Shi, M., Chen, X. X. Antimicrobial peptide evolution in the Asiatic honey bee *Apis cerana cerana*. PLoS One. 2009, 4, e4239.

[17] Alaux, C., Ducloz, F., Crauser, D., Le Conte, Y. Diet effects on honeybee immunocompetence. Biol Lett. 2010, 6, 562-565.

[18] Livak., K. J., Schmittgen, T. D. Analysis of relative gene expression data using real-time quantitative PCR and the 2 (− Delta C (T)) method. Methods. 2001, 25, 402-408.

[19] Yan, H. R., Jia, H. H., Gao, H. R., Guo, X. Q., Xu, B. H. Identification, genomic organization, and oxidative stress response of a Sigma class glutathione S-transferase gene (*AccGSTS1*) in the honey bee, *Apis cerana cerana*. Cell Stress Chaperones. 2013, 18, 415-426.

[20] Burmeister, C., Luërsen, K., Heinick, A., Hussein, A., Domagalski, M., Walter, R. D., Liebau, E. Oxidative stress in *Caenorhabditis elegans*: protective effects of the Omega class glutathione transferase (GSTO-1). FASEB J. 2008, 22, 343-354.

[21] Chen, X. B., Yao, P. B., Chu, X. Q., Hao, L. L., Guo, X. Q., Xu, B. H. Isolation of arginine kinase from *Apis cerana cerana* and its possible involvement in response to adverse stress. Cell Stress Chaperones. 2015, 20, 169-183.

[22] Yao, P. B., Chen, X. B., Yan, Y., Liu, F., Zhang, Y., Guo, X., Xu, B. Glutaredoxin 1, glutaredoxin 2, thioredoxin 1, and thioredoxin peroxidase 3 play important roles in antioxidant defense in *Apis cerana cerana*. Free Radic Biol Med. 2014, 68, 335-346.

[23] Elias-Neto, M., Soares, M. P., Simões, Z. L., Hartfelder, K., Beyond, M. M. Developmental characterization, function and regulation of a Laccase 2 encoding gene in the honey bee, *Apis mellifera* (Hymenoptera, Paine). Insect Biochem Mol. 2010, 40, 241-251.

[24] Huvenne, H., Smagghe, G. Mechanisms of dsRNA uptake in insects and potential of RNAi for pest control: a review. J Insect Physiol. 2010, 56, 227-235.

[25] Leary, S. C. Redox regulation of SCO protein function: controlling copper at a mitochondrial crossroad. Antioxid Redox Signal. 2010, 13, 1403-1416.

[26] Leary, S. C., Cobine, P. A., Kaufman, B. A., et al. The human cytochrome c oxidase assembly factors SCO1 and SCO2 have regulatory roles in the maintenance of cellular copper homeostasis. Cell Metab. 2007, 5, 9-20.

[27] Martínez-Morentin, L., Martínez, L., Piloto, S., et al. Cardiac deficiency of single cytochrome oxidase assembly factor scox induces p53-dependent apoptosis in a *Drosophila* cardiomyopathy model. Hum Mol Genet. 2015, 24 (13), 3608-3622.

[28] Hamza, I., Gitlin, J. D. Copper chaperones for cytochrome c oxidase and human disease. J Bioenerg Biomembr. 2002, 34, 381-388.

[29] Williams, J. C., Sue, C., Banting, G. S., Yang, H., Glerum, D. M., Hendrickson, W. A., Schon, E. A. Crystal structure of human SCO1: implications for redox signaling by a mitochondrial cytochrome c oxidase "assembly" protein. J Biol Chem. 2005, 280 (15), 15202-15211.

[30] Abajian, C., Rosenzweig, A. C. Crystal structure of yeast Sco1. J Biol Inorg Chem. 2006, 11 (4), 459-466.

[31] An, M. I., Choi, C. Y. Activity of antioxidant enzymes and physiological responses in ark shell, *Scapharca broughtonii*, exposed to thermal and osmotic stress: effects on hemolymph and biochemical parameters. Comp Biochem Physiol B: Biochem Mol Biol. 2010, 155, 34-42.

[32] Schauen, M., Hornig-Do, H. T., Schomberg, S., Herrmann, G., Wiesner, R. J. Mitochondrial electron transport chain activity is not involved in ultraviolet A (UVA)-induced cell death. Free Radic Biol Med. 2007, 42, 499-509.

[33] Meng, J. Y., Zhang, C. Y., Lei, C. L. A proteomic analysis of *Helicoverpa armigera* adults after exposure to UV light irradiation. J Insect Physiol. 2010, 56, 405-411.

[34] Park, E. J., Park, K. Induction of reactive oxygen species and apoptosis in BEAS-2B cells by mercuric

chloride. Toxicol in Vitro. 2007, 21, 789-794.

[35] Tang, T., Huang, D. W., Zhou, C. Q., Li, X., Xie, Q. J., Liu, F. S. Molecular cloning and expression patterns of copper/zinc superoxide dismutase and manganese superoxide dismutase in *Musca domestica*. Gene. 2012, 505, 211-220.

[36] Surajudeen, Y. A., Sheu, R. K., Ayokulehin, K. M., Olatunbosun, A. G. Oxidative stress indices in Nigerian pesticide applicators and farmers occupationally exposed to organophosphate pesticides. Int J Appl Basic Med Res. 2014, 4 (Suppl. 1), S37-S40.

[37] Singh, S., Kumar, V., Thakur, S., Banerjee, B. D., Chandna, S., Rautela, R. S. DNA damage and cholinesterase activity in occupational workers exposed to pesticides. Environ Toxicol Pharmacol. 2011, 31, 278-285.

[38] Hernández, A. F., acasaña, M., Gil, F., Rodríguez-Barranco, M., Pla, A., López-Guarnido, O. Evaluation of pesticide-induced oxidative stress from a gene-environment interaction perspective. Toxicology. 2013, 307, 95-102.

[39] Zoysa, M. D., Pushpamali, W. A., Whang, I., Kim, S. J., Lee, J. Mitochondrial thioredoxin-2 from disk abalone (*Haliotis discus discus*): molecular characterization, tissue expression and DNA protection activity of its recombinant protein. Comp Biochem Physiol B: Biochem Mol Biol. 2008, 149, 630-639.

[40] An, S., Zhang, Y., Wang, T., Luo, M., Li, C. Molecular characterization of glutaredoxin 2 from *Ostrinia furnacalis*. Integr Zool. 2013, S1, 30-38.

[41] Morgada, M. N., Abriata, L. A., Cefaro, C., Gajda, K., Banci, L., Vila, A. J. Loop recognition and copper-mediated disulfide reduction underpin metal site assembly of CuA in human cytochrome oxidase. Proc Natl Acad Sci USA. 2015, 112 (38), 11771-11776.

[42] Zhang, Y. Y., Guo, X. L., Liu, Y. L., Liu, F., Wang, H. F., Guo, X. Q., Xu, B. H. Functional and mutational analyses of an Omega-class glutathione *S*-transferase (GSTO2) that is required for reducing oxidative damage in *Apis cerana cerana*. Insect Mol Biol. 2016, 25 (4), 470-486.

[43] Tsuda, M., Ootaka, R., Ohkura, C., Kishita, Y., Seong, K. H., Matsuo, T., Aigaki, T. Loss of Trx-2 enhances oxidative stress-dependent phenotypes in *Drosophila*. FEBS Lett. 2010, 584, 3398-3401.

[44] Moskovitz, J., Oien, D. B. Protein carbonyl and the methionine sulfoxide reductase system. Antioxid Redox Signal. 2010, 12, 405-415.

[45] Banci, L., Bertini, I., Ciofi-Baffoni, V. S., Mangani, S., Martinelli, M., Palumaa, P., Wang, S. A hint for the function of human Sco1 from different structures. Proc Natl Acad Sci USA. 2006, 103, 8595-8600.

[46] Banci, L., Bertini, I., Ciofi-Baffoni, S., Hadjiloi, T., Martinelli, M., Palumaa, P. Mitochondrial copper(I) transfer from Cox17 to Sco1 is coupled to electron transfer. Proc Natl Acad Sci USA. 2008, 105, 6803-6808.

3.3 Characterization of the CDK5 Gene in *Apis cerana cerana* (*AccCDK5*) and a Preliminary Identification of Its Activator Gene, *AccCDK5r1*

Guangdong Zhao, Chen Wang, Hongfang Wang, Lijun Gao,
Zhenguo Liu, Baohua Xu, Xingqi Guo

Abstract: Cyclin-dependent kinase 5 (CDK5) is an unusual CDK whose function has been implicated in protecting the central nervous system (CNS) from oxidative damage. However, there have been few studies of CDK5 in insects. In this study, we identified the AccCDK5 gene from *Apis cerana cerana* and investigated its role in oxidation resistance. We found that AccCDK5 is highly conserved across species and contains conserved features of the CDK5 family. The results of qPCR analysis indicated that AccCDK5 is highly expressed during the larval and pupal stages and in the head and muscle of the adult. We further observed that AccCDK5 is induced by several environmental oxidative stresses. Moreover, the overexpression of the AccCDK5 protein in *E. coli* enhances the resistance of the bacteria to oxidative stress. The activation of CDK5 requires binding to its activator. Therefore, we also identified and cloned cyclin-dependent kinase 5 regulatory subunit 1, which we named AccCDK5r1, from *Apis cerana cerana*. AccCDK5r1 contains a conserved cell localization targeting domain as well as binding and activation sites for CDK5. Yeast two-hybrid analysis demonstrated the interaction between AccCDK5 and AccCDK5r1. The expression patterns of the two genes were similar after stress treatment. Collectively, these results suggest that AccCDK5 plays a pivotal role in the response to oxidative stresses and that AccCDK5r1 is a potential activator of AccCDK5.

Keywords: Cyclin-dependent kinase 5; Cyclin-dependent kinase 5 regulator subunit 1; Oxidative stress; Yeast two-hybrid analysis; *Apis cerana cerana*

Introduction

The Chinese honeybee (*Apis cerana cerana*) is the most widely distributed of all native honeybee species in China. *A. cerana cerana* has many advantages over *A. mellifera*, including the use of sporadic nectariferous plants, a long honey period, and better resistance to cold, mites and disease [1-3]. Chinese honeybees also play a crucial role in ecological balance and the agricultural industry. However, in recent years, due to the introduction of western honeybees, diseases, pesticide abuse, and environmental pollution, the number of Chinese honeybee colonies has plummeted, and the survival of the species has been seriously threatened [4]. Numerous stresses could lead to oxidative stress in the habitat of honeybees, including chemical stresses (pesticides, metal, etc.), physical stresses (radiation, temperature, etc.), and physiological stresses [5-8]. Oxidative stress occurs when the equilibrium between reactive oxygen species (ROS) production and antioxidant defense mechanisms is dysregulated [9]. This results in oxidative damage, as manifested by modifications of cellular lipids, proteins, and DNA [10-12].

Cyclin-dependent kinases (CDKs), a family of serine/threonine kinases, form complexes with cyclins, components that are essential for the kinase activity of CDKs [13]. The functions of CDKs in signal integration to regulate gene transcription and cell division have been clearly established [14]. Cyclin-dependent kinase 5 (CDK5), an unusual member of the CDK family, was discovered in the 1990s and is known as neuronal CDC-2-like kinase [15-17]. CDK5 is a versatile CDK member that regulates many cellular processes, including neuronal migration, actin dynamics, microtubule stability and transport, cell adhesion, axon guidance, synaptic structure and plasticity, and neuronal survival [18]. Dysregulation of CDK5 activity is neurotoxic and may lead to disease, including Alzheimer's disease (AD) and Parkinson's disease (PD) [19-21].

CDK5 regulatory subunit 1 (CDK5r1), encoding the CDK5 activator p35, plays a crucial role in the kinase activity of CDK5 [22]. Activation of CDK5 requires binding to p35, which is a non-cyclin protein that has a short half-life [23, 24]. As an activator of CDK5, p35 contains two domains, including the N-terminal cell localization targeting domain and the CDK5 activation domain. The N-terminal region of p35, known as p10, contains an N-myristoylation consensus sequence (MGXXXS/T) that may be essential for binding to regulatory proteins [25] and that functions in targeting CDK5 to the cell membrane [26]. Furthermore, residues 150-200 of p35 bind CDK5, and residues 279-291 are required for the activation of CDK5 *in vitro* [27]. Because of the higher expression level of CDK5 relative to p35 in neurons, the kinase activity of CDK5 is mainly determined by the expression level of p35. After treatment with neurotoxins, including hydrogen peroxide and glutamate, p35 is degraded via ubiquitin-dependent and ubiquitin-independent pathways to p25, which has a longer half-life than p35 [28-30].

Previous studies have illustrated that the role of CDK5 under oxidative stress is complex. Exposure of neurons to oxidative stress activates calpain, a calcium-dependent protease that converts p35 to p25. This conversion results in the overactivation of CDK5 [31]. Moreover, the increased stability of p25 relative to p35 leads to the inappropriate localization of CDK5 [32]. The dysregulation of CDK5 may affect the cellular antioxidant defense system and could lead to increased oxidative stress [20]. Increased oxidative stress could also cause an increase in ERK activation, and the continuous activation and nuclear localization of ERK could promote neuronal cell death [33-35]. Additionally, it has been reported that the H_2O_2-induced activation of nuclear CDK5 could protect cells from H_2O_2-induced apoptosis via VRK3 phosphorylation, which could lead to the inhibition of ERK activation [36].

In this study, we isolated *AccCDK5* from *A. cerana cerana*. We verified *AccCDK5* expression patterns at different developmental stages and in several tissues. Additionally, we determined the expression patterns of *AccCDK5* in response to several oxidative stressors. We evaluated the antioxidant ability of AccCDK5 overexpressed in *E. coli*. Due to the peculiar relationship between CDK5 and p35, we cloned *AccCDK5r1* and confirmed that *AccCDK5r1* is a homolog of *CDK5r1* in *A. cerana cerana*. Moreover, we obtained evidence for the interaction between AccCDK5 and AccCDK5r1. We also determined the expression patterns of *AccCDK5r1* after several treatments. Based on our results, we speculate that AccCDK5 plays a crucial role in oxidative stress management and that AccCDK5r1 is an activator of AccCDK5.

Methods

Insects and treatments

In this study, Chinese honeybees (*A. cerana cerana*) were obtained from the College of Animal Science and Veterinary Medicine, Shandong Agricultural University (Taian, China). Honeybees of various developmental stages, including the fourth-, the fifth-, and the sixth-instar larvae, white-eyed (Pw), pink-eyed (Pp), and dark-eyed (Pd) pupae, and adult workers (Ad), were obtained. Three samples were analyzed for each stage. The bees were divided into 11 groups, and each group contained 40 randomized individuals. Groups 1-3 were exposed to 4℃, 42℃, or UV (254 nm, 30 MJ/cm^2), respectively. Group 4 was injected with 2 mmol/L H_2O_2 (0.5 μl). Groups 5 and 6 were fed with $CdCl_2$ and $HgCl_2$ (6 mg/L was added to the basic adult diet). Groups 7-10 were treated with pesticides (spirodiclofen, acaricide, acetamiprid, imidacloprid, and phoxim), which were diluted to the final concentration (20 μg/ml was added to food) and the sources of the pesticides and their purity were listed in Table S1. The control groups were fed a basic adult diet, which contained water, 70% powdered sugar, and 30% honey from the source colonies. All bees were maintained in an incubator (at a constant temperature of 34℃ with relative humidity at 70%, in a 24 h dark environment), and three honeybees were sampled randomly from each group at appropriate times. Bees were snap-frozen in liquid nitrogen after treatment and were stored at −70℃.

RNA extraction and cDNA synthesis

By using standard methods for *Apis mellifera* anatomy and dissection as a reference [37], adult workers were dissected into different tissues, including head, epidermis, muscle, midgut, and poison gland. Total RNA was extracted from the samples by using RNAiso Plus (TaKaRa, Japan). The concentration and quality of RNA samples were measured by using a NanoDrop^{TM} 2000/2000c spectrophotometer (NanoDrop products, Wilmington, DE, 19810, USA), and the RNA samples were stored at −70℃. Then, the RNA samples [(1000 ± 200) ng/μl] were reverse transcribed by using 5× All-In-One RT MasterMix (with the AccuRT Genomic DNA Removal Kit) (Applied Biological Materials Inc., Richmond, BC, Canada), which uses oligo dT to prime the reverse transcription. This kit can effectively remove gDNA from RNA samples. Nuclease-free water was used as a negative control in the RT process. The RT procedure was as follows: add the RNA template (up to 2 μg), AccuRT Reaction Mix (4×) (2 μl) and nuclease-free H_2O (up to a total volume of 8 μl) to the tube and incubate at 42℃ for 2 min, then add AccuRT Reaction Stopper (5×) (2 μl), 5× All-In-One RT MasterMix (4 μl) and nuclease-free H_2O (6 μl). The temperatures and times used for the reaction were 10 min at 25℃, 15 min (for qPCR) or 50 min (for PCR) at 42℃, and 5 min at 85℃. The samples were chilled on ice after the RT process and stored at −20℃.

Isolation of the ORF sequence of *AccCDK5*

Recently, genomic sequencing of *A. cerana cerana* has been completed [38]. To clone the ORF sequence of *AccCDK5*, the specific primers AccCDK5-5 and AccCDK5-3 (as shown in Table 1) were designed based on the *AccCDK5* genomic sequence. The primer design method and PCR protocol introduced by a previous study [39] were used. A 25 μl reaction volume was used in the PCR reaction, which contained 2.5 μl *Taq* buffer (TransGen Biotech, Beijing, China), 1 μl dNTP Mixture (Sangon Biotech, Shanghai, China), 1 μl of each primer (10 mmol/L), 1 μl cDNA template, 0.25 μl *Taq* DNA Polymerase (TransGen Biotech, Beijing, China) and 18.25 μl double distilled water. The PCR amplification conditions are as shown in Table 2. The PCR product was purified and ligated into the pEASY-T1 simple vector (TransGen Biotech, Beijing, China) and transformed into Trans1-T1 Phage Resistant Chemically Competent Cells (TransGen Biotech, Beijing, China) for sequencing. The sequencing was carried out by Biosune Biotechnology (Shanghai) Co., Ltd. (Shanghai, China) by using a 3730xl DNA Analyzer (Applied Biosystems, Foster City, CA, USA) with M13 universal sequencing primers (as shown in Table 1).

Table 1 Primers used in this study

Abbreviation	Primer sequence (5'—3')	Description
AccCDK5-5	CGCTAACCACTTTTCATTATTCCGC	cDNA sequence primer, forward
AccCDK5-3	GGTCGTTGCACTACTCGCG	cDNA sequence primer, reverse
AccCDK5r1-5	GCCACCACCACCGCCTCAAC	cDNA sequence primer, forward
AccCDK5r1-3	CGAGGTCGATGCGGAGGGGTC	cDNA sequence primer, reverse
AccCDK5-Y-5	GGATCCATGCAAAAATATGAGAAACTCGAG	Protein expression primer, forward
AccCDK5-Y-3	GTCGACCTGACAACGATCGTTTTTAATGG	Protein expression primer, reverse
AccCDK5-F	CGAACGCCGGTAGACCCTTG	qPCR primer, forward
AccCDK5-R	CCTTGAGCGGGATGATAAAGTGG	qPCR primer, reverse
AccCDK5r1-F	CAACACGCACAACCCGAC	qPCR primer, forward
AccCDK5r1-R	GGAAGTCTCTTAAACGGGTGC	qPCR primer, reverse
AccCDK5-BD-5	CATATGATGCAAAAATATGAGAAACTCGAG	pGBKT7 construction primer, forward
AccCDK5-BD-3	GGATCCCTGACAACGATCGTTTTTAATGG	pGBKT7 construction primer, reverse
AccCDK5r1-AD-5	CATATGATGGGTACCGTGTTGTCGTTC	pGADT7 construction primer, forward

Abbreviation	Primer sequence (5'—3')	Description
AccCDK5r1-AD-3	GAGCTCAGCCGCCTTGGTGGGTAC	pGADT7 construction primer, reverse
β-s	AGAATTGATCCACCAATCCA	Standard control primer, forward
β-x	GGTACCATGCAGCACATATTATTG	Standard control primer, reverse
M13F	TGTAAAACGACGGCCAGT	Universal sequencing primer, forward
M3R	CAGGAAACAGCTATGACC	Universal sequencing primer, forward

Table 2 PCR amplification conditions

Primer pair	Amplification conditions
AccCDK5-5/AccCDK5-3	10 min at 94℃, 40 s at 94℃, 40 s at 53℃, 1 min at 72℃ for 35 cycles, 10 min at 72℃
AccCDK5-Y-5/AccCDK5-Y-3	10 min at 94℃, 40 s at 94℃, 40 s at 50℃, 1 min at 72℃ for 35 cycles, 10 min at 72℃
AccCDK5-BD-5/AccCDK5-BD-3	10 min at 94℃, 40 s at 94℃, 40 s at 50℃, 1 min at 72℃ for 35 cycles, 10 min at 72℃
AccCDK5r1-5/AccCDK5r1-3	10 min at 94℃, 40 s at 94℃, 40 s at 55℃, 90 s at 72℃ for 35 cycles, 10 min at 72℃
AccCDK5r1-AD-5/AccCDK5r1-AD-3	10 min at 94℃, 40 s at 94℃, 40 s at 50℃, 70 s at 72℃ for 35 cycles, 10 min at 72℃

Bioinformatics and phylogenetic analysis of AccCDK5

Conserved domains of AccCDK5 were analyzed by using BLAST by NCBI (http://blast.ncbi.nlm.nih.gov/Blast.cgi). Multiple protein sequence alignments were performed by using DNAMAN software 6.0.3 (Lynnon Biosoft Corporation, San Ramon, CA, USA). The theoretical isoelectric point and molecular weight of AccCDK5 was predicted by using ExPASy (http://web.expasy.org/compute_pi/). The phylogenetic analysis was conducted by using MEGA5.1 software based on the neighbor-joining method.

Real-time quantitative PCR

To determine the expression patterns of *AccCDK5*, real-time quantitative PCR was performed by using the SYBR® PrimeScript™ RT-PCR Kit (TaKaRa, Japan), individual PCR tubes 8-tube strip (clear) (Bio-Rad, Hercules, CA, USA) and the CFX96™ Real-time System (Bio-Rad, Hercules, CA, USA) with specific primers (AccCDK5-F and AccCDK5-R, shown in Table 1) based on the cDNA sequence as described above. The qPCR primers for *AccCDK5* were designed and tested according to previously reported procedures [40, 41]. The *β-actin* gene (GenBank: HM640276) (as shown in Table 1) was selected as a reference gene [42] and was used to normalize the variations in RNA extraction yield and efficiencies of reverse transcription and amplification. The efficiency values and correlation coefficients (R^2) of the qPCR primers are listed in Table S2. We also tested and compared additional reference genes (*ribosomal protein 49* and *tbp-association factor*) by using geNorm software (versions 3.5) (as shown in supplemental Table S3). A 25 μl reaction volume was used for the qPCR reaction, which contained 9.5 μl double distilled water, 2 μl cDNA template, 0.5 μl of each primer

(10 mmol/L), and 12.5 μl SYBR® Premix Ex Taq^{TM}. The qPCR protocol was as follows: 95℃ for 30 s, 40 cycles of 95℃ for 5 s, 55℃ for 15 s and 72℃ for 15 s, and a final melting cycle from 65℃ to 96℃. All of the experimental samples were analyzed in triplicate, and the data from the qPCR were analyzed with the Bio-Rad CFX Manager 3.1 (Bio-Rad, Hercules, CA, USA). The relative expression levels of *AccCDK5* were calculated by using the $2^{-\Delta\Delta C_t}$ comparative CT method [43], and the error bars were calculated by Bio-Rad CFX Manager 3. The mean ± SE from three independent experiments is shown.

The protein expression of AccCDK5 and the antibody preparation

To obtain recombinant AccCDK5 protein, the coding region of *AccCDK5* flanked by *Bam*H I and *Sal* I restriction sites was ligated into the pET-30a (+) vector (Novagen, Darmstadt, Germany). The recombinant plasmid was transformed into *E. coli* BL21 (DE3) (TransGen Biotech, Beijing, China). A positive clone was cultured in LB medium with 50 μg/ml kanamycin at 37℃ overnight. Then, 100-200 μl of the culture was subcultured into 10 ml of fresh LB medium containing 50 μg/ml kanamycin and incubated at 37℃ for 1-2 h until the optical density at 600 nm reached 0.4-0.6. Expression of recombinant AccCDK5 protein was induced by isopropylthio-β-D-galactoside (IPTG) (Cwbiotech, Beijing, China) at a final concentration of 75 μg/ml at 28℃ for 6-8 h. After induction, the bacterial cells were collected by centrifugation at 13,000 r/min at room temperature for 2 min. SDS-PAGE loading buffer was mixed with the bacteria at a 1 ∶ 1 ratio, and the mixture was heated at 100℃ for 10 min. Then, the mixed metaprotein was separated by 12% sodium dodecyl sulfate polyacrylamide gel electrophoresis (SDS-PAGE). The target protein band was excised and macerated with a defined amount of normal saline. The recombinant AccCDK5 protein was injected subcutaneously into white mice to produce antibodies. The mice were injected once a week for four weeks. Four days after the last injection, blood from the mice was collected and maintained at 37℃ for 1 h and then 4℃ for 6 h. The serum was collected by centrifugation at 3,000 r/min at 4℃ for 15 min, after which the serum was aliquoted and stored at −70℃.

Western blotting analysis

Total protein lysate from adult worker bees that were treated with several stressors (4℃, H_2O_2, phoxim, or acaricide) was extracted by using a Tissue Protein Extraction Kit (Cwbiotech, Beijing, China). The total protein lysate was subjected to SDS-PAGE, and the target band was excised and electrotransferred onto a PVDF membrane (Millipore, Bedford, MA, USA) in a semi-dry transfer apparatus. Three replicate blots were produced from each sample. Western blotting analysis was performed according to a previously reported procedure [44]. The anti-AccCDK5 serum was used as the primary antibody at a dilution of 1 ∶ 1,000 (*V/V*). Peroxidase-conjugated goat anti-mouse immunoglobulin G (Dingguo, Beijing, China) was used as the secondary antibody at a dilution of 1 ∶ 2,000 (*V/V*). The development process was performed by using the FDbio-Dura ECL Kit (Fudebio, Hangzhou, China). The Western blotting results were analyzed by using Image-pro plus 6.0 (Media Cybernetics, Rockville, MD, USA).

Disc diffusion assay of recombinant AccCDK5 protein

Disc diffusion assays for cumyl hydroperoxide and $HgCl_2$ were performed by using a method modified from Burmeister et al. [45]. The recombinant AccCDK5 protein was overexpressed in *E. coli* BL21, and the empty pET-30a (+) vector-transformed *E. coli* BL21 cells were used as the control. Approximately 5×10^8 cells were plated on LB-kanamycin agar plates and incubated at 37℃ for 1 h. Then, filter discs (7 mm diameter) soaked in different concentrations of cumyl hydroperoxide or $HgCl_2$ were placed on the surface of the agar. The cells were cultivated at 37℃ for 24 h, after which the inhibition zones around the filter discs were measured. The diameters were measured three times from different angles with a Vernier caliper, and three replicate plates were made for every treatment.

The isolation of the ORF sequence of *AccCDK5r1* and a yeast two-hybrid analysis

Because CDK5 and p35 have been shown to interact, we used the p35 protein sequence of *Homo sapiens* (GenBank: CAA56587.1) as a reference to perform a local BLAST (BLAST-2.3.0+) analysis to search for the p35 homolog in *A. cerana cerana*. From our analysis, an unannotated protein was found, and the specific primers AccCDK5r1-5 and AccCDK5r1-3 (as shown in Table 1) were designed based on this sequence. The PCR product was ligated into a pEASY-T1 vector and transformed into Trans1-T1 Phage Resistant Chemically Competent Cells for sequencing. Bioinformatics and phylogenetic analyses of AccCDK5r1 were performed as described above.

A yeast two-hybrid analysis was performed by using The Matchmaker Gold Yeast Two-Hybrid System (Clontech, Dalian, China) after obtaining the sequences of *AccCDK5* and *AccCDK5r1*. The ORF of *AccCDK5*, flanked by *Nde* I and *Bam*H I restriction sites, was amplified and subcloned into the pGBKT7 vector. Similarly, the ORF of *AccCDK5r1*, flanked by *Nde* I and *Sac* I restriction sites, was ligated into the pGADT7 vector. To obtain positive clones, the plasmids AccCDK5-BD and AccCDK5r1-AD were co-transformed into the Y2H Gold yeast strain and then cultured on selective SD medium (DDO, SD/-Leu/-Trp and QDO, SD/-Ade/-His/-Leu/-Trp). After 3-5 days, the positive clones were subcultured onto QDO medium, to which X-α-Gal (QDO/X) was added for a second round of selection. In this analysis, pGBKT7-53 and pGADT7-T were used as positive controls, whereas pGBKT7-lam and pGADT7-T were used as negative controls.

Expression patterns of *AccCDK5r1* under environmental stress

To determine the expression patterns of *AccCDK5r1* in response to environmental stresses, real-time quantitative PCR was performed as described above.

Primers

The primers used in this study are listed in Table 1. All primers were synthesized by Biosune Biotechnology (ShangHai) Co., Ltd. (Shanghai, China).

Data analysis

The results of the gene expression analyses and the diameters of disc diffusion assay are presented as the mean ± SE from three independent experiments. All data were tested for violations of the assumptions for parametric analyses. Significant differences, labeled with different letters, were determined by Duncan's multiple range tests using the Statistical Analysis System (SAS) version 9.1 software program (SAS Institute, Cary, NC, USA). The same letters indicate that there was no significant difference between the experimental groups, whereas different letters and overlapped letters indicate that there was a significant difference and a non-significant difference between the experimental groups, respectively.

Results

Bioinformatics and phylogenetic analysis of AccCDK5

By using reverse-transcription PCR, *AccCDK5* was cloned from *A. cerana cerana*. The open reading frame (ORF) of *AccCDK5* is 900 bp and encodes a 299-amino acid polypeptide with a predicted molecular weight of 34.05 kDa and an isoelectric point of 7.6. Multiple sequence alignments of several CDK5s from different species revealed that the putative AccCDK5 has high homology with the other CDK5 genes. AccCDK5 shows high identity with AmCDK5 (*Apis mellifera*, NP_001161897.1), DmCDK5 (*Drosophila melanogaster*, GI: 17137070), and HsCDK5 (*Homo sapiens*, GI: 30584911). The high degree of homology indicates that CDK5 is conserved across species. As shown in Fig. 1a, the putative AccCDK5 contained the typical features of CDK5, including Thr14 and Tyr15 sites, an activation loop, and binding sites for non-cyclin regulators.

As shown in Fig. 1b, AccCDK5 has a close evolutionary relationship with AmCDK5 from *Apis mellifera*, which is in agreement with the multiple amino acid sequence alignments of CDK5.

Developmental and tissue-specific expression patterns of AccCDK5

To investigate the expression patterns of *AccCDK5* at different developmental stages and in several tissues, qPCR was used. We found that there were significant differences ($P < 0.001$, $F = 273.76$) among different developmental stages in the relative expression of *AccCDK5*. As shown in Fig. 2a specifically, *AccCDK5* was highly expressed during the larval and pupal stages. During the larval stage, the relative expression of *AccCDK5* increased from L4 to L6 and was the highest at L6. During the pupal stage, *AccCDK5* expression levels in dark-eyed pupae were higher than in the other pupae. The *AccCDK5* expression level during the adult stage was lower than in the other stages. We then analyzed *AccCDK5* expression in different tissues and found that the expression level was higher in head and muscle tissue than in other three tissues (Fig. 2b, $P < 0.001$, $F = 41.87$). The expression pattern was consistent with our expectations given the functions of *AccCDK5*.

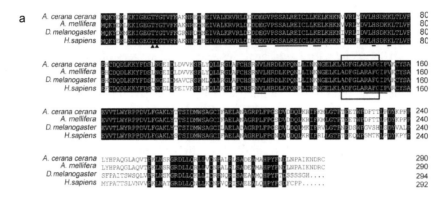

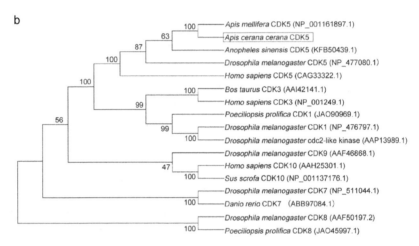

Fig. 1 Characterization of cyclin-dependent kinase 5 (CDK5) from various species. a. Multiple amino acid sequence alignments of AccCDK5 protein sequences with other CDK5 proteins, AmCDK5, DmCDK5, and HsCDK5. Thr14 and Tyr15 are marked by black triangles. The activation loop (T-loop) is boxed. The binding sites of non-cyclin regulators of CDK5 are marked by horizontal lines. b. Phylogenetic analysis of CDK5 from several species. AccCDK5 is boxed.

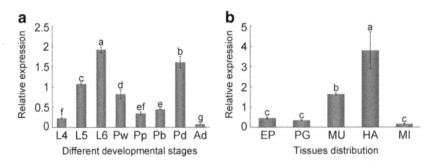

Fig. 2 The relative expression of *AccCDK5* in different developmental stages and tissues. a. Different developmental stages: larval (L4-L6, from the fourth to the sixth instars), pupal (Pw, white-eyed pupae; Pp, pink-eyed pupae; Pb, brown-eyed pupae; and Pd, dark-eyed pupae), and adult workers (Ad). b. Tissue distribution: EP (epidermis), PG (poison gland), MU (muscle), HA (head), and MI (midgut). The error bars represent the mean ± SE from three independent experiments. The letters above the columns represent significant differences ($P < 0.001$) based on Duncan's multiple range tests.

Expression patterns of *AccCDK5* under environmental stresses

To investigate the expression patterns of *AccCDK5* under environmental stresses, adult workers were subjected to ultraviolet light (UV), H_2O_2, 4℃, 42℃, heavy metals ($HgCl_2$, $CdCl_2$), or pesticides (spirodiclofen, acaricide, acetamiprid, imidacloprid, and phoxim) treatments. As shown in Fig. 3, the relative expression of *AccCDK5* was upregulated in the majority of the treatments except the H_2O_2 treatment. Data analysis demonstrated that there were significant differences ($P < 0.001$) between samples from treated bees and the controls and the F statistics were listed in Table S4. As shown in Fig. 3a and b, the relative expression level of *AccCDK5* was increased by 3.7-fold after 0.5 h of exposure to 4℃ and 3.2-fold after 5 h of exposure to 42℃. After UV treatment, the relative expression levels of *AccCDK5* increased by 2-fold at 2 h (Fig. 3c). The H_2O_2 treatment caused a nearly 80% decrease in the relative expression of *AccCDK5* at 30 min (Fig. 3d). The relative expression levels of *AccCDK5* increased by 4.7-fold and 2.4-fold at 3 h and 4.5 h after $HgCl_2$ and $CdCl_2$ treatments, respectively, and then decreased back to the basal levels (Fig. 3e and f). Spirodiclofen treatment changed the mRNA levels of *AccCDK5* only slightly, with a 1.7-fold increase at 1.5 h (Fig. 3g). By contrast, acaricide, acetamiprid, imidacloprid and phoxim treatments increased the relative expression levels of *AccCDK5* by 15.7-fold, 6-fold, 5.9-fold and 10-fold, respectively at different time points (Fig. 3h-k).

Protein expression levels of AccCDK5 under environmental stresses

To study the expression patterns of AccCDK5 in response to environmental stresses, we performed Western blotting analysis. Total protein lysates from *A. cerana cerana* treated with 4℃, H_2O_2, phoxim, or acaricide were probed by using anti-AccCDK5 antibody. The analysis of the densities (as shown in supplemental Table S5-S8 and Fig. S1) indicated that the amounts of protein from each sample were approximately equal. As shown in Fig. 4a ($P < 0.001$, $F = 69.82$) and supplemental Table S3, the levels of AccCDK5 expression were reduced after exposure to 4℃ for 0.5 and 1 h and induced after 2 h and 2.5 h. The expression of AccCDK5 was downregulated after H_2O_2 treatment (Fig. 4b, $P < 0.001$, $F = 67,379.2$, as well as the supplemental Table S4). As shown in Fig. 4c, d and supplemental Table S5 and S6, the levels of expression of AccCDK5 were reduced slightly after phoxim ($P < 0.001$, $F = 270.43$) and acaricide ($P = 0.003$, $F = 15.71$) treatment.

Characterization of recombinant AccCDK5 protein

Next, we determined the protective effects of recombinant AccCDK5 protein by using the disc diffusion method. Our results show that the inhibition zones around the filters soaked with $HgCl_2$ or cumyl hydroperoxide were smaller in diameter on the plates containing *E. coli* overexpressing AccCDK5 than on the control plates (as shown in Fig. 5 and supplemental Table S9 and S10). Analysis of the diameter of the inhibition zones revealed that there were significant differences between *E. coli* that overexpressed AccCDK5 and the controls at the same concentration of $HgCl_2$ ($P < 0.001$, $F = 71.15$) or cumyl hydroperoxide ($P < 0.001$, $F = 94.35$).

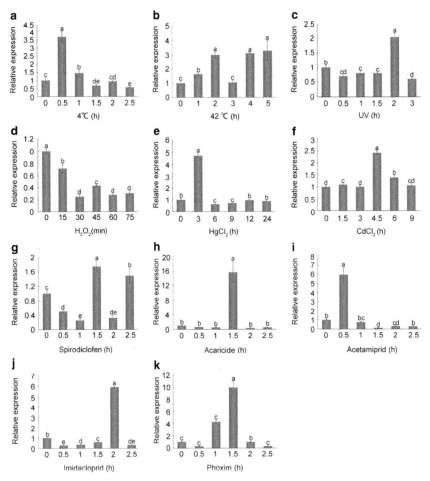

Fig. 3 Expression profiles of *AccCDK5* under environmental stress. Total RNA was extracted from *Apis cerana cerana* treated with various stresses at the indicated time, the treatments include (a) 4℃, (b) 42℃, (c) UV, (d) H_2O_2, (e) $HgCl_2$, (f) $CdCl_2$, (g) spirodiclofen, (h) acaricide, (i) acetamiprid, (j) imidacloprid and (k) phoxim. *β-actin* was used as an internal control. The error bars represent the mean ± SE from three independent experiments. The letters above the columns represent significant differences ($P < 0.001$) based on Duncan's multiple range tests.

Cloning and sequence analysis of *AccCDK5r1*

Therefore, to understand the functional mechanism of AccCDK5, we isolated *AccCDK5r1*, the homolog of mammalian *p35*. The ORF of *AccCDK5r1* is 1,083 bp, and it encodes a 360-amino acid protein. The predicted molecular mass of AccCDK5r1 is 40.42 kDa, whereas its predicted isoelectric point is 9.47. As shown in Fig. 6, multiple sequence alignments suggested that AccCDK5r1 has a high amino acid identity in the region of the N-myristoylation consensus sequence (MGXXXS/T) and the binding and activation site with cyclin-dependent kinase 5 regulator 1 from *Homo sapiens*, *Mus musculus*, and *Danio rerio*. Next, phylogenetic analysis results show that AccCDK5r1 has a close evolutionary relationship with cyclin-dependent kinase 5 activator 1 from *Apis mellifera*.

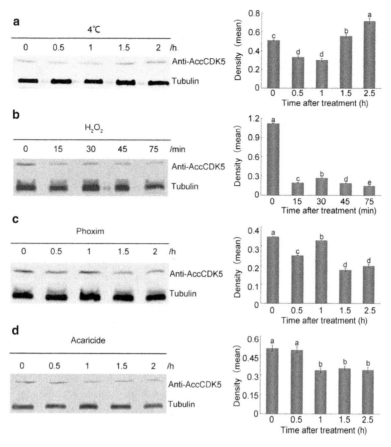

Fig. 4 Western blotting analysis of AccCDK5 changes after (a) 4℃, (b) H_2O_2, (c) phoxim and (d) acaricide treatments. AccCDK5 protein was immunoblotted with anti-AccCDK5. The signal for the binding reaction was visualized with HRP substrate. The error bars represent the mean ± SE from three independent experiments. The letters above the columns represent significant differences based on Duncan's multiple range tests.

Yeast two-hybrid analysis

To investigate the interaction between AccCDK5 and AccCDK5r1, a yeast two-hybrid analysis was performed. As shown in Fig. 7a, cells transformed with the pGBKT7 vector or the AccCDK5-BD plasmid do not grow on SD Medium-His and SD Medium-Ade, which precludes the self-activation of the AccCDK5-BD plasmid. As shown in Fig. 7b, cells transformed with the pGBKT7 vector or the AccCDK5-BD plasmid grew to similar extents, which suggests that AccCDK5-BD is a non-toxic protein. Next, AccCDK5-BD and AccCDK5r1-AD were co-transformed into the Y2H Gold yeast strain. Cells were able to grow on SD Medium-Leu/-Trp plates, which indicates that the plasmids were transformed successfully. The indicated strain was inoculated on plates of SD Medium-Ade-/His-/Leu-/Trp, and cells transformed with AccCDK5-BD and AccCDK5r1-AD grew on SD Medium-Ade/-His/-Leu/-Trp. The positive, interacting clones grew on the SD Medium-Ade/-His/-Leu/-Trp with X-α-gal, which confirmed the interaction between AccCDK5 and AccCDK5r1 (Fig. 7c).

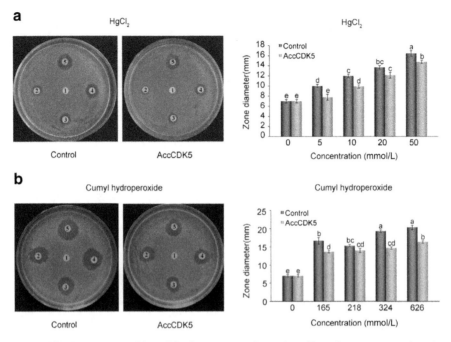

Fig. 5 Disc diffusion assays. Disc diffusion assays by using *E. coli* overexpressing AccCDK5. AccCDK5 was overexpressed in *E. coli*, and bacteria transformed with pET-30a (+) were used as negative controls. Filter discs soaked with different concentrations of $HgCl_2$ or cumyl hydroperoxide were placed on the agar plates. The killing zones around the filters were measured after overnight exposure. The error bars represent the mean ± SE from three independent experiments. The letters above the columns represent significant differences ($P < 0.001$) based on Duncan's multiple range tests.

Expression profiles of AccCDK5r1 under environmental stresses

To explore the relationship between AccCDK5 and AccCDK5r1, we analyzed the expression patterns of *AccCDK5r1* after several stress treatments. Expression of *AccCDK5r1* was induced after stress treatments (Fig. 8). Notably, the expression patterns of *AccCDK5r1* were approximately consistent with those of *AccCDK5* after the same treatment.

Discussion

CDK5 is an atypical member of the CDK family. In the early 1990s, CDK5 was first discovered in bovine brain tissue [46], and since then, great advancements have been made in determining the function and the mechanism of this gene. Activators of CDK5 as well as the activation mechanism of CDK5 and the cellular processes that require CDK5 are distinctly different from those of the other CDK family members [47]. The proper localization and regulation of CDK5 is crucial for its functions, and the absence of CDK5 is lethal [48]. However, the majority of research on CDK5 has focused on humans and other mammals. Few CDK5 studies in insects have been systematically performed.

In this study, we isolated *AccCDK5* from *A. cerana cerana*. The sequence analysis of AccCDK5 revealed that there was high identity with other CDK5s from different species and

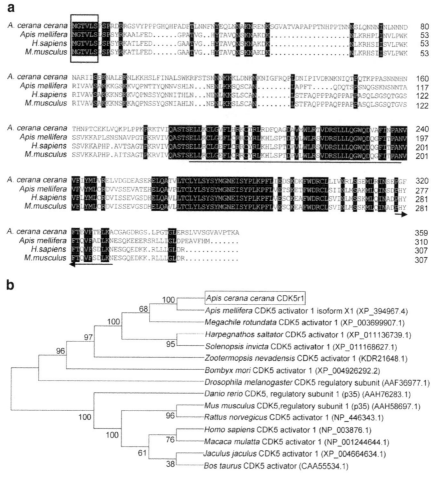

Fig. 6 Characterization of cyclin-dependent kinase 5 regulator 1 (CDK5r1) from different species. a. Multiple amino acid sequence alignments of *Apis cerana cerana* CDK5 regulatory subunit 1 (AccCDK5r1), *Danio rerio* CDK5 regulatory subunit 1 (p35) (AAH76283.1), *Homo sapiens* CDK5 activator 1 (NP_003876.1) and *Mus musculus* CDK5 regulatory subunit 1 (p35) (AAH58697.1). The N-myristoylation consensus sequence (MGXXXS/T) is boxed. CDK5 binding sites are marked by horizontal lines. CDK5 activation sites are marked by arrows. b. Phylogenetic analysis of CDK5r1 from several species. AccCDK5r1 is boxed.

that AccCDK5 contained conserved features of the CDK5 family. These two observations indicated that CDK5 is conserved across species. Phylogenetic analysis showed that AccCDK5 belonged to the CDK5 group and had a close evolutionary relationship with AmCDK5. In summary, we infer that AccCDK5 is a member of the CDK5 family and has functions similar to those of other reported CDK5s.

It has previously been reported that CDK5 kinase activity is indispensable for neurite outgrowth during the course of neuronal differentiation [49]. The central nervous system mass increases exponentially up to 6 d after hatching in *Drosophila melanogaster* [50]. Adult-specific head sensory structures form during the larval and pupal stages in *Drosophila melanogaster* and *Apis mellifera*, and mushroom bodies develop abundantly during the larval stage [51]. Moreover, CDK5 is involved in the regulation of remodeling of mushroom body

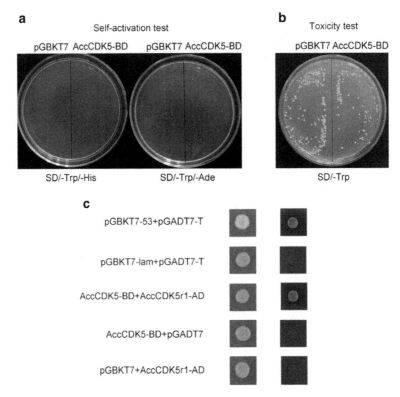

Fig. 7 Yeast two-hybrid analysis of AccCDK5 and AccCDK5r1. a. Self-activation test for AccCDK5-BD. b. toxicity test for AccCDK5-BD. c. AccCDK5-BD and AccCDK5r1-AD fusion constructs were co-transformed into the Y2H Gold yeast strain and grown on DDO and QDO SD media. The positive clones were confirmed on QDO/X SD media.

neurons in *Drosophila melanogaster* [52]. The specific expression pattern of *AccCDK5* in different developmental stages may reflect unique demands in the development processes and nervous system formation. CDK5 is ubiquitously expressed throughout the organism, but the highest kinase activity of CDK5 is detected in post-mitotic neurons [46]. Tissue-specific expression analysis showed that *AccCDK5* is expressed in different tissues but is most highly expressed in the head and muscle, the main locations of the nervous system in bees [37].

There are numerous stresses in the habitat of honeybees, including unsuitable temperatures, UV, H_2O_2, heavy metal and pesticides, which can induce ROS production and eventually lead to oxidative stress [53]. However, the expression patterns of CDK5 under environmental stresses have not been studied extensively in insects. Previous studies demonstrated that downregulation of nestin caused by oxidative stress in neuronal precursor cells led to activation of CDK5 [54], and that CDK5 modified p53 post-translationally and increased p53 stability in response to H_2O_2 [55]. In our study, *AccCDK5* was induced by the exposure of honeybees to 42 ℃, 4 ℃, H_2O_2, UV, $HgCl_2$, $CdCl_2$ and several pesticides.

The normal temperature of a hive is between 33 ℃ and 36 ℃ [56]. High and low temperatures can lead to oxidative stress [57, 58]. Heat shock and cold stress can induce the phosphorylation of Tau, a substrate of CDK5, which is associated with increases in CDK5 [59]. In our study, the transcription levels of *AccCDK5* increased quickly after exposure to 42 and 4 ℃, suggesting that *AccCDK5* is involved in protecting honeybees from ROS damage caused by

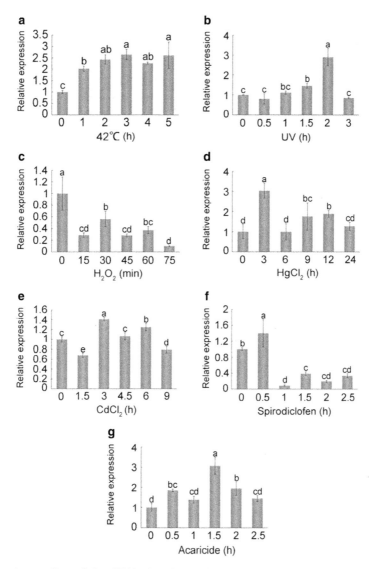

Fig. 8 Expression profiles of *AccCDK5r1* under environmental stresses. Total RNA was extracted from *Apis cerana cerana* treated with various stresses for the indicated amounts of time. The treatments included: (a) 42℃, (b) UV, (c) H_2O_2, (d) $HgCl_2$, (e) $CdCl_2$, (f) spirodiclofen and (g) acaricide. *β-actin* was used as an internal control. The error bars represent the mean ± SE from three independent experiments. The letters above the columns represent significant differences ($P < 0.001$) based on Duncan's multiple range tests.

dramatic temperature fluctuations. UV radiation is absorbed by DNA bases and eventually causes DNA damage and increased ROS production [60, 61]. In addition, CDK5 has been reported to be a mediator of the response to DNA damage [62]. After UV exposure, *AccCDK5* was induced. This effect suggests that *AccCDK5* is related to the DNA damage and ROS response after UV treatment.

Cadmium can induce oxidative stress and, consequently, neuronal death pathways [63]. MeHg-induced hyperphosphorylation of Tau can lead to neuropathological changes in the mouse brain [64]. In our study, *AccCDK5* was induced by the heavy metals $HgCl_2$ and $CdCl_2$.

As mentioned above, phosphorylation of Tau is associated with increases in CDK5. $HgCl_2$ and $CdCl_2$ treatment may lead to hyperphosphorylation of Tau and oxidative stress in the nervous system, thus resulting in increases in *AccCDK5*. The strong oxidizing agent H_2O_2 also causes oxidative damage. Although the transcription and translation levels of AccCDK5 were downregulated, the results of the disc diffusion assay revealed that overexpression of AccCDK5 enhances the resistance of bacteria to oxidative stress, indicating that AccCDK5 may be involved in the response to oxidative stress.

One of the main features of pesticide toxicity is the induction of oxidative stress [65]. Several other studies also showed an increase of oxidative stress after exposure to pesticides, and chronic systemic pesticide exposure results in features of Parkinson's disease, which is connected to the dysregulation of CDK5 [66, 67]. Our results revealed that the expression level of *AccCDK5* was upregulated after pesticide treatments. These results suggest that the ingestion of food containing pesticides may lead to oxidative stress and the dysregulation of AccCDK5. We also detected the downregulation of AccCDK5 at the protein level after acaricide and phoxim treatments, a result that was different from the mRNA expression data. The upregulation of *AccCDK5* at the transcriptional level may be a compensatory mechanism to account for decreased protein levels. For different genes, the protein-per-mRNA ratio is different; additionally, the ratio might change after different treatments. The square of Pearson's correlation coefficient (R^2) between the mRNA and protein concentrations averages 0.09 to 0.46 [68].

Furthermore, the stability of proteins is essential to their structure and function [69]. After exposure to stress, the stability of the AccCDK5 protein may decrease and eventually lead to a decrease in AccCDK5 protein levels. The different concentrations, intensities and mechanism of the stresses (such as heat shock and cold stress, UV, heavy metal, and pesticides) may lead to diverse responses.

Previous studies have demonstrated that most substrates mediate their interaction with CDK5 through p35 and that the regulation of the p35 mRNA level is a major determinant of CDK5 activity. The p35 protein may act as both an activator and an adaptor for CDK5, and this protein plays an indispensable role in CDK5 function [70, 71]. CDK5 and p35 form a complex that participates in cellular processes [72]. Therefore, we cloned the gene from *A. cerana cerana* that codes for p35 and named it *AccCDK5r1*.

The deduced dCdk5a protein, the *Drosophila melanogaster* protein homologous to p35, is much longer than p35 [73]. In our study, the molecular weight of AccCDK5 is greater than 35 kDa, but the high homology of the N-myristoylation consensus sequence (MGXXXS/T) and the CDK5 binding and activation domains, which are required for CDK5 activation, suggests that AccCDK5r1 may play the role of CDK5 activator and adaptor in *A. cerana cerana* [74]. Phylogenetic analysis revealed that AccCDK5r1 has a closer evolutionary relationship with the corresponding proteins in insects than in other species. Based on the sequence and phylogenetic analyses, we propose that *AccCDK5r1* in *A. cerana cerana* is the homolog of *CDK5r1*. Yeast two-hybrid analysis indicates that AccCDK5r1 interacts with AccCDK5, which supports our hypothesis.

The activity of CDK5 is primarily determined by the available amount of CDK5 activator [75, 76]. In this study, the expression patterns of *AccCDK5* and *AccCDK5r1* were similar, which indicates a correlation between *AccCDK5* and *AccCDK5r1*. Notably, *AccCDK5r1* was downregulated after H_2O_2 treatment, and a similar mechanism may be responsible for the downregulation of *AccCDK5*. These results provide further evidence that *AccCDK5r1* is the homolog of *CDK5r1* in *A. cerana cerana*.

Apis cerana cerana, like other insects, encounters multiple environmental stresses. In this study, we employed bioinformatics, phylogenetic analyses, expression profiling, and interaction studies to elucidate the relationship between AccCDK5 and AccCDK5r1. We identified and characterized a potential physiological function of AccCDK5. The results revealed that AccCDK5 and AccCDK5r1 play a role in the response to oxidative stresses. Our work forms the basis for future studies of AccCDK5 and its activator AccCDK5r1.

References

[1] Chen, Y. W., Wang, C. H., An, J., Kaikuang, H. Susceptibility of the Asian honey bee, *Apis cerana*, to American foulbrood, *Paenibacillus larvae larvae*. J Apicult Res. 2015, 39, 169-175.
[2] Radloff, S. E., Hepburn, C., Hepburn, H. R., Fuchs, S., Hadisoesilo, S., Tan, K., Engel, S. M, Kuznetsov, V. Population structure and classification of *Apis cerana*. Apidologie. 2010, 41, 589-601.
[3] Zhao, H. X., Zeng, X. N., Liang, Q., Zhang, X. F., Huang, W. Z., Chen, H. S., Luo, Y. X. Study of the *obp5* gene in *Apis mellifera ligustica* and *Apis cerana cerana*. Genet Mol Res. 2015, 14, 6482-6494.
[4] Yang, G. Harm of introducing the western honeybee *Apis mellifera* L. to the Chinese honeybee *Apis cerana* F. and its ecological impact. Acta Entomologica Sinica. 2005, 48, 401-406.
[5] Kodrík, D., Bednářová, A., Zemanová, M. Krishnan, N. Hormonal regulation of response to oxidative stress in insects-an update. Int J Mol Sci. 2015, 16, 25788-25816.
[6] Meng, J. Y., Zhang, C. Y., Zhu, F., Wang, X. P., Lei, C. L. Ultraviolet light-induced oxidative stress: effects on antioxidant response of *Helicoverpa armigera* adults. J Insect Physiol. 2009, 55, 588-592.
[7] Espín, S., Martínez-López, E., Jiménez, P., María-Mojica, P., García-Fernández, A. J. Effects of heavy metals on biomarkers for oxidative stress in Griffon vulture (*Gyps fulvus*). Environ Res. 2014, 129, 59-68.
[8] An, M. I., Choi, C. Y. Activity of antioxidant enzymes and physiological responses in ark shell, *Scapharca broughtonii*, exposed to thermal and osmotic stress: effects on hemolymph and biochemical parameters. Comparative Biochemistry and Physiology Part B: Biochemistry and Molecular Biology. 2010, 155, 34-42.
[9] Halliwell, B. Reactive oxygen species in living systems: source, biochemistry, and role in human disease. Am J Med. 1991, 91, S14-S22.
[10] Lushchak, V. I. Environmentally induced oxidative stress in aquatic animals. Aquatic Toxicology. 2011, 101, 13-30.
[11] Imlay, J. A., Linn, S. DNA damage and oxygen radical toxicity. Science. 1988, 240, 1302-1309.
[12] Green D. R., Reed, J. C. Mitochondria and Apoptosis. Science. 1998, 281, 1309-1312.
[13] Malumbres, M., Harlow, E., Hunt, T., Hunter, T., Lahti, J. M., Manning, G., Morgan, D. O., Tsai, L. H., Wolgemuth, D. J. Cyclin-dependent kinases: a family portrait. Nat Cell Biol. 2009, 11, 1275-1276.
[14] Morgan, D. O. Cyclin-dependent kinases: engines, clocks, and microprocessors. Annu Rev Cell Dev Bi. 1997, 13, 261-291.
[15] Hellmich, M. R., Pant, H. C., Wada, E. J. F. Neuronal cdc2-like kinase a cdc2-related protein kinase with predominantly neuronal expression. Proc Natl Acad Sci USA. 1992, 89, 10867-10871.
[16] Meyerson, M., Enders, G. H., Wu, C. L., Su, L. K., Gorka, C., Nelson. C., Harlow, E., Tsai, L. H. A family of human cdc2-related protein kinase. Embo J. 1992, 11, 2909-2917.
[17] Lew, J., Beaudette, K., Litwin, C. M. Purification and characterization of a novel proline-directed protein kinase from bovine brain. J Biol Chem. 1992, 267, 13383-13390.

[18] Dhavan, R., Tsai, L. H. A decade of CDK5. Nat Rev Mol Cell Bio. 2001, 2, 749-759.
[19] Giese, K. P. Generation of the Cdk5 activator p25 is a memory mechanism that is affected in early Alzheimer's disease. Front Mol Neurosci. 2014, 7, 36.
[20] Sun, K. H., De Pablo, Y., Vincent, F., Shah, K. Deregulated Cdk5 promotes oxidative stress and mitochondrial dysfunction. J Neurochem. 2008, 107, 265-278.
[21] Zhang, P., Shao, X. Y., Qi, G. J., Chen, Q., Bu, L. L., Chen, L. J., Shi, J, Ming, J, Tian, B. Cdk5-dependent activation of neuronal inflammasomes in Parkinson's disease. Movement Disord. 2016, 31, 366-376.
[22] Moncini, S., Castronovo, P., Murgia, A., Russo, S., Bedeschi, M. F., Lunghi, M.,Selicorni, A, Bonati, M. T., Riva, P, Venturin, M. Functional characterization of CDK5 and CDK5R1 mutations identified in patients with non-syndromic intellectual disability. J Hum Genet. 2016, 61, 283-293.
[23] Tsai, L. H., Delalle. I, Caviness Jr, V. S., Chae, T., Harlow, E. p35 is a neural-specific regulatory subunit of cyclin-dependent kinase 5. Nature. 1994, 371, 419-423.
[24] Tang, D., Yeung, J., Lee, K. Y., Matsushita, M., Matsui, H., Tomizawa, K., Hatase, O., Wang, J. H. An isoform of the neuronal cyclin-dependent kinase 5 (Cdk5) activator. J Biol Chem. 1995, 270, 26897-26903.
[25] Lee, K. Y., Rosales, J. L., Tang, D., Wang, J. H. Interaction of cyclin-dependent kinase 5 (Cdk5) and neuronal Cdk5 activator in bovine brain. J Biol Chem. 1996, 271, 1538-1543.
[26] Minegishi, S., Asada, A., Miyauchi, S., Fuchigami, T., Saito, T. Hisanaga, S. Membrane association facilitates degradation and cleavage of the cyclin-dependent kinase 5 activators p35 and p39. Biochemistry. 2010, 49, 5482-5493.
[27] Poon, R. Y., Lew, J., Hunter, T. Identification of functional domains in the neuronal Cdk5 activator protein. J Biol Chem. 1997, 272, 5703-5708.
[28] Takasugi, T., Minegishi, S., Asada, A., Saito, T., Kawahara, H., Hisanaga, S. Two degradation pathways of the p35 Cdk5 (cyclin-dependent kinase) activation subunit, dependent and independent of ubiquitination. J Biol Chem. 2016, 291, 4649-4657.
[29] Lee M. S., Kwon Y. T., Li, M., Friedlander, R. M. Neurotoxicity induces cleavage of p35 to p25 by calpain. Nature. 2000, 405, 360-364.
[30] Kusakawa, G., Saito, T., Onuki, R., Ishiguro, K., Kishimoto, T., Hisanaga, S. Calpain-dependent proteolytic cleavage of the p35 cyclin-dependent kinase 5 activator to p25. J Biol Chem. 2000, 275, 17166-17172.
[31] Nath, R., Davis, M., Probert, A. W., Kupina, N. C., Ren, X., Schielke, G. P., Wang, K. K. Processing of cdk5 activator p35 to its truncated form (p25) by calpain in acutely injured neuronal cells. Biochem Biophys Res Commun. 2000, 274, 16-21.
[32] Patrick, G. N., Zukerberg, L., Nikolic, M., Dikkes, P., Tsai, L. H. Conversion of p35 to p25 deregulates Cdk5 activity and promotes neurodegeneration. Nature. 1999, 402, 615-622.
[33] Ruffels, J., Griffin, M., Dickenson, J. M. Activation of *ERK1/2*, *JNK* and *PKB* by hydrogen peroxide in human *SH-SY5Y* neuroblastoma cells: role of *ERK1/2* in H_2O_2-induced cell death. Eur J Pharmacol. 2004, 483, 163-173.
[34] Dabrowski, A., Boguslowicz, C., Dabrowska, M., Tribillo, A., G. Reactive oxygen species activate mitogen-activated protein kinases in pancreatic acinar cells. Pancreas. 2000, 21, 376-384.
[35] Stanciu, M., DeFranco, D. B. Prolonged nuclear retention of activated extracellular signal-regulated protein kinase promotes cell death generated by oxidative toxicity or proteasome inhibition in a neuronal cell line. J Biol Chem. 2002, 277, 4010-4017.
[36] Song, H., Kim, W., Choi, J. H., Kim, S. H., Lee, D., Park, C. H., Kim, S., Kim. D. Y., Kim. K. T. Stress-induced nuclear translocation of *CDK5* suppresses neuronal death by downregulating *ERK* activation via *VRK3* phosphorylation. Sci Rep. 2016, 6, 28634.
[37] Carreck, N. L., Andree, M., Brent, C. S., Cox-Foster, D., Dade, H. A., Ellis, J. D., Hatjina, F., Englesdorp, D. V. Standard methods for *Apis mellifera* anatomy and dissection. J Apicult Res. 2013, 52, 1-40.

[38] Park, D., Jung, J. W., Choi, B. S., et al. Uncovering the novel characteristics of Asian honey bee, *Apis cerana*, by whole genome sequencing. Bmc Genomics. 2015, 16, 1.

[39] Templeton, N. S. The polymerase chain reaction. History, methods, and applications. Diagn Mol Pathol. 1992, 1, 58-72.

[40] Giulietti, A., Overbergh, L., Valckx, D., Decallonne, B., Bouillon, R., Mathieu, C. An overview of real-time quantitative PCR: applications to quantify cytokine gene expression. Methods. 2001, 25, 386-401.

[41] Bustin, S. A., Benes, V., Garson, J. A., et al. The MIQE guidelines: minimum information for publication of quantitative real-time PCR experiments. Clin Chem. 2009, 55, 611-622.

[42] Scharlaken, B., De Graaf, D. C., Goossens, K., Brunain, M., Peelman, L. J., Jacobs, F. J. Reference gene selection for insect expression studies using quantitative real-time PCR: the head of the honeybee, *Apis mellifera*, after a bacterial challenge. J Insect Sci. 2008, 8, 1-10.

[43] Livak, K. J., Schmittgen, T. D. Analysis of relative gene expression data using real-time quantitative PCR and the 2 (−Delta Delta C (T)) method. Methods. 2001, 25, 402-408.

[44] Meng, F., Zhang, Y., Liu, F., Guo, X., Xu, B. Characterization and mutational analysis of Omega-class GST (*GSTO1*) from *Apis cerana cerana*, a gene involved in response to oxidative stress. PLoS One. 2014, 9, 93100.

[45] Burmeister, C., Luersen, K., Heinick, A., Hussein, A., Domagalski, M., Walter, R. D., Liebau, E. Oxidative stress in *Caenorhabditis elegans*: protective effects of the Omega class glutathione transferase (*GSTO-1*). The Faseb Journal. 2008, 22, 343-354.

[46] Lew, J., Beaudette, K., Litwin, C. M., Wang, J. H. Purification and characterization of a novel proline-directed protein kinase from bovine brain. J Biol Chem. 1992, 267, 13383-13390.

[47] Cheung, Z. H. The roles of cyclin-dependent kinase 5 in dendrite and synapse development. Biotechnol J. 2007, 2, 949-957.

[48] Trunova, S., Giniger, E. Absence of the Cdk5 activator p35 causes adult-onset neurodegeneration in the central brain of *Drosophila*. Dis Model Mech. 2012, 5, 210-219.

[49] Nikolic, M., Dudek, H., Kwon, Y., Ramos, Y. Tsai, L. The cdk5/p35 kinase is essential for neurite outgrowth during neuronal differentiation. Gene Dev. 1996, 10, 816-825.

[50] Power, M. E. A quantitative study of the growth of the central nerve system of a holometabolous insect, *Drosophila melanogaster*. J Morphol. 1952, 91, 389-411.

[51] Farris, S., Robinson, G., Davis, R. Fahrbach, S. Larval and pupal development of the mushroom bodies in the honey bee, *Apis mellifera*. J Comp Neurol. 1999, 414, 97-113.

[52] Smith-Trunova, S., Prithviraj, R., Spurrier, J., Kuzina, I., Gu, Q., Giniger, E. Cdk5 regulates developmental remodeling of mushroom body neurons in *Drosophila*. Dev Dynam. 2015, 224, 1550-1563.

[53] Lushchak, V. I. Environmentally induced oxidative stress in aquatic animals. Aquatic toxicology. 2011, 101, 13-30.

[54] Sahlgren, C. M., Pallari, H. M., He, T., Chou, Y. H., Goldman, R. D., Eriksson, J. E. A nestin scaffold links Cdk5/35 signaling to oxidant-induced cell death. Embo J. 2006, 25, 4808-4819.

[55] Lee, J. H., Jeong, M. W., Kim, W., Choi, Y. H., Kim, K. T. Cooperative roles of c-Abl and Cdk5 in regulation of p53 in response to oxidative stress. J Biol Chem. 2008, 283, 19826-19835.

[56] Tautz, J., Maier, S., Groh, C., Rossler, W., Brockmann, A. Behavioral performance in adult honey bees is influenced by the temperature experienced during their pupal development. Proceedings of the National Academy of Sciences. 2003, 100, 7343-7347.

[57] Malek, R. L., Sajadi, H., Abraham, J., Grundy, M. A., Gerhard, G. S. The effects of temperature reduction on gene expression and oxidative stress in skeletal muscle from adult zebrafish. Comparative Biochemistry Physiology Part C: Toxicology Pharmacology. 2004, 138, 363-373.

[58] Bagnyukova, T. V., Danyliv, S. I., Zin'ko, O. S., Lushchak, V. I. Heat shock induces oxidative stress in rotan *Perccottus glenii* tissues. J Therm Biol. 2007, 32, 255-260.

[59] Lau, L. F., Seymour, P. A., Sanner, M. A., Schachter, J. B. Cdk5 as a drug target for the treatment of

Alzheimer's disease. J Mol Neurosci. 2002, 19, 267-273.

[60] Gomez-Mendoza, M., Banyasz, A., Douki, T., Markovitsi, D., Ravanat, J. L. Direct oxidative damage of naked DNA generated upon absorption of UV radiation by nucleobases. J Phys Chem Lett. 2016, 7 (19), 3945-3948.

[61] Kottuparambil, S., Shin, W., Brown, M. T., Han, T. UV-B affects photosynthesis, ROS production and motility of the freshwater flagellate, *Euglena agilis* Carter. Aquat Toxicol. 2012, 122-123, 206-213.

[62] Zhu, J., Li, W., Mao, Z. Cdk5: mediator of neuronal development, death and the response to DNA damage. Mech Ageing Dev. 2011, 132, 389-394.

[63] Antoniali, G. Molecular mechanisms of neurodegeneration involving the effect of environmental pollutants on dna repair enzymes. Molecular Signaling and Regulation in Glial Cells. 2014, 44-56.

[64] Fujimura, M., Usuki, F., Sawada, M., Takashima, A. Methylmercury induces neuropathological changes with Tau hyperphosphorylation mainly through the activation of the c-jun-N-terminal kinase pathway in the cerebral cortex, but not in the hippocampus of the mouse brain. Neurotoxicology. 2009, 30, 1000-1007.

[65] Asghari, M. H., Moloudizargari, M., Bahadar, H., Abdollahi, M. A review of the protective effect of melatonin in pesticide-induced toxicity. Expert Opin Drug Met. 2016, 1-10.

[66] Thomaz, J. M., Martins, N. D., Monteiro, D. A., Rantin, F. T., Kalinin, A. L. Cardio-respiratory function and oxidative stress biomarkers in Nile tilapia exposed to the organophosphate insecticide trichlorfon (NEGUVON®). Ecotox Environ Safe. 2009, 72, 1413-1424.

[67] Piner, P., Sevgiler, Y., Uner, N. *In vivo* effects of fenthion on oxidative processes by the modulation of glutathione metabolism in the brain of *Oreochromis niloticus*. Environ Toxicol. 2007, 22, 605-612.

[68] Abreu, R. D. S., Penalva, L. O., Marcotte, E. M., Vogel, C. Global signatures of protein and mRNA expression levels. Mol Biosyst. 2009, 5, 1512-1526.

[69] Becktel, W. J., Schellman, J. A. Protein stability curves. Biopolymers. 1987, 26, 1859-1877.

[70] Lim, A. C. B., Qu, D., Qi, R. Z. Protein-protein interactions in *Cdk5* regulation and function. Neurosignals. 2003, 12, 230-238.

[71] Ross, S., Tienhaara, A., Lee, M. S., Tsai, L. H., Gill, G. C box-binding transcription factors control the neuronal specific transcription of the cyclin-dependent kinase 5 regulator p35. J Biol Chem. 2002, 277, 4455-4464.

[72] Büchner, A., Krumova, P., Ganesan, S., Bähr, M., Eckermann, K., Weishaupt, J. H. Sumoylation of p35 modulates p35/cyclin-dependent kinase (Cdk) 5 complex activity. Neuromol Med. 2015, 17, 12-23.

[73] Ma, E., Haddad, G. A *Drosophila* CDK5alpha-like molecule and its possible role in response to O_2 deprivation. Biochem Biophys Res Commun. 1999, 261, 459-463.

[74] Amin, N. D., Albers, W., Pant, H. C. Cyclin-dependent kinase 5 (cdk5) activation requires interaction with three domains of p35. J Neurosci Res. 2002, 67, 354-362.

[75] Hisanaga, S., Endo, R. Regulation and role of cyclin-dependent kinase activity in neuronal survival and death. J Neurochem. 2010, 115, 1309-1321.

[76] Hisanaga, S., Saito, T. The regulation of cyclin-dependent kinase 5 activity through the metabolism of p35 or p39 Cdk5 activator. Neurosignals. 2003, 12, 221-229.

3.4 Isolation of Carboxylesterase (esterase FE4) from *Apis cerana cerana* and Its Role in Oxidative Resistance during Adverse Environmental Stress

Manli Ma, Haihong Jia, Xuepei Cui, Na Zhai, Hongfang Wang, Xingqi Guo, Baohua Xu

Abstract: Carboxylesterases (CarEs) play vital roles in metabolizing different physiologically important endogenous compounds and in detoxifying various harmful exogenous compounds in insects. Multiple studies of CarEs have focused on pesticide metabolism in insects, while few studies have aimed to identify CarE functions in oxidative resistance, particularly in *Apis cerana cerana*. In this study, we isolated a carboxylesterase gene, esterase FE4, from *Apis cerana cerana* and designated it towards an exploration of its roles as an antioxidant and in detoxification. We investigated *AcceFE4* expression patterns in response to various stressors. A quantitative real-time PCR analysis revealed that *AcceFE4* was upregulated by H_2O_2, imidacloprid, and paraquat, and was downregulated by 4℃, UV radiation, $CdCl_2$, and $HgCl_2$. Additionally, the protein expression of this gene was downregulated at 4℃ and upregulated by H_2O_2. Disc diffusion assays showed that the AcceFE4 recombinant protein-expressing bacteria had a smaller killing zone than the control group with the paraquat, $HgCl_2$ and cumyl hydroperoxide treatments. Moreover, when the gene was knocked down by RNA interference, we observed that multiple oxidant genes (i.e., *AccSOD*, *AccGST*, *AccTrx*, *AccMsrA*, and others) were downregulated in the knockdown samples. Superoxide dismutase (SOD), peroxidase (POD) and catalase (CAT) activity levels were reduced in the knockdown samples relative to the control group. Finally, we measured the enzyme activity of carboxylesterase and found that the enzyme activity was also reduced in the silent samples. Together, these data suggest that *AcceFE4* may be involved in the oxidative resistance response during adverse stress.

Keywords: *Apis cerana cerana*; *AcceFE4*; Oxidation resistance; RNA interference; Enzyme activity; Adverse stresses.

Introduction

The increasing seriousness of environmental pressures, including global warming and the excessive use of pesticides and heavy metal, has intensified threats to the survival of numerous organisms. These environmental stresses can induce the production of reactive oxygen species (ROS) (e.g., superoxide radicals, hydrogen peroxide and hydroxyl radicals), which promote oxidative stress in cells and affect cellular signal transduction in ways that result in cell damage and apoptosis [1, 2]. In aerobic cells, ROS result from metabolic processes that ordinarily maintain a dynamic equilibrium under normal conditions. However, this equilibrium can be disrupted by adverse environmental stresses, which are largely harmful to cellular macromolecules [3-5]. Therefore, ROS detoxification is vital for the organism's survival. Organisms have established complex antioxidant mechanisms to avoid oxidative damage. In the defence against harmful compounds, carboxylesterase functions as an important metabolic enzyme that plays essential roles in the detoxification of various harmful exogenous compounds [6-8].

The CarEs constitute a metabolic enzyme superfamily, the members of which have multiple functions in the metabolic detoxification of pesticides [9], drug resistance [10], juvenile hormone and pheromone degradation [11, 12], and secondary metabolic processes in plants [13]. CarEs are widely distributed in plants, animals, microbes and insects [14]. Based on their sequence similarities and substrate specificities, CarEs can be divided into the following eight classes: α-esterases (ae), β-esterases (be), juvenile hormone esterases (jhe), acetylcholin esterases (AChE), neurotactins (nrt), neuroligins (nlg), gliotactins (gli), and glutactins (glu) [15]. In mammals, CarEs studies have focused on the biotransformation of multiple drugs and the transesterification of xenobiotics, such as insecticides [16, 17]. Insect CarEs are of particular interest because of their contributions to insecticide resistance. Due to this important function, most insect studies have focused on CarEs roles in detoxifying dietary and environmental exogenous compounds, particularly insecticides [6, 7, 9, 10, 18].

Furthermore, the roles of insect CarEs in mediating metabolic detoxification processes and oxidative stress responses have been studied. For example, one study indicated that two carboxylesterases may play significant roles in detoxification of carbaryl and deltamethrin and are most likely involved in the detoxification of chlorpyrifos in *Locusta migratoria* [9]. One study indicated that infection by *Heterorhabditisbeicherriana* induced oxidative stress in *Tenebrio molitor* larvae, which resulted in antioxidant responses: carboxylesterase activities as detoxification enzymes were altered in accordance with SOD, CAT, and GST activity levels [19].

Insect carboxylesterases are rooted in the α-esterase gene cluster, which includes αE7 (also known as E3) from *Lucilia cuprina* (LcαE7), they play important physiological roles in lipid metabolism and are implicated in the detoxification of organophosphorus (OP) insecticides [20]. Montella reported that the metabolic mechanism of carboxylesterase is that the esterases are capable of hydrolysing ester bonds to generate an acid and an alcohol as metabolites [21]. Carboxylesterase is an important detoxifying enzyme that has been implicated in insecticide resistance by its relevant metabolic functions (i.e., catalysis of ester, sulphate, and amino hydrolysis) [22]. These reports have indicated that CarEs may play vital roles in protecting organisms against oxidative damage and oxidative stress. However, the question of whether *AccCarE4* plays effective roles in metabolic detoxification and oxygen free radical scavenging when *A. cerana cerana* is confronted with adverse environmental stress remains unanswered.

A. cerana cerana is the major honeybee species in China; it has specific advantages over other species, including its longer collection period for honey, higher disease resistance, and lower food cost [23]. The balance between regional ecologies and agricultural development previously enabled strong performances by these honeybees [24]. Today, the environmental quality in China is worsening, a condition that is adverse to honeybee survival.

Based on the significant roles of antioxidant processes in insects, we isolated the esterase FE4 gene from *A. cerana cerana* to study its role in oxidative stress resistance. We assessed its expression patterns at different developmental stages and in different tissues. Moreover, the honeybees were exposed to a cold temperature (4 ℃), ultraviolet (UV) radiation, H_2O_2, pesticides (acetamiprid, cyhalothrin and phoxim), herbicide (paraquat) and heavy metals ($HgCl_2$ and $CdCl_2$) to evaluate the *AcceFE4* expression patterns in response to oxidative stress. Meanwhile, at the protein levels, AcceFE4 expression profiles were detected by Western blotting assays. Additional, disc diffusion assays were performed to explore the role of the recombinant AcceFE4 protein *in vitro*. Finally, we used RNA inference technology to knock down *AcceFE4* and to observe the expression levels of other antioxidant genes in the knockdown samples, and we measured enzyme activity of antioxidant enzyme and carboxylesterases. Based on these results, we speculate that *AcceFE4* might play an important role in the oxidative stress response.

Materials and Methods

Insects and treatments

The experimental insects (*A. cerana cerana*) that were used in this study were reared in the artificial honeycomb of Shandong Agricultural University (Taian, China). The honeybees were classified into larvae, pupae, and adults based on age and shape [25]. The fourth(L4)-, the fifth (L5)-instar larvae and the white-eyed (Pw), brown-eyed (Pb), pink-eyed (Pp) and dark-eyed (Pd) pupae were obtained from the hive; the 3-day-old, the 15-day-old and 30-day-old adults were gathered from the honeycomb. The adult honeybees were caged in bee boxes at a constant temperature of 34 ℃ and 80% humidity under a 24 h dark regimen [26]. The adult bees were fed a basic adult diet of water and powdered sugar [27]. The 15- to 30-day post-emergence adult workers were divided into 10 groups, and each group contained 30 honeybees. Each group was exposed to different harmful environmental conditions. Groups 1, 2 were exposed to a cold temperature (4 ℃) and UV light (30 MJ/cm^2) for 0, 0.5, 1.0, 1.5, 2.0, 2.5 h respectively. In addition, the bees in group 3 were injected with 0.5 ml (2 mmol/L) H_2O_2 and treated for 0, 0.25, 0.5, 0.75, 1 h or 1.25 h, while the control group was injected with 0.5 ml phosphate-buffered saline (PBS) (0.5 ml/adult). Groups 4-8 were treated with pesticides (imidacloprid, cyhalothrin, or acetamiprid at a final concentration of 20 mg/ml), and herbicides (phoxim or paraquat at a final concentration of 20 mg/ml) that were diluted in water and added to the food for the experimental groups (treat 0, 0.5, 1.0, 1.5, 2.0, and 2.5 h). Group 9 and 10 were injected with $HgCl_2$ or $CdCl_2$ (0.5 ml of a 2 mmol/L dilution was applied to the thoracic nota of the adult bees), while the control group was injected with PBS (0.5 ml/adult), and the treated bees were collected after treatment at 0, 0.5, 1.0, 1.5, 2.0, 2.5, 3.0 and 3.5 h [28]. The treated bees were flash-frozen in liquid nitrogen and stored at -80 ℃ until further used.

Primer sequences in this study

The primers in this study are listed in Table 1.

Table 1 PCR primers

Abbreviation	Primer sequence (5'—3')	Description
C1	CTTGGTTAACAATGAACAAAC	cDNA sequence primer, forward
C2	TTAAGATAATTTGGATGATAAAAG	cDNA sequence primer, reverse
C3	CATGGTGGCGCATTTGTAGTAGGC	Real-time PCR primer, forward
C4	GGCCCAAGAACATCGTATCAATCC	Real-time PCR primer, reverse
β-s	TTATATGCCAACACTGTCCTTT	Standard control primer, forward
β-x	AGAATTGATCCACCAATCCA	Standard control primer, reverse
C5	GGTACCATGAACAAACAAATAGTAA	Protein expression primer, forward
C6	GTCGACAGATAATTTGGATGATAAAAG	Protein expression primer, reverse
C7	TAATACGACTCACTATAGGGCGAGTCAGCGCTGGAGGAGTT	RNAi primer of AccCarE, forward
C8	TAATACGACTCACTATAGGGCGAGATCTCCGCATCTTCTAA	RNAi primer of AccCarE, reverse
GSTDF	CGAAGGAGAAAACTATGTGGCAG	qPCR primer of AccGSTD, forward
GSTDR	CGTAATCCACCACCTCTATCG	qPCR primer of AccGSTD, reverse
GSTO1F	CCAGAAGTAAAAGGACAAGTTCGT	qPCR primer of AccGSTO1, forward
GSTO1R	CCATTAACATCAACAAGTGCTGGT	qPCR primer of AccGSTO1, reverse
GST4F	CTTCTTAGTTATGGAGGTGTTG	qPCR primer of AccGSTs4, forward
GST4R	GCCATCTGAAATCGTAAAGAG	qPCR primer of AccGSTs4, reverse
Trx1F	GGTTTGAGAATTATACGCACTGC	qPCR primer of AccTrx1, forward
Trx1R	GAGTAAGCATGCGACAAGGAT	qPCR primer of AccTrx1, reverse
Trx2F	GGTTCGGTAGTACTTGTGGAC	qPCR primer of AccTrx2, forward
Trx2R	GGACCACACCACATAGCAAAG	qPCR primer of AccTrx2, reverse
MsrAF	GGTGATTGTTTATTTGGCG	qPCR primer of MsrA, forward
MsrAR	TTTGTATTGCTCTTGTTCACG	qPCR primer of MsrA, reverse
P450F	CGCAAAGAGAATGGGAAGG	qPCR primer of AccCYP4G11, forward
P450R	CTTTTGTGTACGGAGGTGC	qPCR primer of AccCYP4G11, reverse
SOD1F	AAACTATTCAACTTCAAGGACC	qPCR primer of AccSOD1, forward
SOD1R	CACAAGCAAGACGAGCACC	qPCR primer of AccSOD1, reverse
SOD2F	TTGCCATTCAAGGTTCTGGTT	qPCR primer of AccSOD2, forward
SOD2R	GCATGTTCCCAAACATCAATACC	qPCR primer of AccSOD2, reverse
Tpx1F	GGTGGTCTTGGTGAAATGAAC	qPCR primer of AccTpx1, forward
Tpx1R	CTAAACGCAAAGTCTCATCAACAG	qPCR primer of AccTpx1, reverse
Tpx3F	CCTGCACCTGAATTTTCCGG	qPCR primer of AccTpx3, forward
Tpx3R	CTCGGTGTATTAGTCCATGC	qPCR primer of AccTpx3, reverse
CATF	GTCTTGGCCCGAACAATTTG	qPCR primer of AccCAT, forward
CATR	CATTCTCTAGGCCCACCAAA	qPCR primer of AccCAT, reverse

RNA isolation and cDNA synthesis

Trizol reagent (Invitrogen, Carlsbad, CA, USA) was used to extract the total RNA. All RNA samples were treated with Dnase I (Rnasefree) (Promega) to eliminate potential genomic DNA contamination. RNA samples were reverse-transcribed into the first-strand cDNA by using the EasyScript cDNA Synthetic SuperMix (TransGen Biotech, Beijing, China) according to the protocol of the manufacturer, and the cDNA was used as the PCR template for gene cloning and qRT-PCR.

Isolation of the ORF of *AcceFE4*

The open reading frame (ORF) of *AcceFE4* was isolated by PCR amplification with a primer pair (C1 and C2, Table 1) that was designed and synthesised by Biosune Biotechnological Company, Shanghai, China; all primers in this study were obtained from this company. The PCR products were purified and connected with pEASY-T1 vectors (TransGen Biotech, Beijing, China) and were transformed into *Escherichia coli* cells (DH5α) for sequencing.

Bioinformatics analysis and polygenetic tree construction

The active sites and conserved domains of AcceFE4 were obtained from SPDBV_4.10_PC. The AcceFE4 protein sequence was compared with other homologous carboxylesterases from various species that were obtained by using the Basic Local Alignment Search Tool from NCBI (http//:blast.ncbi.nlm.nih.gov/Blast.cgi) [23]. We obtained the *AcceFE4* gene and protein sequences to predict the isoelectric points and molecular weights of the proteins by using DNAMAN version 5.2.2. Phylogenetic trees were constructed based on the homologous sequences of carboxylesterase from the other species with MEGA version 4.1 software by using the neighbour-joining method.

Fluorescence real-time quantitative PCR

qRT-PCR was performed by using the SYBR Premix Ex *Taq* (TaKaRa Dalian, China) and a server on the CFX96™ Real-Time PCR Detection System (Bio-Rad, Hercules, CA, USA) to detect the gene expression changes after various exposures to the adverse stressors. The C3 and C4 specific primers were designed according to the *AcceFE4* sequence. The β-x and β-s primers were designed to amplify the house-keeping gene, *β-actin* (GenBank accession no. HM_640276), which was used as a stable house-keeping gene to normalize the RNA levels [29]. The 20 ml reaction volumes contained 10 ml SYBR *Taq*, 0.5 ml C3 primer, 0.5 ml oC4 primer, 1 ml cDNA template, and 8 ml ddH$_2$O. The amplification conditions were as follows: initial denaturation at 95℃ for 30 s; 41 cycles at 95℃ for 5 s, 55℃ for 15 s and 72℃ for 15 s; with one melting cycle from 65℃ to 95℃ [30]. All experiments were performed in triplicate. The data analyses were obtained from CFX Manager Software version 1.1, and the significant differences between the samples were determined by one-way ANOVA and Duncan's multiple range tests using SPSS software version 17.0 [5].

Expression of recombinant AcceFE4

A bacterial prokaryotic expression vector, pET-30a (+), was used together with transient competent cells to produce the recombinant AcceFE4 protein. The primers (C5 and C6) design were based on the ORF of *AcceFE4* and contained a *Kpn* I or *Sal* I restriction site and were inserted into the pET-30a (+) expression vector, which was previously digested with these restriction sites. The pET-30a (+)-*AcceFE4* expression plasmid was transformed into transient *E. coli* cells. The bacterial solution (10 ml) was added to 10 ml Luria-Bertani (LB) broth that contained 5 ml kanamycin and was incubated at 37℃ for approximately 1.5 h. When the cell density reached 0.4-0.6 (OD_{600}), the recombinant protein was induced with 0.5 mmol/L isopropylthio-β-D-galactoside (IPTG) at 37℃ for 6-8 h. To collect the bacterial cells, the bacterial liquid was centrifuged for 2 min at 13,000 r/min. After centrifugation, the supernatant was discarded, the loading buffer was added and the sample was boiled for 10 min. Protein induction was detected by sodium dodecyl sulphate polyacrylamide gel electrophoresis (SDS-PAGE) with a 12% polyacrylamide gel.

Antibody preparation and Western blotting analysis

After protein induction was assessed by 12% SDS-PAGE, the target protein was excised from the gel and added to 1 ml normal saline (NS) for grinding by using a mortar. White mice received four subcutaneous injections of the products once per week. Blood was collected into a 1.5 ml centrifuge tube, which was placed in a 37℃ water bath for 1 h and then at 4℃ for 6 h. After centrifugation at 3,000 r/min for 15 min, the supernatant antibodies were stored at −80℃.

The protein samples were obtained from the various honeybee treatment groups, and the protein was extracted by using the Tissue Protein Extraction Kit (CWBiotech, Beijing, China). Each protein sample was added to the loading buffer, boiled for 10 min, separated on a 12% SDS-PAGE and blotted onto a polyvinylidine difluoride (PVDF) membrane. The membrane was blocked in 5% (w/V) skim milk (diluted in TBST) for 1 h and incubated overnight at 4℃ with primary antibodies at dilutions of 1 : 500 (V/V). Tubulin (Beyotime, Shanghai, China) was used as the reference antibody for the Western blotting analysis. The membranes were washed three times in TBST, incubated with horseradish peroxidase-conjugated anti-mouse IgG [1:2,000 (V/V)] (Dingguo, Beijing, China) for 2 h at room temperature, and washed three times [5]. Finally, the results were observed by using the SuperSignal® West Pico Trial Kit (Thermo Scientific Pierce, IL, USA), based on the instructions provided in the FDbio-Dura ECL Kit (FDbio Science Biotech, Hangzhou, China).

Disc diffusion assay

The disc diffusion assay was performed according to Liu et al. [5]. The recombinant AcceFE4 was expressed in transient *E. coli* cells. The bacterial culture (5×10^8 cell/ml) was seeded onto LB kanamycin agar plates and incubated at 37℃ for 1 h. Cells with the pET30a (+) empty vector were used as the control. The filter discs (6 mm diameter) were placed onto the surfaces of the agarose medium plates, and the filters were soaked with different concentration of hazardous substances, including $HgCl_2$ (0, 10, 20, 40, and 80 mmol/L), paraquat (0, 10, 50, 100, and 200 mmol/L) and cumene hydroperoxide (0, 10, 50, 100, and 200 mmol/L). The cells were grown for 12 h at 37℃.

dsRNA (dseFE4-RNA and dsGFP-RNA) synthesis and injection

We selected a 507 bp fragment that was part of the ORF of *AcceFE4* and designed a primer pair (C7 and C8, Table 1) according to the target fragment; T7 polymerase promoter sequences were included at the start of the *AcceFE4* forward primer (C7) and at the end of the *AcceFE4* reverse primer (C8). PCR products were purified and transformed into *Escherichia coli* cells (DH5α) for sequencing. Briefly, the purified products were used as templates for dsRNA synthesis with the T7 RiboMAX™ Express RNAi System (Promega, USA) according to the instructions of the manufacturer. The green fluorescent protein gene (GFP, GenBank accession number: U87974) was used as the control [31]. The genes in *Apis cerana cerana* doesn't share homology with dsRNA-GFP, thus the RNAi response will not be triggered by dsRNA-GFP in the body of *Apis cerana cerana*. The synthesised dsRNA samples were adjusted to final concentrations of 6 mg/ml. The adult honeybees were injected with 0.5 ml dseFE4-RNA or 0.5 ml dsGFP-RNA into the thoracic nota by using Microsyringes [9]. In addition, 0.5 ml water was injected with bees as the control group. The samples were flash-frozen in liquid nitrogen at 12 and 36 h post-injection and were stored at −80℃ until further used.

mRNA level analysis of antioxidant genes after the knockdown of *AcceFE4*

To assess the transcriptional levels of antioxidant genes after the knockdown of *AcceFE4*, we designed primers (Table 1) for these genes, which included *AccSOD1* (GenBank ID: JN700517), *AccSOD2* (GenBank ID: JN637476), *AccGSTD* (GenBank ID: JF798573), *AccGSTs4* (GenBank ID: JN008721), *AccGSTO1* (GenBank ID: KF496073), *AccTpx1* (GenBank ID: HM641254), *AccTpx3* (GenBank ID: JX456217), *AccTrx1* (GenBank ID: JX844651), *AccTrx2* (GenBank ID: JX844649), *AccCAT* (GenBank ID: KF765424), *AccMsrA* (GenBank ID: HQ219724), and *AccCYP4G11* (GenBank ID: KC243984) [2]. qRT-PCR was performed to analyze these genes after *AcceFE4* was knocked down.

Enzymatic activities in the AcceFE4 RNAi samples

We extracted protein samples from whole-bee adults 12, 24 and 36 h after they had been injected with dseFE4-RNA. The total protein levels were quantified with the BCA Protein Assay Kit (Nanjing Jiancheng Bioengineering Institute, Nanjing, China). We measured the activities of SOD, POD, CAT and CarEs enzyme by using SOD, POD, CAT and CarEs kits, respectively, according to the instructions of the manufacturer [32]; these kits were all obtained from the Nanjing Jiancheng Bioengineering Institute.

Results

Isolation and characterisation of *AcceFE4*

To determine whether *AcceFE4* played a role in oxidation resistance in honeybees, we isolated the ORF of *eFE4* in *A. cerana cerana*. The ORF of *AcceFE4* consisted of 1,584 bp and encoded a 527-amino acid protein with a predicted molecular mass of 58 kDa and a theoretical isoelectric point of 7.07. The amino acid sequence for AcceFE4 was compared with other homologous sequences, including *Apis mellifera* carboxylesterase, *Melipona quadrifasciata* esterase FE4, *Habropoda laboriosa* esterase FE4, and *Eufriesea mexicana* esterase E4, and is shown in Fig. 1. The high homology suggested that the CarEs family was highly conserved between these species and the common CarEs characteristics were associated with a catalytic triad consisting of Ser-Glu-His and a nucleophilic group that surrounded the active site serine residue (GXSXG) (Fig. 1). To further explore the evolutionary relationships between the CarEs from the different insect species, we constructed a phylogenetic CarEs tree. As shown in Fig. 2, we assembled four branches of CarEs, and *AcceFE4* was classified as α-esterase.

Quantitative real-time PCR of *AcceFE4* at different developmental stages and its tissue distribution

The *AcceFE4* expression profile was assessed via qRT-PCR at the different honeybee developmental stages of L4, L5, Pw, Pb, Pp, Pd, A3, A15 and A30. As shown in Fig. 3A, *AcceFE4* expression was the highest in the Pp phase. In the larvae, the expression levels in L5 were higher than in L4. In the pupae, the expression levels in pink phase pupae were higher than in the other pupal phases, while expression levels were at their lowest in the Pd pupae developmental stage. The expression levels in 3-day-old adult workers were higher than in other adult bees. Total RNA was collected from the head, epidermis, midgut, muscle, and venom gland. As shown in Fig. 3B, their expression patterns indicated that the *AcceFE4* transcripts were most abundant in the epidermis, followed by the venom gland.

Transcriptional expression profiles of *AcceFE4* in response to various stressors

We assessed the *AcceFE4* transcriptional levels in *A. cerana cerana* in response to various stressor treatments by qRT-PCR; the untreated honeybees were used as the control group for qRT-PCR normalization. As shown in Fig. 4, the transcription levels were clearly downregulated at 4°C and with the $HgCl_2$, $CdCl_2$ and phoxim treatments. Fig. 4A shows that the transcription level continued to decline and reached its minimum at 2.5 h of the 4°C treatment. Fig. 4I and J shows that with the $HgCl_2$ and $CdCl_2$ treatments, the *AcceFE4* expression levels continued to noticeably decline, particularly at 3.5 h. Fig. 4C shows that the expression levels at 15, 45 and 60 min were higher relative to the control group during the H_2O_2 treatment. We also analyzed the expression levels after treatments with UV radiation

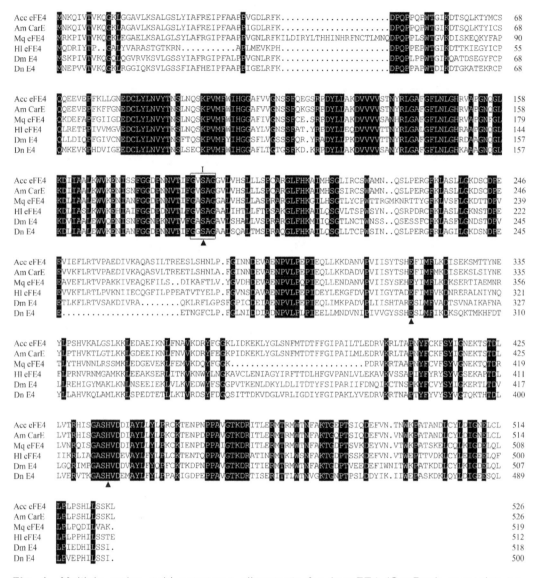

Fig. 1 Multiple amino acid sequence alignments for AcceFE4 (GenBank accession no. XM_017049526.1) from *A. cerana cerana*, AmCarE (GenBank accession no. NP_001136081.1) from *A. mellifera*, Mq eFE4 (GenBank accession no. KOX71850.1) from *Melipona quadrifasciata*, Hl eFE4 (GenBank accession no. KOC67535.1) from *Habropoda laboriosa*, Dm E4 (GenBank accession no. OAD58521.1) from *Eufriesea mexicana*, and Dn E4 (GenBank accession no. KZC12051.1) from *Dufourea novaeangliae*. Conserved domains are shown in black. The Ser-Glu-His catalytic triads are indicated with black triangles. Domain I is the nucleophilic elbow that surrounds the serine residue in the active site (GXSXG).

(Fig. 4B), the expression profile of *AcceFE4* was reduced at 0.5, 1.0 and 1.5 h than the control group. After bees treated for imidacloprid and acetamiprid, *AcceFE4* expression levels were induced as showed in Fig. 4D and H. As shown in Fig. 4F the expression of the gene was upregulated at 1.0, 1.5 and 2.0 h than the control group. After treatment with cyhalothrin condition. In Fig. 4G, the expression of *AcceFE4* was upregulated by paraquat treatment.

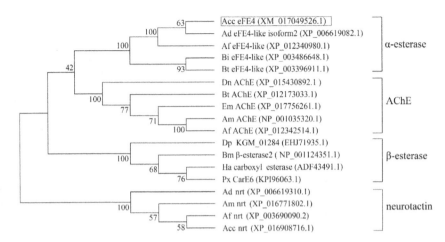

Fig. 2 Phylogenetic analysis of the CarEs homologous sequences from various species by using the neighbour-joining method with bootstrap values of 500 replicates. Four main groups of the CarE superfamily are shown, and AcceFE4 is boxed. The sequences were obtained from NCBI database.

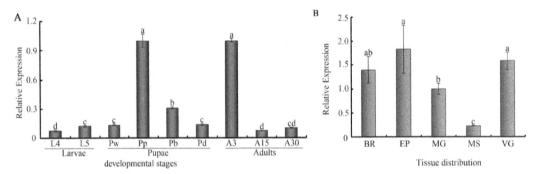

Fig. 3 A. *AcceFE4* transcription expression levels during the different developmental stages. *AcceFE4* expression was measured at multiple developmental stages, including the larvae (L4 and L5) and pupae (Pw, Pp, Pb, Pd) stages, and in adult workers (A3, A15 and A30). The data are presented as the means ± SE of three replicates. The letters above the bar indicate significant differences ($P < 0.005$) as determined by Duncan's multiple range tests using SPSS software version 17.0. B. *AcceFE4* expression profiles in different tissues from *A. cerana cerana*. *AcceFE4* expression was detected in epidermis (EP), venom gland (VG), muscle (MS), brain (BR), and midgut (MG). The data are presented as the means ± SE of three independent experiments. The letters above the columns represent significant differences ($P < 0.05$) based on one-way ANOVA and Duncan's multiple range tests by using SPSS software version 17.0.

Expression of the recombinant AcceFE4 protein

As shown in Fig. 5, the induced protein was clearly apparent on the albumen glue. The second to the fifth samples were the induced protein; the first sample was the control group, which did not receive IPTG. The target protein was approximately 58 kDa.

Western blotting analysis

To further assess the role of eEF4 in the oxidative response in *A. cerana cerana* and to

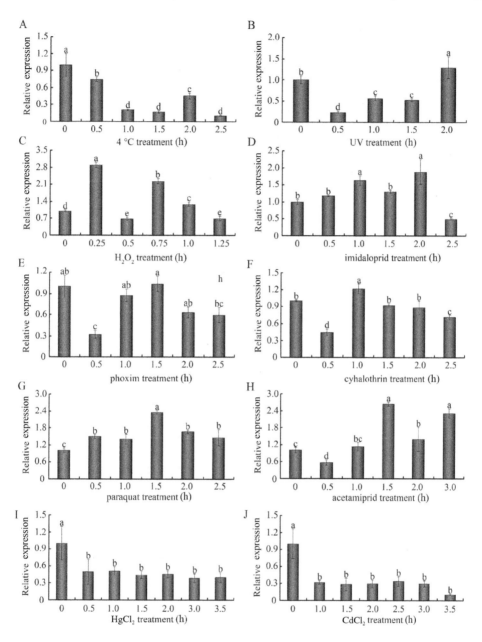

Fig. 4 *AcceFE4* expression patterns in response to abiotic stress. Total RNA was extracted from *A. cerana cerana* after treatments with various stressors. The adult worker bees were exposed to the following conditions: 4℃, UV, H_2O_2, imidacloprid, phoxim, cyhalothrin, paraquat, acetamiprid, $HgCl_2$ or $CdCl_2$. The data are presented as the means ± SE of three replicates. The letters above the bar indicate significant differences ($P < 0.005$) as determined by Duncan's multiple range tests using SPSS software version 17.0.

verify the results of the transcription expression profiles for *AcceFE4*, we analyzed protein expression in response to adverse stress by Western blotting. The anti-AcceFE4 antibody was used to detect the *AcceFE4* protein, and tubulin was used as the normalization reference. As shown in Fig. 6A, protein expression was consistent with the transcriptional level at 4℃, and

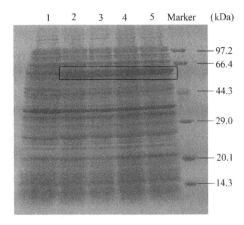

Fig. 5 Recombinant AcceFE4 protein expression. The SDS-PAGE gel lanes are as follows: lane 1, recombinant AcceFE4 expression without IPTG; lanes 2-5, recombinant AcceFE4 expression with IPTG.

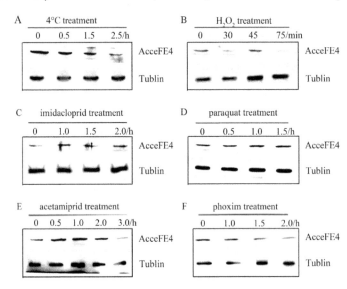

Fig. 6 Western blotting analysis of AcceFE4 expression levels after treatments with various abiotic stressors. Total protein was extracted from adult bees that were treated with various stressors, including 4℃, H_2O_2, imidacloprid, paraquat, acetamiprid and phoxim. Tubulin was used as a reference protein for the Western blotting analysis.

AcceFE4 expression was inhibited; its lowest expression level occurred at 2.5 h after treatment. The expression level was induced at 45 min after the treatment with H_2O_2, as shown in Fig. 6B. The AcceFE4 protein was reduced with the phoxim treatment, particularly at 2 h. In summary, *AcceFE4* expression at the protein level was consistent with its expression at the mRNA level.

AcceFE4 assessment by disc diffusion under various conditions

The disc diffusion method was used to explore the role of the recombinant AcceFE4 protein. The overexpressed recombinant AcceFE4 protein was exposed to various stressors, including cumyl hydroperoxide, paraquat and $HgCl_2$. As shown in Fig. 7, the bacteria that expressed the recombinant protein had a smaller killing zone than the pET-30a (+) empty vector control, and the results indicated that the recombinant protein contained antioxidant activity.

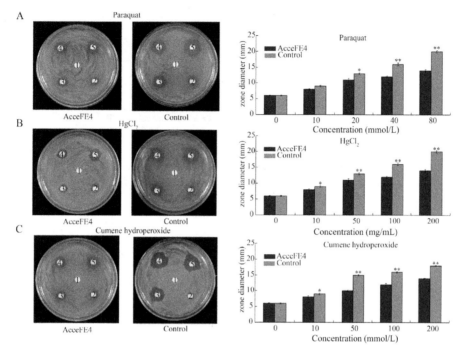

Fig. 7 Disc diffusion assays to detect the resistance of AcceFE4-overexpressing bacterial cells to paraquat, cumyl hydroperoxide and $HgCl_2$. The halo diameters of the killing zones were compared by using histograms. A. The paraquat concentrations are 0, 10, 20, 40, and 80 mmol/L. B. The $HgCl_2$ concentrations are 0, 10, 50, 100, and 200 mg/ml. C. The cumyl hydroperoxide concentrations are 0, 10, 50, 100, and 200 mmol/L. Independent-sample t tests were used for comparisons of baseline characteristics between the treatment and control groups. Significant differences are indicated with two asterisks at $P < 0.05$.

Knockdown of AcceFE4 in *A. cerana cerana*

The RNAi method was used to confirm the antioxidant function of *AcceFE4* in *A. cerana cerana*. As shown in Fig. 8, lower *AcceFE4* transcript expression quantities were observed in the dseFE4-RNA- injected adults at 12 and 36 h post-injection, and the silencing was efficient. Additionally, the expression assessment of the control adult group, which was injected with dsGFP-RNA and injected with water, showed minimal expression changes, indicating that the dsGFP-RNA and water injections did not influence *AcceFE4* expression [33].

Transcription levels of antioxidant genes after the knockdown of *AcceFE4*

To further investigate the role of *AcceFE4* against oxidative damage, we analyzed the transcription levels of other antioxidant genes. The qRT-PCR results in Fig. 9 show that *AccSOD1*, *AccGSTD*, *AccGSTs4*, *AccGSTO1*, *AccTpx1*, *AccTpx3*, *AccTrx1*, *AccTrx2*, *AccCAT*, *AccMsrA*, and *AccCYP4G11* were reduced when *AcceFE4* was knocked down at 36 h.

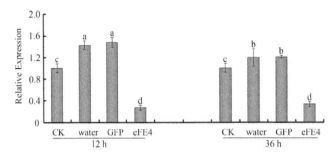

Fig. 8 Silencing efficiency of *AcceFE4* after the adult bees were injected with 3 mg dsRNA. Groups CK (no injection), water and dsGFP were used as the controls. Total RNA was extracted and detected by qRT-PCR at 12 and 36 h after injection. *β-actin* was used as an internal reference gene. The data are presented as the means ± SE of three replicates. The letters above the bar indicate significant differences ($P < 0.005$) as determined by Duncan's multiple range tests using SPSS software version 17.0.

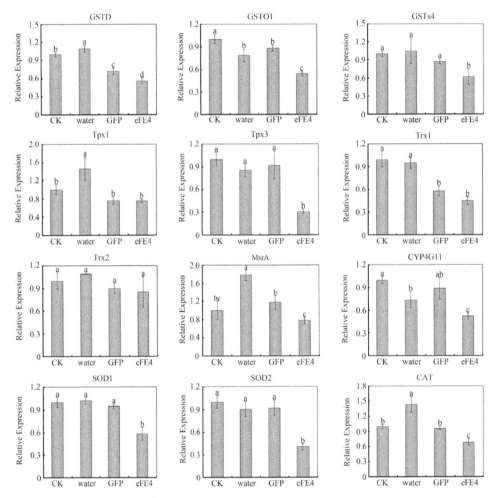

Fig. 9 Expression levels of other antioxidant genes (*GSTD, GSTO, Trx1, Trx2, MsrA, CYP4G11, SOD1, SOD2, Tpx1, Tpx3, CAT* and *GSTs4*) following the knockdown of *AcceFE4* at 36 h were analyzed by qRT-PCR. The *β-actin* gene was used as the internal control. The data are presented as the means ± SE of three replicates. The letters above the bar indicate significant differences ($P < 0.005$) as determined by Duncan's multiple range tests using SPSS software version 17.0.

Determination of enzymatic activities after the knockdown of *AcceFE4*

To confirm the oxidation resistance of *AcceFE4* in *A. cerana cerana* at the protein level, we measured the enzymatic activities of the knockdown samples (at 36 h). Fig. 10 B, C, D indicated that the CAT, SOD, POD activities in the honeybees in which *AcceFE4* was knocked down were reduced relative to the control honeybees, while the change in the control group was minimal. As shown in Fig. 10A, the enzyme activity of carboxylesterases was reduced in the silent samples than in the control group.

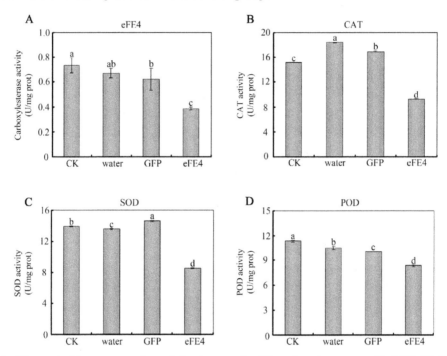

Fig. 10 The antioxidant enzymatic activity changes for (A) carboxylesterase, (B) CAT, (C) SOD, (D) POD when *AcceFE4* was knocked down at 36 h. Total protein was extracted from adult bees 36 h after the injections. Groups CK (no injection), water and dsGFP were used as the controls. The data are presented as the means ± SE of three replicates. The letters above the bar indicate significant differences ($P < 0.005$) as determined by Duncan's multiple range tests using SPSS software version 17.0.

Disscussion

Many previous studies indicated that carboxylesterases (CarEs), cytochrome P450 monooxygenases (P450) and glutathione *S*-transferases (GSTs) are the major metabolic detoxification enzymes that are involved in detoxification of insecticides and in resistance development in insect populations [34-37]. And these enzymes may play significant roles in protecting organisms from adverse environmental stress. However, many studies have reported that P450 and GSTs play important roles in oxidative stress [27, 38-40]. Thus, we predict that CarE may protect against oxidative stress. And we have conducted a series of experiments to prove our prediction. Thus, we guessed that *AcceFE4* may participate in oxidative stress and metabolic detoxification process in *A. cerana cerana*.

In this work, we imitated several types of ROS damage that *A. cerana cerana* may encounter during their lifespan and explored the resulting changes in eFE4 expression at the mRNA and protein levels to look for evidence of their antioxidant functionality environmental stresses. We first identified an esterase FE4 from *A. cerana cerana* and named it *AcceFE4*. Sequence analysis indicated that *AcceFE4* shared a characteristic α/β hydrolase structure, including a catalytic triad consisting of Ser-Glu-His and a GQSAG consensus sequence, suggesting that *AcceFE4* was biologically active [9]. Additionally, phylogenetic analysis revealed an *AcceFE4* clade that was implicated in dietary detoxification. Taken together, these results indicated that *AcceFE4* was a member of the carboxylesterase family.

To better understand the role of *AcceFE4* might play, we analyzed *AcceFE4* mRNA expression at multiple developmental stages and its tissue distribution. The expression of developmental stages revealed that the *AcceFE4* mRNA expression was higher in the pupal stages than in the adult stages. Because ROS cause oxidant damage (particularly in the fast-growing organisms) due to high oxygen levels [39], we speculated that *AcceFE4* may be played a vital role in preventing ROS-mediated damage during the pupae growth stages and would be essential for the development of the honeybee. Indeed, several studies have demonstrated that midgut is the first barrier of xenobiotics in peroral toxicity [34-37]. It has been reported that metabolic process occurred in midgut tissue [41]. Similarly, the expression level of *AcceFE4* in tissues showed that highly expressed in midgut. Taken together, these data indicate that *AcceFE4* may play important roles in preventing adverse damage during different developmental stages and in different tissues.

In order to explore the function of *AcceFE4*, we imitated adverse life-threatening environmental conditions that could induce ROS-mediated damage and pose threats to the honeybee over the course of its life; these conditions included cold temperature, H_2O_2, UV radiation [42, 43]. Temperature is an abiotic environmental factor that causes physiological changes in organisms [44, 45], and ROS are key mediators of cold-induced apoptosis [46]. In this study, we found that when the honeybees were exposed to 4℃, the *AcceFE4* mRNA levels were downregulated. Cold stress can induce hepatocytes and liver endothelial cells to undergo apoptosis due to ROS-mediated damage in mammals [47]. Low temperatures decrease metabolic rates and increase oxygen radical formation during flight in honeybees [48]. However, after H_2O_2 treatment, the gene was highly induced. H_2O_2 is a typical oxidant that causes oxidative damage and can induce ROS elevation, which results in cell death by oxidative stress [49, 50]. Furthermore, there is also evidence that UV is a harmful environmental stressor that can produce oxidative stress in honeybees and can cause oxidative stress through ROS production, which damages membrane lipids and proteins [51, 52]. The qRT-PCR results showed that *AcceFE4* was repressed during the UV exposure, which indicated that *AcceFE4* may play different roles in UV resistance and may be involved in different signal transduction processes [45]. Moreover, heavy metals, such as $HgCl_2$ and $CdCl_2$, can enhance intracellular ROS formation, which promoting cellular oxidative stress [53, 54]. We found that the gene was repressed after the $HgCl_2$ and $CdCl_2$ treatments. These findings indicated that *AcceFE4* was responsive to heavy metal and played a significant role in the physiology of the honeybee.

Taken together, these results showed that *AcceFE4* was responsive to adverse and oxidative stressors, indicating that *AcceFE4* was important in the life of *A. cerana cerana*.

Nowadays, the mechanism of carboxylesterases insecticides metabolic detoxification has been studied very clear, however few studies demonstrated that antioxidative mechanism about carboxylesterase in insects [55]. For instance, as an important detoxifying enzyme, carboxylesterase has been implicated in insecticide resistance with its relevant metabolic functions, such as catalyzing hydrolysis of ester, sulphate, and amid [56]. Studies have shown that CarE is important in the hydrolysis of a broad range of endogenous and xenobiotic ester-containing compounds, such as OPs, carbamates, and pyrethroids [21]. Insecticides are significant threats to a honeybee's life and can induce the oxidation of lipid biomembranes, leading to impaired biochemical and physiological functions [57]. In this work, we imitated adverse conditions that bees may encounter in their life, such as, phoxim, paraquat, acaricide, cyhalothrin and so on. Nearly all treatments caused a change of *AcceFE4* expression profiles. For example, imidacloprid treatment promoted increased *AcceFE4* expression. Ge et al. reported that imidacloprid can induce oxidative stress and DNA damage in zebrafish [58]. We suggest that *AcceFE4* may play different roles in the honeybee. Additionally, *AcceFE4* transcription levels were consistent with its protein expression levels. These data confirmed our speculation that eFE4 may be involved in metabolic detoxification and adverse stress responses in *A. cerana cerana*.

Additionally, in the disc diffusion assay, we imitated three types of environmental stress to explore the function of *AcceFE4*. When the paraquat, $HgCl_2$ and cumene hydroperoxide treatments were used, we observed that the killing zones of the *E. coli* that overexpressed AcceFE4 were smaller than those of the control bacteria. All together with previous results indicate that the bacterial growth may be medicated by the overexpression of AcceFE4. And we speculate that AcceFE4 overexpression may increase ROS resistance in bacterial cells and protect *A. cerana cerana* from bacterial damage.

RNAi technology has become a promising investigative strategy for understanding gene roles, particularly for insects where transgenic technology is insufficient [59]. We silenced the gene by microinjecting dseFE4-RNA into *A. cerana cerana*. The controllable dose and minimal invasiveness make the microinjection method widely useful for diverse insect types [31, 60, 61]. Here, we knocked down *AcceFE4* by using RNAi technology to further explore *AcceFE4* functions. In our research, we found that the expression levels of *AccSOD*, *AccTrx*, *AccMsrA*, *AccTpx*, *AccGST*, *AccCAT* and other genes were reduced when *AcceFE4* was silenced. However, these genes have been reported that they play important roles in the oxidative stress [30, 33, 45 ,62, 63], which indicating that *AcceFE4* may participate in the oxidative process. Organism possess a variety of antioxidant systems for resisting oxidative damage, including the glutathione *S*-transferases (GSTs), superoxide dismutases (SODs), catalase (CAT), thioredoxin peroxidase (Tpx), methionine sulfoxide reductase A (MsrA), and thioredoxin (Trx) systems, which consist of multifunctional proteins and enzymes. The Omega-class glutathione *S*-transferase (GSTO2) is required for reducing oxidative damage in *A. cerana cerana* [2]. Tsuda has reported that the loss of Trx-2 enhances oxidative stress-dependent phenotypes in *Drosophila*, which are hypersusceptible to paraquat and exhibit high levels of expression of antioxidant genes, such as those encoding SOD, CAT, and glutathione synthetase [64]. Moreover, it has been reported that the MsrA system may mediate methionine sulfoxide modifications and that absence of the MsrA enzyme may result in elevated levels of protein carbonyls in yeast and mice [65].Thus, the antioxidant genes expression changed after the knockdown of *AcceFE4*, indicating that *AcceFE4* may be response to the oxidative stress. Antioxidant enzymes (CAT, POD and SOD) play important roles in scavenging H_2O_2 [39, 45]. When their activity levels were downregulated by the knockdown of *AcceFE4*, which may be result in the change of H_2O_2 levels. This finding provided further confirmation of the involvement of *AcceFE4* in the antioxidant process. In our study, the enzyme activities decreased in the knockdown protein samples compared to the control groups. Moreover, the declining activity of CarE indicated that *AcceFE4* may play important roles in response to oxidative damage. Based on these findings, we speculate that *AcceFE4* may play important roles in oxidative stress resistance when the bees encountered the adverse environmental stresses in their life.

In conclusion, we have isolated a metabolism and detoxification enzyme, eFE4, from *A. cerana cerana*, and we have studied the transcript and protein expression levels of *AcceFE4* in response to various environmental stressors. *AcceFE4* was involved in the oxidative process and protected the organism against ROS mediated damage. Additionally, after successful knockdown of *AcceFE4*, the declines in the levels of multiple antioxidant genes and the reduced enzymatic activities of CAT, SOD, and POD confirmed our results. This article provides a better understanding of the detailed mechanisms that underlie the *AcceFE4* response to environmental stress and will be useful for studying CarE in the organism. However, these data address only the possibility that *AcceFE4* plays roles in adverse stress responses. To better understand the characteristics of *AcceFE4*, further studies that validate the function of gene will be necessary.

References

[1] Simon, H. U., Haj-Yehia, A., Levi-Schaffer, F. Role of reactive oxygen species (ROS) in apoptosis induction, Apoptosis. 2000, 5 (5), 415-418.

[2] Zhang, Y. Y., Guo, X. L., Liu, Y. L., Liu, F., Wang, H. F., Guo, X. Q., Xu, B. H. Functional and mutational analyses of an omega-class glutathione *S*-transferase (GSTO2) that is required for reducing oxidative damage in *Apis cerana cerana*. Insect Mol Biol. 2016, 25 (4), 470-486.

[3] Brennan, L. J., Keddie, B. A., Braig, H. R., Harris, H. L. The endosymbiont *Wolbachia pipientis* induces the expression of host antioxidant proteins in an *Aedes albopictus* cell line. PLoS One. 2008, 3 (5), e2083.

[4] Zhang, Y., Yan, H., Lu, W., Li, Y., Guo, X., Xu, B. A novel Omega-class glutathione *S*-transferase gene in *Apis cerana cerana*: molecular characterisation of *GSTO2* and its protective effects in oxidative stress. Cell Stress Chaperones. 2013, 18 (4), 503-516.

[5] Liu, S. C., Liu, F., Jia, H. H., Yan, Y. Y., Wang, H. F., Guo, X. Q., Xu, B. H. A glutathione *S*-transferase gene associated with antioxidant properties isolated from *Apis cerana cerana*. Naturwissenschaften. 2016, 103 (5-6), 43.

[6] Enayati, A. A., Ranson, H., Hemingway, J. Insect glutathione transferases and insecticide resistance. Insect Mol Biol. 2005, 14 (1), 3-8.

[7] Kontogiannatos, D., Michail, X., Kourti, A. Molecular characterization of an ecdysteroid inducible carboxylesterase with GQSCG motif in the corn borer, *Sesamia nonagrioides*. J Insect Physiol. 2011, 57 (7), 1000-1009.

[8] Liu, S., Gong, Z. J., Rao, X. J., Li, M. Y., Li, S. G. Identification of putative carboxylesterase and glutathione S-transferase genes from the antennae of the *Chilo suppressalis* (Lepidoptera: Pyralidae). J Insect Sci. 2015, 15 (1).

[9] Zhang, J., Li, D., Ge, P., Yang, M., Guo, Y., Zhu, K. Y., Ma, E., Zhang, J. Z. RNA interference revealed the roles of two carboxylesterase genes in insecticide detoxification in *Locusta migratoria*. Chemosphere. 2013, 93 (6), 1207-1215.

[10] Cui, F., Qu, H., Cong, J., Liu, X. L., Qiao, C. L. Do mosquitoes acquire organophosphate resistance by functional changes in carboxylesterases? FASEB J. 2007, 21 (13), 3584-3591.

[11] Kamita, S. G., Hammock, B. D. Juvenile hormone esterase: biochemistry and structure. J Pestic Sci. 2010, 35 (3), 265-274.

[12] Vogt, R. G., Riddiford, L. M., Prestwich, G. D. Kinetic properties of a sex pheromone-degrading enzyme: the sensillar esterase of *Antheraea polyphemus*. Proc Natl Acad Sci USA. 1985, 82 (24), 8827-8831.

[13] Dong, J. F., Zhang, J. H., Wang, C. Z. Effects of plant allelochemicals on nutritional utilization and detoxication enzyme activities in two *Helicoverpa* species. Acta Entomol Sin. 2002, 45 (3), 296-300.

[14] Walker, C. H., Mackness, M. I. Esterases: problems of identification and classification. Biochem Pharmacol. 1983, 32 (22), 3265-3269.

[15] Ranson, H., Claudianos, C., Ortelli, F., Abgrall, C., Hemingway, J., Sharakhova, M. V., Unger, M. F., Collins, F. H., Feyereisen R. Evolution of supergene families associated with insecticide resistance. Science. 2002, 298 (5591), 179-181.

[16] Imai, T. Human carboxylesterase isozymes: catalytic properties and rational drug design. Drug Metab Pharmacokinet. 2006, 21 (3), 173-185.

[17] Ahmad, S., Forgash, A. J. Nonoxidative enzymes in the metabolism of insecticides. Drug Metab Rev. 1976, 5 (1), 141-164.

[18] Li, X., Schuler, M. A., Berenbaum, M. R. Molecular mechanisms of metabolic resistance to synthetic and natural xenobiotics. Annu Rev Entomol. 2007, 52, 231-253.

[19] Li, X., Liu, Q., Lewis, E. E., Tarasco, E. Activity changes of antioxidant and detoxifying enzymes in *Tenebrio molitor* (Coleoptera: Tenebrionidae) larvae infected by the entomopathogenic nematode *Heterorhabditis beicherriana* (Rhabditida: Heterorhabditidae). Parasitol Res. 2016, 115(12), 4485-4494..

[20] Jackson, C. J., Liu, J. W., Carr, P. D., et al. Structure and function of an insect α-carboxylesterase (αEsterase7) associated with insecticide resistance. Proc Natl Acad Sci USA. 2013, 110 (25), 10177-10182.

[21] Montella, I. R., Schama, R., Valle, D. The classification of esterases: an important gene family involved in insecticide resistance —a review. Mem Inst Oswaldo Cruz. 2012, 107 (4), 437-449.

[22] Zhang, J., Zhang, J., Yang, M., Jia, Q., Guo, Y., Ma, E., Zhu, K. Y. Genomics-based approaches to screening carboxylesterase-like genes potentially involved in malathion resistance in oriental migratory locust (*Locusta migratoria manilensis*). Pest Manag Sci. 2011, 67 (2) 183-190.

[23] Chen, X., Yao, P., Chu, X., Hao, L., Guo, X., Xu, B. Isolation of arginine kinase from *Apis cerana cerana* and its possible involvement in response to adverse stress. Cell Stress Chaperones. 2015, 20 (1), 169-183.

[24] Yang, G. Harm of introducing the western honeybee *Apis mellifera* L. to the Chinese honeybee *Apis cerana* F. and its ecological impact. Acta Entomol Sin. 2005, 3, 401-406.

[25] Meng, F., Zhang, Y., Liu, F., Guo, X., Xu, B. Characterization and mutational analysis of Omega-class GST (GSTO1) from *Apis cerana cerana*, a gene involved in response to oxidative stress. PLoS One. 2014, 9 (3), e93100.

[26] Alaux, C., Ducloz, F., Crauser, D., Le Conte, Y. Diet effects on honeybee immunocompetence. Biol Lett. 2010, 6 (4), 562-565.

[27] Zhu, M., Zhang, W., Liu, F., Chen, X., Li, H., Xu, B. Characterization of an *Apis cerana cerana* cytochrome P450 gene (*AccCYP336A1*) and its roles in oxidative stresses responses, Gene. 2016, 584

[28] Li, G., Jia, H., Wang, H., Yan, Y., Guo, X., Sun, Q., Xu, B. A typical RNA-binding protein gene (*AccRBM11*) in *Apis cerana cerana*: characterization of *AccRBM11* and its possible involvement in development and stress responses. Cell Stress Chaperones. 21 (6), 1005-1019.

[29] Yan, Y., Zhang, Y., Huaxia, Y., Wang, X., Yao, P., Guo, X., Xu, B. Identification and characterisation of a novel 1-Cys thioredoxin peroxidase gene (*AccTpx5*) from *Apis cerana cerana*. Comp Biochem Physiol B: Biochem Mol Biol. 2014, 172-173, 39-48.

[30] Jia, H., Sun, R., Shi, W., Yan, Y., Li, H., Guo, X., Xu, B. Characterization of a mitochondrial manganese superoxide dismutase gene from *Apis cerana cerana* and its role in oxidative stress. J Insect Physiol. 2014, 60, 68-79.

[31] Elias-Neto, M., Soares, M. P., Simoes, Z. L., Hartfelder, K., Bitondi, M. M. Developmental characterization, function and regulation of a Laccase2 encoding gene in the honey bee, *Apis mellifera* (Hymenoptera, Apinae). Insect Biochem Mol Biol. 2010, 40 (3), 241-251.

[32] Lu, R., L., Wang S., Xing G., et al. Effects of acrylonitrile on antioxidant status of different brain regions in rats. Neurochem Int. 2009, 55 (7), 552-557.

[33] Yao, P., Chen, X., Yan, Y., Liu, F., Zhang, Y., Guo, X., Xu, B. Glutaredoxin 1, glutaredoxin 2, thioredoxin 1, and thioredoxin peroxidase 3 play important roles in antioxidant defense in *Apis cerana cerana*. Free Radic Biol Med. 2014, 68, 335-346.

[34] Ahmad, S., Duval, D. L., Weinhold, L. C., Pardini, R. S. Cabbage looper antioxidant enzymes: tissue specificity. Insect Biochem. 1991, 21 (5), 563-572.

[35] Cohen, M. B., Schuler, M. A., Berenbaum, M. R. A host-inducible cytochrome P450 from a host-specific caterpillar: molecular cloning and evolution. Proc Natl Acad Sci USA. 1992, 89 (22), 10920-10924.

[36] Snyder, M. J., Stevens, J. L., Andersen, J. F., Feyereisen, R. Expression of cytochrome P450 genes of the CYP4 family in midgut and fat body of the *tobacco hornworm, Manduca sexta*. Arch Biochem Biophys. 1995, 321 (1), 13-20.

[37] Yu, Q. Y., Lu, C., Li, W. L., Xiang, Z. H., Zhang, Z. Annotation and expression of carboxylesterases in the silkworm, *Bombyx mori*. BMC Genomics. 2009, 10, 553.

[38] Shi, W., Sun, J., Xu, B., Li, H. Molecular characterization and oxidative stress response of a cytochrome P450 Gene (CYP4G11) from *Apis cerana cerana*. Z. Naturforsch. C. 2013, 68, 509-521.

[39] Corona, M., Robinson, G. E. Genes of the antioxidant system of the honey bee: annotation and phylogeny. Insect Mol Biol. 2006, 15 (5) 687-701.

[40] Aurori, C. M., Buttstedt, A., Dezmirean, D. S., Mărghitaș, L. A., Moritz, R. F., Erler, S. What is the main driver of ageing in long-lived winter honeybees: antioxidant enzymes, innate immunity, or vitellogenin? J Gerontol A Biol Sci Med Sci. 2013, 69 (6) 633-639.

[41] Akpan, B. E., Okorie, T. G. Allometric growth and performance of the gastric caeca of *Zonocerus variegatus* (L.) (Orthoptera: Pyrgomorphidae). Acta Entomol Sin. 2003, 46, 558-566.

[42] Lushchak, V. I. Environmentally induced oxidative stress in aquatic animals. Aquat Toxicol. 2011, 101 (1), 13-30.

[43] Kottuparambil, S., Shin, W., Brown, M. T., Han, T. UV-B affects photosynthesis, ROS production and motility of the freshwater flagellate, *Euglena agilis* Carter. Aquat Toxicol. 2012, 122-123, 206-213.

[44] An, M. I., Choi, C. Y. Activity of antioxidant enzymes and physiological responses in ark shell, *Scapharca broughtonii*, exposed to thermal and osmotic stress: effects on hemolymph and biochemical parameters. Comp Biochem Physiol B: Biochem Mol Biol. 2010, 155 (1), 34-42.

[45] Yao, P., Lu, W., Meng, F., Wang, X., Xu, B., Guo, X. Molecular cloning, expression and oxidative stress response of a mitochondrial thioredoxin peroxidase gene (*AccTpx-3*) from *Apis cerana cerana*. J Insect Physiol. 2013, 59 (3), 273-282.

[46] Rauen, U., Polzar, B., Stephan, H., Mannherz, H. G., De Groot, H. Cold-induced apoptosis in cultured hepatocytes and liver endothelial cells: mediation by reactive oxygen species. FASEB J. 1999, 13 (1),

155-168.

[47] Harari, P. M., Fuller, D. J., Gerner, E. W., Heat shock stimulates polyamine oxidation by two distinct mechanisms in mammalian cell cultures. Int J Radiat Oncol Biol Phys. 1989, 16 (2), 451-457.

[48] Harrison, J. F., Fewell, J. H. Environmental and genetic influences on flight metabolic rate in the honey bee, *Apis mellifera*. Comp Biochem Physiol A: Mol Integr Physiol. 2002, 133 (2), 323-333.

[49] Goldshmit, Y., Erlich, S., Pinkas-Kramarski, R. Neuregulin rescues PC12-ErbB4 cells from cell death induced by H_2O_2. Regulation of reactive oxygen species levels by phosphatidylinositol 3-kinase. J Biol Chem. 2001, 276 (49), 46379-46385.

[50] Casini, A. F., Ferrali, M., Pompella, A., Maellaro, E., Comporti, M. Lipid peroxidation and cellular damage in extrahepatic tissues of bromobenzene-intoxicated mice. Am J Pathol. 1986, 123 (3), 520-531.

[51] Ravanat, J. L., Douki, T., Cadet, J. Direct and indirect effects of UV radiation on DNA and its components. J Photochem Photobiol B. 2001, 63 (1-3), 88-102.

[52] Cadet, J., Sage, E., Douki, T. Ultraviolet radiation-mediated damage to cellular DNA. Mutat Res. 2005, 571 (1-2), 3-17.

[53] Park, E. J., Park, K. Induction of reactive oxygen species and apoptosis in BEAS-2B cells by mercuric chloride. Toxicol In Vitro. 2007, 21 (5), 789-794.

[54] Tang, T., Huang, D. W., Zhou, C. Q., Li, X., Xie, Q. J., Liu, F. S. Molecular cloning and expression patterns of copper/zinc superoxide dismutase and manganese superoxide dismutase in *Musca domestica*. Gene. 2012, 505 (2), 211-220.

[55] Li, H. M., Buczkowski, G., Mittapalli, O., Xie, J., Wu, J., Westerman, R., Schemerhorn, B. J., Murdock, L. L., Pittendrigh, B. R. Pittendrigh, Transcriptomic profiles of *Drosophila melanogaster* third instar larval midgut and responses to oxidative stress. Insect Mol Biol. 2008, 325-339.

[56] Zhang, B., Helen, H. S., Wang, J., Liu, H. Performance and enzyme activity of beet armyworm *Spodoptera exigua* (Hübner) (Lepidoptera: Noctuidae) under various nutritional conditions. Agric Sci China. 2011, 10, 737-746.

[57] Narendra, M., Bhatracharyulu, N. C., Padmavathi, P., Varadacharyulu, N. C. Prallethrin induced biochemical changes in erythrocyte membrane and red cell osmotic haemolysis in human volunteers. Chemosphere. 2007, 67 (6), 1065-1071

[58] Ge, W., Yan, S., Wang, J., Zhu, L., Chen, A., Wang, J. Oxidative stress and DNA damage induced by imidacloprid in zebrafish (*Danio rerio*). J Agric Food Chem. 2015, 63(6), 1856-1862.

[59] Huvenne, H., Smagghe, G. Mechanisms of dsRNA uptake in insects and potential of RNAi for pest control: a review. J Insect Physiol. 2010, 56 (3), 227-235.

[60] Liu, S., Ding, Z., Zhang, C., Yang, B., Liu, Z. Gene knockdown by intro-thoracic injection of double-stranded RNA in the brown planthopper, *Nilaparvata lugens*. Insect Biochem Mol Biol. 2010, 40 (9), 666-671.

[61] Futahashi, R., Tanaka, K., Matsuura, Y., Tanahashi, M., Kikuchi, Y., Fukatsu, T., Laccase 2 is required for cuticular pigmentation in stinkbugs. Insect Biochem Mol Biol. 2011, 41 (3), 191-196.

[62] Gong, Z., Guo, X., Xu, B. Molecular cloning, characterisation and expression of methionine sulfoxide reductase A gene from *Apis cerana cerana*. Apidologie. 2011, 43 (2), 182-194.

[63] Yan, H., Jia, H., Gao, H., Guo, X., Xu, B. Identification, genomic organization, and oxidative stress response of a Sigma class glutathione S-transferase gene (*AccGSTS1*) in the honey bee, *Apis cerana cerana*. Cell stress Chaperones. 2012, 18, 415-426.

[64] Tsuda, M., Ootaka, R., Ohkura, G., kishita, Y., Seong, K. H., Matsuo T., Aigaki, T. Loss of Trx-2 enhances oxidative stress-dependent phenotypes in *Drosophila*. FEBS Lett. 2010, 584, 3398-3401.

[65] Moskovitz, J., Oien, D. B. Protein carbonyl and the methionine sulfoxide reductase system. Antioxid Redox Signal. 2010, 12, 405-415.

3.5 Characterization of a Decapentapletic Gene (*AccDpp*) from *Apis cerana cerana* and Its Possible Involvement in Development and Response to Oxidative Stress

Guilin Li, Hang Zhao, Hongfang Wang, Xulei Guo, Xingqi Guo, Qinghua Sun, Baohua Xu

Abstract: To tolerate many acute and chronic oxidative stress-producing agents that exist in the environment, organisms have evolved many classes of signal transduction pathways, including the transforming growth factor β (TGF-β) signal pathway. Decapentapletic gene (*Dpp*) belongs to the TGF-β superfamily, and studies on *Dpp* have mainly focused on its role in the regulation of development. No study has investigated the response of *Dpp* to oxidative pressure in any organism, including *Apis cerana cerana* (*A. cerana cerana*). In this study, we identified a *Dpp* gene from *A. cerana cerana* named *AccDpp*. The 5'-flanking region of *AccDpp* had many transcription factor binding sites that relevant to development and stress response. *AccDpp* was expressed at all stages of *A. cerana cerana*, with its the highest expression in 15-day worker bees. The mRNA level of *AccDpp* was higher in the poison gland and midgut than in other tissues. Furthermore, the transcription of *AccDpp* could be repressed by 4℃ and UV, but induced by other treatments, according to our qRT-PCR analysis. It is worth noting that the expression level of AccDpp protein was increased after a certain time when *A. cerana cerana* was subjected to all simulative oxidative stresses, a finding that was not completely consistent with the result from qRT-PCR. It is interesting that recombinant AccDpp restrained the growth of *Escherichia coli*, a function that might account for the role of the antimicrobial peptides of AccDpp. In conclusion, these results provide evidence that *AccDpp* might be implicated in the regulation of development and the response of oxidative pressure. The findings may lay a theoretical foundation for further genetic studies of *Dpp*.

Keywords: *Apis cerana cerana*; Decapentapletic gene; Growth and development; Stress response; Expression pattern; Western blotting analysis

Introduction

Apis cerana cerana is a well-known subspecies of oriental bees. Compared with *Apis*

mellifera (*A. mellifera*), *A. cerana cerana* has a strong resistance to mites, acute sense of smell, and can forage the nectar and pollen of wide range of flowers, including wild plants. These advantages are irreplaceable by *A. mellifera* [1-3]. However, recently, excessive uses of pesticides and the existence pollutants, climate change with extreme heat and cold, ultraviolet radiation, and heavy metal in the environment, which can lead to the generation of reactive oxygen species (ROS), cause serious harm to the survival of honeybees [4-7].

ROS homeostasis and signalling are essential to the organisms, but their exact function remains a mystery. Hydrogen peroxide (H_2O_2), hydroxyl radical (HO·), and superoxide anion (O^{2-}), are generated endogenously or exogenously by ROS. A low concentration of ROS is essential to the organism, and accumulating evidence has suggested that ROS can serve as pivotal signalling molecules to participate in the regulation of various cellular functions, including cell growth, proliferation, survival and the immune response [8, 9]. However, excess ROS result in various disease states, including cancer, aging, diabetes, and neurodegeneration, and are implicated in the damage of macromolecules, such as lipids, protein, and nucleic acids. Generally, oxidative stress occurs when antioxidant defence mechanisms are compromised or antioxidant protection is overwhelmed by a high level of ROS. Redox-sensitive signalling proteins can be modified by oxidative stress, which leads to aberrant cell signalling [10]. As signalling molecules, ROS may be connected to many signal pathways by targeting transduction or specific metabolic cellular components, which may execute and initiate the program of cell apoptotic death. Transforming growth factor β (TGF-β) signal transduction can also be activated by ROS.

The TGF-β superfamily was first discovered by Roberts [11] and successfully extracted in human blood by Assoian [12]. TGF-β superfamily members include TGF-βs, activins, bone morphogenetic protein (BMP), inhibins, and growth differentiation factor (GDF). There are many subtypes in the TGF-β superfamily. In mammals, the subtypes mainly include TGF-β1, TGF-β2, and TGF-β3, which are expressed in various tissues and have different expression levels [13]. TGF-β plays an essential role in signaling pathways that control metazoan cell growth, differentiation, and participates in the formation of tissues and organs, as well as in the immune response of the body [14, 15].

Previous studies have also indicated that TGF-β signalling could mediate ROS production and control redox [16]. The increase in ROS contributes to TGF-β-induced cell apoptosis in cirrhotic hepatocytes [17], while the inhibition of ROS lowers the susceptibility of the cell to TGF-β-induced apoptosis. High levels of TGF-β, IL-6, and IL-1 can induce a ROS-mediated response to DNA damage [18]. Chen et al. (2012) proposed that the signalling pathway involved in ROS could increase TGF-β expression, resulting in increased ROS production and promoting TGF-β-dependent fibrogenesis [19]. TGF-β elicited cell apoptosis in mouse foetal hepatocytes via an oxidative process [20]. The expression of catalase and MnSOD can be regulated by TGFβ in airway smooth muscle cells, leading to the change of ROS levels in the body [21].Michaeloudes et al. (2011) suggested that TGF-β upregulated the level of *Nox4*, possibly generating H_2O_2 and ultimately contributing to the increase of ROS. Renal autoregulation could be impaired by TGF-β through the generation of ROS [22]. ROS can be induced by TGF-β and then activate *p38* [23-25]. TGF-β participates in mediating transglutaminase 2 activation in the oxidative stress response, causing protein aggregation [26].

The expression of catalase and MnSOD can be regulated by TGF-β in airway smooth muscle cells, leading to the change of ROS levels in the body [26]. However, in *A. cerana cerana*, the TGF-β signalling pathway has not been studied.

Dpp (decapentapletic gene), similar to *BMP2* and *BMP4* of vertebrates, is a member of the TGF-β superfamily in insects and is a secreted molecule. It was first found and studied more clearly in *Drosophila melanogaster* (*D. melanogaster*). In *Drosophila* embryonic development, *Dpp* is one type of segment polarity gene that belongs to the group of zygotic genes. Dpp is a morphogen in the process of insect development and guides cell growth, differentiation and senescence in a dosage-dependent manner [27-29]. Many previous genetic analyses have demonstrated that *Dpp* also played a crucial role in many developmental events through positional information in the intercellular signalling pathway [30, 31]. Ninov et al. found that Dpp signalling pathways can directly regulate cell motility and retraction [32]. There are many genes in the signalling pathway of Dpp, such as *Dpp*, *Put*, *Tkv*, *Mad*, *Med*, *Shn*, and *Brk*, among which *Dpp* is located the most upstream. The Dpp signalling pathway has been implicated in many developmental processes and can both activate and repress gene transcription [33]. Studies of *Dpp* have mainly focused on its role in growth and development. Although *Dpp* is a member of the TGF-β superfamily, whether *Dpp* is related to ROS remain unknown.

A. cerana cerana plays a critical role in the development of honey industry and maintains the ecological balance. Though its genome information had been uncovered in 2015 [34], its was not be released. In addition, to date, only 195 mRNA sequences of *A. cerana cerana* have been submitted in the NCBI database. Thus, it is essential to obtain more information concerning gene expression for the study of the function and biological mechanisms of Chinese bees. To our knowledge, the role of *Dpp* in *A. cerana cerana* has not been studied. In this paper, we isolated and characterized the *Dpp* gene from *A. cerana cerana* and detected its expression profile in different tissues, at different development stages, and under various oxidative stresses at the mRNA and protein levels. So far, this is the first report concerning the relationship between the *Dpp* gene and oxidative stress.

Materials and Methods

Experimental insects and various treatments

The insects (*A. cerana cerana*) used in this work were reared in the artificial beehives of Shandong Agricultural University (Taian, China). In general, each colony has one queen to lay eggs, which has completed mating and will stay in the hive all the time, unless swarming or flying fled. Honeybees of different developmental stages were classified based on the criteria of previous reports [35]. The egg (Eg), 1-day to 7-day larvae (L1-L7), pre-pupal phase pupae (Po), pupae [white-eyed (Pw), pink-eyed (Pp), brown-eyed (Pb) and dark-eyed (Pd) pupae], and 1-day worker bees (A1) were collected directly from the hive, while adult honeybees [15-day worker bees (A15), and 30-day worker bees (A30)] were collected at the entrance of the hive by marking 1-day worker bees with paint 15 and 30 days earlier. The 15-day worker bees were divided into ten groups ($n=40$/group) and kept at 34°C under standard conditions as described by Alaux et al. (2010) [36]. Each group was treated with various stress conditions (Table S1), which could be involved in oxidative stress [4-7], and the control groups (untreated 15-day worker bees) were incubated at 34℃ and fed with normal food. Bees that were injected with phosphate buffered saline (PBS) (0.5 μl/worker) were the injection controls of group injected with H_2O_2. Methomyl, vitamin C (VC), $HgCl_2$ and $CdCl_2$ were dissolved in water, and acaricide, cyhalothrin and paraquat were diluted by water. The honeybees in the above experiments were collected at the indicated time. To analyze tissue-specific expression, different tissues of the 15-day worker bees, including the leg, wing, muscle, midgut, haemolymph, rectum, poison gland, honey sac, antennae and epidermis, were dissected on ice. All of the specimens were flash-frozen in liquid nitrogen and stored at −70℃ until they were used. Each experiment was performed in triplicate.

Extraction of total RNA, synthesis of cDNA and genomic DNA preparation

Total RNA from *A. cerana cerana* was extracted and cDNA was synthesized by using Trizol reagent (Invitrogen, Carlsbad, CA, USA) and an EasyScript First-Strand cDNA Synthesis SuperMix (TransGen Biotech, Beijing, China), respectively, as per the protocol of the manufacturers. For expression profile analysis of *AccDpp* at different development stages and under different types of abiotic stresses, whole honeybee was used to extract RNA, and for the analysis of the expression patterns of *AccDpp* at different development, the RNA was extracted from different tissues. The extraction of genomic DNA was performed according to the instructions offered by the EasyPure Genomic DNA Extraction Kit (TransGen Biotech, Beijing, China).

Primers and amplification conditions

The primer pairs used in this study are listed in Table S2 and were synthesized by Sangon Biotechnological Company (Shanghai, China). All of the polymerase chain reaction (PCR) amplification procedures are listed in Table S3.

Cloning of the full-length cDNA, 5′-flanking region, and genomic sequence of *AccDpp*

Acquisition of the full-length cDNA, 5′-flanking region, and genomic sequence of *AccDpp* was carried out as described by Chen et al. (2015) [37].

Bioinformatics analysis

The MatInspector database (http://www.cbrc.jp/research/db/TFSEARCH.html) was used to predict the putative transcription factor binding sites (TFBs) of the *AccDpp* promoter. The GC content of the gene was predicted by the DNASTAR program (version 7.01). NCBI servers (http://blast.ncbi.nlm.nih.gov/Blast.cgi) were used to select the homologous sequence of *AccDpp* and to predict the conserved domain of *Dpp* from different species. DNAMAN version 5.22 (Lynnon Biosoft, Quebec, Canada) and the ProtParam tool (http://www.*expasy*.

ch/*tools/protparam*.html) were used to determine the physical and chemical properties of *AccDpp*. Molecular Evolutionary Genetics Analysis (MEGA version 4.1) was chosen to generate the phylogenetic tree. The prediction of antimicrobial peptides and signal peptide of AccDpp was performed by using the antimicrobial peptide database and Signalp 4.1 Server, separately. The online software SWISS-MODEL was used to build the possible three-dimensional structure of AccDpp, and SPDBV version 4.1 was chosen to analyze the three-dimensional structure of AccDpp.

Fluorescent real-time quantitative PCR

Fluorescent real-time quantitative PCR (qRT-PCR) was carried out according to the protocol of Zhang et al. (2013) to check the mRNA expression profile of *AccDpp*. The expression of *AccDpp* was normalized by *β-actin* (GenBank accession number: HM-640276), which is stably expressed [38-42]. Untreated 15-day worker bees were used as controls.

Protein expression, purification and preparation of anti-AccDpp

The stop codon and signal peptide-less open reading frame (ORF) of *AccDpp* with the *Kpn* I and *Sac* I restriction sites were cloned into the expression vector pET-30a(+) (Novagen, Madison, WI) and was transformed into Transetta (DE3) chemically competent cells (*Escherichia coli*, TransGen Biotech, Beijing, China). The induction and purification of recombinant AccDpp were performed based on previous reports [38]. The preparation of antibodies was performed according to the procedure of Meng et al. (2014) with some modification [43]. In brief, the target protein was separated by 12% SDS-PAGE. The SDS-PAGE albumin glue that contained the target protein was cut and ground with moderate benzylpenicillin sodium for injection (Lukang Pharmaceutical, Jining, China) and sodium chloride injection (0.9%) (Cisen Pharmaceutical, Jining, China). The ground sample was used to inject white mice (Taibang, Taian, China).

Western blotting analysis

The total protein of *A. cerana cerana* was extracted and quantified according to the protocol provided by a tissue protein extraction kit (ComWin Biotech, Beijing, China) and a total protein assay kit (using a standard BCA method; ComWin Biotech, Beijing, China), respectively. After equal amounts of the protein of each sample were separated by 12% SDS-PAGE, they were electrotransferred onto a PVDF membrane (ComWin Biotech, Beijing, China) by using the wet transfer method. Then, membrane rinsed with 10 ml TBST buffer solution containing 0.5 g of Difco^{TM} Skim Milk (Solarbio, Beijing, China). The primary antibodies (anti-AccDpp polyclonal antibody; 1∶100 dilution) were used to incubate the membrane at 4℃ overnight. After rinsing in TBST three times, the secondary antibodies (peroxidase-conjugated goat anti-mouse immunoglobulin G; Jingguo Changsheng Biotechnology, Beijing, China) at a dilution of 1∶2,000 (*V/V*) were used to probe the membrane. Finally, the membrane was washed with TBST. The results of antigen-antibody binding were detected by using the SuperSignal™ West Dura Extended Duration Substrate (Thermo Fisher Scientific, Shanghai, China).

Disc diffusion assay

Escherichia coli cells overexpressing AccDpp and with the pET30a(+) vector were grown in LB-kanamycin agar plates and incubated at 37℃ for 45 min. Next, the agar plates were covered with five filter discs (6 mm in diameter), which were soaked with 2 μl of various concentrations of reagents. The reagents contained $HgCl_2$ (0, 5, 10, 20, and 70 mg/ml), paraquat (0, 50, 200, 300, and 500 mmol/L), $CdCl_2$ (0, 300, 500, 700, and 900 mmol/L), and cumene hydroperoxide (0, 12.5, 25, 50, and 100 mmol/L). $HgCl_2$ and $CdCl_2$ were dissolved by water. Paraquat and cumene hydroperoxide were diluted by water and absolute ethyl alcohol, respectively. The treated cells were cultured overnight at 37℃.

GenBank accession properties of the genes used in this paper

Many genes were used to perform bioinformatics analysis. Their species name and GeneBank accession number are listed in Table S4.

Results

Characterization of *AccDpp*

The full-length cDNA of *AccDpp* (GenBank accession number: KT750952) is 1,652 bp, with a 1,104 bp open reading frame (ORF) that encodes 390 amino acids. The amino acid sequence of AccDpp contains a signal peptide with 23 amino acids (Fig. 1). Thus, the mature protein of AccDpp only contains 367 amino acids, and is a secretory protein. The molecular weight and theoretical pI of mature AccDpp was 41.38 kDa and 9.65, respectively. The *AccDpp* gene is flanked by a 167 bp 5′ untranslated region (5′ UTR) and a 312 bp 3′ UTR (Fig. 1). In the 3′ UTR of *AccDpp*, a typical polyadenylation signal sequence (AATAA) existed.

Fig. 2A revealed that the C-terminus of *Dpp* of different species was highly conserved, while the N-terminus was not. The TGF-β-propeptide domain and TGF-β domain of *Dpp* in various species were predicted by the NCBI Conserved Domain Database. The results showed that the TGF-β-propeptide domain and TGF-β domain existed in the N-terminus and C-terminus of the Dpp protein, respectively (Fig. 2B). The TGF-β-propeptide is known as a latency-associated peptide (LAP) in TGF-β. LAP is a homodimer that is disulphide linked to the TGF-β binding protein. The TGF-β domain is a multifunctional peptide that controls proliferation, differentiation, and other functions in many cell types. These two conserved domains may decide the functions of AccDPP in *A. cerana cerana*.

A neighbour-joining phylogenetic tree was generated by MEGA 4.1 to explore the evolutionary relationships of Dpp among different species, and the result revealed that AccDpp was more closely related to AmDpp than other species (Fig. 3A). Fig 3B showed the possible three-dimensional structure of AccDpp that may contribute to better understanding of the role

```
AACGGAGTTCAACATCGATGTATCGATATTAAAGACTCCCGGGGGCGAGGTTTTGAGACAATAGTTTCTGAGTGAGGTTACATGGACTC  89
ATTGAAATGTCGTGGCTCCTCAAACAACGAACTTACGATTAGCTCAGGGAAGCAACATCCGTGAGCAACGACATCACAATGCGCATGTTG 179
                                                                                M  R  M  L    4
CTGCCACTGAGATTGGTTTTCGTCGCAGCAGGTTTGCTGCTGGCAACTACCAAAGCAAATGTTCATTTGCATCTACATTCTGACACTGCC 269
 L  P  L  R  L  V  F  V  A  A  G  L  L  L  A  T  T  K  A  N  V  H  L  H  S  D  T  A        34
AAACATTTTGCCATTGGTGAACGTGAACAAGCAATTGCTGGAATGGAAGCTAGCCTTTTATCTCTTCTTGGTTTTGCTAAGAGGCCAAAA 359
 K  H  F  A  I  G  E  R  E  Q  A  I  A  G  M  E  A  S  L  L  S  L  L  G  F  A  K  R  P  K   64
CCACAAGGTCCTGCTTATGTTCCAGAATCATTAAAAAAACTTTTTATCAAACAAAATACAATTGGTACAGCTGATATTGCAAAACCAGGA 449
 P  Q  G  P  A  Y  V  P  E  S  L  K  K  L  F  I  K  Q  N  T  I  G  T  A  D  I  A  K  P  G   94
ATACATGCCAGATCTGCTAATACAGTGCGTTCCTTTTCTCATGTTGAGAGTAAGGTGGATCATAAATTTCAATCTCCAAATCGCTTTCGT 539
 I  H  A  R  S  A  N  T  V  R  S  F  S  H  V  E  S  K  V  D  H  K  F  Q  S  P  N  R  F  R  124
CTTTCTTTTGACTTAAGTAGTATTCCTTCAGGAGAAAAATTACAAGCTGCAGAATTAAGTCTTTCTCGCGTACCTATTTTAAATCAAGAT 629
 L  S  F  D  L  S  S  I  P  S  G  E  K  L  Q  A  A  E  L  S  L  S  R  V  P  I  L  N  Q  D  154
GATTCCAATCCAAAATTGGGTAGAATACTTATATATGATATTTTACGTCCTGGTGCAAAAGGATATAGTTCACCATTATTACGTTTAATT 719
 D  S  N  P  K  L  G  R  I  L  I  Y  D  I  L  R  P  G  A  K  G  Y  S  S  P  L  L  R  L  I  184
GACAGTAAATTAATAAATACTAGAAAAAATGGCACAATTAGCCTAGATGTTCATCCTGCTGTAGAAAGATGGATAAAGGATCCCAAAAAT 809
 D  S  K  L  I  N  T  R  K  N  G  T  I  S  L  D  V  H  P  A  V  E  R  W  I  K  D  P  K  N  214
AATCATGGACTTTTGGTGCATGTACGTGGAGTTAAACAGGAGCATGTACGTTTGAAAAGAAACACAAATGAAAGAAATGATACATGGGTT 899
 N  H  G  L  L  V  H  V  R  G  V  K  Q  E  H  V  R  L  K  R  N  T  N  E  R  N  D  T  W  V  244
GTTAATCGCCCAATGTTATTTACTTATACAGATGATGGCAGATATAAAATGTCTTCTGCAAGTCAAATAATGGATCGCCGTGCTAGAAGA 989
 V  N  R  P  M  L  F  T  Y  T  D  D  G  R  Y  K  M  S  S  A  S  Q  I  M  D  R  R  A  R  R  274
GCAGCACTTCGAAAAAATCGTCGAAAAGATGGACGTGAGAATTGTAGGCGACATCCGCTTTATGTTGACTTTGCTGATGTTGGTTGGAAT 1079
 A  A  L  R  K  N  R  R  K  D  G  R  E  N  C  R  R  H  P  L  Y  V  D  F  A  D  V  G  W  N  304
GATTGGATTGTTGCTCCCCCTGGTTATGATGCCTTTTATTGTCATGGTGATTGTCCCTTTCCATTAGCAGATCATTTAAATTCTACCAAT 1169
 D  W  I  V  A  P  P  G  Y  D  A  F  Y  C  H  G  D  C  P  F  P  L  A  D  H  L  N  S  T  N  334
CATGCAATTGTTCAAAACTTGGTTTATTCAACAAAACCAAGCATGGTGCCTAAAGCTTGCTGTGTACCTACTGCTCTTAGTTCAATATCC 1259
 H  A  I  V  Q  N  L  V  Y  S  T  K  P  S  M  V  P  K  A  C  C  V  P  T  A  L  S  S  I  S  364
ATGCTTTATCTCGATGAAGAAATAAAGTAGTATTGAAAAATTATCAAGACATGTCTGTCCTTGGATGTGGATGTCGTTAACTATTCATT 1349
 M  L  Y  L  D  E  E  N  K  V  V  L  K  N  Y  Q  D  M  S  V  L  G  C  G  C  R  *            390
AATTGGTTCAACAAAGTATCCTAACAGTGAAAATGCATGTGCAATAAAAATGAAAATAGATAAAAAGGTATTAAAAGATAAGAAACAAA  1439
GAACTAAGATTTCAATAATGCCATAAATAATGTTATAAATAATGTTTTGAAGATGTTGAATTGTTTATTAAATATTTACACAATTGAGGC 1529
ATTCTTGGAGGATTTTATAAATTTATTATGTATGAAAATTTTACTTATATGATTAATTGTTAAAAATTCAGATAATTCTGACTGCCTAAC 1619
TTATTTTTTATTCACAAAATAAAATAAAATCTG                                                         1652
```

Fig. 1 The cDNA sequence of *AccDpp* and its amino acid sequence. The top line shows the nucleotide sequence of *AccDpp*, and the second line shows the deduced amino acid sequence. The start codon (ATG) and stop codon (TAG) are boxed. The polyadenylation signal (AATAA) sequence is marked by an oval. The underlined region indicates the signal peptide, and the shaded amino acid sequence denotes the predicted antimicrobial peptide. The sequence was deposited in GenBank, and the GenBank accession number: KT750952.

of AccDpp. The subunit of AccDPP had more β folds (15) than α helices (6) (Fig. 3B and Table S5), that may be related to the function of AccDpp.

The analysis of genetic sequence structure of *AccDpp*

A 4,066 bp sequence of *AccDpp* was isolated to study its genomic feature and included three introns and three exons (GenBank accession number: KT750953). It is interesting, a long intron was located inside the 5′ UTR of *AccDpp*. Both in this long intron and in the 5′ UTR contained

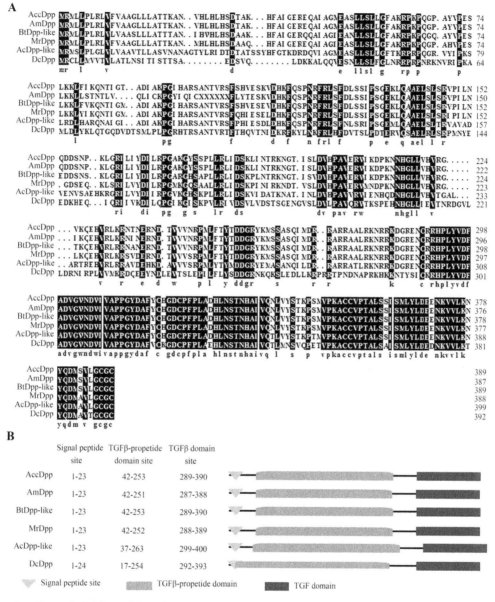

Fig. 2 Characterization of Dpp from various species. The amino acid sequences of Dpp were all downloaded from the NCBI database (S4 Table). A. Alignment of the deduced AccDpp protein sequence with other Dpp proteins. B. Conserved domain of Dpp. The conserved domains are marked by different shapes.

many putative transcription factor binding sites (TFBs) (Fig. S1), including fifty-three CdxA, fourteen CF2-II, ten HSF, one NIT2, and one BR-C. Thus, this intron and 5′ UTR might be involved in the regulation of transcription of *AccDpp*. The GC content of the exons of *AccDpp* was higher than that of its intron (Table S6), its similar with other *Dpp* genes. The size and GC content of Dpp exons from different species had a higher homology than its introns (Fig. 4 and Table S6), representing the conservation and variability of the same gene during evolutionary periods.

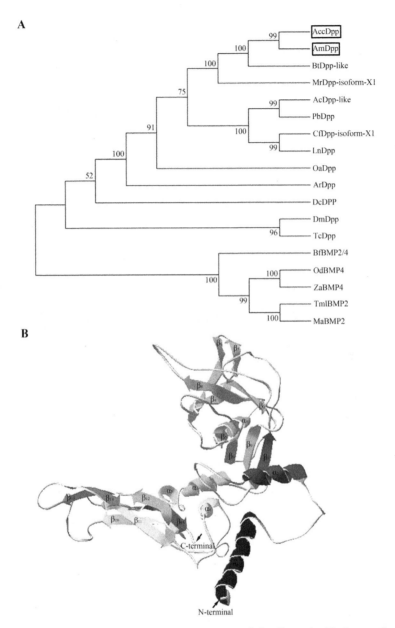

Fig. 3 Phylogenetic analysis and the tertiary structure of AccDpp. A. Phylogenetic analysis of AccDpp from different species. The species source of the above analysis is listed in S4 Table. B. The tertiary structure of AccDpp. Helices, sheets, and coils are presented in different colours. (For color version, please sweep QR Code in the back cover)

Putative transcription factor binding sites on the *AccDpp* promoter

A 1,571 bp promoter sequence (GenBank accession number: KT750953) was isolated to investigate the organization of regulatory regions of *AccDpp*. As shown in Fig. 5, fifty-three CdxA, seven CF2-II, six HSF, three NIT2, and one BR-C were identified in the promoter of *AccDpp*. Heat shock transcription factor (HSF) can respond to heat shock and contributes to

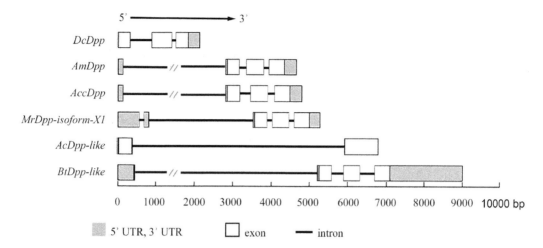

Fig. 4 Schematic representation of the DNA structures of *Dpps*. The legend presents the pattern of untranslated regions, introns, and exons. Some introns of *Dpp* have been partially abridged in this figure for their partial sequence. The length of the genomic DNA of *Dpps* is loaded from the NCBI database, and their GeneBank accession numbers are listed in Table S4.

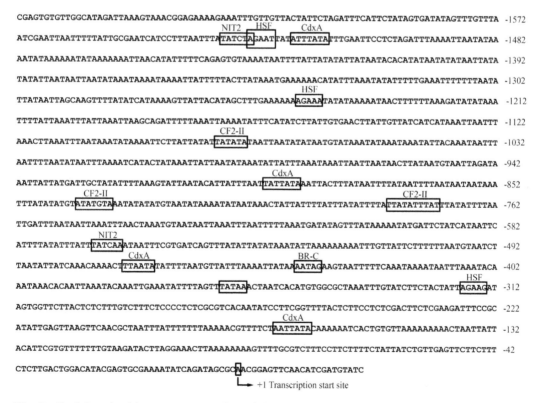

Fig. 5 Partial nucleotide sequences and prediction transcription factor binding sites in the promoter region of *AccDpp*. The transcription start site and putative transcription factor binding sites mentioned in this paper are marked with arrows and boxes, respectively. The sequence was deposited in GenBank, and the GenBank accession number: KT750953.

building a cytoprotective state of the cell [44]. CdxA, CF2-II, NIT2, and BR-C are associated with embryo or tissue development [45-48]. The results indicated that *AccDpp* might participate in organismal growth and various environmental stress responses.

Expression and characterization of recombinant AccDpp

AccDpp was overexpressed in Transetta (DE3) with two His-tags and separated by SDS-PAGE. The recombinant protein had a molecular mass of 47.98 kDa, approximately 41.38 kDa attributed to AccDpp and approximately 6.60 kDa to cleavable N- and C-terminal His-tags (Fig. 6). It is interesting that only when the signal peptide of AccDPP was removed could recombinant AccDpp be induced by IPTG. The signal peptide can guide the nascent polypeptide chain across the endoplasmic reticulum; however, *E. coli* does not have an endoplasmic reticulum. Moreover, the recombinant AccDpp was almost insoluble in Transetta cells, and the HisTrap™ FF column could not purify the recombinant AccDpp (data not shown).

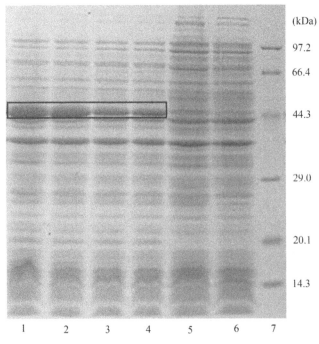

Fig. 6 Expression of AccDpp in Transetta (DE3) chemically competent cells. Recombinant AccDpp was separated by SDS-PAGE, and then stained with Coomassie Brilliant blue. Lane 1-4, expression of AccDpp after IPTG induction for 4, 5, 6, and 7 h. Lane 5 and Lane 6, non-induced of recombinant AccDpp and induced overexpression of pET-30a (+) vector for 7 h. The box shows the site of recombinant AccDpp.

Temporal and spatial expression of *AccDpp* and its protein

Various developmental stages and tissue expression profiles of *AccDpp* were investigated by qRT-PCR. *AccDpp* had expression at all stages and was highly expressed in 15-day adult bees (Fig. 7A). The stages L3, L5, Pp, Pb, and A15 were selected to carry out Western blotting to detect their protein levels. Anti-AccDpp was used to detect AccDpp. The results showed that the content of AccDpp protein was higher in stage L3 and L5, followed by stage A15, Pp, and Pb (Fig. 7D), a result that was not completely consistent with the qRT-PCR findings. *AccDpp* was expressed in all of the selected tissues, while the highest expression level in the poison gland, followed by the midgut (Fig. 7B). Western blotting was also performed to explore the AccDpp content in different tissues. As shown in Fig. 7C, the expression of AccDpp was higher in the poison gland than in the epidermis, rectum, and midgut. These data revealed that *AccDpp* might be related to development and growth in *A. cerana cerana*.

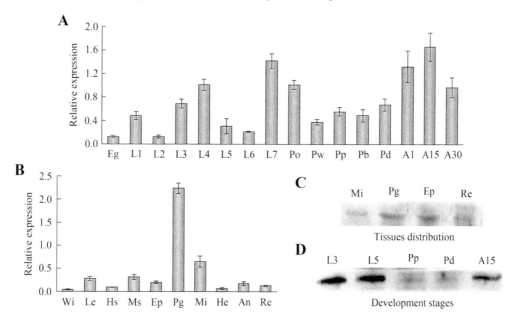

Fig. 7 Expression profile of *AccDpp* and Western blotting analysis of AccDpp at different developmental stages and different tissues. A and B. the mRNA level of *AccDpp* at different developmental stages: egg (Eg), 1-day to 7-day larvae (L1-L7), pre-pupal phase pupae (Po), pupae [white-eyed (Pw), pink-eyed (Pp), brown-eyed (Pb) and dark-eyed (Pd) pupae], 1-day worker bees (A1), 15-day worker bees (A15), and 30-day worker bees (A30), and different tissues: leg (Le), wing (Wi), muscle (Ms), midgut (Mi), haemolymph (He), rectum (Re), poison gland (Pg), honey sac (Hs), antennae (An) and epidermis (Ep), respectively. The data are the mean ± SE of three independent experiments. The letters above the columns suggest significant differences ($P < 0.0001$) according to Duncan's multiple range tests. C, and D. the expression level of AccDpp protein at different tissues and stages of development, separately.

Expression patterns of *AccDpp* under different types of abiotic stresses

Although the expression level of AccDpp protein was higher in the L3 and L5 stages than in the A15 stage (Fig. 7D), the larvae were not easily bred, and its quantity was less. Moreover, the mRNA level of the A15 stage was higher than that of the other stages (Fig. 7A). Therefore, the 15-day worker bees were selected to be treated with 4 ℃, 44 ℃, H_2O_2, UV, VC, acaricide, cyhalothrin, paraquat, methomyl, $HgCl_2$, and $CdCl_2$. As shown in Fig. 8A and Fig. 8B, the mRNA level of *AccDpp* was induced and repressed after 44 and 4 °C treatment, respectively, and reached maximums and minimums at 1 and 5 h, separately. When the 15-day worker bees were exposed to methomyl, acaricide, cyhalothrin, and paraquat, the transcription levels of *AccDpp* were all upregulated and accumulated to their highest level at 0.5, 4, 0.5, and 2 h (Fig. 8C-F). Under stress using with H_2O_2 and VC, the mRNA expression of *AccDpp* was slightly increased at 1 and 3 h (Fig. 8G, H), respectively, compared with the control. Conversely, the expression of *AccDpp* was reduced after UV treatment (Fig. 8I). When the 15-day worker bees were fed with food containing $CdCl_2$ and $HgCl_2$, the transcription levels of *AccDpp* were increased 4.67-fold and 11.74-fold compared to untreated honeybees, respectively, although the mRNA levels of *AccDpp* were gradually downregulated over time (Fig. 8J, K). The above results indicated that *AccDpp* might participate in a stress response.

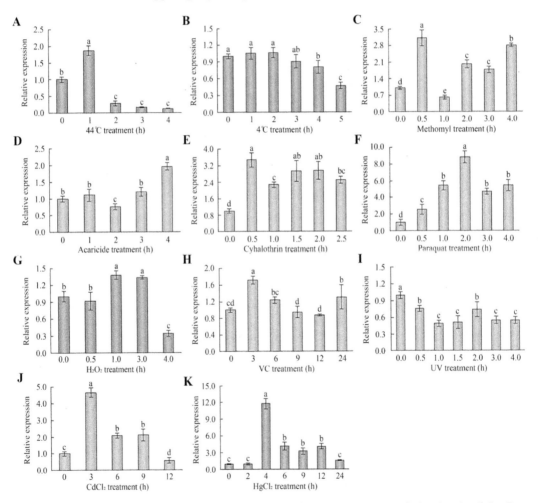

Fig. 8 Expression of *AccDpp* under different stress conditions. The transcription levels of *AccDpp* were analyzed via qRT-PCR. Untreated 15-day worker bees and the *β-actin* gene were used as controls and an internal control, separately. The data are the mean ± SE of three independent experiments. Significant differences ($P < 0.001$) were represented by different letters on the bar based on Duncan's multiple range tests.

Western blotting analysis of AccDpp under abiotic stress conditions

Further studies (Western blotting analysis) aimed at exploring the level of AccDpp under the condition of abiotic stress. On the whole, the amount of AccDpp was increased to a certain degree after exposure to all of the stressful agents (Fig. 9), although the time and extent of induction period showed some differences with the expression patterns of *AccDpp* under the same stress conditions. After being subjected to 44 ℃ for 1, 3, and 4 h, the level of AccDpp reached a peak at 1 h (Fig. 9A), which consistent with the result of qRT-PCR. As shown in Fig. 9I, in contrast to the mRNA level of *AccDpp*, AccDpp accumulated at 3 h when 15-day bees were subjected to 4 ℃. Following exposure to methomyl, acaricide, cyhalothrin, and paraquat, the expression level of AccDpp increased at different times (Fig. 9B-D and J). VC, UV and H_2O_2 potently enhanced the amount of *AccDpp* protein at 6, 4.0, and 1.5 h (Fig. 9E, F and K), respectively. $CdCl_2$ and $HgCl_2$ treatment caused noticeable increase in the protein level of AccDpp (Fig. 9G, H). These findings revealed that AccDpp might play a pivotal role when *A. cerana cerana* was subjected to stress stimuli.

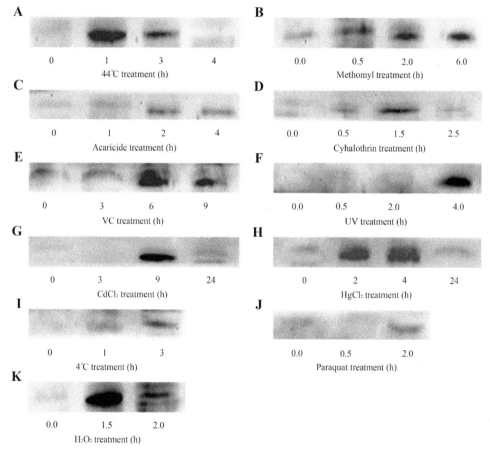

Fig. 9 Western blotting analysis of AccDpp. 15-day-old adult bees were exposed to 44 ℃ (A), methomyl (B), acaricide (C), cyhalothrin (D), VC (E), UV (F), $CdCl_2$ (G), $HgCl_2$ (H), 4 ℃ (I), paraquat (J), and H_2O_2 (K). An equivalent concentration of extracted protein was loaded in every lane under the same treatment conditions.

Disc fusion assay of recombinant AccDpp.

Recombinant AccDpp protein was exposed to four reagents to provide further evidence that AccDpp was related to the stress response. *E. coli* with pET-30a (+) vector used as the control. The results showed that the killing zones were larger around the filters on the plates with cells overexpressing AccDpp than around the filters of the control plates (Fig. 10), a finding that was

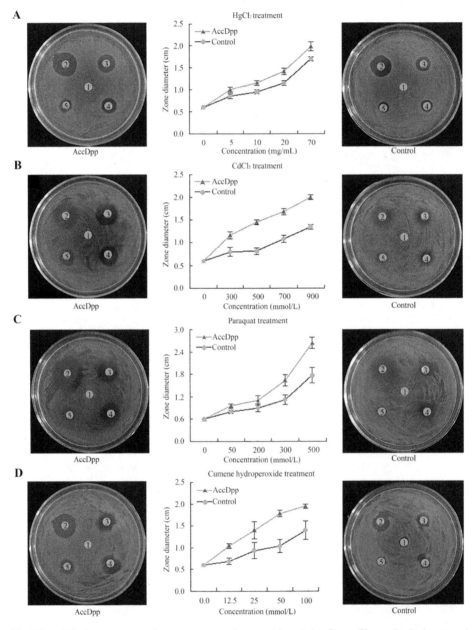

Fig. 10 Disc diffusion assays of overexpressed recombinant AccDpp. The selected reagents are $HgCl_2$, $CdCl_2$, paraquat, and cumene hydroperoxide. The numbers on the filter discs from 2-5 represent the concentration of reagents from small to large, and the number 1 indicates the control. The data are the mean ± SE of three independent experiments.

opposite to the expected results. This suggested that AccDpp might have antimicrobial activity. The antimicrobial peptide database was used to predict the antimicrobial activity of the peptide in AccDpp. Most of active antimicrobial peptides have a net charge between +3 to +8. However, peptide that having a neutral charge may also have antimicrobial activity. The predictable results showed that there were at least seven antimicrobial peptides in the sequence of AccDpp (Fig. 1). These data indicated that AccDpp was likely to have antibacterial activity.

Discussion

The TGF-β superfamily is associated not only with the growth, differentiation, and apoptosis of cells but also with ROS. *Dpp* belongs to the TGF-β superfamily. Studies in model organisms have suggested that *Dpp* notably contributes to the body axis decision and the development of appendages [49, 50]. However, few reports have discussed the role of *Dpp* in the ROS response in insects.

In this paper, we used the Chinese bee as an experimental insect and successfully isolated *Dpp* gene (*AccDpp*). The ORF of *AccDpp* encoded 390 amino acids, which included a signal peptide consisting of 23 amino acids (Fig. 1). The sequence of the C-terminus of *Dpp* from different species was highly conserved compared with that of the N-terminus (Fig. 2A). The TGF-β domain of *Dpp* from different species consisted of 101 amino acids (Fig. 2A, B). The results of Fig. 2A and Fig. 2B suggested that the TGF-β domain and TGF-β-propeptide domain might decide the conservation and diversity of function of Dpp protein among various species during the course of evolution, respectively. Phylogenetic analysis showed that AccDpp presented the closest evolutionary relationships with AmDpp (Fig. 3A). The sequence identity of AccDpp and AmDpp can reach 92.82%. Such a high sequence identity is also present in other genes of *A. cerana cerana* [51-55] and *A. mellifera*, and some protein sequences are even exactly the same. However, the traits and characteristics of these two bees are very different, possibly due to the environment and minor differences between the genes.

Additionally, a 1,571 bp promoter sequence of *AccDpp* was cloned. Sequence analysis showed that there are many transcription factor binding sites (TFBs) in the promoter (Fig. 5) that play a role in development and the stress response. It is worth mentioning that an intron more than 2,000 bp sequence presented inside the 5′ UTR of *AccDpp*. Many TFBs existed in this intron and the 5′ UTR of *AccDpp* (Fig. S1), which may also control the transcription of *AccDpp* as its promoter. Such a long intron also exists in the coding region of *Dpp* of other species (Fig. 4 and Table S6). Recent evidence had shown that the expression of *Dpp* was extremely complicated and could be regulated by the adjustment of the 5′ and 3′ coding region, and a 50 kb intron (this intron interrupted the protein coding region) in *D. melanogaster* [56, 57]. This long intron can also exist inside the 5′ UTR of *Dpp* in other species. The different positions of it may be the result of the evolution of species. Research had demonstrated that *Dpp* played a role in developmental processes [33] and was expressed at all of the development stages of *Polyrhachis vicina*, *D. melanogaster*, and *Bombyx mori*. qRT-PCR analysis showed that *AccDpp* was expressed from the egg to adult and had the highest transcription level at the A15 stage in *A. cerana cerana*, suggesting that *AccDpp* participated in the development of the Chinese bee (Fig. 7A), which was not consistent with the Western blotting result (Fig. 7D). The same result was obtained in the expression pattern of *AccDpp* in different tissues (Fig. 7B, C). That may be due to the particular body needs of *AccDpp* at the mRNA and protein levels. The poison gland, midgut, and epidermis are associated with self-defence, protection from oxidative damage and exogenous substance detoxification [58], and the stabilization of physical as well as stress response [59], respectively. The tissue-specific expression of *AccDpp* indicated it may have protective activity against the impairment of environmental stress and xenobiotics.

The above results prompted us to explore the role of *AccDpp* under oxidative stress conditions. Abrashev et al. (2008) suggested that heat shock could induce the antioxidant response and oxidative stress [60]. A decrease in the temperature leads to the transcription and translation of many genes, including genes that are induced following ROS induction. The antioxidative and metabolic systems were changed after exposure to cold stress in rats [61]. ROS can also be induced by the accumulation of toxic pesticides, resulting in oxidative injury in the living body [62, 63]. For example, the early embryonic development of amphibians was seriously affected by the widespread use of paraquat, which could induce ROS generation. The processes related to cell aging were intensely affected by adding pesticides in the culture of yeast *Saccharomyces cerevisiae* [64], likely because pesticides induce oxidative lesions by stimulating the production of free radicals. UV irradiation provokes ROS formation, leading to the activation of complex signalling pathways, such as mitogen activated protein kinase (MAPK) and nuclear factor kappa-β (NF-κβ) pathways, finally causing cellular death [4, 65]. H_2O_2 is one of the three major types of ROS, resulting directly from the action of oxidase enzymes or from the dismutation of superoxide anion radicals [66]. Experimental evidence had indicated that DNA oxidative lesions and mutation can be induced by cadmium, which could influence cell proliferation, differentiation, and apoptosis and might be associated with carcinogenesis [6, 67]. Several studies had indicated that mercury played a role in the generation of oxygen radicals [68, 69]. As an antioxidant, vitamin C (VC) can mitigate oxidative stress [70]; however, VC can also cause oxidative damage of DNA [71]. Thus, we can see that heat, cold, pesticide, heavy metal, UV, H_2O_2, and VC are all related to oxidative stress. So we selected 4℃, 44℃, acaricide, cyhalothrin, paraquat, methomyl, $HgCl_2$, $CdCl_2$, VC, UV and H_2O_2 to simulate oxidative stress conditions to treat *A. cerana cerana* and test the response of *AccDpp*.

AccDpp expression might be related with temperature (4℃) and UV stress (Fig. 8B, 8I), but not enough to prevent the translation of *AccDpp* (Fig. 9I, 9F). The transcription levels of *AccDpp* were elevated after exposure to 44℃ (Fig. 8A), methomyl (Fig. 8C), acaricide (Fig. 8D), cyhalothrin (Fig. 8E), H_2O_2 (Fig. 8G), and $CdCl_2$ (Fig. 8J) to a certain degree, although its comparative expression profile varied in response to different conditions, suggesting that *AccDpp* might be relevant to the oxidative stress response. VC treatment increased the mRNA level of *AccDpp* (Fig. 8H). We speculated that the dose of VC was sufficient to induce *AccDpp* to participate in the reaction of ROS. Moreover, the transcription levels of *AccDpp* were significantly enhanced by paraquat and $HgCl_2$ (Fig. 8F, 8K), indicating that paraquat and $HgCl_2$ were more conducive to the translation of *AccDpp*. Our transcriptional analysis of *AccDpp* suggested that *AccDpp* might play a role in oxidative stress.

Furthermore, Western blotting was performed to explore the protein level of *AccDpp* when

A. cerana cerana was subjected to other oxidative pressures, including 44 ℃, methomyl, acaricide, cyhalothrin, VC, UV, $CdCl_2$, $HgCl_2$, and H_2O_2. The findings indicated that AccDpp expression was enhanced under these conditions compared with the untreated group. The extent of induction of AccDpp was more obvious under 44 ℃, H_2O_2, VC, UV, $HgCl_2$, and $CdCl_2$ conditions. It is noteworthy that the induced degree and time point of *AccDpp* showed a sensible difference at the mRNA and protein levels. Although mRNA and its corresponding protein both exist in the cell, only the protein plays a role. ROS could increase TGF-β expression, and TGF-β could mediate the production of ROS [16-18, 20]. Thus, we suggest that AccDpp is implicated in the oxidative stress response.

Concerning the protein levels of *AccDpp* that were not consistent with its transcriptional patterns, the following explanations should be considered. First, the increased level of AccDpp could be a result of the accumulation of protein. Although the transcription of *AccDpp* was repressed, the already existing mRNA could continue to be translated. Second, the simulated environmental stress regulated the transcription and translation of *AccDpp* through different signal transduction pathways. Third, that is a result of post-transcriptional regulation. A recent paper reported that, although the *invE* mRNA was readily detectable, the expression of its protein was tightly repressed. Mitobe et al. reported that RNA-binding protein Hfq was involved in the regulation of *invE* gene expression through post-transcriptional regulation [72]. Last but not least, there are several RNAs involved in mRNA transcription and translation, such as miRNAs and circRNAs. For example, miRNAs are implicated in the regulation of many pivotal processes of enamel maturation by affecting mRNA translation and stability in rat incisors [73]. Many studies had demonstrated that circRNAs could regulate the splicing, transcription, post-transcription, and activation of protein [74, 75]. The difference in the expression profiles of *AccDpp* and AccDpp may due to their regulation by miRNAs and circRNAs.

To evaluate whether recombinant AccDpp has an antioxidant function in *E. coli* cells, disc diffusion assays were performed. However, the findings showed that the killing zones were not smaller around the filters on the plates with cells overexpressing recombinant AccDpp than around the filters on the control plates (Fig. 10). A previous study reported that recombinant arginine kinases of *A. cerana cerana* could inhibit the growth of bacteria, which because of the antimicrobial peptide in arginine kinase protein [37]. The antimicrobial peptides not only had broad-spectrum antibacterial activity but also high antibacterial activity [76]. Therefore, we considered the antibacterial activity of the antimicrobial peptides of AccDpp led to the result of the disc diffusion assay experiment. The antimicrobial peptides play an important role in the humoral immune defence [77]. The TGF-β superfamily can participate in the immune response of organisms [14, 15]. The antimicrobial peptides of AccDpp may cause AccDpp to participate in the immune response.

Collectively, these results provided evidence that *AccDpp* might play a role in the development of *A. cerana cerana* and oxidative stress response. Findings of this present reported will be conducive to studying the development of Chinese bees and other insects. This will provide a foundational knowledge to explore and understand the TGF-β signal transduction pathway in the future.

References

[1]   Peng, Y. S., Fang, Y., Xu, S., Ge, L. The resistance mechanism of the Asian honey bee, *Apis cerana* Fabr., to an ectoparasitic mite, *Varroa jacobsoni* Oudemans. J Invertebr Pathol. 1987, 49 (1), 54-60.

[2]   Cheng, S. L. Special management of *Apis cerana cerana*. The Apicultural Science in China. 2001, 1, 488-512.

[3]   Oldroyd, B. P., Wongsiri, S. Asian honey bees: biology, conservation, and human interactions. Harvard University Press, United States. 2009.

[4]   Boileau, T. W. M., Bray, T. M., Bomser, J. A. Ultraviolet radiation modulates nuclear factor kappa β activation in human lens epithelial cells. J Biochem Mol Toxicol. 2003, 17 (2), 108-113.

[5]   Thomas, C. D., Cameron, A., Green, R. E., , et al. Extinction risk from climate change. Nature. 2004, 427 (6970), 145-148.

[6]   Matés, J. M., Segura, J. A., Alonso, F. J., Márquez, J. Roles of dioxins and heavy metals in cancer and neurological diseases using ROS-mediated mechanisms. Free Radic Biol Med. 2010, 49 (9), 1328-1341.

[7]   Kerr, J. T., Pindar, A., Galpern, P., et al. Climate change impacts on bumblebees converge across continents. Science. 2015, 349 (6244), 177-180.

[8]   Lal, M. A., Brismar, H., Eklöf, A. C., Aperia, A. Role of oxidative stress in advanced glycation end product-induced mesangial cell activation. Kidney Int. 2002, 61 (6), 2006-2014.

[9]   Ray, P. D., Huang, B. W., Tsuji, Y. Reactive oxygen species (ROS) homeostasis and redox regulation in cellular signaling. Cell Signal. 2012, 24 (5), 981-990.

[10] Finkel, T., Holbrook, N. J. Oxidants, oxidative stress and the biology of ageing. Nature. 2000, 408 (6809), 239-247.

[11] Roberts, A. B., Anzano, M. A., Lamb, L. C., Smith, J. M., Sporn, M. B. New class of transforming growth factors potentiated by epidermal growth factor: isolation from non-neoplastic tissues. Proc Natl Acad Sci USA. 1981, 78 (9), 5339-5343.

[12] Assoian, R. K., Komoriya, A., Meyers, C. A., Miller, D. M., Sporn, M. B. Transforming growth factor-beta in human platelets. Identification of a major storage site, purification, and characterization. J Biol Chem. 1983, 258 (11), 7155-7160.

[13] Massagué J. TGF-beta signal transduction. Annu Rev Biochem. 1998, 67, 753-791.

[14] Massagué, J. How cells read TGF-β signals. Nat Rev Mol Cell Biol. 2000, 1 (3), 169-178.

[15] Moustakas, A., Pardali, K., Gaal, A., Heldin, C. H. Mechanisms of TGF-β signaling in regulation of cell growth and differentiation. Immunol Lett. 2002, 82 (1), 85-91.

[16] Samarakoon, R., Overstreet, J. M., Higgins, P. J. TGF-β signaling in tissue fibrosis: redox controls, target genes and therapeutic opportunities. Cell Signal. 2013, 25 (1), 264-268.

[17] Black, D., Bird, M. A., Samson, C. M., et al. Primary cirrhotic hepatocytes resist TGFβ-induced apoptosis through a ROS-dependent mechanism. J Hepatol. 2004, 40 (6), 942-951.

[18] Hubackova, S., Krejcikova, K., Bartek, J., Hodny, Z. IL1-and TGFβ-Nox4 signaling, oxidative stress and DNA damage response are shared features of replicative, oncogene-induced, and drug-induced paracrine 'bystander senescence'. Aging (Albany NY). 2012, 4 (12), 932.

[19] Chen, J., Chen, J. K., Nagai, K., Plieth, D., Tan, M., Lee, T. C., Threadgill, D. W., Neilson, E. G., Harris, R. C. EGFR signaling promotes TGFβ-dependent renal fibrosis. J Am Soc Nephrol. 2012, 23 (2), 215-224.

[20] Sánchez, A., Álvarez, A. M., Benito, M., Fabregat, I. Apoptosis induced by transforming growth factor in fetal hepatocyte primary cultures involvement of reactive oxygen intermediates. J Biol Chem. 1996, 271 (13), 7416-7422.

[21] Michaeloudes, C., Sukkar, M. B., Khorasani, N. M., Bhavsar, P. K., Chung, K. F. TGF-β regulates *Nox4*, *MnSOD* and catalase expression, and *IL-6* release in airway smooth muscle cells. Am J Physiol Lung Cell Mol Physiol. 2011, 300 (2), L295-L304.

[22] Sharma, K., Cook, A., Smith, M., Valancius, C., Inscho, E. W. TGF-β impairs renal autoregulation via generation of ROS. Am J Physiol Renal Physiol. 2005, 288 (5), F1069-F1077.

[23] Chiu, C., Maddock, D. A., Zhang, Q., Souza, K. P., Townsend, A. R., Wan, Y. TGF-β-induced p38 activation is mediated by Rac1-regulated generation of reactive oxygen species in cultured human keratinocytes. Int J Mol Med. 2001, 8 (3), 251-255.

[24] Hanafusa, H., Ninomiya-Tsuji, J., Masuyama, N., Nishita, M., Fujisawa, J. I., Shibuya, H., Matsumoto, K., Nishida, E. Involvement of the p38 mitogen-activated protein kinase pathway in transforming growth factor-β-induced gene expression. J Biol Chem. 1999, 274 (38), 27161-27167.

[25] Nebreda, A. R., Porras, A. p38 MAP kinases: beyond the stress response. Trends Biochem Sci. 2000, 25 (6), 257-260.

[26] Shin, D. M., Jeon, J. H., Kim, C. W., et al. TGFβ mediates activation of transglutaminase 2 in response to oxidative stress that leads to protein aggregation. FASEB J. 2008, 22 (7), 2498-2507.

[27] Padgett, R. W., Wozney, J. M., Gelbart, W. M. Human *BMP* sequences can confer normal dorsal-ventral patterning in the *Drosophila* embryo. Proc Natl Acad Sci USA. 1993, 90 (7), 2905-2909.

[28] Lecuit, T., Brook, W. J., Ng, M., Calleja, M., Sun, H., Cohen, S. M. Two distinct mechanisms for long-range patterning by Decapentaplegic in the *Drosophila* wing. Nature. 1996, 381 (6581), 387-393.

[29] Nellen, D., Burke, R., Struhl, G., Basler, K. Direct and long-range action of a DPP morphogen gradient. Cell. 1996, 85 (3), 357-368.

[30] Irish, V., Gelbart, W. M. The decapentaplegic gene is required for dorsal-ventral patterning of the *Drosophila* embryo. Genes Dev. 1987, 1, 868-879.

[31] Wharton, K. A., Ray, R. P., Gelbart, W. M. An activity gradient of decapentaplegic is necessary for the specification of dorsal pattern elements in the *Drosophila* embryo. Development. 1993, 117 (2), 807-822.

[32] Ninov, N., Menezes-Cabral, S., Prat-Rojo, C., Manjón, C., Weiss, A., Pyrowolakis, G., Affolter, M., Martín-Blanco, E. Dpp signaling directs cell motility and invasiveness during epithelial morphogenesis. Curr Biol. 2010, 20 (6), 513-520.

[33] Affolter, M., Marty, T., Vigano, M. A., Jaźwińska, A. Nuclear interpretation of Dpp signaling in *Drosophila*. EMBO J. 2001, 20 (13), 3298-3305.

[34] Park, D., Jung, J. W., Choi, B. S., et al. Uncovering the novel characteristics of Asian honey bee, *Apis cerana*, by whole genome sequencing. BMC Genomics. 2015, 16 (1), 1.

[35] Michelette, E. D. F., Soares, A. E. E. Characterization of preimaginal developmental stages in Africanized honey bee workers (*Apis mellifera* L.). Apidologie. 1993, 24 (4), 431-440.

[36] Alaux, C. Ducloz, F., Crauser, D., Le Conte, Y. Diet effects on honeybee immunocompetence. Biol Lett. 2010, 6(4), 562-565.

[37] Chen, X., Yao, P., Chu, X., Hao, L., Guo, X., Xu, B. Isolation of arginine kinase from *Apis cerana cerana* and its possible involvement in response to adverse stress. Cell Stress Chaperones. 2015, 20 (1), 169-183.

[38] Zhang, Y., Yan, H., Lu, W., Li, Y., Guo, X., Xu, B. A novel Omega-class glutathione *S*-transferase gene in *Apis cerana cerana*: molecular characterisation of *GSTO2* and its protective effects in oxidative stress. Cell Stress Chaperones. 2013, 18 (4), 503-516.

[39] Li, Y., Zhang, L., Kang, M., Guo, X., Xu, B. *AccERK2*, a map kinase gene from *Apis cerana cerana*, plays roles in stress responses, developmental processes, and the nervous system. Arch Insect Biochem Physiol. 2012, 79 (3), 121-134.

[40] Lourenço, A. P., Mackert, A., dos Santos Cristino, A., Simões, Z. L. P. Validation of reference genes for gene expression studies in the honey bee, *Apis mellifera*, by quantitative real-time RT-PCR. Apidologie. 2008, 39 (3), 372-385.

[41] Scharlaken, B., De Graaf, D. C., Goossens, K., Peelman, L. J., Jacobs, F. J. Differential gene expression in the honeybee head after a bacterial challenge. Dev Comp Immunol. 2008, 32 (8), 883-889.

[42] Umasuthan, N., Revathy, K. S., Lee, Y., Whang, I., Choi, C. Y., Lee, J. A novel molluscan sigma-like glutathione *S*-transferase from *Manila clam, Ruditapes philippinarum*: cloning, characterization and

transcriptional profiling. Comp Biochem Physiol C: Toxicol Pharmacol. 2012, 155 (4), 539-550.

[43] Meng, F., Zhang, Y., Liu, F., Guo, X., Xu, B. Characterization and mutational analysis of Omega-class GST (*GSTO1*) from *Apis cerana cerana*, a gene involved in response to oxidative stress. PLoS One. 2014, 9 (3), e93100.

[44] Santoro, M. G. Heat shock factors and the control of the stress response. Biochem Pharmacol. 2000, 59 (1), 55-63.

[45] Ericsson, A., Kotarsky, K., Svensson, M., Sigvardsson, M., Agace, W. Functional characterization of the *CCL25* promoter in small intestinal epithelial cells suggests a regulatory role for caudal-related homeobox (Cdx) transcription factors. J Immunol. 2006, 176 (6), 3642-3651.

[46] Fu, Y. H., Marzluf, G. A. Characterization of *nit-2*, the major nitrogen regulatory gene of *Neurospora crassa*. Mol Cell Biol. 1987, 7 (5), 1691-1696.

[47] Gogos, J. A., Hsu, T., Bolton, J., Kafatos, F. C. Sequence discrimination by alternatively spliced isoforms of a DNA binding zinc finger domain. Science. 1992, 257 (5078), 1951-1955.

[48] Spokony, R. F., Restifo, L. L. Anciently duplicated Broad Complex exons have distinct temporal functions during tissue morphogenesis. Dev Genes Evol. 2007, 217 (7), 499-513.

[49] Niwa, N., Inoue, Y., Nozawa, A., Saito, M., Misumi, Y., Ohuchi, H., Yoshioka, H., Noji, S. Correlation of diversity of leg morphology in *Gryllus bimaculatus* (cricket) with divergence in *dpp* expression pattern during leg development. Development. 2000, 127 (20), 4373-4381.

[50] Ferguson, E. L., Anderson, K. V. Decapentaplegic acts as a morphogen to organize dorsal-ventral pattern in the *Drosophila* embryo. Cell. 1992, 71 (3), 451-461.

[51] Imjongjirak, C., Klinbunga, S., Sittipraneed, S. Cloning, expression and genomic organization of genes encoding major royal jelly protein 1 and 2 of the honey bee (*Apis cerana*). J Biochem Mol Biol. 2005, 38 (1), 49-57.

[52] Li, J. H., Zhang, C. X., Shen, L. R., Tang, Z. H., Cheng, J. A. Expression and regulation of phospholipase A2 in venom gland of the Chinese honeybee, *Apis cerana cerana*. Arch Insect Biochem Physiol. 2005, 60 (1), 1-12.

[53] Shi, J., Lua, S., Du, N., Liu, X., Song, J. Identification, recombinant production and structural characterization of four silk proteins from the Asiatic honeybee *Apis cerana*. Biomaterials. 2008, 29 (18), 2820-2828.

[54] Srisuparbh, D., Klinbunga, S., Wongsiri, S., Sittipraneed, S. Isolation and characterization of major royal jelly cDNAs and proteins of the honey bee (*Apis cerana*). J Biochem Mol Biol. 2003, 36 (6), 572-579.

[55] Zhao, H., Gao, P., Du, H., Ma, W., Tian, S., Jiang, Y. Molecular characterization and differential expression of two duplicated odorant receptor genes, *AcerOr1* and *AcerOr3*, in *Apis cerana cerana*. J Genet. 2014, 93 (1), 53-61.

[56] St Johnston, R. D., Hoffmann, F. M., Blackman, R. K., Segal, D., Grimaila, R., Padgett, R. W., Irick, H. A., Gelbart, W. M. Molecular organization of the decapentaplegic gene in *Drosophila melanogaster*. Genes Dev. 1990, 4 (7), 1114-1127.

[57] Blackman, R. K., Sanicola, M., Raftery, L. A., Gillevet, T, Gelbart, W. M. An extensive 3′ cis-regulatory region directs the imaginal disk expression of decapentaplegic, a member of the TGF-beta family in *Drosophila*. Development. 1991, 111 (3), 657-666.

[58] Enayati, A. A., Ranson, H., Hemingway, J. Insect glutathione transferases and insecticide resistance. Insect Mol Biol. 2005, 14 (1), 3-8.

[59] Marionnet, C., Bernerd, F., Dumas, A., et al. Modulation of gene expression induced in human epidermis by environmental stress *in vivo*. J Invest Dermatol. 2003, 121 (6), 1447-1458.

[60] Abrashev, R. I., Pashova, S. B., Stefanova, L. N., Vassilev, S. V., Dolashka-Angelova, P. A., Angelova, M. B. Heat-shock-induced oxidative stress and antioxidant response in *Aspergillus niger* 26. Can J Microbiol. 2008, 54 (12), 977-983.

[61] Yuksel, S., Dilek, A., Ozfer, Y. Antioxidative and metabolic responses to extended cold exposure in rats. Acta Biol Hung. 2008, 59 (1), 57-66.

[62] Banerjee, B. D., Seth, V., Ahmed, R. S. Pesticide-induced oxidative stress: perspective and trends. Rev Environ Health. 2001, 16 (1), 1-40.

[63] Soltaninejad, K., Abdollahi, M. Current opinion on the science of organophosphate pesticides and toxic stress: a systematic review. Med Sci Monit. 2009, 15 (3), RA75-RA90.

[64] Owsiak, A., Marchel, M., Zyracka, E. Impact of pesticides used in the culture of the vine on the viability of the yeast *Saccharomyces cerevisiae* wine in chronological aging. J Microbiol Biotechnol Food Sci. 2015, 4, 48.

[65] Jiang, Q., Zhou, C., Healey, S., Chu, W., Kouttab, N., Bi, Z., Wan, Y. UV radiation down-regulates *Dsg-2* via Rac/NADPH oxidase-mediated generation of ROS in human lens epithelial cells. Int J Mol Med. 2006, 18 (2), 381-387.

[66] Agarwal, A., Gupta, S., Sekhon, L., Shah, R. Redox considerations in female reproductive function and assisted reproduction: from molecular mechanisms to health implications. Antioxid Redox Signal. 2008, 10 (8), 1375-1404.

[67] Filipič, M., Fatur, T., Vudrag, M. Molecular mechanisms of cadmium induced mutagenicity. Hum Exp Toxicol. 2006, 25 (2), 67-77.

[68] Lee, Y. W., Ha, M. S., Kim, Y. K. Role of reactive oxygen species and glutathione in inorganic mercury-induced injury in human glioma cells. Neurochem Res. 2001, 26 (11), 1187-1193.

[69] Wiggers, G. A., Pecanha, F. M., Briones, A. M., Perez-Giron, J. V., Miguel, M., Vassallo, D. V., Cachofeiro, V., Alonso, M. J., Salaices, M. Low mercury concentrations cause oxidative stress and endothelial dysfunction in conductance and resistance arteries. Am J Physiol Heart Circ Physiol. 2008, 295 (3), H1033-H1043.

[70] Chen, Y., Luo, G., Yuan, J., Wang, Y., Yang, X., Wang, X., Li, G., Liu, Z., Zhong, N. Vitamin C mitigates oxidative stress and tumor necrosis factor-alpha in severe community-acquired pneumonia and LPS-induced macrophages. Mediators Inflamm. 2014, 2014, 1-11.

[71] Lee, S. H., Oe, T., Blair, I. A. Vitamin C-induced decomposition of lipid hydroperoxides to endogenous genotoxins. Science. 2001, 292 (5524), 2083-2086.

[72] Mitobe, J., Morita-Ishihara, T., Ishihama, A., Watanabe, H. Involvement of RNA-binding protein Hfq in the osmotic-response regulation of *invE* gene expression in *Shigella sonnei*. BMC Microbiol. 2009, 9 (1), 110.

[73] Yin, K., Hacia, J. G., Zhong, Z., Paine, M. L. Genome-wide analysis of miRNA and mRNA transcriptomes during amelogenesis. BMC Genomics. 2014, 15 (1), 998.

[74] Memczak, S., Jens, M., Elefsinioti, A., et al. Circular RNAs are a large class of animal RNAs with regulatory potency. Nature. 2013, 495 (7441), 333-338.

[75] Ashwal-Fluss, R., Meyer, M., Pamudurti, N. R., et al. circRNA biogenesis competes with pre-mRNA splicing. Mol Cell. 2014, 56 (1), 55-66.

[76] Brogden, K. A. Antimicrobial peptides: pore formers or metabolic inhibitors in bacteria? Nat Rev Microbiol. 2005, 3 (3), 238-250.

[77] Brown, S. E., Howard, A., Kasprzak, A. B., Gordon, K. H., East, P. D. A peptidomics study reveals the impressive antimicrobial peptide arsenal of the wax moth *Galleria mellonella*. Insect Biochem Mol Biol. 2009, 39 (11), 792-800.

3.6 Molecular Cloning, Expression and Oxidative Stress Response of the Vitellogenin Gene (*AccVg*) from *Apis cerana cerana*

Weixing Zhang, Zhenguo Liu, Ming Zhu, Lanting Ma, Ying Wang, Hongfang Wang, Xingqi Guo, Baohua Xu

Abstract: Vitellogenin (Vg) is a yolk precursor protein in most oviparous females. However, Vg has not been studied in the *Apis cerana cerana*. In this work, the Vg gene of the *A. cerana cerana* has been cloned and sequenced. The gene codes for a protein consisting of 1,770 amino acids in seven exons with a predicted molecular mass and isoelectric point of 200 kDa and 6.46, respectively. Additionally, an 807 bp 5′-flanking region was isolated, and potential transcription factor binding sites associated with development and stress response were identified. Quantitative real-time PCR analysis showed that *AccVg* is highly expressed in pupae during different developmental stages. In addition, the expression of *AccVg* could be induced by cold (4, 16℃), $CdCl_2$, and pesticide treatments. Real-time quantitative PCR (RT-qPCR) and Western blotting analysis indicated that *AccVg* transcription was induced by various abiotic stresses. Western blotting was used to measure the expression levels of AccVg protein. Taken together, these results suggest that *AccVg* most likely plays essential roles in antioxidant defence and that it may be of critical importance to honeybee survival.

Keywords: Vitellogenin; *Apis cerana cerana*; RNA interference; Oxidative stress response; RT-PCR

Introduction

In China, *Apis cerana cerana* is commonly found and plays an important role in the balance of regional ecologies and agricultural economics as a pollinator of flowering plants. In honeybees, worker-worker division of labour is based on a process of behavioural maturation; for the first 2-3 weeks of adult life, workers perform tasks inside the hive, such as brood care and food storage, and as they become older, they progress to tasks outside, including foraging for pollen and nectar [1].

Organisms are exposed to a series of unfavourable environmental stressors on a nearly constant basis, including temperature change, heavy metal, pesticides, and ultraviolet (UV) radiation, all of which are believed to induce the formation of reactive oxygen species (ROS) [2, 3]. The generation of increased ROS and the production of oxidative damage are associated with induced thermal stress [4]. ROS can be generated by all oxygenic organisms during aerobic metabolism [5, 6]. Therefore, understanding the antioxidant system of the honeybee and its ROS defence mechanisms has become a crucial issue for this industry.

Large lipid transfer proteins (LLTP) have multiple roles in animals, such as lipid transporters, inflammation suppressors, immunomodulators, and blood coagulators. Vg is one of the ancient forms of these proteins, with an estimated 700 million-year history [7]. Vgs are mostly produced in the liver (vertebrates), hepatopancreas (crustaceans) or fat body (invertebrates), secreted into the blood, and taken up by targets via receptor-mediated endocytosis [8-10]. Honeybee Vg, known as a conserved yolk precursor protein, is a 180 kDa glycolipoprotein synthesized in fat body cells and released to the haemolymph (the insect blood) [11, 12]. In egg-laying queens, Vg constitutes over 50% of total haemolymph proteins and is taken up by developing oocytes [13]. In the haemolymph of nurse bees, Vg comprise 30%-50% of total proteins, and impacts both antioxidant and immune function [14-16]. Additionally, Vg is present in the haemolymph of young worker bees, as well as in drones where it may account for up to 5% of total haemolymph proteins [17]. Vg has an antioxidant function in workers [15]. Knockdown of Vg expression in workers causes decreased resistance to oxidative stress. The Vg titre positively correlates with the oxidative stress tolerance of the honeybee because the antioxidant property of Vg shields other haemolymph molecules from reactive oxygen species (ROS) [15]. Additionally, a causal link between honeybee vitellogenin activity and lifespan is supported by the zinc-binding capacity of the protein product [14], which also suggests an antioxidant function [18-20].

The vitellogenins have been studied extensively in a wide variety of animals, both vertebrates and invertebrates, including insects. In the present study, we isolated and characterized the Vg in *A. cerana cerana* (*AccVg*). Additionally, we analyzed the expression of Vg during different developmental stages and under various abiotic stress conditions. Finally, we investigated the activity of *AccVg* under various stress conditions, including cold, insecticides and heavy metal exposure.

Materials and Methods

Biological specimens and treatments

The Chinese honeybees, *A. cerana cerana*, maintained at the experimental apiary of Shandong Agricultural University (Taian, Shandong Province, China), were used in the experiments following different treatments. Newly emerged workers within an 8 h period were tagged with Opalith numbered plates (Graze, Germany) at emergence from hives. When the tagged workers were 15 days these were called foragers. They were captured and randomly divided into 6 groups, every groups with three independent biological samples of 150 individuals, caged in wooden cage during the treatment which were kept in an incubator at a constant temperature (31℃) and humidity (60%), and fed with pollen-and-sucrose solutions [21].

For the environmental stress analysis, the worker bees were treated with cold (4℃) or heat (42℃) for the indicated times. For the pesticide treatments, cyfluthrin, phoxim and paraquat were diluted to a final concentration of 20 mg/L. The diluted liquid (200 μl) was sprayed into gauze and the gauze was put in the wooden cage. For the heavy metal treatment, worker bees were fed with pollen-and-sucrose solutions containing $CdCl_2$ (1.0 μg/ml). The control group was fed only the pollen-and-sucrose solution. The bees were kept in the cages and exposed to the different treatments for up to four hours. For different developmental stages and tissues, untreated larvae and adult workers were harvested for gene expression analysis. A total of 60 fifteen-day-old workers were anesthetized on ice, and the brains, thoraces, and guts were dissected. Three individual bees were dissected for each stage. All of the bees were flash-frozen in liquid nitrogen at an appropriate time and stored at −80℃ for RNA or DNA extraction.

RNA extraction, cDNA synthesis, and genomic DNA isolation

Total RNA was extracted by using the RNAiso Plus (TaKaRa, Dalian, China) according to the instructions of the manufacturer, followed by RNase-free DNase I treatment. The single-stranded cDNA was synthesized by using Transcript® All-in-One First-Strand cDNA Synthesis SuperMix for qPCR (TransGen Biotech, Beijing, China) at 42℃ for 15 min, followed by 85℃ for 15 s. Genomic DNA was isolated from newly emerged workers by using an EasyPure Genomic DNA Extraction Kit according to the instructions of the manufacturer (TransGen Biotech).

Primers

The primers used in the present study were listed in Table 1.

Table 1 PCR primers in this study

Abbreviation	Primer sequence (5′→3′)	Description
VG1	ATGTTGCTACTTCTAACGCTT	cDNA sequence primer, forward
VG2	CGTAGAGGCTGTTGTCGTGT	cDNA sequence primer, reverse
VG3	CAACACCGCCGCCACCTTAT	cDNA sequence primer, forward
VG4	TTGCTGTTCGCCTGCTTGTCCG	cDNA sequence primer, reverse
VG5	GAGGAAATGTGGGAACTGAT	cDNA sequence primer, forward
VG6	ACGAAAGAAAGCGACAAAGT	cDNA sequence primer, reverse
5P1	CGGGCATCATACAAACCACAT	5′ RACE reverse primer, outer
5P2	TGGTGGTCGGTGAACAAACT	5′ RACE reverse primer, inner
AAP	GGCCACGCGTCGACTAGTAC(G)$_{14}$	Abridged anchor primer
AUAP	GGCCACGCGTCGACTAGTAC	Abridged universal amplification primer
3P1	AAGGAGACCGACGACAAGA	3′ RACE forward primer, outer
3P2	AAGAACGAGGCCGCGATGAA	3′ RACE forward primer, inner
B26	GACTCTAGACGACATCGA(T)$_{18}$	3′ RACE universal primer, outer
B25	GACTCTAGACGACATCGA	3′ RACE universal primer, inner

Continued

Abbreviation	Primer sequence (5'→3')	Description
N1	ATGTTGCTACTTCTAACGCTT	Genomic sequence primer, forward
N2	CGTAGAGGCTGTTGTCGTGT	Genomic sequence primer, reverse
N3	CAACACCGCCGCCACCTTAT	Genomic sequence primer, forward
N4	TTGCTGTTCGCCTGCTTGTCCG	Genomic sequence primer, reverse
N5	GAGGAAATGTGGGAACTGAT	Genomic sequence primer, forward
N6	ACGAAAGAAAGCGACAAAGT	Genomic sequence primer, reverse
IF1	GGTCTGAGCAGTTGCACGCTG	Inverse PCR forward primer, outer
IF2	TCTAAACGAAACAAAGCGGCA	Inverse PCR reverse primer, outer
IR1	TCGCAATCTTCATCGTTGTA	Inverse PCR forward primer, inner
IR2	GTTCAAGTCCTCGGTGATAG	Inverse PCR reverse primer, inner
P1	AAAGCGGGGTACAGGATG	Promoter special primer, forward
P2	GGTGAGCAACGCCTTTA	Promoter special primer, reverse
VG-1	GGATCCTTGGAAAGCTATGACACAG (*Bam*H I)	Protein expression primer, forward
VG-2	CTCGAGCTGGCTCTGGCTAACTTTACC (*Xho* I)	Protein expression primer, reverse
RT1	TCGGTGACTACGAGCAGGAT	Real-time PCR primer, forward
RT2	GGGCAGAGGTTTGGACACT	Real-time PCR primer, reverse
β-s	CCGTGATTTGACTGACTACCT	Standard control primer, forward
β-x	AGTTGCCATTTCCTGTTC	Standard control primer, reverse

Isolation of the full-length cDNA and the genomic sequence of *AccVg*

To obtain the fragment of cDNA of *AccVg*, special primers (VG1/VG2, VG3/VG4, VG5/VG6) were designed and synthesized (Shanghai Sangon Biotechnological Company, Shanghai, China) by the conserved regions of the Vg from *A. mellifera*, *Nasonia vitripennis* and *Locusta migratoria*. Fig. S1 shows the position of primers in Vg genes. The cDNA of newly emerged workers was used for cloning the full-length cDNA. On the basis of the sequence of the cloned conservative fragments, specific primers (3N1, 3N2, 5N1, 5N2) were designed and synthesized for 3'- and 5'- rapid amplification of cDNA end (RACE). All the PCR products were purified by using a gel extraction kit (Solarbio, Beijing, China). The PCR products were ligated into pEASY-T3 vector (TransGen Biotech) and transformed into *Escherichia coli* strain DH5α for sequencing.

To isolate the *AccVg* gene, three pairs of specific primers (N1/N2, N3/N4 and N5/N6) were designed based on the *AccVg* cDNA sequence. The PCR reaction uses 1 μl genomic DNA as template, 0.25 μl *Tap* DNA polymerase, 2.5 μl PCR buffer, 1 μl dNTP, 1 μl forward primer and reverse primer, and 18.25 μl ddH$_2$O in a 25 μl volume with the following cycling conditions: initial denaturation program (94 ℃ for 10 min), followed by 35 cycles of 94 ℃ for 40 s 55 ℃ for 40 s and 72 ℃ for 120 s. Finally, the PCR products were separated, purified, ligated into pEASY-T3 vector, and transformed into *E. coli* DH5α for sequencing.

Amplification of the 5'-flanking region of *AccVg*

The 5'-flanking region of *AccVg* was obtained by using the inverse polymerase chain reaction (I-PCR). The genomic DNA was digested with the restriction endonuclease *Nsi* I at 37℃ overnight, self-ligated to form circles by using T4 DNA ligase (TaKaRa, Dalian, China), and used as a template for I-PCR. Two pairs of primers (IF1/IR1, IF2/IR2) were designed and synthesized based on the genomic DNA sequence. Finally, the sequence of the 5'-flanking region of *AccVg* was used to design specific primers (P1/P2) for direct PCR amplification from genomic DNA. The PCR reaction uses 1 μl genomic DNA as template, 1 μl dNTP, 1 μl forward primer and reverse primer, 0.25 μl *Tap* DNA polymerase, 2.5 μl PCR buffer, and 18.25 μl ddH$_2$O in a 25 μl volume with the following cycling conditions: initial denaturation program (94℃ for 10 min), followed by 35 cycles of 94℃ for 40 s 50℃ for 40 s and 72℃ for 90 s. All of the PCR products was purified, ligated into the pMD19-T vector and sequenced.

Bioinformatic and phylogenetic analyses

The theoretical isoelectric point and molecular weight of AccVg were predicted by using ExPASy online software (http://web.expasy.org/compute_pi/). The sequences of *AccVg* homologous with other species were retrieved by using the NCBI server (http://blast.ncbi.nlm.nih.gov/Blast.cgi) and aligned by using DNAMAN version 5.2.2 (Lynnon Biosoft, Quebec, Canada). The phylogenetic analysis was carried out by using the Molecular Evolutionary Genetics Analysis (MEGA version 4.1) (http://www.megasoftware.net/) based upon the neighbour-joining method. The SWISS-MODEL (http://swissmodel.expasy.org/) online software was used to predict the tertiary structure of AccVg. The TFSEARCH website (http://www.cbrc.jp/research/db/TFSEARCH.html) was used to predict the transcription factor binding sites.

Production of recombinant AccVg and antibody preparation

By using the primers VG-1/VG-2 (Table 1), a 1,128 bp cDNA fragment was obtained (a part of the whole cDNA, 920-1295 amino acids), cloned into the expression vector pET-28a (+), and the recombinant plasmid pET-28a (+)-*AccVg* was transformed into *E. coli* BL21 (ED3), which was grown at 37℃ in 10 ml Luria-Bertani (LB) broth supplemented with 40 μg/ml kanamycin. When the cell density reached 0.6 (OD$_{600}$), isopropylthio-β-D-galactoside (IPTG) was added to a 0.5 mmol/L final concentration to induce the expression of the recombinant protein. Cells were centrifuged at 6,500 g at 4℃ for 10 min, after incubation at 37℃ for 4 h. The pellet was resuspended in 10 ml lysis buffer, sonicated and centrifuged again at 4℃ at 13,000 g for 15 min. The recombinant protein was purified by using the HisTrap^{TM} FF column (GE Healthcare, Uppsala, Sweden) according to the instructions of the manufacturer. Finally, the purified protein was separated by sodium dodecyl sulphate polyacrylamide gel electrophoresis (SDS-PAGE) on 7.5% acrylamide gels. The purified protein was more than 95% pure by SDS-PAGE gel analysis. The purified protein was injected subcutaneously into New Zealand White male rabbits weighing 2-3 kg for generation of antibodies [22].

Quantitative RT-PCR

To determine the expression patterns of the *AccVg* in *A. cerana cerana*, quantitative RT-PCR was used. Amplification of the *β-actin* transcript (Gene ID: LOC108003299) was used as a sample control [23]. The special primers RT1/RT2 and the *actin* gene primers actin-f/ actin-r are listed in Table 1. The quantitative RT-PCR was carried out by using the TransStart® Tip Green qPCR SuperMix (TransGen Biotech) in a 20 μl volume on a CFX96™ Real-time System (Bio-Rad) with the following cycling conditions: denaturation program (94℃ for 30 s), followed by 40 cycles of 94℃ for 5 s 55℃ for 15 s and 72℃ for 10 s. At the end of the amplification phase, a melting curve analysis was performed with a minimum of 60℃ to a maximum of 94℃ and 0.2℃ increases for every 15 s to test the specificity and identity of the qRT-PCR products. Negative controls without the addition of RT enzyme were run to check for contamination by genomic DNA. To check reproducibility, each SYBR green assay was performed in triplicate, and for each data point, we analysed three independent biological samples. The relative transcript quantities of the *Vg* and *actin* genes were calculated by using the comparative C_t method (Applied Biosystems, User Bulletin 2) [24, 25]. The statistical analysis of the differences was determined by Duncan's multiple range test by using Statistical Analysis System (SAS) version 9.1 software.

Western blotting analysis

Whole-body proteins were extracted by using RIPA Lysis Buffer (Beyotime, Jingsu, China) according to the instructions of the manufacturer and quantified with BCA Protein Assay Kit (Beyotime). Soluble proteins were separated by SDS-PAGE on 7.5% acrylamide gels, transferred to a polyvinylidene fluoride (PVDF) membrane (Millipore, Immobilon 0.22 μm), and processed for Western blotting. Membranes were treated with the anti-AccVg serum against AccVg protein raised in rabbits, used at a dilution of 1∶500. The goat anti-rabbit IgG, HRP Conjugated (CWBIO, China) was used as the secondary antibody at a 1∶2,000 dilution. Proteins were visualized by using chemiluminescent peroxidase substrate (Beyotime Biotechnology, Shanghai, China), and then the blots were quantified by using a Fusion Fx system (UVP, America) and the analysis software FusionCapt Advance System (UVP, America).

Statistical analysis

All data are reported as the mean ± standard error (SE), unless otherwise stated. The Statistical Analysis System (SAS) version 9.1 was used for data analysis. Differences between the treatment groups were analyzed by using the Duncan's multiple range test and analysis of variance (ANOVA) to identify homogenous groups of treatments. $P < 0.05$ were considered statistically significant. The different letters above bars denote significant differences between the treatment groups.

Results

Molecular characterization of *AccVg* and features of the recombinant AccVg protein

The full-length cDNA sequence of *AccVg* (GenBank accession number: kp398512) was 5,411 bp, including 5,313 bp in the open reading frame, 26 bp in the 5′-untranslated region and 72 bp in the 3′-untranslated region. The ORF encodes a polypeptide of 1,770 amino acids with a calculated molecular mass of 200 kDa and a theoretical isoelectric point (pI) of 6.46, respectively (Fig. S2). Multiple sequence alignment revealed high amino acid identity (89.05%) between *AmVg* and *AccVg* (Fig. S3).

To investigate the evolutionary relationship between AccVg and other Vgs from different species, a phylogenetic tree of Vgs was constructed. Of the eight insect species for which Vgs were examined, AccVg was categorized into the Hymenoptera. As shown in Fig. 1, phylogenetic analysis revealed that AccVg was more closely related to AmVg than to homologues in other species. The SWISS-MODEL was used to predict and reconstruct the three-dimensional structure (Fig. S4).

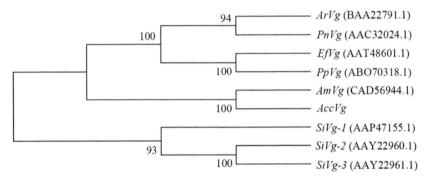

Fig. 1 Phylogenetic relationships on the basis of the amino acid sequences between Vgs from different insect species.

The molecular mass of AccVg protein is so large that preparation of recombinant plasmid pET-28a (+)-AccVg was difficult. We expressed a portion of the protein (920-1295 aa) to prepare antibodies against AccVg. SDS-PAGE analysis showed that the recombinant protein was soluble and had a molecular mass of approximately 50 kDa (Fig. 2). The entire Western blotting of the specific antibody against AccVg is shown in Fig. S5.

Genomic structure and the characterization of the 5′-flanking region of *AccVg*

By cloning and assembling cDNA fragments and intron, we obtained the genomic sequence of *AccVg*. To better understand the organization of the regulatory elements, primers (IF1/IF2 and IR1/IR2) were used to obtain the 5′-flanking region of *AccERR*, by using inverse PCR (I-PCR). Briefly, total genomic DNA was digested with the restriction endonuclease *Nsi* I at 37°C overnight, self-ligated to form circles by using T4 DNA ligase (TaKaRa, Dalian, China), and used as a template for I-PCR. The genomic DNA sequence of *AccVg* contains 7,064 bp, which includes seven exons that are separated by six introns with high AT content and flanked by the 5′ splice donor GT and the 3′ splice acceptor AG signals. The introns have typical eukaryotic characteristics with high AT content.

To further analyze the relationship of *AccVg* with other *Vg*s, the genomic structures of *Encarsia formosa Vg*, *A. mellifera Vg*, *Toxorhynchites amboinensis Vg* and *Anopheles aegypti Vg* were compared, as shown in Fig. 3. The analysis showed that *EfVg*, *AmVg* and *AccVg* contain seven exons and possess the longest stretch of sequence similarity, whereas *TaVg* and

AaVg contained only three exons. The numbers, sizes, and relative positions of all the introns of *Vg*s in different species are different.

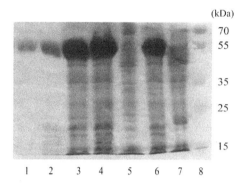

Fig. 2 An SDS-PAGE analysis of the recombinant AccVg protein. Lane 8, protein marker; Line 1 and 2, purified recombinant protein; Line 3 and 4, suspension of recombinant part of AccVg, respectively; Line 6 and 7, induced and non-induced overexpression of pET-28a (+)-*AccVg* in BL21, respectively; Line 5, induced overexpression of pET-28a (+) in BL21.

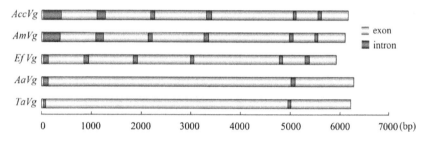

Fig. 3 The DNA structures of Vgs. The length extrons and introns of genomic DNA of *AccVg*, *AmVg* (NC_007073), *EfVg* (AY553878), *AaVg* (AY691327), *TaVg* (AY691326) are shown according to the scale below.

To better understand the molecular mechanism of the transcriptional regulation of *AccVg*, a 833 bp 5'-flanking fragment was isolated. A series of putative *cis*-acting elements was predicted by using TFSEARCH software. A series of putative *cis*-acting elements was also predicted after analysing the promoter sequence of *AccVg*. Some heat shock elements, which are the binding sites for heat shock factor (HSF), were also found [26]. Several important *cis*-acting elements involved in embryo or organ development, including Broad-Complex (BR-C), NIT2, and CdxA were identified [27]. Additionally, the ovo homolog-like transcription factor (TF) binding site involved in epidermis differentiation was predicted. The CAAT-box, which is critical for transcription, was found in the 5'-flanking region of the promoter. Other putative transcription factors are shown in Fig. 4. These putative transcription factor binding sites imply that *AccVg* may be subject to developmental regulation.

Developmental and tissue expression profiles of *AccVg*

Vg expression has been discovered not only in adults but also in younger stages and late pupae in *A. mellifera* [13, 23, 28-30]. Thus, we were interested in studying whether *Vg* is expressed in the developmental stages (larvae, pupae, adults) and in various tissues of

```
AAAGCGGGGTACAGGATGTAAATATCTAGAATTCAAAGTTTAACATTGAAAATATCATTGCCGTGAACT           -739
     Ovo homolog-like transcription factor              Ovo homolog-like transcription factor
CATGACACTCGTGTTATTTTATATATCATATTTTTTAATCAATCGTAATATTTACGAATGAAAATCTCG            -670
                                          CAAT box        Ovo homolog-like transcription factor
           Cdx A
TGTTTACTTGCATTTATATTAACGATAAATTTTTTTATATCATATTCATCATATTGTATCATTATCAC            -601
                NIT 2           CAAT box
ATTACGAATATTTCGATGATTTTGTATCAAGATACTGATCAATGTATCAGAGAGATTTTTTTTTTTTA            -532
ATTATTAAAAGTAATCATAAAATTAAAATAAAATCTTAATTGTATCTTAATCATAGTAATTATTAAAAA           -463
TTATAAAATATTTTAAATTAAAATCTAGGGGTGAAAATTCAAAGTGGTAAGCAAAGATTAACTTTAAGA           -394
TTAAAAATGCAACTTTTAAACTATTTGCTAAAAATAAATTTTTCACTTTTTTATGTAAAATTATGTTAT           -325
                                                  CAAT box     HSF
GTAATTTTATACTTCTCGATTATATATCGACCTTGTAATCGCATCTTAAAAATCAATTTAAAGAAAACA           -256
             CAAT box                  Cdx A
GATATTATTATCACGCATTCGAAGACAATAATATTGAATCATTTATAGTTACATGTTTTTTCAAAAGA            -187
                                            HSF
TATTATTATTAGCTGTTGTTACGCCAAACACAGATTATAGACATGCTAGAAACGAGCGGAGAGATAAAA           -118
              BR-C                    CAAT box       HSF
TAAAAAAGATAAAATAAAAAAAAATAGACATAATAACGTGTTTCAATCTAGCAAAAGAGAAAAATGCTG            -49
                                            +1
ATTTAGCATAAATAAGATATAAATTGCTAAGCCGAATCAAATGCATCGTTACTTTCTCGAAAGTTGTCT           +21
                                                   └─► Transcription start
TCAACATGTTGCT                                                                    +34
     └─► Translation start
```

Fig. 4 The characterization of 5'-flanking region in *AccVg*. Predicted transcription factor binding sites are boxed. The transcription start site and translation start site are marked with an arrowhead.

A. cerana cerana, by using total RNA for qPCR analysis and protein isolates in Western blotting. In workers, *AccVg* mRNA expression slowly rose until reaching a maximum at the third-instar larval, decreased in prepupae, became practically undetectable in pupae, and increased again in newly emerged workers (Fig. 5A) ($F = 99.16$; $P < 0.05$). The expression of *AccVg* in different tissues is shown in Fig. 5B. In terms of various tissues, the highest amount of *AccVg* was detected in the abdominal epidermis because the epidermis contains more fat body cells that are known to produce abundant vitellogenin, more than other tissues. The lowest quantity transcript was detected in the brain ($F = 101.13$; $P < 0.05$).

Expression patterns of *AccVg* under different environmental stresses

Previous studies have indicated that the expression of *Vg* could be induced by abiotic stress. To better understand whether *AccVg* is involved in various abiotic stress responses, the relative expression of *AccVg* under cold (4℃), heat (42℃), heavy metal and insecticides (cyfluthrin, phoxim, paraquat) exposure was determined by qPCR. Following the cold (4℃) treatment, the transcription level of *AccVg* was slightly elevated (Fig. 6A) ($F = 6.10$; $P < 0.05$), but a significant alteration of *AccVg* expression was observed after the heat (42℃) treatment (Fig. 6B) ($F = 155.88$; $P < 0.05$). As shown in Fig. 6C, expression of *AccVg* was enhanced by heavy metal ($CdCl_2$) treatment and reached the highest transcription level at 9 h of exposure ($F = 42.39$; $P < 0.05$). With regard to the insecticide treatments, *AccVg* was induced by cyfluthrin, phoxim, and paraquat, and reached the highest transcription level at 4, 2 and 2 h, respectively. The results shown are relative to control levels at time 0, nonetheless, we also checked the levels of *AccVg* expression in control bees during the duration of the treatment period to make sure that there was no change other than the one induced by the treatments. These results suggested that *AccVg* may be involved in abiotic stress response.

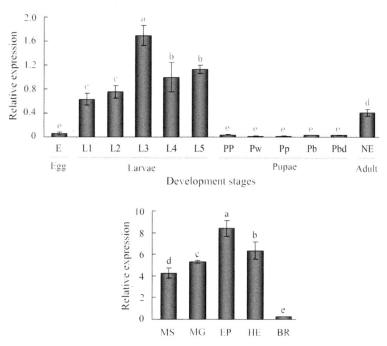

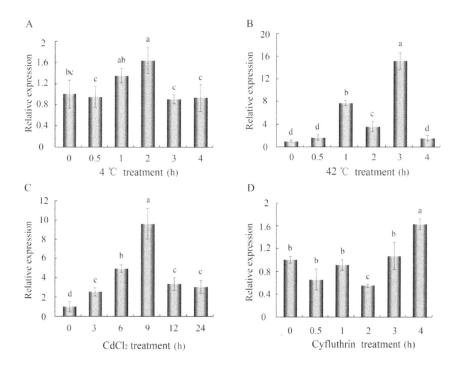

Fig. 5 *Vg* expression during different developmental stages and in tissues of *A. cerana cerana* workers. A. Quantitative RT-PCR analysis of *Vg*. L1, L2, L3, L4, L5=larval stage; PP, Pw, Pp, Pb, Pbd=pupal phase (PP, prepupae; Pw, white-eyed pupae, unpigmented cuticle, Pp, pink-eyed pupae, unpigmented cuticle; Pb, brown-eyed pupae, unpigmented cuticle; Pbd, brown-eyed pupae, dark pigmented cuticle), and adults (NE: newly emerged workers) according to Michelette and Soares [31]. B. *AccVg* expression profiles in the brain (BR), epidermis (EP), muscle (MS), hemolymph (HP), midgut (MG).

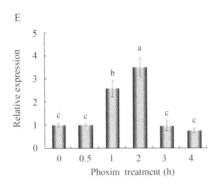 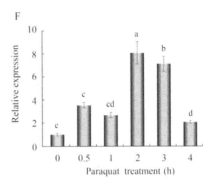

Fig. 6 The transcription levels of *AccVg* under different stress condition. These conditions include cold (4℃) (A), heat (42℃) (B), CdCl$_2$ (C), cyfluthrin (D), phoxim (E), paraquat (F). The RT-qPCR results for the treatments were normalized to bees without any stress treatment (0 h) for (A-F). The different letters above the columns indicate significant differences ($P < 0.05$) according to Duncan's multiple range tests.

Western blotting analysis

To gain a comprehensive understanding of the molecular mechanisms underlying the effects of abiotic stress, the protein levels of AccVg were examined by Western blotting analysis. An anti-AccVg antibody was generated and used to detect AccVg. As shown in Fig. 7, when workers were exposed to heat (42℃), CdCl$_2$, phoxim or paraquat, the expression of AccVg was significantly increased. In terms of mRNA, the high mRNA levels seen in figure correspond to low Vg amounts detected in the Western blotting, which suggests the existence of mechanisms operating on the translatability and/or stability of *Vg* mRNA.

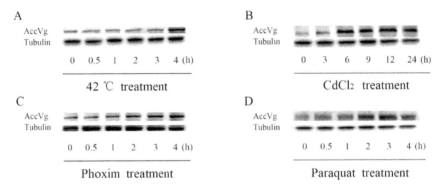

Fig. 7 Western blotting analysis of Vg in different treatments of *A. cerana cerana* workers. SDS-PAGE (7.5%) was performed with 80 ng of total proteins from workers. Images are representative of three replicates.

Discussion

Previous studies demonstrated that Vg plays an important role in energy metabolism and contributes to growth and development in invertebrates. A growing number of studies have been conducted on the roles of Vg in response to adverse stress. However, no studies had systematically imitated the variety of stresses that animals may experience during their life. In this study, the honey bees were foragers obtained from the comb, which were always subjected to the outside environment. Several stresses that honey bees may encounter during their lifespan were imitated, and the resulting changes were explored.

The phylogenetic tree constructed in this study indicated that the Vg amino acid sequence has been conserved throughout evolution and that AccVg is in the typical Vg cluster, sharing the highest degree of homology with Vg of *A. mellifera*. Taken together, these results indicate that the gene we isolated is a bona fide *Vg* homologue.

The timing and level of gene expression are determined by the promoter region, both of which depended on the role the gene plays. To predict the putative roles of *AccVg*, an 807 bp 5′-flanking region was cloned and putative binding sites for transcription factors were predicted. The region contained a large number of putative binding sites for transcription factors involved in development and the response to environmental stress, such as a BR-C binding site and an ovo homolog-like TF binding site, that are involved in the early stages of tissue development and growth. This result prompted us to study the putative roles of *AccVg*. To achieve this goal, we analyzed the transcription levels of the gene encoding *AccVg* during all development stages, in different tissues and under various abiotic stresses. Stage-specific expression analysis indicated that the *AccVg* gene is expressed throughout the life cycle of bees and that the highest accumulation of *AccVg* was detected in the third larval instar (Fig. 5A). *AccVg* expression was still very high in adults, although lower than in the third-instar larval. At the egg and pupal stages, *AccVg* expression was the lowest. This finding suggests that during the larval stages, the function of *AccVg* is to control the rate of development, whereas during the pupal stages, the expression of *AccVg* decreased. The stage-specific expression of *AccVg* indicated that *AccVg* might play roles when the bees were exposed to adverse stress.

Additionally, the mRNA level of *AccVg* was also characterized after various abiotic stress treatments (Fig. 6). When the honeybees were treated with $CdCl_2$, *AccVg* expression was slightly increased and reached the highest level at 9 h; extension of the treatment time resulted in decreased expression. Many heavy metals, including $CdCl_2$ [32], can induce the formation of endogenous ROS, which include superoxide anions, hydrogen peroxide, and hydroxyl radicals. Endogenously generated ROS should be kept in balance, as high ROS concentration can damage DNA, protein, and lipids [33]. In this study, *AccVg* expression increased rapidly, peaking at 9 h, and then decreased in response to heavy metal exposure. The increased *AccVg* expression following $CdCl_2$ exposure could enhance the tolerance of *Apis cerana cerana* to oxidative stress. Considering our result, we hypothesized that heavy metal (such as $CdCl_2$) could cause stress and induce the expression of *AccVg* through distinct pathways.

A large number of putative HSF binding sites was identified in the *AccVg* 5′-flanking region that are likely to participate in heat-induced transcriptional activation. Western blotting analysis of AccVg after 42 °C treatment confirmed the result, suggesting that *AccVg* might play a role in response to extreme temperature. Pesticides were the main threat to a honeybee's life. Pesticide treatments, which induce the formation of reactive oxygen species, resulted in an increase in the expression of *AccVg* (Fig. 6) [34]. Bus et al. [35] noted that paraquat induced oxidative stress by redox cycling with various cellular diaphorases and oxygen to produce superoxide radicals. Our data suggest that Vg activity protects the bee against oxidative damage induced by pesticides. *AccVg* is preferentially oxidized, which is a property that is indicative of an antioxidant function [20, 36, 37]. With regard to protein expression, the Western blotting results also indicated the response of AccVg to various oxidative stresses (Fig. 7). All the results demonstrated that *AccVg* would be induced and activated after abiotic stress, which indicated that AccVg quite possibly plays a significant role in the response of *A. cerana cerana* to adverse environmental stresses. Though some of these stressors also cause increases of gene expression in *A. cerana cerana*, the transcription levels and patterns are quite different from those of the AccVg protein levels.

In conclusion, we identified the *Vg* gene in *A. cerana cerana*, cloned the 5'-flanking region and predicted putative binding sites for transcription factors. Next, the patterns of *Vg* expression during different developmental stages and under different adverse stresses were examined at the transcriptional level. Then, the expression patterns at the protein level were examined to confirm these results. All of these results raise the possibility that *AccVg* plays a role in the response to adverse stress. To further understand the characteristics of *AccVg*, additional studies that directly address the functional significance of this gene and its products should be carried out.

References

[1] Winston, M. The Biology of the Honey Bee. Harvard University Press, United States. 1987, 35 (35), 316-318.

[2] Lushchak, V. I. Environmentally induced oxidative stress in aquatic animals. Aquat Toxicol. 2011, 101, 13-30.

[3] Kottuparambil, S., Shin, W., Brown, M. T., Han, T. UV-B affects photosynthesis, ROS production and motility of the freshwater flagellate, *Euglena agilis* Carter. Aquat Toxicol. 2012, 122, 206-213.

[4] Yang, L. H., Huang, H., Wang, J. J. Antioxidant responses of citrus red mite, *Panonychus citri* (McGregor) (Acari: Tetranychidae), exposed to thermal stress. J Insect Physiol. 2010, 56, 1871-1876.

[5] Dröge, W. Free radicals in the physiological control of cell function. Physiol Rev. 2002, 82, 7-95.

[6] Balaban, R. S., Nemoto, S., Finkel, T. Mitochondria, oxidants, and aging. Cell. 2005, 120, 483-495.

[7] Hayward, A., Takahashi, T., Bendena, W. G., Tobe, S. S., Hui, J. H. Comparative genomic and phylogenetic analysis of vitellogenin and other large lipid transfer proteins in metazoans. Febs Lett. 2010, 584, 1273-1278.

[8] Spieth, J., Nettleton, M., Zucker-Aprison, E., Lea, K., Blumenthal, T. Vitellogenin motifs conserved in nematodes and vertebrates. J Mol Evol. 1991, 32, 429-438.

[9] Tufail, M., Takeda, M. Molecular characteristics of insect vitellogenins. J Insect Physiol. 2008, 54, 1447-1458.

[10] Tseng, D. Y., Chen, Y. N., Kou, G. H., Lo, C. F., Kuo, C. M. Hepatopancreas is the extraovarian site of vitellogenin synthesis in black tiger shrimp, *Penaeus monodon*. Comp Biochem Physiol A: Mol Integr Physiol. 2001, 129, 1909-1917.

[11] Wheeler, D. E., Kawooya, J. K. Purification and characterization of honey bee vitellogenin. Arch Insect Biochem. 1990, 14, 253-267.

[12] Fleig, R. Role of the follicle cells for yolk uptake in ovarian follicles of the honey bee *Apis mellifera* L. (Hymenoptera: Apidae). Int J Insect Morphol. 1995, 24 (95), 427-433.

[13] Hartfelder, K., Engels, W. Social insect polymorphism: hormonal regulation of plasticity in development and reproduction in the honeybee. Curr Top Dev Biol. 1998, 40, 45-77.

[14] Amdam, G. V., Simões, Z. L., Hagen, A., Norberg, K., Schrøder, K., Mikkelsen, Ø., Kirkwood, T. B., Omholt, S. W. Hormonal control of the yolk precursor vitellogenin regulates immune function and longevity in honeybees. Exp Gerontol. 2004, 39, 767-773.

[15] Seehuus, S. C., Norberg, K., Gimsa, U., Krekling, T., Amdam, G. V. Reproductive protein protects functionally sterile honey bee workers from oxidative stress. Proc Natl Acad Sci USA. 2006, 103, 962-967.

[16] Corona, M., Velarde, R. A., Remolina, S., Moran-Lauter, A., Wang, Y., Hughes, K. A., Robinson, G. E. Vitellogenin, juvenile hormone, insulin signaling, and queen honey bee longevity. Proc Natl Acad Sci USA. 2007, 104, 7128-7133.

[17] Trenczek, T., Zillikens, A., Engels, W. Developmental patterns of vitellogenin haemolymph titre and rate of synthesis in adult drone honey bees (*Apis mellifera*). J Insect Physiol. 1989, 35, 475-481.

[18] Amdam, G. V., Omholt, S. W. The regulatory anatomy of honeybee lifespan. J Theor Biol. 2002, 216, 209-228.

[19] Goto, S., Nakamura, A., Radak, Z., Nakamoto, H., Takahashi, R., Yasuda, K., Sakuraiac, Y., Ishiic, N. Carbonylated proteins in aging and exercise: immunoblot approaches. Mech Ageing Dev. 1999, 107, 245-253.

[20] Quinlan, G. J., Martin, G. S., Evans, T. W. Albumin: biochemical properties and therapeutic potential. Hepatology. 2005, 41, 1211-1219.

[21] Pham-Delègue, M. H., Girard, C., Métayer, M. L. Long-term effects of soybean protease inhibitors on digestive enzymes, survival and learning abilities of honeybees. Entomol Exp Appl. 2000, 95 (1), 21-29.

[22] Bitondi, M., Simoes, Z. P. The relationship between level of pollen in the diet, vitellogenin and juvenile hormone titres in Africanized *Apis mellifera* workers. J Apicult Res. 1996, 35, 27-36.

[23] Guidugli, K. R., Piulachs, M., Bellés, X. Vitellogenin expression in queen ovaries and in larvae of both sexes of *Apis mellifera*. Arch Insect Biochem. 2005, 59 (4), 211-218.

[24] Livak, K. J., Schmittgen T. D. Analysis of relative gene expression data using real-time quantitative PCR and the 2(−Delta Delta C (T)) Method. J Virol Methods. 2001, 25 (4), 402-408.

[25] Tang-Feldman, Y. J., Wojtowicz, A., Lochhead, G. R. Use of quantitative real-time PCR (qRT-PCR) to measure cytokine transcription and viral load in murine cytomegalovirus infection. J Virol Methods. 2006, 131 (2), 122-129.

[26] Ding, Y., Hawkes, N., Meredith, J., Eggleston, P., Hemingway, J., Ranson, H. Characterization of the promoters of Epsilon glutathione transferases in the mosquito *Anopheles gambiae* and their response to oxidative stress. Biochem J. 2005, 387, 879-888.

[27] Zhang, Y., Yan, H., Lu, W., Li, Y., Guo, X., Xu, B. A novel Omega-class glutathione S-transferase gene in *Apis cerana cerana*: molecular characterisation of GSTO2 and its protective effects in oxidative stress. Cell Stress Chaperon. 2013, 18, 503-516.

[28] Engels, W., Kaatz, H., Zillikens, A. Honey bee reproduction: vitellogenin and caste-specific regulation of fertility. Advances in Invertebrate Reproduction 1990, 5, 495-502.

[29] Barchuk, A. R., Bitondi, M. M., Simões, Z. L. Effects of juvenile hormone and ecdysone on the timing of vitellogenin appearance in hemolymph of queen and worker pupae of *Apis mellifera*. J Insect Sci. 2002, 2 (1), 1.

[30] Piulachs, M. D., Guidugli, K. R., Barchuk, A. R. The vitellogenin of the honey bee, *Apis mellifera*: structural analysis of the cDNA and expression studies. Insect Biochem Mole. 2003, 33 (4), 459-465.

[31] Michelette, E. R. D. F., Soares, A. E. E. Characterization of preimaginal developmental stages in Africanized honey bee workers (*Apis mellifera* L). Apidologie. 1993, 24 (4), 431-440.

[32] Liu, J., Qu, W., Kadiiska, M. B. Role of oxidative stress in cadmium toxicity and carcinogenesis. Toxicol Appl Pharm. 2009, 238, 209-214.

[33] Narendra, M., Bhatracharyulu, N., Padmavathi, P., Varadacharyulu, N. Prallethrin induced biochemical changes in erythrocyte membrane and red cell osmotic haemolysis in human volunteers. Chemosphere. 2007, 67, 1065-1071.

[34] Day, B. J., Patel, M., Calavetta, L., Chang, L. Y., Stamler, J. S. A mechanism of paraquat toxicity involving nitric oxide synthase. Proc Natl Acad Sci USA. 1999, 96, 12760-12765.

[35] Bus, J. S., Aust, S. D., Gibson, J. E. Superoxide- and singlet oxygen-catalyzed lipid peroxidation as a possible mechanism for paraquat (methyl viologen) toxicity. Biochem Biophys Res Commun. 1974, 58, 749-755.

[36] Salo, D. C., Lin, S. W., Pacifici, R. E. Superoxide dismutase is preferentially degraded by a proteolytic system from red blood cells following oxidative modification by hydrogen peroxide. Free Radical Bio Med. 1998, 5 (5-6), 335-339.

[37] Andrus, P. K., Fleck, T. J., Gurney, M. E., Hall, E. D. Protein oxidative damage in a transgenic mouse model of familial amyotrophic lateral sclerosis. J Neurochem. 1998, 71, 2041-2048.

4

The Molecular Mechanisms of Growth and Development of *Apis cerana cerana*

The Chinese honeybee (*Apis cerana cerana*) is a eusocial insect and has an important role in the pollination of flowering crops at low temperatures. Global warming, excessive pesticide use and competition with western honeybees are causing *A. cerana cerana* to become an endangered species. Organisms undergo multiple morphological changes, including organ formation, throughout the process of development, and these developmental changes are regulated by changes in gene expression.

In the past 10 years of research, we have isolated more than 50 genes from Chinese honeybees. The results show that most genes and their encoded proteins are evolutionarily conserved, but there are still differences. Structural conservatism determines the similarity of gene function, and structural differences may be caused by the diversity of species in the process of evolution. In addition, by analyzing the characteristics of growth and tissue distribution, we found that some genes play an important role in the growth and development of bees.

The cuticle is the protective shield of insects and is constructed of many cuticle proteins and other components, such as chitin filaments, which have different temporal and spatial patterns. The insect cuticle is renewed during each molting cycle when the new cuticle is synthesized by the underlying epidermis and the old cuticle is shed. We isolated and characterized a CPR gene from *A. cerana cerana*. In this study, we evaluated the expression of this gene in different developmental stages and tissues and investigated the changes in the *AccCPR1* expression patterns in response to ecdysteroid. The results showed that the *AccCPR1* transcript was present predominantly in the head, thoracic and abdominal epidermis in 1-day-old adult workers. The *AccCPR1* transcript was present as the ecdysteroid titer decreased after reaching a peak. When larvae were reared with different concentrations of 20E, the mRNA expression of *AccCPR1* was repressed. Exposure of the thoracic integument of the pupae to different concentration of 20E also repressed *AccCPR1* expression, which recovered after the removal of 20E. Our study contributes to the knowledge regarding the roles of typical CPR genes from insects in cuticle protein gene structure and regulation and suggests that the *AccCPR1* gene might be involved in the ecdysteroid response.

TGF-β activated kinase-1 (TAK1) is essential in the regulation of organism growth and differentiation and plays a pivotal role in developmental processes in many species. A number of studies have demonstrated that mutation or deletion of the *TAK1* gene causes adverse effects on development. Recently, it has been demonstrated that *TAK1* acts as an essential regulator in cartilage morphogenesis, vascular development, keratinocyte apoptosis and cell shape control during embryonic development. In this study, we characterized the *TAK1* gene in *A. cerana cerana*, analyzed mRNA expression after treatment with toxic substances, and assessed protein localization in larvae. Promoter analysis of *AccTAK1* revealed the presence of transcription factor binding sites related to early development. Real-time quantitative PCR and immunohistochemical experiments revealed that *AccTAK1* was expressed at high levels in the fourth-instar larvae, primarily in the abdomen, in the intestinal wall cells of the midgut and in the secretory cells of the salivary glands. In addition, *AccTAK1* expression in the fourth-instar larvae could be dramatically induced by treatment with pesticides and organic solvents. These observations suggest that *AccTAK1* may be involved in the regulation of early development in the larval salivary gland and midgut.

The honeybee is a highly eusocial insect that exhibits several striking social behaviors including age-related labor division, symbolic dance language and a precise learning system. These complex social behaviors should be associated with certain coordinated systems in the brain. Ribosomal proteins (RPs) play pivotal roles in developmental regulation. The loss or mutation of ribosomal protein L11 (RPL11) induces various developmental defects. We isolated a single-copy gene, *AccRPL11*, and characterized its connection to brain maturation. *AccRPL11* expression was highly concentrated in the adult brain and significantly induced by abiotic stresses such as pesticides and heavy metal. Immunofluorescence assays demonstrated that *AccRPL11* was localized to the medulla, lobula and surrounding tissues of the esophagus in the brain. The post-transcriptional knockdown of *AccRPL11* gene expression resulted in a severe decrease in the expression levels of other brain development-related genes, namely, *p38*, *ERK2*, *CacyBP* and *CREB*. Immunofluorescence signal attenuation was also observed in *AccRPL11*-rich regions of the brain in dsAccRPL11-injected honeybees. These results suggested that the *AccRPL11* gene may have a function in brain maturation during the adult phase.

Similar to vertebrate adipose tissue and liver, insect lipid droplets (LDs) are prominent cellular organelles participating in a number of biological processes, such as energy metabolism and nutrient storage, by mediating cellular lipolysis. PAT family proteins are evolutionary highly conserved lipid droplet-associated proteins that are involved in lipid metabolism and transport, intracellular trafficking and signaling. Lipid storage droplet 1 (LSD-1), a PAT family protein located around lipid droplets in insects, is intimately linked to lipid droplet formation and lipid metabolism. *AccLSD-1* was expressed ubiquitously from feeding larval to adult stages, and its expression level was the highest at the brown-eyed pupae (Pb) stage. Sequence analysis indicated that the central region of the LSD-1 protein has significant sequence similarity and that a typical *LSD-1* gene is composed of 8 exons and 7 introns. The first intron of *AccLSD-1*, including several PPARγ-response elements (PPREs), is located in the 5' UTR. The effect of CLA, Rosi and the combination on *AccLSD-1* expression indicated that 1% CLA and 0.5 mg/ml Rosi were a suitable diets for rearing adult workers in the laboratory. *AccLSD-1* was downregulated by CLA but upregulated by Rosi. Furthermore, the combination of CLA and Rosi remarkably rescued the suppression of *AccLSD-1* expression by CLA alone. These results suggested that *AccLSD-1* is associated with *A. cerana cerana* development, especially during pupal metamorphosis, and could be regulated by CLA or Rosi possibly by activating PPARγ.

4.1 Characterization of the Response to Ecdysteroid of a Novel Cuticle Protein R&R Gene in the Honeybee, *Apis cerana cerana*

Rujiang Sun, Yuanying Zhang, Baohua Xu

Abstract: Genes encoding cuticle proteins are helpful subjects to study the molecular mechanisms of insect molting and metamorphosis. In this study, we isolated and characterized a novel cuticle protein R&R gene, referred to as *AccCPR1*, from *Apis cerana cerana*. The open reading frame of *AccCPR1* has a length of 573 nt and encodes a protein of 190 amino acids that contains a chitin binding region and is a typical cuticle R&R-2 protein. Five putative E74 binding sites and four BR-C binding sites were predicted in the 5'-flanking region, which suggests a potential function in molting and metamorphosis. RT-qPCR showed that *AccCPR1* transcript occurred as the ecdysteroid titer decreased after reaching a peak, which suggests *AccCPR1* expression requires a "pulse" regimen of ecdysteroids. This hypothesis was tested by using different experimental strategies. When larvae were reared with different concentrations of 20E in their diet, the ecdysteroid peak repressed *AccCPR1* expression. Exposure of the thoracic integument of the pupae *in vitro* to different concentration of 20E repressed *AccCPR1* expression, which recovered after the removal of 20E. These results suggest that *AccCPR1* is a typical cuticle R&R-2 protein that plays an important role in the development, and an ecdysteroid pulse is critical for high *AccCPR1* gene expression.

Keywords: *AccCPR1* gene; *Apis cerana cerana*; RT-qPCR; Ecdysteroids; 20E

Introduction

The cuticle is the protective shield of insects and is constructed of many cuticle proteins and other components, such as chitin filaments, which have different temporal and spatial patterns [1]. The main components of the cuticle are chitin fibers embedded in a matrix of various proteins [2]. Although chitin is a simple polysaccharide, it is present in cuticle in various percent, and it is suggested that these differences are partly responsible for the properties of the individual proteins and the degree of secondary sclerotization [3, 4].

However, even with this variety, some common characteristics can be found in cuticle proteins. The CPR family (named after the Rebers and Riddiford Consensus), which contains the R&R consensus sequence first recognized by Rebers and Riddiford [5], is by far the largest CP (cuticular proteins) family and is found in every examined species of arthropod [6]. The extended version of the R&R consensus sequence is recognized as pfam00379 in the Pfam database, and the pfam motif is responsible for chitin binding in the arthropod cuticle [7, 8]. This extended, conserved R&R consensus sequence most likely interacts with chitin through a β-sheet structure [9]. Three distinct forms of the extended R&R consensus sequence have been recognized and named by Andersen [10, 11]: RR-1, RR-2 and RR-3. RR-1 proteins have been isolated from soft (flexible) cuticle, whereas RR-2 proteins have been associated with hard (rigid) cuticle [10]. Very few cuticular protein sequences contain RR-3 [11].

The insect cuticle is renewed during each molting cycle when the new cuticle is synthesized by the underlying epidermis and the old cuticle is shed. The bulk of chitinous cuticle (procuticle) is deposited while the titers of ecdysteroids are declining during a molting cycle, which suggests a particular mode of regulation [12]. Several studies have shown that some cuticle protein genes expression may increase at different molting stages whereas titers of ecdysteroids decrease after reaching high levels, i.e., presented as a pulse [1, 12, 13]. Moreover, a few cuticle protein genes are induced by the continuous presence of moderate levels of 20E, which suggests that some cuticle genes are not regulated by a typical pulse of 20E [14, 15]. Therefore, cuticular genes, which are regulated by ecdysteroids in a stage- and tissue-specific manner, should be perfect subjects to study the mechanism of insect molting and metamorphosis during insect development. Recently, several insect whole genome sequences have been made available. Over 700 putative cuticle protein genes in different CP families in the whole genome data of six species were obtained because of their similarity to a small number of authentic CP sequences [6]. There have been numerous studies concerning CPs in Diptera, Orthoptera, Lepidoptera, Coleoptera and Hemiptera. The CPs of different families share different common motifs and features in conserved regions [6] and have different developmental expression patterns [16].

The Chinese honeybee, *Apis cerana cerana*, is an important indigenous species that plays an essential role in the balance of regional ecologies and agricultural economic development as the pollinator of flowering plants [17]. To date, no CP gene has been identified in the Chinese honeybee. In this study, we isolated and characterized a CPR gene, described as *AccCPR1*, from *A. cerana cerana*, and evaluated its expression in different developmental stage and tissues. We investigated the changes in the *AccCPR1* expression patterns of the response to ecdysteroid. Our study contributes to the knowledge regarding typical *CPR* gene from insects on cuticle protein gene structure and regulation. Our findings suggest that the *AccCPR1* gene may be involved in the ecdysteroid response.

Materials and Methods

Bees and treatments with 20E

The larvae, pupae and adults of Chinese honeybees (*A. cerana cerana*), which were maintained in artificial beehives at the experimental apiary of Shandong Agricultural University, China, were identified according to the criteria established by Michelette and

Soares [18]. They were collected at the fourth (L4)- and the fifth (L5)-instar larvae; pharate pupae (PP) stage; successive pharate adult stages, including the following phases that show white (Pw), pink (Pp), dark pink (Pdp) and brown (Pb) eyes, and then the light (Pbl)- and dark (Pbd)-pigmented pharate adult stages. Adult workers of known age were obtained by marking newly emerged bees with paint and collecting them after 1, 10 and 30 days. Larvae, pupae and adults were dissected, adhering muscle and fat body were removed as much as possible and entire integument was used in individual samples for each developmental period. Various tissues from 1-day-old adult workers, including the head, thoracic and abdominal epidermis, muscle, midgut and Malpighian tubules, were freshly collected.

A. cerana cerana larvae were collectively reared in 24-well plates (Costar, Corning Incorporated, USA) and placed into a desiccator with a constant temperature (34℃) and constant humidity (70%) that was maintained by using a 10% glycerol solution inside the desiccator. The larvae were fed according to the method established by Silva et al. [19]. Once the larvae reached their thirdinstar, they were randomly transferred to artificial diets containing different concentrations of 20E: 1, 0.1, 0.01 and 0.001 μg/ml of the diet. One 24-well plate was filled with 48 larvae (two larvae/well); all wells in a plate received the same concentration of 20E. Ethanol, which was the 20E solvent, was used in the control plates. The final concentration of ethanol was 0.1 mg/ml. The larvae that were still alive after 96 h of feeding on the treatment diets were collected in groups.

Based on the experimental design of Soares et al. [20], the thoracic integument of newly-ecdysed pupae (Pw phase) was incubated in 5 ml of commercially available Grace's culture medium (Invitrogen, Carlsbad, CA, USA), which is prepared specifically for insect tissues. The integument was incubated in the presence of 1, 0.1 or 0.001 μg 20E/ml culture medium for a period of 3, 6 or 18 h. During this process, some of the 6 h-incubated integuments were washed for 6 or 12 h in hormone-free medium. Ethanol, which was the 20E solvent, was used in the control incubations. All honeybees in this study were frozen in liquid nitrogen at the indicated time points and stored at −80℃ for RNA extraction.

Primers

The sequences of the primers used for cloning in this study are listed in Table 1. The amplification conditions and products are shown in Table 2.

Table 1 Primers used in this study

Abbreviation	Primer sequence (5′ → 3′)	Description
MP1	CTTTCATTATCAAGCAGGACG	cDNA sequence primer, forward
MP2	GTTCCTCGGCGATTTTTTG	cDNA sequence primer, reverse
5P1	CGTCCTGCTTGATAATGAAAG	5′ RACE reverse primer, outer
5P2	TAAGAACCTCGAATATGTCCAG	5′ RACE reverse primer, inner
AAP	GGCCACGCGTCGACTAGTAC(G)$_{14}$	Abridged anchor primer
AUAP	GGCCACGCGTCGACTAGTAC	Abridged universal amplification primer
3P1	TCCAATCAATCTTCCACAAG	3′ RACE forward primer, outer
3P2	GAGAGAGCAAAGGACAAAC	3′ RACE forward primer, inner

Continued

Abbreviation	Primer sequence (5' → 3')	Description
B26	GACTCTAGACGACATCGA(T)$_{18}$	3' RACE universal primer, outer
B25	GACTCTAGACGACATCGA	3' RACE universal primer, inner
TP1	CGAAAGTATCTACACGAAGAG	Full-length cDNA primer, forward
TP2	CACAAATTAGTACGCATGATCG	Full-length cDNA primer, reverse
TN1	CGAAAGTATCTACACGAAGAG	Genomic sequence primer, forward
TN2	TCCAGTACCATCTCCATTTTC	Genomic sequence primer, reverse
TN3	AGAAAATGGAGATGGTACTGG	Genomic sequence primer, forward
TN4	CACAAATTAGTACGCATGATCG	Genomic sequence primer, reverse
QP1	GAGAGAGCAAAGGACAAAC	Inverse PCR forward primer, outer
QP2	CACGAACGACAAAGAAATTAC	Inverse PCR reverse primer, outer
QP3	GGAAACTGAACGATTAGCC	Inverse PCR forward primer, inner
QP4	CTCTTCGTGTAGATACTTTCG	Inverse PCR reverse primer, inner
QS1	TGACTCTTGATCTATGGCTTC	Promoter special primer, forward
QS2	GAGAGTTTGCCAGTGAAAG	Promoter special primer, reverse
WQP1	ACCACAAATAACAGATGCAGAG	Promoter special primer, forward
WQP2	AGGGCTAATCAGAATAAATGCG	Promoter special primer, reverse
RTP1	CTTTCATTATCAAGCAGGACG	Real-time PCR primer, forward
RTP2	AAGATGCGTGAAATCCAT	Real-time PCR primer, reverse
β-s	GTTTTCCCATCTATCGTCGG	Standard control primer, forward
β-x	TTTTCTCCATATCATCCCAG	Standard control primer, reverse

Table 2 PCR amplification conditions

Primer pairs	PCR amplification conditions
MP1/MP2	94℃ for 5 min, 35 cycles of 94℃ for 40 s, 47℃ for 40 s and 72℃ for 40 s, then 72℃ for 10 min
5P1/AAP	94℃ for 5 min, 32 cycles of 94℃ for 40 s, 46℃ for 40 s and 72℃ for 30 s, then 72℃ for 5 min
5P2/AUAP	94℃ for 5 min, 35 cycles of 94℃ for 40 s, 47℃ for 40 s and 72℃ for 30 s, then 72℃ for 10 min
3P1/B26	94℃ for 5 min, 32 cycles of 94℃ for 40 s, 46℃ for 40 s and 72℃ for 30 s, then 72℃ for 5 min
3P2/B25	94℃ for 5 min, 35 cycles of 94℃ for 40 s, 46℃ for 40 s and 72℃ for 30 s, then 72℃ for 10 min
TP1/TP2	94℃ for 5 min, 35 cycles of 94℃ for 40 s, 47℃ for 40 s and 72℃ for 1 min, then 72℃ for 10 min
TN1/TN2	94℃ for 5 min, 35 cycles of 94℃ for 40 s, 47℃ for 40 s and 72℃ for 1 min, then 72℃ for 10 min
TN3/TN4	94℃ for 5 min, 35 cycles of 94℃ for 40 s, 47℃ for 40 s and 72℃ for 1 min, then 72℃ for 10 min
QP1/QP2	94℃ for 5 min, 32 cycles of 94℃ for 40 s, 45℃ for 40 s and 72℃ for 2 min, then 72℃ for 10 min
QP3/QP4	94℃ for 5 min, 35 cycles of 94℃ for 40 s, 46℃ for 40 s and 72℃ for 2 min, then 72℃ for 10 min
QS1/QS2	94℃ for 5 min, 35 cycles of 94℃ for 40 s, 47℃ for 40 s and 72℃ for 1 min 30 s, then 72℃ for 10 min
WQP1/WQP2	94℃ for 5 min, 35 cycles of 94℃ for 40 s, 47℃ for 40 s and 72℃ for 2 min 30 s, then 72℃ for 10 min

RNA extraction, cDNA synthesis and DNA preparation

Total RNA was extracted from the integument of one-day-old adult workers by using Trizol Reagent (Invitrogen, Carlsbad, CA, USA), according to the protocol of the manufacturer.

Total RNA was digested by RNase-free DNase Ⅰ (Promega, Madison, WI, USA) to remove potential DNA contamination. The first-strand cDNA was synthesized by using a reverse transcriptase system (TransGen Biotech, Beijing, China), and it was purified and polyadenylated as described by Wang et al. [21]. Genomic DNA was prepared from the fifth-instar larvae by using the EasyPure Genomic DNA Extraction Kit (TransGen Biotech) following the protocol of the manufacturer and used as template in PCR reactions.

Isolation of the full-length cDNA of *AccCPR1*

To obtain the internal fragment of cDNA of *AccCPR1*, primers MP1 and MP2 (Table 1) were designed and synthesized (Shanghai Sangon Biotechnological Company, China) from the conserved regions of the CPRs from several insects. To clone the full-length cDNA, the gene-specific primers 5P1/5P2 and 3P1/3P2 (Table 1) were designed and synthesized based on the previously obtained partial *AccCPR1* sequence and used for 5′ rapid amplification of cDNA end (RACE) and 3′ RACE, respectively. For 3′ RACE, the first round of PCR was performed by using the primers 3P1 and B26 (Table 1). The 50-fold diluted PCR product was then used for the second round of amplification with the primers 3P2 and B25 (Table 1). 5′ RACE was performed by using the purified cDNA. The product of the first round, which was performed with the primers 5P1 and AAP (Table 1), was diluted 20-fold and used for the second round of amplification with the primers 5P2 and AUAP (Table 1). Finally, the accurate full-length cDNA sequence was obtained by PCR by using primers TP1 and TP2 (Table 1). These primers were designed and synthesized on the basis of the putative full-length cDNA that was deduced by comparing and aligning the above three fragments by using DNAMAN software 5.2.2 (Lynnon Biosoft, QC, Canada). All PCR amplification conditions are shown in Table 2. All cloned sequences in this study were obtained as follows: the PCR products were purified by using a gel extraction kit (TaKaRa, Japan), ligated to the pEASY-T3 vector (TransGen Biotech), transformed into the *Escherichia coli* strain DH5α and then sequenced.

Amplification of the genomic sequence and the 5′-flanking region of *AccCPR1*

Two pairs of primers (TN1/TN2 and TN3/TN4) were designed and synthesized from the full-length cDNA of *AccCPR1*. Two fragments of 969 and 757 nt were obtained from genomic DNA by a PCR protocol (Table 2). The PCR products were purified, cloned into pMD18-T and then transformed into competent *E. coli* DH5a cells for sequencing.

The total genomic DNA used as the template was digested by restriction endonuclease *Mbo* Ⅰ at 37℃ overnight and then ligated to the vectors by using T4 DNA ligase (TaKaRa, Japan). The first PCR was performed by using primers QP1 and QP2 and the next PCR was performed by using the primers QP3 and QP4. To further verify the validity of the sequence, the specific primers WQP1 and WQP2, which were designed based on the obtained promoter sequence, were directly used to amplify the 5′-flanking region of *AccCPR1* from *A. cerana cerana* genomic DNA. The cloning and sequencing strategies were the same as described above. The primers and reaction conditions are provided in Table 1, Table 2, respectively.

Bioinformatics

Homologous CPR sequences from other insects were retrieved from GenBank at NCBI and aligned by using DNAMAN version 5.2.2 software (Lynnon Biosoft Company, USA). The isoelectric point (pI) and molecular mass (M_W) were computed by using PeptideMass (http://us.expasy.org/tools/peptide-mass.html). Signal peptides in AccCPR1 were detected by using the SignalP 4.0 Server (http://www.cbs.dtu.dk/services/SignalP/). The secondary structure was predicted by using the PSIPRED Protein Structure Prediction Server (http://bioinf.cs.ucl.ac.uk/psipred/). Phylogenetic analysis was performed to check the evolutionary relationships among sequences of various origins by using the neighbor-joining (NJ) method in MEGA version 4.0 software. The MatInspector database (http://www.cbrc.jp/research/db/TFSEARCH.html) was used to search for transcription factor binding sites in the 5′-flanking region.

Real-time qPCR analysis

To quantitatively compare the expression levels of *AccCPR1* mRNA in the experiments, a real-time quantitative RT-PCR assay was performed by using a CFX 96™ Real-time System (Bio-Rad, USA). The method of extracting total RNA and reverse transcription was the same as described above. One set of specific primers, RTP1 and RTP2 (Table 1), for *AccCPR1* and one set of primers, β-s and β-x (Table 1), for the *A. cerana cerana β-actin* gene (GenBank accession number: XM_640276) were designed and synthesized. Real-time qPCR was performed by using SYBR® Premix Ex *Taq* (TaKaRa, Japan) in 25 μl on a CFX 96™ Real-time System (Bio-Rad, USA). The qPCR amplification conditions were as follows: initial denaturation at 95 ℃ for 30 s; 41 cycles at 95 ℃ for 5 s, 53 ℃ for 15 s and 72 ℃ for 15 s; and a single melting cycle from 65 ℃ to 95 ℃. Three independent samples were analyzed, and each sample was analyzed three times. The linear relationship, amplification efficiency, and data analysis were conducted by using CFX Manager Software version 1.1. *AccCPR1* levels were normalized to the corresponding control *β-actin* levels. The $2^{-\Delta\Delta C_t}$ comparative C_t method was used to calculate the relative expression levels of *AccCPR1* [22, 23]. Overall differences in the abundance of *AccCPR1* transcripts were determined by one-way ANOVA analysis, based on a post hoc Tukey test by using Statistical Analysis System (SAS) version 9.1 (Version 8e, SAS Institute, USA).

Results

Identification and molecular properties of the cDNA sequence of *AccCPR1*

A novel *CPR* gene, referred to *AccCPR1*, was identified from the *A. cerana cerana* cuticle by using the RT-PCR and RACE-PCR methods. The full-length cDNA (GenBank accession number: JQ282798), which has a length of 822 nt, contained an open reading frame (ORF) of 573 nt encoding a 190-amino acid polypeptide with a predicted molecular mass of 21.3 kDa and an isoelectric point of 6.50. The non-coding regions consisted of a 51 nt in the 5′

untranslated terminal region (UTR) and 198 nt in the 3′ UTR.

The NCBI conserved domain (CD) search revealed that the chitin_bind_4 domain of the Arthropoda chitin_bind_4 superfamily of CPRs was present in the deduced amino acid sequence of AccCPR1, which suggests that the *AccCPR1* gene belongs to the CPR family. In addition, the extended version of the R&R consensus sequence in the Pfam database was pf00379, chitin_bind_4. Hence, the deduced amino acid sequence of AccCPR1 potentially containing this R&R consensus sequence was submitted to Pfam at http://pfam.sanger.ac.uk/search. We also found that the AccCPR1 protein sequence belonged to the chitin_bind_4 family, which is described as insect cuticle proteins. Moreover, the deduced amino acid sequence of AccCPR1 was recognized in the Prosite database (http://www.expasy.ch/prosite) as a chitin-binding R&R-2 cuticle protein (PS5115), and the alignment from the 35th to the 107th amino acid was a chitin-binding region (Fig. 1A). The analysis of the secondary structure indicated that β-sheets occurred frequently (42.5%) throughout the structure of the region, whereas there were none in other regions, which is consistent with previous reports that stated

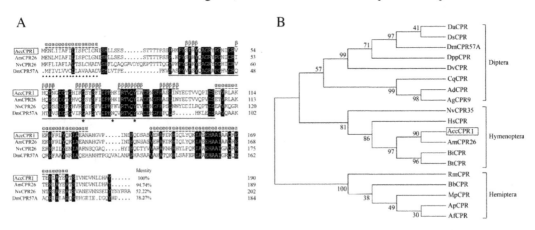

Fig. 1 Multiple amino acid alignment and phylogenetic analysis. A. Comparison of the deduced amino acid sequences of AccCPR1 (GenBank accession number: JQ282798) from *A. cerana cerana*, AmCPR26 (XP 001122183) from *Apis mellifera*, DmCPR57A (NP 611489) from *Drosophila melanogaster* and NvCPR26 (NP 001166278) from *N. vitripennis*. Identical amino acid residues are shaded in blue The degree of identity relative to AccCPR1 is shown at the end of each sequence. The full underline indicates the conserved chitin binding domain in CPR. The amino acids conserved in the strict R&R consensus: G-x8-G-x6-(Y/F) are marked by *. The arrowheads ▲ indicate the signal peptide domain in CPR. The secondary structure assignment of the *AccCPR1* gene is shown above the sequence. B. A phylogenetic tree generated by MEGA version 4.0 by using CPRs from various species, including *A. cerana cerana* (GenBank accession number: JQ282798), *Bombus impatiens* (XP 003487232), *Anopheles darlingi* (EFR19535), *A. mellifera* (XP 001122183), *Drosophila virilis* (XP 002049596), *N. vitripennis* (NP 001166278), *Drosophila ananassae* (XP 001959005), *Drosophila sechellia* (XP 002034700), *Anopheles gambiae* (XP 003436223), *Acyrthosiphon pisum* (NP 001165717), *Rhopalosiphum maidis* (AAO63550), *Aphis fabae* (AAO63551), *Myzus persicae* (AAL29466), *Bombus terrestris* (XP 003399157), *Brevicoryne brassicae* (AAO63548), *Drosophila pseudoobscura pseudoobscura* (XP 001361173), *Culex quinquefasciatus* (XP 001844392), *Drosophila melanogaster* (NP 611489) and *Harpegnathos saltator* (EFN83555).

that the secondary structure of the CPR conserved motif is most likely rich in β-sheets [9, 24]. The strict R&R consensus: G-x8-G-x6-(Y/F) was found in this region. The predicted amino acid sequence of AccCPR1 exhibited 94.74%, 52.22%, and 38.27% similarity to the CPRs from *Apis mellifera* (XP_001122183), *Nasonia vitripennis* (NP_001166278) and *Drosophila melanogaslter* (NP_611489), respectively. Moreover, as shown in Fig. 1A, there were signal peptides at the amino terminal of the CPRs that are involved in protein transfer across the membrane.

To study the evolutionary relationships among different CPRs from various insects, a phylogenetic tree was constructed by using MEGA (version 4.0) software, and all of the amino acid sequences used were obtained from GenBank. As showed in Fig. 1B, the insect CPRs can be classified into 3 clusters. The results indicate that AccCPR1 has a greater level of sequence identity with Hymenoptera CPRs than with Diptera or Hemiptera CPRs.

Cloning and structural analysis of the *AccCPR1* genomic sequence

To further analyze the structure of the *AccCPR1* gene, we also isolated and characterized the genomic DNA sequence. The genomic DNA sequence of *AccCPR1* (GenBank accession number: JQ282799) consisted of 1,833 nt, and four exons and three introns were found (Fig. 2). All of the introns had typical eukaryotic characteristics with high AT content and conserved 5'-GT and AG-3' sequences at 5' and 3' splice sites, respectively. An alignment of exons and introns was then performed by using the CPR sequences from *A. cerana cerana*, *A. mellifera*, *Anopheles gambiae* and *D. mojavensis*. This analysis revealed that *AccCPR1* and *AmCPR26* contained three introns, whereas *AgCPR26* and *DmCPR57A* contained two introns. A signal peptide of 17 amino acids starting at the initiator methionine of the N-terminal amino acid was identified by using Signal P4.0 Server (http://www.cbs.dtu.dk/services/SignalP/). Moreover, a comparison of the genomic revealed that in *AccCPR1* genomic sequence, the first intron interrupts the signal peptide after five coding amino acids, as also observed in *A. mellifera* (NC_007070), *A. gambiae* (NT_078265) and *D. melanogaster* (NP_611489) (Fig. 2).

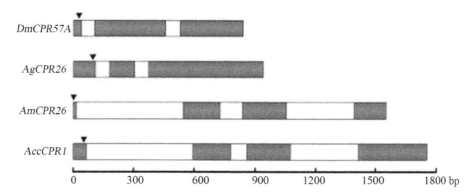

Fig. 2 Schematic representation of cuticle protein R&R (CPR) DNA structures. The lengths of the exons and introns of genomic DNA from *Drosophila mojavensis* (NP 611489), *Anopheles gambiae* (XP 003436223), *Apis mellifera* (XP 001122183) and *Apis cerana cerana* are indicated according to the scale below. Light grey and black are used to highlight introns and exons, respectively. The translational initiation codons (ATG) are marked by triangular symbol.

Characterization of the 5′-flanking region of AccCPR1

To identify regulatory elements upstream of the *AccCPR1* gene, a 1,522 nt DNA fragment upstream of the translation start site was obtained from the *A. cerana cerana* genome by inverse PCR (I-PCR). Several putative transcription factor binding sites in the 5′-flanking region were then predicted by using the web software program MatInspector. Five E74 and four BR-C binding sites were detected. In addition, a TATA-box, which represents the putative core promoter element upstream of the transcription start site, was found (Fig. 3).

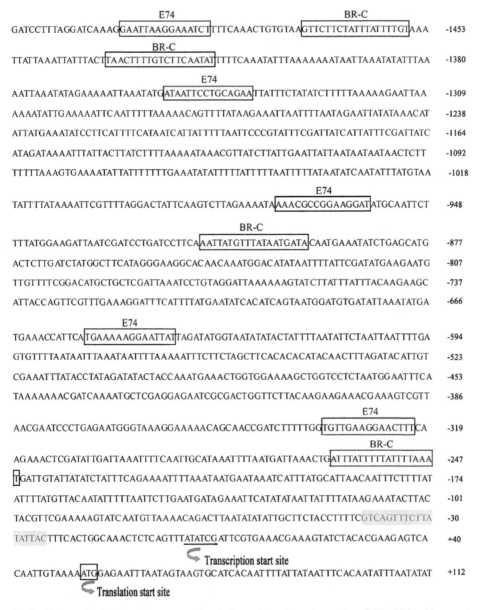

Fig. 3 Partial nucleotide sequences and putative *cis*-acting elements of the 5′-flanking region of *AccCPR1*. The transcription factor binding sites mentioned in the text are boxed. The transcription start site and translation start site are both indicated with arrowheads.

Expression of *AccCPR1* during different developmental stages and in different tissues

To determine the expression of *AccCPR1* at specific developmental stages, we first obtained a general overview of the expression pattern of *AccCPR1* in various developmental stages of worker bees by real-time PCR (qPCR). As shown in Fig. 4A, the expression of *AccCPR1* markedly differed among different developmental stages. The expression of *AccCPR1* was very low in the fourth-instar larval phase (L4) and then increased sharply during the fifth-instar larval (L5) and prepupal (PP) phases. The expression dramatically decreased during early pupal phases, and reached a very low level at the pink-eye pupal phase (Pp), which coincided with the onset of pharate adult development. Subsequently, the amount of *AccCPR1* transcripts showed a steep increase again and reached a relatively high level during the intermediate and later pharate adults (Pdp, Pb, Pbl, Pbm and Pbd) and in newly emerged (1 d) adults, which was in accordance with the deposition of the cuticle. The expression of *AccCPR1* then decreased to very low levels gradually in nurse workers (10 days old) and foragers (30 days old).

We also compared the developmental profile of *AccCPR1* expression with the hemolymph ecdysteroid titer (see in Fig. 4A). The transition from a low to a high level of *AccCPR1* transcripts (from the L5 to PP phases and from Pp pupae to pharate adults) occurred as the ecdysteroid titer was decreasing after a peak that triggered the onset of the pharate pupal and pharate adult stages.

To investigate whether the expression of *AccCPR1* occurred only in the epidermis, various tissues were extracted from 1-day-old adult workers. Tissue-specific expression analysis indicated that a high amount of transcript was present predominantly in the head and thoracic and abdominal epidermis (Fig. 4B). The transcription levels were moderate in the midgut and very low in the muscle and Malpighian tubules. These results indicate that *AccCPR1* may function in the development and differentiation of the cuticle and gut lining.

Effect of 20E on *AccCPR1* expression

Based on the hypothesis that *AccCPR1* gene expression is regulated by the titer of ecdysteroids, we reared the larvae with different concentrations of 20E to analyze the effect of 20E on *AccCPR1* expression. As shown in Fig. 4C, compared to the control (0 μg/ml), the expression of *AccCPR1* after rearing with 20E (1 μg/ml) was very low. The transcription of *AccCPR1* after rearing with 20E (0.001 μg/ml) and that of the control groups, i.e., groups treated with ethanol alone (the 20E solvent), was almost identical. Moreover, the expression of *AccCPR1* decreased dramatically as the concentration of 20E increased in the diet. Thus, high 20E titers may retard *AccCPR1* expression; however, a low concentration or the absence of 20E may permit expression of *AccCPR1*.

To test this hypothesis further, we analyzed the effect of 20E on *AccCPR1* expression by incubating integument pieces of Pw pupae with different 20E titers. As shown in Fig. 4D, the expression of AccCPR1 incubated in the presence of 1 μg 20E/ml culture medium was significantly lower after 3, 6, or 18 h of incubation than at 0 h. However, in the integuments

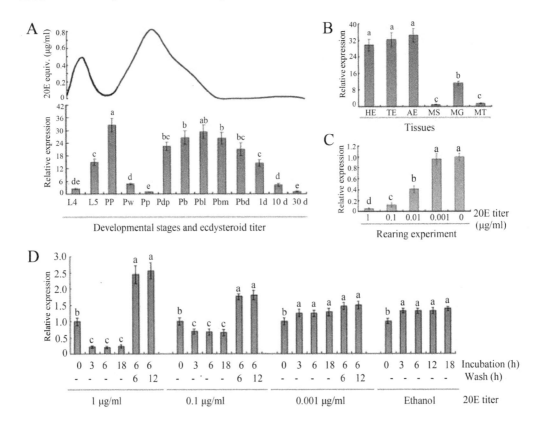

Fig. 4 Expression profiles of *AccCPR1* determined by real-time quantitative PCR. A. Column diagram indicated the expression patterns of *AccCPR1* in whole developmental stages. L4 and L5 are the fourth and the fifth larval instars, and PP is prepupae pupae. Pw, Pp, Pdp and Pb are unpigmented pupae with white, pink, dark pink and brown eyes. Pbl, Pbm and Pbd are pupae with a light-, medium- or dark-pigmented thorax, and 1-day-old, 10-day-old and 30-day-old workers represent the newly emerged, nurse and forager bees, respectively. Line chart presented the ecdysteroid titers that redrawn from Rachinsky et al., Pinto et al. and Hartfelder et al. [25-27]. B. Tissue distribution of *AccCPR1*. Total RNA was extracted from the dissected head (HE), thoracic epidermis (TE), abdominal epidermis (AE), muscle (MS), midgut (MG) and Malpighian tubules (MT) of newly emerged bees. C. Effect of dietary 20E on *AccCPR1* gene expression. In comparison with the non-supplemented diet (0 µg/ml; control group), various 20E-supplemented diets (1, 0.1, 0.01, 0.001 µg/ml) significantly affect *AccCPR1* gene expression. D. Levels of *AccCPR1* transcripts in the thoracic integument of newly ecdysed pupae incubated *in vitro* for 3, 6 or 18 h in the presence were subsequently washed for 6 or 12 h, totaling 12 or 18 h of incubation. Control incubations in the presence of ethanol (the 20E solvent) were performed for 3, 6, 12, 18 h and 0 h corresponds to newly dissected, non-incubated integuments (T0). The *A. cerana cerana β-actin* gene was applied as an internal standard control. The histograms indicate the relative expression levels. Data represent the means ± SE of three independent experiments. The different letters above the columns indicate significant differences ($P < 0.01$) as determined by using one-way ANOVA.

incubated for 6 h following 6 h or 12 h of washing in hormone-free medium, the expression increased significantly to a relatively high level. The expression level after 12 h of washing was higher than that after 6 h of washing, but the difference was not significant. The level of *AccCPR1* transcripts after treatment with 0.1 µg 20E/ml culture medium was lower compared than that at 0 h, and after washing, the expression increased but the difference was not as much as treatment with 1 μg 20E/ml. Treatment with 0.001 μg 20E/ml was found to have a slight effect on *AccCPR1* expression, as the expression after 3, 6 or 18 h of incubation increased minimally and washing in hormone-free medium did not cause a noticeable effect. The levels of *AccCPR1* transcripts after incubation with ethanol only (the 20E solvent) at every stage were slightly higher than that at 0 h of treatment, and the expression levels were approximately equal after incubation. These incubation experiments showed that a high ecdysteroid titer (1 μg 20E/ml) inhibited *AccCPR1* expression and that the removal of the hormone from the incubation medium disinhibited the expression, which is similar to the expression pattern of *AccCPR1* in *vivo* (see Fig. 4A). Similar results were observed during treatment with a low concentration (0.1 μg 20E/ml); however, the variation of the amplitude was relatively small. The amount of *AccCPR1* transcripts was slightly higher when incubated in 0.001 μg 20E/ml than the baseline at 0 h, which can be easily explained if we consider that the titer of 20E (0.001 μg/ml) is lower than the residual ecdysteroid level in the 0 h integuments, which thus permits an increase in the level of *AccCPR1* transcripts. Similar results after ethanol treatment suggested that residual ecdysteroids in the newly dissected integuments (0 h) are diluted during incubation in ethanol (the 20E solvent) and that the decline of 20E concentration permits *AccCPR1* gene expression.

Discussion

Research on CPRs is mainly concentrated in model insect species, such as *D. melanogaster*, whereas the expression of CPRs in the Chinese honeybee (*A. cerana cerana*) has not been previously reported. To investigate the features of honeybee CPRs and their possible involvement in the response to ecdysteroids, we identified a cuticle R&R protein from *A. cerana cerana*. An effective method for naming CP genes is to begin each name with a genus/species abbreviation of 3-4 letters followed by the CP family name and then the number of the gene in that CP family. Ideally, the CP genes should be numbered based on their orders on chromosomes, but annotation generally precedes complete assembly of a genome [6]. As we are not sure the order of the gene on chromosomes of *A. cerana cerana*, we named it *AccCPR1* for it is the first CPR gene discovered in the species.

Based on the deduced amino acid sequence of AccCPR1, several characteristics were found: (1) A chitin-binding domain was found in the amino acid sequence of AccCPR1, and a bioinformatics analysis revealed that *AccCPR1* may belong to the chitin-binding R&R-2 type of cuticle protein [7]. Multiple sequence alignment indicated that the deduced amino acid sequence of AccCPR1 shared higher similarity with CPRs from other insects in the chitin binding region than in other regions. (2) β-sheets occurred frequently throughout the chitin binding region, which is an optimal location for interaction with chitin [9]. (3) The strict R&R consensus: G-x8-G-x6-(Y/F) remains almost invariant in CPR proteins [6]. AccCPR1 and its orthologs are among the very few CPR proteins that are exceptions to a component of the strict R&R consensus. In their sequences and in the orthologs, the bolded and underlined G is replaced by something else (H in the *A. cerana cerana* and *Apis mellifera*, F in the *N. vitripennis*, N in the *Drosophila melanogaster*).

A typical characteristic of *CPR* genomic sequences in insects is that the first intron interrupts the signal peptide, normally after four or five coding amino acids [28-30]. The first intron of *AccCPR1* after five coding amino acids may play a crucial role in the regulation of gene expression. In this study, we detected five E74 binding sites and four putative BR-C binding sites in the 5′-flanking region of *AccCPR1*. E74, which is a known nuclear receptor transcription factor, is coordinately induced when the ecdysteroid titer is low, which implies the existence of temporal signals in *Drosophila* [31]. BR-C, which controls a tissue-specific response to the steroid hormone ecdysone at the onset of metamorphosis [32], is the most critical element in *CPR* gene transcription. Previous studies on the *Bombyx mori WCP10* promoter revealed that BmBR-C plays a regulatory role in the transcription of *BMWCP10*, and BmBR-C contributes greatly to the expression of *BMWCP10* induced by 20E [33].

A developmental profile revealed that expression of *AccCPR1* requires a "pulse" regimen of ecdysteroids, i.e., an original stimulation by elevated concentrations of ecdysteroid, followed by a decline to basal levels during pharate adult stages, and the expression is maintained at relatively high levels in newly emerged honeybees. This expression "pulse" pattern has been observed in *A. mellifera* [20], *B. mori* [34, 35], *D. melanogaster* [36] and *Tenebrio molitor* [37]. The expression profile of *AccCPR1* in different tissues of newly emerged honeybees showed that it was expressed highly in the head and thoracic and abdominal epidermis and moderately in the midgut. Therefore, *AccCPR1* constructs the cuticle in all regions of the body and may be used in the midgut lining to form the peritrophic matrices lining the midgut epithelium [38]. According to FlyBase, the *D. melanogaster* ortholog *DmCPR57A* is not expressed in the midgut, although it is expressed in the hindgut. Since *D. melanogaster* and *A. cerana cerana* belong to different groups of arthropods, their expression profiles may not be the same. The spatial and temporal expression profile demonstrated that *AccCPR1* may play an important role in cuticle development and differentiation.

Studies on morphological and chemical properties have suggested that there are pronounced differences between the outer layer, which is deposited before ecdysis (pre-ecdysial cuticle, exocuticle), and the inner layer, which is deposited after ecdysis (post-ecdysial cuticle, endocuticle) [2]. The pre-ecdysial expression of *AccCPR1* indicated that the corresponding protein may be deposited in the exocuticle. However, the transcription levels of *AccCPR1* increased abruptly in the earlier pharate adult phases and were maintained during the process of tanning and sclerotization of the adult cuticle, which suggests that the AccCPR1 protein may be part of the endocuticle. If a gene transcript appears first in a pharate stage and persists into the next stage, it is difficult to tell whether it contributes to the pre-ecdysial cuticle or the post-ecdysial cuticle [6]. Andersen suggested that RR-2 proteins were predominantly from hard cuticles and RR-1 proteins were from soft cuticles [11]. In this study, *AccCPR1* maintained high expression levels during pharate adult stages when the cuticle undergoes tanning and intensifies sclerotization, which indicates that this gene encodes a cuticle protein associated with hard cuticles. The result agrees with presence of the RR-2 consensus sequence in the AccCPR1 protein.

The requirement of an ecdysteroid pulse for the expression of cuticle protein genes has been demonstrated *in vitro* by using transient ecdysteroid exposure [39-41]. In this study, the amount of *AccCPR1* transcripts increased when the ecdysteroid titer was declining after a peak during the pharate pupal stage and earlier Pp pharate adult stage. This expression pattern shows that high *AccCPR1* gene expression is triggered by a pulse regimen of ecdysteroids. We conducted two experiments *in vitro* to test this hypothesis. We performed rearing experiments to observe the effect of 20E on the level of *AccCPR1* transcripts in the integument. We selected the larval-pharate pupae for these experiments because the larvae develop normally outside the hive in incubators when appropriate food, temperature and relative humidity are provided. The rearing experiments suggested that the transcription levels of *AccCPR1* were decreased when the larvae were fed a high concentration of 20E (1 μg/ml). We hypothesize that pharate pupae had high ecdysone titers *in vivo* after being fed with 20E (1 μg/ml) during the larval stage. Potential histological effects of the conditions of the rearing experiment require further examination.

The tanning of the cuticle in the late pharate adult phases (Pbl, Pbm and Pbd phases) starts in the thoracic integument and continues to the abdominal integument because the prothoracic gland is the main source of ecdysteroids that allows the pupal peak to occur at the Pp stage in the honeybee [42]. In the 20E incubation experiment, we collected the thoracic integument in the early pupae (Pw stage) to ensure that the ecdysteroid titer was increasing and that the *AccCPR1* gene transcript *in vivo* would be then expressed as the ecdysteroid titer decreased after its peak. The incubation experiments suggested that *AccCPR1* expression was inhibited in the integument that was exposed to a high concentration of 20E (1 μg/ml). Subsequently, the amount of transcript recovered when the hormone was removed from the incubation medium, which suggests that a pulse of ecdysteroids is required to trigger *AccCPR1* expression.

In summary, this study identified the structural characteristics and functional implications of a novel *AccCPR1* gene and thus demonstrated that the *AccCPR1* gene encodes a RR-2 cuticle protein. *AccCPR1* was expressed highly in pharate pupae and newly emerged bees, which suggests that the AccCPR1 protein is needed for cuticle renewal and differentiation. As demonstrated by rearing and incubation experiments, the levels of *AccCPR1* transcripts are modulated by the titer of ecdysteroids.

References

[1] Willis, J. H. Metamorphosis of the cuticle, its proteins, and their genes. *In*: Gilbert, L. I., Tana, J. R., Atkinson, B. G. Cell Biology: A Series of Monographs: Meta-morphosis. Postembryonic Reprogramming of Gene Expression in Amphibian and Insect Cells. Academic Press, San Diego. 1996, 253-282.

[2] Neville, A. C. Biology of the Arthropod Cuticle. Springer, Berlin. 1975.

[3] Andersen, S. O., Højrup, P., Roepstorff, P. Insect cuticular proteins. Insect Biochem Mol Biol. 1995, 25, 153-176.

[4] Andersen, S. O., Peter, M. G., Roepstorff, P. Cuticular sclerotization in insects. Comp Biochem Physiol B: Biochem Mol Bio. 1996, 113, 689-705.

[5] Rebers, J. F., Riddiford, L. M. Structure and expression of a *Manduca sexta* larval cuticle gene homologous to *Drosophila* cuticle genes. J Mol Biol. 1988, 203, 411-423.

[6] Willis, J. H., 2010. Structural cuticular proteins from arthropods: Annotation, nomenclature, and sequence characteristics in the genomics era. Insect Biochem Mol Biol. 40, 189-204.

[7] Rebers, J. E., Willis, J. H. A conserved domain in arthropod cuticular proteins binds chitin. Insect Biochem Mol Biol. 2001, 31, 1083-1093.

[8] Iconomidou, V. A., Willis, J. H., Hamodrakas, S. J. Unique features of the structural model of 'hard' cuticle proteins: implications for chitin-protein interactions and cross-linking in cuticle. Insect Biochem Mol Biol. 2005. 35, 553-560.

[9] Iconomidou, V. A., Willis, J. H., Hamodrakas, S. J. Is beta-pleated sheet the molecu-lar conformation which dictates formation of helicoidal cuticle? Insect Biochem Mol Biol. 1999, 29, 285-292.

[10] Andersen, S. O. Amino acid sequence studies of endocuticular proteins from the desert locust, *Schistocerca gregaria*. Insect Biochem Mol Biol. 1998, 28, 421-434.

[11] Andersen, S. O. Studies on proteins in post-ecdysial nymphal cuticle of locust, *Locusta migratoria*, and cockroach, *Blaberus craniifer*. Insect Biochem Mol Biol. 2000, 30, 569-577.

[12] Riddiford, L. M., Hiruma, K., Zhou, X., Nelson, C. A. Insights into the molecular basis of the hormonal control of molting and metamorphosis from *Manduca sexta* and *Drosophila melanogaster*. Insect Biochem Mol Biol. 2003, 33, 1327-1338.

[13] Mathelin, J., Quennedey, B., Bouhin, H., Delachambre, J. Characterization of two new cuticular genes specifically expressed during the post-ecdysial molting period in *Tenebrio molitor*. Gene. 1998, 211, 351-359.

[14] Horodyski, F. M., Riddiford, L. M. Expression and hormonal control of a new larval cuticular multigene family at the onset of metamorphosis of the tobacco hornworm. Dev Biol. 1989, 132, 292-303.

[15] Bouhin, H., Braquart, C., Charles, J. P., Quennedey, B., Delachambre, J. Nucleotide sequence of an adult-specific cuticular protein gene from the beetle *Tenebrio molitor*: effects of 20-hydroxyecdysone on mRNA accumulation. Insect Mol Biol. 1993, 2, 81-88.

[16] Charles, J. P. The regulation of expression of insect cuticle protein genes. Insect Biochem Mol Biol. 2010, 40, 205-213.

[17] Yang, G. Harm of introducing the western honeybee *Apis mellifera* L. to the Chinese honeybee *Apis cerana* F. and its ecological impact. Acta Entomol Sin. 2005, 3, 401-406.

[18] Michelette, E. R. F., Soares, A. E. E. Characterization of preimaginal developmental stages in Africanized honey bee workers (*Apis mellifera* L). Apidologie. 1993, 24, 431-440.

[19] Silva, I. C., Message, D., Cruz, C. D., Campos, L. A. O., Sousa-Majer, M. J. Rearing Africanized honey bee (*Apis mellifera* L.) brood under laboratory conditions. Genet Mol Res. 2008, 8, 623-629.

[20] Soares, M. P., Elias-Neto, M., Simoes, Z. L., Bitondi, M. M. 2007. A cuticle protein gene in the honeybee: expression during development and in relation to the ecdysteroid titer. Insect Biochem Mol Biol. 37, 1272-1282.

[21] Wang, J., Wang, X., Liu, C., Zhang, J., Zhu, C., Guo, X. 2008. The NgAOX1a gene cloned from *Nicotiana glutinosa* is implicated in the response to abiotic and biotic stresses. Biosci Rep. 28, 259-266.

[22] Livak, K. J., Schmittgen, T. D. Analysis of relative gene expression data using real-time quantitative PCR and the 2 (−DeltaC(T)) method. Methods. 2001, 25, 402-408.

[23] Bustin, S. A., Beaulieu, J., Huggett, J., Jaggi, R., Kibenge, F. S., Olsvik, P. A., Penning, L. C., Toegel, S. MIQE précis: Practical implementation of minimum standard guidelines for fluorescence-based quantitative real-time PCR experiments. BMC Mol Biol. 2010, 11, 74.

[24] Hamodrakas, S. J., Willis, J. H., Iconomidou, V. A. A structural model of the chitin-binding domain of cuticle proteins. Insect Biochem Mol Biol. 2002, 32, 1577-1583.

[25] Rachinsky, A., Strambi, C., Strambi, A., Hartfelder, K. Caste and metamorphosis—hemolymph titers of juvenile hormone and ecdysteroids in last instar honeybee larvae. Gen Comp Endocrinol. 1990, 79, 31-38.

[26] Pinto, L. Z., Hartfelder, K., Bitondi, M. M. G., Simoes, Z. L. P. Ecdysteroid titer in pupae of highly social bees related to distinct modes of caste development. J Insect Physiol. 2002, 48, 783-790.

[27] Hartfelder, K., Bitondi, M. M. G., Santana, W. C., Simões, Z. L. P. Ecdysteroid titer and reproduction

in queens and workers of the honey bee and of a stingless bee: loss of ecdysteroid function at increasing levels of sociality? Insect Biochem Mol Biol. 2002, 32, 211-216.

[28] Binger, L. C., Willis, J. H. Identification of the cDNA, gene and promoter for a major protein from flexible cuticles of the giant silkmoth *Hyalophora cecropia*. Insect Biochem Mol Biol. 1994, 24, 989-1000.

[29] Charles, J. P., Chihara, C., Nejad, S., Riddiford, L. M. A cluster of cuticle protein genes of *Drosophila melanogaster* at 65A: sequence, structure and evolution. Genetics. 1997, 147, 1213-1224.

[30] Lemoine, A., Mathelin, J., Braquart-Varnier, C., Everaerts, C., Delachambre, J. A functional analysis of ACP-20, an adult-specific cuticular protein gene from the beetle *Tenebrio*: role of an intronic sequence in transcriptional activation during the late metamorphic period. Insect Mol Biol. 2004, 13, 481-493.

[31] Sullivan, A. A., Thummel, C. S. Temporal profiles of nuclear receptor gene expression reveal coordinate transcriptional responses during *Drosophila* development. Mol Endocrinol. 2003, 17, 2125-2137.

[32] Kalm, L. V., Crossgrove, K., Seggern, D. V., Guild, G. M., Beckendorf, S. K. The Broad-Complex directly controls a tissue-specific response to the steroid hormone ecdysone at the onset of *Drosophila* metamorphosis. EMBO J. 1994, 13, 3505-3516.

[33] Wang, H. B., Iwanaga, M., Kawasaki, H. Activation of BMWCP10 promoter and regulation by BR-C Z2 in wing disc of *Bombyx mori*. Insect Biochem Mol Biol. 2009, 39, 615-623.

[34] Zhong, Y. S., Mita, K., Shimada, T., Kawasaki, H. Glycine-rich protein genes, which encode a major component of the cuticle, have different developmental profiles from other cuticle protein genes in *Bombyx mori*. Insect Biochem Mol Biol. 2006, 36, 99-110.

[35] Nita, M., Wang, H. B., Zhong, Y. S., Mita, K., Iwanaga, M., Kawasaki, H. Analysis of ecdysone-pulse responsive region of BMWCP2 in wing disc of *Bombyx mori*. Comp Biochem Physiol B: Biochem Mol Biol. 2009, 153, 101-108.

[36] Kawasaki, H., Hirose, S., Ueda, H. BetaFTZ-F1 dependent and independent activation of Edg78E, a pupal cuticle gene, during the early metamorphic period in *Drosophila melanogaster*. Dev Growth Differ. 2002, 44, 419-425.

[37] Rondot, I., Quennedey, B., Delachambre, J. Structure, organization and expression of two clustered cuticle protein genes during the metamorphosis of an insect, *Tenebrio molitor*. Eur J Biochem. 1998, 254, 304-312.

[38] Merzendorfer, H., Zimoch, L. Chitin metabolism in insects: structure, function and regulation of chitin synthases and chitinases. J Exp Biol. 2003, 206, 4393-4412.

[39] Fechtel, K., Natzle, J. E., Brown, E. E., Fristrom, J. W. Prepupal differentiation of *Drosophila* imaginal discs: identification of four genes whose transcripts accumulate in response to a pulse of 20-hydroxyecdysone. Genetics. 1988, 120, 465-474.

[40] Suzuki, Y., Matsuoka, T., Iimura, Y., Fujiwara, H. Ecdysteroiddependent expression of a novel cuticle protein gene BMCPG1 in the silkworm, *Bombyx mori*. Insect Biochem Mol Biol. 2002, 32, 599-607.

[41] Noji, T., Ote, M., Takeda, M., Mita, K., Shimada, T., Kawasaki, H. Isolation and comparison of different ecdysoneresponsive cuticle protein genes in wing discs of *Bombyx mori*. Insect Biochem Mol Biol. 2003, 33, 671-679.

[42] Hartfelder, K. Structure and function of the prothoracic gland in honey bee (*Apis mellifera* L.) development. Invertebr Reprod Dev. 1993, 23(1), 16.

4.2 Characterization of the TAK1 Gene in *Apis cerana cerana* (*AccTAK1*) and Its Involvement in the Regulation of Tissue-specific Development

Fei Meng, Mingjiang Kang, Li Liu, Lu Luo, Baohua Xu, Xingqi Guo

Abstract: TGF-β activated kinase-1 (TAK1) plays a pivotal role in developmental processes in many species. Previous research has mainly focused on the function of TAK1 in model organisms, and little is known about the function of TAK1 in Hymenoptera insects. Here, we isolated and characterized the TAK1 gene from *Apis cerana cerana* (*AccTAK1*). Promoter analysis of *AccTAK1* revealed the presence of transcription factor binding sites related to early development. Real-time quantitative PCR and immunohistochemistry experiments revealed that *AccTAK1* was expressed at high levels in the fourth-instar larvae, primarily in the abdomen, in the intestinal wall cells of the midgut and in the secretory cells of the salivary glands. In addition, *AccTAK1* expression in the fourth-instar larvae could be dramatically induced by treatment with pesticides and organic solvents. These observations suggest that *AccTAK1* may be involved in the regulation of early development in the larval salivary gland and midgut.

Keywords: *AccTAK1*; *Apis cerana cerana*; Development characterization; Expression pattern; Immunohistochemistry

Introduction

Organisms undergo multiple morphological changes, including organ formation, throughout the process of development, and these developmental changes are regulated by changes in gene expression. Extensive studies have elucidated some of the genes, such as those in the mitogen activated protein kinase (MAPK) pathway, that are required for development [1-3].

MAPK cascades are composed of three interlinked protein kinases, MAPKKKs, MAPKKs and MAPKs, which play important roles in development and in environmental responses [1]. A member of the MAPKKK superfamily, transforming growth factor-β activated kinase-1 (TAK1), was first identified as a signaling molecule in TGF-β and BMP signaling pathways [4]. TAK1 belongs to the mixed-lineage kinase (MLK) family that includes ASK1, MLK3 and MUK [5]. TAK1 contains a conserved kinase domain at its N-terminus, and its glycine-rich loop (G-x-G-x-x-G) and hinge region (MEYAEGGS) are necessary for gene activation [6].

Recently, it has been demonstrated that TAK1 acts as an essential regulator in cartilage morphogenesis, vascular development, keratinocyte apoptosis and cell shape control during embryonic development [2, 3, 7, 8]. However, these roles of TAK1 in developmental regulation vary from species to species, even among tissues [9]. In *Xenopus* development, overexpression of *xTAK1* in early embryos induced ventral mesoderm apoptosis [10]. Overexpression of *dTAK1* in *Drosophila* caused strong apoptosis in the visual system, leading to a small eye phenotype [11]. Additionally, TAK1 has diverse biological functions in mouse development. Mutation studies in the mouse indicated that TAK1 plays important roles in vascular development and smooth muscle formation [3]. TAK1-deficient mice exhibit severe defects in cartilage development, in addition to chondrodysplasia and joint abnormalities [2]. The keratinocyte-specific depletion of TAK1 in mice caused skin wrinkling and apoptosis [7].

The toxic substance has become a factor influencing development. It was reported that pesticides remaining in plants could influence the normal development of the digestive system and glands in insects, causing apoptosis during the early development stages [12]. The Asian honeybee (*Apis cerana cerana*) is a eusocial insect and has important roles in the pollination of flowing crops at low temperatures [13]. Global warming, excessive pesticide use and competition with western honeybees are causing *A. cerana cerana* to become an endangered species [14]. Thus far, no reports have been published regarding TAK1 in the honeybee. Given that TAK1 plays crucial roles in development, our interest focused on identifying the *TAK1* gene whose function correlates with development in *A. cerana cerana*. In this study, we characterized the *TAK1* gene in *A. cerana cerana*, analyzed mRNA expression after treatment with toxic substances, and assessed protein localization in larvae.

Materials and Methods

Honeybees, artificial feeding and tissue preparation

The colonies of *A. cerana cerana* were maintained in the experimental apiary of Shandong Agricultural University (Taian, China). Larvae, pupae and adult bees were identified by age, eye color and cuticle pigmentation. The fourth-instar larvae were transferred to polyethylene cell cups and reared on an artificial diet in an incubator with 96% relative humidity at 34℃. Pesticides (imidacloprid, cyhalothrin, acetamiprid, and dimethoate) and organic solvents (methanol, alcohol) were added to the normal diet in the experimental group, and the final concentrations of the additives are displayed in Supplementary Table 1. Various tissues from the fourth-instar larvae, including the head, thorax, abdomen, muscle, epidermis, Malpighian tubules, midgut and salivary gland, were freshly collected, frozen in liquid nitrogen and stored at −80℃.

Primers and amplification conditions

Primers and PCR amplification conditions are listed in Supplementary Table S2 and Supplementary Table S3, respectively.

Cloning of the cDNA of *AccTAK1*

Total RNA was extracted with Trizol (Invitrogen, Carlsbad, CA, USA), and the first-strand cDNA was subsequently synthesized by using reverse transcriptase (TransGen Biotech, Beijing, China) according to the protocol of the manufacturer. An internal conserved fragment was obtained by reverse transcription-PCR by using the primers MP1 and MP2, which were designed based on the sequences of TAK1 from other species. Then, specific primers (5W/5N and 3W/3N) were generated for the 5' rapid amplification of cDNA end (RACE) and 3' RACE, respectively. According to the deduced full-length cDNA, specific primers, QP1 and QP2, were designed to amplify the full-length cDNA of TAK1 in *A. cerana cerana* (designated as *AccTAK1*). The PCR product was purified, cloned into the pEASY-T3 vector (TransGen Biotech, Beijing, China), and sequenced.

Predictions of the theoretical isoelectric point and molecular weight were performed with PeptideMass (http://us.expasy.org/tools/peptide-mass.html). Homologous TAK1 sequences in other species were retrieved from the NCBI server and aligned by using DNAMAN version 5.2.2 (Lynnon Biosoft, Quebec, Canada). Phylogenetic analysis was carried out by the Molecular Evolutionary Genetics Analysis (MEGA version 4.1) software by using the neighbor-joining method.

Isolation of genomic DNA and the promoter of *AccTAK1*

Genomic DNA was extracted from whole bees by using the EasyPure Genomic DNA Extraction Kit (TransGen Biotech, Beijing, China). Several specific primers were designed and used to isolate the genomic DNA of *AccTAK1*. For promoter amplification, genomic DNA was completely digested with *Xba* I at 37°C overnight, ligated with T4 DNA ligase (TaKaRa, Dalian, China) and used as the template. The promoter region was obtained by using inverse-PCR (I-PCR) with specific primers QDS1/ QDX1 and QDS2/QDX2. Ultimately, primers MQS and MQX were used to verify the promoter sequences. Furthermore, the TFSEARCH database (http://www.cbrc.jp/research/db/TFSEARCH.html) was used to analyze the promoter region of *AccTAK1*.

Protein expression, purification and preparation of anti-AccTAK1 serum

To construct the expression vector, primers (YHS/YHX) were designed that incorporated a *Sac* I site in the forward primer and a *Hin*dIII site in the reverse primer. After PCR, the fragment digested with *Sac* I and *Hin*dIII was inserted into the pET-30a (+) vector excised with the same restriction endonucleases, and the pET-30a (+)-AccTAK1 plasmid was transformed into BL21 (DE3). The cultures were induced with 0.1 mmol/L IPTG and centrifuged at 5,000 g at 4°C for 10 min. Then, the pellet was resuspended in lysis buffer, sonicated and centrifuged again. Finally, both the supernatant and pellet solubilized in PBS were analyzed by SDS-PAGE. The target protein was purified by using the MagneHis™ Protein Purification System (Promega, Madison, WI, USA). The purified protein was injected subcutaneously into white mice for generation of antibodies as described by Meng et al. [24].

Immunohistochemical analysis in larvae

The second- and the fourth-instar larvae were fixed in 4% paraformaldehyde overnight. The fixed samples were washed, dehydrated through ethanol in a series of ascending concentrations, cleared in xylene and finally embedded in paraffin. Vertical sections (10 μm) were prepared and mounted onto poly-L-lysine-coated glass slides. Dewaxed sections were blocked with 10% normal goat serum for 30 min and incubated overnight with the anti-*AccTAK1* antibody (1 ∶ 500) at 4℃. On the next day, after washing three times in PBS, the sections were incubated with fluorescein isothiocyanate (FITC)-conjugated goat anti-rat IgG (1 ∶ 100, Beyotime, Jiangsu, China) at 37℃ for 1 h in the dark. For further study, the sections were blocked again and incubated with the anti-AccTAK1 antibody. Subsequently, the sections were incubated with tetramethylrhodamine isothiocyanate (TRITC)-conjugated goat anti-rat IgG (1 ∶ 50, Beyotime, Jiangsu, China) for 1 h and counterstained with 4,6-diamidino-2-phenylindole (DAPI, 1 ∶ 1,000, Sigma, Saint Louis, MO, USA). Finally, sections were observed by laser confocal microscopy (LSM 510 META, Carl Zeiss AG, Germany) at 408, 488, and 561 nm, to observe DAPI, FITC and TRITC, respectively. The omission of the anti-AccTAK1 antibody in the reaction served as a negative control.

Real-time quantitative PCR

Real-time quantitative PCR (qPCR) was carried out by using the SYBR® PrimeScript™ RT-PCR Kit (TaKaRa, Dalian, China) in a 25 μl volume on a CFX96™ Real-time System (Bio-Rad) with the following conditions: 95℃ for 30 s, 40 cycles of amplification (95℃ for 5 s, 53℃ for 15 s, and 72℃ for 15 s) and a melting cycle from 65℃ to 95℃. All samples were analyzed in triplicate. The linear relationship, amplification efficiency and data analysis were conducted by using CFX Manager Software version 1.1. The analysis of significant difference was determined by Duncan's multiple range test using Statistical Analysis System (SAS) version 9.1 software.

Results

Cloning and characterization of the full-length cDNA of *AccTAK1*

The cDNA sequence of *AccTAK1* (GenBank accession no. GU295942) is 1,891 bp in length and contains a 117 bp 5′ untranslated region (UTR), a 127 bp 3′ UTR and a 1,647 bp open reading frame (ORF) that encodes a peptide of 548 amino acids with a predicted molecular mass of 62.8 kDa and a pI of 6.08. A multiple sequence alignment showed that the N-terminal region of AccTAK1 shares high homology with other species and that the AccTAK1 kinase domain is similar to the typical TAK1 kinase domain. The glycine-rich loop, with the sequence G-x-G-x-x-G, and the hinge region, containing the residues MEYAEGGS, were also identified (Fig. 1A). Furthermore, two regions at the C-terminal tail, in which some short consensus regions were present, were absent in *Apis mellifera* TAK1 (AmTAK1) compared to *AccTAK1*.

To investigate the evolutionary relationship of AccTAK1 to other species and MLK family members, a neighbor-joining phylogenetic tree was generated by MEGA 4.1 (Fig. 1B). The results revealed four groups, and the *AccTAK1* protein was categorized into group I, where it clustered with AmTAK1.

Identification of the genomic structure of *AccTAK1*

The full-length genomic sequence of *AccTAK1* (GenBank accession no. GU295943) was 9,511 bp, contained nine exons and was similar to *AmTAK1* (Supplement Fig. 1). Surprisingly, part of the sixth and the ninth exons that are present in *AccTAK1* are absent in *AmTAK1*, leading to missing amino acids in AmTAK1, as shown in Fig. 1A.

To study the organization of the regulatory region of *AccTAK1*, a 1,118 bp fragment (GenBank accession no. GU295943) upstream of the transcription start site was isolated by I-PCR. Several putative transcription factor binding sites (TFBS) were identified in the promoter region, including BR-C, which is involved in the tissue-specific regulation of salivary gland development [15], as well as CF2-II, DFD and HB, which are necessary for early development [16-18]. In addition, the basic TATA-box and CAAT-box were also identified in the promoter region. The positions and functions of putative TFBS are shown in Table 1. These data imply that the *AccTAK1* gene may be involved in the early development of specific tissues.

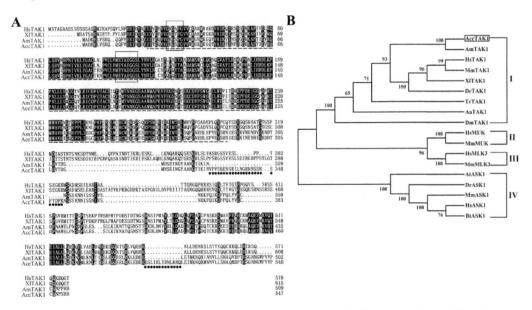

Fig. 1 Multiple amino acid alignment and phylogenetic analysis. A. Identical amino acid residues in this alignment are shaded black. The dashed line shows the conserved kinase domain in TAK1. The glycine-rich loop (G-x-G-x-x-G) and the hinge region (MEYAEGGS) are boxed. The amino acid residues marked with dots denote the differences between AccTAK1 and AmTAK1. B. A phylogenetic tree was constructed from different species with the neighbor-joining method by using MEGA 4.1 software. Four main groups are shown and AccTAK1 is boxed. All gene messages of various species are listed in Supplement Table 4.

Table 1 Putative transcription factors, binding sites, positions, and possible functions of the 5'-flanking region of *AccTAK1*

Transcription factor	Putative binding site and position	Function
BR-C	TAAATAAATAAAA: −546 TTAATAAATAAAA: −1037	Tissue-specific response to salivary gland morphogenesis and regulation of the third-instar and early prepupal development
CF2-II	GTATATATA: −395, −418	DNA binding zinc finger domain and participation in regulation of distinct sets of genes during embryo development
DFD	TTTCGAATTAATTAAA: −1075	Cell development and organ formation in embryo development
HB	TTATAAAAAA: −334 ATATAAAAAA: −365	Regulation of gene expression at the level of transcription and tissue development in early stages

Expression profiles of *AccTAK1* at different development stages and in various tissues

To identify the expression pattern of *AccTAK1* at different developmental stages, qPCR was carried out. Stage-specific expression profiling showed that the amount of *AccTAK1* mRNA increased rapidly from the second-instar larvae to the fourth-instar larval stage, decreased from the fourth-instar larvae to the white eyes pupae, increased slightly in pink eyes pupae, and gradually dropped at the adult stage. The highest expression level was detected in the fourth-instar larvae, and almost no significant differences were found between pupae and adult stages (Fig. 2A). To investigate the reason for the highest expression during developmental stages, various tissues were extracted from the fourth-instar larvae, and tissue-specific expression analysis indicated that a high amount of transcript was detected predominantly in the abdomen, midgut and larval salivary gland (Fig. 2B). There were no significant differences in transcription levels among the thorax, head, muscle, epidermis and Malpighian tubules. These results strongly indicate that *AccTAK1* may function in the early development of specific tissues in the abdomen.

Expression profiles of *AccTAK1* after feeding different pesticides and organic solvents

To determine whether *AccTAK1* is involved in the development of specific tissues in the abdomen, artificial diets containing different pesticides or organic solvents were used to feed the fourth-instar larvae, and then qPCR was used to analyze the expression of *AccTAK1*. As shown in Fig. 2C, after toxic substance treatments, the expression of *AccTAK1* was significantly induced ($P < 0.01$), except after imidacloprid treatments ($P = 0.0216$). Expression levels after organic solvent treatments were stronger compared to pesticide treatments. The mRNA level was 9.07-fold and 8.59-fold higher than that in normal larvae for the methanol and alcohol treatments, respectively. These results reveal that the expression of *AccTAK1* may be involved in abiotic stress, and they further suggest that the expression of *AccTAK1* is related to the development of specific tissues in the abdomen.

Tissue localization of AccTAK1 in the larval stage

To determine the tissue-specific localization of AccTAK1, paraffin sections from the second- and the fourth-instar larvae were subjected to immunohistochemical analysis by using

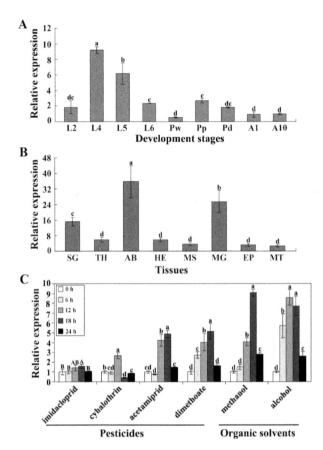

Fig. 2 Expression profile of *AccTAK1* by real-time quantitative PCR. A. *AccTAK1* mRNA expression at different developmental stages. Total RNA was isolated from larvae (L2-the second instars, L4-the fourth instars, L5-the fifth instars and L6-the sixth instars), pupae (Pw-white eyes pupae, Pp-pink eyes pupae and Pd-dark eyes pupae) and adults (A1-one day post-emergence adults and A10-ten days post-emergence adults). B. Tissues distribution of *AccTAK1*. Total RNA was extracted from dissected salivary gland (SG), thorax (TH), abdomen (AB), head (HE), muscle (MS), midgut (MG), epidermis (EP) and Malpighian tubules (MT) of the fourth-instar larvae. C. Expression of *AccTAK1* after treatment with different pesticides or organic solvents by feeding the fourth-instar larvae. Samples were collected at 6, 12, 18 and 24 h after different feeding treatments. Vertical bars represent the mean ± SE (n = 3). Means labeled with different lowercase letters were highly significantly different ($P < 0.01$), and means labeled with different capital letters were significantly different ($P < 0.05$), as determined by Duncan's multiple range test using SAS version 9.1 software.

an anti-AccTAK1 antibody. The results showed that AccTAK1 is ubiquitously expressed in the second-instar larvae (data not shown). However, the TAK1 expression in the fourth-instar larvae became less widespread, and specific tissues with higher expression were clearly seen (Fig. 3). Expression in the intestinal wall cells of the developing midgut and the secretory cells in the salivary gland were prominent. These changes in tissue distribution may be caused by the formation of specific tissues during post-embryonic development. The localization in the larval salivary gland and intestinal wall also imply that the *AccTAK1* gene plays an important role in the development of specific tissues.

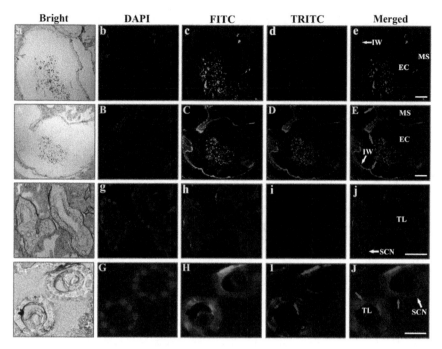

Fig. 3 Immunohistochemical localization of *AccTAK1* in the fourth-instar larvae. a-j are negative controls for A-J, respectively. a-e and A-E. Midgut; scale bar, 120 μm. f-j and F-J. Salivary gland; scale bar, 30 μm. EC, enteric cavity; IW, intestinal wall; TL, tubule lumen; SCN, secretory cell nucleus. Merged images are a combination of DAPI, FITC, and TRITC staining. An obvious reaction is marked by red arrows. (For color version, please sweep QR Code in the back cover)

Discussion

TAK1 is essential in the regulation of organism growth and differentiation. A number of studies have demonstrated that mutation or deletion of the *TAK1* gene causes adverse effects on development [2, 3]. In this study, the *AccTAK1* gene was isolated from *A. cerana cerana*. The deduced amino acid sequences of AccTAK1 showed high homology in the N-terminal region with TAK1 in other species (Fig. 1A). In human *TAK1*, two alternative exons are located in the C-terminal region and their use determines the four splicing variants, TAK1-a, TAK1-b, TAK1-c, and TAK1-d, which perform unique functions in tissue-specific development [19]. We presume that the difference between *AccTAK1* and *AmTAK1* in two domains at the C-terminal region may be caused by splicing variants, and this difference suggests that the role of *AccTAK1* may be different from others.

We isolated the promoter region of *AccTAK1*, and sequence analysis revealed the existence of binding sites for CF2-II, DFD, HB and BR-C, which are necessary for early development (Table 1), suggesting that the *AccTAK1* gene functions in early development similar to TAK1 from other species [10, 20]. Interestingly, BR-C also participates in the tissue-specific response to salivary gland morphogenesis [15]. The *AccTAK1* gene may be related with the early development of specific tissues, which may vary from TAK1 in other species.

To confirm this hypothesis, we examined the expression pattern of *AccTAK1* at different developmental stages. The results showed that *AccTAK1* gene transcripts were detected predominantly in the fourth- and the fifth-instar larvae (Fig. 2A), suggesting that the *AccTAK1* gene plays an important role during the growth of the fourth- and the fifth-instar larvae. However, the expression level was not constant from the fourth-instar larvae to the fifth-instar larvae, and we speculate that this difference may be related to the generation or degeneration of specific tissues at particular developmental stages. Tissue-specific expression analysis indicated that a high amount of *AccTAK1* was present in the abdomen, midgut and larval salivary gland (Fig. 2B), which is inconsistent with some early suggestions that TAK1 is mainly expressed in the head and nervous system [10]. This difference may be a result of the life habits of the honeybee. In the larval stage, the abdomen is the most active tissue. The midgut is also a vital tissue in the abdomen for food digestion and nutrient absorption. The larval salivary gland, also known as the silk gland, is located ahead of the abdomen which play important roles in silk production and pupae protection [21, 22]. Notably, Silva-Zacarin et al. [22] demonstrated that silk formation stopped at the end of the fifth-instar in Hymenoptera. We speculate that this is the reason for the lower expression level in the fifth-instar larvae compared to the fourth-instar larvae. In addition, these results suggest that *AccTAK1* is involved in larval salivary gland morphogenesis, which is consistent with the function of BR-C that participates in the tissue-specific response to salivary gland development.

Moreover, we analyzed the expression of *AccTAK1* after treatment with pesticides or alcohol vapors. The results showed that *AccTAK1* expression was induced by toxic substances, suggesting that *AccTAK1* could be involved in responses to abiotic stresses. The regression of salivary gland cells was related to starvation and alcohol vapors [23]. Excess pesticides could influence the normal development of the digestive system and glands in insects, triggering an alteration in secretory and absorptive functions [12]. Feeding artificial diets containing toxic substances to the fourth-instar larvae destroyed function in the midgut and salivary gland, and induced the expression of *AccTAK1*. These results also imply that *AccTAK1* may function in the development of the larval salivary gland and midgut. Likewise, immunohistochemistry showed that AccTAK1 in the fourth-instar larvae was specifically localized in the secretory cells of the larval salivary gland and intestinal wall cells of the midgut (Fig. 3). This result provides evidence that *AccTAK1* may be involved in alimentation and transport.

In conclusion, analysis of *AccTAK1* by mRNA expression, protein localization and promoter analysis strongly suggests that the gene plays important roles in the developmental regulation of the larval salivary gland and midgut. The further characterization of this newly identified gene in early development will help to improve the survival rate of honeybee larvae.

References

[1] Rodriguez, M. C., Petersen, M., Mundy, J. Mitogen-activated protein kinase signaling in plants. Annu Rev Plant Biol. 2010, 61, 621-649.

[2] Shim, J. H., Greenblatt, M. B., Xie, M., Schneider, M. D., Zou, W., Zhai, B., Gygi, S., Glimcher, L. H. TAK1 is an essential regulator of BMP signaling in cartilage. EMBO J. 2009, 28, 2028-2041.

[3] Jadrich, J. L., O'Connor, M. B., Coucouvanis, E. The TGFβ activated kinase TAK1 regulates vascular development *in vivo*. Development. 2006, 133, 1529-1541.

[4] Yamaguchi, K., Shirakabe, K., Shibuya, H., Irie, K., Oishi, I., Ueno, N., Taniguchi, T., Nishida, E., Matsumoto, K. Identification of a member of the MAPKKK family as a potential mediator of TGF-beta signal transduction. Science. 1995, 270, 2008-2011.

[5] Gotoh, I., Adachi, M., Nishida, M. Identifica-tion and characterization of a novel MAP kinase kinase kinase. MLTK. J Biol Chem. 2001, 276, 4276-4286.

[6] Brown, K., Vial, S. C., Dedi, N., Long, J. M., Dunster, N. J., Cheetham, G. M. Structural basis for the interaction of TAK1 kinase with its activating protein TAB1. J Mol Biol. 2005, 354, 1013-1020.

[7] Sayama, K., Hanakawa, Y., Nagai, H., et al. Transforming growth factor-β-activated kinase 1 is essential for differentiation and the prevention of apoptosis in epidermis. J Biol Chem. 2006, 281, 22013-22020.

[8] Mihaly, J., Kockel, L., Gaengel, K., Weber, U., Bohmann, D., Mlodzik, M. The role of the *Drosophila* TAK homologue dTAK during development. Mech Dev. 2001, 102, 67-79.

[9] Delaney, J. R., Mlodzik, M. TGF-beta activated kinase-1: new insights into the diverse roles of TAK1 in development and immunity. Cell Cycle. 2006, 5, 2852-2855.

[10] Shibuya, H., Iwata, H., Masuyama, N., Gotoh, Y., Yamaguchi, K., Irie, K., Matsumoto, K., Nishida, E., Ueno, N. Role of TAK1 and TAB1 in BMP signaling in early *Xenopus* development. EMBO J. 1998, 17, 1019-1028.

[11] Takatsu, Y., Nakamura, M., Stapleton, M., Danos, M. C., Matsumoto, K., O'Connor, M. B., Shibuya, H., Ueno, N. TAK1 participates in c-Jun N-terminal kinase signaling during *Drosophila* development. Mol Cell Biol. 2000, 20, 3015-3026.

[12] Franco, R., Sánchez-Olea, R., Reyes-Reyes, E. M., Panayiotidis, M. I. Environmental toxicity, oxidative stress and apoptosis: ménage à trois. Mutat Res. 2009, 674, 3-22.

[13] Li, H. L., Zhang, Y. L., Gao, Q. K., Cheng, J. A., Lou, B. G. Molecular identification of cDNA, immunolocalization, and expression of a putative odorant-binding protein from an Asian honey bee, *Apis cerana cerana*. J Chem Ecol. 2008, 34, 1593-1601.

[14] Xu, P., Shi, M., Chen, X. X. Antimicrobial peptide evolution in the Asiatic honey bee *Apis cerana*. PLoS One. 2009, 4, e4239.

[15] Von Kalm, L., Crossgrove, K., Von Seggern, D., Guild, G. M., Beckendorf, S. K. The Broad-Complex directly controls a tissue-specific response to the steroid hormone ecdysone at the onset of *Drosophila* metamorphosis. EMBO J. 1994, 13, 3505-3516.

[16] Gogos, J. A., Hsu, T., Bolton, J., Kafatos, F. C. Sequence discrimination by alternatively spliced isoforms of a DNA binding zinc finger domain. Science. 1992, 257, 1951-1955.

[17] Rørth, P. Specification of C/EBP function during *Drosophila* development by the bZIP basic region. Science. 1994, 266, 1878-1881.

[18] Stanojević, D., Hoey, T., Levine, M. Sequence-specific DNA-binding activities of the gap proteins encoded by hunchback and Krüppel in *Drosophila*. Nature. 1989, 341, 331-335.

[19] Dempsey, C. E., Sakurai, H., Sugita, T., Guesdon, F. Alternative splicing and gene structure of the transforming growth factor β-activated kinase 1. Biochim Biophys Acta. 2000, 1517, 46-52.

[20] Jadrich, J. L., O'Connor, M. B., Coucouvanis, E. Expression of TAK1, a mediator of TGF-β and BMP signaling, during mouse embryonic development. Gene Expr Patterns. 2003, 3, 131-134.

[21] Silva-Zacarin, E. C., Taboga, S. R., Silva de Moraes, R. L. Nuclear alterations associated to programmed cell death in larval salivary glands of *Apis mellifera* (Hymenoptera: Apidae). Micron. 2008, 39, 117-127.

[22] Silva-Zacarin, E. C., Silva de Moraes, R. L., Taboga, S. R. Silk formation mechanisms in the larval salivary glands of *Apis mellifera* (Hymenoptera: Apidae). J Biosci. 2003, 28, 753-764.

[23] Sehnal, F., Akai, H. Insect silk glands: their types, development and function, and effects of environmental factors and morphogenetic hormones on them. Int J Insect Morphol Embryol. 1989, 19, 79-132.

[24] Meng, F., Zhang, L., Kang, M., Guo, X., Xu, B. Molecular characterization, immunohistochemical localization and expression of a ribosomal protein L17 gene from *Apis cerana cerana*. Arch Insect Biochem Physiol. 2010, 75, 121-138.

4.3 Cloning, Structural Characterization and Expression Analysis of a Novel Lipid Storage Droplet Protein-1 (LSD-1) Gene in Chinese Honeybee (*Apis cerana cerana*)

Li Liu, Zhihong Gong, Xingqi Guo, Baohua Xu

Abstract: Lipid storage droplet 1 (LSD-1), a PAT family protein located around lipid droplets in insects, is intimately linked to lipid droplets formation and lipid metabolism. Conjugated linoleic acid (CLA) and rosiglitazone (Rosi) have previously been shown to modulate the expression of several PAT family proteins through peroxisome proliferator-activated receptor-γ (PPARγ). In the present study, we isolated and characterized a novel *LSD-1* gene, referred to *AccLSD-1*, from Chinese honeybee (*Apis cerana cerana*). Sequence analysis indicated that the central region of LSD-1 protein had significant sequence similarity and a typical *LSD-1* gene was composed of 8 exons and 7 introns. Interestingly, the first intron of *AccLSD-1* including several PPARγ-response elements (PPREs) was located in the 5′ UTR. Analysis of 5′-flanking region of *AccLSD-1* revealed a number of putative *cis*-acting elements, including three PPREs. Quantitative real-time PCR showed that *AccLSD-1* expressed ubiquitously from feeding larva to adult, and its expression level was the highest at brown-eyed pupae (Pb) stage. The effect of CLA, Rosi and combination on *AccLSD-1* expressions indicated 1% CLA and 0.5 mg/ml Rosi were considered as the suitable diets for rearing adult workers in laboratory, and *AccLSD-1* was downregulated by CLA whereas upregulated by Rosi. Furthermore, the combination of CLA and Rosi remarkly rescued the suppression of *AccLSD-1* expression by CLA alone. These results suggest that *AccLSD-1* is associated with *A. cerana cerana* development, especially during pupal metamorphosis, and can be regulated by CLA or Rosi possibly via activating PPARγ.

Keywords: Lipid storage droplet 1; *Apis cerana cerana*; Quantitative real-time PCR; PPARγ; Conjugated linoleic acid; Rosiglitazone

Introduction

Like vertebrate adipose tissue and liver, the insect lipid droplets (LDs) are prominent cellular organelles participating in a number of biological processes, such as energy metabolism and nutrient storage, by mediating cellular lipolysis [1, 2]. They are consisted of a hydrophobic core of neutral lipids, predominantly triacylglycerols (TGs) or cholesteryl esters

(CEs), surrounding by a monolayer of phospholipids and lipid droplets associated proteins [3]. As the best-understood lipid droplets associated proteins, PAT family proteins have been received attention as the crucial role in coordinating LDs motility and regulating LDs function in organismal energy homeostasis [3, 4]. It has been demonstrated that PAT family proteins widespreadly exist in organisms from prokaryotes to eukaryotes and have significant sequence similarity, especially in N-terminal sequences [3, 5]. In mammal, five PAT family proteins including perilipin, adipose differentiation-related protein (ADRP), tail-interacting protein of 47 kDa (TIP47), S3-12, and OXPAT have been independently identified, cloned, and characterized by a number of previous studies [6, 7]. However, only two PAT proteins, lipid storage droplet protein-1 and -2 (LSD-1 and LSD-2), are expressed by insect genomes [6, 8]. Recent observations have suggested that LSD-1 and LSD-2, the same as the mammal PAT family proteins, were shown to attach to lipid droplets [8]. Moreover, LSD-1 functioned as active players in the control of lipolysis and LSD-2 is linked to storage of neutral lipids [6]. Most of the knowledge of the lipolytic process was acutely regulated by adipokinetic hormone (AKH), a G protein-coupled receptor that activates cAMP-dependent protein kinase A (PKA). Subsequently, LSD-1 phosphorylated by PKA functions as regulators of the rate of lipolysis by modulating the activity of TGL which is the main lipase in insect fat body [1, 8, 9]. Arrese et al. have proposed that phosphorylation of LSD-1 promotes a 1.7-fold activation of TGL [9]. However, unlike perilipin A which has a dual function on TG metabolism, LSD-1 is not essential to shield TG from cytosolic lipases [10]. Thus, a full understanding of lipid homeostasis will require unraveling the regulatory mechanism of LSD-1.

In mammal, several PAT family proteins such as perilipin, S3-12 and OXPAT/PAT-1 are directly controlled by peroxisome proliferator-activated receptor-γ (PPARγ), which is a member of the PPAR subfamily of nuclear hormone receptors [11, 12]. In vertebrates, PPARγ is essential in adipose tissue for glucose uptake and TG accumulation, and can be activated by conjugated linoleic acid (CLA) and rosiglitazone (Rosi) [13]. The activation of PPARγ heterodimerizes with the retinoid X receptor α (RXRα) and regulate energetic metabolism processes by the activation of peroxisome proliferator response element (PPRE) in the promoter region of adipogenic target genes, such as *perilipin*, *adiponectin*, and *adipocyte fatty acid binding protein* (αFABP/αP2) [14-16]. Previous studies have demonstrated that CLA has significant favorable effect on carcinogenesis, diabetes, atherosclerosis, immune function, and body composition in mammal [14]. Moreover, dietary CLA can enhance the quality of several animal foods such as poultry, pork and fish, and CLA can also be accumulated in the tissues of houseflies and silkworms [17, 18]. It has been reported that CLA is involved in the regulation of above functions associated with the activation of PPARγ. For example, isomer-specific effect of CLA on gene expression in human adipose tissue has been reported to be genotype dependent and at least in part mediated by PPARγ [14]. The *trans*-10, *cis*-12 CLA causes cell delipidation *in vitro* by downregulation of PPARγ [19]. The *trans*-10, *cis*-12 CLA isomer specifically decreases both lipogenesis and adipogenesis by downregulating the expression of PPARγ [14]. Rosi, commonly used to treat type II diabetes, has beneficial effects on weight gain, insulin resistance, inflammation, fibrosis, and angiogenesis [20]. Treatment with Rosi increases perilipin expression in subcutaneous fat, subsequently resulting in lipid storage in subcutaneous fat, fat redistribution and insulin sensitization[21]. In adipose, the improvement of adipocyte function and insulin resistance by Rosi may be relevant to the activation of PPARγ [22].

It is reported that CLA and Rosi are capable of affecting *perilipin* gene expression via PPARγ [6, 21], and PAT family proteins LSD-1 and perilipin A appears highly similar in function and structure [1]. However, there is little information about the effect CLA and Rosi on PAT family proteins in insects. In addition, the structure and biochemical properties of PAT family proteins are well described both in mammals and model organisms, but no detailed data of the PAT family proteins has been reported in *Apis cerana cerana*, the important economic insect of China. In this study, we isolated a novel *LSD-1* gene belonging to a member of PAT family proteins, defined as *AccLSD-1*, and investigated the effects of Rosi and CLA on *AccLSD-1* expression in adult workers of *A. cerana cerana*. Our results indicated that *AccLSD-1* appeared to be ubiquitously expressed on whole developmental stages, especially in pupal metamorphosis when *AccLSD-1* was transcribed in a parabola fashion and reached the highest level at brown-eyed pupae (Pb) stage. Moreover, the transcripts of *AccLSD-1* could be upregulated by Rosi and downregulated by CLA. The congenerous application of CLA and Rosi showed obvious alteration of the *AccLSD-1* expression that was significantly higher than that of CLA alone group. These results indicated that *AccLSD-1* might be involved in pupal development, and can be regulated by CLA and Rosi possibly via activating PPARγ.

Materials and Methods

Materials

The Chinese honeybee, *A. cerana cerana*, was maintained at Shandong Agricultural University (Shandong, China). The conjugated linoleic acid (CLA), purchased from Qingdao Auhai biotech Co., Ltd. (Shandong, China), was composed of 36.2% c9, t11 CLA, 39.2% t10, c12 CLA, 5.0% palmitic acid, 2.1% stearic acid, 10.9% oleic acid, 1.5% linoleic acid and 5.1% others. Rosiglitazone was obtained from Zhengzhou Lion Biological Technology Co., Ltd. (Henan, China).

Primers

The sequences of primers used for cloning in the present study were listed in Table 1. The amplification conditions and products were shown in Supplementary Table S1.

Table 1 The primers used for PCR amplification

Primer	Primer sequence (5' → 3')	Description
For the cloning of the full-length cDNA		
MP1	GACGACATTCTGATTTGCCGC	Gene special primer, forward
MP2	CTGCCAAGATTGCGAGGAGT	Gene special primer, reverse
3P1	GGAACAAGGAACGGAATGC	3' RACE forward primer, primary
3P2	GCCAACCACTGTAGAGCAAT	3' RACE forward primer, nested
B26	GACTCTAGACGACATCGA(T)$_{18}$	Universal primer, primary
B25	GACTCTAGACGACATCGA	Universal primer, nested
5P1	TGTCTACTGCAACATCCGC	5' RACE reverse primer, primary

Primer	Primer sequence (5' → 3')	Description
5P2	TGTGGCGGAAGATGAACTG	5' RACE reverse primer, nested
AAP	GGCCACGCGTCGACTAGTAC(G)$_{14}$	Abridged anchor primer
AUAP	GGCCACGCGTCGACTAGTAC	Abridged universal amplification primer
ZP1	CCGTTCGCACCTTGAGC	Full-length cDNA primer, forward
ZP2	TGGATGAAACTCACGCGG	Full-length cDNA primer, reverse
For the cloning of the genomic and 5'-flanking region		
GP1	TCCAGTCGGCCGTTCGC	Genomic sequence primer, forward
GP2	ATCGTACACATTGCCAGCTATG	Genomic sequence primer, reverse
GP3	CATAGCTGGCAATGTGTACGAT	Genomic sequence primer, forward
GP4	TGTCTACTGCAACATCCGC	Genomic sequence primer, reverse
GP5	GCCTGTTCTGATGCGTGC	Genomic sequence primer, forward
GP6	GCATTCCGTTCCTTGTTCC	Genomic sequence primer, reverse
GP7	GGAACAAGGAACGGAATGC	Genomic sequence primer, forward
GP8	TGGATGAAACTCACGCGG	Genomic sequence primer, reverse
QDP1	CGAGAGAAGTACAACCTTG	I-PCR forward primer, outer
QDP2	GTTTCATTCGATTACTATGTG	I-PCR forward primer, inner
QDP3	TTATGAAAATCTACGATCG	I-PCR reverse primer, outer
QDP4	TAATCGTGTTGGATAGTTGATC	I-PCR reverse primer, inner
WQP1	CATATGGGTGTTTGATGAG	Promoter specific primer, forward
WQP2	GATAGCGAAGTGTGGATTG	Promoter specific primer, reverse
Primers used for quantitative real-time PCR		
RTP1	CGAATCATTTATTGGAGC	Real-time PCR primer, forward
RTP2	GACCTGTGGTTATTGTGG	Real-time PCR primer, reverse
β-action-s	GTTTTCCCATCTATCGTCGG	Standard control primer, forward
β-action-x	TTTTCTCCATATCATCCCAG	Standard control primer, reverse

Amplification of the full-length cDNA of *AccLSD-1*

Total RNA was extracted from the abdominal fat body tissues of adult workers by using Trizol reagent (Invitrogen, USA) and then digested with RNase-free DNase I to remove the genomic DNA. Single-strand cDNA was synthesized by using reverse transcriptase system (TransGen Biotech, China) and it was purified and polyadenylated as described by Wang et al. [23].

By using primers MP1 and MP2 (Table 1) designed by the conserved regions of the *LSD-1* genes from *Apis mellifera*, *Drosophila melanogaster*, *Bombyx mori* and *Manduca sexta*, an internal conservative fragment of approximately 856 bp was amplified under the PCR protocol (Supplementary Table S1). To clone the full-length cDNA, the gene specific primers 5P1/5P2 and 3P1/3P2 (Table 1) were designed and synthesized based on the already obtained partial sequence of *AccLSD-1*, and used for the 5′ rapid amplification of cDNA end (RACE) and 3′ RACE, respectively. For 3′ RACE, the first round of PCR was performed with the primers 3P1 and B26 (Table 1). Then the 10-fold diluted PCR product was used for the second round of amplification with the primers 3P2 and B25 (Table 1). The 5′ RACE was performed by using the purified cDNA. The product of the first round which was performed with the primers 5P1 and AAP (Table 1) was diluted 20-fold and used for the second round of amplification with the primers 5P2 and AUAP (Table 1). The complete coding region of *AccLSD-1* was amplified by using gene specific primers ZP1 and ZP2 (Table 1). All PCR products were cloned into the pMD18-T vector (TaKaRa, Japan) and sequenced by Shanghai Biosune Biotechnology Company (Shanghai, China).

The cloning of the genomic DNA of *AccLSD-1*

Genomic DNA was prepared from the fifth-instar feeding-stage larvae by using EasyPure Genomic DNA Extraction Kit (TransGen Biotech, China) following the protocol of the manufacturer and used as template in PCR reactions.

The *AccLSD-1* genomic sequence was subdivided into four segments for cloning due to its extremely large size, and amplified from genomic DNA by PCR protocol (Supplementary Table S1). The primers used for cloning genomic DNA were designed based on the cDNA sequence and provided in Table 1. The cloning and sequencing strategies were as the same as described above.

Sequence analysis of *AccLSD-1*

The deduced amino acid sequence was obtained by using the DNA-protein translation tool at Expasy (http://ca.expasy.org/tools/dna.html) and aligned with those of several insects by using a DNAMAN software 5.2.2 (Lynnon Biosoft Company). The Molecular Evolution Genetics Analysis (MEGA) version 4.0 was used to construct a phylogenetic tree by using the neighbor-joining method. The isoelectric point (pI) and molecular weight (M_W) were computed by PeptideMass (http://us.expasy.org/tools/peptide-mass.html). LSD-1 phosphorylation sites were predicted by using NetPhosk 1.0 (http://www.cbs.dtu.dk/services/NetPhosK) and the Protscale program (http://www.expasy.ch/tools/protscale.html) was used for structural prediction.

Cloning of the 5′-flanking region of *AccLSD-1*

In order to isolate the 5′-flanking region of *AccLSD-1*, we designed gene specific primers QDP1/QDP3 and QDP2/QDP4 (Table 1) according to 5′ end of the genomic fragment and amplified the 5′-flanking region of *AccLSD-1* by using the inverse polymerase chain reaction (I-PCR). Total genomic DNA (approximate 5 μg) used as template was digested completely with *Nde* I at 37°C overnight, and then ligated to the circles by using T4 DNA ligase (TaKaRa, Japan). The cloning and sequencing strategies were as the same as described above. To further verify the validity of the sequence, the specific primers WQP1 and WQP2 designed based on the obtained promoter sequence were directly used to amplify 5'-flanking region of *AccLSD-1* from *A. cerana cerana* genomic DNA. The transcription factor (TF) binding site predicting tool (http://tfbind.ims.u-tokyo.ac.jp) was used to analyze the putative TF binding site in the 5'-flanking region of *AccLSD-1* as described by Tsunoda and Takagi [24].

Quantitative real-time PCR (qPCR) analysis

To determine the expression profile of *AccLSD-1* mRNA in different developmental stages, three larval instars (L3, L4, L5), prepupal (PP) phase, successive pupal stages (Pw, Pp, Pb, Pbl, Pbm, Pbd) and adult workers of known age were collected and staged according to Bitondi et al. [25]. Quantitative real-time PCR (qPCR) was performed by using *AccLSD-1* gene specific primers RTP1 and RTP2 (Table 1) and *A. cerana cerana β-actin* gene (GenBank accession number: HM640276) primers β-action-s and β-action-x (Table 1) by CFX96™ Real-time System (Bio-Rad). The first-strand cDNA samples from partial larvae (the third- to the fifth-instar larvae), prepupae (PP), pupae (Pw, Pp, and Pb), and adult workers (1st and 10th day) were synthesized as described above and diluted (1 : 10, V/V) in water as the temple for qPCR. The total 25 μl reaction system, mainly consisted of 2 μl cDNA (about 20 ng), 12.5 μl SYBR® Premix Ex *Taq*™(TaKaRa, Japan) and 0.5 μl each primer (10 μmol/L), was carried out in triplicate under the thermal cycling: preincubation at 95℃ for 30 s, 40 cycles of 95℃ for 10 s, 53℃ for 20 s, and 72℃ for 15 s and the melting cycle from 65℃ to 95℃. The relative quantification of *AccLSD-1* transcript was calculated by using the comparative C_t method [26].

Rearing the adult workers of *A. cerana cerana* in laboratory

Two experiments were performed. In the first experiment, the effects of various concentrations of CLA or Rosi diets on the expression pattern of *AccLSD-1* were investigated to determine the properly amount of CLA or Rosi to rear adult workers of *A. cerana cerana*. Adult workers were reared as described by Alaux et al. [27]. In brief, 1-day-old bees were divided into three groups and reared in cages placed into an incubator at 32℃ with 70% relative humidity (RH). Group 1 served as control group was fed on the basic adult diet (BAD) containing candy (30% honey from the source colonies, 70% powdered sugar) and water. Group 2 was fed the mixture of BAD and CLA diets of different concentrations (0.5%, 1%, 5%, and 10% CLA). The mixture of BAD and various Rosi diets with final concentration of 0.5, 1.0, 5.0, 10 mg/ml were fed as the food for group 3, respectively. The second experiment was conducted to detect the effect on the expression of *AccLSD-1* by applying CLA and Rosi together. Adult workers were reared as described in the first experiment, except that they were respectively fed with 1% CLA diets, 0.5 mg/ml Rosi and their combination diets (1% CLA and 0.5 mg/ml Rosi) on the first day of experiment.

All diet treatments were independently replicated three times with approximately 30 individual adults. Diets were provided for adult workers ad libitum. Adult workers of each group were collected at 10 days after the start of the experiment to detect the *AccLSD-1* gene expression pattern by qPCR as described above. Statistical analysis was determined by using SAS software V8e (SAS Institute, Cary, NC, USA).

Results

Isolation and sequence analysis of cDNA sequence of *AccLSD-1*

Through RT-PCR and RACE-PCR methods, a novel *LSD-1* gene, referred to *AccLSD-1*, was cloned from *A. cerana cerana*. The full-length cDNA (GenBank accession number: GU722328), which was 1,313 bp in length, contained an open reading frame (ORF) of 1,161 bp encoding a 386-amino acid polypeptide with a theoretical molecular weight of 42.6 kDa and isoelectric point of 10.03. The non-coding regions, cloned by using the RACE-PCR, were consisted of 40 bp 5' untranslated terminal region (UTR) and 112 bp 3' UTR with the poly (A) tail.

Multiple sequence alignment based on the deduced LSD-1 proteins of *A. cerana cerana* and other five insects showed that the LSD-1 proteins had significant sequence similarity across insects. Moreover, the extensive similarities of the central 63 amino acids (aa) region (216-278 aa) suggested that it should be the spot involved in lipid binding, which was the same as described by Arrese et al. [10]. But the PAT domain, a highly conserved 100 aa region near N-terminal sequence of PAT proteins, had low sequence similarity among these alignment proteins of insects (Supplementary Fig. S1).

The PKA-phosphorylation sites of AccLSD-1 predicted by Netphosk 1.0 contained ten serine residues (ser11, 23, 24, 45, 47, 52, 299, 327, 338, 356) and two threonine residues (thr243, 257) (Supplementary Fig. S1). The LSD-1 proteins of *M. sexta* and *D. melanogaster* contained five serine PKA sites and six PKA-phosphorylation sites, respectively [8, 10]. However, there was no consistent agreement on the conserved PKA sites involved in activation of lipolysis among insects.

Phylogenetic analysis of the identified PAT family protein sequences from insects and vertebrates indicated that AccLSD-1 had high homology with those of other insects, which was consistent with the result of sequence alignment analysis. However, the relationship between insects and vertebrates was more distant (Supplementary Fig. S2).

Hydropathy analysis of AccLSD-1 polypeptide suggested the highest hydrophobic region existed between residues 216-235, which further supported the conserved central region was the binding site for LDs (Supplementary Fig. S3).

Cloning and structural analysis of the genomic sequence of *AccLSD-1*

In order to further analyze the structure of the *AccLSD-1* gene, we also isolated and characterized the genomic DNA sequence. The comparison of the genomic sequence and cDNA sequence of *AccLSD-1* gene revealed that *AccLSD-1* genomic DNA (GenBank accession number: GU722332) spanning 5,224 bp in length contained 8 exons and 7 introns (Fig. 1). All introns had the eukaryote typical characterization with high AT contents and conserved 5'-GT and AG-3' sequence at 5' and 3' splice sites respectively.

The comparison of the genomic sequence of *AccLSD-1* and other insect *LSD-1s* revealed that they all have the same number of introns. In spite of *AmLSD-1* containing 7 exons and 6 introns, there was an intron located in the 5' UTR according to the published *A. mellifera* genome (data not shown). Furthermore, the position and length of *AccLSD-1* introns were similar to *AmLSD-1* but significantly different from *TcLSD-1* and *DmLSD-1*. Notably, the first intron of *AccLSD-1*, approximately 1,413 bp, was located in the 5' UTR(Fig. 1). These results suggested that the first intron may play an important role in transcription of *AccLSD-1*.

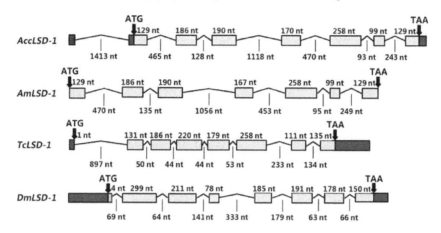

Fig. 1 Exon-intron organization of *AccLSD-1* and its homologues, including *AmLSD-1*, *TcLSD-1* and *DmLSD-1*. Exons, introns and untranslated regions (UTR) are shown with light gray boxes, lines and dark gray boxes, respectively. The initiation and termination codons of each gene are marked with arrows. Length of the exons and introns is indicated by number of nucleotides (nt). The sequences used are as follows: *AccLSD-1* (*A. cerana cerana*, GU_722332), *AmLSD-1* (*A. mellifera*, NC_007078), *TcLSD-1* (*T. castaneum*, NC_007417), *DmLSD-1* (*D. melanogaster*, NT_033777).

Characterization of the 5'-flanking region of *AccLSD-1*

To identify upstream regulatory elements of *AccLSD-1* gene, a 1,231 bp fragment of the 5'-flanking region was generated from the *A. cerana cerana* genome by I-PCR. Then a number of putative *cis*-acting elements in the 5'-flanking region were predicted by using the TFBIND software. The results indicated that in addition to the typical TATA-box and CAAT-box, the binding sites of PPARγ and CAAT/enhancer binding protein α (C/EBPα), which are the major and determining adipogenic transcription factors, were found in this region. As shown in Fig. 2, three putative PPREs were located at −148 to −167, −411 to −430 and −1146 to −1165, respectively. In addition, several important transcription factors such as heat shock factors (HSF) and T-cell factor (data not shown) required for regulating various environmental stresses were predicted.

Expression pattern of *AccLSD-1* in different developmental stages

To detect the expression levels of *AccLSD-1* at an appropriate developmental stage, we firstly performed a general overview on the expression pattern of *AccLSD-1* in the whole developmental stages. As shown in Fig. 3A, (Pw-Pb) and decreased again at progressive body pigmentation stages (Pbl-Pbd). Interestingly, the highest level of *AccLSD-1* was detected at Pb stage (Fig. 3B). After the adult molt, the expression of *AccLSD-1* had a low level at the first day of adult stage and maintained a high level at the 10th day of the adult stage at least (Fig. 3A). We

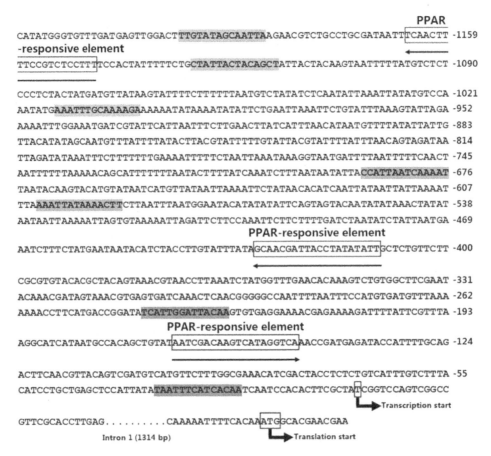

Fig. 2 The nucleotide sequence and putative transcription factor-binding sites of the 5′-flanking region of *AccLSD-1*. The transcription start site and translation start site are marked with the arrowheads. Boxed sequences and arrows indicate three putative PPAR-responsive elements found by the TFBIND software and their direction, respectively. Partial CAAT/enhancer binding protein α (C/EBPα) binding sites are highlighted in gray.

further investigated the expression pattern of *AccLSD-1* in the whole adult stage (Fig. 3C). The observation suggested that the expression of *AccLSD-1* exhibited much stronger in youth workers (25-day-old) than childhood (2-day-old) and old age ones (50-day-old). Given these results, we presumed that *AccLSD-1* might be involved in all developmental stages, especially in the pupal molt. The developmental stages with a high level of *AccLSD-1* expression included the fifth-instar larvae, Pb, and 10-day-old adult stages. In this study, we chose the 10-day-old adult stage as the appropriate spot for determination of the gene expression.

Effect of CLA, Rosi, and their combination on *AccLSD-1* mRNA expression

Based on the results above, we determined to analyze the effect of CLA or Rosi on *AccLSD-1* expression at 10-day old adult stage. Dietary CLA had been shown to decrease the expression of *AccLSD-1* in Fig. 4. The expression of *AccLSD-1* was decreased dramatically along with the increasing concentration of CLA diets (Fig. 4A). Treatment of Rosi alone had been found to progressively improve the expression of *AccLSD-1* as the increasing concentration from 0.5

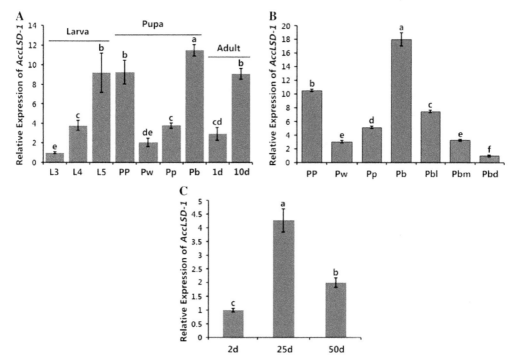

Fig. 3 Quantitative real-time PCR analysis of *AccLSD-1* expression profiles during development. The *A. cerana cerana* β-actin gene is applied as internal standard control. Different letters above the bars indicate significant difference as determined by using one-way ANOVA analysis ($P < 0.05$). Data represent the means ± SE ($n = 3$). A. Expression patterns of *AccLSD-1* in whole developmental stage. L3, L4, L5, the 3rd-, the 4th- and the 5th-instar larval phases; PP, prepupal phase; Pw, Pp, Pb, unpigmented pupae of white-, pink- or brown-eyed. B. Expression patterns of *AccLSD-1* during pupal stage. Pbl, Pbm, Pbd, light, medium and dark pigmented thorax pupae. C. Expression patterns of *AccLSD-1* during adult stage. 2-day-old, 25-day-old, 50-day-old workers represent the childhood, youth and old age workers, respectively.

to 1.0 mg/ml. However, it was not affected significantly when the concentration of Rosi diets was higher than 1.0 mg/ml. The amount of *AccLSD-1* expression was almost five-fold higher in 1.0 mg/ml Rosi group comparable to control group (BAD) (Fig. 4B). Considering that the high concentration of CLA or Rosi had significant influence on the food intake and survival rate of adult workers, we chose the 1% CLA and 0.5 mg/ml Rosi as the suitable diets for rearing adult workers of *A. cerana cerana* in the laboratory. Subsequently, we examined the combined effect of 1% CLA and 0.5 mg/ml Rosi on the *AccLSD-1* expression of adult workers. Compared with the control group, CLA alone suppressed *AccLSD-1* transcription significantly. Nevertheless, when combined with Rosi treatment, the mixed group significantly attenuated the suppression of *AccLSD-1* mRNA levels by CLA alone and the transcription of *AccLSD-1* was 2.2-fold higher in the mixed group than the control group (Fig. 4C).

Discussion

PAT family proteins are evolutionary highly conserved lipid droplet-associated proteins that are involved in lipid metabolism and transport, intracellular trafficking and signaling [6]. In

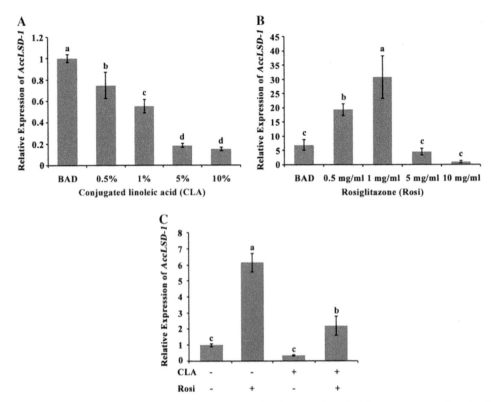

Fig. 4 Effects of dietary CLA, Rosi and their combination on *AccLSD-1* gene expression. A, B. In comparison with BAD group (control group), various CLA diets (0.5%, 1%, 5%, and 10% CLA) and Rosi diets (final concentration: 0.5, 1.0, 5.0, 10 mg/ml) significantly affect *AccLSD-1* gene expression. C. Quantitative real-time PCR for relative expression of *AccLSD-1* mRNA in bees fed with (+) or without (-) 1% CLA or 0.5 mg/ml Rosi diets. The combination of CLA and Rosi completely rescued the suppression of *AccLSD-1* mRNA levels by CLA alone. Different letters above the bars indicate significant difference as determined by using one-way ANOVA analysis ($P < 0.05$). Data represent the means ± SE ($n = 3$).

the present study, we isolated and characterized a member of the PAT family proteins, AccLSD-1, from *A. cerana cerana* for the first time.

Based on the sequence and structure of LSD-1, several interesting characterizations were found: (1) Significant sequence similarity of LSD-1 proteins had not been found in structural motif, "PAT domain", but in the central region of a predicting binding site for LDs. Previous observations in *M. sexta* and *D. melanogaster* have suggested that LSD-1 adopts a putative α-helical conformation to bind to the lipid droplet surface and contains several unfold regions which are likely to connect with other proteins, such as LSD-2 and TGL [8]. However, the precise molecular mechanism that underlies the interaction between LSD-1 and LSD-2 on LDs is yet unknown. In addition, three proteins including CG2254, CG1112 and CG10691 have been identified to locate on the surface of lipid droplet [4]. Therefore, there may be more proteins associated with LDs, and LSD-1 is potentially involved in specific protein-protein interactions. (2) A typical characterization of *LSD-1* in insects was composed of 8 exons and 7 introns. Notably, the first intron of *AccLSD-1* was located in the 5' UTR, which is different from that of other insect *LSD-1* genes but is similar to the mammal *perilipin* gene [28].

Sequence analysis on the first intron of *AccLSD-1* revealed that it contained three PPREs (results not shown), which is involved in the transactivation of the adipogenic target genes by PPARγ. Previous observation has shown that the Acyl-CoA-binding Protein (ACBP) is activated by PPARγ through an intronic PPRE [29]. These results imply that the first intron is likely to regulate the *AccLSD-1* expression through interacting with PPARγ. (3) Sequence analysis on the *AccLSD-1* promoter revealed a number of adipogenic transcription factors and stress response elements that may be involved in multiple signalling pathways in *A. cerana cerana* lipid metabolism and stress responses. Interestingly, three putative PPREs identified in the promoter region of *AccLSD-1* is similar to that of mammal PAT family protein [30, 31], suggesting that some conservation with respect to the regulatory mechanisms of PAT family protein exists between PAT family protein in mammals and in insects.

PZAT family proteins play fundamental roles in lipid homeostasis and lipid-droplet biology [6]. Evidence from *in vivo* and *in vitro* studies has shown that the phosphorylation of LSD-1, highly similar to perilipin A, seems to be a strong correlation with the lipolysis [1, 8]. Lipid metabolism in the lipid droplets is essential for insect development and metamorphosis [2]. In insects, lipid from the diet is digested and absorbed by midgut, and subsequently either been stored in fat body or utilized for flight muscle and ovary, which is determined by several factors, such as development stage, nutritional state, sex, and migratory flight [2, 32]. Therefore, we determined the *AccLSD-1* mRNA expression during lipid metabolism in different development stages in the study. The results revealed that *AccLSD-1* constitutively expressed throughout feeding larval, pupal and adult stages, which had also been reported in *M. sexta* [10]. Interestingly, there was an abrupt decrease in the quantity of transcripts in both larval-pupal transition and pupal-adult transition. Moreover, *AccLSD-1* was transcribed in a parabola fashion rather than persistently reduced trend in nonfeeding pupal stage. The expression patterns of *AccLSD-1* indicated that this gene may be associated with the successive metamorphosis of *A. cerana cerana*, especially during pupal development. Previous studies in *Bombyx* suggest that fat body cells are dissociated to individuals and progressively underwent programmed cell death (PCD) including autophagy and apoptosis during pupal metamorphosis. Meanwhile, progenitor cells of adult fat body, which can promote the growth and differentiation of adult tissues, are gradually generated in pupal stage [2]. Therefore, the special expression pattern of *AccLSD-1* in pupal stage of honeybees may be connected with the change of fat body cells and progenitor cells. Additionally, the study in *M. sexta* reveals that LSD-1 shows the highest mRNA level in the feeding larvae [10]. However, *AccLSD-1* in *A. cerana cerana* exhibited the highest mRNA level at Pb stage. In particular, Pb stage is the spot for transformation from eye pigmentation to body pigmentation during pupal metamorphosis [33]. Thus we deduced that *AccLSD-1* may be involved in pupal body pigmentation. Besides, it is notable that the expression level of *AccLSD-1* is gradually elevated during the stage of feeding larval and adult stage when they could be fed by nurse bees or foraging by themselves normally. In *Drosophila*, the expression level of *LSD-1* in adult is downregulated in response to starvation [6, 34]. Therefore, we hypothesized that the transcriptional levels of *LSD-1* were positively correlated with nutritional status and the rate of lipolysis. These discoveries provide new knowledge for further understanding the function of LSD-1. However, much more functional information of insect LSD-1s remains to be explored.

Since there is an intriguing correlation between the activation of LSD-1 and lipid metabolism, it will be particularly important to understand the regulatory mechanisms of LSD-1. Although significant similarity between the regulation of lipolysis in mammals and in insects [6], no data is yet known to account for the mechanisms. In mammal adipose tissue, PPARγ is certified to be an important mediator on the activation of PAT family protein [11]. Numerous regulators involved in modulating the transcriptional activation of PPARγ include natural or synthetic ligands [13] and multiple signaling pathways such as mitogen-activated protein kinase (MAPK), c-Jun N-terminal kinase (JNK), cyclic adenosine monophosphate (cAMP) and Akt/protein kinase B (PKB) signaling pathways [35-37]. As the PPARγ-specific ligands, CLA (predominantly *trans*-10, *cis*-12 CLA) and Rosi have been demonstrated to induce the expression of PPARγ [19, 22] and modulate expression of genes involved in neutral lipid metabolism acting through PPARγ [14, 22]. Consistent with previous reports, we showed that feeding with either CLA or Rosi alone all caused a significant change in mRNA levels of *AccLSD-1* in a dose-dependent manner, suggesting that the *AccLSD-1* might be transcriptionally regulated by PPARγ. Besides, three PPREs identified in the *AccLSD-1* promoter region further suggested that *AccLSD-1* is a novel PPARγ target gene. Whether CLA or Rosi induces the transcription of the *AccLSD-1* through PPARγ, and whether the regulation of *AccLSD-1* depends on the interaction of PPARγ with the promoter or intronic PPRE of *AccLSD-1* remains to be elucidated.

Previously, it was demonstrated that CLA and Rosi have been demonstrated to play adverse roles in adipose mass, weight maintain and insulin resistance [22, 38]. Moreover, the combination of CLA with Rosi can attenuate the suppression of adipose abundant genes such as *lipoprotein lipase* (*LPL*) and *adiponectin receptor* (*AdipoR2*) by CLA alone [22]. Our findings agree with previous reports showing supplementation of both Rosi and CLA significantly reduced suppression in expression of *AccLSD-1* induced by dietary CLA alone. It is possible that there is some difference in the regulatory mechanism of CLA and Rosi on *AccLSD-1* expression. Additionally, the results showed that the transcripts of *AccLSD-1* could be dramatically upregulated by Rosi and downregulated by CLA, respectively. The expression pattern of *AccLSD-1* also indicated that CLA and Rosi had different potential functions and regulatory mechanisms on lipid metabolism. In mammal, there are enormous reports about the potential mechanisms of CLA on modulating adipogenic gene but little about Rosi. *Trans*-10, *cis*-12 CLA has been certified to be directly linked to the suppression of PPARγ activity by phosphorylating PPARγ via ERK [13] and suppresses adipogenic gene expression and metabolism by attenuating PPARγ DNA-binding affinity or transcriptional activation via nuclear factor κB (NF-κB) and MAPK [13, 39]. However, the mechanisms by which CLA or Rosi inducing the *AccLSD-1* expression are still unknown.

In summary, this study identified the structural characterization and functional implication of a novel *AccLSD-1* gene, showing that *AccLSD-1* functions as a considerable modulator on *A. cerana cerana* development, especially in the pupal metamorphosis, and may be regulated by PPARγ. The results pave a way for a subsequent functional analysis and regulatory mechanism of *AccLSD-1*.

References

[1] Patel, R. T., Soulages, J. L., Hariharasundaram, B., Arrese, E. L. Activation of the lipid droplet controls the rate of lipolysis of triglycerides in the insect fat body. J Biol Chem. 2005, 280 (24), 22624-22631.

[2] Liu,Y., Liu, H., Liu, S., Wang, S., Jiang, R. J., Li, S. Hormonal and nutritional regulation of insect fat body development and function. Arch of Insect Biochem & Physiol. 2009, 71 (1), 16-30.

[3] Martin, S., Parton, R. G. Lipid droplets: a unified view of a dynamic organelle. Nat Rev Mol Cell Biol. 2006, 7 (5), 373-378.

[4] Beller, M., Riedel, D., Jansch, L., Dieterich, G., Wehland, J., Jackle, H., Kühnlein, R. P. Characterization of the *Drosophila* lipid droplet subproteome. Mol & Cell Proteomics. 2006, 5 (6), 1082-1094.

[5] Miura, S., Gan, J. W., Brzostowski, J., Parisi, M. J., Schultz, C. J., Londos, C., Oliver, B., Kimmel, A. R. Functional conservation for lipid storage droplet association among Perilipin, ADRP, and TIP47 (PAT)-related proteins in mammals, *Drosophila*, and *Dictyostelium*. J Biol Chem. 2002, 277 (35), 32253-32257.

[6] Bickel, P. E., Tansey, J. T., Welte, M. A. PAT proteins, an ancient family of lipid droplet proteins that regulate cellular lipid stores. Biochim Biophys Acta. 2009, 1791, (6), 419-440.

[7] Kimmel, A. R., Brasaemle, D. L., McAndrews-Hill, M., Sztalryd, C., Londos, C. Adoption of PERILIPIN as a unifying nomenclature for the mammalian PAT-family of intracellular, lipid storage droplet proteins. J Lipid Res. 2010, 51 (3), 468-471.

[8] Arrese, E. L., Rivera, L., Hamada, M., Mirza, S., Hartson, S. D., Weintraub, S., Soulages, J. L. Function and structure of lipid storage droplet protein 1 studied in lipoprotein complexes. Arch Biochem Biophys. 2008, 473 (1), 42-47.

[9] Arrese, E. L., Howard, A. D., Patel, R. T., Rimoldi, O. J., Soulages, J. L. Mobilization of lipid stores in *Manduca sexta*: cDNA cloning and developmental expression of fat body triglyceride lipase, TGL. Insect Biochem Mol Biol. 2010, 40 (2), 91-99.

[10] Arrese, E. L., Mirza, S., Rivera, L., Howard, A. D., Chetty, P. S., Soulages, J. L. Expression of lipid storage droplet protein-1 may define the role of AKH as a lipid mobilizing hormone in *Manduca sexta*. Insect Biochem Mol Biol. 2008, 38 (11), 993-1000.

[11] Dalen, K. T., Schoonjans, K., Ulven, S. M., Weedon-Fekjaer, M. S., Bentzen, T. G., Koutnikova, H., Auwerx, J., Nebb, H. I. Adipose tissue expression of the lipid droplet-associating proteins S3-12 and perilipin is controlled by peroxisome proliferator-activated receptor-γ. Diabetes. 2004, 53 (5), 1243-1252.

[12] Wolins, N. E., Quaynor, B. K., Skinner, J. R., et al. OXPAT/PAT-1 is a PPAR-induced lipid droplet protein that promotes fatty acid utilization. Diabetes. 2006, 55 (12), 3418-3428.

[13] Kennedy, A., Chung, S., LaPoint, K., Fabiyi, O., McIntosh, M. K. *Trans-10, cis-12* conjugated linoleic acid antagonizes ligand-dependent PPARγ activity in primary cultures of human adipocytes. J Nutr. 2008, 138 (3), 455-461.

[14] Brown, J. M., Boysen, M. S., Jensen, S. S., Morrison, R. F., Storkson, J., Lea-Currie, R. Isomer-specific regulation of metabolism and PPARγ signaling by CLA in human preadipocytes. J Lipid Res. 2003, 44 (7), 1287-1300.

[15] Yajima, H., Kobayashi, Y., Kanaya, T., Horino, Y. Identification of peroxisome-proliferator responsive element in the mouse HSL gene. Biochem Biophys Res Commun. 2007, 352 (2), 526-531.

[16] Rypka, M., Vesely, J. Evidence for differential effects of glucose and cycloheximide on mRNA levels of peroxisome proliferator-activated receptor- (PPAR-) machinery members: superinduction of PPAR-γ1 and -γ2 mRNAs. Acta Biochim Pol. 2010, 57 (2), 209-215.

[17] Park, C. G., Kim, S. J., Ha, Y. L. Dietary conjugated linoleic acid (CLA) in house fly, *Musca domestica*, with no adverse effects on development. J Asia-Pacific Entomol. 2000, 3 (2), 59-64.

[18] Park, C. G., Park, G. B., Kim, Y. S., Kim, S. J., Min, D. B., Ha, Y. L. Production of silkworms with conjugated linoleic acid (CLA) incorporated into their lipids by dietary CLA. J Agric Food Chem.

2006, 54 (18), 6572-6577.

[19] Stachowska, E., Baskiewicz, M., Marchlewicz, M., Czuprynska, K., Kaczmarczyk, M., Wiszniewska, B., Chlubek, D. Conjugated linoleic acids regulate concentration triacylglycerol and cholesterol of macrophages/foam cells by modulating of CD36 expression. Acta Biochim Pol. 2017, 57 (3), 379-384.

[20] Sandoval, P., Loureiro, J., Gonzalez-Mateo, G., et al. PPAR-γ agonist rosiglitazone protects peritoneal membrane from dialysis fluid-induced damage. Lab Invest. 2010, 90 (10), 1517-1532.

[21] Kim, H. J., Jung, T. W., Kang, E. S., Kim, D. J., Ahn, C. W., Lee, K. W., Lee, H. C., Cha, B. S. Depot-specific regulation of perilipin by rosiglitazone in a diabetic animal model. Metabolism. 2007, 56 (5), 676-685.

[22] Liu, L. F., Purushotham, A., Wendel, A. A., Belury, M. A. Combined effects of rosiglitazone and conjugated linoleic acid on adiposity, insulin sensitivity, and hepatic steatosis in high-fat-fed mice. Am J Physiol Gastrointest Liver Physiol. 2007, 292 (6), G1671-G1682.

[23] Wang, J., Wang, X., Liu, C., Zhang, J., Zhu, C., Guo, X. The *NgAOX1α* gene cloned from *Nicotiana glutinosa* is implicated in the response to abiotic and biotic stresses. Biosci Rep. 2008, 28 (5), 259-266.

[24] Tsunoda, T., Takagi, T. Estimating transcription factor bindability on DNA. Bioinformatics. 1999, 15 (7-8), 622-630.

[25] Bitondi, M. M., Nascimento, A. M., Cunha, A. D., Guidugli, K. R., Nunes, F. M., Simoes, Z. L. Characterization and expression of the *Hex 110* gene encoding a glutamine-rich hexamerin in the honey bee, *Apis mellifera*. Arch Insect Biochem Physiol. 2006, 63 (2), 57-72.

[26] Chen, Y. P., Higgins, J. A., Feldlaufer, M. F. Quantitative realtime reverse transcription-PCR analysis of deformed wing virus infection in the honeybee (*Apis mellifera* L.). Appl Environ Microbiol. 2005, 71 (1), 436-441.

[27] Alaux, C., Ducloz, F., Le Crauser, D., Conte, Y. Diet effects on honeybee immunocompetence. Biol Lett. 2010, 6 (4), 562-565.

[28] Lu, X., Gruia-Gray, J., Copeland, N. G., Gilbert, D. J., Jenkins, N. A., Londos, C. The murine perilipin gene: the lipid droplet-associated perilipins derive from tissue-specific, mRNA splice variants and define a gene family of ancient origin. Mamm Genome. 2001, 12 (9), 741-749.

[29] Helledie, T., Grontved, L., Jensen, S. S., et al. The gene encoding the Acyl-CoA-binding protein is activated by peroxisome proliferator-activated receptor γ through an intronic response element functionally conserved between humans and rodents. J Biol Chem. 2002, 277 (30), 26821-26830.

[30] Girnun, G. D., Domann, F. E., Moore, S. A., Robbins, M. E. Identification of a functional peroxisome proliferator-activated receptor response element in the rat catalase promoter. Mol Endocrinol. 2002, 16 (12), 2793-2801.

[31] Nagai, S., Shimizu, C., Umetsu, M., Taniguchi, S., Endo, M., Miyoshi, H., Yoshioka, N., Kubo, M., Koike, T. Identification of a functional peroxisome proliferator-activated receptor responsive element within the murine perilipin gene. Endocrinology. 2004,145 (5), 2346-2356.

[32] Arrese, E. L., Patel, R. T., Soulages, J. L. The main triglyceride-lipase from the insect fat body is an active phospholipase A1: identification and characterization. J Lipid Res. 2006, 47 (12), 2656-2667.

[33] Michelette, E. R. F., Soares, A. E. E. Characterization of preimaginal developmental stages in Africanized honey bee workers (*Apis mellifera* L.). Apidologie. 1993, 24 (4), 431-440.

[34] Gronke, S., Mildner, A., Fellert, S., Tennagels, N., Petry, S., Müller, G., Jäckle, H., Kühnlein, R. P. Brummer lipase is an evolutionary conserved fat storage regulator in *Drosophila*. Cell Metab. 2005, 1 (15), 323-330.

[35] Kim, S. P., Ha, J. M., Yun, S. J., Kim, E. K., Chung, S. W., Hong, K. W., Kim, C. D., Bae, S. S. Transcriptional activation of peroxisome proliferator-activated receptor-γ requires activation of both protein kinase A and Akt during adipocyte differentiation. Biochem Biophys Res Commun. 2010, 399 (1), 55-59.

[36] Hu, E., Kim, J. B., Sarraf, P., Spiegelman, B. M. Inhibition of adipogenesis through MAP

kinase-mediated phosphorylation of PPARγ. Science. 1996, 274 (5295), 2100-2103.

[37] Camp, H. S., Tafuri, S. R., Leff, T. c-Jun N-terminal kinase phosphorylates peroxisome proliferator-activated receptor-gamma1 and negatively regulates its transcriptional activity. Endocrinology. 1999, 140 (1), 392-397.

[38] Herrmann, J., Rubin, D., Hasler, R., et al. Isomer-specific effects of CLA on gene expression in human adipose tissue depending on PPARγ2 P12A polymorphism: a double blind, randomized, controlled cross-over study. Lipids Health Dis. 2009, 8 (1), 35.

[39] Chung, S., Brown, J. M., Provo, J. N., Hopkins, R., McIntosh, M. K. Conjugated linoleic acid promotes human adipocyte insulin resistance through NFκB-dependent cytokine production. J Biol Chem. 2005, 280 (46), 38445-38456.

4.4 Ribosomal Protein L11 is Related to Brain Maturation During the Adult Phase in *Apis cerana cerana* (Hymenoptera, Apidae)

Fei Meng, Wenjing Lu, Feifei Yu, Mingjiang Kang, Xingqi Guo, Baohua Xu

Abstract: Ribosomal proteins (RPs) play pivotal roles in developmental regulation. The loss or mutation of ribosomal protein L11 (*RPL11*) induces various developmental defects. However, few RPs have been functionally characterized in *Apis cerana cerana*. In this study, we isolated a single copy gene, *AccRPL11*, and characterized its connection to brain maturation. *AccRPL11* expression was highly concentrated in the adult brain and was significantly induced by abiotic stresses such as pesticides and heavy metal. Immunofluorescence assays demonstrated that AccRPL11 was localized to the medulla, lobula and surrounding tissues of esophagus in the brain. The post-transcriptional knockdown of *AccRPL11* gene expression resulted in a severe decrease in adult brain than in other tissues. The expression levels of other brain development-related genes, *p38*, *ERK2*, *CacyBP* and *CREB*, were also reduced. Immunofluorescence signal attenuation was also observed in *AccRPL11*-rich regions of the brain in dsAccRPL11-injected honeybees. Taken together, these results suggest that AccRPL11 may be functional in brain maturation in honeybee adults.
Keywords: *AccRPL11*; *Apis cerana cerana*; Knockdown; Brain maturation; Immunofluorescence

Introduction

The honeybee is a highly eusocial insect that exhibits several striking social behaviors including age-related labor division, symbolic dance language and a precise learning system [1, 2]. These complex social behaviors should be associated with certain coordinated systems in the brain. The changes in mushroom bodies are associated with division labor in adult honeybee colonies [3]. Recently, the expression of genes related to brain development has also been implicated in behavioral maturation [4-6].

There are many genes associated with development in eukaryotes such as the mitogen-activated protein kinase (MAPK) family and the ribosomal protein (RP) family of genes, which participate in growth at different developmental stages. The ribosome is an essential, highly complex ribonucleoprotein organelle that is responsible for translating mRNA into protein. It is composed of a large (60S) and a small (40S) subunit. In eukaryotes, the large subunit of a mature ribosome is composed of 28S, 5.8S and 5S rRNA and approximately 47 RPs [7]. Some RPs are believed to have critical biological functions in the regulation of development, sexual differentiation and extraribosomal transcription and translation [8-10]. The knockout mutation of *RPL24* in mice causes skeletal deformities including a kinked tail, a ventral midline spot and white hind feet [11]. Adult flies that are heterozygous for *RPL5* and *RPL38* have larger wings than wild-type flies [12].

Ribosomal protein L11 (RPL11) is a component of the large 60S ribosomal subunit and plays several important roles in organisms. The protein consists of two major domains: a compact C-terminal domain (CTD) and a tightly folded N-terminal domain (NTD), which is necessary for peptide release [13]. Previous studies in mammalian cell lines have indicated that RPL11 and the tumor suppressor protein p53 have close functional links with each other. It has also been shown that RPL11, in its role as a regulator of p53, can bind directly to MDM2, inducing the activation of p53. The activation results in cell cycle arrest and DNA repair and developmental abnormalities [14-17]. Recent studies in humans and animal models, such as zebrafish, have shown that the defects caused by *RPL11* mutation or knockdown are mediated via a p53-dependent apoptotic response. For example, the loss of *RPL11* in zebrafish affects embryonic brain development, leading to p53-induced lethality [18]. A zebrafish *RPL11* mutant induces defects in hematopoietic stem cells, which result in the unique disease phenotype of Diamond-Blackfan anemia (DBA) [19]. The *RPL11* mutation in humans is also associated with production of DBA and the accompanying hematopoiesis abnormality can be caused by either p53 family members or p53-independent pathways [20]. These data confirm that the *RPL11* gene has a critical role in organ development.

The abnormal expression of *RPL11* clearly results in tissue-specific abnormalities in a variety of species; however, even within a species, *RPL11* displays different expression patterns. In *Arabidopsis thaliana*, the high expression level of one *RPL11* gene was detected only in the vasculature and cotyledons, whereas the expression of a second *RPL11* gene was restricted to the pollen, root stele and developing anthers [21, 22]. Studies on RP genes have been focused predominantly on such model organisms as mice, humans and zebrafish. Consequently, information about other important species is lacking, especially in insects. The honeybee has become an excellent insect model for the study of social behavior due to its unique process of behavioral maturation. The Asian honeybee (*Apis cerana cerana*) is an important indigenous species that is necessary for low temperature pollination and ecological diversity. Because RPs play different roles in development, we were interested in identifying the role of the *RPL11* gene in *A. cerana cerana*. In this study, we identified a *RPL11* gene in *A. cerana cerana* (*AccRPL11*), predicted putative transcription factor binding sites (TFBS) in its promoter region and investigated its expression pattern. Based on previous evidence, we presumed that the *AccRPL11* could play a role in the brain, and this hypothesis was experimentally examined in the brain by using immunofluorescence localization. RNAi experiments were also performed to investigate the impact of *AccRPL11* depletion on the brain.

Material and Methods

Insects

Colonies of *A. cerana cerana* at different developmental stages were obtained from hives maintained at the experimental apiary of Shandong Agricultural University, Taian, China. One-day-old adults were collected soon after they emerged from the honeycomb and transferred to a hand-feeding box. The adults were fed a diet of water and pollen and maintained in an incubator with 60% relative humidity at 34℃ under a 24 h dark regimen.

Isolation of the *AccRPL11* gene

Total RNA was extracted by using Trizol reagent (Invitrogen, Carlsbad, CA, USA), and the first-strand cDNA was synthesized by using reverse transcriptase (TransGen Biotech, Beijing, China). Genomic DNA was isolated from the entire body of the adult honeybee by using the EasyPure Genomic DNA Extraction Kit (TransGen Biotech, Beijing, China).

Primers RP1 and RP2 were designed based on conserved regions of *RPL11* genes from several species and were used to obtain an internal fragment of *RPL11* by PCR. By using the sequence of amplified fragment, two pairs of primers, 5P1/2 and 3P1/2, were subsequently generated and employed for 5' RACE and 3' RACE, respectively. Another pair of primers, Q1 and Q2, was synthesized to verify the sequence of deduced full-length cDNA of *AccRPL11*. Several specific primers were designed and used to isolate *AccRPL11* genomic DNA. The genomic DNA was completely digested with *Eco*R I and self-ligated by using T4 DNA ligase (TakaRa, Dalian, China) to produce a template for promoter amplification by using inverse PCR (I-PCR). Two primer pairs (QDZS1/2 and QDZX1/2) were used to generate a fragment containing the *AccRPL11* promoter, and the primers QDZQ1 and QDZQ2 were used to verify its sequence. The primers and their sequences are listed in Supplementary Table S1.

Bioinformatic analysis

Homologous RPL11 protein sequences from various species were retrieved from the NCBI database and aligned by using DNAMAN version 5.2.2 (Lynnon Biosoft, Quebec, Canada). Conserved domains in RPL11 were predicted by using the Conserved Domain Database available at NCBI (http://www.ncbi.nlm.nih.gov/Structure/cdd/cdd.shtml). The molecular mass and isoelectric point (pI) of relevant protein was calculated by using PeptideMass (http://web.expasy.org/cgi-bin/compute_pi/pi_tool). Phylogenetic analysis was performed by using the neighbor-joining method with the Molecular Evolutionary Genetic Analysis (MEGA) software version 4.0 [23]. Several transcription factor binding sites in the 5'-flanking region of *AccRPL11* were identified by using the TFSEARCH database (http://www.cbrc.jp/research/db/ TFSEARCH).

Genomic and mRNA hybridization

For Southern hybridization, 30 μg of genomic DNA of *A. cerana cerana* was completely digested by using three restriction endonucleases (*Eco*R I , *Eco*R V and *Pvu* II) for 48 h at

37℃. After precipitation with ethanol, the DNA was separated on 1% agarose gel and transferred to a Hybond-N + nylon membrane (Amersham, Pharmacia, UK) by using capillary blotting. For Northern blotting, 20 μg of total RNA extracted was separated on 1% agarose-formaldehyde gel and transferred to a Hybond-N + nylon membrane. A highly specific probe radiolabeled with [α^{32}P] was synthesized with primers SB1 and SB2 by using the Primer-a-Gene Labeling System (Promega, Madison, WI, USA). Southern blotting and Northern blotting hybridizations were performed as described previously [24]. Briefly, membranes were prehybridized overnight and then hybridized for 48 h at 65℃ with a gene-specific probe. After hybridization, the membranes were rinsed twice in 2× SSC and 0.2% SDS followed by 0.2× SSC and 0.2% SDS for 10 min each at 42℃. The radioactive signal was visualized by using a FLA-7000 phosphorimager (Fujifilm).

Protein expression and purification of recombinant AccRPL11 protein

Primers (YH1/YH2) were designed and incorporated to *Sac* I and *Hin*dIII, respectively. The amplified fragment was inserted into the expression vector pET-30a (+) excised with *Sac* I and *Hin*dIII. The recombinant plasmid pET-30a (+)-AccRPL11 was transformed into *Escherichia coli* BL21 (DE3) for subsequent protein expression studies. The induction, expression and protein purification of AccRPL11 were performed as described previously [25]. The concentration of the purified protein was determined with Coomassie Brilliant Blue G-250 by using bovine serum albumin as a standard. For polyclonal antibody production, purified AccRPL11 protein was subcutaneously injected into white mice.

Preparation of dsAccRPL11 RNA

Primers R1 and R2 harboring T7 polymerase promoter sequences at their 5′ ends were designed for the synthesis of *AccRPL11* double-stranded RNA (ds*AccRPL11*). The resulting PCR product was purified by using the Gel Extraction Kit (Solarbio, Beijing, China), and purified PCR product was used to synthesize the *dsAccRPL11* by using RiboMAX™ T7 Large Scale RNA Production Systems (Promega, Madison, WI, USA). The ds*AccRPL11* RNA preparation was digested by using DNase I, precipitated with isopropanol and redissolved in RNase-free water. The two strands of RNA were annealed by heating the RNA to 95℃ for 1 min and then cooling it gradually to room temperature. As a control, we also synthesized dsRNA from the green fluorescent protein gene (GFP gene) [26]. The concentration of dsRNA was determined by using spectrophotometry.

RNA interference (RNAi)

Newly emerged honeybee adults were collected for the RNAi experiments. Prior to injection, adults were placed on ice for 5 min to decrease their body pressure, and individuals showing signs of hemolymph leakage were discarded after injection. Each adult was injected between the first and the second abdominal segments with a microsyringe containing 10 μl (5 μg) of ds*AccRPL11* diluted 1 : 5 in RNase-free water. At the same time, an equivalent amount of ds*GFP* and equivalent volume of RNase-free water were also injected as injection controls.

All of the adults were then maintained in an incubator with 60% relative humidity at 34℃ under a 24 h dark regimen. During the incubation period, healthy adults were sampled at appropriate times, and various tissues were dissected to determine their RNAi status.

Immunofluorescence analysis

Brain tissue dissected from the normal honeybee adults and dsRNA-injected adults at 7 days postinjection were fixed overnight in 4% paraformaldehyde at 4℃. The next day, the tissue samples were washed thoroughly, dehydrated, cleared, dipped in wax and, finally, embedded in paraffin. Slices of 10 μm thickness were produced and mounted onto poly-L-lysine-coated glass slides. Dewaxed sections were blocked for 30 min in 10% normal goat serum at room temperature and incubated with the anti-AccRPL11 polyclonal antibody (1 ∶ 50 dilution) at 4℃ overnight. The sections were then washed in PBS three times and incubated with fluorescein isothiocyanate (FITC)-conjugated goat anti-mouse IgG (1 ∶ 100 dilution, Beyotime Biotech, Jiangsu, China) in the dark at 37℃ for 1 h. Finally, the sections were rinsed with PBS and visualized for immunoreactivity at 488 nm under a confocal laser microscope (LSM510 META, Carl Zeiss AG, Germany). Samples that were not treated with the anti-AccRPL11 antibody served as negative controls.

Semiquantitative RT-PCR

The primer pairs P1/P2, E1/E2, C1/C2, C3/C4 and β1/β2 were used to amplify fragments of *p38* (GenBank ID: GU321334), *ERK2* (GenBank ID: GU321335), *CacyBP* (GenBank ID: GU722329), *CREB* (GenBank ID: XM_623113.3) and *β-actin* (GenBank ID: XM640276), respectively. *β-actin* gene was used as a loading control for quantification purposes. The signal intensities [mean signal intensity (INT) per square millimeter] of the bands were determined by using the Quantity-One™ Image Analysis software (Bio-Rad, Hercules, CA, USA).

Real-time quantitative PCR (RT-qPCR)

Real-time qPCR was performed in 25 μl reaction volume with the SYBR® PrimeScript™ RT-PCR Kit (TaKaRa, Dalian, China) and a CFX96™ Real-Time PCR Detection System (Bio-Rad). The amplification reaction protocol was as follows: initial denaturation at 95℃ for 30 s; 40 cycles at 95℃ for 5 s, 56℃ for 15 s and 72℃ for 15 s; and a single melting cycle from 65℃ to 95℃. All samples were run in triplicate. The relative expression levels of *AccRPL11* transcripts were determined by using the $2^{-\Delta\Delta C_t}$ method. Data were analyzed by Duncan's multiple range test using Statistical Analysis System (SAS) version 9.1 software.

Results

Characterization of AccRPL11

We isolated the full-length cDNA sequence of *RPL11* from *A. cerana cerana* (termed

AccRPL11) by using a combination of reverse transcription PCR (RT-PCR) and rapid amplification of cDNA end (RACE). The cDNA sequence of *AccRPL11* (GenBank ID: HQ828077) was 675 bp long and consisted of a 13 bp 5' untranslated region (UTR), a 62 bp 3' UTR and an open reading frame (ORF) of 600 bp. The predicted molecular mass and pI of the putative full-length protein was 22.844 kDa and 10.08, respectively. Multiple sequence alignment revealed high amino acid identity (72.86%-92.46%) between AccRPL11 and RPL11 sequences from other species including *Zea mays*, *Bombyx mori*, *Drosophila melanogaster*, *Homo sapiens*, *Mus musculus* and *Nasonia vitripennis*. These results suggest that RPs are conserved in the evolution of different organisms. In addition, five typical RPL11 characteristic regions were identified (Fig. 1a). Interestingly, these domains were not concentrated in the N- and C-terminal regions, which is contrary to what was observed in a previous study of RPL11 from *Chironomus riparius* [27]. A phylogenetic tree showed a clear division between the RPL11 proteins of the Insecta, Arachnoidea and Mammalia classes (Fig. 1b). AccRPL11 was located in a higher evolutionary position in Insecta and was closely related to the RPL11 protein from *A. mellifera*. The results showed that the genes may encode protein of similar protein, which in turn implies similar function.

We performed Southern blotting to investigate the copy number of *AccRPL11* in the *A. cerana cerana* genome. Genomic DNA extracted from honeybee adults was individually digested with *Eco*R I, *Eco*R V and *Pvu* II, and a fragment from the 5' end of the *AccRPL11* cDNA was synthesized as a probe. As shown in Fig. 1c, only a single discrete band was detected in the lanes of the Southern blotting containing the three restriction digests, implying that there is a single gene for *RPL11* in the *A. cerana cerana* genome.

Genomic structure of *AccRPL11*

We amplified the *AccRPL11* gene from genomic DNA to study the structure of its genomic locus. The complete genomic sequence of *AccRPL11* (GenBank ID: JN699055) was 1,959 bp long and, similar to other RPL11 sequences, consisted of four introns. There was gene structure similarity between *AccRPL11* and other species (Supplementary Fig. 1). We found that the lengths of the 5' UTRs in multiple species were short. In some cases, they were no longer than 6 bp, suggesting that they may perform similar roles. Surprisingly, the last three exons of *NvRPL11*, *AmRPL11* and *AccRPL11* were identical in length. The results strongly suggest that there is a highly conserved gene structure for *RPL11* in Hymenoptera.

Identification of the 5'-flanking region of AccRPL11

Next, we isolated the 5'-flanking region of *AccRPL11* (GenBank ID: JN699055) to study the organization of the regulatory region of the gene. Several putative TFBS were predicted by using the web-based software program TFSEARCH. Four CdxA-, two NIT-2-, two HSF- and one SOX-5-binding sites were identified in the promoter region. A binding motif for Croc, a brain development-related transcriptional activator, was also found (Fig. 2) [28].

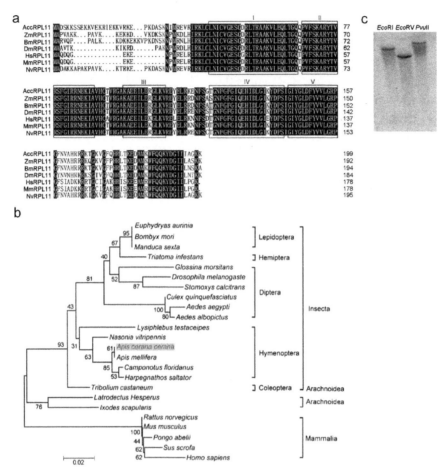

Fig. 1 Characterization of AccRPL11. a. Multiple amino acid sequence alignment of the AccRPL11 from *A. cerana cerana*, ZmRPL11 from *Zea mays*, BmRPL11 from *Bombyx mori*, DmRPL11 from *Drosophila melanogaster*, HsRPL11 from *Homo sapiens*, MmRPL11 from *Mus musculus* and NvRPL11 from *Nasonia vitripennis*. Identical amino acid residues are shaded in blue The five conserved RPL11 sequence motifs are boxed. b. A phylogenetic tree of RPL11 protein from several species was constructed by using the neighbor-joining method. Three classes (Insecta, Arachnoidea and Mammalia) and five categories (Lepidoptera, Hemiptera, Diptera, Hymenoptera and Coleoptera) are shown on the tree. The scale bar represents 0.02 amino acid substitutions per site. Bootstrap values, provided at each node, are an indication of the reliability of each branch within the phylogenetic tree. *A. cerana cerana* is highlighted in gray. Gene sequence information is provided in Supplementary Table 2. c. Southern blotting analysis of *AccRPL11* in *A. cerana cerana* genome. Genomic DNA is completely digested with three restriction enzymes, *Eco*R I, *Eco*R V and *Pvu* II and only a single discrete band is detected by hybridization in each lane.

Protein purification and Western blotting of AccRPL11

We examined the expression of the AccRPL11 protein to understand its function in *A. cerana cerana*. The complete ORF of *AccRPL11* (excluding the stop codon) was inserted into the pET-30a (+) vector and transformed into *E. coli* BL21 as a recombinant histidine fusion protein. The overexpression of AccRPL11 was induced with 0.1 mmol/L IPTG for 6 h.

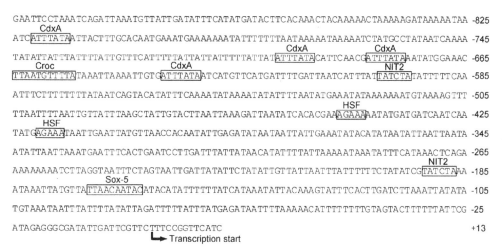

Fig. 2 Characterization of the 5'-flanking region in *AccRPL11*. Predicted transcription factor binding sites are boxed and the transcription start site is marked with an arrowhead.

SDS-PAGE analysis of the crude cell extract showed an AccRPL11 protein with a molecular mass of approximately 25 kDa (Supplementary Fig. 2a). The protein was purified and the final concentration of the protein was approximately 0.5 mg/ml. The appropriate amounts of purified proteins were subcutaneously injected into white mice to generate polyclonal antibodies. To evaluate the sensitivity of the mouse anti-AccRPL11 polyclonal antibody, the induced AccRPL11 and purified AccRPL11 were subjected to Western blotting analysis. The result showed only a single band of approximately 25 kDa in both cases, which indicates that the anti-AccRPL11 antibody is reasonably specific (Supplementary Fig. 2b).

Developmental regulation and tissue distribution of AccRPL11

Total RNA was extracted from various developmental stages of the honeybee, and Northern blotting was performed to examine the developmental regulation of *AccRPL11* expression. RNA extracts from every developmental stage showed a hybridization signal. The signal decreased slightly at the pupal stage and gradually reached a maximum in adults (Fig. 3a), suggesting that *AccRPL11* plays essential roles in mediating growth. To determine whether gender had any effect on *AccRPL11* expression, total RNA extracted from drones and worker bees was examined by using Northern blotting (Fig. 3b). The analysis indicated no significant difference in *AccRPL11* expression between males and females. Samples were also extracted from different tissues of adults to investigate the tissue distribution of *AccRPL11*. As shown in Fig. 3b, *AccRPL11* transcripts were observed in all of the tissues examined, with the highest concentration detected in the brain. Thus, *RPL11* may play a role in the brain of *A. cerana cerana*.

Pesticides and heavy metal can damage the brain system and induce cancer and neurodegenerative diseases [29]. Previous studies reported that acute cadmium could drastically inhibit rRNA transcription [30]. To determine whether the role of *AccRPL11* associated with the brain, three types of pesticides and heavy metal were fed to honeybee adults, and RNA was subsequently extracted for qPCR. The results showed that all treatments increased

AccRPL11 expression; however, the heavy metal induced greater increases in transcription levels than the pesticides (Fig. 3c, d). Therefore, these results suggest that brain stimulus under pesticides and heavy metal could affect the expression of *AccRPL11*.

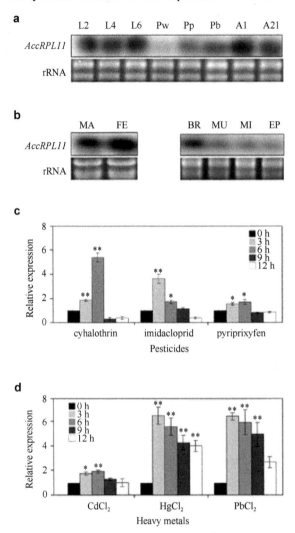

Fig. 3 Expression profile of *AccRPL11*. a. Northern blotting analysis of *AccRPL11* expression at different developmental stages. Total RNA was isolated from larvae (L2, the second instars, L4, the fourth instars and L6, the sixth instars), pupae (Pw, white eyes pupae, Pp, pink eyes pupae and Pb, black eyes pupae) and adults (A1, 1-day post-emergence adults and A21, 21-day post-emergence adults). b. Northern blotting analysis of *AccRPL11* expression in male and female honeybees and in various tissues. Total RNA was extracted from drones (MA), working bees (FE), brain (BR), muscle (MU), midgut (MI) and epidermis (EP). Ethidium bromide stained rRNA was used as a loading control. c. qPCR analysis of *AccRPL11* expression after treatment with different pesticides. d. qPCR analysis of *AccRPL11* expression after treatment with different heavy metal. Samples were collected at 3, 6, 9 and 12 h after treatment. The *AccRPL11* expression level was normalized against a wild type control (adults on a normal diet). Vertical bars represent mean ± SE (n=3). Two asterisks and asterisk refer to a highly significant difference (P<0.01) and a significant difference (P<0.05), respectively, as determined by Duncan's multiple range test using SAS version 9.1 software.

RNAi-mediated silencing of *AccRPL11* gene

We performed RNAi experiments to determine the role of *AccRPL11* in the brain. One-day post-emergence adults were injected with 5 μg ds*AccRPL11*, and brains were dissected at 1, 5, 9 and 13 days postinjection. Equivalent amounts of ds*GFP* or RNase-free water were also injected as controls. Northern blotting showed almost no difference in the transcription levels of *AccRPL11* between the control groups and the 2-day-old adults. However, lower transcription levels were observed in the ds*AccRPL11*-injected adults at 5 and 9 days postinjection. Interestingly, The transcription levels of *AccRPL11* were increased in brains from the 14-day-old ds*AccRPL11*-injected adults (Fig. 4a). The silencing effect of ds*AccRPL11* on *AccRPL11* protein levels was also assessed by using an ELISA. As expected, protein levels decreased approximately three-fold in adults sampled at 5 and 9 days postinjection compared to those from the first day after injection. However, reduced protein levels were no longer apparent at 13 days postinjection (Fig. 4b). These data strongly suggest that the silencing of the *AccRPL11* gene was successfully achieved in the brain of *A. cerana cerana*, but the inhibition was reversed by 13 days after the injection of ds*AccRPL11*. Therefore, we selected adults at 7 days postinjection to further analyze *AccRPL11* expression in the brain.

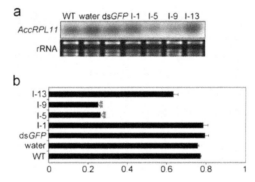

Fig. 4 Effects of dsRNA injection on the expression of *AccRPL11* and other related genes. a. Northern blotting analysis of *AccRPL11* mRNA expression in dsRNA-injected adults. Total RNA was isolated from the brains of noninjected (WT), RNase-free water-injected (water) and *dsGFP*-injected (*dsGFP*) bees. RNA was also isolated from *dsAccRPL11*-injected adult bees at 1 day (I-1), 5 days (I-5), 9 days (I-9) and 13 days (I-13) postinjection. Injections were performed on 1-day post-emergence adults. b. ELISA analysis of AccRPL11 protein expression in dsRNA-injected adults.

Phenotypes caused by RNAi of *AccRPL11* gene in the brain

To investigate the function of *AccRPL11* in the brain, qPCR experiments showed that the expression level of *AccRPL11* in 7-day postinjection adults was more severely decreased in the brain than in other tissues (Fig. 5a). Furthermore, the expression levels of four other brain and nerve development-related genes, *p38*, *ERK2*, *CacyBP* and *CREB*, also showed a significant difference in these honeybees relative to the controls (Fig. 5b).

Furthermore, paraffin sections from the wild-type and ds*AccRPL11*-injected adult honeybee brains were subjected to immunofluorescence analysis by using the anti-AccRPL11 polyclonal antibody and fluorescein-conjugated goat anti-mouse IgG. As shown in Fig. 5c, AccRPL11 immunoreactivity was widely but not evenly distributed in different sensory and motor neuropils. The prosencephalon was more intensely stained than the mesencephalon and metencephalon. A positive reaction was only detected in the surrounding tissues of esophagus of the metencephalon. In contrast, AccRPL11 was ubiquitously expressed in the optic lobes, especially in the medulla and lobula, but expression was low in the mushroom bodies. However, in the ds*AccRPL11*-injected honeybee, the optic lobes and surrounding tissues of esophagus contained lower amounts of AccRPL11, and the localization of the protein was similar to that of the control group (Fig. 5d). The results suggest that AccRPL11 is likely to fulfill a function in brain maturation during adult phase.

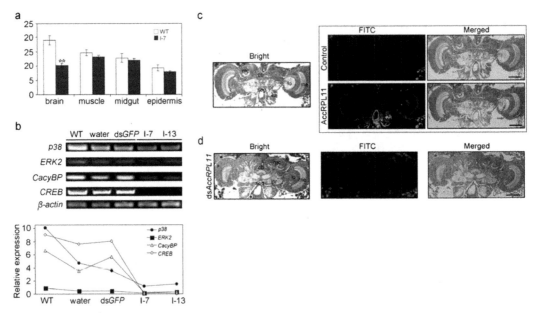

Fig. 5 RNAi phenotypes in the brain of *A. cerana cerana* caused by ds*AccRPL11*. a qPCR analysis of *AccRPL11* expression in various tissues of noninjected (WT) and ds*AccRPL11*-injected adults at 7 days postinjection (I-7). Vertical bars represent the mean ± SE (n=3). Two asterisks refer to a highly significant difference ($P<0.01$), as determined by Duncan's multiple range test using SAS version 9.1 software. b. RT-PCR analysis of the expression of the *A. cerana cerana p38*, *ERK2* and *CacyBP* genes and the *A. mellifera CREB* gene in controls (WT, water and ds*GFP*) and ds*AccRPL11*-injected adults (I-7 and I-13). The *β-actin* gene was used as an internal control. The accompanying polygram shows the relative expression levels of the genes. The signal intensities (INT/mm^2) of the bands were measured by using Quantity-One™ Image Analysis software (Bio-Rad, Hercules, CA, USA). c. Immunofluorescence localization of AccRPL11 in a transverse section of the brain from wild-type *A. cerana cerana* adults. An anti-AccRPL11 polyclonal antibody was used as the primary antibody, and FITC-conjugated goat anti-mouse IgG was employed as the secondary antibody. Major domains in the brain are indicated, and immunofluorescence staining is indicated with red arrows. Scale bar, 500 μm. d. Immunofluorescence analysis in a transverse section of the brain from ds*AccRPL11*-injected adults at 7 days postinjection. Positive staining is barely visible in any region where AccRPL11 would normally be localized. The major domains in the brain are indicated as follows: cb, central body; lo, lobula; lpr, lateral protocerebrum; mc, median calyx of the mushroom body; me, medulla and sog, subesophageal ganglion. Scale bar, 500 μm.

Discussion

Ribosomal protein L11 is a major inhibitor of the MDM2-p53 feedback loop and is of key importance for organism development. Mutations or deletions within RPL11 induce a variety of developmental defects [18, 19, 20]. Although RPs are highly evolutionarily conserved, they are unlike other house-keeping genes because the defects caused by RP gene knockdown vary from species to species [7, 11, 12]. To gain new insight into *RPL11* function in other important species, we identified and characterized an *RPL11* gene from *A. cerana cerana*.

The N-terminal and CTDs of RPL11 are the most important functional domains of the protein. The N-terminus is necessary for peptide release and translational accuracy [13]. In *C. riparius*, a high homology of RPL11 was found in the N-terminal region compared with other insects [27]. However, we did not obtain the same result in our multiple sequence alignment. In this study, five conserved domains, characteristic of RPL11, were identified in AccRPL11, but they were not located in the N- or C-terminal region of the protein (Fig. 1a). The conservation supports our assertion that AccRPL11 is an integral member of the RPL11 subfamily; however, the localization discrepancy suggests that the role of *AccRPL11* may be different from the *C. riparius* RPL11 protein.

Previous studies have shown that the two *RPL11* genes in *A. thaliana* display different expression patterns [21, 22]. In *A. cerana cerana*, our Northern blotting analysis showed that *RPL11* was expressed throughout development and that the highest transcription levels were found in adults, not only in early development (Fig. 3a). The result is inconsistent with previous work indicating that the expression of *RPL11* is significantly higher in the embryonic stage than in other stages of development in Diptera insects [27]. The differences may be caused by the use of different species: Hymenoptera is not only one of the largest orders of insects but also is one in which species exhibit highly social behavior. The differential expression pattern during specific stages of development should reveal a distinct function for *RPL11*. Both the highest expression levels being found in the brain (Fig. 3b) and the brain development-related transcriptional activator Croc (Fig. 2) suggest that *AccRPL11* may play a role in the brain. Pesticides and heavy metal have been clearly shown to alter neurological and immunological functions and to induce cancer and neurodegenerative diseases [29]. The significant increase in *AccRPL11* expression after pesticides and heavy metal treatments (Fig. 3c, d) may provide further evidence to our hypothesis that *AccRPL11* is involved in brain function.

RNAi-mediated silencing of the *AccRPL11* gene was performed to validate this hypothesis. In our study, the injection of ds*AccRPL11* in honeybee adults had a negative effect on *AccRPL11* expression in the brain than in other tissues (Fig. 5a). Indeed, the loss of *RPL11* can induce developmental abnormalities in the zebrafish brain, causing defects such as an enlarged prosencephalon and a malformed metencephalon [18]. *ERK2* and *p38* have critical roles in the central nervous system [31, 32], and *CacyBP* is an important neuronal protein [33]. Furthermore, the disruption of *CREB* function in the brain has been shown to cause neurodegeneration [34]. The expression of these brain development-related genes (*ERK2*, *p38*, *CacyBP* and *CREB*) gradually decreased over time after honeybees were injected with ds*AccRPL11* in the RNAi experiments (Fig. 5b). This result implied AccRPL11 may play a role in the brain of adult worker honeybees.

The honeybee, a social insect, possesses a complex brain with a highly regulated neuronal network that is thought to influence its highly organized social behavioral patterns [35]. In our immunofluorescence experiments, the presence of AccRPL11 in the mushroom bodies, medulla, lobula and surrounding tissues of esophagus was not apparent after the *AccRPL11* gene was silenced (Fig. 5c, d). The honeybee brain is divided into three regions, the prosencephalon, mesencephalon and metencephalon. The prosencephalon, which includes the mushroom bodies, optic lobes and center body, primarily receives and processes visual and major neuronal signals [36, 37]. Color discrimination in foraging is associated with links between the visual and memory systems [36]. In the optic lobes, the lateral lamina receives brightness stimuli and transports the signals to the medulla, which preserved and amplified the color divergence in the lobula [36, 38, 39]. The attenuation of the immunofluorescence signal in the medulla and lobula in *AccRPL11*- knockdown honeybees suggests that ds*AccRPL11* may interfere with the visual system maturation in brain. In addition, the localization of AccRPL11 in the surrounding tissues of esophagus in the metencephalon did change in ds*AccRPL11*- injected adults (Fig. 5d). The subesophageal ganglion in the honeybee receives gustatory sensory information from sensory projections and motor stimuli from motor neurons. After processing, the subesophageal ganglion returns these neuronal signals to their respective endpoints [40]. The localization and attenuation of AccRPL11 to the surrounding tissues of esophagus might indicate that AccRPL11 may function in a role in motor system maturation, which is necessary for foraging, nursing and other behaviors as well as visual system.

Consequently, our results on the higher expression levels of AccRPL11 in the brain, the ds*AccRPL11*-mediated depletion of brain transcripts, a decrease in the brain development related gene expression and immunofluorescence attenuation in *AccRPL11* knockdown honeybees implied that the *AccRPL11* gene may have a function in brain maturation during adult phase.

References

[1] Woyciechowski, M., Moroń, D. Life expectancy and onset of foraging in the honeybee (*Apis mellifera*). Insect Soc. 2009, 56 (2), 193-201.

[2] Dyer, F. C. The biology of the dance language. Annu Rev Entomol. 2002, 47 (1), 917-949.

[3] Withers, G. S., Fahrbach, S. E., Robinson, G. E. Selective neuroanatomical plasticity and division of labour in the honeybee. Nature. 1993, 364 (6434), 238-240.

[4] Page Jr, R. E., Scheiner, R., Erber, J., Amdam, G. V. The development and evolution of division of labor and foraging specialization in a social insect (*Apis mellifera* L.). Curr Top Dev Biol. 2006, 74 (6), 253-286.

[5] Nelson, C. M., Ihle, K. E., Fondrk, M. K., Page, R. E., Amdam, G. V. The gene vitellogenin has multiple coordinating effects on social organization. PLoS Biol. 2007, 5 (3), e62.

[6] Marco Antonio, D. S., Guidugli-Lazzarini, K. R., do Nascimento, A. M., Simões, Z. L., Hartfelder, K. RNAi-mediated silencing of vitellogenin gene function turns honeybee (*Apis mellifera*) workers into extremely precocious foragers. Naturwissenschaften. 2008, 95 (10), 953-961.

[7] Kenmochi, N. Ribosome's and ribosomal proteins. *In*: Cooper D. Nature encyclopedia of the human genome. 2003, 5, 77-82.

[8] Byrne, M. E. A role for the ribosome in development. Trends Plant Sci. 2009, 14 (9), 512-519.
[9] Duncan, K. A., Jimenez, P., Carruth, L. L. The selective estrogen receptor-alpha coactivator, RPL7, and sexual differentiation of the songbird brain. Psychoneuroendocrinology. 2009, Suppl.1, S30-S38.
[10] Lindström, M. S. Emerging functions of ribosomal proteins in gene-specific transcription and translation. Biochem Biophys Res Commun. 2009, 379 (2), 167-170.
[11] Oliver, E. R., Saunders, T. L., Tarlé, S. A., Glaser, T. Ribosomal protein L24 defect in belly spot and tail (Bst), a mouse Minute. Development. 2004, 131 (16), 3907-3920.
[12] Marygold, S. J., Coelho, C. M., Leevers, S. J. Genetic analysis of RpL38 and RpL5, two minute genes located in the centric heterochromatin of chromosome 2 of *Drosophila melanogaster*. Genetics. 2005, 169 (2), 683-695.
[13] Bouakaz, L., Bouaka, E., Murgola, E. J., Ehrenberg, M., Sanyal, S. The role of ribosomal protein L11 in class I release factor- mediated translation termination and translational accuracy. J Biol Chem. 2006, 281 (7), 4548-4556.
[14] Vogelstein, B., Lane, D., Levine, A. J. Surfing the p53 network. Nature. 2000, 408 (6810), 307-310.
[15] Zhang, Y., Wolf, G. W., Bhat, K., Jin, A., Allio, T., Burkhart, W. A., Xiong, Y. Ribosomal protein L11 negatively regulates oncoprotein MDM2 and mediates a p53-dependent ribosomal-stress checkpoint pathway. Mol Cell Biol. 2003, 23 (23), 8902-8912.
[16] Bhat, K. P., Itahana, K., Jin, A., Zhang, Y. Essential role of ribosomal protein L11 in mediating growth inhibition-induced p53 activation. EMBO J. 2004, 23 (12), 2402-2412.
[17] Murayama, E., Kissa, K., Zapata, A., Mordelet, E., Briolat, V., Lin, H. F., Handin, R. I. Herbomel, P. Tracing hematopoietic precursor migration to successive hematopoietic organs during zebrafish development. Immunity. 2006, 25 (6), 963-975.
[18] Chakraborty, A., Uechi, T., Higa, S., Torihara, H., Kenmochi, N. Loss of ribosomal protein L11 affects zebrafish embryonic development through a p53-dependent apoptotic response. PLoS One. 2009, 4 (1), e4152.
[19] Danilova, N., Sakamoto, K. M., Lin, S. Ribosomal protein L11 mutation in zebrafish leads to haematopoietic and metabolic defects. Br J Haematol. 2011, 152 (2), 217-228.
[20] Quarello, P., Garelli, E., Carando, A., et al. Diamond-Blackfan anemia: genotype-phenotype correlations in Italian patients with RPL5 and RPL11 mutations. Haematologica. 2010, 95 (2), 206-213.
[21] Williams, M. E., Sussex, I. M. Developmental regulation of ribosomal protein L16 genes in *Arabidopsis thaliana*. Plant J. 1995, 8 (1), 65-76.
[22] Barakat, A., Szick-Miranda, K., Chang, I. F., Guyot, R., Blanc, G., Cooke, R., Delseny, M., Bailey-Serres, J. The organization of cytoplasmic ribosomal protein genes in the *Arabidopsis* genome. Plant Physiol. 2001, 127 (2), 398-415.
[23] Tamura, K., Dudley, J., Nei, M., Kumar, S. MEGA4: molecular evolutionary genetics analysis (MEGA) software version 4.0. Mol Biol Evol. 2007, 24 (8), 1596-1599.
[24] Wang, M., Kang, M., Guo, X., Xu, B. Identification and characterization of two phospholipid hydroperoxide glutathione peroxidase genes from *Apis cerana cerana*. Comp Biochem Physiol C: Toxicol Pharmacol. 2010, 152 (1), 75-83.
[25] Meng, F., Zhang, L., Kang, M., Guo, X., Xu, B. Molecular characterization, immunohistochemical localization and expression of a ribosomal protein L17 gene from *Apis cerana cerana*. Arch Insect Biochem Physiol. 2010, 75 (2), 121-138.
[26] Elias-Neto, M., Soares, M. P., Simões, Z. L., Hartfelder, K., Beyond, M. M. Developmental characterization, function and regulation of a Laccase2 encoding gene in the honey bee, *Apis mellifera* (Hymenoptera, Paine). Insect Biochem Mol Biol. 2010, 40 (3), 241-251.
[27] Martínez-Guitarte, J. L., Planelló, R., Morcillo, G. Characterization and expression during development and under environmental stress of the genes encoding ribosomal proteins L11 and L13 in *Chironomus riparius*. Comp Biochem Physiol B: Biochem Mol Biol. 2007, 147 (4), 590-596.
[28] Jeffrey, P. L., Capes-Davis, A., Dunn, J. M., Tolhurst, O., Seeto, G., Hannan, A. J., Lin, S. L.

CROC-4: a novel brain specific transcriptional activator of c-fos expressed from proliferation through to maturation of multiple neuronal cell types. Mol Cell Neurosci. 2000, 16 (3), 185-196.

[29] Franco, R., Sánchez-Olea, R., Reyes-Reyes, E. M., Panayiotidis, M. I. Environmental toxicity, oxidative stress and apoptosis: ménage à trios. Mutat Res. 2009, 674 (1-2), 3-22.

[30] Planelló, R., Martínez-Guitarte, J. L., Morcillo, G. Ribosomal genes as early targets of cadmium-induced toxicity in *Chironomus riparius* larvae. Sci Total Environ. 2007, 373 (1), 113-121.

[31] Guo, S. W., Liu, M. G., Long, Y. L., et al. Region-or state-related differences in expression and activation of extracellular signal-regulated kinases (ERKs) in naïve and pain-experiencing rats. BMC Neurosci. 2007, 8, 53.

[32] Oliveira, C. S., Rigon, A. P., Leal, R. B., Rossi, F. M. The activation of ERK1/2 and p38 mitogen-activated protein kinases is dynamically regulated in the developing rat visual system. Int J Dev Neurosci. 2008, 26 (3-4), 355-362.

[33] Zhai, H., Shi, Y., Jin, H., et al. Expression of calcyclin-binding protein/Siah-1 interacting protein in normal and malignant human tissues: an immunohistochemical survey. J Histochem Cytochem. 2008, 56 (8), 765-772.

[34] Mantamadiotis, T., Lemberger, T., Bleckmann, S. C., et al. Disruption of CREB function in brain leads to neurodegeneration. Nat Genet. 2002, 31 (1), 47-54.

[35] Ament, S. A., Corona, M., Pollock, H. S., Robinson, G. E. Insulin signaling is involved in the regulation of worker division of labor in honey bee colonies. Proc Natl Acad Sci USA. 2008, 105 (11), 4226-4231.

[36] Dyer, A. G., Paulk, A. C., Reser, D. H. Color processing in complex environments: insights from the visual system of bees. Proc Biol Sci. 2011, 278 (1707), 952-959.

[37] Sinakevitch, I., Mustard, J. A., Smith, B. H. Distribution of the octopamine receptor AmOA1 in the honey bee brain. PLoS One. 2011, 6 (1), e14536.

[38] Brandt, R., Rohlfing, T., Rybak, J., Krofczik, S., Maye, A., Westerhoff, M., Hege, H. C., Menzel, R. Three-dimensional average-shape atlas of the honeybee brain and its applications. J Comp Neurol. 2005, 492 (1), 1-19.

[39] Paulk, A. C., Gronenberg, W. Higher order visual input to the mushroom bodies in the bee, *Bombus impatiens*. Arthropod Struct Dev. 2008, 37 (6), 443-458.

[40] Schröter, U., Menzel, R. A new ascending sensory tract to the calyces of the honeybee mushroom body, the subesophageal-calycal tract. J Comp Neurol. 2003, 465 (2), 168-178.